D0745704

Confronting Climate Change is a guide to the risks, dilemmas, and opportunities of the emerging political era, in which the impacts of a prospective global warming could affect all regional, public, and even individual decisions. Written by a renowned group of scientists, political analysts, and economists, all with direct experience in climate change related deliberations, *Confronting Climate Change* is a survey of the best available answers to three vital questions:

what do we know so far about the foreseeable dangers of climate change?
how reliable is our knowledge?
what are the most rewarding ways to respond?

CONFRONTING CLIMATE CHANGE

Risks, Implications and Responses

CONFRONTING CLIMATE CHANGE

Risks, Implications and Responses

EDITED BY
IRVING M. MINTZER

ASSISTANT EDITORS
Art Kleiner and Amber Leonard

PRODUCTION EDITOR
Arno Rosemarin

ASSISTANT PRODUCTION EDITOR
Heli Pohjolainen

Stockholm Environment Institute

CAMBRIDGE
UNIVERSITY PRESS

Published by the Press Syndicate of the University of Cambridge
The Pitt Building, Trumpington Street, Cambridge CB2 1RP
40 West 20th Street, New York, NY 10011-4211, USA
10 Stamford Road, Oakleigh, Victoria 3166, Australia

First published 1992

Printed in Great Britain at the University Press, Cambridge

A catalogue record for this book is available from the British Library

Library of Congress cataloguing in publication data

ISBN 0 521 42091 1 hardback
ISBN 0 521 42109 8 paperback

Cover illustration: Courtesy of USEPA/Bruce Presentations

Contents

III Energy Use and Technology

IV Economics and the Role of Institutions

V Equity Considerations and Future Negotiations

Foreword

It is not only the non-specialist, the man and woman in the street and the ordinary person who finds "climate change" and "global warming" a fascinating yet difficult topic. In most societies some tenuous link to our agricultural origins ensures that the weather is a frequent feature of conversation. But weather is not climate - even if it results from it. Conflicting signs, different emphasis placed on the many strands of evidence, new knowledge and different propensities to be optimistic or pessimistic all lead to difficulties in identifying the "signal from the noise," in recognizing trends in global climate change - in discerning evidence of a real climate warming effect.

Even scientists, trained in the scientific method are, from time to time periodically perplexed. Many physicists, chemists and those used to working at the "chemical" end of biology feel a need to have more evidence, more measurement, more research. At home with the process of inductive reasoning, hypothesis establishment and direct experimental procedures, any consensus view on climate change presents some problems due to the range of uncertainties. The whole climate change issue is, however, much more susceptible to approaches based on deductive reasoning, where information is assembled and interpretations made on the basis of the best available evidence so that a "working hypothesis" or explanation is produced, involving a minimum of assumptions. There is nothing new or "unscientific" in this approach. Agricultural scientists, and others, are used to working from sample estimates, frequency distributions and probabilities; the whole of the Earth's geological record, and the evolutionary basis of biology, has been interpreted in this way. Wait for the definitive experiment and you wait for ever.

In the area of climate change and climate change prediction there is only one definitive experiment possible, and that is a rather long-term one. It may be prudent to make some well-chosen responses before we are certain "beyond reasonable doubt." And fortunately, there is much accumulating evidence and the possibility of climate simulation through General Circulation Models of, not only increased sophistication but also of improved realism. Of course, uncertainty is still the name of the game but we should not fall into the trap of making the mistake that could be characterized by adapting a well-known remark of Edmund Burke — nobody makes a greater mistake than he who thinks he knows nothing because he knows so little!

Of course there is a need for more information, further research and continued assessment of the evidence, the effects and the possible policy and management responses. It is in relation to this need for a continued updating of the assessment that this volume has been produced. It has drawn on the expertise — and thoughtfulness — of the international community of professionals concerned with climate change issues. It also attempts, by the editorial commentary that accompanies each chapter, to evolve a synthesis as well as a synopsis. It does not take up an advocacy stance, but seeks to expose the issues and inform the reader. In this it is a continuation of a programme element of the Stockholm Environment Institute that has focused for some years on, and contributed to, responses to potential man-induced climatic modification.

The impact of potential climate change is a challenge to national and international planners and policy makers. Equally it challenges industry, commerce and all elements of the local or wider community. It is for these people that the book is written.

M.J. Chadwick
Director
STOCKHOLM ENVIRONMENT INSTITUTE
Stockholm, Sweden

Acknowledgements

Many people have contributed generously to this book. The work has been greatly enhanced by their efforts and we wish to express our deep appreciation for their help. Gordon T. Goodman, Chairman of the Board of the Stockholm Environment Institute, gave the incisive and compassionate vision which framed the foundation of this assessment. Mike Chadwick and Lars Kristoferson, the Director and Deputy Director of the Institute respectively, gave strength and direction to our pursuit of that vision. Bert Bolin, Roald Sagdeev, Sir Crispin Tickell, Shridath Ramphal, Robert Frosch, R.K. Pachauri, Patricia Close, Lourival do Carmo Monaco and Ali Mazrui helped us to structure the vision and to focus our attention on the "big picture." Gerald Leach, Georgii Golitsyn, Paul Crutzen, David Hall, Roberto Sanson Mizrahi, Hans Oeschger, Konrad von Moltke, Måns Lönnroth, Ophelia Mascarenhas, Gilbert White, Roger Rainbow, Erik Belfrage and Alvaro Umana acted as a planning and review committee for this effort. They provided constructive comments on the plans for the project and critical reviews that challenged and focused our efforts. John Holdren, Anne and Paul Ehrlich, Roger Revelle and Alan Miller offered intellectual guidance and encouragement that helped us navigate the mists of scientific uncertainty, political turmoil and human fallibility. More than fifty reviewers provided constructive criticism and creative feedback on the early drafts of the chapters of this book. Their names appear on the following pages.

We are especially indebted to those at SEI in Stockholm who aided us in the practical tasks of producing a book involving forty-four coauthors on four continents working in ten time zones. Without their dedicated, continuing and professional efforts, our task could not have been completed. Absolutely critical to our success have been the countless hours spent by our Production Editors, Arno Rosemarin and Heli Pohjolainen who were responsible for the book's layout, graphics, final editing and typesetting. We also thank Solveig Nilsson for her extra efforts keying in changes from the corrected and reedited pageproofs and Krister Svärd, the Institute's librarian for backup throughout this project.

Among those working with us in the United States, we would like to take special note of the critically important contributions to this book made by several people. Patricia Feuerstein examined the web of ideas that spans these chapters and created an index that will guide each reader to the issues and concepts he or she wishes to explore in depth. Shawn Armstrong, David Blaivos, Gwen Anderson, Paul Carroll, Janis Dutton, Donna Kapsides, Anuriti Sud, Judy Webb and Vince Schaper helped us to organize and implement the tasks we faced in ways too numerous to mention. And we are, of course, unrelentingly grateful for the wealth of advice on production problems that was provided to us by Marilyn Powell in our periodic moments of extreme need. All our efforts would have been of little import without the help and dedication of our coauthors. But even these would have come to little except for the tireless, resolute and good humored efforts of their spouses, staff and assistants, who took our phone calls at inconvenient hours, unscrambled the confusing messages and tracked down our colleagues across oceans and continents to answer "just one more question" before we could put the book to bed.

With all this help, our task was eased and simplified. Despite the help, some errors may remain within these pages. The responsibility for all of these is solely ours.

Irving M. Mintzer
STOCKHOLM ENVIRONMENT INSTITUTE
Box 2142, 103 14 Stockholm, Sweden
April,1992

List of Reviewers

Sharad P. Adhikary, Department of Hydrology and Meteorology, Kathmandu, Nepal.
Gilbert Arum, KENGO, Nairobi, Kenya.
Richard E. Benedick, World Wildlife Fund, Washington, DC, USA.
John T. Blake, Jamaica Meteorological Service, Kingston, Jamaica.
Deborah Bleviss, International Institute for Energy Conservation, Washington, DC, USA.
Peter Brewer, Monterey Bay Aquarium and Research Institute, Monterey, CA, USA.
Chris Burnup, Business Council of Australia, Melbourne, Australia.
Adinath Chatterjee, CESC Limited, Calcutta, India.
Karn Chiranond, Department of Treaties and Legal Affairs, Ministry of Foreign Affairs, Bangkok, Thailand.
B.V. Chitnis, TATA Consulting Engineers, Bombay, India.
Paul J. Crutzen, Max Planck Institute for Air Chemistry, Mainz, Germany.
C. Dasgupta, Ministry of External Affairs, New Delhi, India.
Roger Dower, World Resources Institute, Washington, DC, USA.
Hadi Dowlabadi, Carnegie Mellon University, Pittsburgh, PA, USA.
Kerry Emanuel, Massachusetts Institute of Technology, Cambridge, MA, USA.
Malin Falkenmark, NFR, Stockholm, Sweden.
Robert Friedman, Congress of the United States Office of Technology Assessment Washington, DC, USA.
Axel Friedrich, Umveltbundesamt, Berlin, Germany.
Joseph Gabut, Department of Foreign Affairs, Papua New Guinea.
L. Danny Harvey, Department of Geography, University of Toronto, Toronto, Canada.
Hillard Huntington, Energy Modelling Forum, Stanford University, Palo Alto, CA, USA.
Tariq Osman Hyder, Ministry of Foreign Affairs, Islamabad, Pakistan.
Ivar Isaksen, University of Oslo, Oslo, Norway.
David Jhirad, US Agency for International Development, Washington, DC, USA.
Mohamed Khalil, African Center for Technology Studies, Nairobi, Kenya.
Dan Lashof, Natural Resources Defence Council, Washington, DC, USA.
Steve Leatherman, Department of Geography, University of Maryland, College Park, MD, USA.
Jeremy Leggett, Greenpeace International, London, UK.
Michael MacCracken, Lawrence Livermore Laboratory, Livermore, CA, USA.
Cecilia MacKenna, Ministry of External Relations, Santiago, Chile.
Abdullahi Majeed, Department of Meteorology, Republic of Maldives.
Bayani Mercado, Department of Foreign Affairs, Pasay City, Philippines.
Aloke Mookherjea, Fläkt India Limited, Calcutta, India.
Frank Muller, Center for Global Change, University of Maryland, College Park, MD, USA.
Fernando Novillo-Soravia, Argentine Mission in Geneva, Geneva, Switzerland.
Catharina Nystedt, ABB Fläkt, Stockholm, Sweden.
Hans Oeschger, Physikalisches Institute, University of Bern, Bern, Switzerland.
R.K. Pachauri, TATA Energy Research Institute, New Delhi, India.
Atiq Rahman, Bangladesh Institute for Advanced Studies, Dhaka, Bangladesh.
Roger Rainbow, Shell International Petroleum Company, Shell Centre, London, England.
Roger Revelle (Deceased, 1991), Scripps Institute of Oceanography, University of California, San Diego, CA, USA.
Richard Richels, Electric Power Research Institute, Palo Alto, CA, USA.
Alan Robock, Department of Meteorology, University of Maryland, College Park, MD, USA.
Annie Roncerel, Climate Network Europe, Brussels, Belgium.
Norman Rosenberg, Resources for the Future, Washington, DC, USA.
Cynthia Rosensweig, NASA Goddard Institute for Space Studies, New York, NY, USA.
Juan Salazar-Sancisi, Ministry of Foreign Affairs, Quito, Ecuador.
Robert Schiffer, NASA Headquarters, Washington, DC, USA.
Kirk Smith, Environment and Policy Institute, East West Center, Honolulu, HI, USA.

Aca Sugandhy, Ministry of Population and Environment, Jakarta Pusat, Indonesia.
Tang Cheng-Yuan, Ministry of Foreign Affairs, Beijing, China.
Peter Thacher, World Resources Institute, Washington, DC, USA.
Tyler Volk, Department of Applied Sciences, New York University, New York, NY, USA.
Arthur Westing, Putney, VT, USA.
Pamela Wexler, Center for Global Change, University of Maryland, College Park, MD, USA.
Gilbert White, Natural Hazards Research Center, Boulder, CO, USA.
Montague Yudelman, World Wildlife Fund, Washington, DC, USA.
Derwood Zaelke, Center for International Environmental Law, Washington, DC, USA.

CHAPTER 1

Living in a Warming World

Irving M. Mintzer

Human activities are changing the composition of our atmosphere at an unprecedented rate. If current trends continue, our planet could face a climatic shock unlike anything experienced in the last 10,000 years. It would not be felt as an immediate blow, that is a shift from status quo to catastrophe. The climate cannot disappear like an endangered species. Nor can it explode like a runaway reactor. Nevertheless, the risks of rapid climate change — rapid by geologic and climatic standards — are rising rapidly in our time. In this context, "shock" is an appropriate term. It describes the impact that the resulting set of changes may have on human economies and natural ecosystems.

Climate change is not a new phenomenon. Earth's climate has changed before, many times in the last two billion years. But something important is different this time.

1 A Change is in the Air

The woman gathering fuelwood in the Sudan senses a difference. She survived 2 extremely dry years and then, in August 1988, saw a year's rain fall in three days. The British coal miner suspects that something is askew in the picture outside his kitchen window. He was shocked as he saw the "100-year storm" blow across Britain twice in 5 years, tearing up trees that he had regarded as a permanent feature of the landscape. The American, Australian, and African farmers who have seen deep droughts and big rains crush their crops again and again in the last 5 years, can feel a difference in the soil and smell a difference in the air. The Bangladeshi boatman, who saw the "once-in-a-century" typhoon surge out of the Bay of Bengal twice in 20 years — washing across the alluvial delta of the Ganges-Bhramaputra — senses that the weather has changed since his childhood. The South Pacific scientist who studies the ecology of coral reefs knows that something significant has disturbed the seabed; she sees large masses of dying coral. And the Swiss ski resort owners, who waited through two long winters without sufficient snow on which to ski, all know something is different.

But what is it? Is it just an unusual string of random, independent events? Is it a passing phase, a simple stochastic variation in a complex non-linear system? Or is the collective experience of many people in varied walks of life an early indicator of a more profound and lasting change, a major shift in global and regional climates?

2 Mechanics of the Heat Trap

Many of the world's best physical, chemical, and biological scientists continue to puzzle over these questions. Working together with economists and political scientists under the aegis of the Intergovernmental Panel on Climate Change (IPCC), the leading scientists from more than sixty countries have developed long-term scenarios, have used complex computer models to run simulation experiments and expanded their study of climate to past eons and other planets (IPCC, 1990 and 1992).[1] In the process, they have developed a better understanding of the global climate system and of the forces that cause climate to change.

Global climate, the long-term statistical average of millions of daily weather events, is a tapestry composed of many

[1] Sponsored by the World Meteorological Organization (WMO) and the United Nations Environment Programme (UNEP), the IPCC prepared a three-volume assessment in 1990 summarizing the state of the art on climate modelling, climate impacts, and response strategies (IPCC, 1990). In 1992, the IPCC completed an updated report and a supplement to that work (IPCC, 1992).

threads. Each of the regional climate patterns observed in cities, towns, and rural villages offers a glimpse of one of the threads, a sign of some deeper climatic pattern. Each reflects a combination of complex local interactions between the atmosphere, the oceans and the biota. Over time, these threads interweave. Trends in temperature, precipitation, soil moisture, and numerous other factors combine into highly variable, localized events to create the regional conditions that give each locale its special character.

Solar radiation fuels the global climate machine. Changes in the distribution of sunlight falling on the Earth and heat emitted from it are the principal driving forces that determine the character of global climate. The magnitude of these energy flows is affected by shifts in the Earth's orbit around the sun, increases or decreases in cloud cover, transformation of the land surfaces of the continents, variations in ocean currents, and changes in the composition of the atmosphere.

The average annual temperature of the planet is following an upward trend. In the last century, the mean surface temperature has increased by about 0.5- 0.7 °C. Seven of the eleven warmest years in the last hundred have occurred in the last decade. Last year, 1991, was the warmest year of our instrumental temperature record and the winter of 1991-92 follows the pattern.

It is not just the global averages that seem to be changing. Each person experiences the global climate through the variability of local weather events. From many anecdotal observations, the weather seems to be more variable now than it has been in the past. And some weather data also suggest that regional climates became less stable in the 1980s. In many areas, the frequency and severity of extreme weather events seem to be increasing. But the changes recorded so far are still within the statistical ranges of natural variability. From a mathematical point of view, nothing definitive can yet be proven about future climate change — but something makes the changes feel different this time.

During the last five years, as the international scientific assessment proceeded within the IPCC and international cooperative research advanced under the International Geosphere-Biosphere Program (IGBP), scientists have learned a great deal about the dynamics of the climate system. But despite all this groundbreaking work, they still cannot simply explain the apparent increase in extreme weather events. They cannot say, for sure, whether a major long-term climate change is under way. Nor can they say with confidence precisely where, or when, or how severely the regional impacts of these future changes will be felt.

They can, however, say some important things about the changes that are now taking place. Human activities are changing the composition and behaviour of the atmosphere at an unprecedented rate. And pollutants from a wide range of human activities — including energy use, industrial production, agriculture, forestry, and land use changes — are increasing the global atmospheric concentration of certain heat-trapping gases.

The most dangerous of these trace gases include carbon dioxide, methane, nitrous oxide, and the synthetic compounds called chlorofluorocarbons (CFCs). Because of their atomic structure, these gases are transparent to incoming solar radiation. Most of the sunlight passes through them and is not absorbed. But the same gases absorb and re-emit light at longer wavelengths — such as the thermal infra-red radiation that is released naturally and continuously from the earth's surface. When these heat-trapping gases release the energy they have absorbed, they re-emit it in all directions. The re-emitted radiation carries most of the heat upward, out of the atmosphere; but some is re-emitted downward, warming air, land and water below.

In effect, these gases act like a blanket, trapping heat close to the surface that would otherwise escape through the atmosphere to outer space. This process is commonly called "the greenhouse effect" because it reminds some observers of the heat-trapping effect of the glass walls in a horticultural greenhouse.

The greenhouse effect is neither new nor due solely to human activities. It is a natural component of the Earth's geophysical balance and has been occurring for the last two billion years. For thousands of millennia, natural background concentrations of greenhouse gases (principally water vapour and carbon dioxide, CO_2) trapped sufficient heat near the surface to raise our planet's average temperature by about 33 °C above what it would otherwise have been. This process increased the surface temperature from -18 to +15°C. The warmer temperature allowed water to exist on the surface as a liquid — rather than as ice — and to become the medium for biological evolution of life.

But in the last century, this natural background process, "the greenhouse effect", has become the "greenhouse problem". Human activities have steadily increased the concentrations of various heat-trapping gases, enhancing the warming effect. The world's atmospheric scientists, while they may disagree on the details, have in fact reached a consensus about the global implications of this buildup (IPCC, 1990 and 1992). If current trends in the emissions of greenhouse gases continue, the surface will warm by about 0.3 °C per decade. By sometime around the middle of the next century, the cumulative warming effect will raise the average surface temperature of our planet somewhere between 1.5 and 4.5 °C above the natural "background" temperature which existed before the beginning of the industrial revolution in the eighteenth century (IPCC, 1990).[2]

Scientists understand the main outlines of these changes at the global level but many important (and persistent) uncertainties remain. Little can now be said with confidence

---

[2] A warming of 1.5-4.5 °C is equivalent to the increase in temperature that would be expected if the pre-industrial concentration of CO_2 alone were doubled while the concentration of all other heat-trapping gases remained at the background, pre-industrial level — a benchmark used frequently in climate modeling experiments.

about the response of regional climates to the global buildup of heat-trapping gases. Current understanding of the feedback processes that link the atmosphere, oceans and biota — a set of closely coupled, non-linear systems — is rudimentary at best. Little is now known about the character or location of any thresholds of non-linearity in the responses of these systems to future stresses. But we can be certain that the future will contain ample surprises — much as the discovery of the Antarctic Ozone Hole was a surprise, an unforeseen threshold of non-linearity in the response of the upper atmosphere (the stratosphere) to increasing concentrations of chlorine and bromine.

A change in global temperature induced by human activities of one, two or even five degrees does not sound like it would make much difference, especially since it is superimposed on a natural process that has already heated the planet's surface by more than 30 °C. Those 30 degrees represent a vast difference — the difference between our warm, hospitable planet and a lifeless ball of ice. And because of paleoclimatic temperature records, we know enough to be reasonably certain that even small additional changes in the average planetary temperature can produce dramatic changes in climate. For example:

- A difference of 1 °C in average global temperature is all that separates today's equable (i.e. warm) climate from that of the Little Ice Age. During this cold period that lasted from the 14th to the 17th Century, traditional crops failed frequently in Europe. On at least several occasions, the Baltic Sea froze, allowing people to walk, skate, or sled from continental Europe to Scandinavia.

- A worldwide increase of 2 °C above today's level would push average global temperature beyond anything experienced in the last 10,000 years. At no time during the period of written human history have people faced such conditions as these. The shifts in temperature will vary from location to location and could cause crop zone boundaries to shift. Humans could certainly survive the climate of a 2-degree rise - we have done so before. But we have no written or cultural records with which to learn from the successful (and unsuccessful) adaptations of our ancestors.

- A warming of 5 °C from the present level would make the average global temperature hotter than at any time in the last three million years. During those previous hotter periods, there was no polar ice cap in the Northern Hemisphere; the sea level was as much as 75 metres higher than it is now. Tropical and subtropical regions extended as far north as Canada and England.

3 Effects of a Greenhouse Warming

If current trends continue, climatic conditions will change more quickly in the next few decades than they have in the last several millennia. Some of the effects — such as the increases in average temperature and average sea level — will be felt worldwide and can be predicted today with confidence. Initially, some areas may even seem to benefit from these shifts. Other changes will be principally regional in character. The specific timing and severity of these regional impacts cannot now be predicted with confidence.

The impacts of climate change on sea level, weather-related disasters, fresh water resources, food production, and population, are described below. Chapters in this volume illustrate the physical impacts of climate change and explore the effects of these impacts on relations between nations. These include chapters by Richard A. Warrick and Atiq A. Rahman ("Future Sea Level Rise: Environmental and Socio-Political Considerations," Chapter 7), Martin L. Parry and M.S. Swaminathan ("Effects of Climate Change on Food Production," Chapter 8), Peter H. Gleick (Effects of Climate Change on Shared Fresh Water Resources," Chapter 9), James K. Mitchell and Neil J. Ericksen ("Effects of Climate Change on Weather-Related Disasters," Chapter 10), and Nathan Keyfitz ("The Effect of Changing Climate on Population," Chapter 11) .

The clearest and most widely discussed global impact of a greenhouse warming will be an increase in average sea level. Two processes will contribute to the rise in mean sea level. An atmospheric warming of several degrees would warm the upper layer of the ocean, causing it to expand in volume like the liquid mercury in a hospital thermometer. A greenhouse warming would also melt some of the snow and ice lodged in high mountain glaciers. This would increase the runoff to streams and rivers, with the resulting meltwater ultimately finding its way into the oceans. The combined effects of thermal expansion and the melting of mountain glaciers are projected to raise average global sea level by 20-100 cm (with a best guess of about 60 cm) during the next century. If, as some preliminary evidence suggests, snow accumulation increases in the Antarctic, however, the average sea-level rise will tend to be at the low end of the expected range.

The zones of greatest vulnerability to sea level rise are the flat, heavily populated alluvial deltas of the world's great river systems and the low-lying areas of many island states. Preliminary studies of the Nile Delta, for example, suggest that a sea level rise of about 100 cm could flood an area that now houses about 15% of Egypt's population and produces approximately the same proportion of the country's food.

However, as Richard Warrick and Atiq Rahman point out in Chapter 7, local effects of sea level rise will vary markedly. These effects will be determined less by the average global change and more by local factors that include the extent of subsidence or uplift along a specific stretch of coast, the condition and response of biotic systems in the affected areas, and the efforts of human societies to protect coastal structures.

The rapidity of regional climate changes, as well as their magnitude, will determine the extent of the damage that results. The expected rate of change due to the continuing buildup of greenhouse gases causes scientists to assess whether rapid climate change could disrupt the stability of national economies and natural ecosystems. Because regional climates will continue to change throughout the next century, humans and other species attempting to adapt to the new conditions will always be "shooting at a moving target," coping with climate conditions that never seem to settle into any permanent, stable, equilibrium condition.

Irrespective of the magnitude of the *average* global warming, temperature changes due to greenhouse gas buildup will be unevenly distributed around the globe. The areas at high latitudes — closer to the poles — are expected, for example, to experience a warming 2-3 times the global average. At low latitudes — closer to the equator — temperatures are expected to rise only 50-75% of the global average increase. Thus, a global warming due to greenhouse gas buildup would shrink the temperature gradient between the cold polar regions and the warm tropics — the natural temperature differential that fuels the thermodynamic engine of the global weather machine.

Climatologists believe that, if this temperature gradient shrinks, it could dramatically alter the patterns of air and ocean currents that determine regional climates. Global warming could cause the patterns of the jetstream in the atmosphere or warm currents in the oceans to shift, changing the climate dynamics that give each geographic location its ecological and cultural identity. For instance, if the Gulf Stream, that river of warm ocean water that now travels from the coast of Florida to the coast of Norway, were to move westward 200 km — away from the European continent — into the North Atlantic, we could see a world in which Europe got hotter, on average, but Great Britain got colder and wetter.

Not only would the average weather conditions be changed by global warming, but the frequency of large storms and extreme weather events could be altered as well. The evidence linking such effects to an enhanced greenhouse effect is inconclusive at present, with the exception of some model results that suggest a likely increase in the number of extreme rainstorms (Houghton, J.T., G.J. Jenkins and J.J. Ephraums, ed., 1990). In other words, no one has proved, and possibly no one *can* prove, that the extreme weather events so widely reported in the last few years have any direct relationship to global warming or greenhouse gases. But these storms and floods exemplify the kinds of events that may occur with increasing frequency in the decades ahead. And they give policy-makers, investors, and citizens a sense of how well our existing institutions are prepared to respond to even small effects of climate change.

Mitchell and Ericksen (Chapter 10) review the historical data on weather-related disasters. In the future as in the past, they argue, it is the poor — both within and among countries — who will experience the largest damages from weather-related disasters, as measured by the percentage of annual income lost in these events. Therefore, Mitchell and Ericksen conclude, those concerned with the impacts of climate change should now join forces with those traditionally concerned with disaster relief to insure that systematic management strategies are implemented. It makes sense to prepare for, and to minimize, future damages from such events — whether they result from "natural hazards" and random events, or are the consequence of a human-induced greenhouse warming.

Even if a greenhouse warming does not increase the frequency and severity of storms, it is likely to alter the timing, duration, and distribution of rain and snowfall. As Gleick (Chapter 9) observes, there is little reason for confidence in the ability of current models to predict the regional distribution of rainfall in a warmer world. For example, we cannot predict today which regions or river systems will experience water shortages most acutely. But some general conclusions can be drawn. Several models suggest that mid-continent, mid-latitude areas would be drier, especially in summer. The most reliable conclusion is simply that precipitation, runoff, and soil moisture — all critical variables in areas which depend on rain-fed agriculture — will be quite different in the future from what they are today. Internationally shared resources of clean, potable water will be stretched to cover larger irrigated areas and serve increasingly thirsty populations.

During a century when world population (and food demand) is expected to more than double, rapid changes in temperature and precipitation patterns could have some important negative effects on food production. Already 80% of the world's potential arable land has been broken open by the plough. Even if agricultural technology, enhanced by modern chemistry and biotechnology, achieves dramatic increases in yields per hectare, the amount and location of lands suitable for traditional agricultural practices may shrink (or at least change) dramatically, especially if current trends in greenhouse gas buildup continue. Parry and Swaminathan (Chapter 8) note that there will be gains in regional output as well as losses, although it is impossible to determine with certainty which areas will receive the benefits of changing climate and which will suffer the losses. Some types of plants will thrive while others suffer; and species other than commercial cultivars will be affected. Weeds and agricultural pests may profit more from the changes than will the local crop species. Perhaps the most disturbing effect on agricultural systems is the pattern of change itself: farmers will have to adapt to circumstances which could stay in continuous and unpredictable flux for several decades.

The problems of population growth and climate change are highly interactive. The more rapidly population increases, the more difficult it will be to deal with the effects of rapid climate change. All people will be affected to some extent; but in general, the greatest damages will most likely be visited on the poor — those with the fewest options for adapting quickly to altered climates. Among nations, the developing countries are particularly vulnerable to the impacts of climate change. This is especially so for those with broad, flat coastal plains, those which are economically dependent on agriculture, and those which are currently protected from open-ocean storms by coral reefs and will become less so if these reefs deteriorate. Within countries, the poorest of the poor — those with little or no land and often dependent on subsistence agriculture — may be seriously at risk.

As Keyfitz suggests (Chapter 11), greenhouse gas-related environmental damage may add immense new urgency to the already great pressure for migration of people from poor to wealthy nations. This urgency may stem from extreme weather events and sea level rise (which could make some inhabited areas uninhabitable), or from the alteration of traditional growing conditions. The advanced industrial societies will be buffered somewhat, especially against the early impacts of climate change, by their wealth and by the technological options that wealth provides. Ultimately, however, the problems of climate change (and their solutions) will transcend national boundaries. To the extent that climate change induces human migrations or food and resource shortages, all nations share the effects.

Thus, we see that the effects of a rapid climate change are likely to be widely distributed and strongly felt. If current trends continue, the impacts of rapid climate change will affect the distribution of natural species as well as national rates of economic growth. It is tempting to speculate which regions might be the winners and which the losers in this great game of weather roulette. But we do not know the timing, severity, and extent of any impacts on specific locales — including the distribution of changes in precipitation, agricultural output, and the frequency of extreme weather events. The impacts of these changes could increase intra-regional and inter-regional tensions, enhancing the existing prospects for conflicts between States. On balance, no areas can safely assume that they will necessarily be advantaged by climate change.

4 Are We Approaching a Catastrophic Climate Change?

It is fashionable to focus public and scientific attention on the effects of greenhouse gas buildup, as though climate change were the only (or at least the most important) environmental problem facing the world today. But the risks of rapid climate change do not exist in isolation, encapsulated into some closed compartment of the lower atmosphere. P.J. Crutzen and G.S. Golitsyn ("Linkages Between Global Warming, Ozone Depletion, Acid Deposition and Other Aspects of Global Environmental Change," Chapter 2) draw our attention to the linkages between global warming due to the greenhouse effect, stratospheric ozone depletion, acid deposition, and other aspects of global environmental change taking place today.

Like many processes in nature, the characteristics of climate — both global and regional — are the result of complex interactions between several closely coupled, non-linear systems. These systems include the dynamic, circulating fluxes of the ocean and atmosphere and the complex web of interacting species that make up the terrestrial and marine biota. Greenhouse gas buildup, ozone depletion, and acid deposition all occur in the same dilute, low-temperature reaction vessel — the atmosphere. The changes which ensue from each of these processes naturally and unavoidably affect each other.

Non-linearity in these systems means that the dimensions of an effect are not necessarily proportional to the size of the stimulus that changes the system. If a significant force — like the heat-trapping effect of increasing greenhouse emissions — is doubled, the effect on climate may not double. Rather, that force will combine with other factors, producing an effect that may appear small at first, until the force reaches a certain threshold — after which the effect may increase suddenly and dramatically. Although the feedback mechanisms which govern and couple these processes are not fully understood at this time, climate scientists now know that specific thresholds of non-linearity exist. Crossing those thresholds may transform the entire climate system into a new and quite different state. Several of the chapters in this volume, including Martin I. Hoffert ("Climate Sensitivity, Climate Feedbacks, and Policy Implications", Chapter 3), Hans Oeschger and Irving M. Mintzer ("Lessons from the Ice Cores: Rapid Climate Changes During the Last 160,000 Years," Chapter 4), and Michael B. McElroy ("Changes in Climates of the Past: Lessons for the Future," Chapter 5) explore aspects of these potential transitions, whose historic counterparts are evident in the long-term geologic record of past climate changes. We do not now know what, if anything, will push the global climate system across one of these thresholds into a very different climate regime. But by studying the historical and geological evidence of past changes, we learn which mechanisms *could* come into play in the future. Data from ice cores taken in Central Greenland and Antarctica suggest that climates long past may have shifted by as much as 5 °C in periods as short as a few centuries.

Today's human-induced changes in the climate system already represent stresses *equal in magnitude* to those associated with major glacial-interglacial transitions in the past. Some (as yet unspecified) combination of the feedback processes visible and active in Earth's geologic past could come together again to promote a rapid future climate change. This is, of course, much more likely if current trends in the emissions of greenhouse gases continue into the next century.

Is the Earth's climate already changing due to the buildup of greenhouse gases? T.M.L. Wigley, G.I. Pearman and P.M. Kelly ("Indices and Indicators of Climate Change: Issues of Detection, Validation, and Climate Sensitivity," Chapter 6) assess whether a statistically significant climate warming can be identified in the historical record of the last two centuries. Basing their analysis on the best available records of global temperature change, they find a continued pattern of variability from year to year. These variations take place within the larger pattern of a clear upward trend in temperature. But the size of the observed temperature rise is still statistically within the range of natural variability, as deduced from the temperature records of the last few centuries.

Therefore, we cannot prove nor disprove the hypothesis that the observed increase is due to the concurrent buildup of greenhouse gases. Wigley et al. conclude that we must continue to monitor global temperature, both on land and in the sea, in order to establish (with a high degree of statistical confidence) whether or not a global warming due to greenhouse gas buildup is indeed in progress. They suggest that it may be important to begin monitoring other variables as well, and to increase the geographic extent and quality of the temperature-reporting network, if we are to find the "fingerprint" of climate change in current observations.

The analysis of Wigley et al. leaves us with uncertainty about the severity of the crisis. But the continued buildup of greenhouse gases is a very risky business, with large and mostly negative impacts to be expected if rapid climate change occurs. As we continue to increase the atmospheric concentration of radiatively active trace gases, we are, in effect, haphazardly twisting the dials on the complex machine of our global climate. In that context, must we wait until there are demonstrable and painful damages all around us before we intervene to manage the risks of rapid climate change? It is argued here that the world need *not* wait for the worst damages to occur; we can hedge our bets, sustaining the prospects for economic development while limiting the risks of rapid climate change.

5 The Sustainable Energy Approach: Hedging our Bets on Climate Change

In every country of the world, economically important human activities lead to emissions of greenhouse gases. For some activities — like the cultivation of paddy rice in flooded soils — there is no practical alternative to the current methods which produce these emissions. But in other areas of human activity — ranging from the manufacture of industrial chemicals, metals and fertilizer, to the production and use of electricity, to the provision of passenger and freight transport — there is a clear potential for hedging our bets on climate change. By carefully selecting among technically feasible and cost-effective investment alternatives in light of their greenhouse gas emissions impacts, it is possible to reduce the risks of rapid climate change while promoting the prospects for sustained and equitable development. In this volume, we focus on energy because energy-related activities (including extraction, mobilization, and use of fuels for electricity, transportation, manufacturing, cooking and heat) contribute more than any other factors to the risks of rapid climate change. Essays by John Holdren ("The Energy Predicament in Perspective," Chapter 12), David Jhirad and Irving M. Mintzer ("Electricity: Technological Opportunities and Management Challenges to Achieving A Low-Emissions Future," Chapter 13) and Jayant Sathaye and Michael Walsh ("Transportation in Developing Nations: Managing the Institutional and Technological Transition to a Low-Emissions Future," Chapter 14) explore the range of cost-effective, emissions-reducing energy options.

The choices made today among technically feasible and cost-effective options for energy supply and use will significantly affect the rate of future emissions growth. Consider that carbon dioxide is the most important greenhouse gas emission — accounting, in itself, for about half of the annual global increase in the greenhouse effect. About 70-90% of the carbon dioxide emitted each year from human activities comes from energy use — specifically, from the combustion of fossil fuels. (Deforestation and land use changes account for most of the other 10-30%.) Energy use also contributes significantly to the buildup of other greenhouse gases, including methane, nitrous oxide and tropospheric ozone.

Future energy choices must be evaluated in the context of current patterns of energy use. As Holdren observes, the current global pattern of energy use is neither economically rational, ecologically sustainable, nor socially equitable. The rich industrialized countries, which represent about 25% of the world's population, consume about two-thirds of the primary energy. In the process, they release more than 50% of the total greenhouse gases. Industrial countries produce nearly 75% of the fossil-fuel derived emissions of CO_2, and almost 60% of total carbon dioxide emissions. These proportions may be about to change, however, because total energy use in developing countries is expected to grow rapidly in the decades ahead. Developing country populations are increasing rapidly, their end uses are shifting from biomass to commercial fossil fuels, pressure for electrification is increasing within them and domestic demand for expanded mobility through the use of motorized transport is growing. From a traditional development perspective, these trends are desirable and offer economic benefits. But in the context of global climate change, these trends represent an dangerous potential. Unless systematic steps are taken to control emissions growth in these countries — and, more importantly, in industrialized countries — the quantity of greenhouse gases emitted annually to the atmosphere could increase substantially.

In all countries, the pattern of energy use depends on two key factors: the choice of technologies and the efficiency with which these technologies are implemented. Choices among alternative energy investments made during the next decade — for example, in the electricity supply and transportation sectors — will affect the rates of growth for energy, greenhouse gas emissions and gross national product throughout the 21st Century. Some industrial countries (e.g. Japan and Germany) are reaping benefits while seeking to reduce greenhouse gas emissions from energy supply and use. By introducing efficiency improving and renewable energy technologies, these countries have increased their effectiveness as energy users and developed new technologies for export. The same process must be encouraged in other industrialized countries and could also take place in developing countries. Companies engaging in co-development of these new technologies with partners in developing countries are more likely to capture a significant share of many new and evolving markets. As Germany did in the 1950s and 1960s, and Japan did in the 1960s and 1970s, newly industrializing countries that encourage such partnerships may expect in the future to leapfrog beyond some of the castoff, outmoded and emissions-intensive technologies that were relied upon by countries that achieved industrialization in earlier periods.

Transportation is a major source of energy-related greenhouse gas emissions. Reducing the emissions from transportation, especially during a period of increasing demands for mobility, presents a difficult challenge. This challenge must be met first in industrial countries, but eventually in developing countries as well — where it will be especially difficult to limit emissions increases. Sathaye and Walsh (Chapter 14) analyse the difficulties and hopes for the transport sector, particularly concerning motor vehicles. They note that substantial opportunities exist for reducing the energy intensity of cars, trucks, buses, and airplanes. In addition, a number of new "tailpipe" technologies are emerging for decreasing emissions per unit of fuel consumed. However, while a combination of new vehicle designs, tailpipe controls, and traffic management schemes may be sufficient to stabilize vehicle-related emissions in some industrial countries, these measures will not be enough to keep greenhouse gas emissions from rising in the transport sector of the developing world. Exploding demand for motorized vehicles, increases in vehicle miles travelled per year, and worsening traffic congestion, will combine to raise emissions from the transport sector in the decades ahead.

Sathaye and Walsh outline a comprehensive strategy to keep the increases in greenhouse gas emissions to a minimum. If new and creative partnerships can be established that prevent developing countries from becoming a dumping ground for inefficient vehicles, their strategy will offer benefits to virtually all stakeholders. It increases the profit potential of the companies willing to develop the market, expands citizen mobility by offering them a less congested transport system and minimizes damage to the environment. But it requires a commitment to investment in transport infrastructure and urban planning — a commitment that may be hard to mobilize in the cash-strapped societies of the developing world.

The problem in the electricity sector is similar in character to the transport problem, but the structure of the solution is different. Jhirad and Mintzer (Chapter 13) survey the institutional challenges and the technological opportunities for minimizing the emissions of greenhouse gases from the electricity supply sector. In the industrial countries, electricity demand will increase in the commercial/industrial sector during the next two decades as more and more manufacturing processes are electrified. For example, demand will increase in the residential sector as the introduction of computers, telefax machines, stereos, and air conditioning equipment increase the "plug load" on the utility system. In developing countries, the components of load growth are driven both by the need to provide basic services and by the desire to support luxury appliances. Lighting loads will grow rapidly as rural villages are electrified. Urban residences, commercial buildings and industrial facilities will all experience growing demands for electricity. All of these factors suggest an increase in greenhouse gas emissions from electricity production.

Even without the global warming problem, meeting the projected increases in electricity demand in developing countries will be extremely difficult. Jhirad and Mintzer note that utility companies in these countries face a triple bind — declining technical and financial performance, limited access to external capital and increasingly stringent demands for environmental protection. Many technological options exist for improving the efficiency of electricity end-uses and reducing the rate of emissions per unit of electricity produced. These technologies include more efficient lighting and better electric motor drives on the end-use side, and solar, wind, biomass, fuel cells and advanced combustion systems on the electricity supply side of the equation. But a number of significant market failures and institutional obstacles limit the ability of suppliers and consumers to implement the most economically and technically efficient solutions.

Market-oriented measures and institutional reforms will be needed simultaneously, both to reduce the rate of emissions growth and to sustain the prospects for local economic development. Several kinds of institutions will be pivotal to the success of such a strategy in the electricity sector. The multi-lateral banks and other development assistance institutions have a key role. Only if these institutions can provide additional funds, particularly to invest in local capacity building, is there any hope for achieving the joint objectives of economic development and emissions control.

In addition, no effort by the banks and development assistance agencies will be sufficient without price and policy reforms. Utilities must be reoriented toward efficiency — both financial and technical. To accomplish this, it will be necessary to recover the full cost of electricity production and use. New blood, new policies, and new management strategies will be needed to overcome persistent market failures in the electricity sector. Utility management will play a vital role in a world threatened by rapid climate change.

However, reform of public and governmental institutions is not enough. Secondarily, and not unlike the situation in the transport sector, new management strategies will be required in the private sector. Small and major manufacturing corporations in the electric machinery supply industry, and producers of energy end-use devices, will compete for market shares in a large and rapidly growing market in the developing world. The multinational enterprises that can commit now to the joint development of advanced, environmentally sound technologies — through partnerships with enterprises in developing countries — will only be able to capture a substantial share of these rapidly expanding markets if external costs to the environment are internalized. Those companies which make it their business to be environmentally responsible as well as technologically advanced are likely to earn significant profits from these growing markets — re-establishing customer loyalty and, in the process, making a significant contribution to global stability. This, then, is one way that responsible businesses can hedge the bet on climate change. Corporations that choose to develop advanced, efficiency-improving, and emissions-reducing technologies will improve their long-run prospects for survival, whether or not the world is now approaching a radical change in climate.

6 Do the Risks of Rapid Climate Change Justify the Costs of Early Response?

Both new technologies and greater efficiency, while they require an up-front investment, are more cost-effective than the traditional means of providing energy. But to capture that competitive advantage requires a shift in accounting procedures, levelling the playing field by incorporating the environmental costs of energy into the price of fuels. This would realign the choices among technologies in a more ecologically and economically rational manner, promoting economic development while minimizing the risks of rapid climate change.

While all the remedies discussed so far will entail real and substantial costs, there are inherent macro-economic advantages for societies that implement them. Jochem and Hohmeyer suggest in Chapter 15 ("The Economics of Near-Term Reductions in Greenhouse Gases") what they call the "rational use of energy:" a process of selecting energy technologies to minimize the total costs of energy supply and use, considering all financial, environmental, and social costs. Even for a country such as Germany — already well advanced in its programme of industrialization, in which large investments have already been made to pick the easy fruits of energy conservation — there are substantial economic benefits to be gained from further investments in efficiency-improving technologies. Contrary to what some analysts in the US have suggested, these investments will promote economic development, increase domestic employment, improve the national balance of trade by encouraging high technology exports and environmental and social costs not now accounted for in traditional economic analyses.

Successful national strategies to make the pattern of energy use more economically rational can take a variety of forms. They might include measures to (1) "get the prices right" for electricity and fuels (through taxes or incentives that reward efficiency), (2) increase the flow of information about cost-effective energy options, (3) encourage investments in renewable energy technologies and efficiency-improving devices and (4) set high performance standards for widely used energy end-use devices. Rather than being a burden on the economy, Jochem and Hohmeyer conclude that these types of measures would stimulate a high-quality of economic development. As other writers (including Harvard Business School's "competitive advantage" pioneer, Michael Porter) have also suggested, these measures will also increase national competitiveness while enhancing the economic and technical efficiency of the domestic economy. This general group of strategies has been called the "No Regrets" approach, because it is composed of measures that will deliver demonstrable benefits even if rapid climate change does *not* take place.

Some sceptics disagree, charging that the results cited by Jochem and Hohmeyer cannot be generalized beyond the German case. These analysts argue for what has become known as the "Wait and See" approach. They urge governments and corporations to avoid the potential risks of early investments in advanced technologies. Too many uncertainties persist in climate science, they claim; regional impacts are too unpredictable; and we are too ignorant of future economic effects. Thus, the "Wait and See" analysts argue that any shift away from the current pattern of energy and economic growth would impose heavy economic costs on human societies. And, they say, some premature investments may not even generate significant environmental benefits, in part because of the unexpected effects of still-poorly-understood ecological linkages.

These two strategies — "No Regrets" and "Wait and See" — epitomize the debate over policy responses to climate change. But which offers a safer, more promising direction for the world's governments? In Chapter 16, R.K. Pachauri and Mala Damodaran (" 'Wait and See' versus 'No Regrets': Comparing the Costs of Economic Strategies") analyse the results from a variety of economic models to contrast the costs and potential benefits of both approaches. They start with a critique of several of the most widely publicized global economic models that have been recently applied to comparing these two strategies. Then, based on a careful review of the work of Nordhaus, Manne and Richels, and Peck and Teisberg, the authors conclude that under either the "Wait and See" or the "No Regrets" strategy, the costs of a greenhouse warming will not be evenly distributed. Although the precise distribution of these costs is uncertain, with the "Wait and See" strategy, the distribution of costs will be independent of the actions of individual stakeholders (including individual nations). On the other hand, under a "No Regrets" approach, the measures taken will offer benefits to each stakeholder group in direct proportion to its members' investments of time, money, effort, and technology.

When coupled with increased research to reduce the remaining uncertainties in the science of climate change, the "No Regrets" strategy offers substantial and immediate benefits. This strategy offers potentially affected parties a measure of control over their own destiny and a mechanism for hedging against uncertain future risks. For any stakeholder that implements it, this strategy increases the likelihood of capturing private gains as well as public benefits — whether the planet is about to face a major climate change or only a continuing period of high year-to-year variability.

But can these advantages be formalized, broadly distributed, and somehow made more permanent? A number of analysts have argued that cementing these gains for the long term requires substantial reform of existing institutions.

7 Institutional Challenges in a Warming World

Global warming due to the buildup of greenhouse gases is not an isolated problem. It is currently the most prominent, and over the long term probably the most significant, of a new class of environmental problems. These problems are closely linked to each other, and likely to appear with increasing frequency in the decades ahead. Unlike the predominantly local character of earlier environmental issues, these global problems share several common characteristics that will make it significantly more difficult for individuals, corporations, or governments to develop successful response strategies in the future.

Continuing scientific uncertainty will cloud our knowledge of the direct causes, feedback processes and regional effects of these problems. Long lag times will persist between the recognition of the symptoms of a problem, and the observation of any ameliorating effects that result from a successful response strategy. And those who benefit from the activities that contribute to the cumulative risks will remain unable to identify or compensate those who bear the costs of the damages. Dealing with these complexities will create new challenges for existing institutions. In some cases, these complexities have already begun to stimulate the growing perception of a need for new institutions.

Since 1972, when United Nations Conference on the Human Environment was held in Stockholm, Sweden, there have been many international meetings devoted to environmental issues. In the follow-up to the 1972 Stockholm Conference, the United Nations established the UN Environment Programme (UNEP) to coordinate international efforts to protect the global environment. Under UNEP's aegis, several major international treaties have been negotiated to protect the environment. The Vienna Convention (1985) and the subsequent Montreal Protocol on Substances that Deplete the Ozone Layer (1987) represent the first instances in which a collaborative effort of governments, international organizations, corporations, and non-governmental organizations has led to a negotiated agreement to reduce a major risk of environmental damage before the worst consequences have been realized.

Now, the legal precedents arising from those international agreements are about to be developed further. Kilaparti Ramakrishna and Oran R. Young ("International Organizations in a Warming World: Building a Global Climate Regime," Chapter 17) observe that the concurrent efforts of the Intergovernmental Panel on Climate Change and the Intergovernmental Negotiating Committee for a Framework Convention on Climate Change will eventually lead to a new international agreement to control the risk of greenhouse gas buildup.

Even if a framework convention is successfully agreed and signed, Ramakrishna and Young note that the larger challenge remains — to implement a flexible, cooperative, long-term regime to manage global emissions of greenhouse gases and to minimize the damages due to rapid climate change. Having studied the historical precedents and the mandates of existing institutions, they conclude that the dimensions of a global climate regime are so broad, the potential conflicts between stakeholders so complex, and the challenges of monitoring and enforcement so daunting, that no existing institution will be fully adaptable to the necessary task of implementation. Despite the inherent difficulties of forming and funding new institutions, the proposal by Ramakrishna and Young will cause many participants to rethink their positions as the negotiations draw closer to culmination.

Ramakrishna and Young, among others, offer a compelling argument that only a separate climate institution can maintain authority and jurisdiction, given the multidisciplinary challenges of climate change management. This carefully considered approach is a radical alternative to the present direction of negotiations. For example, it may prove controversial for the World Bank, which currently manages international climate finance through its Global Environmental Facility (GEF). The World Bank seeks an agreement that would vest all international environment funds under the control of the Bank's Board of Executive Directors, to be managed by the staff and the administering body of the GEF. Some may see conflicts between proposals for a new climate institution and the interests of the United Nations Development Programme, UNEP and the World Meteorological Organization. Such jurisdictional disagreements do not necessarily mean that the proposal will be stillborn, but it does suggest that there is likely to be a long and difficult labour before the birth.

Whatever institutional solution does emerge, it will have to take into account the growing influence of a critically important group of actors, often ignored but essential to success: non-governmental organizations (NGOs). In Chapter 18, Navroz K. Dubash and Michael Oppenheimer ("Modifying the Mandates of Existing Institutions: Non-Governmental Organizations") note that NGOs have been a key force in resolving environmental problems for the last several decades. Particularly since 1972, the technical competence and the political agendas of these organizations have broadened considerably. In industrialized countries, environmental NGOs have strengthened their base of scientific and technical competence while moving their horizons outward from local to international environmental problems. In developing countries, development-oriented NGOs are increasingly concerned with environmental problems. They recognize that their role is crucial to the evolution of balanced, equitable, and efficient strategies for sustainable development.

In the last decade, NGOs in the industrialized North and the developing South have sought to make common cause, each sensitizing their governments to the concerns of the other. NGOs have become a vital link between North and South, helping to keep development priorities on the agenda in international environmental negotiations. These organizations still face the challenge of finding ways to maintain attention simultaneously on environmental problems and development issues — both at the local and national level — while continuing to place questions of distributional equity on the agenda of international policy-making.

Another group of vitally important actors — multinational corporations — has been largely silent as the drama of climate negotiations has unfolded on the international stage. Their silence, however, does not imply that they are necessarily bit players. Corporations involved in energy supply and distribution, chemicals and metals manufacturing, packaging and paper production, and transportation will make investment decisions during the next decade that are likely to shape the trajectory of greenhouse gas emissions far into the next century. While often listening attentively to the current dialogue, many senior executives in these enterprises have been content to observe, rather than participate fully in the negotiations process. Only a few courageous corporate leaders — including the members of the Business Council for Sustainable Development and the International Environmental Bureau of the International Chamber of Commerce — have sought to shape their own investment decisions in light of the international concerns about the risks of rapid climate change.

Peter Schwartz, Napier Collyns, Ken Hamik and Joseph Henri ("Modifying the Mandates of Existing Institutions: Corporations," Chapter 19) analyse the emerging international movement toward corporate environmentalism, finding something quite unexpected. Corporate environmentalism is *not* an altruistic response to corporate guilt, a public relations ploy, or the pet project of a few radical industrialists. Instead, corporate environmentalism is a discipline for adding administrative and financial value to a company. It also reflects the steely-eyed realization by responsible executives that the enterprises they lead will be dramatically affected by the policy decisions made in these international negotiations. They know that it is not just the weather that is changing, but the international business climate.

Schwartz et al., outline a market-driven, long-run programme that can help a firm evolve through these turbulent times, achieving sustainable management of their own resources, as well as greenhouse gas emissions. This programme does not require a stream of charitable contributions to NGOs; it is based instead on finding a sound strategy for hedging against the risks of an uncertain future, sharpening the competitive edge of the firm in its principal markets, developing customer loyalty, and improving the long-run profitability of the firm.

As a postscript to the argument outlined in Chapter 19, Art Kleiner ("The Lesson of Continuous Improvement") observes a number of important synergisms between corporate environmentalism and the statistically oriented management theories of "the quality movement." Leading Japanese, American, and European corporations have captured substantial economic benefits — including increased market share and enhanced customer loyalty — through such customer- and community-oriented practices as continuous improvement and waste minimization. Kleiner observes that the precepts of the quality movement reinforce the central tenets of aggressive, forward-looking corporate environmentalism: rigorous, periodic self-examination; the development and refinement of new technologies; and the commitment to structural change in the production process that systematically reduces the number of product defects and makes the repetition of mistakes more difficult.

As Schwartz et al. conclude, those firms willing to confront uncertainty and make a commitment to quality develop their own improved leadership. By carrying this capacity for leadership into the negotiations of an international agreement on climate change, responsible corporate leaders can help shape a more practical and successful convention.

Another set of actors cannot be ignored in the process of reforming climate-change-oriented institutions. These are the international trade and tariff negotiators, whose agreements establish the channels through which greenhouse gas-reducing technologies are shared. Konrad von Moltke ("International Trade, Technology Transfer, and Climate Change," Chapter 20) explores the linkages, contradictions, and synergisms between the two separate sets of negotiations — over environment and trade, respectively — under way this year. The current round of negotiations on the General Agreement on Tariffs and Trade (GATT) — known as the Uruguay Round — is often considered unconnected to the climate talks. However, the GATT negotiations will be a key forum for international debate — not just on trade, but also including several issues of central importance to the climate negotiations. Decisions made by governments in the GATT forum will affect both the investment options available to corporations and the opportunities for technology transfer in support of sustainable development programmes. Key issues include the treatment of intellectual property rights, specialized subsidies, and non-tariff trade barriers.

It is possible, concludes von Moltke, to achieve the separate objectives of both environmental and trade negotiations simultaneously. But this solution requires supplementing the current international trade goal of economic efficiency with an additional criterion — efficiency in the use of resources, especially those that are key to the risks of climate change. Conversely, if attention is not paid to the implicit linkages between economic and resource efficiencies, attempts to achieve the goals of trade policy could undermine international efforts to protect the environment. Capturing the complementary benefits requires what some will no doubt view as a heretical change in the calculus of economic growth. It will be necessary to identify and apply consistent methods for internalizing the environmental costs of energy use, so that they are systematically incorporated into the price of fuels. It is also vital to rationalize the management of state-mandated subsidies.

And even this may not be enough. Von Moltke suggests making structural changes in the interface between trade and environmental regimes. Specialized institutions for dispute mediation will be necessary in each domain, along with cooperative cross-jurisdictional approaches that maintain the flexibility to handle each individual case within the regime that is most central to its substance. For example, something as simple as the dissemination of a new refrigerator using ozone-friendly, CFC-substitutes may require agreements among tariff, patent, and environmental protection authorities. If this type of institutional evolution is feasible, it may be possible to protect the environment as well as intellectual property rights, to facilitate technology transfer in addition to free trade, and to promote equitable and sustainable growth.

8 Crafting a Fair Bargain

Several nagging questions remain as the world moves slowly toward a negotiated international agreement limiting the risks of rapid climate change and minimizing unavoidable damages. What is a fair bargain among all the affected stakeholders in both industrial and developing nations? How will responsibility for current and past contributions to future risks be assigned? How should the targets for future reductions be set? How will the rights to emit greenhouse gases into the shared atmosphere, and the burden of costs from the damages caused by those gases, be allocated henceforth?

Michael Grubb, James Sebenius, Antonio Magalhaes and Susan Subak ("Sharing the Burden," Chapter 21) offer a sound and rational basis for addressing these questions. Beginning with "the facts" about current and past emissions, noting the weaknesses and limitations in the existing data, they compare and evaluate a range of rationales for assigning responsibility and allocating emissions rights. Their analysis brings the unspoken biases inherent in different accounting systems to light, and it illuminates the effects that each bias would have on the calculation of who pays and who receives compensation.

The recent negotiations under the INC demonstrate that there is no magic formula or single plan that is guaranteed to please all the parties. Grubb et al. do not promise any quick fix; like the rest of us, they expect the negotiations to be protracted, highly political and difficult. But these authors suggest an accounting system with the potential to be both politically practical, technically feasible and fair. Their proposal involves a flexible, evolving mixture of allocation criteria for future emissions rights, based on a combination of population and current emissions. Such a system — incorporating tradeable emissions permits that would expire after a specified period — could offer an incentive for all countries to limit future emissions growth. It would also stimulate sufficient resource transfers to enable developing countries to adopt more advanced and efficient technologies than they otherwise might, and it would provide a basis for reconsideration and revision of the overall scheme as new information emerged.

William A. Nitze, Alan S. Miller and Peter H. Sand ("Shaping Institutions to Build New Partnerships: Lessons from the Past and a Vision for the Future," Chapter 23) have a clear vision of the sort of cooperative process through which allocation proposals (and other climate-related proposals) could best be considered. Nitze et al. view the Conference of the Parties to a Climate Convention as a vehicle for strengthening existing institutions and, as circumstances require, establishing new ones. Deliberations might first take the form of a convention with short-term goals for stabilizing national emissions of greenhouse gases; but from there, a comprehensive protocol could evolve for reducing global emissions over the long term. Noting that achievement of this goal will require national strategies and action plans, they encourage the involvement of non-governmental organizations at the regional, national, and international levels. An inclusive, evolutionary process would inherently tend to encourage the goals set forth in other chapters of this book: stimulating cost-effective investments in energy efficiency, catalysing large-scale resource transfers from North to South, reducing the amount of waste and materials expended in producing material well-being and improving the performance of existing institutions. This all sounds very well but how are we to get from here to there?

One of the most important steps is to get a better shared understanding of the views held by each of the key stakeholders in the negotiations. Much has already been written and said about the perceptions of the climate problem in the advanced market economies of the OECD. But there is little general understanding of the point of view held by those leaders and thinkers of the developing world who have considered the climate problem in detail.

Tariq Osman Hyder ("Climate Negotiations: the North-South Perspective," Chapter 22) places the current climate negotiations plainly in the context of a more general North-South dialogue. As a civil servant in Pakistan, a spokesperson for the Group of 77 (G-77) in the management meetings of the Global Environmental Facility, and as his country's spokesperson in the Intergovernmental Negotiating Committee for a Framework Convention on Climate Change, Hyder is uniquely qualified to provide this clear and provocative point of view.

As Hyder observes, the current negotiations principally involve three key "players," each with distinct interests: the United States, the rest of the industrialized countries, and the developing world. Despite occasional differences of tactical approach, the first two share a common long-term interest quite different from the interests of the developing countries. Hyder notes that the current global recession — a backdrop for the climate talks — puts the industrialized countries in the driver's seat for these and other global negotiations. Pressing its advantage, the North has demanded various concessions from the South in exchange for offers of "aid" and technology transfer. The pattern continues in the climate talks.

The South, for its part, seeks to give highest priority to questions of long-term economic development. In this context, Hyder urges against attacking the symptoms of climate change, or jumping on the climate change bandwagon because it is the most fashionable problem of the day. Rather, it is preferable that international negotiations focus on the underlying economic causes of the current environmental situation.

Hyder appreciates why environmental issues in general — and the risks of rapid climate change in particular — have taken an important place on the policy agenda of the North. But he urges us to recognize the issues that take precedence in the South: stable and assured access to world markets with fair terms of trade, opportunities to attract new capital for necessary investments, integration of scientific and technical advances into national development plans, and additional finances to promote the shared goal of sustainable development. Just as the neglect of environmental concerns leads to economic problems, the neglect of the economic concerns of developing countries will lead to environmental damage — as a consequence of poverty, population increase, waste, warfare, careless use of resources, and the adoption of inferior technologies. The interrelatedness of the world no longer allows countries in one part of the globe to profit at the expense of those elsewhere.

Although all of the objectives of developing countries cannot be achieved through the climate negotiations, Hyder suggests that these talks may be the right vehicle for establishing several principles of mutual respect that could be the basis for all successful future negotiations. These principles include: equity, sovereignty, the right to development, the need for sustainable development, recognition of the special circumstances of developing countries, and a commitment to international cooperation. If continued and sustained goodwill allows for agreement on these basic principles, the prospects for achieving a successful, effective, and practical climate convention are very good.

9 What's Missing in this Picture?

As this discussion suggests, the risks of rapid global climate change in our world — like the risks of global war in an earlier era — are too complex and too important to be left to the specialists. The causes and the impacts of global climate change touch every aspect of human society and every natural ecosystem from the poles to the equator. No country or species is guaranteed immunity from the effects. No region or group can be safely assured of escaping future damages.

The risks of unrestrained buildup of greenhouse gases are well-documented in the works of the IPCC and in the reports of many national investigatory commissions. But the potential benefits of climate change are less well-understood. More must be learned about these. Perhaps the most important missing element in our picture of a world transformed by greenhouse gas buildup is the recognition that this critical threat to human societies can also be perceived as an incentive for bold cooperation in efforts to build an equitable and sustainable pattern of economic development. Pursuing the "No Regrets" Strategy could transform international relations. We might leave behind the obsolete Cold-War perspective of a struggle to the death among lifelong enemies, and gradually adopt a 21st century vision of mutually-reinforcing economic competition in a world with many centres of power. But if this new world is to survive, competition must be guided both by a shared sense of compassion and fairness toward the poor and a sustained concern for the quality of the environment.

If you come away from this volume with one understanding, it should be of the importance of linkages: the interconnections between the risks of rapid climate change and so many other problems of central concern to national governments, corporate leaders, non-governmental organizations, and individual citizens. Climate change is inexorably linked to ozone depletion, acid deposition and urban pollution, deforestation and loss of biological diversity, and desertification. It is intimately tied to the most vital economic undertakings of our time — energy production and use, transportation, agriculture, forestry, building construction, industry and manufacturing, and so on. It affects (and is affected by) the fundamental concerns of human society: population growth, urban density and planning, management (and mismanagement) of institutions, and the quality of life for individuals and families. It is no accident that environmental matters, and particularly global warming, have captured the attention of schoolchildren around the planet. What we decide now is shaping their world.

Finally, consideration of climate change is necessarily involved in the great policy negotiations of our time — not only through new environmental treaties, but also through trade negotiations, water rights disputes, and international security debates. The implementation of a framework convention on climate change could affect all these issues.

Certainly, dealing with the risks of rapid climate change would be easier if the risks themselves were well characterized and carefully quantified. Today, neither is completely possible. Future rates of emissions growth can only be estimated. The regional impacts of climate change can not be predicted with confidence. The best efforts of the international scientific community — embodied in billions of dollars of research — will leave major uncertainties unresolved for decades to come. No reasonable amount of additional research support would be enough to resolve all these uncertainties before the end of the century. We can not know how rapidly these conditions will change in the future, or how long the processes of change will continue. We cannot know if there will be some stable future equilibrium climate, or if the planet will oscillate rapidly between widely different, temporarily stable (i.e. meta-stable) climate states — as the ice-core evidence suggests it did in the geologic past.

Heretofore, scientific uncertainties have allowed governments to assume that climate change dangers were unimportant, exaggerated or premature. Such nonchalance can no longer be maintained without sobering political and environmental consequences. While the effects cannot be definitively tied to greenhouse gas emissions, the circumstantial evidence is great and the risks of further danger are severe. How then are governments, corporations, and individuals to respond? Science offers us no quick fix or magic solution. We face instead a problem of risk management. How societies deal with risk is a political and a financial problem, more than it is an engineering question.

Our purpose in bringing together the opinions represented in this volume is to broaden the discussion of global climate change and link it firmly with current international discussions of politics and investment. We seek, in particular, to draw into the debate — and into the international climate negotiations — two groups who are able to be highly effective in the process now under way. These two groups are: (1) senior officials in the economics, finance, trade and development ministries of developing countries; and (2) senior, responsible business executives in multinational corporations — especially those managers who contemplate making major investments in long-lived facilities that will release significant quantities of greenhouse gas emissions during the coming decades.

In highlighting the linkages, the uncertainties, the geopolitical implications, and equity considerations integral to consideration of the climate problem, we seek not to overwhelm or frighten these groups (or others) by exaggerating the risks, but to challenge them by emphasizing the complexities and the linkages to other issues of concern today. The challenge we offer to everybody is simple and stark: get into the game now. Contribute to the process in an active way, before a new bargain is struck. If you do not participate with an open mind and full spirit, you may expect a regime to emerge that is first flaccid, then draconian, cementing inequities in global trade. It may constrain economic development first in the South and then in the North, and obstruct technology transfer in both directions. In the end, you will have to learn to live with an outcome that will probably be flawed, cumbersome, unbalanced, bureaucratic and impractical.

If, and only if, broad-based participation by all stakeholder groups can be stimulated quickly, are we likely to achieve a systematic, comprehensive, balanced, and pragmatic response strategy that promotes the prospects for sustainable development, promotes equity, and minimizes the irreversible environmental damages of abrupt climate change. We offer this volume as a spur to that increased participation.

In conclusion, we want to emphasize that the international process of managing the risks of rapid climate change is not just an exercise in damage control. It offers an important — and in some ways unique — opportunity: to use the threat of global environmental change as a vehicle for expanding international cooperation — on scientific as well as trade issues — and as an incentive for the development of the advanced, more efficient, and less polluting technologies that can propel humankind forward into the 21st Century. If the human race embraces the challenges which this opportunity presents — enthusiastically, energetically and with good courage — then we may truly be on the path to a sustainable world.

References

Houghton, J.T., G.J. Jenkins and J.J. Ephraums, 1990: *Climate Change: the IPCC Scientific Assessment*, Cambridge University Press, Cambridge, UK.

IPCC (Intergovernmental Panel on Climate Change), 1990: *Assessment Report of the Intergovernmental Panel on Climate Change*, World Meteorological Organization/United Nations Environment Programme, Geneva, Switzerland.

IPCC (Intergovernmental Panel on Climate Change), 1992: *Update and Assessment Report of the Intergovernmental Panel on Climate Change*, World Meteorological Organization/United Nations Environment Programme, Geneva, Switzerland.

A Thumbnail Sketch of this Book

The chapters which comprise this volume are divided into five groups.

The first group highlights the uncertainties in the current scientific understanding of the climate system, and the implications of these uncertainties for the sense of political urgency surrounding the climate problem. These chapters explore the feedbacks and linkages between the ocean, atmosphere, and terrestrial biota, whose actions could push the overall system from a period of slow and gradual climate change into a period of accelerated, abrupt and unpredictable reactions to the stress of an altered atmospheric balance. Several chapters focus on the lessons that can be learned about the mechanisms of rapid future climate change from the study of natural archives of past climate change — including air trapped in ice cores, fossil remains buried in seabed sediments and other paleoclimatic data. The section concludes with a review of what we can deduce about current climate change from the historical temperature record, and what we might learn about future rates of change from monitoring other sorts of indicators.

The second group of chapters reviews the impacts of rapid climate change and their geopolitical implications. Rapid climate change may exacerbate the intra- or inter-regional tensions caused by other problems: sea level rise, food shortage, dwindling fresh-water resources, and weather-related disasters. The last chapter in this section explores the relationships between climate change and population growth in the next century.

The third set of chapters focuses attention on one key contributor to greenhouse gas emissions — the supply and use of energy. The first chapter in this section sets the global energy problem in perspective. The remaining chapters highlight the technological opportunities and the institutional challenges to achieving a low-emissions future, especially in the two sectors where the demand for fuel is sure to grow consistently during the decades ahead — transportation and electricity.

The fourth section focuses on economic questions and institutional issues. It begins with a chapter that explores the economic impacts of a strategy to achieve near-term emissions reductions in an advanced industrial economy — one which has already achieved substantial improvements in efficiency during the last two decades. The next chapter analyses the challenges of comparing two very different policy response strategies — "No Regrets" vs. "Wait and See." A third chapter explores the linkages between negotiations on global trade, climate change and technology transfer. Next comes a trio of chapters which all address the same question — but about different subjects. Each asks: How can existing institutions evolve to meet most successfully the special challenges of problems like the risks of rapid climate change? The first asks this question about international agencies, the second about non-governmental organizations and the third about corporations.

The last section of this volume addresses the question of how to get an equitable bargain from the current negotiations. The first chapter reviews various systems for assigning responsibility for past emissions and allocating the rights to future emissions. The second chapter highlights the principal obstacles to reaching a cooperative solution, from the point of view of the developing countries, and outlines a set of principles that could be the basis for overcoming these obstacles. The final chapter proposes a flexible approach to the remaining negotiations that could lead the way to a cooperative resolution in the near-term debate.

CHAPTER 2

Linkages between Global Warming, Ozone Depletion, Acid Deposition and Other Aspects of Global Environmental Change

Paul J. Crutzen and Georgii S. Golitsyn

Editor's Introduction

Environmental problems do not occur in isolation from each other; in fact there is often an ironically cruel symmetry about them. For example, carbon dioxide (CO_2) is the indispensable feedstock for life on Earth and the amount of CO_2 in the atmosphere regulates the process known as "the greenhouse effect." This effect is a fundamental geophysical mechanism that warms the lower atmosphere and thus allows life to exist on the surface. The buildup of CO_2 could thus be beneficial, stimulating plant growth. However, the very rapid pace of growth of atmospheric CO_2 (and other greenhouse gases), occurring at present by many human activities, are expected to upset the stability of traditional climates and dramatically alter the conditions that support already strained human societies and natural ecosystems.

Environmental problems can not be separated from human activities. Chlorofluorocarbons (CFCs) are an entirely man-made family of industrial gases. For more than three decades, these inexpensive and versatile compounds — which are neither toxic, nor corrosive, nor explosive — were touted as the "miracle drugs" of the chemical industry. Now we know that they are an important cause of stratospheric ozone depletion and also contribute to the greenhouse effect. As a consequence of our expanding knowledge of their environmental effects, the traditional (and most dangerous) varieties of these compounds are the focus of an internationally agreed phaseout programme.

And finally, environmental activities are inextricably woven into a complex tapestry of interactive atmospheric and biological mechananisms. These natural feedback processes have no regard for international conventions, legal boundaries, or political ideologies. When the oxidation products of exhaust gases from fossil fuel combustion mix with water in the lower atmosphere, acid fallout results—depositing acidic salts and liquids at locations hundreds or even thousands of kilometres from the site of fuel combustion. This acidic deposition interacts with other pollutants in air, water, and soil to damage trees and aquatic ecosystems. If acid fallout reduces the ability of forests to carry out photosynthesis or, similarly, if ozone depletion leads to the death of marine phytoplankton that perform photosynthesis in the ocean, the ability of the biota to remove carbon dioxide from the air is reduced and global warming accelerated by the enhanced greenhouse effect. Because these sorts of complex linkages and feedback mechanisms are so vital to our understanding of global environmental change, we open this book with a guide to understanding them. For Paul Crutzen and Georgii Golitsyn, individual environmental issues are "symptoms" of deeper problems. Often these problems are linked directly to human activities and national policies. There are many cases where two seemingly unrelated pollution problems, or several widely separated disruptions of natural ecosystems, can be traced to a single economically important activity — such as fossil fuel use or deforestation. Most important, environmental problems are often closely linked by the effects of narrowly (mis)conceived policies which, while trying to make one symptom less severe, often make several problems worse.

Professor Paul Crutzen, a Dutch atmospheric chemist directing the Air Chemistry Laboratory of the Max Planck Institute in Mainz, Germany, and Academician Georgii Golitsyn, an atmospheric physicist directing the Institute of Atmospheric Physics of the Russian Academy of Sciences in Moscow, have used scientific cooperation to create a benevolent linkage of their own, bridging one of the major political gaps of the Twentieth Century—that between Germany and Russia. Their collaboration on this chapter highlights the technical complexities of atmospheric problems and the challenges inherent in the search for comprehensive policy solutions.

- I.M.M.

1 The Context of Global Environmental Change

Evidence for environmental changes can be seen everywhere. We are all too familiar with it. Fewer birds and frogs can be heard in our forests and wetlands, and fewer butterflies seen in our meadows than only a few decades ago. Landscapes change as forests are cut for timber or to make way for agricultural fields; orchards are cleared for expanding industries and cities. Tropical forests are cleared for pastures which become wastelands due to overgrazing. Examples of such changes vary from place to place but have much in common throughout the world. The effects of these changes can be seen on local, regional and global scales — they affect human health, the functioning of ecosystems, atmospheric chemical composition, stratospheric ozone and the Earth's climate. But the causes of these changes, at root, are only two: the growth of human populations, and the careless growth of technology.

In this context, the task of the scientist is to observe, discover and anticipate these disturbances, to quantify them, to identify their causes, and increasingly to present policy makers with options for the containment and reversal of serious environmental damage. This is particularly challenging because of the range of problems which now involve two or more nations, or, in some cases all nations of the world. No human-decreed boundaries are observed by tropospheric ozone, sulphuric and nitric acid (the ingredients of acid rain), or the reactive chlorine and nitrogen oxide radicals that destroy stratospheric ozone. In these cases the medium of transfer is often the atmosphere. The atmosphere is, however, not only a medium of transfer; chemical reactions take place there as the pollutants combine, often involving solar radiation as the energy source. Those reactions become, in effect, secondary sources of pollution, which have potentially even more environmental significance than the direct emissions.

1.1 The importance of linkages

Many human actions have not just one but several environmental consequences. For instance, burning of fossil fuel produces not only a major greenhouse gas (carbon dioxide), but also nitrogen oxides and sulphur dioxide (especially, coal and many sorts of oil); and it introduces metals into the atmosphere. Hence, acid rains and intoxication of soils and water. The chlorofluorocarbons not only destroy ozone in the stratosphere but are powerful greenhouse gases (on a per-molecule basis). These kind of examples are well known, but now more complex interconnections are surfacing.

We will describe several of these atmospheric chemical combinations in this article; they are examples of the important, but often unexpected *linkages* between environmental maladies. We use the word "linkages" to emphasize the fact that seemingly unrelated human practices actually have important chemical or physical relationships. This adds a level of complexity to environmental protection efforts. It means that measures meant to solve one problem can unexpectedly exacerbate another; well-intentioned policies can produce the opposite of their desired results. The following four examples demonstrate the prevalence of the "linkage effect":

- **The high chimney solution:** After the Second World War, to alleviate severe local air pollution problems from industrial emissions, high chimneys were built in many European and American cities. The consequence was, however, that the pollutants — mainly sulphur dioxide (SO_2), nitric oxides (NO_x), particulates, and heavy metals — were transported over much longer distances, changing local problems into regional and trans-boundary problems. In some cases, this allowed the emissions to reach new ecosystems which were insufficiently buffered against them.

- **The reductions paradox:** Today, to counteract the regional buildup of tropospheric ozone and alleviate the acid rain problem, environmental officials often call for reductions in emissions of nitric oxide (NO) and sulphur dioxide (SO_2). However, reducing NO will also lead (as we shall see shortly) to reduced hydroxyl (OH) concentrations in the atmosphere. Because hydroxyl reacts with most atmospheric gases, both of natural and anthropogenic origin, the lowering of OH concentrations will lead to higher concentrations of many trace gases, the greenhouse gas methane (CH_4) being an important example. Reducing SO_2 will reduce sulphate particle concentrations. The positive side-effects of this are healthier air and a clearer view through the atmosphere (better visibility); but on the other hand, it will also lead to less reflection ("backscattering") of solar radiation to space, thus removing the "masking effect" of these pollutant releases and increasing the warming of the lower atmosphere (Charlson et al., 1990 and Wigley, 1991). Because sulphate particles are also probably the main source of cloud condensation nuclei, lower SO_2 emissions may lead to fewer cloud droplets, causing still less backscattering of solar radiation to space, and maybe reduced general cloudiness as well — which itself may mean greater climate change.[1]

[1] Although the SO_2-climate feedback is still speculative and in need of quantification, the depicted chain of events appears quite feasible (IPCC, 1990).

- **The hydrocarbon conundrum:** To abate photochemical smog formation, American environmental planners sought ways to reduce emissions of reactive hydrocarbons (mostly from auto and smokestack emissions) in the urban areas of the US. Despite this effort, in many cases, especiallyn rural areas, no significant reductions were observed in ozone concentrations (McKeen et al., 1991). We now believe these policies had underestimated the importance of the fact that vegetation also emits hydrocarbons naturally. Current policies thus lean more toward reducing NO (nitric oxide) emissions. Although small when compared to those of hydrocarbons, NO and NO_2 are most critical as catalysts in the production of smog ozone. As a byproduct, however, the new policy will also lead to lower concentrations of hydroxyl radicals, and thus increase the concentrations and the lifetime of methane in the atmosphere.

- **The methylchloroform surprise:** Prior to 1970, until it was discovered that they contribute to photochemical smog formation and lead to carcinogenic products, tri- and tetrachloroethylene (C_2HCl_3 and C_2Cl_4) were extensively used for dry-cleaning purposes. Subsequently these chemicals were phased out. As a replacement product, the much less reactive compound trichloromethane (CH_3CCl_3, more commonly called methylchloroform), was introduced into the market. Unexpectedly, a substantial fraction of the methylchloroform emissions reaches the stratosphere, contributing to ozone depletion. The compound is now included in the list of ozone-depleting substances to be phased out by the year 2000 under the London Amendment to the Montreal Protocol on Substances that Deplete the Ozone Layer. When methylchloroform was first introduced, the importance of OH radicals in cleaning the atmosphere had not even been proposed. Unknowingly, a significant local pollution problem was converted into a global hazard.

1.2 The need for comprehensive solutions

Thus, simple acts do not produce simple effects — either in creating problems or trying to fix them. However, the moral lesson of linkages is of course not that we should abandon environmental protection efforts in despair; but that we should act with as much attention as possible to the interrelationships and dynamic processes of the total environment. Thus, we plead in this chapter for comprehensive approaches, and against piecemeal "solutions," which too often lead to the substitution of one problem by another, possibly more severe one. We can also warn of some possible pitfalls, and we emphasize the need for vigorous research on critical environmental processes. Much of the scientific discussion that follows is based on an excellent review by the Intergovernmental Panel on Climate Change (IPCC, 1990).

2 Global Warming and its Environmental Effects

2.1 Temperature changes

During the 1980s, the painstaking work of reconstructing mean global temperature changes, from the records of local weather observations, was more-or-less completed. As a result (IPCC, 1990), we now know that since 1860, the global mean surface air temperature has risen by about 0.5 °C. This rise in temperatures has not taken place at a constant rate. Noticeable temperature increases occurred between 1910 and 1940, and since the early 1970s. (See Figure 1 for a chart of the changes.) In the years between, temperatures did not

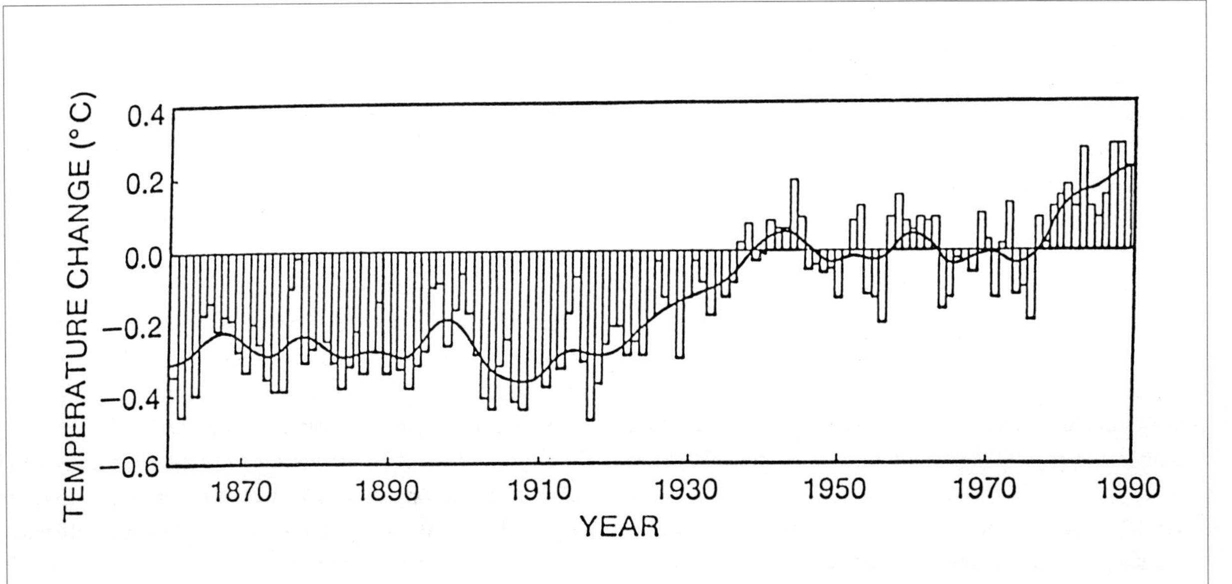

Figure 1. Global-mean combined land-air and sea surface temperatures from 1861 through1989, plotted relative to the average (0.0) for the years 1951 through 1980. Note that the rise in temperatures has not taken place at a consistent rate; noticeable increases occurred between 1910 and 1940, and since the early 1970s. Source: IPCC, 1990.

change much; but the 1980s have produced the six warmest years on record, and 1990 was far (about 0.2 °C) warmer than the three previous warmest years (1981, 1988 and 1989).

Based on paleoclimatic information from warmer epochs and climate models, researchers have concluded that the expected warming will eventually be much more marked at higher latitudes and during winter. Although temperature records in Siberia may appear consistent with such a change, records in North America do not agree with it. Furthermore, the transition to a new climate state may initially not reflect the expected changes, due to the inertia of the climate system — in particular the ocean circulation and cryospheric processes. Climate researchers also agree that the observed increase of mean global surface temperature is caused, at least partly, by the observed increase in concentrations of the greenhouse gases in particular: carbon dioxide (CO_2), methane (CH_4), nitrous oxide (N_2O), the chlorofluorocarbon gases ($CFCl_3$ and CF_2Cl_2), and low-stratospheric and tropospheric ozone (whose effect on global warming is rather complex, as we will discuss later in this article).

It is clear that, without countermeasures, the observed changes in the composition of the atmosphere as a result of man's activity will continue, with adverse direct effects on the biosphere and substantial risks.[2]

Even under the most favourable of assumptions, the global average temperature increase estimated by the Intergovernmental Panel for Climate Change (IPCC, 1990) for the next century is at least 1 °C. That may seem rather minor but, in fact, the rate is much faster than the most rapid previous rate of warming in the present geological era — which has been deduced at about 5 °C during the few thousand year transition between the last glacial and the present interglacial periods (NAS, 1983). The maximum temperature increase which is considered credible for the next century in the IPCC scenarios, 5.5 °C, would put the world into climate conditions approaching those that may have existed during the Cretaceous, some hundred million years ago, when the continent-ocean distribution was much different from the present. Thus, there is a significant risk for creating, within only a few generations, "another world" to which many ecosystems and a vastly expanded human population may not be able to adapt.

2.2 Precipitation and ocean levels

Often precipitation is of even greater importance than temperature in its impact, especially on agriculture. Assembling statistics on this is very difficult because precipitation varies even more than temperatures, both in time and space. The overall effect on soil moisture content and thus plant productivity in space and time is of particular importance.

A third important, large scale factor is the increase in the level of the world oceans (as discussed in Warrick and Rahman, 1992, this volume). Reconstruction of the historic record for the last 100 years shows that the mean sea level has increased by about 15 cm (IPCC, 1990). Only about half of this can be explained by the melting of mountain glaciers, and the thermal expansion of water in the warmer oceans, indicating uncertainties in present scientific knowledge. Estimates for the middle of the next century predict a further sea level rise by 20 to 70 cm, depending on which climate scenario and model is used (IPCC, 1990).

For many low countries, especially in South Asia, this constitutes an increasing threat with potentially disastrous consequences, tragically demonstrated by the many hundred thousands of deaths resulting from the April 1991 storm surge and flood in Bangladesh. Although not proven to be connected to climate change, the threat for a greater frequency of such events may increase due to sea level rise and the expected, more frequent tropical cyclones resulting from increases in sea surface temperatures.

3 The Role of Feedback Mechanisms

We can best understand the interrelationships between these factors through their many positive and negative feedbacks. A feedback mechanism is a process in which a force that affects a system is itself made stronger or weaker by the reaction of that system. In this chapter, we use the concept of feedbacks as aids in building a growing understanding of how a particular action may beneficially, or adversely, affect the entire system.

3.1 Positive feedback: effects of warming on snow and ice

For example, consider how temperature change affects regions covered in snow and ice. Warmer temperatures melt the snow and ice cover, which means that the darker land surface underneath is revealed; this darker surface absorbs more solar energy, causing still further temperature increases. In this example of *positive* feedback, the warming temperature "reinforces" itself. The temperature change, as a result, will be more dramatic than anyone would ordinarily expect. This is the main reason for the stronger-than-average climate warming at high latitudes. In general, when positive feedback takes place, it exacerbates the effect of the new force, and makes it stronger.

3.2 Negative feedback: cloud formation

Sometimes, however, there is a negative feedback effect. It undermines a force exerted on the environment. The result may be to move the system in the opposite direction from what would be expected if only positive or amplifying feedbacks were present.

An example of negative feedback is provided by processes that increase the extent of cloud cover. When the Earth

[2] We will not repeat the arguments regarding the detection of the greenhouse warming versus natural climate variability. (It still cannot be excluded that the observed temperature rise may be due to this.) For a discussion, the reader is referred to the first part of the report "Scientific Assessment of the Climate Change" (IPCC, 1990), or to Wigley et al., (1992; this volume).

surface grows warmer, more water evaporates. That means that more clouds are formed; those clouds, in turn, reflect more solar radiation back into space than the darker ground beneath them would reflect. As a result, the Earth surface grows cooler. The realization of this effect can be seen in the fact that the total cloud amount in any region, northern *or* southern hemisphere, is larger in summer than in winter.

It should, however, be noted that cloud processes can cause both positive and negative feedbacks, with the latter being potentially stronger. High and thus cold clouds can also enhance the greenhouse effect, as they trap upward going thermal radiation from the warmer Earth surface and lower atmosphere, while emitting less energy to space — a positive feedback. The opposite is the case for low clouds.

4 Projections for Future Climate

The climate of the next few decades to centuries depends primarily on three factors:

- the levels of future greenhouse gas emissions,

- the chemical fate of these gases in the atmosphere,

- the response (feedbacks) of the total climate system (land, sea, ice and snow) and biosphere (system of Earth life forms) to increasing greenhouse gas forcing.

From our horizon today, it is hard to see which of the above factors is most critical. However, we can only influence the first one directly and the last one indirectly. Some have recommended larger scale geo-engineering efforts to directly manipulate climate — e.g. by enhancing the sulphate levels in the stratosphere. However, such efforts may spur feedback processes with unexpected and maybe deleterious effects, such as additional stratospheric ozone depletion.

4.1 Effects of business as usual

Despite the inherent uncertainties of climate projections[3], we have enough information to make some definite assertions. We know that if society proceeds in "the business as usual" mode of operation, the equivalent doubling of CO_2 would be reached in less than 40 years from now (IPCC, 1990). (That figure includes the direct warming effect — i.e. radiative forcing — on the surface temperature from all greenhouse gases.)

The projections for future climate are done now by using results of climate models and using data on past warmer epochs (assuming, with some justification from reconstructions, that at least zonal distributions of changes are similar for these epochs). Admittedly, both methods of estimating climate change — from paleoclimatic evidence and from climate models — have their deficiencies and uncertainties. The latter arise mostly from inadequate description of cloud processes and their interaction with the Earth radiation field (e.g. most models produce a maximum cloud amount in winter - contrary to the observations). Because of the lag effect, due to heat uptake in ocean waters and incomplete representation of ocean heat transfer, the estimated temperature increases of 1.5 to 5.5 °C may be delayed by 20-30 years or more. Nevertheless, as we have already noted, even the slowest expected temperature increase rates of our era are very fast on geological time scales.

Even more important for agriculture, forestry, fisheries, and water resources are predictions of the geographical and seasonal distribution of changes in surface temperatures, precipitation, soil moisture content, winds, clouds, frequency of droughts and floods, and length of the frost-free season. Here, something can also be learned from paleoclimatic studies and maybe from present records, but most of these questions must await future research for answers.

The very warm winters that have been experienced so frequently in the Eurasian middle latitudes during the decade of the 1980s may be the first signs of the kind of winter weather that may be more typical of the next century. The rise in the level of the Caspian Sea level and the rise in the level of the oceans in general, the more frequent and intense hurricanes at the Eastern parts of Eurasian and American continents, and the melting of permafrost may be heralds of future climate changes. Alternatively these changes may be just normal variations. Only time will tell for sure.

5 Chemical Changes in the Atmosphere

Although the permanent constituents Nitrogen (N_2), Oxygen (O_2) and Argon (Ar) make up more than 99.9% of the volume of the "dry" atmosphere, they play no dominant role in its chemistry and are also only of little significance for the climate of the Earth. Their atmospheric abundances are outside the control of human activities.

Of much greater importance to both climate and chemistry are the highly variable concentrations of water (H_2O), and several much less abundant gases which have always been present in the atmosphere: carbon dioxide (CO_2), methane (CH_4), nitrous oxide (N_2O) and ozone (O_3). Also important is a suite of gases which are entirely man-made: the chlorofluorocarbons $CFCl_3$, CF_2Cl_2, $C_2F_3Cl_3$, and other industrial chlorocarbon gases such as CCl_4, CH_3CCl_3 etc. Relatively recent observations have shown that all of these gases have been increasing due to natural causes and human (agricultural and industrial) activities.

5.1 Chemical changes over time

Natural composition changes over the last 160,000 years were detected by analysing ancient air bubbles contained in ice cores drilled at the Vostok station of the USSR in Antarctica and several sites in Greenland. The records

[3] There will always be uncertainties about our projections for future climate and state of the environment, but research to be carried out in the International Geosphere-Biosphere and the World Climate Programs is aimed at reducing them.

Table 1. Summary of key greenhouse gases influenced by human activities[a]

Parameter	CO_2 (carbon dioxide)	CH_4 (methane)	CFC-11 (trichloro-fluoro-methane)	CFC-12 (dichloro-fluoro-methane)	N_2O (nitrous oxide)
Late glacial (15,000 yrs BP)	195 ppmv	0.35 ppmv	0	0	244 ppbv
Pre-industrial atmospheric concentration (1750-1800)	280 ppmv	0.8 ppmv	0	0	288 ppbv
Current atmospheric concentration (1990)[b]	353 ppmv	1.72 ppmv	280 pptv	484 pptv	310 ppbv
Current rate of annual atmospheric accumulation	1.8 ppmv	0.015 ppmv	9.5 pptv	17 pptv	0.8 ppbv
Relative radiative forcing	1	27	12400	15800	210
Global warming potential (GWP)	1	2	7	6000	150
Atmospheric lifetime[c] in years	100-200	10	65	130	150

ppmv = parts per million, by volume
ppbv = parts per billion, by volume;a billion here means thousand million (10^9)
pptv = parts per trillion, by volume; a trillion here denotes a million times a million (10^{12})
[a] Ozone has not been included in the table because of lack of precise data.
[b,c] IPCC (1990)

analyzed by French and Swiss scientists show that during the ice ages the atmosphere contained about 200 parts per million by volume (ppmv) of CO_2 and 400 parts per billion by volume (ppbv) of CH_4, increasing to 280 ppmv and 700 ppbv respectively during the (subsequent) interglacial ages. There are indications of a similar behavior of N_2O. It is difficult to say whether atmospheric chemical changes influenced temperature shifts, or vice versa. In fact, it is most likely that both processes were affected by a number of positive biospheric feedbacks. This raises considerable concern for the stability of present climate: a warmer climate may well reinforce the release of even more greenhouse gases from the biosphere. Unfortunately, there is nothing in the climate record which suggests that during the current geological era the biosphere would react with negative feedbacks (in the fashion of the Gaia hypothesis) to stabilize climate.

Following the establishment of the present interglacial some 10,000 years ago, the ice core records tell that a rather stable atmospheric composition emerged. It lasted, locked in, until about 300 years ago. The observed rise of the greenhouse gases CO_2, CH_4 and N_2O during the past centuries, which accelerated during the present century, thus clearly points to the impact of human activities. The relevant records are assembled in Table 1. They show growths in atmospheric CO_2, CH_4, and N_2O concentrations by 25%, 110% and 8%, respectively, since the middle of the 18th Century until 1990. The entirely anthropogenic trace gases $CFCl_3$ and CF_2Cl_2 showed increases from 0 to 280 pptv and 480 pptv, respectively.

5.1.1 Growth in CO_2

The initial growth in atmospheric CO_2 was mainly due to deforestation and agricultural pioneering activities, especially in North America, leading to partial oxidation of soil organic matter. Since the 1960s, the fossil fuel combustion source of CO_2 has surpassed the biospheric CO_2 source. However, due to the growth in tropical deforestation, especially during the 1980s, the biospheric source has remained significant, equal to maybe 20-40% of the fossil fuel source. It is estimated (IPCC, 1990; p. 13) that during the period 1850-1986 almost 200 Gigatonne of carbon (200 Gt; one Gt = 10^{15} grams C) had been released to the atmosphere as CO_2 by fossil fuel burning, and about 120 Gt by various agricultural activities. Of this a little more than 40% has remained in the atmosphere, the so-called "airborne fraction." Most of the remaining carbon has probably been taken up by the oceans, but current estimates of oceanic uptake appear too low to give a satisfactory balance (see Table 2). Therefore,

Table 2. Summary of the atmospheric CO_2 budget

	Gigatonnes Carbon per year (GtC/yr)
Emissions from fossil fuels into the atmosphere	5.4±0.5
Emissions from deforestation and land use	1.6±1.0
Accumulation in the atmosphere	3.4±0.2
Uptake by the ocean	2.0±0.8
Net imbalance	**1.6±1.4**

Source: IPCC (1990)

either the oceanic sink or the terrestrial (and coastal) carbon sink is larger than hitherto estimated (IPCC, 1990).

5.1.2 Growth in CH_4

Methane is produced biologically during the decay of organic matter under anoxic (i.e. oxygen-free) conditions. Emissions from wetlands are its main natural source. Large emissions from rice fields, landfills and the stomachs of ruminating animals (especially cattle) are, however, strongly perturbing the natural cycle of CH_4 in the atmosphere. In addition, methane also escapes as associated gas from coal mines and oil wells, and from natural gas production and distribution systems. The total CH_4 load in the atmosphere is of the order of 4 Gt or 4000 teragrams (1 teragram = 10^{12} grams). There, it tends to react with hydroxyl (OH) molecules, leading to the production of CO and CO_2. Because hydroxyl reacts in this way with several different compounds, this gas (which is only present in minute quantities in the atmosphere) is dubbed "the detergent of the atmosphere".

Atmospheric CH_4 may be increasing not only because of growing emissions, but also because its loss from the atmosphere by reaction with OH may become less efficient, due to higher consumption of OH by other reactions. The most important of these are reactions with CH_4 and CO in the atmosphere.

It is possible to obtain estimates of the oxidation rates of CH_4 in the atmosphere, and thus of the release rates of CH_4 at the Earth's surface. From knowledge of the global distribution and the industrial sources of methylchloroform (CH_3CCl_3), another compound which is removed following reaction with OH, the average concentration of OH in the atmosphere can be calculated. This calculation provides a means for estimating the atmospheric loss of CH_4. This quantity is equal to about 420 teragrams (Tg)/year. In addition, about 30 Tg CH_4/yr are lost through stratospheric reactions other than with OH. Since methane is also oxidized by soil microorganisms and is increasing at a rate of almost 1% per year to balance the mass flows, scientists conclude that the total release of methane to the atmosphere must be about 500 Tg/year. About 20% of the atmospheric methane is free of radiocarbon ^{14}C, possibly pointing to a source of 100 Tg/year due to release from the fossil fuel sector. The current atmospheric budget of CH_4 is shown in Table 3.

Table 3. Estimated sources and sinks of methane

	Annual Release ($Tg\ CH_4$)	Range ($Tg\ CH_4$)
Source		
Natural wetlands		
(bogs, swamps, tundra, etc)	115	100 - 200
Rice paddies	110	25 - 170
Enteric fermentation (animals)	80	65 - 100
Gas drilling, venting, transmission	45	25 - 50
Biomass burning	40	20 - 80
Termites	40	10 - 100
Landfills	40	20 - 70
Coal mining	35	19 - 50
Oceans	10	5 - 20
Freshwaters	5	1 - 25
CH_4 Hydrate destabilization	5	0 - 100
Sink		
Removal by soils	30	15 - 45
Reaction with OH in the atmosphere	500	400 - 600
Atmospheric increase	**44**	**40 - 48**

Sources: IPCC (1990), Cicerone and Oremland (1988)

5.1.3 Growth in nitrous oxide

Nitrous oxide (N_2O), whose atmospheric growth rate is approaching 0.3% per year, is a gas with a relatively long atmospheric residence time of 150-200 years. It is destroyed in the stratosphere by photochemical reactions which partly lead to the production of nitric oxide (NO). This very important process is the main source of stratospheric NO, which in turn serves as a catalyst in ozone destruction reactions and thus regulates the amount of ozone in the stratosphere. Thus, as is well known, stratospheric ozone regulates the penetration of solar ultraviolet radiation (of wavelengths shorter than 310 nm) to the Earth's surface, a factor which has considerable importance for the biosphere (Crutzen, 1971; WMO, 1985).

Most of the release of N_2O from the Earth's surface probably occurs as a consequence of microbiological processes (nitrification and denitrification) in waters and in soils. Its rate of increase is thus influenced by land use disturbances, agricultural activity and nitrogen fertilizer application. From these connections, we see a remarkable linkage in the chain of feedbacks between biospheric processes (a tree clearing in Nigeria, or a fertilizer spread in Ohio) and fundamental chemical processes, affecting ozone in the stratosphere.

Until only a few years ago, scientific knowledge about the sources and sinks of N_2O were thought to be in rather good shape (WMO, 1985), with most atmospheric increase believed due to emissions from coal and oil combustion and biomass burning. Since then it has been shown that earlier measurements were incorrect, so that combustion processes are now believed to be only a minor contribution (Muzio and Kramlich, 1988). The atmospheric N_2O budget still (as shown in Table 4) presents us with major uncertainties.

5.1.4 Chlorofluorocarbons

Finally, we must discuss the chlorofluorocarbon (CFC) gases. Like N_2O, these gases do not react in the troposphere and are mainly destroyed in the stratosphere by solar ultraviolet radiation. In the process of destroying CFCs, chlorine (Cl) atoms and chlorine monoxide radicals (ClO) are produced which, on a molecule by molecule basis, are almost an order of magnitude more powerful than NO and NO_2 in destroying ozone by catalytic reactions. Because of the very rapid rise in their atmospheric concentrations during the past three decades, the chlorine that was carried by the

Table 4 . Estimated sources and sinks of nitrous oxide (N_2O)

Source		Range (TgN per year)
Oceans		1.4-2.6
Soils	(tropical forests)	2.2-3.7
	(temperate forests)	0.7-1.5
Combustion	0.1-0.3	
Biomass burning		0.02-2.2
Fertilizer (including ground-water)		0.01-2.2
Total:		**4.4-10.5**
Sink		
Removal by soils		?
Photolysis in the stratosphere		7.0-13.0
Atmospheric increase		**3.0-4.5**

Source: IPCC (1990)

chlorofluorocarbon gases to the stratosphere has caused major depletions of stratospheric ozone, especially over Antarctica.

5.2 *Changes in the balance of greenhouse gases*

5.2.1 *Greenhouse gas effects*

CO_2, CH_4, N_2O and the chlorofluorocarbon gases are also major greenhouse gases, which due to their propensity to absorb terrestrial heat radiation and increasing atmospheric concentrations, effect a warming of the Earth's surface. Since pre-industrial times the resulting radiative forcing, or direct warming effect, resulting from the increase in these gases, has amounted to 2.5 watts per square metre (W/m^2). This may be compared to a 4.3 W/m^2 forcing which, according to radiative calculations, would take place with a doubling of the atmospheric CO_2 content from pre-industrial levels. That, in turn, would suggest an equilibrium temperature increase of 1.5-4.5 °C (IPCC, 1990). Thus the current rise in greenhouse gases may already have committed the planet to a temperature rise by 0.9-2.6 °C, only about 0.5 °C of which has been realized.

The difference between the actual temperature rise so far, and the temperature rise to which the planet is committed, is partly due to the absorption of heat into the oceans. It may also indicate the presence of negative feedback in the climate system — e.g. due to cloud-related climate cooling effects. At present clouds exert a net cooling effect on climate (Raval and Ramanathan, 1989). Whether clouds will moderate the radiative warming caused by growing concentrations of greenhouse gases, also in the future, is unknown. An additional negative feedback on climate warming may result from the increased sulfate loading, leading to stronger backscattering (outward deflection) of solar radiation.

5.2.2 *Greater influence of less abundant greenhouse gases*

The important contributions by gases other than CO_2 to the greenhouse effect, which until quite recently were neglected, are noteworthy. According to the IPCC (1990), for the decade of the 1980s the greenhouse forcing contribution by CO_2 merely accounted for 55% of the total. The remaining 45% was due to the chlorofluorocarbon gases (24%), methane (15%) and nitrous oxide (6%).

The importance of these minor greenhouse gases (CH_4, N_2O and the chlorofluorocarbons) is quite remarkable, considering that their absolute concentrations (and rates of increase), are small compared to those of CO_2. The reason for this is that, due to the already relatively large amounts of CO_2 in the atmosphere, the absorption of infrared radiation by CO_2 in several wavelength regions is already complete. For the other, much less abundant, greenhouse gases, which absorb at other wavelengths, this saturation effect has not yet occurred.

Applying formulas for saturation[4] from the IPCC report (1990), one finds that the addition of one mole of CH_4 to the atmosphere has about a 21 times larger heating effect than the addition of one mole of CO_2. For N_2O, $CFCl_3$ and CF_2Cl_2, the corresponding relative radiative forcing factors are 206, 12,400 and 15,800, respectively.

5.3 *Calculation of global warming potential for trace gases*

In order to compare the radiative heating over time, one must take into account the different atmospheric residence times of greenhouse gases. This leads to the definition of the (relative) Global Warming Potential (GWP) of a trace gas, compared to that of CO_2.

As the atmospheric lifetime of CH_4 is only about 10 years, compared to about 120 years for CO_2, its GWP, comparing one mole of emission of each, would be only about 3. For N_2O, $CFCl_3$ and CF_2Cl_2, with atmospheric residence times of about 150, 60 and 130 years, respectively, the GWP factors would be about 150, 6,000 and 17,000.

The correct calculation and application of GWP is a complicated, but important, issue, which most likely is going to play a major role in future international negotiations. The issue is complicated by the fact that CH_4, N_2O and the chlorofluorocarbon gases are also of great importance for the chemistry of the atmosphere. For example, they affect stratospheric and tropospheric ozone, and stratospheric water vapour. Indirectly, this will also have an effect on climate through both positive and negative feedbacks, according to the IPCC (1990), leading in the case of CH_4 to more than a doubling of its GWP. This estimate is, however, too high (Lelieveld and Crutzen, 1992).

---

[4] The potential sensitivity of climate to greenhouse gas emissions can be expressed in terms of additionally trapped infrared energy (F) in the atmosphere, resulting from the addition of one unit (e.g. one mole, on one part per billion by volume) of a trace gas to its atmospheric abundance X:

$$f_i = \delta F/\delta X_i$$

Formulas of these factors for CO_2, CH_4, N_2O and the chlorofluorocarbon gases have been given by the IPCC (1990).

6 Changes in tropospheric (lower atmospheric) chemistry

6.1 Tropospheric ozone

Although only about 10% of all ozone is located in the troposphere, this portion is nevertheless of the greatest importance for the chemistry of the atmosphere. The reason is that solar ultraviolet radiation causes the decomposition (photolysis) of ozone into energetic oxygen (O) atoms. These, in turn, react with water vapour to produce hydroxyl (OH) the "detergent" of the atmosphere. Hydroxyl, as we have already noted, reacts with almost all gases that are emitted by natural processes and anthropogenic activities in the atmosphere.

Thus, although ultraviolet radiation is damaging to life and ozone is a toxic gas in large quantities, both are also of substantial importance for keeping the atmosphere "clean." Hence the need for policies which ultimately maintain a safe level of total-column ozone, preserving a stable concentration in the upper atmosphere (the stratosphere) and minimizing the buildup of ozone in the troposphere.

6.2 Other tropospheric gases

The chemistry of the troposphere is changing due to strong anthropogenic emissions, which rival or surpass those caused by natural processes. Of large importance for the overall chemical functioning of the troposphere are the emissions of NO, CH_4 and CO, because they determine changes in the tropospheric concentrations of ozone, and thus of hydroxyl.

While CH_4 and CO are largely removed by reaction with OH, they are also the main reactants of OH in most of the troposphere. Thus, because CH_4 concentrations have been increasing worldwide by almost 1% per year during the past decades, and there are indications that CO concentrations are likewise increasing at similar rates — at least in the northern hemisphere (Zander et al, 1989; Dianov-Klokov et al., 1989; Golitsyn et al., 1990) — a lowering of the concentrations of OH radicals may be expected.

However, increased inputs of NO_x into the atmosphere, mainly by fossil combustion processes at temperate latitudes and biomass burning in the tropics, will work in the opposite direction — leading to higher concentrations of ozone and OH. As most NO input occurs in the northern hemisphere, the expected ozone increase should occur mainly there.[5] Table 5 shows estimated annual NO emissions, and their sources.

Because ozone is an intensely toxic gas, which also interferes with the photosynthesis of vegetation, the rise in tropospheric ozone causes crop and health damage, whose abatement for the US alone has been estimated at several thousand million dollars annually (OTA, 1990). Ozone may also contribute to the widespread forest damage which has been reported in several regions of North America and Europe.

Photochemical ozone production, however, is observed not only in the industrial world. It appears in the tropics during the dry season as well, owing to the use of fires in forest-clearing and dry savanna grass-clearing activities. Consequently, high surface ozone volume mixing ratios, approaching those measured in the industrial regions of the northern hemisphere, are frequently measured in the tropics and subtropics during the dry season (Crutzen and Andreae, 1990; Fishman et al., 1990). Current estimates are that 2-5 $\times$ 10^{15} g of biomass carbon are annually burned, releasing NO_x and reactive hydrocarbons — exactly those ingredients that most immediately add to photochemical ozone production (Crutzen and Andreae, 1990).

6.3 The potential for rapidly increasing tropospheric ozone

Increasing emissions of CO and CH_4 would tend to lower OH concentrations, while those of NO have the opposite effect. As the atmospheric lifetime of NO_x is much shorter than that of CH_4 and CO, the former effect may be more important and a plausible deduction is that global average OH concentrations, may go down. Whether this is indeed the case is not known. According to theoretical model estimates (Valentin, 1990), only about 60% of the atmospheric increase in methane may be due to increasing emissions, the remainder due to a lowering of global average OH concentrations. This also

Table 5. Estimated emissions of nitric oxide (NO)

Anthropogenic emissions:	Range (TgN per year)
Industrial	13-29
Aircraft	0.3
Biomass burning	3-7
Agriculture (fertilizers)	0.5-1
Natural emissions:	
Soil exhalations	5-15
Lightning	2-10
Marine	0.15
Flow from stratosphere	0.5

Sources: WMO (1985).

[5] Indeed, measurements made at a number of locations in the northern hemisphere, including several clean air sites, indicate rates of surface ozone concentration increases of about 1% per year (IPCC, 1990; Crutzen, 1988; Bojkov, 1988). This trend also follows from the oldest records of reliable surface ozone measurements from the end of the last century — which were made at the Montsouris observatory site, previously located at the outskirts of Paris. When only those measurements from wind directions are considered that did not come from Paris, the reported ozone volume mixing ratios were generally just below 10 ppbv (Volz and Kley, 1988). Typical ozone surface volume mixing ratios which are measured at various sites in Europe nowadays are at least 2-3 times higher. In addition, very large excursions in ozone concentrations, exceeding 100 ppbv, are commonly reported during photochemical summer smog episodes in many regions of the industrialized northern hemisphere.

implies that global average OH concentrations may decrease by 0.3% per year. Similar results were obtained by other researchers (Isaksen and Hov, 1987; Thompson and Cicerone, 1986). Because almost all gases that are emitted into the atmosphere are removed by reaction with OH, this reduction in the oxidizing power of the atmosphere may be of considerable long-term significance.

In general, it may be expected that, owing to increasing CH_4 and CO concentrations, OH radical concentrations are decreasing in the more pristine atmospheric environments which contain relatively little NO, while increasing in the more polluted NO-rich environments. This would cause a gradual shift in atmospheric photochemical oxidation intensity from clean, mostly tropical oceanic to more polluted

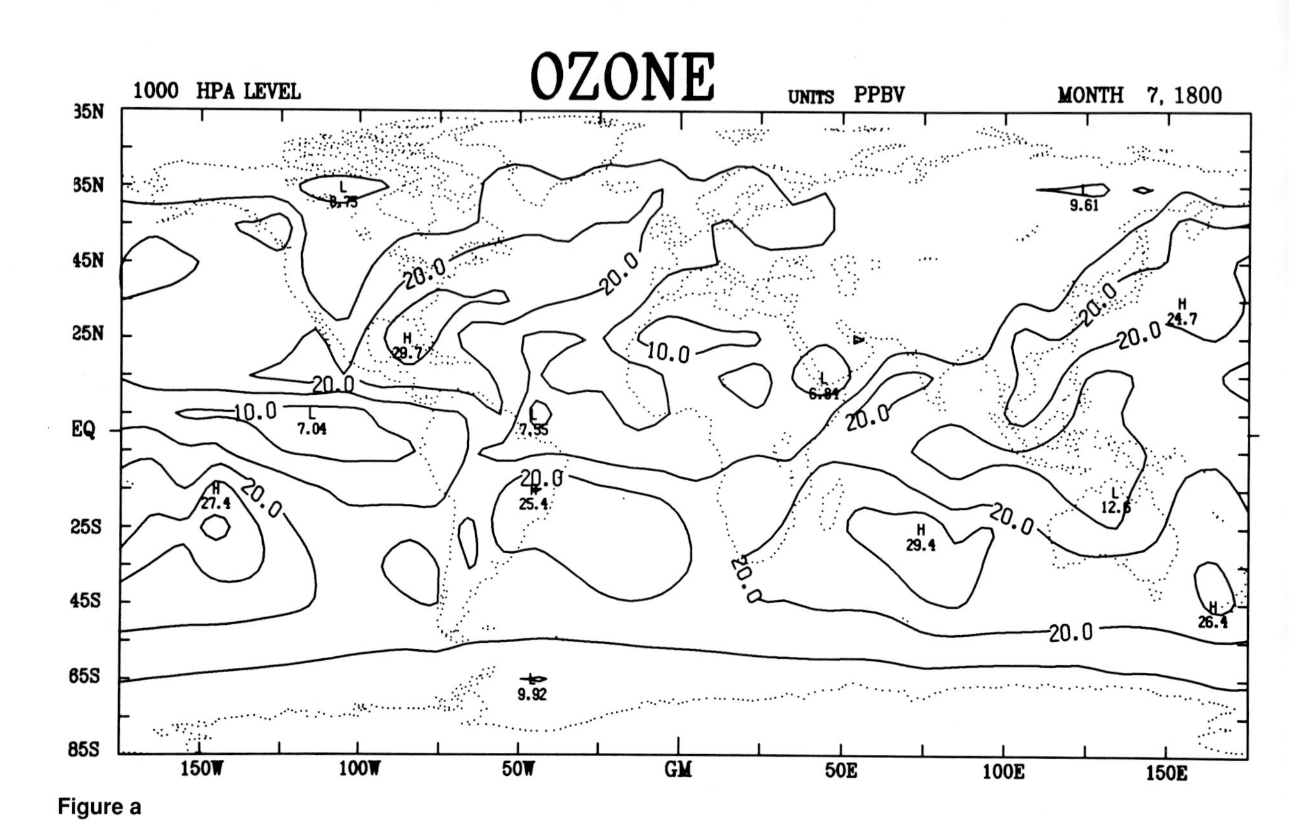

Figure a

Figure b

Figure 2a, b, c and d. Surface ozone concentrations may have more than doubled over large regions of the northern hemisphere during the last 100 years.

Panel a shows model calculations of the surface volume mixing ratio of ozone, plotted against a map of the world, for a pre-industrial period: July, 1800. Panel b shows the calculated ozone levels, plotted against the same map, during the month of July, 1980.

Panels c and d compare the same two months (July 1800 and July 1980), showing meridional cross sections of ozone volume mixing ratios. *HPA* (the Y-axis) represents the atmospheric pressure in units of hPa (hectopascal), which depends upon temperature and altitude. Source: Crutzen and Zimmermann, 1991.

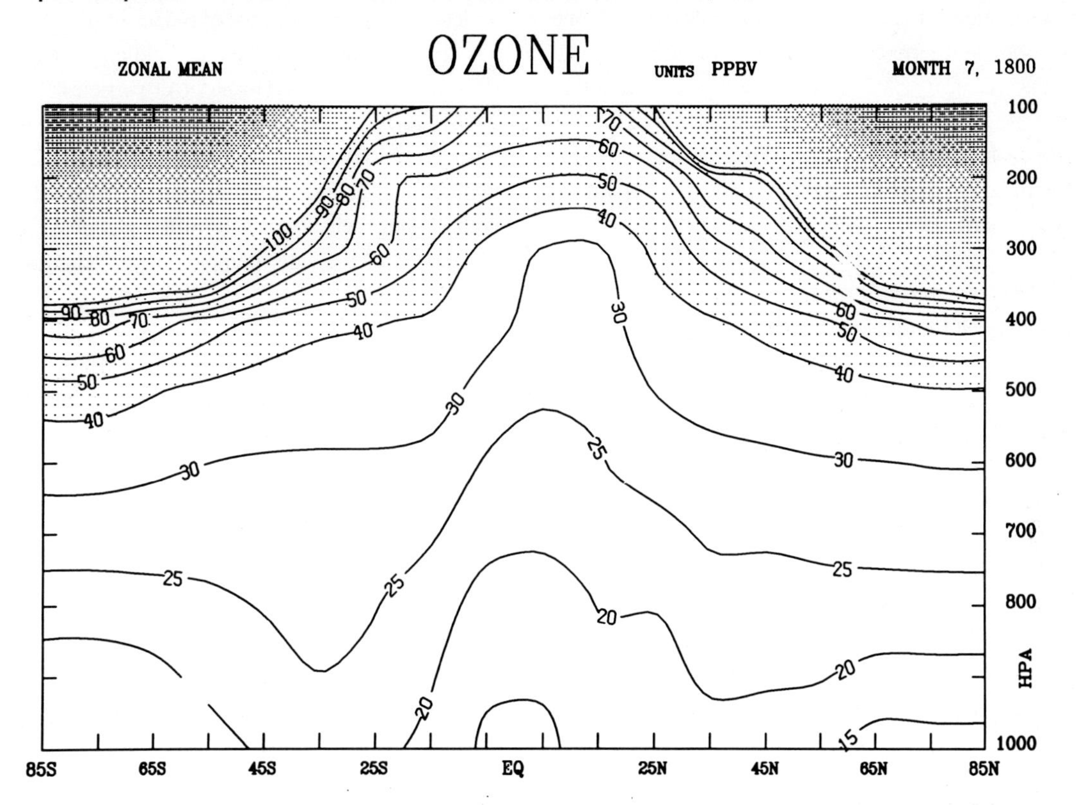

Figure c

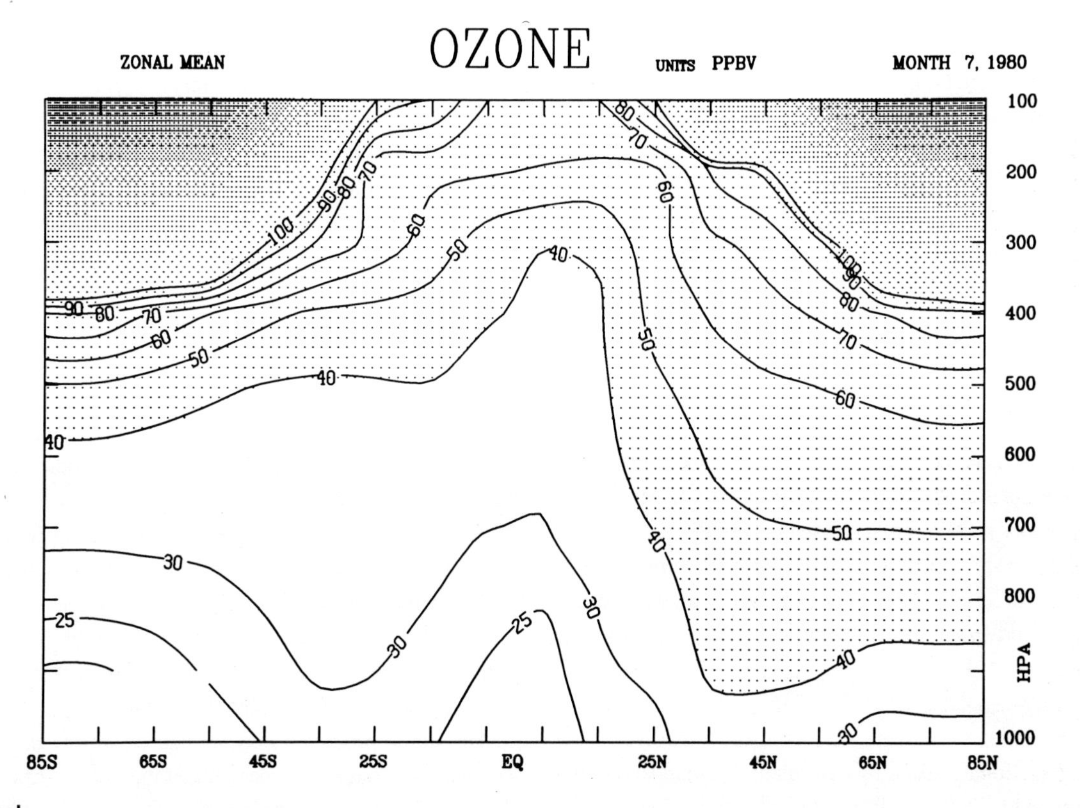

Figure d

continental environments, leading to higher ozone concentrations in the latter. With industrial growth in the developing world, and the resulting increase in NO emission, larger parts of the continental tropics are likely to experience higher ozone concentrations in the future — with possible adverse consequences for the biosphere.

The potential for large ozone increases, mainly due to anthropogenic NO emissions, is also clearly indicated by model calculations. Figure 2, showing calculated ozone concentrations with a model of transportation, emissions and atmospheric chemistry, shows that more than a doubling in surface ozone concentrations as well as increases in the rest of the troposphere may have occurred over large regions of the northern hemisphere during the last 100 years (Crutzen and Zimmermann, 1991).

6.4 Nitric oxide and acid rain

Anthropogenic NO emissions are not only of importance because of their role in catalytic ozone-forming processes, but also because, together with SO_2 emissions, they strongly determine the acidity of precipitation. Thus, for several thousands of kilometres around industrial regions, "acid rain" has had, and continues to have, a strong impact on the health of the environment — causing lake acidification, fish death, and forest damage.

In the industrial world, regulatory measures are now gradually bringing this environmental hazard under control, but the approaching industrialization in the developing world, coupled with the use of "cheap" sulphur-rich coal in many developing countries as a source of energy, may well lead to even more severe environmental disturbances than have been noted in the industrialized world (Rodhe and Herrera, 1988).

7 Stratospheric Ozone Changes

Nowhere has the deleterious global impact of human activities been more clearly demonstrated than in the stratosphere. Since the end of the 1970s, drastic ozone depletions have been recorded during the spring months of September to November over the Antarctic continent. (Although it is a toxic pollutant at the surface, in the upper atmosphere ozone is an essential shield against the dangers of ultraviolet radiation from the sun.) Despite some quasi-biannual variation, the ozone depletions have been steadily growing to reach more than 50% of the total ozone column, wiping out virtually all ozone between 15 and 22 km altitude (WMO, 1988; Farman et al., 1985; Hofmann et al., 1987). Moreover, the ozone loss is no longer restricted only to the Antarctic. It is affecting the temperate zone of the northern hemisphere as well (Stolarski et. al., 1991), approaching 1% per year during the winter and spring months during the 1980s, as shown in Figure 3.

From atmospheric observations, the cause for the ozone depletions has now been clearly identified. It involves catalytic destruction of ozone by chlorine (Cl) atoms and chlorine monoxide (ClO) radicals, the photochemical breakdown products of the industrial chlorofluorocarbon gases in the stratosphere. An entirely unexpected chain of events, involving several positive feedbacks, is responsible for the particularly damaging activity of the chlorine gases during these conditions.

7.1 The chain of feedbacks involves the following:

- naturally cold winter and early springtime temperatures, which promote the removal of oxides of nitrogen from the gas phase by freezing them into solid nitric acid trihydrate (HNO_3, $3H_2O$), referred to as "NAT particles" (Toon et al., 1986; Crutzen and Arnold, 1986). The presence of gaseous nitrogen oxides would tend to trap inorganic chlorine as hydrochloric acid and chlorine nitrate, which do not react with ozone; their absence from the gas phase favors the conversion of the latter compounds into chemically reactive Cl and ClO;

- the production of molecular chlorine (Cl_2) by the reaction of hydrochloric acid (HCl) and chlorine nitrate ($ClONO_2$) on the surface of the trihydrate particles (Molina et al., 1987; Tolbert et al., 1987).

- the dissociation of Cl_2 by solar ultraviolet radiation, producing Cl atoms;

- the establishment of an ozone destruction cycle, involving a reaction of ClO with itself, which is particularly effective in the lower stratosphere. The efficiency of this process is proportional to the square of the stratospheric chlorine content, and thus growing by about 7% per year (Molina and Molina, 1987).

This intricate interplay between reactive chlorine and nitrogen is remarkable — especially the protective role of the oxides of nitrogen against ozone destruction by the chlorine species. It is particularly remarkable in view of the fact that the oxides of nitrogen themselves, under natural circumstances, are largely responsible for controlling ozone concentrations above about 25 km.

7.2 Implications of ozone destruction

For five to ten years after the industrial production of chlorofluorocarbon gases is curtailed (by the end of this century, according to international agreement), the chlorine content of the atmosphere will continue to increase due to their slow diffusion into the stratosphere, thus aggravating the ozone loss. Further damage to the ozone layer by chlorine gases, possibly in combination with bromine gases, must therefore unfortunately be expected until the year 2010. Therefore, a return of the stratosphere will start, but because of the slow breakdown of the chlorofluorocarbon gases the healing process will last a hundred years.

Additional problems may be caused by the continuing increase by about 0.25% per year of the reservoir of atmospheric N_2O, the precursor of the NO_x catalysts. For the future,

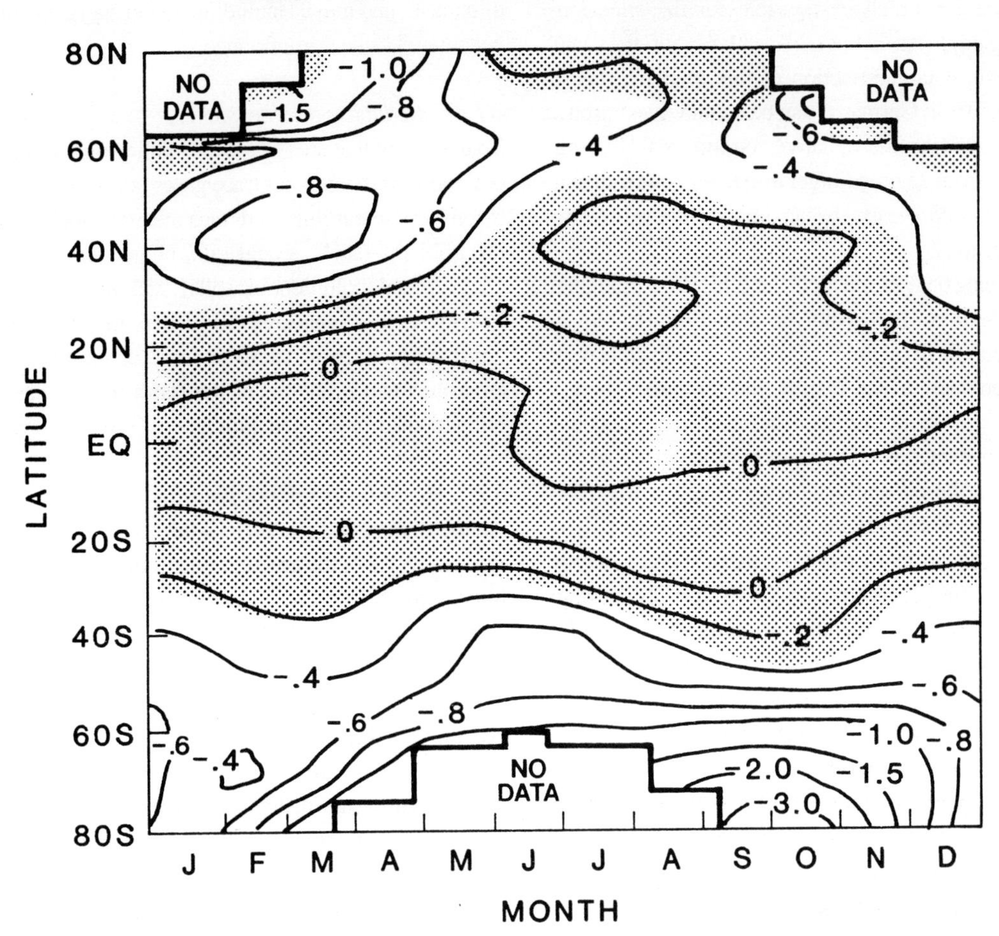

Figure 3. Trends in total column ozone depletion, expressed in per cent per year, observed by the TOMS total ozone satellite instrument during the period from November, 1978 to May, 1990. White areas represent latitude-months where depletion of -0.4% or more was observed. Source: Stolarski et al., 1991.

the emissions of NO and H_2O by hypersonic advanced aircraft, which are in their planning stages in the US, Europe and Japan, may also become of concern, because they may lead to further ozone depletions, through reactions with NO and NO_2 as catalysts in the middle and upper stratosphere (Johnston et. al., 1989). In addition, following reaction of $ClONO_2$ and HCl on the NAT particles, aircraft emissions of NO_x and H_2O may also promote "ozone hole" reactions in the cold winter and springtime — due to the increasing likelihood of formation of NAT particles (Peter et al., 1991).

8 What kind of policy should we thus aim for?

As our discussion has shown, the changes in climate, atmospheric chemistry and the environmental conditions are proceeding side by side, and are closely interrelated. The problem complex is summarized by the term "Global Change": The worldwide environmental system has been disturbed in so many ways, that simple cures and issue-by-issue approaches in many cases are no longer satisfactory solutions to its problems. Thus in the future, environmental policy must increasingly involve the scientific community.

8.1 Basic chemical guidelines

Although scientific knowledge is not always available, some general guidelines can nevertheless be given. To stabilize climate and global atmospheric chemistry, reductions in anthropogenic emissions are necessary for:

CH_4 by about 15%;
CO_2 by about 60%;
N_2O by about 75% (IPCC, 1990).

8.1.1 Reducing methane

Reduction measures for CH_4 should especially involve reductions in releases from the fossil fuel energy sector, augmented by efforts to cut back also on emissions from landfills. More justification for this policy comes from the fact that about 20% of the atmospheric CH_4 content is ^{14}C-free, or "old" methane (Wahlen, 1989).

Such actions are especially called for in the industrial nations, which are largely responsible for the emissions. Coal mining produces about 35% (±10%) of the "old" methane. A much larger fraction of this methane than at present (only 20% in Germany) can be captured and used as an energy source. As the radiative forcing of CH_4 in the atmosphere is about 25 times larger than that of one molecule of CO_2, this would clearly be an important measure for climate protection.

The remaining 65% of old methane is likewise thought to come from the fossil fuel sector, in particular escaping as "associated" gas from oil production sites, and from leaks in natural gas production and distribution systems. There is much controversy about this issue, because some "accounting" studies conducted in OECD countries indicate that CH_4 emissions from the oil and gas sectors together may only contribute at most 20% of the old methane source. It is, therefore, speculated that leaks from these sources may be substantially larger in Eastern Europe and the USSR. If so, a substantial source of CH_4 could be eliminated.

Any tightening of CH_4 releases from the fossil fuel sector will not only reduce the growth of CH_4, but also of atmospheric CO_2, to which CH_4 is oxidized in the atmosphere. As the climate forcing by one molecule of CH_4 is about 21 times larger than that of one molecule of CO_2 in the atmosphere, these are important climate protection measures. The same applies to CH_4 escape from landfills, currently estimated to be about 10% of the total anthropogenic CH_4 source. With landfills, it is possible to burn most of the methane and use it as an energy source. As most organic material which is dumped in landfills is derived from plants, this is a solar energy source, which does not even effect a net increase of CO_2 in the atmosphere.

8.1.2 Reducing carbon dioxide

CO_2 emission reductions clearly require energy conservation measures and the development of alternative, renewable energy sources which are not based on fossil fuels. But while one may be somewhat optimistic about achieving the necessary reductions of enough methane releases to stabilize atmospheric concentrations, the necessary reductions of CO_2 emissions will be much harder to accomplish. Thus far, initiatives have been announced by some nations to achieve reductions by up to 25% by the year 2005.

In the short term, that is for the next decades, CO_2 emission reductions must come especially from energy savings (better home insulation, introduction of passive and active solar heating, tightening of CH_4 leaks, use of power plant waste heat, improved transportation systems, and introduction of energy-saving devices). Although shifts from coal to natural gas for energy production will lead to less CO_2 emission, this is most likely only a temporary measure, because natural gas reserves are insufficient to sustain future energy needs and use of coal may become inevitable, especially for the developing world. For longer time horizons, the introduction of non-CO_2-producing energy sources — preferably solar energy systems, but maybe also new generations of nuclear, and maybe including fusion energy plants — will become of critical importance.

8.1.3 Reducing trace gases

To the extent that energy savings policy will also reduce the emissions of many other trace gases (such as SO_2, NO_x, CO, reactive hydrocarbons, and trace metals), other environmental problems (such as acid rain, photochemical smog, and heavy metal pollution of soils) will be relaxed as well. Energy savings measures thus constitute the most comprehensive approach to alleviate some of the most important environmental problems of the Earth. In any case, it is an obligation to future generations.

8.2 Measures for developing nations

Although the industrial nations have so far been responsible for about 85% of the greenhouse gas increases, scenarios for future climate forcing will increasingly depend on the developing world. It will be of the greatest importance to introduce fossil energy-saving measures there as soon as possible, because early introductions will be particularly effective — providing both energy cost savings and preventing environmental damage in tropical and subtropical ecosystems. Due to lesser buffering capacity (less organic matter in the soil), these ecosystems are particularly vulnerable to disturbances.

In these countries, emissions of CO_2 and CH_4 are currently closely related to agricultural activities. An important atmospheric source of CO_2 is tropical deforestation, contributing 10-40% of the fossil fuel source (IPCC, 1990). It appears that the present rate of deforestation may only be substantially reduced if the rate of population growth can be curtailed, and productivity in existing tropical agriculture strongly increased. The latter will require further substantial research and financial investments. Future developments in agriculture, e.g. the use of fertilizers, should (as much as possible) aim at reducing greenhouse gas emissions (e.g. N_2O).

Atmospheric CH_4 releases in the developing world come to a substantial degree from rice fields and cattle holdings (IPCC, 1990; Cicerone and Oremland, 1988). *Prima facie* it appears that these emissions can only grow in the future, due to increasing populations. Here, the hope is for more effective production methods and maybe the introduction of agricultural practices that reduce the emissions of CH_4 from rice fields. Care must be taken, however, that these do not cause enhanced emissions of N_2O to the atmosphere. Research in this area has only just begun.

8.3 Reforestation

This brings us to other aspects of the role of agriculture and forestry in "Global Change," which likewise contribute in significant ways to the growth of greenhouse gases in the atmosphere. We have already mentioned the release of CO_2

due to deforestation and oxidation of organic matter in agricultural soils. As a countermeasure, large reforestation activities have been proposed in some countries. For instance, in 1990 the US announced that they would plant one billion trees in each subsequent year. However, even this massive reforestation project would only be equivalent to a 5% reduction of the annual US carbon dioxide emissions.

In addition, there may be many pitfalls. Forty years ago, the Soviet Union embarked on a vigorous programme of planting forest strips to protect crops. First the strips were rather successful, but although their maintenance required only minimal attention, with time even this was lacking, and most reforested land is now in rather bad condition.

The method can, however, work. Some hundred-year-old tree plantings have shown a very positive influence on crops, on soil water content, and on biodiversity. The success of reforestation programmes will thus depend on great care and knowledge. The risk for failures may be particularly large in the tropics, where monocultures have been vulnerable to environmental degradation. The best method for reforestation in the tropics may well be to leave the degraded areas alone, and have them return to their natural conditions on their own or with some well-designed assistance. Field research projects are of great importance in order to prevent expensive failure.

Finally, we would like to point to a potentially ominous environmental side effect of reforestation: increased emissions of N_2O, in particular if reforestation is accompanied by extensive use of nitrogen fertilizer. The risk for increased N_2O releases may be particularly great in tropical forests, which in their natural form are already major sources of this gas (Keller et al., 1986). The processes which are mainly responsible for the atmospheric N_2O increase are largely unknown, emphasizing a strong need for intensive research efforts. These may be particularly important in the tropics, as observations in the background atmosphere indicate that most N_2O is produced at low latitudes (Prinn et al., 1990).

8.4 Agriculture

Although the present rate of increase of N_2O in the atmosphere of 0.2-0.3% per year may seem rather modest, this rate may increase in the future (there are indeed some signs of this in the measurement records). The terrestrial nitrogen cycle is already disturbed by the application of nitrogen fertilizer in agriculture, at a rate approaching 100 Teragrams of nitrogen per year — already equal to the natural nitrogen fixation rate, and growing at the rapid rate of about 4% per year. This, combined with the mushrooming need for food production in developing nations, could well set the stage for some major problems if no more suitable agricultural methods are developed which do not require the input of such large quantities of new nitrogen.

The large additions of fixed nitrogen to agricultural soils may not only cause increased N_2O release, but lead to major eutrophication problems in rivers, lakes, and coastal areas, and high nitrate levels in ground water. Furthermore, high yield cultivation systems are not only characterized by excessive inputs of N-fertilizer, but also by massive fossil fuel energy "subsidies" during production, packaging and transportation to the consumer. It will therefore be an important challenge to develop less wasteful, less nitrogen-and fossil-fuel-intensive alternative methods of agriculture.

8.5 Burning of plant matter

It is revealing that while "highly developed" agriculture needs so much energy and fixed nitrogen, in the developing world both are wasted in large quantities by the burning of plant matter during the dry season. According to estimates by Crutzen and Andreae (1990), 2 to 5 billion tonnes of biomass carbon (2-5 x 10^{15} g C/yr) are burned each year, consisting of firewood, dry savanna grass, agricultural wastes and dead vegetation cut for deforestation activities.

Although only the loss of carbon from deforestation activities is a net source of atmospheric carbon dioxide — the rest is merely recycled carbon — the sheer amount of wasted renewable energy is hair-raising. It is larger than the total commercial energy consumption in the developing world. Furthermore, many other trace gases — in particular, NO, CO, CH_4, and reactive hydrocarbons — are released to the atmosphere as well, in quantities comparable to those arising from industrial processes. The production of methane by biomass burning may constitute as much as 10% of its total source (Quay et al., 1991). It is thus a major factor in greenhouse warming. In addition, the production of smoke particles contributes a large fraction to the total particulate matter loading in the atmosphere. Because the smoke particles serve as effective cloud condensation nuclei, the smoke emissions from biomass burning may be as important to the Earth radiation balance as the sulphur dioxide emissions from coal and oil burning (Crutzen and Andreae, 1990).

A particularly interesting finding was that about 40% of the nitrogen originally present in the plant material before burning was converted to molecular nitrogen (in a process which we have called pyrodenitrification), thus constituting a substantial loss of valuable nutrient nitrogen in savannas and farm lands in the tropics and subtropics (Kuhlbusch et al., 1991). As other N-containing gases and smoke particles are emitted by the fires as well, and deposited elsewhere, the overall nitrogen budget of tropical ecosystems that are frequently exposed to biomass burning may be substantially negative with potential major long-term consequences for their productivity. As very little scientific study has been devoted to N cycling in tropical ecosystems, including pathways of fertilizer N, the possibility of significantly increased releases of N_2O from tropical ecosystems should be explored in order to design agricultural methods which prevent major environmental problems in the future.

9 Conclusion

We have shown that many environmental problems of soils, waters, climate and atmosphere are closely interlinked and should not be treated in isolation from each other. As many problems are of regional or global dimensions, their solution will require international participation and solidarity as well as sacrifices at the personal, local and national levels. To resolve environmental problems, in the future we should search for comprehensive solutions — avoiding the quick-fixes which have been applied so often in the past, but which frequently have substituted one problem for another. This will require much more basic and policy-oriented science, and close interactions between decision makers and scientists. We therefore, again, make a plea for support of the core projects of the IGBP (International Geosphere Biosphere Programme).

However, even with the best science available, we can never be certain that our "predictions" about the future are correct. However, this can go either way, leading to possible overpredictions and underestimations of potential environmental effects. A case in point for the latter is the gross underestimation of stratospheric ozone depletions by models due to the use of chlorofluorocarbon gases prior to the discovery of the "ozone hole." Although stratospheric ozone depletions had been predicted to occur, they were initally estimated to be of the order of 1% per decade. In fact what has been observed are ozone depletions many times larger for extensive regions at high and mid latitudes. With this sad experience in fresh memory, we should be extremely cautious in assuming that e.g. future climate chamges will be only a fraction as bad as current climate models predict.

Finally, at some point in the future (the sooner the better), humankind must reflect on what really makes a human happy and what living conditions we ought to bequeath to our children, grandchildren and generations thereafter. It is clear that we are at a critical turning point in history, and that we must face the challenge. If we realize that the marks of humankind on the environment are clearly discernable on all scales — in the air, soils and waters — and that each individual's actions has not only local, but global implications, there may still be great potential left for all of us to design a society which is in better harmony with the rich environment which Nature presented to us at our birth.

Can't see the eons for the years: The Caspian Sea level and long-term climate variability

Sometimes the puzzles of climatic cause and effect are especially difficult to untangle, because they require us to notice a pattern that may go back millions of years.

Consider the case of the Caspian Sea, the largest inner continental closed basin in the world, with an area close to 400,000 km^2.

On the east coast of the Sea, there is a dam which was constructed in the early 1980s, to section off the Bay of Kara-Bogaz-Gol. Its purpose is to keep Caspian Sea water from draining out into this secondary bay, whose level is a few meters lower than the Caspian water level. Before the dam, the bay had regularly drained from the Sea several cubic kilometers per year, through a narrow strait.

The dam builders had argued that their project was needed to save the Caspian Sea, whose water level had fallen 1.7 m in the decade of 1930s, and continued to fall (albeit more slowly, by another 1 m between 1940 and 1977). In addition to the dam, its dam proponents had also sought to divert several northern rivers into the Volga, which flows into the Caspian Sea. Otherwise, they said, the sea would continue to shrink.

Fortunately, the river diversion faced such strong public opposition that it never materialized. For since 1978, the water level of the Caspian Sea has begun to rise dramatically. By 1992 it had risen almost 2 meters. The rise can be explained by the combined action of increased river runoff, increased precipitation and decreased evaporation (Golitsyn and Panini, 1989), which in its turn is partly due to weaker winds (Golitsyn et al., 1990). About 10 per cent of the rise is directly caused by that dam.

Of course, one can not say for sure that this rise is due to greenhouse gas warming; it may reflect some regional climatic variations. Nevertheless, the present rapid rise of sea level is consistent with what we know from paleoclimatic reconstructions of warmer periods. Both during the Holocene Optimum (6,000 years ago) and the Eemian (125,000 years ago) the Caspian sea level was several meters higher than now. Presumably, it will one day reach that level again.

This case convincingly shows the price of incorrect knowledge of long-term climate variations, which were totally neglected here. Today the rise of the sea already causes a great number of problems to coastal cities, ports, and oil and gas excavation and distribution facilities. A further rise in sea level would aggravate the problem. (Other dramatic examples of the effects of natural climate variability include the Sahel drought and the "dust bowl" of the 1930s in the US)

References

Bojkov, R.D., 1988: Ozone changes at the surface and in the free troposphere. *Proceedings of the NATO Advanced Research Workshop on Regional and Global Ozone and its Environmental Consequences,* I.S.A. Isaksen, ed., Reidel-Dordrecht, Holland, p. 83-96.

Charlson, R., J. Langner and H. Rodhe, 1990: Sulphate aerosol and climate. *Nature,* **348**, 22.

Cicerone, R. and R. Oremland, 1988: Biogeochemical aspects of atmospheric methane. *Global Biogeochem. Cycles,* **2**, 299-327.

Crutzen, P. J., 1971: Ozone production rates in an oxygen, hydrogen, nitrogen oxide atmosphere. *J. Geophys. Res.,* **76**, 7311-7327.

Crutzen, P.J., 1988: Tropospheric ozone: an overview. *Proceedings of the NATO Advanced Research Workshop on Regional and Global Ozone and its Environmental Consequences,* I.S.A. Isaksen ed., Reidel-Dordrecht, Holland, pp. 3-23.

Crutzen, P.J. and M.O. Andreae, 1990: Biomass burning in the tropics: impact on atmospheric chemistry and biogeochemical cycles. *Science,* **250**, 1669-1678.

Crutzen, P.J. and F. Arnold, 1986: Nitric acid cloud formation in the cold Antarctic stratosphere: a major cause for the springtime "ozone hole." *Nature,* **324,** 651-655.

Crutzen, P.J. and P.H. Zimmermann, 1991: The changing photochemistry of the troposphere. *Tellus,* **43AB**. 136-151.

Dianov-Klokov, V.I., L.N. Yurganev, L. N. Grechko and A.V. Dzhola, 1989: Spectroscopic measurements of atmoshperic carbon monoxide and methane. T. Latitudinal distribution, *J. Atmos. Chem,* **8**, 139-151.

Farman, J.C., B.G. Gardiner and J.D. Shanklin, 1985: Large losses of ozone in Antarctica reveal seasonal ClO_x/NO_x interaction. *Nature,* **315**, 207-210.

Fishman, J., C.E. Watson, J. C. Larsen and J.A. Logan, 1990: Distribution of tropospheric ozone determined from satellite data. *J. Geophys. Res.,* **95** (D4), 3599-3617.

Golitsyn, G.S. and G. N. Panini, 1989: Contemporary changes of the Caspian Sea level. *Soviet Meteorology and Hydrology,* No. 1, pp. 57-64, Moscow, USSR.

Golitsyn, G.S., A.V. Dzuba, A. G. Osipian and G. N. Panini, 1990: Regional climate changes and their impact on the Caspian Sea level rise. *Proc. of the USSR Academy of Sciences* (Duklady), **313**(5), 1224-1227, Moscow, USSR.

Hofmann, D. J., J.W. Harder, S.R. Rolf and J.M. Rosen, 1987: Balloon-borne observations of the development and vertical structure of the Antarctic ozone hole in 1986. *Nature,* **326**, 59-62.

IPCC (Intergovernmental Panel on Climate Change), 1990: *Climate Change: The IPCC Scientific Assessment.* Houghton, J.T., G.J. Jenkins and J.J. Ephraums, eds.; Cambridge University Press, 365 pp.

Isaksen, I.S.A. and O. Hov, 1987: Calculation of trends in the tropospheric concentration of O_3, OH, CH_4 and NO_x. *Tellus,* **39B**, 271-285.

Johnston, H. S., D.E. Kinnison and D.J. Wuebbles, 1989: Nitrogen oxides from high-altitude aircraft: an update of potential effects on 0_3. *J. Geophys. Res.,* **94**, 16351-16363.

Keller, M., W.A. Kaplan and S.C. Wofsy, 1986: Emissions of N_2O, CH_4 and CO_2 from tropical soils. *J. Geophys. Res.,* **91**, 11791-11802.

Kuhlbusch, T.A., J.M. Lobert, P.J. Crutzen and P. Warneck, 1991: Molecular nitrogen emissions from denitrification during biomass burning. *Nature,* **351**, 135-137.

Lelieveld, J. and P.J. Crutzen, 1992: Influence of atmospheric chemical feedbacks on greenhouse warming by methane. *Nature,* **355**, 339-342.

McKeen, S.A., E.-Y. Hsie and S.C. Liu, 1991: A study of the dependence of rural ozone on ozone precursors in the Eastern United States, *J. Geophys. Res.,* **96**, 15377-15394.

Molina, L.T. and M.J. Molina, 1987: Production of Cl_2O_2 from the self-reaction of the ClO radical, *J. Phys. Chem.,* **91**, 433-436.

Molina, M.J., T.-L. Tso, L.T. Molina and F.C.-Y. Wang, 1987: Antarctic stratospheric chemistry of chlorine nitrate, hydrogen chloride, and ice: release of active chlorine: *Science,* **238**, 1253-1257.

Muzio, L.J. and J.C. Kramlich, 1988: An artifact in the measurement of N_2O from combustion sources. *Geophys. Res. Lett.*, **15**, 1369-1372.

NAS (National Academy of Sciences), 1983: *Changing Climate,* National Academy Press, Washington, DC, 496 pp.

OTA (Office of Technology Assessment), 1990: Congress of the United States. *Catching our Breath.* Next Steps for Reducing Urban Ozone. US Congress, Washington, DC.

Peter, Th., C. Bruhl and P.J. Crutzen, 1991: Increase in the PSC-formation probability caused by high-flying aircraft.*Geophys. Res. Lett.,* **18**, 1465-1468.

Prinn, R., D. Cunnold, R. Rasmussen, P. Simmonds, F. Alyea, A. Crawford, P. Fraser and R. Rosen, 1990: Atmospheric emissions and trends of nitrous oxide deduced from 10 years of ALE-GAGE data. *J. Geophys. Res.,* **95**, 18369-18386.

Quay, P.D., S.L. King, J.M. Landsdown and D.O. Wibur, 1991: Carbon isotopic composition of atmospheric CH_4: fossil and biomass burning source strengths. *Global Biogeochemical Cycles,* **5**, 25-48.

Raval, A. and V. Ramanathan, 1989: Observational determination of the greenhouse effect. *Nature,* **342**, 758-761.

Rodhe, H. and R. Herrera (eds.), 1988: *Acidification in Tropical Countries,* SCOPE 36, Wiley-Chichester, 405 pp.

Stolarski, R.S., P. Bloomfield and R.D. McPeters, and J.R. Herman, 1991: Total ozone trends deduced from Nimbus 7 TOMS data. *Geophys. Res. Lett.*, **18**, 1015-1018.

Thompson, A.M. and R.J. Cicerone, 1986: Possible perturbations to atmospheric CO, CH_4 and OH. *J. Geophys. Res.,* **91**, 10853-10864.

Tolbert, M.A., M.J. Rossi, R. Malhotra and D.M. Golden, 1987: Reaction of chlorine nitrate with hydrogen chloride and water at Antarctic stratospheric temperatures. *Science,* **238**, 1258-1260.

Toon, O.B., P. Hamill, R.P. Turco and J. Pinto, 1986: Condensation of HNO_3 and HCl in the winter polar stratosphere. *Geophys. Res. Lett.,* **13**, 1284-1287.

Valentin, K.M., 1990: Numerical modeling of the climatological and anthropogenic influences on the chemical composition of the troposphere since the last glacial maximum, PhD dissertation, Johannes Gutenberg-University, Mainz.

Volz, A. and D. Kley, 1988: Ozone measurements made in the 19th Century: an evaluation of the Montsouris series. *Nature,* **332**, 240-242.

Wahlen, M., N. Takata, R. Henry, B. Deck, J. Zeglen, J.S. Vogel, J. Southon, A. Shemesh, R. Fairbanks and W. Broecker, 1989. Carbon-14 methane sources and in atmospheric methane: the contribution from fossil fuel carbon. *Science,* **245**, 286-290.

Wigley, T.M.L., 1991: Could reducing fossil-fuel emissions cause global warming? *Nature,* **349**, 503-506.

WMO (World Meteorological Organization), 1985: Atmospheric Ozone 1985, Global Ozone Research and Monitoring Project, Report No. 16, Geneva.

WMO, 1988: Report of the International Ozone Trends Panel, Global Ozone Research and Monitoring Project, Report No. 18, Geneva.

Zander, R., P. Demoulin, D.H. Ehhalt, U. Schmidt and C.P. Rinsland, 1989: Secular increase of the total vertical column abundance of carbon monoxide above Central Europe since 1950. *J. Geophys. Res.*, **94**, 11021-11028.

CHAPTER 3

Climate Sensitivity, Climate Feedbacks and Policy Implications

Martin I. Hoffert

Editor's Introduction

The study of climatic problems includes much more than the atmosphere. In this chapter, Martin Hoffert looks back over the Earth's history to explore the natural archives of paleoclimate data, using modern techniques of isotopic and oceanographic analysis. He combines these analyses with computer modelling studies, which were designed to estimate the sensitivity of global climate to changes in atmospheric composition, the position of the Earth's orbit, and circulation of the world ocean. In the process, as one of the paper's reviewers (geophysicist Tyler Volk) observes, Hoffert spins a complex web, providing us a rapid and enthralling tour of Ice Ages, the planetary evolution of Earth, Mars, and Venus, atmospheric physics, soil chemistry, and ocean circulation. And he explains the strengths, weaknesses, and fundamental limits of climate models in terms that are readily penetrable by the lay reader.

Because its range is the widest, this chapter is the longest in the book. But it also expands directly on the discussion of linkages in the previous chapter, to show the interconnections among the major components of our planetary system — air, ocean, land, and biota. Hoffert notes that everything on the surface of the planet — from the smallest dust particle to the largest iceberg — absorbs and re-emits the radiation that comes to us from the sun. And he explains how each of these components interacts with Earth's lifeforms to affect the overall radiative balance of the planet with its surroundings.

Woven throughout this chapter, you will find a thread of respect for the fundamental uncertainties in our understanding of natural processes, and the intellectual challenge facing geophysicists, geochemists, and other scientists who try to model future effects of human activities on climate. None the less, this chapter concludes with a strong plea for political action to reduce the risks of rapid climate change while we carefully monitor the state of the planet and continue to pursue basic and applied research in planetary science.

Martin Hoffert is a professor at, and former chairman of, the Department of Applied Science at New York University — one of the pre-eminent interdisciplinary graduate-level science programmes in the United States. He is best known for his work in geochemistry and oceanography and for his path- breaking applications of small, transparent, and understandable models to large complex geophysical processes.

Syukuro Manabe, one of the few true masters of the three-dimensional general circulation models of the atmosphere, caps off Hoffert's presentation, describing a potent new international research strategy for meeting the challenge of uncertainty head-on. *(See page 51)* *Manabe urges us to continue and intensify carefully targeted research on planetary dynamics, linking three current lines of attack: (1) process studies of atmospheric dynamics that can improve scientific knowledge of cloud feedbacks and land surface processes; (2) monitoring of the natural environment with both in situ measurements and observations from satellites; and (3) diagnostic studies to promote model validation, testing, and impact assessment. With this three-pronged approach, Manabe argues, the scientific community can provide additional policy-relevant data on the prospects for rapid climate change. This additional data can help national leaders and business executives to sustain the momentum of continued economic development in a world where future climate is uncertain and, potentially, quite different from that of the recent past.*

- I. M. M.

1 Climate Feedbacks and Uncertainty

What limits our capability to project global warming into the future? It is not an inability to compute the direct amount of radiative heating by greenhouse gases. Rather, it is our poor understanding of climate feedbacks. These are processes of the climate system, that amplify or diminish any of the climatic changes from direct global warming (Manowitz, 1990). For example, cloud-radiative feedback (also described by Crutzen and Golitsyn, 1992, this volume) is the process in which greenhouse gas emissions lead not just to rising temperatures, but also to changes in the distribution and characteristics of clouds over the planet. Clouds scatter incident solar radiation back to space (cloud albedo), but they also tend to retain heat radiated from the surface below (cloud greenhouse). As a result, changes in cloud distributions induced by greenhouse gas emissions can promote both warming and cooling, in different ways, simultaneously. It is not yet clear which of these effects prevails when the climate changes.

A feedback is positive if it amplifies a warming or cooling trend, negative if it diminishes the same warming or cooling trend. Positive feedbacks promote large climate changes, negative ones promote climate stability.

Our understanding of the cloud feedback process is incomplete enough that, by itself, it produces a factor-of-three uncertainty in future temperature changes. The statistical term "factor-of-three" refers to a range of prediction so uncertain that the top limit is three times the bottom limit. A typical climate model might, for example, be used to forecast a set of potential global temperature rises. When the feedback effects of cloud cover are built in to the model, they add a "factor-of-three" ambiguity. As a result, we can be no more specific than to predict a rise of between 1.5 and 4.5° Celsius for an atmospheric CO_2 doubling (Houghton et al., 1990). The cloud feedback problem is presently under intense study, partly in hopes of rendering more accurate forecasts; but it is by no means the only "wild card" that can affect global warming.

In this chapter, I will review a number of wide-ranging considerations bearing on feedback effects — first covering the physical and chemical processes underlying greenhouse calculations, then discussing feedbacks in the climatological record of Earth, Mars, and Venus, and focusing ultimately on certain potentially large wild card feedbacks from the oceans and biosphere. Let it be said up front that there are speculative aspects to this review. But there are also very practical reasons to explore these possibilities.

International bodies like the Intergovernmental Panel on Climate Change (IPCC) have projected dramatic temperature rise and other climate changes from anthropogenic greenhouse gases over the next century (Houghton et al., 1990). Some critics emphasize uncertainties of the climate models on which these projections are based, and conclude that global warming is likely to have been overestimated (Marshall Institute, 1989; Lindzen, 1990; Ellsaesser, 1990). That may be. But in fairness, because of the unpredictability of feedback processes, uncertainties in climate models are as likely to *underestimate* future global warming as to overestimate them (Schneider, 1990).

1.1 Lessons from the climatic past

In modelling the future, it is instructive to look to our climatic past for guidance (MacCracken et al., 1990). Even without a fossil-fuel greenhouse, the present world climate is, from the perspective of the past few million years, anomalously warm — an interglacial hiatus from which, in the absence of human intervention, we are destined to descend to another ice age. The climate history of the Earth over the past 1.6 million years, the so-called Pleistocene period, is one of cycles between glacial (cooler) and interglacial (warmer) periods. Ice volume and temperature fluctuations, with periodicities of 100,000, 41,000 and 23,000 years, are seen in the geologic record — and are thought to arise from variations in the Earth's orbit (Crowley and North, 1991).

But one needs to go back millions, even tens of millions, of years to find atmospheric CO_2 levels (and corresponding periods of global warmth) comparable to what humanity can produce in coming centuries. Figure 1, showing a range of prehistoric climate reconstructions, suggests that comparable levels last existed in the mid-Pliocene era, about three million years ago. As a result of various non-equilibrium effects — different feedbacks coming into play over different time scales — simple "paleo-analogues" of global warming may not exist (Crowley, 1990). If current climate and carbon cycle models are even approximately correct, the present era of human history can bequeath on to future generations an anomalous "super-interglacial" period — an unprecedented hot spell persistent enough to overcome the next few glacial cycles. For reasons discussed subsequently, humanity's combustion of the fossil fuel reserve could alter the global climate over time scales comparable to the lifetime of the human species. Those who see *Homo sapiens* as somehow apart from natural processes might well conclude we are at "the end of nature" (McKibben, 1989).

1.2 Implications of uncertainty on climate policy

As we move toward uncharted climatic territory, it seems all the more important to understand relevant feedback mechanisms that may have operated on Earth and on other Earth-like planets. It may become rapidly necessary to make climate policy in the face of substantial uncertainty. The stakes are too high to rule out plausible climate change processes for which there is physical evidence.

Indeed, recent environmental history tells us that surprises can be expected in the precise way that any human-triggered environmental perturbation unfolds. Atmospheric photochemistry models were predicting ozone depletion from chlorofluorocarbon (CFC) emissions well before they were measured. But the specifics of the "ozone hole" were unanticipated. Local ozone depletion have proven much more extreme than the gradual decreases that had been predicted by models. The Antarctica ozone hole is now understood to depend on chemical reactions catalysed in polar stratospheric clouds, reactions which had not been included in models. Although the implications of ozone depletion are still hotly debated, they were perceived as serious enough in the post-ozone-hole world to rapidly adopt a global CFC regulatory treaty — the Montreal Protocols. Though controversial at first, rollbacks in CFC emissions have proven easier to implement than some had originally thought. Reductions in the emissions of CO_2, the major greenhouse gas, may prove more challenging because at least 85% of global energy consumption is presently from fossil fuels (Hammond, 1990; Holdren, 1992, this volume).

Could some climatic feedback from the oceans or biosphere trigger a response to greenhouse warming for which the world is ill-prepared? A broad perspective, including consideration of climatic long shots, seems only prudent if one seeks to minimize "unpleasant surprises in the greenhouse" (Broecker, 1987).

2 The astrophysics of the Greenhouse Effect

2.1 The validity of greenhouse models

When global climate is constant, conservation of energy dictates that the solar energy absorbed by the atmosphere, clouds and surface must equal the energy radiated back from the Earth's atmosphere, clouds, and surface to cold space. This is the fundamental energy balance on which climate modelling is based. Necessarily — due to the complexities of molecular absorption and emission of radiation — the computer codes from which the solar and infrared components of radiation are computed are too often intimidating to the uninitiated. Atmospheric radiation is a discipline where the modelling art looks mysterious even to scientifically knowledgeable researchers in adjacent fields.

This does not mean radiative heating rates computed from greenhouse gases are a significant source of error or uncertainty. Radiation models developed and intercompared by a dedicated research community over decades, with the help of satellite data, are believed accurate to within a few percent.

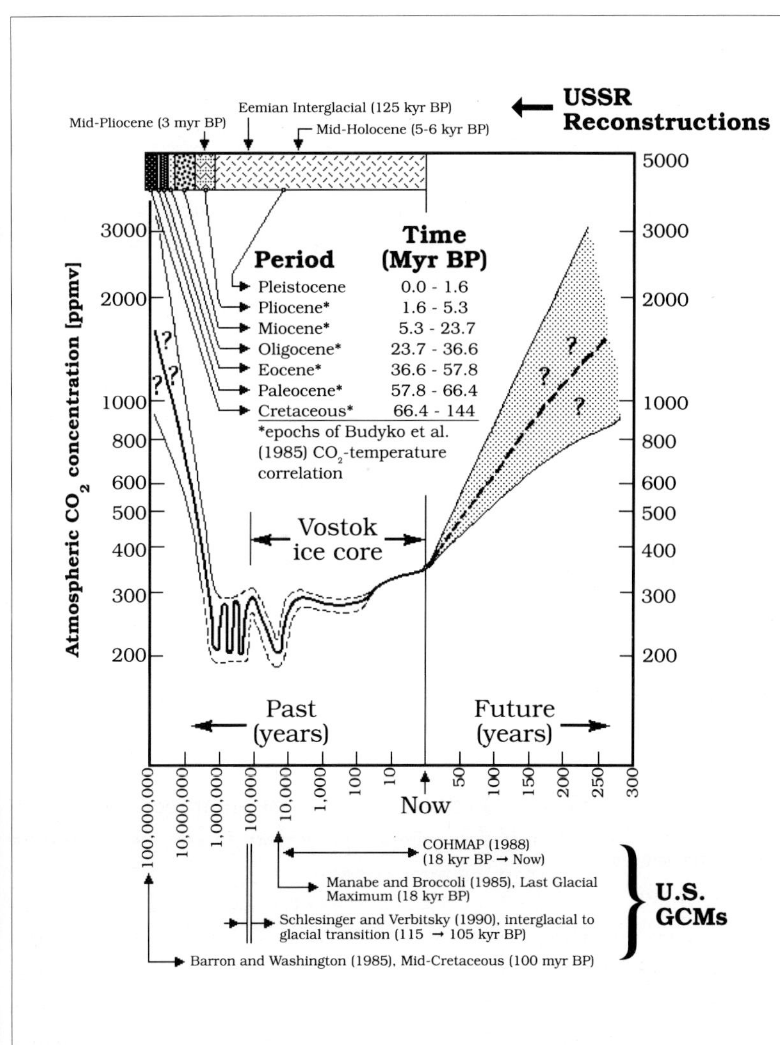

Figure 1. Variations of atmospheric carbon dioxide concentration, shown over time: from 100 million years in the past to 300 years into the future. The upper scale (top left corner) shows only past paleoclimates as reconstructed by Soviet researchers, keyed against a table showing the name and duration of each period. (MacCracken et al., 1990). The lower scale shows periods simulated by general circulation models, both for the past (to the left of "Now") and for the future (shown to the right of "Now"). Note that while the future scale is linear, the past recedes exponentially.

It is believed that the paleoclimate variations, were partly caused by the greenhouse effect, from changes in carbon dioxide levels. Human emissions of fossil fuel carbon dioxide in the coming decades and centuries could produce levels of greenhouse warming comparable to those produced millions of years ago.

2.2 The Earth as an absorber and emitter of radiation

Simple theories of the Earth's radiative behaviour compare the Earth to the astrophysical concept of a "black body". Like the black hole of astrophysics, a black body absorbs all radiation which contacts it. Unlike a black hole, it is also a perfect emitter of radiant energy. The amount of electromagnetic radiation which black bodies emit is proportional to the fourth power of their temperature.[1]

The greater the temperature of a black body, the shorter the wavelength of the radiant energy which it emits. For example, stars are very nearly black bodies; which is why hot stars are blue (short wavelength), cold stars red (long wavelength) and our yellow sun is somewhere in between. It is possible to express the Earth's energy loss as equivalent to that of a black body. Most of this radiation leaving the Earth is in the infrared part of the spectrum (between the wavelengths of 8 and 30 microns). These wavelengths of radiation are emitted and absorbed by greenhouse gases, including water vapour, carbon dioxide (CO_2), ozone (O_3), methane (CH_4), and nitrous oxide (N_2O) and the chlorofluorocarbons.

That greenhouse gases warm the surface of the Earth and other planets is a well-established scientific fact. A classic proof that the greenhouse effect is taking place on Earth is that the present global mean surface temperature, about 288 Kelvin or 15° Celsius, is 33 °C warmer than the Earth's black-body temperature (255K).[2] The additional thirty-three degrees Celsius of surface warmth is due to the absorption and re-emission of radiation by atmospheric water vapour and (to a lesser extent) by carbon dioxide and ozone. What is, to some extent, still uncertain is the amount and timing of global warming from humanity's greenhouse gas emissions.

2.3 Greenhouse gas molecules and the shrinking infrared "window"

An orbiting observer who could "see" in thermal infrared could not see the Earth's surface very well because the view would be blocked by water vapour, carbon dioxide and ozone — trace gases which we can see through, but which absorb infrared radiation. But the atmosphere's infrared opacity has "windows" of transparency at some infrared wavelength bands — that let some wavelengths of radiation through. Increases in the amount of greenhouse gases produce global warming because they further block infrared windows. Some gases, because of the strength and distribution of their absorption spectrum, are more effective blockers than others (Table 1). For example, molecule for molecule an increase in chlorofluorocarbon-12 concentration produces almost sixteen thousand times the greenhouse heating of carbon dioxide.

This comparison does not take into account the lifetime of various greenhouse gas molecules, and their integrated effect over time. For example, CO_2 will almost certainly continue to be the major greenhouse gas, because there are so many molecules of CO_2 being added, and they persist so long in the atmosphere.

2.4 The link between surface temperature and incoming radiation levels

It is also critical to the greenhouse effect that in the troposphere — the first 11 kilometres of atmosphere above the surface, where clouds and weather systems exist — the temperature drops off with altitude about 6.5 °C for every kilometre higher. The altitude from which radiation escapes from the Earth to space (called the radiation height), is currently about five kilometers above the surface — roughly halfway up the troposphere. When they escape, the infrared photons carry with them, in their distribution of wavelengths (or "temperature"), a signature of the atmosphere's radiation height. As more greenhouse gases are added, the atmosphere gets optically thicker — it radiates infrared from higher, colder altitudes. This creates an imbalance between the absorbed solar radiation and emitted longwave radiation. The climate system restores the balance by raising the surface temperature to a new value. This process is the greenhouse effect.

Because the rate at which temperature drops off with altitude tends to remain constant, it is possible to find a height at which the absorbed solar energy equals the black body radiation to space. This is called the effective radiation height. The temperature at the radiation height is coupled to the Earth's surface temperature — as one changes, so does the other. The coupling between the two temperatures — that of the surface and that of the radiation height — results from a process known as "convective adjustment," and allows us to make a first estimate of the climate sensitivity.

In analyses of climate change, it is important to distinguish between radiative forcing and surface temperature

Table 1. Radiative heating of anthropogenic greenhouse gases relative to carbon dioxide per molecule added to the atmosphere

Gas	Relative radiative heating
Carbon dioxide (CO_2)	1
Methane (CH_4)	21
Nitrous oxide (N_2O)	206
Chlorofluorocarbon-11 (CFC-11)	12,400
Chlorofluorocarbon-12 (CFC-12)	15,800

(After MacCracken et al., 1990).

[1] Per unit area, the rate of energy loss from a black body at temperature T is σT^4, where σ is the *Stefan-Boltzmann constant* (≈ 5.67 x 10^{-8} W m^{-2} K^{-4}).

[2] T_{eff} is the effective black body temperature, with S_o the solar energy per unit area falling on a plane perpendicular to the sun's rays at the top of the atmosphere (the solar constant = 1368 watts per square metre (W m^{-2}) and the fraction of solar radiation reflected back to space (the planetary albedo ≈30%). A global energy balance gives: $T_{eff} = [S_o(1-\alpha)/(4\sigma)]$ ≈255 K (-18 °C).

change. Radiative forcing refers to the rate at which heat is added to or removed from the climate system — by changes in incident sunlight, planetary albedo (reflectivity), or greenhouse gas concentrations. Climate *sensitivity* is the ratio of the temperature change to the radiative forcing. Gaining a better understanding of the sensitivity of the Earth's climate is a fundamental goal of climate change research.

If, for example, the atmosphere were heated by an additional four watts per square metre — the heating predicted by radiative models for a CO_2 doubling since the beginning of the industrial period — we would expect the surface to warm eventually by about one degree Celsius.[3] This is because the climate sensitivity based on black-body cooling alone would correspond to a surface warming of about 1° Celsius. However, this warming could be either increased by positive feedback, or decreased by negative feedback. The realized global warming will also be affected by the heat capacity of the oceans, which would tend to postpone the temperature rise somewhat into the future.

A more general discussion will be given presently, linking climate sensitivity, feedbacks and transient climate change on decadal to century time scales. But in keeping with the broad objectives of this survey, it is useful to first discuss how climate feedbacks can operate over much longer time scales — and not only on Earth.

3 The Evolution of Climate

Greenhouse gases seem to have profoundly affected the climates of at least three "terrestrial" planets: Venus, Earth and Mars. These all have surface temperatures that are warmer than they would be without the greenhouse effect (Table 2), and all have very likely experienced evolutionary histories in which the greenhouse gas content of their atmospheres has varied (Kasting et al., 1988).

3.1 The vanished greenhouse of Mars

At present, Mars is so cold (218K, or -55 °C) that, even with its large seasonal and daily temperature fluctuations, liquid water does not normally exist. Such low temperatures mean there can be very little water vapour in the atmosphere, though appreciable amounts of water as permafrost are believed to exist beneath the surface and in permanent ice caps. (Overlying the water ice caps at both poles are "seasonal" ice caps of solid carbon dioxide, which condense from the atmosphere in winter and return as a gas to the atmosphere in summer.) Given this deep freeze, water vapour feedback on Mars today is so weak that even with twice the surface pressure of CO_2 as the Earth, Mars produces less than half the greenhouse warming. This is not too surprising, as most of the 33 degrees of greenhouse warming on Earth comes from water vapour.

What is surprising is that images of the Martian surface, returned by Mariner and Viking spacecraft, show evidence of ancient riverbeds — channels cut into the surface long ago, perhaps billions of years ago, by running water — complete with tributaries, meandering paths, and other properties of rivers on Earth (Figure 2). This suggests Mars was once much warmer, warm enough to sustain liquid water (requiring a surface temperature greater than 273 Kelvin). The most plausible explanation of this prior warmth is a strong greenhouse warming from an earlier, much denser CO_2 atmosphere. Because Mars is smaller than the Earth, it cooled earlier and is tectonically inactive at present. We may surmise that early in its history, like the Earth, it released volatile gases, including carbon dioxide and water. The highest mountain in the solar system is an inactive volcano on Mars: Olympus Mons, taller than Mount Everest.

As Mars cooled, degassing of CO_2 (injection to the atmosphere from fissures in Martian surface rock) slowed. At the same time, carbon dioxide may have been drawn out of the atmosphere and converted to calcium carbonate rocks in the crust — a process that could continue so long as the temperature remained above the freezing point of water.

[3] In this work, we call the energy radiated to space per unit surface area per unit surface temperature change the radiative damping coefficient, λ. The climate sensitivity, λ^{-1}, is inversely proportional to the radiative damping coefficient. A first approximation of λ, including the convective adjustment but neglecting all other atmospheric feedbacks, is the blackbody cooling coefficient, $\lambda^* = 4\sigma T_{eff}^3 \approx 4.0\ Wm^{-2}\ K^{-1}$.

Table 2. Variation of effective and surface temperature on three planets showing the influence of greenhouse gases

Planet	Atmosphere pressure [atm]	Greenhouse gases	Orbit [AU]	Solar constant [W m²]	Albedo [%]	Effective temp [K]	Surface temp [K]	Greenhouse warming [°C]
Venus	90	CO_2	0.723	2620	76	229	750	521
Earth	1	H_2O, CO_2	1.000	1368	30	255	288	33
Mars	0.006	CO_2	1.524	589	25	210	218	8

Figure 2. Images of the Martian surface, returned by the Viking spacecraft, show evidence of ancient river beds, complete with tributaries, meandering paths, and other properties of rivers. This suggests that Mars was once much warmer, warm enough to sustain liquid water. The most plausible explanation is a strong greenhouse warming from a much denser CO_2 atmosphere early in the planet's history. Source: Based on Nasa Viking images assembled by the US Geological Survey - USGS Miscellaneous Investigations Series.

(Liquid water is necessary to precipitate calcium carbonate from carbon dioxide gas.) Presumably, the process stopped when the planet cooled to the freezing point of water. This scenario may explain why Mars' atmospheric pressure is so close to the triple point — the point where solid, liquid and vapour phases of water coexist. A Martian greenhouse more than three-and-a-half billion years ago is the most plausible hypothesis for its apparently wetter climate early on. The critical evidence confirming this hypothesis would be Mars rock samples showing carbonate content consistent with absorption of a prior thick carbon dioxide atmosphere. Such samples will hopefully be obtained and analysed in future missions to Mars.

3.1.1 Earth compared to Mars

Evidently, Earth averted the deep freeze of Mars. This is at least partially because the Earth's crust has remained geologically active. Unlike Mars, the CO_2 in Earth's carbonate rock is recycled to the atmosphere through a combination of chemical reactions and plate tectonics. It takes 100 million years for molten rock to upwell at the boundaries of tectonic plates under the sea, spread across the ocean floor and, finally, plunge back down into the deep mantle. Each of these stages releases CO_2 into the atmosphere or swallows it into the rock shell of the Earth again — a cycle which has taken place many times over since the Earth formed. The carbon rock cycle is a geochemical conveyor belt partly driven by radioactive heat released by the mantle. During the 4.6 billion years of planetary evolution there has been a gradual slowing of this conveyor belt as the Earth cooled, and a reduction in the amount of carbon dioxide in the Earth's atmosphere. Fortunately for life on Earth, the conveyor belt has not altogether stopped, as it has, apparently, on Mars.

3.2 The runaway greenhouse of Venus

An example of a positive feedback gone wild is the runaway greenhouse of Venus, where vaporized greenhouse gases apparently caused progressively more warming until all surface volatiles, including water and carbon dioxide, were driven to the gaseous phase. Venus today is an uninhabitable hell with a dense carbon dioxide atmosphere and a surface temperature of 750 Kelvin (477 °C) — hot enough to melt lead. In the unlikely event they ever existed, Venus' liquid water oceans have vanished without a trace. Even the water vapour that might, under more favourable conditions, have condensed to oceans on Venus is gone — the hydrogen lost to space, the oxygen re-combined in rocks at the surface.

3.2.1 Earth compared to Venus

Fortunately again, a life-threatening runaway greenhouse effect like that of Venus, but caused by humanity's continued fossil fuel carbon combustions, is unlikely on Earth. Burning all the oil, natural gas, coal and shale could not heat the atmosphere enough to volatilize the carbonate rocks to carbon dioxide or boil the oceans.

But we ought not to become too complacent about avoiding a Venus-like runaway greenhouse. Major, long-term changes in the Earth's climate lasting thousands of years and beyond are likely if humanity consumes a significant fraction of its recoverable fossil fuel reserve. The inadvertent introduction of greenhouse gases to the Earth's atmosphere since the worldwide industrial revolution in amounts large enough to affect global climate has been called a "grand geophysical experiment" (Revelle and Suess, 1957). For better or worse, our descendants will experience the climatic impacts of this experiment.

3.3 Evolution of the Earth's greenhouse effect

Stars brighten as they evolve. The sun, for example, has brightened by 30% since the solar system formed. All things being equal, a simple energy balance indicates that the Earth's surface would have been below the freezing point of water at the time that life evolved — a biological impossibility known as the faint-young-sun paradox (Kasting et al., 1988). The greenhouse effect of an early atmosphere much richer in CO_2 can be invoked to resolve this paradox, with geologic evidence to support it. But, in the face of increasing solar luminosity, how was the planetary "thermostat" regulated in the eons after life had evolved? Some stabilizing climatic feedback process is strongly indicated.

Walker et al. (1981) were the first to propose that, on geologic time scales, climate is stabilized by factors affecting the rate at which calcium silicate rocks are geochemically *weathered* (that, is, converted to calcium carbonate by reaction with atmospheric carbon dioxide). Recall that weathering may also have played a significant role in removing Mars' early CO_2 atmosphere. The rate at which CO_2 is removed from the Earth's atmosphere by rock weathering increases as temperature increases for two reasons: (1) rainfall and runoff increase, carrying more carbonate to the ocean; and (2) the respiration of soil organisms increases.

The second mechanism — by releasing more carbon dioxide to the soil, which diffuses to the atmosphere — initially produces a positive feedback: Warmer surface temperatures promote the release of more carbon dioxide, which in turn promotes warmer temperatures, and so on. But in the long run, higher temperatures will increase weathering from rainfall — a negative feedback. The warming of surface temperatures would promote, through weathering, a drain of CO_2 out of the atmosphere, which would drive CO_2 levels down. That, in turn, would create a cooling trend opposed to the initial warming. The cooling trend would eventually slow down the drain of CO_2, until it reached an equilibrium. Thus, the long-term effect of geochemical weathering is to stabilize global temperature — a negative climate feedback.

3.4 The role of life in greenhouse evolution

Although many geochemists believe carbon dioxide variations over geologic time can be explained abiotically, they are increasingly challenged by theories in which life plays a more or less important role. Even the geochemical weathering rate feedback is now understood to be affected by living organisms in many different ways (Volk, 1987, 1989).

The most fiercely controversial idea in this arena may be the Gaia hypothesis of James Lovelock and Lynn Margulis (Lovelock, 1988). This hypothesis — some think "metaphor" is a better term — holds that living organisms on Earth actively regulate atmospheric composition and climate in the face of challenges like the increasing luminosity of the sun. Specific Gaian mechanisms that have been invoked are the emission of greenhouse gases and/or reflective cloud-producing gases (Charlson et al., 1987) by organisms to produce a planetary homeostasis. Though a compelling case has by no means been made for the "strong" — planetary homeostasis — version of Gaia, most earth scientists would accept a "weak" version — that biological process and feedbacks affect global climate. We will examine biological processes and biological feedbacks on climate in more detail presently.

3.5 Continental drift

The gradual reduction in the CO_2 content of the atmosphere, as the Earth cooled, was probably modulated by the detailed dynamics of continental drift. Such modulations are recorded as changes in sea floor spreading rates (measured by magnetic field reversals relative to sea-floor spreading centers like the mid-Atlantic ridge). There is also evidence from variation in carbon isotopes and organic carbon burial rates that over the last 600 million years — the Phanerozoic eon — atmospheric CO_2 rose and fell by large factors (Budyko et al., 1987; Berner, 1990). On these time scales, supercontinents break up and reassemble, and ice caps come and go in a flash.

4 Lessons from Climate Evolution

4.1 Evaluating the case for pessimism: "return to Cretaceous"

One hundred million years ago — the mid-Cretaceous when dinosaurs roamed the Earth — was a time of great warmth. The Earth was about 10 °C hotter and the poles ice-free, very possibly in response to greenhouse warming from an atmosphere four to ten times richer in carbon dioxide (Barron and Washington, 1985). We have only a crude idea of what planetary ecosystems looked like in those days, or of how well a human technological civilization would have fared. Interestingly, the recoverable fossil fuel reserve contains enough carbon to raise atmospheric CO_2 to mid-Cretaceous levels (Figure 1). While paleoclimatic analogies have problems, it is not unreasonable to think of the Cretaceous climate as a "worst case" greenhouse warming scenario, if humanity burns the fossil fuel reserve to depletion.

4.2 Evaluating the case for optimism: "return to Eden"

Given that large variations in atmospheric carbon dioxide and climate have occurred over Earth history — all the while within boundaries habitable to life — it is not obvious that humanity's greenhouse will necessarily be "bad". Some Soviet climatologists have recently used a paleoclimate analogue approach to project a more favourable world climate in the latter half of the 21st century, as the result of continued fossil fuel burning (Budyko, 1991; Budyko and Izrael, 1991). Climate model projections, coupled with the prospect of indefinite climate change, suggest that any regional "winners" in global warming will be transient at best, while other regions would suffer reduced agricultural productivity due to increased drought frequency, with continuous changes. Budyko and his colleagues, on the other hand, see a warmer climate as one with wetter continents everywhere. Combined with increased crop fertilization from higher CO_2 levels, this warmer climate would provide a bonanza of global agricultural productivity — perhaps enough, they suggest, to avert the starvation of billions in the much more populous world of the 21st Century.

pid climate change as well, and about the need for a constructive response.[4] At this point, even the executive branch of the US government (which tends to be quite conservative on global warming) has concluded in its National Energy Strategy that "there is sufficient credible scientific concern to start acting to curb the buildup of so-called greenhouse gases" (White House, 1991).

I will argue that even if the long-term climatic impacts prove to be beneficial — and it is by no means clear that they will — a rationale exists for limiting the rate of global warming based on the rate ecosystems can respond to climatic change. I also want to show that wild card climatic feedbacks can affect these warming rates.

4.3 Evaluating the case for climate engineering

Profound technological developments are sometimes stimulated by accidental discoveries. One outcome of our species' inadvertent effect on planetary climate could be the ability to engineer climatic change. Proposals have already been made to compensate for global warming by geo-engineering — seeding the stratosphere with reflective aerosol particles, deflecting incident solar radiation with space-based mirrors, sequestering atmospheric CO_2 by fertilizing polar zone plankton and other ingenious manipulations (NAS, 1991). These might be considered humanity's intervention in the climate system to provide negative feedbacks. Apart from their technological feasibility, such capabilities raise a host of legal and ethical questions that are only beginning to be addressed.

Opponents of geo-engineering argue that such approaches pre-suppose a much better knowledge and understanding of the climate system than all but the most confident of scientists are prepared to claim. Moreover, these measures would not address the underlying problem at its scource: they won't do anything to reduce co-related problems such as acid rain,

[4] The US National Energy Strategy concludes that "there is sufficient credible scientific concern to start acting to curb the buildup of so-called greenhouse gases" (White House, 1991).

photochemical pollutants, and urban air pollution, which solutions based on reducing fossil fuel use would address. Finally, they are likely to require continuous action which, if interrupted for any reason, could lead to disaster if the greenhouse heating, which these solutions are designed to mask, were allowed to grow indefinitely.

If humanity can make major alterations in the Earth's climate without really trying, what could be done in a more technologically sophisticated age to change the inhospitable climates of neighbouring planets? Terraforming — the creation of Earth-like habitats on other planets — has received serious scientific attention in recent years (McKay et al., 1991). An earlier NASA study, prior to analysis of Viking mission data, assumed enough greenhouse warming potential in the CO_2 Mars polar caps that a modest reduction in their albedo (the amount of radiation they reflect) could raise planetary temperatures above freezing. This could be achieved, the study concluded, by covering the polar caps with low-albedo dust or growing dark plants on them. (Averner and MacElroy, 1976). We now know that the "dry ice" caps at Mars' poles are almost entirely volatilized and re-condensed each Mars year, and that the NASA terraformers overestimated the CO_2 mass of these caps available for greenhouse warming by many orders of magnitude.

More recent ideas include releasing carbon dioxide in surface rocks. McKay (1987) estimates an amount equivalent to one Earth atmosphere of pressure could be released in 200 years by diverting 1% of the incident solar energy to this purpose. But again, it is not yet known how much carbon dioxide exists either absorbed or in carbonate rocks (Fogg, 1989).

Allaby and Lovelock (1984) in their speculative novel, *The Greening of Mars,* explored the creation of a warmer, wetter and generally more habitable Mars by colonists seeding its atmosphere with chlorofluorocarbons. In their novel, CFCs are shipped from Earth as payload in formerly nuclear-tipped missiles left over from the Cold War — though the authors significantly underestimated the mass of CFCs it would take. Even with "free" interplanetary transportation, it might be more cost-effective to make chloroflorcarbons from material available on Mars. Also, though CFCs are certainly efficient greenhouse gases (Table 1), they are not necessarily optimum. The infrared absorption of CFCs is, after all, an incidental property of an artificial molecule developed for refrigeration and aerosol spray can applications. The best approach might be to manufacture on Mars new greenhouse molecules engineered to strongly and broadly absorb infrared radiation at very low concentrations.

Terraforming today is a not yet entirely respectable concept that serves mainly to provide plot opportunities for science fiction. But in the same sense that genetic engineering has the potential to alter the pace and direction of "natural" biological evolution, so geo-engineering has the potential to accelerate and divert natural climate evolution. Given sustainable sources of solar or fusion power, and a long enough term commitment, the feasibility of geo-engineering planetary atmospheres is real enough. Such schemes could profitably exploit climate feedbacks from greenhouse gases liberated from planetary surfaces, particularly the water vapour feedback from permafrost driven from Mars' regolith and ice caps. But before that happens, humanity will have faced the challenge of the "greenhouse century" — the next hundred years of greenhouse-gas-induced climate change on Earth.

5 Sensitivity, Feedbacks and Climatic Transients

There are important relationships between radiative feedback, the rate of global warming, and climate sensitivity. To get a feeling for these relationships, it is helpful to consider processes that come into play when the atmosphere is heated by some direct radiative forcing.

For our purposes, it is irrelevant whether this forcing comes from a change in the level of solar radiation, a change in planetary albedo (reflectiveness), or from the addition of greenhouse gases — so long as it comes from outside the climate system (Houghton et al., 1990).

5.1 Heat flux to the surface

Climatic changes are due to both external and internal factors. The internal factors can be either autonomous (self-generated) or they can be responses or feedbacks to some externally imposed change. An example of an autonomous internal change is the so-called El Niño — a dramatic change in the wind and temperature system of the equatorial Pacific, that occurs over two- to five-year intervals, and impacts the global climate. This variation is believed to arise from an interaction within the atmosphere-ocean system, rather than a response to radiative forcing imposed from outside. Climate feedbacks are triggered over various time scales, ranging from "fast" meteorological processes to "slow" geologic processes (some of which were discussed in the preceding section). We can calculate the heating of the planetary surface from the following formula that includes both external radiative forcing and internal feedbacks:

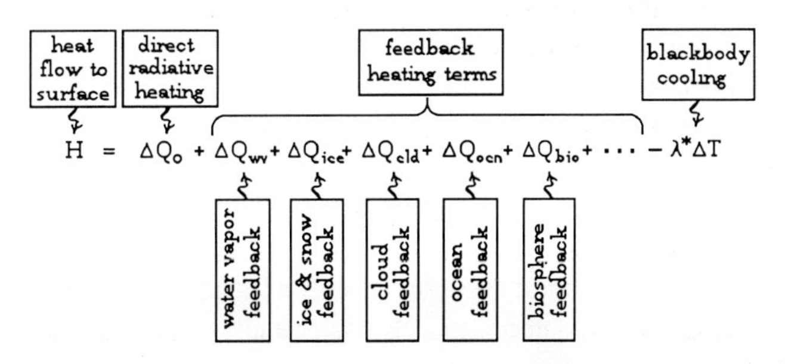

Perhaps the best-verified fast feedback is due to water vapour — a greenhouse gas whose concentration is largely controlled by surface temperature. When the surface warms, evaporation and precipitation both increase for the world as a whole; the water vapour content of the atmosphere goes up, and the greenhouse effect from this additional water vapour produces an incremental radiative forcing (ΔQ_{wv}). Satellite studies confirm that this water vapour feedback roughly doubles the direct radiative heating within a few months (Raval and Ramanathan, 1989). Twenty-five years ago, Manabe and Wetherald (1967) obtained a similar conclusion by the use of a radiative-convective atmospheric model. A warmer surface also normally results in less snow and ice at high latitudes, which means lower planetary albedo and an increased radiative forcing (ΔQ_{ice}). The result is more net radiant heat input to the climate system. Temperature-ice albedo feedback has both fast-reacting (sea ice and land snow) and slowly reacting (glacial ice) components.

A potentially large but presently poorly understood fast radiative feedback is due to clouds (ΔQ_{cld}). This is often broken down into a number of sub-processes. (For more about cloud feedback, see Crutzen and Golitsyn, 1992; this volume.) Other radiative feedbacks are heat fluxes from greenhouse gases or reflective aerosols driven from the oceans (ΔQ_{ocn}) and the biosphere (ΔQ_{bio}).

This leads to the following simple equation for the net heat flux to the Earth's surface:

$$H = \Delta Q_0 - \lambda \Delta T, \text{ where}$$

$$\lambda = \lambda^* - \left(\frac{\Delta Q}{\Delta T}\right)_{wv} - \left(\frac{\Delta Q}{\Delta T}\right)_{ice} - \left(\frac{\Delta Q}{\Delta T}\right)_{cld} - \left(\frac{\Delta Q}{\Delta T}\right)_{ocn} - \left(\frac{\Delta Q}{\Delta T}\right)_{bio} \cdots [2]$$

is the radiative damping coefficient. Notice that the basic blackbody dampingcoefficient λ^* is reduced if the feedback heating rates, the $(\Delta Q/\Delta T)_x$ terms, are positive. This is called positive feedback. When feedbacks are positive, the radiative damping coefficient decreases, and the climate sensitivity increases. Conversely, a negative feedback tends to decrease the climate sensitivity.

Under hypothetical steady state conditions, after a sufficient length of time subsequent to the imposition of ΔQ_0, the net heat flux H will vanish. The global warming would then depend only on the change in direct radiative forcing (ΔQ_0) and the radiative damping coefficient ($\Delta T = \Delta Q_0/\lambda$).

5.2 *Model predictions of climate feedbacks*

It is possible to use climate models to estimate the radiative damping coefficient and temperature changes for different feedback processes (Schlesinger, 1985; Mitchell, 1989). Typical values derived from current climate models are indicated in Table 3.

Of course, these estimates are only as good as the underlying feedback process models. Major studies are under way to determine the sensitivity of general circulation models (GCMs) to their own internal representations of feedback processes. This work is being coordinated by the Program for Climate Model Diagnosis and Intercomparison (PCMDI) at Lawrence Livermore National Laboratory. This work will help to explain why different models estimate a different climate sensitivity for the same radiative forcing. Although intercomparisons of climate models may seem more preoccupied with the meta-universe of simulations than with the Earth System itself, they are essential if we are to understand how feedbacks work.

5.3 *Models as laboratories of alternatives*

In the preceding section, I discussed a number of useful things that have been learned about feedbacks from the climatic history of the Earth and Earth-like planets. But paleoclimatologists who study ancient climates are at best passive observers. They cannot go back in a time machine to intentionally experiment with, or change, climate history. As humanity conducts its Grand Geophysical Experiment with greenhouse gases, numerical simulations are perhaps the only way to conduct parallel "laboratory experiments."

It is an article of faith that useful results will emerge from these simulations before substantial climate changes in the real world take place. Hopefully the results will arrive soon enough to allow policy to make a difference.

5.4 *Model uncertainties and cloud radiative feedback*

Most of the variability between global climate models can be traced to differences in cloud radiative feedback. This is illustrated in Table 3. When analyzing a CO_2 doubling since the industrial era, the predictions range from 1.5 ° to 4.8 °C. If model-to-model differences in climate sensitivity were entirely due to differences in the way cloud radiative feedback processes are depicted, we would expect the planetary cooling rate to be correlated by the feedback parameter. Figure 3, derived from the intercomparison by Cess et al. (1989) of 14 GCMs, shows this is very nearly the case.[5]

Notice in Figure 3 that different global climate models predicted different cloud radiative feedbacks — ranging from slightly negative to strongly positive. This dramatically illustrates how uncertainties in cloud radiative feedback translate into uncertainties in estimates of global climate sensitivity. Unfortunately we cannot say from an examination of the models alone, which values of cloud radiative feedback are the most realistic. Observational studies using satellites and selected field measurements are underway to resolve this issue. The main point we can make is that radiative feedbacks can strongly impact equilibrium climate sensitivity.

5.5 *Models and the time scale of climate response*

The atmospheric build up of greenhouse gases is occurring on time scales that range from decades to centuries. If the response of global climate to continually increasing greenhouse gas concentrations were immediate, it could be calculated from equilibrium atmospheric general circulation models (GCMs). There are many such models, and as we have seen, they exhibit quite different climate sensitivities. How-

---

[5] (The symbol λ is defined by these authors as the gain — the reciprocal of the planetary cooling rate defined here. Hopefully there will be no confusion on this point if the differences in symbology are kept in mind.)

Table 3. Effects of "fast" radiative feedbacks on planetary cooling and equilibrium climate response to carbon dioxide doubling.

Feedback	Radiative heating per unit temperature rise [$W\ m^{-2}\ °C^{-1}$]	cumulative radiative damping coefficient, λ [$W\ m^{-2}\ °C^{-1}$]	cumulative change in equilibrium temp., ΔT [°C][a]
Blackbody cooling	-4.0	-4.0	1.1
Water vapour[b]	1.4	2.6	1.7
Sea ice/land snow[c]	0.4	2.2	2.0
Cloud[d]	-0.8 to 1.3	3.0 to 0.9	1.5 to 4.8
Oceans, biosphere	?	?	?

[a] Computed from $\Delta T = \Delta Q_0/\lambda$, where $\Delta Q_0 = 4.4\ W\ m^{-2}$ for a CO_2 doubling.
[b] Sum of water vapour greenhouse, lapse rate and baroclinic instability feedbacks.
[c] Omits glacial ice feedback.
[d] Sum of cloud cover, altitude, formation, liquid water and radiative property feedbacks.

ever, the response of the climate system will not be immediate but will be delayed by the time it takes to warm the oceans, an effect called thermal inertia. One can imagine, on entering a cold house, a lag from the moment one turns up the thermostat, until the room actually becomes warm. This is a similar phenomenon.

A complicating factor is that the atmosphere and the oceans come to equilibrium on different time scales — about a month for the atmosphere versus a thousand years for the oceans (Hoffert and Flannery, 1985). This delayed response is called the transient climate change, and to calculate it, it is necessary to employ a coupled atmosphere-ocean climate model. There are fewer such models than atmospheric GCMs, and they contain additional uncertainties, including the very uncertain cloud-radiative feedback.

Because of the long and varied time scales of climate responses, the sheer computational requirements for simulating the coupled atmosphere/ocean system can become enormous. Perhaps future generations of parallel-processing supercomputers will routinely provide detailed transient climate projections for various locales and seasons. Indeed, a comprehensive analysis of the impact of global climate change requires such detailed information.

But at this point in time, it has proven useful in comparative assessments of climate response to radiative forcing, to use simpler ocean climate models, of the type proposed by Hoffert et al. (1980). Such a model, for example, was used in the recent IPCC assessment of changes in potential greenhouse gas concentration, and their effects over the next century (Houghton et al., 1990).

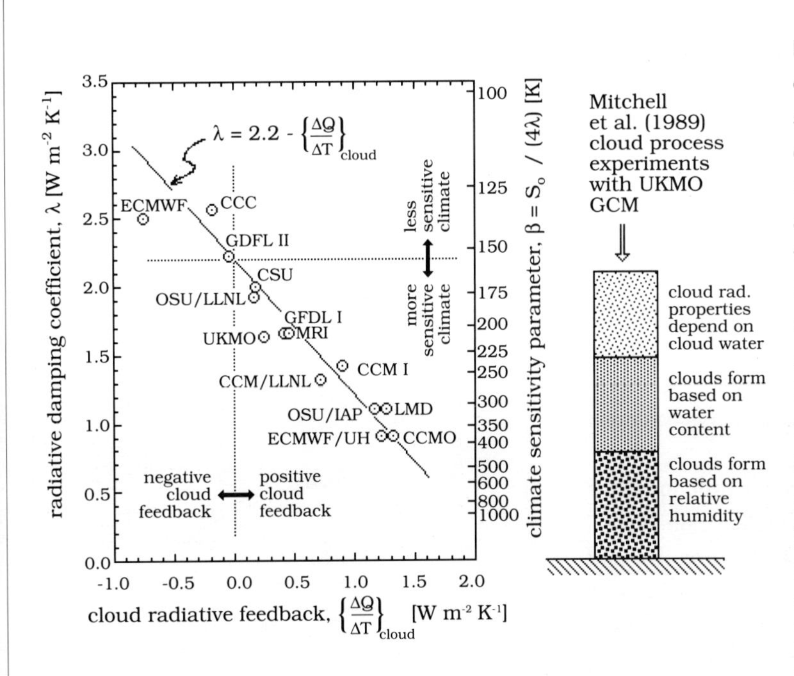

Figure 3. The effect of cloud processes on climate sensitivity. The left panel shows the radiative damping coefficient (left vertical axis) and climate sensitivity parameter β (right vertical axis) as a function of cloud radiative feedback $(\Delta Q/\Delta T)_{cld}$ in 14 general circulation models. GCM sensitivities and acronym labels are those given in the GCM inter-comparison by Cess et al. (1989). Notice that different global climate models predict different cloud radiative feedbacks — ranging from slightly negative to strongly positive (see also Table 3, "cloud"). The bar graph on the right, derived from Mitchell et al. (1989), shows the effect of changing cloud process models (hence cloud feedback) on the radiative damping coefficient and climate sensitivity.

6 Deriving Climate Sensitivity from Paleoclimate Data

Uncertainty about climate sensitivity has had great political impact; it has raised difficult questions about the importance of implementing policies to reduce the risks of rapid climate change. To address some of these issues, Curt Covey of Lawrence Livermore National Laboratory and I undertook a study of whether global climate sensitivity and transient climate change could be derived from paleoclimatic data, as an alternative to general circulation modelling (Hoffert and Covey, 1991). The methodology requires reconstructing both the radiative forcing and the temperature responses of selected ancient climates. Such reconstructions are based on proxy data — including such diverse archives of climate information as tree rings, lake sediments, ice cores, pollen, loess, ocean cores, corals, paleosols, geomorphic features and sedimentary rocks. The systematic reconstruction of climate and atmospheric composition from these proxy records is both an art and a science (see Box).

One useful principle of empirical science is that interpolation is safer than extrapolation, so our first step was to look at the sensitivity implied by climates very different from our own. Presumably, the future climate will fall between the present climate and climatic extremes observed over geologic history.

One hundred million years ago, in the middle of the Cretaceous Era, the Earth was about 10 °C warmer than today. Though the sun was slightly dimmer, this was more than offset by the greater sunlight absorption by a darker, more ocean-covered planet. Substantial greenhouse heating was provided by high concentrations of atmospheric carbon dioxide, six to ten times higher than present levels. At the other extreme, we considered the deep freeze of the Last Glacial Maximum (LGM), 18,000 years ago.

An orbital explanation for interglacial cycling was first proposed by Milutan Milankovitch (1920). He held that cyclic changes in seasonal solar radiation, caused by periodic variation in the Earth's orbit, drives climate fluctuation on glacial-interglacial timescales ranging from 10,000 to 100,000 years. However, the annual mean insolation — the average amount of sunlight incident on the Earth over the year — remains virtually constant. It is now strongly suspected that what causes the Ice Ages is the persistence of snow cover from year to year and changes in concentration of greenhouse gases resulting from the seasonal and latitude changes in sunlight. In a sense, the changes in the Earth's orbit are thought to be the pacemaker of Ice Ages, though they do not themselves cause the large climate changes.

The record of these changes — particularly the changes in CO_2 and CH_4 concentration — can be observed by studying polar ice cores. The best evidence comes from the 160,000-year-long record in the ice drilled at Vostok station on the Antarctic plateau. When combined with data from Greenland ice cores, these records reveal much about the last global-scale glacial-interglacial cycle (Chappellaz et al., 1990). These data suggest that, during the coldest glacial periods, sulphate aerosol particles were more abundant and the atmospheric concentration of CO_2 was 80 ppm below pre-industrial levels.

Most significant, for our present concerns, is the sensitivity of climate to these changed conditions — both at the Last

Figure 4. Two historical curves show a similar relationship between temperature change and latitude, when scaled to the radiative forcing of the time — one during the Cretaceous period, 100 million years BP, and the other during the Last Glacial Maximum (LGM), 18,000 years BP.

The stippled region (above) shows the range of Cretaceous climate response as the Earth grew warmer. This was obtained by subtracting the present zonal mean temperature distribution from that of 100 million years BP and averaging the two hemispheres. The LGM curve (below) is synthesized from sea surface temperature changes derived in the CLIMAP (1976) study (small open circles) and from air temperature on the Antarctic plateau, recorded in the Vostok ice core deuterium isotope record (small solid circle).

In both the Cretaceous and LGM reconstructions, the solid curves are derived from a hypothetical "universal" zonal temperature response, scaled to the positive and negative radiative forcing reconstructed for 100 million and 18 thousand years ago. In both cases, the global mean planetary cooling rate was 2.2 W m^{-2} K^{-1}, and the CO_2 doubling sensitivity was 2.0 °C, including the effect of cloud feedback.

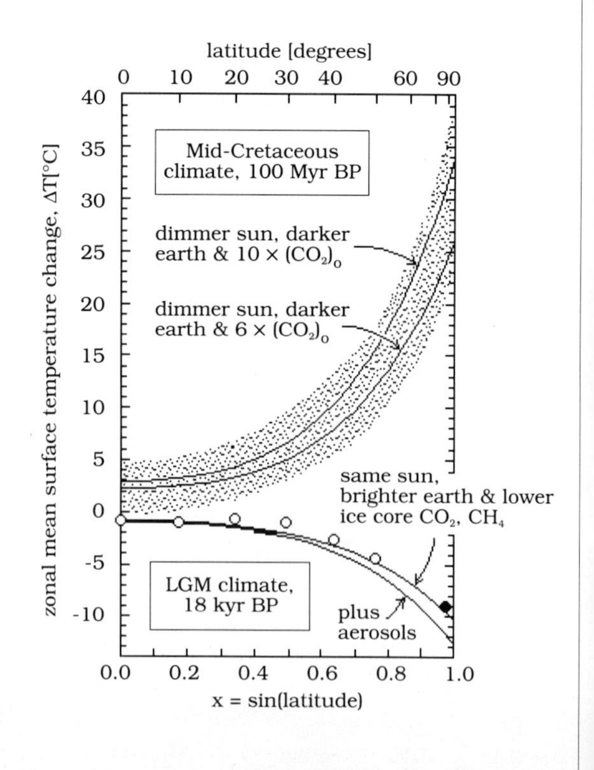

Glacial Maximum and during the mid-Cretaceous. As illustrated in Figure 4, both the cooling of 18,000 years ago and the warming of 100 million years ago showed the same pattern of temperature change versus latitude, and very nearly the same climate sensitivity. Though one period was cold and the other warm, the global average temperature change and the distribution of temperature with latitude were quite similar — when scaled to the radiative forcing of those times. Both periods exhibited relatively weak temperature response in the tropics, compared to large high-latitude amplification of warming or cooling. We were impressed that both the very hot and very cold paleoclimates could be recovered with the same assumption about the sensitivity of global climate to a doubling of atmospheric CO_2, i.e. approximately 2 °C. This is also roughly equivalent to the sensitivity of clear sky atmospheric columns today, based on satellite measurements (Hoffert and Covey, 1991). The implication of this is that while the net effect of cloud-related feedback processes is small — only about 10% or less of total sensitivity — the positive feedback effect of water vapour is quite significant. Water vapour increases climate sensitivity to doubled CO_2, from about 1.1 °C to 2 °C.

In light of the possibility of adaptation, the impact of climate change on natural ecosystems may depend more on the rate of climate change than on the ultimate temperature change. To understand the effects of future increases in greenhouse gas concentrations on observed rates of change in the next century, a number of additional factors must be considered. Ocean thermal inertia, for example, could slow the observable rate of temperature increase caused by any greenhouse gas buildup.

To explore the implications of paleocalibration for 21st Century warming rates, we re-ran the IPCC ocean climate model with the Business as Usual scenario, assuming a climate sensitivity to doubled CO_2 of 2.0 °C. Not surprisingly, we found warming rates of about 0.3 °C per decade over most of the 21st Century. This would produce a climatic warming wave, sweeping poleward at about 6 kilometres per year, on average for the Earth as a whole. However, climatic changes would be much smaller in the tropics, and larger at the high latitudes, so this velocity of the warming wave represents a characteristic value for temperature latitudes. This is faster by a factor of three to thirty than the rate tree species could follow moving climate zones by spreading their seeds (Shugart et al,, 1986). And this does not account for barriers to poleward migration from urbanization and agriculture. Of course, humanity could intervene to accelerate the migration of desirable tree species, using techniques akin to forest management, but the threat to the survival of natural ecosystems — what some have called "the end of nature" — is a major issue for environmental ethics and economics. If, as Vellinga and Swart (1991) have proposed, global warming rates beyond a "green" limit of 1 °C per century are risky, our work suggests that the IPCC "Business as Usual" scenario would almost certainly pose a severe problem in the 21st Century.

We can now say with some confidence that if climate sensitivity were as sluggish as some critics suggest, it is unlikely that the Earth would have experienced the Cretaceous hothouse and Pleistocene chills. Our paleoclimate-derived sensitivity, while not so large as that of some climate models, is large enough to worry about. Those disinclined to

Paleoclimates and the Future Climate

Despite some excellent large scale studies like CLIMAP and COHMAP, which have attempted to reconstruct world climate since the Last Glacial Maximum, paleoclimatologists have tended to emphasize local as opposed to global climate reconstructions. To project anthropogenic global warming, Budyko and Izrael (1991) proposed the "paleo-analogue" method. They employed reconstructed spatial distributions of temperature and precipitation patterns of "warm" analogue periods. They also presented maps of temperature and precipitation for their periods: the Holocene optimum (6000 years ago), the Eemian interglacial (120,000 years ago) and the Pliocene optimum (three million years ago).

There has been some reluctance by the climate community and the IPCC to accept these results, because the raw data and transfer functions used to reconstruct these periods are not well characterized in the literature and because the radiative forcing of their Holocene, Eemian and Pliocene periods is not well known. Budyko and Izrael derive their global climate sensitivity from other periods. Crowley (1990) has argued that, because of differences in the distribution of radiative forcing, there may be no warm period that is a satisfactory analogue for future climate.

Moreover, Crowley observes that because future temperatures may be increasing at a very high rate — as much as 2-4 °C per century, we will have a very unique combination of warm atmospheres and polar ice sheets. These are conditions very different from the pre-Pleistocene warm periods. General circulation modellers have tended to view paleoclimate reconstructions as sources of surface boundary conditions — sea surface temperatures and ice sheet locations — rather than test cases of climate sensitivity (Street-Perrott, 1991). All of these factors suggest that the potential of paleoclimate reconstructions to reduce the uncertainty in climate sensitivity has been insufficiently exploited, although a number of legitimate objections must still be addressed.

consider global warming a problem on the grounds that global climate model results are inherently uncertain might want to re-examine their positions in light of these results. Independent of complex computer models, there is good evidence of risk from continued greenhouse gas emissions. If certain wild card feedbacks not yet included in our paleo-calibration are activated, then warming rates would increase even more, and the risk to planetary ecosytems would have to be faced even earlier.

7 Biosphere and Ocean Feedbacks

What are some of these additional "wild card" feedbacks that have not yet found their way into global climate models? The Earth's biosphere has the potential to induce climatic feedbacks during global warming by changing the reflectivity of the Earth's surface or cloud cover (Charlson et al., 1987) or by releasing additional quantities of greenhouse gases (Lashof, 1989). Ocean circulation affects the distribution of carbon between the atmosphere and oceans, as well as the rate at which heat is transferred from the relatively warm surface layer to the cold depth. Both these processes can have unpredictable feedback effects on the rate of climate change.

7.1 Potential feedbacks from the terrestrial biosphere

So long as the "standing crop" of biomass remains constant, virtually all the carbon dioxide removed from the atmosphere through photosynthesis is regenerated by respiration, mostly as a result of the oxidation by bacteria of dead organic matter.[6] Of every 10,000 carbon atoms cycled from atmosphere to life forms and back this way, a few escape oxidation. They are buried in anoxic environments and transformed into an organic rock called kerogen. For every ten thousand carbon atoms in kerogen, a few are transformed into recoverable fossil fuels. In other words, the rate of fossil fuel formation is only a very small fraction of the rate of the primary productivity of organic carbon. But, because the process has been going on for hundreds of millions of years, the fossil fuel carbon reservoir is now large compared with the amount currently in the atmosphere. None the less, if human energy consumption continues to grow at its present rate, the global stock of fossil fuels would be burned to depletion in a few hundred years.

Table 4 presents current estimates of the fossil fuel reserve, given in units of both energy and carbon. The table shows that conventional fossil fuel resources contain about 2500 gigatonnes (Gt) of C. This estimate could easily double to 5000 Gt if shale oil were included.[7] If all this carbon were to remain in the atmosphere as CO_2, it would equal about 2400 parts per million by volume (ppmv), or six times present atmospheric levels. Table 5 indicates the amount of carbon in other reservoirs of the Earth System.

One well-documented feedback mechanism affecting atmospheric carbon dioxide is deforestation — either the direct cutting back of the standing crop of trees for agriculture, or inadvertently, through a forest ecosystems' inability to survive climate change. The problem of determining the effect of deforestation is complicated by regrowth of trees and vegetation, which takes some of the CO_2 out of the atmosphere that deforestation puts in. Despite the well-documented loss of tropical rainforests, it has been difficult to ascertain from limited sampling whether the world's forests are a net source or sink of carbon dioxide today. This underscores the difficulty of projecting whether terrestrial ecosystems will be sources or sinks of carbon in the future.

[6] Photosynthesis is the fundamental energy-gathering process of life: sunlight + carbon dioxide + water → organic carbon + oxygen. This occurs mainly in the leaves of terrestrial plants and in microscopic blue green algae in the ocean. The rate of organic carbon fixation by photosynthesis is called the primary productivity.

[7] The unit 1 Gt C equals 1 gigatonne carbon or 10^{15} grams of carbon, independent of the chemical compound it is incorporated in. If the carbon is in the form of atmospheric carbon dioxide, then 1 Gt C is equivalent to 0.473 parts per million by volume (ppmv) of CO_2.

Table 4. Energy and carbon content of proven fossil fuel and U-235 reserves.

Energy source[a]	Energy content [10^{21} J]	Carbon content [Gt C]
Hard coal	77.3	1900
Soft coal	17.7	440
Oil	5.2	104
Natural gas	4.2	57
Uranium (recoverable at < $130 kg)	1.4	0
Totals	**105.8**	**2501**

[a] Hammond, 1990, p.320, excluding shale derivatives

Table 5. Carbon in the earth system (after MacCracken et al., 1990).

Carbon reservoir	Carbon content [Gt C = 10^{15} g C]
Atmosphere (present value)	740
Terrestrial biota	560
Detritus and labile humus	1,400
Stable humus and peat	700
Marine biota	2
Dead marine organic matter	1,800
Inorganic dissolved carbon in oceans	35,000
Organic carbon rocks (see Table 4 for fossil fuels)	12,000,000
Carbonate rocks	94,000,000

Figure 5. A model developed by Jenkinson et al. (1991) predicts that, if the global warming rate estimated by paleo-calibration occurs (3 °C per century), about 60 gigatonnes of carbon would be released from the soil during the next 60 years. This figure shows the model's estimates of carbon released as CO_2 from soil organic matter, from now to the year 2050.

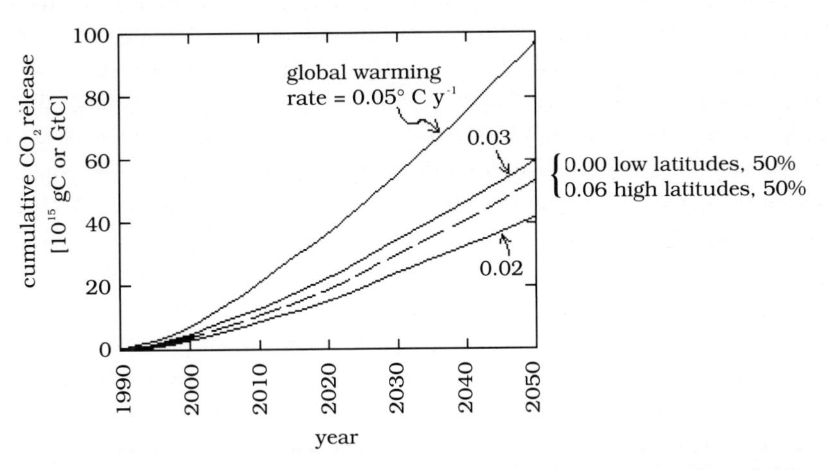

Solid curves represent net carbon loss from the global stock of soil organic carbon, for 17 life zones excluding wetlands, at three different latitude warming rates: 0.02 °C y^{-1} (bottom), 0.03 (middle), and 0.05 (top). The dashed curve is net carbon loss, as a result of a 0.03 °C y^{-1} average warming rate, divided unequally between the zones, with a rise of 0.06 °C y^{-1} in high- latitude zones (tundra, boreal desert, cool desert, cool temperature steppe, moist boreal forest, wet boreal forest, and cool temperate forest, which together contain 50% of the total non-wetlands soil organic carbon), and no temperature rise in the ten remaining life zones, again excluding wetlands.

7.2 *Soil feedbacks*

Although most of the public's concern about the biota has focused on the release of carbon through deforestation, the first meter of soil contains even more carbon than do all the world's forests. Because of the size of the soil carbon reservoir, changes in the organic carbon content of soil could be even more important than changes in the standing crop of trees. Table 5 shows that 1400 Gt C is tied up as organic detritus and labile humus, typically in the uppermost meter of soil. This is twice as much carbon as is now contained in the atmosphere. One effect of global warming could be to accelerate respiration and decomposition of soil organic matter, thereby releasing more CO_2 to the atmosphere and further enhancing the warming trend. As Figure 5 illustrates, a model developed by Jenkinson et al. (1991) predicts that, if the global warming rate estimated by paleo-calibration occurs (3 °C per century), about 60 gigatonnes of carbon would be released from the soil during the next 60 years.

On the 100-year time scale, increased soil respiration would act as a modest positive climate feedback, reinforcing the rate of global warming. But on the longest geologic time scales, an increase in soil respiration could become a negative feedback, as the rate of atmospheric CO_2 removal to sedimentary carbonates by calcium-silicate weathering increases.

7.3 *Potential feedbacks from ocean life*

The amount of carbon in the marine biota is less than 1% that of the land biosphere, though its primary productivity is comparable. The marine biosphere "runs faster" than the land biota, with organic carbon lifetimes of plankton measured in days as opposed to decades for trees. Despite the small biomass of the marine organisms, changes in their productivity can be invoked as a possible explanation for variations of atmospheric CO_2 of 170 Gt C (80 ppmv) recorded between glacial to interglacial periods in polar cores at 170 Gt C (Chappellaz et al.,1990).

Although the entire living marine biosphere contains only 2 Gt C of biomass, plankton can exert considerable leverage on the distribution of the surrounding 35,000 Gt of inorganic carbon dissolved in the ocean (Table 5). The combination of photosynthesis near the surface and the respiration of fecal pellets falling through the water column drives a "biological pump" that moves carbon from the warm surface layer into the deep ocean. During periods of stable climate, this biological pump maintains a vertical gradient in dissolved inorganic carbon (Volk and Hoffert, 1985). A change in ocean productivity affects the distribution of this dissolved carbon, as well as the concentration of carbon dioxide at the surface, and hence the atmospheric CO_2 concentration that is in equilibrium with the surface ocean.[8]

[8] Mix (1989) used data on planktonic foraminifera species in modern and ice-age Atlantic sediments to assess spatial patterns of changes in marine productivity. These changes, if extrapolated to the global ocean, support models in which a significant portion of CO_2 changes are driven by variations in productivity. This is an interesting finding, but incomplete for purposes of predicting future CO_2 feedbacks on climate. Some other mechanism, like changes in sea level or ocean circulation, is needed to explain how the intensity of the ocean's biological pump changes when the climate changes.

One of the factors controlling the biological pump is the supply of essential nutrients to marine organisms. Over most of the ocean surface, phosphate and nitrate are the "biolimiting" nutrients, virtually depleted in most surface waters where phytoplankton use all there is. Because they are rich in nutrients, high-latitude waters surrounding Antarctica continent are relatively high in productivity. But the productivity would be even higher if dissolved phosphate nutrients near the surface were fully utilized by phytoplankton. In the Southern Ocean, Martin et al. (1990) suggest that another element, iron, is biolimiting. They hypothesize that significant amounts of atmospheric carbon might be sequestered by fertilizing the Southern Ocean with supertanker loads of iron filings. However, Peng and Broecker (1991) conclude, on the basis of model calculations, that even if iron fertilization worked perfectly it could not significantly reduce atmospheric CO_2 concentrations. Joos et al. (1991) did find, on the basis of their model calculations, that atmospheric CO_2 concentrations could be lowered by up to 190-227 Gt C (90-107 ppmv) by the year 2100, for the IPCC Business as Usual scenario. These authors believe, however, that such large biotic uptake is unlikely in practice — again, in part, because of feedbacks. For example, organic carbon produced in the surface waters of Antarctica would be oxidized at more equatorward latitudes, returning CO_2 to the water column and eventually to the atmosphere.

A possible link between ozone depletion and climate change is a CFC-generated ozone hole around Antarctica — increasing ultraviolet radiation and killing plankton and other marine organisms in the food chain in the Southern Ocean. Without these organisms, the global level of net primary productivity would fall, CO_2 would build up more rapidly in the atmosphere, and the warming effect would increase. Another link arises from the fact that a CO_2 increase leads to a cooling of the stratosphere at the same time that it warms the surface and the lower atmosphere. The Antarctic ozone hole is indirectly caused by the occurrence of stratospheric water-nitric acid clouds, which form only at temperatures below about -80 °C (Solomon, 1990). By increasing the region over which these clouds can form, the CO_2 could very well exacerbate stratospheric ozone depletion. These are only two of a host of potential feedbacks between biogeochemical cycles and climate change.

7.4 Ocean circulation

One of the most intriguing wild cards is the possibility of changes in ocean circulation. The hemispheric asymmetry of the present continents produces a dramatic difference in the way deep water forms in the two hemispheres. In contrast to the "open" Southern Ocean around Antarctica, the main site of seasonal sea ice formation in the Northern Hemisphere is the Arctic ocean, a landlocked water body virtually inacces-

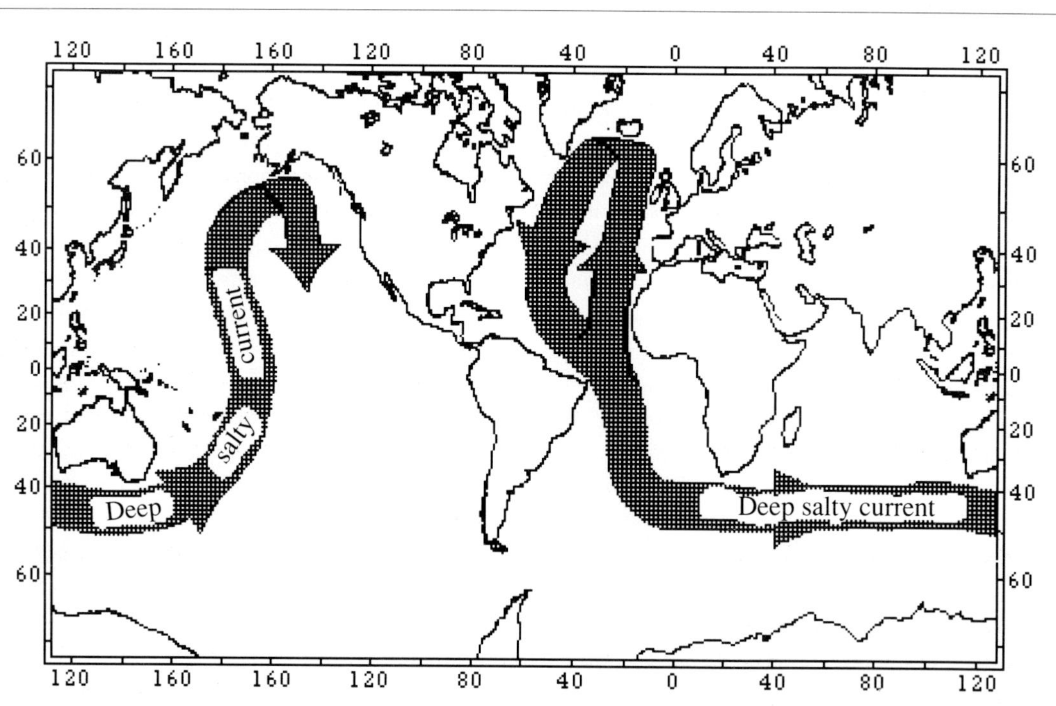

Figure 6. A mechanism operating in today's ocean, which may create rapid climate shifts. Salt-laden North Atlantic Deep Water (NADW) flows down the length of the Atlantic, around Africa through the Southern Indian Ocean, and finally northward in the deep Pacific. This acts as a large-scale salt transport system, compensating for the transport of water vapour through the atmosphere from the Atlantic to the Pacific. This system may be self-stabilizing. Records from ice and sediment cores suggest that it was disrupted in glacial times and replaced by an alternate mode of operation. Source: Adapted from Broecker, 1987. Reprinted by permission from Nature, vol. 328, pp. 123-126. Copyright (c) 1987, MacMillan Magazines Ltd.

sible to the North Pacific and communicating weakly (if at all) with the North Atlantic. It is presently believed that the source of North Atlantic Deep Water (NADW) is the permanent density-stratified layer of sea water of the Iceland and Greenland Seas. This is a region of high salinity, created because evaporation removes more moisture there than is replaced by local precipitation.

This salty North Atlantic water enters the Atlantic through the Norwegian sea; as it moves southward down the Atlantic it flows beneath the high-salinity (but warmer) Mediterranean outflow, and never does attain the density needed to sink to the bottom of the Atlantic. But on reaching the Southern Ocean, the NADW mixes to the surface and is cooled in the Antarctic Circumpolar Current, where it becomes part of the source water for the densest, coldest and deepest bottom water that feeds all the world's oceans: Antarctic Bottom Water (AABW).

Figure 6 illustrates the paths of deep, dense salty currents in the oceans. Broecker et al. (1985) speculated that these currents have two alternate modes of operation, one in which NADW formation is strong and another in which it is weak or absent.[9] A temporary shut-off of NADW may have occurred during the so-called Younger Dryas period, 10,800 to 10,000 years ago, when fresh water from the Mississippi may have been diverted to the North Atlantic, punctuating the gradual warming since the Last Glacial Maximum with a short-lived global cooling. This was followed by a very rapid rewarming and return to the slow emergence from the Last Glacial Maximum. It has also been suggested that the North Atlantic was flooded with fresh water, via the St. Lawrence River's outlet, by an abrupt drainage of Lake Agassiz, a giant mid-continental lake of meltwater. What is known, from Greenland ice core records, is that there was a local temperature rise as large as 7 °C over 50 years, and a 50% increase in rainfall over 20 years, during this Younger Dryas Termination (Dansgaard et al., 1989). This has prompted speculation that a future shut-down of NADW could occur during the next century's global warming by greenhouse gases, possibly threatening Europe with localized cooling in the midst of a more general global warming (Calvin, 1991). In addition to such direct climatic effects, rapid changes in ocean circulation could trigger changes in ocean chemistry that feedback on global climate.

The implication of Broecker's hypothesis is that rapid changes in ocean circulation can trigger changes in ocean chemistry that feed back on climate (Broecker and Peng, 1986, 1989; Boyle, 1988). Such circulation-driven changes in ocean carbon chemistry (or in the biological productivity of the ocean) may explain the rapid increases of atmospheric carbon dioxide and methane that are observed in ice cores and seem to occur during glacial terminations.

We do not understand the dynamics of glacial-interglacial transitions well enough yet to make predictions about deep water circulation feedbacks. We need to know much more about the natural variability of the oceans. The most sophisticated current atmospheric-ocean climate model, run for 100 simulated years with realistic build ups of CO_2, does not suddenly flip its NADW circulation (Manabe et al., 1991). But this does not prove that the real ocean will not experience such a change. There are indications that climate variability may be partly due to inherently unpredictable chaotic-dynamic interactions between the atmosphere and the oceans (Gaffin et al., 1986). The best we can say at this point is that an abrupt change in ocean circulation during the next century is an important wild card that cannot be ruled out.

Carbon dioxide is not the only greenhouse gas that can be driven to the atmosphere by global warming. No one has definitively explained the large rate of methane increase currently in progress — but some feedback processes may be involved. Gas clathrates are crystalline inclusions of water (hydrates) bound with light hydrocarbons — primarily methane. The crystals are found in abundance beneath Arctic permafrost and along the continental shelf below the ocean floor. Many authors have hypothesized a climatic feedback that would involve methane being degassed from high-latitude methane hydrate reservoirs as a consequence of global warming (Bell, 1982; Khalil and Rasmussen, 1988; Lashof, 1989; MacDonald, 1990). No observational evidence yet supports this hypothesis. But, depending on the actual rate of methane release if it occurred, this could be a significant positive climate feedback.

On the other hand, data from Antarctic ice cores indicate that atmospheric methane concentration varied by a factor of two in remarkably close association with glacial-interglacial climatic changes during the past 160,000 years (Chappellaz et al., 1990). These changes could have involved methane hydrates (Nisbet, 1990), or other methane sources such as natural wetlands (Fung et al., 1991). In any case, the geological record strongly suggests that feedbacks between climate and methane sources or sinks will affect methane concentration as the climate warms. Observational studies currently in progress will hopefully help to determine the potential importance of climate-methane feedbacks, which are likely to involve methane sources both from hydrates and from wetlands.

8 The Dilemma for Policy Makers

8.1 *Lessons of the IPCC scenarios*

In 1990, the Intergovernmental Panel on Climate Change (IPCC) sought to assess the effect of various greenhouse emission policies on climate change. Because of publication time constraints, Houghton et al. (1990) developed a set of

[9] These speculations were supported by Manabe and Stouffer (1988), who found two stable equilibria in time integrations of a coupled atmosphere-ocean climate model. These had identical boundary conditions but different initial conditions. In one of these solutions, a fairly realistic interhemispheric thermohaline circulation was maintained, in the other it was absent. The solution is quite sensitive to the salt water versus fresh water balance in the North Atlantic.

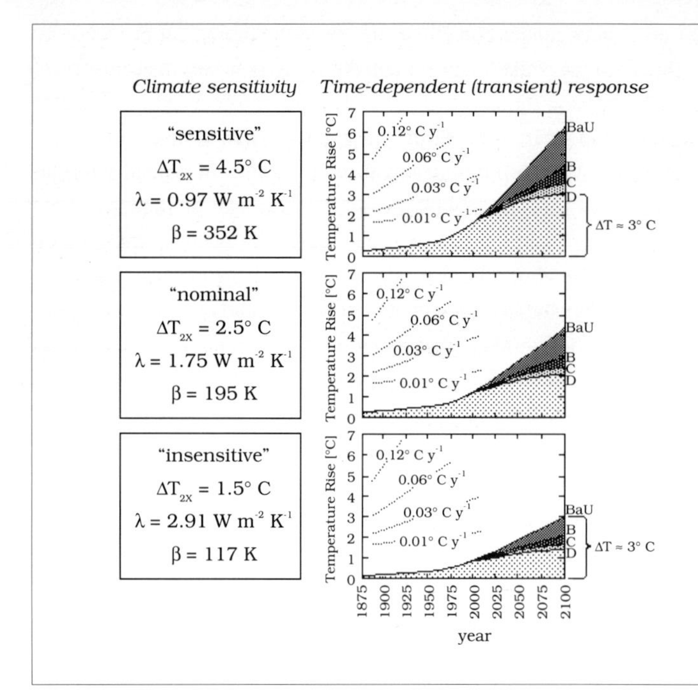

Figure 7. Twelve predictions of the global mean surface temperature response, projected to the year 2100. The left column shows three equilibrium climate sensitivities. For each sensitivity, the right column shows temperature changes under each of four assumptions regarding future greenhouse gas concentrations in the atmosphere. The four assumptions are the IPCC BaU (Business as Usual), B, C, and D scenarios (adapted from Houghton et al., 1990).This figure shows the importance of climate sensitivity to policy decisions in the face of uncertainty. If sensitivity is high (upper panel), then draconian rollbacks are needed to keep global warming below 3 °C by the year 2100. If climate sensitivity is low (lower panel), then the same three- degree warming will occur for Business as Usual.

scenarios which projected the evolution of greenhouse gas concentrations in the atmosphere over the 21st Century, and the corresponding radiative forcing.

A more policy-relevant approach would have been to develop scenarios based on the chain of causality from economic development to emissions to concentrations and finally to global warming. This is possible in principle, because there is a causal link between the rate of greenhouse gas emissions over time, and the concentration history of greenhouse gases in the atmosphere that results from these emissions. For example, the CO_2 concentrations versus time can be computed from global carbon-cycle models, which account for the distribution of carbon in the atmosphere, in the ocean, and in terrestrial and marine biota as CO_2 is emitted from fossil fuel burning and possibly deforestation. Hopefully, as part of the ongoing IPCC process, comprehensive systems models will emerge which permit analysis of all the links in the causal chain. This would make it easier to separate the factors which are well known from the factors which are uncertain.

None the less, there is much to be learned about the impact of climatic feedbacks and climate sensitivity on global warming policy options from the responses of an atmosphere-ocean-climate model to the four IPCC "scenarios" (Figure 7).

In an approximate sense, the first scenario, "Business as Usual (BaU)," portrays a 21st Century world in which greenhouse gas emissions increase as they will if fossil fuels remain the dominant world energy source, and population growth continues as in current projections, particularly in the third world. "BaU" is meant to represent an average projection, though it could significantly underestimate emissions from developing countries if they fuel the economic growth they aspire to with relatively abundant (and cheap) local coal reserves (Table 4). Given the energy and technology transfer policies of the developed countries, there would be a strong economic incentive for the developing countries to exploit relatively inexpensive coal reserves.

Scenarios "B" through "D" represent progressively stronger greenhouse gas emissions controls, with "B" corresponding approximately to a freeze in greenhouse gas emissions at 1990 levels, and "D" a draconian rollback in which fossil fuels are cut back by 60-80% during the next 50 years (Krause et al., 1989; Houghton et al., 1990, Annex).

Figure 7 shows the projected temperature rise of all four scenarios, for each of three different climate sensitivities. It is not surprising that the greatest warming takes place under Business as Usual with the highest sensitivity; and the smallest warming (although still more than 1 °C in the next century) is for the draconian rollback ("D") scenario with the least sensitive climate. But economic considerations make intermediate combinations of these results more interesting. For example, if climate sensitivity is high (upper panel), then draconian rollbacks are needed to keep global warming *below* three degrees Celsius by the year 2100. If climate sensitivity is near its lower uncertainty boundary (lower panel), then the same three degree warming will occur for Business as Usual. Though the global warming is approximately the same, the costs to implement a mitigation policy are vastly different.

8.2 Feedback uncertainties in policy making

The central dilemma of global warming policy is decision making in the face of large uncertainties. Some critics of global warming policy have emphasized the uncertainty in GCMs, and further argued that global warming will be near the lower end of these projections, implying thereby that corrective actions are premature. On the other hand, these projections are as likely to be underestimates as overestimates, and may be too conservative if the wild card feedbacks discussed in this chapter become important in the next century.

The problem is somewhat analogous to that faced by a patient whose physician has informed him that his condition requires surgery. Though the studies of the impact of this surgery are somewhat equivocal, to not make a decision is a decision in itself of major significance to his survival.

The central question for the policy maker is: What are allowable rates of greenhouse gas emissions that would maintain climate change below some "threshold of acceptability"? We cannot answer this question yet, but we can specify the ingredients of an answer. We need to know the climate sensitivity (including cloud feedbacks so poorly characterized by GCMs); how ocean heat storage links climate sensitivity to global warming rates; and how feedbacks from the biosphere and oceans, along with possibly compensatory cooling from sulphate aerosols, would affect the ultimate climate response. We also need better criteria specifying what is meant by "unacceptable" warming and rates of warming. This brings into play a panoply of philosophical, ethical, political, environmental, and ecological perspectives.

The decision making process is also complicated by fundamental cultural differences between scientists and policy makers that can render communication on science policy nonproductive. For professional reasons, the scientist is guarded, sometimes hedging his or her conclusions in with so many caveats that they are virtually useless to the policy maker. Politics, the art of the possible, requires decision making in the face of uncertainty. This volume attempts to bridge the gap between science and policy. And in this chapter, I have focused attention on how climatic feedbacks could impact global warming policy decisions. It is clear from this survey that the connections are many, and the ramifications virtually endless. Some possible "wild card" feedbacks or other significant factors have almost certainly been left out because no one has thought of them yet.

8.3 Charting the route of acceptable action

None the less, there is light at the end of the tunnel. We have found that a semi-empirical approach to estimating climate sensitivity can be quite useful. For example, the paleoclimatic

Triad Strategy for Improving Climate Prediction

Syukuro Manabe

Unfortunately, the projections of future climate change made by the Intergovernmental Panel on Climate Change (IPCC; Houghton et al., 1990) are subject to large uncertainties. These uncertainties reflect our inability to model the various processes that control any future climate change. It is therefore necessary to improve various components of climate models, such as the representation of cloud feedback and land surface processes.

In addition, it is essential to carefully assess the model's predictions of future climate change in light of the results obtained by monitoring actual climate changes and the factors causing these changes. Comparing the predicted and observed climate will eventually enhance our confidence in model prediction.

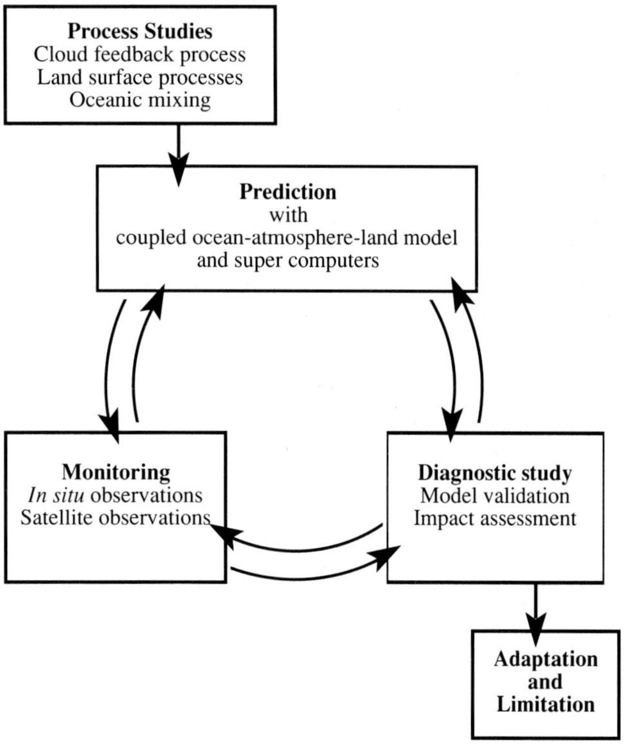

A comprehensive triad strategy for continuously improving the art and science of climate prediction is illustrated by the flow diagram in Figure 8. This strategy involves coordinating three concurrent activities. The first involves the monitoring of the coupled ocean-atmosphere-land surface system by *in situ* and remote sensing. The second applies state-of-the-art models to the prediction of future climate change. The third activity incorporates in-depth analysis of predicted changes in conjunction with the observations of the climate system.

The insights gained from this coordinated strategy would be indispensable — not only for enhancing our confidence in model prediction of future climate change, but also for adapting effectively to altered climate conditions and for reducing the rate of buildup in atmospheric concentration of greenhouse gases (see also Wigley, 1992; this volume).

Process Studies
Cloud feedback process
Land surface processes
Oceanic mixing

Prediction
with
coupled ocean-atmosphere-land model
and super computers

Monitoring
In situ observations
Satellite observations

Diagnostic study
Model validation
Impact assessment

Adaptation and Limitation

record may provide an alternative path to general circulation models around the uncertainties of cloud-radiative feedback. Most experts in this field would probably agree that a resolution of the cloud feedback issue by more realistic cloud models — tested against observations and ultimately incorporated as parameterizations in general circulation climate models — is the most scientifically acceptable approach to resolving this issue. But that may be decades away. Satellite observations, too, could be very helpful, but the Earth Observing System is also decades away from operational status. In the meantime, the paleoclimatic record awaits analysis. We believe that its implications may be reliable enough to use as a basis for political decisions. Preliminary results with this method are encouraging enough to pursue it much more vigorously (Oeschger and Mintzer, 1992; this volume).

A common near-term goal for an international climate convention is a global emission rate freeze (Krause et al., 1989). A committee of the US National Academy of Sciences (NAS, 1991), studying this problem advocated a "no regrets" strategy — a response focused on zero- or negative-cost options like improved energy efficiency and fuel switching (Pachauri, 1992; this volume). Despite lingering criticisms on grounds of climate model uncertainties (Michaels, 1991), "No regrets" is generally perceived as an economically painless, if not particularly farsighted, policy. Indeed in 1990, even without a carbon agreement, annual global CO_2 emissions, at 5.803 gigatonnes of carbon (Gt C), were down slightly from 5.813 Gt C in 1989 (Wald, 1991) — a *de facto* emissions freeze.

Over the next few years, an emissions freeze based on energy conservation, efficiency and fuel switching is technologically feasible (Holdren, 1992; this volume). Given the pressures of population growth and economic development, it will become increasingly difficult to implement in the coming decades without new non-fossil energy technology, particularly as the already committed population growth of the next century makes itself felt. An important implication of the causal link between emissions and concentrations is that an emission rate freeze results in continually increasing atmospheric carbon dioxide concentrations, radiative heating and global warming. All of these continue to rise to some extent throughout the next century in all four IPCC scenarios (Figure 7), though the amount of warming clearly decreases as more stringent emissions controls are applied. If, for example, stabilization of greenhouse gas levels were needed, a worst-case scenario something close to a fossil-fuel phaseout in the next 50 years could become necessary. Very little contingency planning has gone into the transition from a fossil fueled to a solar/fusion powered world implied by scenario D. This is a further example of how climate modelling and systems analysis can identify key problems for policy makers.

In the immediate future, first-order policy questions will continue to be impacted by the uncertainties of climate feedbacks. Their resolution may be an acid test of the fledgling science of Global Change. I have argued in this chapter that semi-empirical methods, based on the history of the Earth's climate imprinted in the geological record, can provide a useful approach to this problem. Perhaps those policy makers unconvinced of the reality of global warming by GCM model predictions might reassess their positions based on the paleoclimate record.

Acknowledgments

It is a pleasure to thank C. Covey, L.D.D. Harvey, M.C. MacCracken, S. Manabe, K. Taylor and T. Volk for insightful contributions to this paper. The help of H. Jones in transmitting edited manuscripts via electronic mail was also invaluable. Research on the climate sensitivity, cloud feedback, and ocean modelling at New York University, some of which is reported here, was supported by the US Department of Energy, Carbon Dioxide Research Program, Office of Health and Environmental Research, and the National Institute for Global Environmental Change.

References

Allaby, M. and J. Lovelock, 1984: *The Greening of Mars.* New York, St. Martin's Press.

Averner, M.M. and R.D. MacElroy, eds., 1976: *On the Habitability of Mars: An Approach to Planetary Ecosynthesis.* NASA **SP-414**, Washington, DC, National Aeronautics and Space Administration.

Barron, E.J. and W.M. Washington, 1985: Warm Cretaceous Climates: High atmospheric CO_2 as a plausible mechanism. In *The Carbon Cycle and Atmospheric CO_2: Natural Variations Archean to Present*, E.T. Sundquist and W.S. Broecker, eds, *Geophys. Mong* **32**, 546-553, Washington, DC, American Geophysical Union.

Bell, P.R., 1982: Methane hydrate and the carbon dioxide question. In *Carbon Dioxide Review:* 1982. W.C. Clark, ed., 401-406, New York, Oxford University Press.

Berner, R.A., 1990: Atmospheric carbon dioxide levels over Phanerozoic time. *Science,* **249**, 1382-1386.

Boyle, E.A., 1988: The role of vertical chemical fractionation in controlling late Quaternary atmospheric carbon dioxide. *J. Geophys. Res.,* **93C**, 15,701-15,714.

Broecker, W.S., 1987: Unpleasant surprises in the greenhouse? *Nature,* **328**, 123-126.

Broecker, W.S. and T.-H. Peng, 1986: Carbon cycle, 1985: glacial to interglacial changes in the operation of the global carbon cycle. *Radiocarbon,* **28**, 309-327.

Broecker, W.S. and T.-H. Peng, 1989: The cause of the glacial to interglacial atmospheric CO_2 change. *Global Biogeochemical Cycles,* 3, 215-239.

Broecker, W.S., D.M. Peteet and D. Rind, 1985: Does the ocean-atmosphere system have more than one stable mode of operation? *Nature,* **328**, 123-126.

Budyko, M.I.,ed., 1991: Climate of the Future. Manuscript submitted by USSR delegation of Working Group VIII of US-USSR Agreement on Protection of the Environment for review as a possible supplement to MacCracken et al. (1990). US

National Climate Program Office, Suite 518, 1825 Connecticut Ave, N.W., Washington, DC.

Budyko, M.I. and Y.A. Izrael, 1991: *Anthropogenic Climatic Change*, pp. 277-318, Tucson, University of Arizona Press.

Budyko, M.I., A.B. Ronov and A.L. Yanshin, 1987: *History of the Earth's Atmosphere*. New York, Springer-Verlag.

Calvin, W.H., 1991: Greenhouse and Icehouse. In *Whole Earth Review*, **73**, Winter 1991, 106-111.

Cess, R.D., G.L. Potter, J.P. Blanchet, G.J. Boer, S.J. Ghan, J.T. Kiehl, H.Le Treut, Z.-X. Li, X.-Z. Liang, J.F.B. Mitchell, J.-J. Morchette, D.A. Randell, M.R. Riches, E. Roeckner, U. Schlese, A. Slingo, K.E. Taylor, W.M. Washington, T.R. Wetherald and I. Yagai, 1989: Interpretation of cloud-climate feedback as produced by 14 atmospheric general circulation models. *Science*, **245**, 513-516.

Chappellaz, J., J.M. Barnola, D. Raynaud, Y.S. Korotkevich and C. Lorius, 1990: Ice-core record of atmospheric methane over the past 160,000 years. *Nature*, **345**, 127-131.

Charlson, R.J., J.E. Lovelock, M.O.Andreae and S.G. Warren, 1987: Ocean phytoplankton, atmospheric sulfur, cloud albedo and climate. *Nature*, **326**, 655-661.

CLIMAP Project Members, 1976: The surface of the ice age earth. *Science*, **191**, 1131-1137.

Crowley, T.J., 1990: Are there any satisfactory geologic analogs for a future greenhouse warming? *Journal of Climate*, **3**, 1282-1292.

Crowley, T.J. and G. North, 1991: *Paleoclimatology*. New York, Oxford University Press.

Dansgaard, W., J.W.C. White and S.J. Johnsen, 1989: The Abrupt Termination of the Younger Dryas Climate Event. *Nature*, **339**, 532-534.

Ellsaesser, H.W., 1990: A different view of the climatic effect of CO_2 — updated. *Atmósfera*, **3**, 3-29.

Fogg, M.J., 1989: The creation of an artificial dense Martian atmosphere: a major obstacle to the terraforming of Mars. *J. Brit. Interplanet. Soc.*, **42**, 577-582.

Fung, I., J. John, J. Lerner, E. Matthews, M. Prather, L.P. Steele and P.J. Fraser, 1991: Three-dimensional model synthesis of the global methane cycle. *JGR*, **96**, 13, 13033-13065.

Gaffin, S.R., M. I. Hoffert and T. Volk, 1986: Nonlinear coupling between surface temperature and ocean upwelling as an agent in historical climate variations. *J. Geophys. Res.*, **91**, 3944-3950.

Hammond, A.L., ed., 1990: *World Resources 1990-91: A Guide to the Global Environment*, New York, Oxford University Press.

Hoffert, M.I. and C. Covey, 1991: Projecting 21st Century greenhouse warming from paleoclimate data and ocean models. Submitted to *Nature*.

Hoffert, M.I. and B.F. Flannery, 1985: Model projections of the time-dependent response to increasing carbon dioxide. In *Projecting the Climatic effects of Increasing Carbon Dioxide*. M.C. MacCracken and F.M. Luther, eds., DOE/ER-0237, 149-190, Washington, DC, US Department of Energy, Atmospheric and Climate Research Division.

Hoffert, M.I., A.J. Callegari and C.-T. Hsieh, 1980: The role of deep sea heat storage in the secular response to climatic forcing. *J. Geophys. Res.*, **85** (C11), 6667-6679.

Houghton, J.T., G. Jenkins and J.J. Ephraums, 1990: *Climate Change: The IPCC Scientific Assessment*. New York, Cambridge University Press.

Jenkinson, D.S., D.E. Adams and A. Wild, 1991: Model estimates of CO_2 emissions from soil in response to global warming. *Nature*, **351**, 304-306.

Joos, F., J.L. Sarmiento and U. Siegenthaler, 1991: Estimates of the effect of Southern Ocean iron fertilization on atmospheric CO_2 concentrations. *Nature*, **349**, 772-775.

Kasting, J.F., O.B. Toon and J.B.Pollack, 1988: How climate evolved on the terrestrial planets. *Scientific American*, February 1988, 90-97.

Khalil, M.A.K. and R.A. Rasmussen, 1988: Climate-induced feedbacks for the global cycles of methane and nitrous oxide. *Tellus*, **41B**, 554-559.

Krause, F., W. Bach and J. Koomey, 1989: How much fossil fuel can still be burned? In *Energy Policy in the Greenhouse*, Vol. 1, Chapter I.4, International Project for Sustainable Energy Paths (IPSEP), El Cerrito, California.

Lashof, D.A., 1989: The dynamic greenhouse: Feedback processes that may influence future concentrations of atmospheric trace gases and climatic change. *Climatic Change*, **14**, 213-242.

Lindzen, R.S., 1990: Some coolness concerning global warming. *Bull. Am. Meter. Soc.*, **71**, 288-299.

Lovelock, J., 1988: *The Ages of Gaia: A Biography of Our Living Earth*. New York, W.W. Norton & Company.

MacCracken, M.C., A.D. Hecht, M.I. Budyko and Y.A. Izrael, eds., 1990: *Prospects for Future Climate: A Special US/USSR Report on Climate and Climate Change*. Chelsea, Michigan, Lewis Publishers.

MacDonald, G.J., 1990: Role of methane clathrates in past and future climates. *Climatic Change*, **16**, 247-281.

Manabe, S. and R.J. Stouffer, 1988: Two stable equilibria of a coupled ocean-atmosphere model. *J. Climate*, **1**, 841-866.

Manabe, S. and R.T. Wetherald, 1967: Thermal equilibrium of the atmosphere with a given distribution of relative humidity. *J. Atmos. Sci.*, **24**, 241-259.

Manabe, S., R.J. Stouffer, M.J, Spelman and K. Bryan, 1991: Transient responses of a coupled ocean-atmosphere model to gradual changes of atmospheric CO_2. Part I: Annual mean response. *J. Climate*, **4**, 785-818.

Manowitz, B., ed., 1990: Global Climate Feedbacks, *Proceedings of the Brookhaven National Laboratory Workshop* June 3-6, 1990, Washington, DC, US Department of Energy, Atmospheric and Climate Research Division.

Marshall Institute, 1989: *Scientific Perspectives on the Greenhouse Problem*. Washington DC, George C. Marshall Institute.

Martin, J.H., S.E. Fitzwater and R.M. Gordon, 1990: Iron deficiency limits phytoplankton growth in Antarctic waters. *Global Biogeochemical Cycles*, **4**, 5-12.

McKay, C.P., 1987: Terraforming: Making an Earth of Mars. *The Planetary Report*, **VII**(6), 26-27.

McKay, C.P., O. B. Toon and J. S. Kasting, 1991: Making Mars habitable. *Nature*, **352**, pp. 489-496.

McKibben, W., 1989: *The End of Nature*. New York, Random House.

Michaels, P., 1991: Global warming: The new National Academy of Sciences report, *Cato Review of Business and Government*, Summer 1991, 20-23.

Milankovitch, M., 1920: *Theorie Mathematique des Phenomenes Thermiques Produits par la Radiation Solaire*, Paris, Gauthier-Villars, Paris.

Mitchell, J.F.B., 1989: The "greenhouse" effect and climatic change. *Rev. Geophys.*, **27**, 115-139.

Mitchell, J.F.B., C.A. Senior and W.J.Ingram, 1989: CO_2 and climate: a missing feedback? *Nature*, **341**, 132-134.

Mix, A.C., 1989: Influence of productivity variations on long-term atmospheric CO_2. *Nature*, **337**, 541-544.

NAS (National Academy of Sciences), 1991: *Policy Implications of Greenhouse Warming*. Washington, DC, National Academy Press.

Nisbet, E.G., 1990: The end of the ice age. *Can. J. Earth Sci.*, **27**, 148-157.

Peng, T.H. and W.S. Broecker, 1991: Dynamical limitations on the Antarctic iron fertilization strategy. *Nature*, **349**, 227-229.

Ramanathan, V. and W. Collins, 1991: Thermodynamic regulation of ocean warming by cirrus clouds deduced from observations of the 1987 El Niño. *Nature,* **351**, 27-32.

Raval and V. Ramanathan, 1989: Observational determination of the greenhouse effect. *Nature,* **342**, 758-761.

Revelle, R.R. and H.E. Suess, 1957: Carbon dioxide exchange between atmosphere and ocean and the question of an increase of atmospheric CO_2 during past and present decades. *Tellus,* **9**, 18-27.

Schlesinger, M.E., 1985: Analysis of results from energy balance and radiative-convective models. In *Projecting the Climatic Effects of Increasing Carbon Dioxide,* DOE/ER-0237, 281-319, M.C. MacCracken and F.M. Luther, eds., Washington, DC., US Department of Energy, Atmospheric and Climate Research Division.

Schneider, S.H., 1990: The global warming debate heats up: an analysis and perspective. *Bull. Am. Meter. Soc.,* **71**, 1292-1304.

Shugart, H.H., M. Ya. Antonovsky, P.G. Jarvis and A.P. Stanford, 1986: CO_2, climatic change and forest ecosystems. In *The Greenhouse Effect, Climatic Change and Ecosystems,* 475-521, B. Bolin, B.R. Döös, J. Jager, and R.A. Warrick, eds., SCOPE 29, 475-521, New York, Wiley.

Solomon, S., 1990: Progress towards a quantitative understanding of Antarctic ozone depletion. *Nature,* **347**, 347-354.

Street-Perrott, F.A., 1991: General circulation (GCM) modelling of paleoclimates: a critique. *The Holocene,* **1**, 74-80.

Vellinga, P. and R. Swart 1991: The greenhouse marathon: A proposal for a global strategy. *Climatic Change,* **18**, viii-xii.

Volk, T. and M.I. Hoffert, 1985: Ocean carbon pumps: analysis of relative strengths and efficiencies of in ocean-driven atmospheric CO_2 changes. In *The Carbon Cycle and Atmospheric CO_2:* Natural Variations Archean to Present, E.T. Sundquist and W.S. Broecker, eds., *Geophys. Monogr. Ser.,* **32**, 91-110, Washington, DC, American Geophysical Union.

Volk, T., 1987: Feedbacks between weathering and atmospheric CO_2 over the last 100 million years. *Am. J. Sci.,* **287**, 763-779.

Volk, T., 1989: Rise of angiosperms as a factor in long-term climatic cooling. *Geology,* **17**, 107-110.

Wald, M.W., 1991: Carbon dioxide emissions dropped in 1990, ecologists say. *New York Times International,* Sunday, Dec. 8.

Walker, J.C.G., P.B. Hays and J.F. Kasting, 1981: A negative feedback mechanism for the long-term stabilization of Earth's surface temperature. *J. Geophys. Res.,* **86**, 9776-9782.

White House, 1991: *National Energy Strategy: Executive Summary,* First Edition 1991/1992, February 1991, Washington, DC.

CHAPTER 4

Lessons from the Ice Cores: Rapid Climate Changes During the Last 160,000 Years

Hans Oeschger and Irving M. Mintzer

Editor's Introduction

As snow accumulates in the perpetually cold areas of Greenland and Antarctica, air collects in small hollow pockets within the ice. Over time, the ice closes over and the air is permanently trapped beneath the surface. Advanced techniques of isotopic analysis—similar to those discussed by Michael McElroy in the following chapter — reveal important information about the chemical composition of the atmosphere and the average surface temperatures, at the time when the ice closed over the air pockets. Comparing ice core samples taken in Greenland and the Antarctic, Hans Oeschger and Irving Mintzer report the evidence, just emerging now, of large and rapid climate fluctuations in the past, and suggest that the dynamic mechanisms that drove the planet through the great changes of the distant past could be repeated in the not-too-distant future.

For example, ice cores from the Camp Century and Dye 3 sites in Greenland indicate that, during the last major glacial to inter-glacial transition approximately 13,000 years ago, climate in the North Atlantic region appears to have fluctuated between two very different, meta-stable states. Analysis of oxygen isotopes in marine sediments suggests that one state supported a quite mild climate, approximately 5 °C warmer, on average, than the other. The transitions between these two states may have taken place in periods lasting less than 100 years.

As part of the analysis of ice core records, Oeschger and others have recently applied a new mathematical technique called a deconvolution algorithm. Application of this technique to the ice cores reveals the history of carbon dioxide concentrations over very long periods. Combining this information with the known history of fossil-fuel derived CO_2 emissions reveals the time trajectory of CO_2 emissions from the biota. This analysis suggests that during the last decades, some of the increases in CO_2 in the atmosphere may have been offset by a "fertilization effect"—a response that enhances the ability of land plants and marine phytoplankton to absorb carbon dioxide. This fertilization effect appears to have been enough, in itself, to offset CO_2 increases from deforestation and land use changes—so far.

Hans Oeschger, Professor of Physics at the University of Bern, Switzerland and Irving Mintzer, Coordinator of the Climate Programme at the Stockholm Environment Institute, interpret the results of the deconvolution analysis to evaluate the effects of small changes in the rate of growth of fossil fuel-related emissions on the atmospheric concentration of carbon dioxide. Oeschger is widely recognized as one of the principal developers of ice-core techniques and a seminal thinker on the application of these data to the understanding of the climatic history of the planet. Based on an analysis of the effects of the price-driven decline in emissions growth that occurred after oil price shocks of the 1970s, they reach a hopeful conclusion: that even small changes made in the near term can significantly reduce the risks of rapid climate change in the next century.

- I. M. M.

1 Introduction

In the coldest places on Earth, the snow never melts. In these glacial environments, each year's new snows fall on the previous year's accumulation. Layers of ice build up on the surface of the glaciers. In the process, air is trapped in tiny, hollow pockets within the ice.

The air trapped within these pockets becomes a continuous, long-term record of the contents and behaviour of our atmosphere. Within the last decade, the development of new analytic techniques has converted this glacial record into a natural archive, from which many kinds of useful information could be gleaned. In some parts of Greenland and the Antarctic, this continuous record spans a period of more than 100,000 years. The longest such record analysed to date — a record covering 160,000 years — is taken from ice core samples drawn at Vostok Station, Antarctica. And new ice cores are now being drilled in Central Greenland that could produce records as long as 200,000 years.

These new cores will do more, however, than just extend the current record. They will also provide high-resolution data about rapid climate changes of the more recent geologic past. But even in the meantime, new methods of mathematical and isotopic analysis are emerging that produce new insights about rapid climate transitions, using the data gleaned from the world's longest ice cores to summarize essential findings on the CO_2/climate relationship.

These studies analyse certain parameters from the ice core samples, such as the CO_2 content of occluded air (air trapped within the ice cores), and the ratio between isotopes of oxygen (^{18}O and ^{16}O) in water molecules. As we shall see, this ratio between ^{18}O and ^{16}O (referred to as the index quantity, $\delta^{18}O$) can be used as a proxy measure with which to deduce the surface temperature at the time that the ice was laid down on the glacier. To a first approximation, the higher the value of $\delta^{18}O$, the warmer the indicated temperature at the time the ice was formed.

In this chapter, we will attempt several things:

(1) We will discuss reconstructions of the trajectory of atmospheric carbon dioxide (CO_2) concentration, from the last glacial period to the present era. This is important because strong and rapid climate changes occurred during this period, possibly accompanied by changes in the concentration of CO_2 and other greenhouse gases.

(2) We will summarize recent work, discussing reconstructions of the pre-industrial level of CO_2 in the atmosphere. Determining a pre-industrial baseline is important because it allows us to suggest a background level of atmospheric CO_2 during the current warm interglacial period - an equilibrium level, equivalent to what atmospheric concentrations of greenhouse gases would be *without* the heavy use of fossil fuels by Northern, industrialized societies.

(3) We will present recent reconstructions of the buildup of CO_2 from the beginning of the industrial revolution to 1958, the beginning of precise measurements at Mauna Loa, Hawaii. This history provides valuable insights into the mechanisms through which carbon was exchanged among the atmosphere, ocean, and biosphere, during a period in which the rate of emissions of CO_2 and other greenhouse gases exceeded the rate of natural removal of these trace substances.

(4) Some of the most recent results from ice core studies and atmospheric measurements will be analysed by models of how excess CO_2 could be partitioned among the principal planetary reservoirs. Drawing on the ice core data, we will examine the links between historical greenhouse gas concentrations and past average surface temperatures to estimate the potential greenhouse gas feedback into the climate system today. These correlations offer important new insights into the future risks of rapid climate change due to greenhouse gas buildup.

2 Maintaining the Balance: Natural Controls on the Atmospheric Concentration of Greenhouse Gases

Over millions of years, the principal processes controlling atmospheric CO_2 concentrations have been geologic in nature. The extent of volcanic activity determines carbon dioxide emission levels. The geologic weathering of mountains determines the CO_2 removal rate.

By contrast, on a time scale of decades to centuries, the atmospheric concentration of CO_2 is determined by more dynamic processes — processes that control the flux of CO_2 exchanged between the atmosphere and the terrestrial biota (the amalgamated aggregate of all Earth life forms), and between the atmosphere and the ocean surface. The net change in atmospheric CO_2 concentration is the result of small differences in the magnitude of these large movements of CO_2. The current flux cycling from the atmosphere to the biota and back is about 100 gigatonnes (Gt) of carbon per year (100 Gt C/yr). The simultaneous flux between the atmosphere and the warm, upper mixed layer of the ocean is approximately 90 Gt C/yr.

The human inputs of CO_2 are a one-way flux to the atmosphere, due principally to burning fossil fuel and deforestation, and to other land use conversions. Compared to the natural exchange fluxes, human inputs of CO_2 are relatively small. At current rates of fossil fuel use, the weight of carbon released as CO_2 from fuel combustion is approximately 5.9 Gt C/yr - 1/20 of the non-human flux. The rate of emissions from deforestation is estimated with less precision, and lower confidence, to be approximately 2 Gt C/yr.

Given the small differences between these large fluxes of CO_2 in and out of the atmosphere, one might legitimately ask whether human-induced emissions could have much effect on the atmospheric level of CO_2 and on global climate. The best available information comes from the 160,000 year Vostok ice core. And the evidence from this core indicates a strong correlation between atmospheric CO_2 and global climate. During periods of stable climate, the atmospheric concentration of CO_2 was essentially constant. By contrast, during periods of major climatic change, the CO_2 concentration reflects adjustments in the flux of carbon among the principal reservoirs. Furthermore, measurements of the air trapped in the Vostok core suggest that, over the long periods represented in this core sample, a buildup of greenhouse gases occurred essentially in phase with the temperature increase. The observed declines in greenhouse gas concentrations generally followed the cooling observed in the isotopic record of global temperature.

3 Atmospheric CO_2: Looking at the Long-Term Record

The atmospheric CO_2 concentration has gone both up and down during the long geologic history of the Earth. Analyses of the ice cores taken from glacial areas in Greenland and the Antarctic suggest that, throughout the past glacial-interglacial cycle, the atmospheric concentration of CO_2 has included brief episodes of relatively rapid climate change (on the order of a degree per century), scattered among lengthy periods of more gradual change. The principal driver of climatic oscillations has been the position of the Earth's orbit around the sun and an ever-changing, interactive relationship between the climate system and the circulation of the oceans.

3.1 Rapid climatic oscillations during the last glacial period

Relatively slow climatic variations associated with the glacial-interglacial transition have been documented in the Antarctic ice-core record and in marine sediments. These natural archives support a theory originally put forth by Milankovitch, of long-term climatic change driven by changes in the Earth's orbit relative to the sun.

But recent analyses of ice cores taken from the Camp Century and Dye 3 sites in Greenland indicate that much more rapid oscillations of temperature and greenhouse gas concentrations have occurred during the last glacial period, from 80,000 to 30,000 years ago (Dansgaard et al., 1982). These changes are illustrated in Figure 1.

During the glacial era from 40,000 to 30,000 years before the present era (BP), for example, evidence from the Greenland ice cores show that CO_2 concentration fluctuated rapidly, by as much as 50 parts per million (ppmv) over periods as short as 1000 years (Stauffer et al., 1984). The ^{18}O record from these cores suggests that the temperature in Greenland changed by c. 5 °C during these episodes and that the concentration of dust in the atmosphere also fluctuated dramatically (Dansgaard et al., 1984; Langway et al., 1984). Indications of the changes between cold and mild climates can be seen in the record of the cosmogenic beryllium isotope (^{10}Be) taken from these cores as well. In fact, all the ice core parameters from these samples support the idea that during parts of the last glacial period, the Earth System may have oscillated rapidly between two quasi-stable states with very different climate regimes (Oeschger et al., 1985).

Broecker et al. (1985) hypothesized that the main driving force for these climatic oscillations was a change in the circulation of ocean water in the North Atlantic. They noted that when water vapour is exported from one ocean drainage basin to another, salt is left behind by evaporation, increasing the salinity and the density of the remaining mass of surface water and enhancing the process of deep-water formation. This, in turn, is presumed to affect ocean circulation, sea-surface temperature, and the amount of heat available to

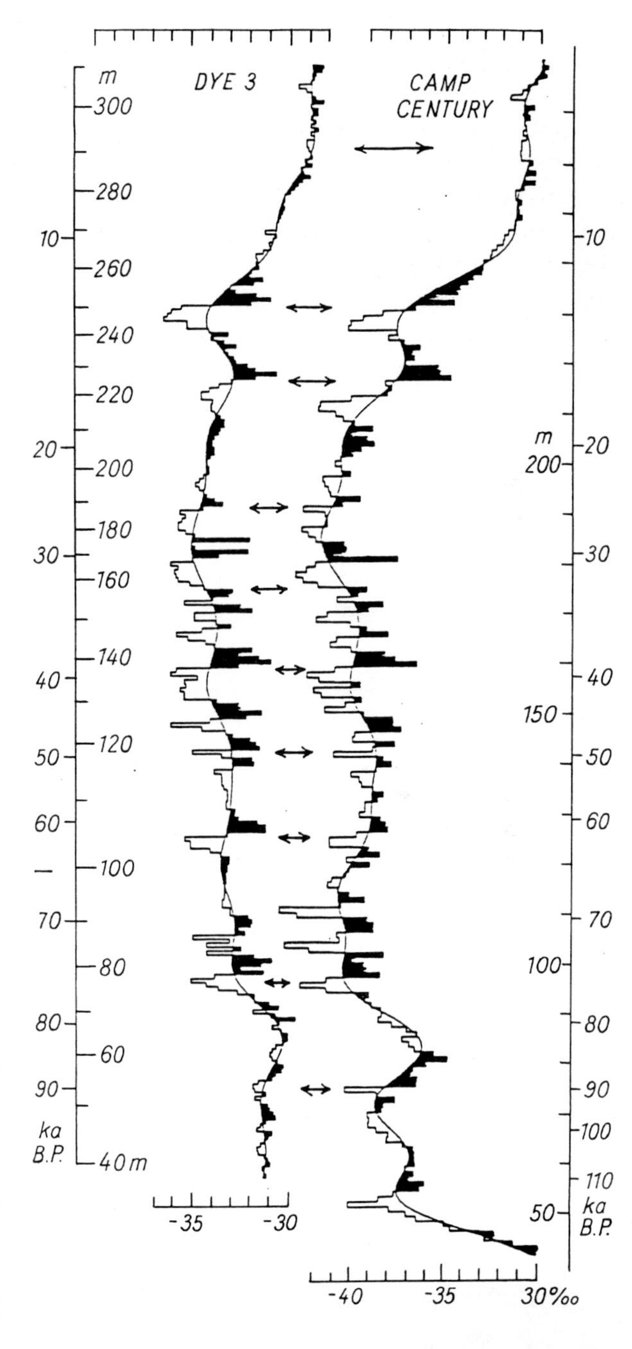

Figure 1. Rapid fluctuations of temperature during the last glacial period, from 80,000 to 30,000 years ago. These are derived from d^{18}O profiles for two Greenland ice cores, and show rapid bi-modal fluctuations (Dansgaard et al., 1982). Initial measurements on the same two cores (Camp Century and Dye 3) indicate that CO_2 concentrations also varied between two states. Reprinted by permission from Science, vol. 218, pp 1273-1277, copyright (c) by AAAS.

warm the air currents that heat the North Atlantic region, especially in Europe.

The short duration of the bimodal North Atlantic oscillations offer support for this hypothesis; the observed oscillations preclude a mechanism based on changes in the Earth's orbit, and suggest a change in some process internal to the Earth system. If that hypothesis is correct, the interaction results from complex feedback mechanisms (Broecker et al., 1985). Changes in water vapour transport can lead, for example, to changes in both the rate and pattern of ocean circulation (Broecker and Denton, 1989; see also McElroy, 1992; this volume).

However, a series of detailed CO_2 measurements of samples from Byrd Station, Antarctica, should have revealed similarly rapid CO_2 variations. The Byrd station cores showed no such signal of rapid variations (although there were small fluctuations of $\delta^{18}O$ and CO_2). One possible explanation for this discrepancy between hemispheric records focuses on the observation that ice in the Antarctic region takes relatively longer to close over and trap air than does ice formed in surfaces of the Greenland glacier. This hypothesis suggests that the Antarctic samples have a tendency to smooth out the larger, more rapid CO_2 variations observed in the Greenland cores. Drilling of a new, high resolution core in the Antarctic may provide the data necessary to test this hypothesis (Neftel et al., 1988). The CO_2 and CH_4 signals in Antarctic cores are expected to be approximately in phase with the observations made in Greenland, because the atmosphere is well mixed for these gases on a yearly time scale.

Regarding the $\delta^{18}O$ temperature signal, it is possible, however, that the transfer of the North Atlantic climate signal to the Antarctic may have resulted in a smoothing of the signal. This could imply that the smaller number of $\delta^{18}O$ oscillations observed in the Vostok core represent clusters of events that can be observed separately in the Greenland cores. (See Figure 2.)

For the last 160,000 years, the interactions between air, ocean, and biota have reflected these complex and dynamic relationships. Some data suggest, for example, that CO_2 concentrations have actually gone down during warm as well as cold periods. Neftel et al. (1988) suggest that at the beginning of the warm Holocene era there was another surprising and rapid change in atmospheric composition. Their analysis indicates that in a warming period of a few thousand years, atmospheric CO_2 concentrations *fell* by about 30 ppmv, from 280 ppmv to 250 ppmv.

This apparent increase in the rate of extraction of CO_2 from the atmosphere-ocean system may be explained by rapidly increasing uptake among the terrestrial biota. Oeschger (1991) suggests that this seemingly counter-intuitive change in concentration occurred at this particular time because, sustained by a warming climate, soils and plants began to then grow back over the continents, re-covering

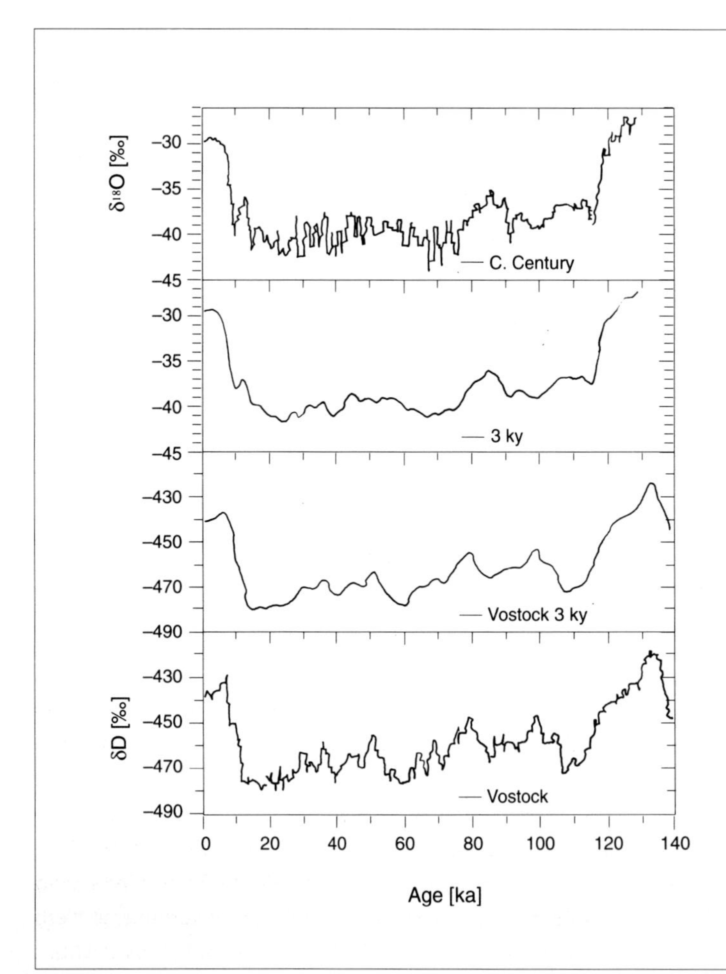

Figure 2. Comparison of stable isotope records from Greenland (Camp Century) and Antarctica (Vostok) dD profiles. In the Camp Century record, during the period from 80,000 BP until 30,000 BP (top panel), transitions between a mild and a cold climate are indicated. Filtering the raw data with a 3,000 year filter (middle two panels) improves the impression of similarity of the two records. The lower frequency variations in the Byrd Station record during this period may reflect clusters of the events observed in the Greenland cores. Sources: Dansgaard et al., 1982 (Greenland) and Jouzel et al., 1987 (Vostok).

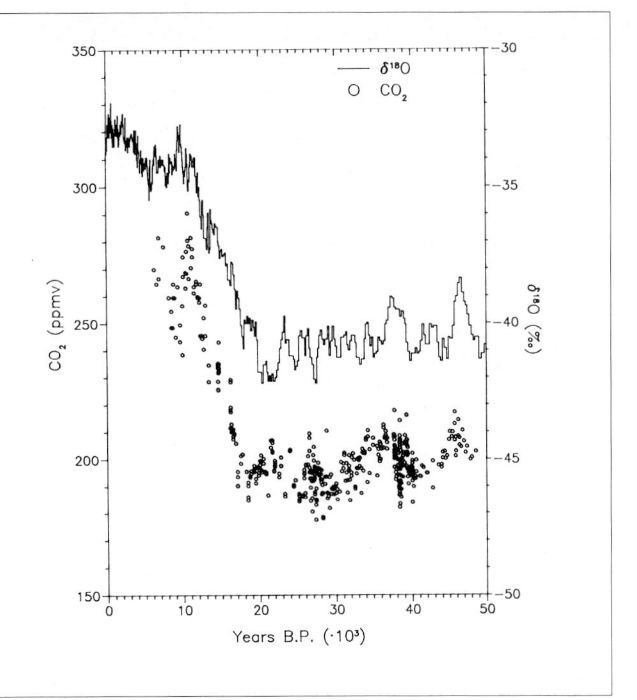

Figure 3. Variations in CO_2 and $\delta^{18}O$ at Byrd Station, Antarctica, for the last 50,000 years (Neftel et al., 1988, and additional unpublished data). The 10% drop in CO_2 at the end of the last deglaciation (10 ky BP) may have been caused by vegetative regrowth following the melting of continental ice sheets.

areas which had been covered by ice for the millennia of the preceding glacial period.

3.2 Rapid climate change between the Holocene period and the last glacial epoch

A particularly large change in atmospheric composition took place about 10,000 years ago, in the transition from the last warm period (the Holocene) to the last glacial epoch. In this period, atmospheric CO_2 declined by about 90 ppmv. The ice core data show that not only CO_2 but also methane (CH_4) concentrations decreased at this time (e.g. Lorius, 1989). The declining concentration of CO_2, when combined with the concurrent decline in the atmospheric concentration of CH_4, caused a decrease in the radiative forcing of the climate system. The decline in greenhouse gas-induced warming was approximately 2 watts per square metre (W/m^2). A climate model study by Broccoli and Manabe (1987) suggests that this decline in greenhouse forcing contributed significantly to the climatic coupling between the hemispheres during the inter-glacial to glacial transition.

Analyses of $\delta^{18}O$ and CO_2 from samples of this period taken at Byrd Station, Antarctica are shown in Figure 3 (Neftel et al., 1988). The decline in $\delta^{18}O$ about 10,000 years ago indicates the climatic transition to the last glacial period. The CO_2 observations for the concurrent period show a parallel decrease from 280 to 190 ppmv. The shifts of CO_2 and $\delta^{18}O$ might well have occurred in phase with each other.

The magnitude and speed of the change in temperature and atmospheric composition during the last inter-glacial to glacial transition is important to us today. For comparison, we note that in the last two centuries the concentration of atmospheric CO_2 has increased by an equivalent amount. The buildup of greenhouse gases due to human activities during this recent period has committed the planet to a roughly equal change in radiative forcing. Of course, this does not mean that the more recent emissions will necessarily cause a change in climate of equal magnitude to that of the last inter-glacial to glacial transition.

However, the disruption of natural ecosystems and the damages to national economies that result from any large change in climate will be determined not just by the magnitude of the change but by the speed as well. The geologic change that occurred 10,000 years ago was spread over a period of a few millennia; the more recent change has occurred over a period of only a few centuries. This rapid rate of change suggests that it may be more difficult for ecosystems and societies to respond smoothly to any resulting future change in climate.

3.3 Atmospheric CO_2: observations of the recent past

The ice core data also provide additional, convincing evidence about the more recent changes in the atmosphere. Using an ice core from Siple Station, Antarctica, drilled by a US-Swiss team, it was possible to obtain a clear and consistent reconstruction of the trajectory of atmospheric CO_2 buildup during the last two centuries (Neftel et al., 1985; Friedli et al., 1986). Figure 4 illustrates the observed changes in atmospheric CO_2 concentration since 1750. CO_2 has a characteristically long residence time in the atmosphere; thus, sufficient atmospheric mixing occurs that the measurements taken in the Antarctic region are reliable representations of the global concentration of this gas.

The observations made in the most recent (i.e. youngest) samples from the Siple Station core overlap with the period of precise measurements made since 1958 at Mauna Loa. The results show good correspondence in the period of overlap between the two observational records. More specifically, the analysis of the Siple core shows that the global concentration of CO_2 in the atmosphere was approximately 280 ppmv in about AD 1800. The atmospheric concentration

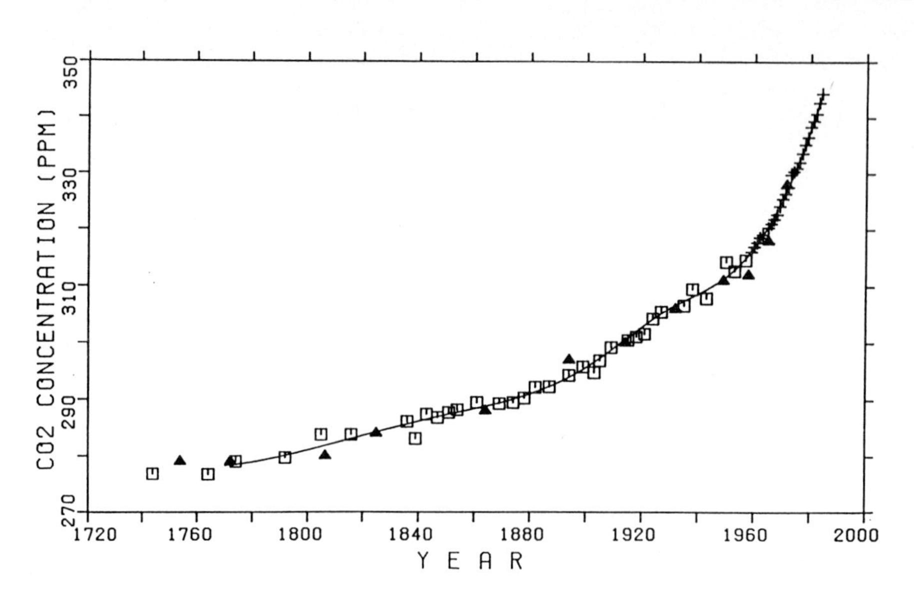

Figure 4. Atmospheric CO_2 increase in the past 250 years, as indicated by measurements on air trapped in ice from Siple Station, Antarctica (squares; Neftel et al., 1985:Friedli et al., 1986) and by direct atmospheric measurements at Mauna Loa, Hawaii (triangles: Keeling et al., 1989).

began to increase at about this time and increased significantly in the 19th Century. During the 20th Century, the growth rate increased, reaching approximately 355 ppmv today. The current rate of growth in atmospheric CO_2 concentration is approximately 0.5% per year or about 1.8 ppmv per annum.

4 The Deconvolution Analysis: Human-Induced Emissions of CO_2 During the Industrial Era

How does this buildup in concentration compare with the pattern of known emissions of CO_2 from human activities? Do the current models provide adequate explanations for the partitioning of excess carbon among the principle reservoirs? These are the key questions that determine our confidence in the ability of the models to explain the changes observed in the past and to predict the fate of future emissions of CO_2.

4.1 Estimating the amount of carbon stored in the atmosphere

If, due to carbon emissions, atmospheric CO_2 were increased, the CO_2 flux out of the atmosphere into the oceans and into the biota (due to fertilization) would exceed the return from these reservoirs into the atmosphere. Thus, excess CO_2 would be taken up by the oceans or the biota, or both.

Thus, it is important to analyse the flow of CO_2 into each of these reservoirs. The net flux of CO_2 into the ocean is estimated as the product of a fixed gas transfer coefficient between sea water and air, multiplied by the difference in CO_2 partial pressure between the atmosphere and ocean. The partial pressure of CO_2 in the surface mixed layer of the ocean is determined by the rate at which CO_2 is transported from the warm mixed layer at the surface to the cold, deep layers of the ocean.

Radioisotope analysis using ^{14}C indicates that most of the carbon that enters the surface ocean as CO_2 is absorbed into the top several hundred metres of the ocean, passing downwards by slow processes of diffusion. In a few exceptional areas, such as parts of the North Atlantic, more rapid and deeper turnover occurs. Observations on the sea bottom of radioactive tritium produced by atmospheric explosions of nuclear weapons suggests that in these areas, a process of rapid convection (sometimes referred to as thermohaline circulation), moves CO_2 rapidly into the ocean depths.

The constant rain of organic material — both fecal material and dead organisms — does not contribute significantly to sequestering excess carbon on a time scale of decades. The level of activity by marine biota is controlled by other factors (e.g. light, temperature, and limiting nutrients) and is not sensitive to small changes in the concentration of CO_2 at the surface.

To simulate the uptake of excess CO_2 in the atmosphere by the ocean, models have been developed which describe the processes of CO_2 transfer at the ocean surface, and the subsequent transport of CO_2 to greater depths. These models enable us to estimate how a pulse input of CO_2 into the atmosphere would be distributed in the atmosphere-ocean system over time. The total increase in atmospheric CO_2 is then the sum of the fractions of each year's CO_2 emissions that remain in the atmosphere. These annual fractions can be derived from the pulse input response function. This math-

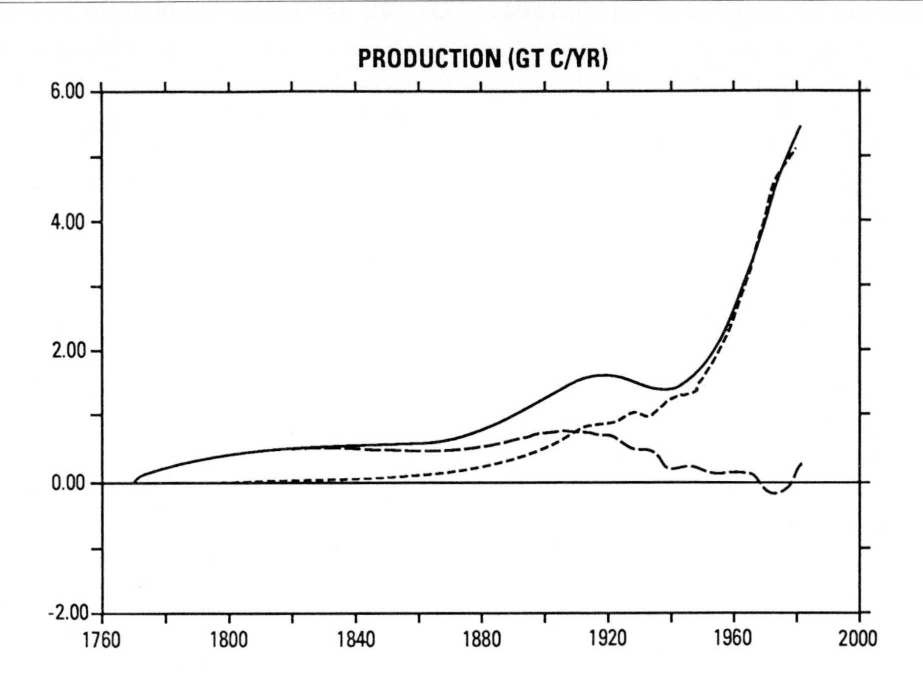

Figure 5. CO_2 production rates over the last two hundred years, obtained by deconvolving the measured CO_2 increase using the box-diffusion model. The solid line shows total carbon dioxide production. The short dashed line shows observed fossil fuel input. The long dashed line represents the difference, which equals the net biospheric CO_2 input. Sometime between 1910 and 1920, fossil fuel production of CO_2 overtook biospheric production; by 1960, fossil fuel production dominated total carbon dioxide production. Source: Siegenthaler and Oeschger, 1987.

ematical procedure is the convolution of the input history and the pulse input function.

From the ice core studies and the atmospheric measurements, one knows the history of atmospheric CO_2 increases. By applying the inverse mathematical procedure - the deconvolution from the trajectory of CO_2 increases, one can calculate the complete history of human-induced CO_2 releases.

4.2 Total human-induced release of CO_2

The total annual emissions of CO_2 from human activity is the sum of emissions due to fossil fuel combustion — and net emissions from the biota that are induced by human activities. Figure 5 illustrates one reconstruction of the trajectory of total CO_2 emissions induced by human activities during the last two centuries (Siegenthaler and Oeschger, 1987). The total human-induced emissions of CO_2 were relatively constant from 1780 to 1860. This was followed by a slow increase through 1910. An upward "bump" in the rate occurred between 1910 and 1920. CO_2 emissions continued to rise steadily until 1940, when the rate of emissions increased dramatically. Since 1940, the overall pattern of human-induced emissions of CO_2 has closely followed the historic expansion in fossil fuel consumption.

4.3 Fossil fuel-related emissions of CO_2

All fossil fuel combustion results in the oxidation of carbon to CO_2, but not all fossil fuels contribute equally to historical or present emissions.[1] Taken together, the release of CO_2 from all fossil fuel burning have produced about 5.4±0.5 Gt of C per year during the last decade (Houghton et al. 1990).

It is surprising how closely the deconvolved total CO_2 production (Siegenthaler and Oeschger 1987) agrees with the estimates of production and emissions from fossil fuel combustion from Marland (1989) and Rotty and Masters (1985) reported of each country to the United Nations.

The short-dashed line in Figure 5 illustrates the trajectory of fossil-fuel emissions since approximately 1850. During this period, fossil-fuel derived emissions of CO_2 have increased from less than 0.1 Gt of C per year in the middle of the 19th Century to over 5.4 Gt of C per year in the last decades of the 20th Century.

4.4 Biospheric release of CO_2

The principal activities responsible for human-induced emissions of CO_2 from the biota are deforestation and changes in land use patterns. Deforestation releases CO_2 through processes of rapid oxidation (burning of standing forests) and

[1] Delivering one thermal kWh from coal releases about twice as much CO_2 as producing the same amount of energy from natural gas. Producing one thermal kWh from oil produces about 50% more CO_2 than does releasing the same amount of heat from natural gas. Depending on how they are synthesized, burning synthetic fuels from coal or oil shale will release about three to seven times as much CO_2 as burning an equivalent amount of natural gas.

slow oxidation (bacterial decomposition of forest residues). Changing land use patterns expose carbon in the soil to the forces of erosion by wind and water. Some of the mobilized carbon is oxidized in the air to CO_2. The best recent estimates suggest that the rate of gross CO_2 emissions from deforestation and land use changes during the 1980s is approximately 1.6 ± 1.0 Gt of C per year (Houghton et al., 1990).

The rate of carbon uptake by terrestrial biota can respond to an increase in atmospheric CO_2. All else being equal, increased atmospheric concentrations of CO_2 induce a fertilization affect in the terrestrial biosphere. Because other simultaneous changes in precipitation, runoff, and soil moisture also effect net primary productivity of the biosphere, the global impact of the fertilization effect is very hard to estimate quantitatively. It is preferable, therefore, to combine the increased uptake of CO_2 due to fertilization, and the release of biotic CO_2 due to deforestation and other land use changes, into a single quantity — the net biospheric production. The long-dashed line in Figure 5 illustrates the difference between the total deconvoluted production and the estimated fossil fuel emissions. This difference is referred to as the net biotic emissions of CO_2.

4.5 *Monitoring the excess: the net imbalance of CO_2 among the reservoirs*

As noted above, the net addition of carbon to the atmosphere from fossil fuel burning has been about 5.4 Gt of C per year during the last decade. The estimated emissions from deforestation and land use changes are about 1.6 Gt of C per year. This suggests a total influx of carbon to the atmosphere that is equivalent to about 7.0 Gt per year. The estimated rate of oceanic uptake of CO_2 is approximately 2.0 ± 0.8 Gt of C per year. As a result of these differences in the flux of carbon between the principal reservoirs, excess carbon is building up in the atmosphere.

But the observed rate of increase for the decade of the 1980s, 1.6 ppmv per year corresponding to 3.4 Gt of C per year, is insufficient to explain the fate of the total excess in carbon emissions. The net imbalance in the carbon budget is of the same order of magnitude as the estimated emissions from deforestation and land use. This suggests that the impact to date of the fertilization effect on the terrestrial biota may be approximately equal to the impact of deforestation and land use changes.

This hypothesis remains very difficult to prove, given the complexity of the Earth system and our currently incomplete view of how it operates. Other possible explanations of the so-called "missing sink" for carbon may prove to explain better the redistribution of carbon among the reservoirs over the last decade.

4.6 *A note of caution about the results from the deconvolution analysis*

The fact that the analysis of the CO_2 production function (based on the deconvolution method) produces strikingly similar results to that obtained by other analysts using fuel-use data does not necessarily prove that the approach is completely correct. The coincidence of results could be a fortuitous artifact of some quirk in the method; or it could reveal some basic insights about the workings of the climate system.

One simple explanation might be that the carbon cycle model used for this analysis rather accurately describes the complex dynamics of CO_2 uptake by the oceans, even if the growth rate in emissions increases by a factor of two. This assumes that the atmosphere-ocean system functioned in a more or less constant manner during the period of observation, and that the fossil fuel use data was estimated with high accuracy.

But these results imply that net emissions from the biota were insignificant; this is troubling and hard to believe. If biotic release were at the high end of the IPCC estimate (2.6 Gt of C per year), it suggests that the fertilization effect of CO_2 just cancelled out the biotic release due to deforestation and land use conversion despite large shifts in land use. It would be easier to accept the hypothesis that there was minimal difference between the uptake of CO_2 due to enhanced fertilization and the gross emissions from deforestation and land use change, if the absolute sizes of these CO_2 fluxes were small. As yet, there is no reasonably satisfactory explanation for this.

5 The Grand Geophysical Experiment

In 1957, Roger Revelle remarked that, through its massive increase in the emissions of carbon dioxide and other greenhouse gases, the human race was performing a grand, unplanned geophysical experiment on the Earth's climate system (Revelle and Suess, 1957). Contrary to all good scientific practice, there is no "control" environment against which to measure the effects of this specific stress on the system. The results of the experiment are still not known and its significance not widely understood. Even now, the response of the climate system to the large-scale injections of trace gases can not be predicted with accuracy. But there is good reason to believe that the system is very sensitive to even small changes in the rate of growth in annual emissions.

In particular, we note that because fossil fuel use contributes such a large fraction of total annual emissions and these emissions are quite sensitive to abrupt price changes, some elements of the planet's response can be gauged by investigating the effects of the oil price shocks of 1973-74. During the 30 years prior to 1973, the price of oil declined in constant dollar terms and the rate of fossil fuel-derived emissions grew steadily. For most of this period, CO_2 emissions grew at a rate of about 4.4% per year.

The oil crises of the 1970s shocked the global economy, spurred widespread efforts to improve the efficiency of energy use, and caused a significant reduction in the rate of growth in CO_2 emissions. From 1970 to 1990, fossil fuel-derived emissions increased from 4.7 to a little less than 6 Gt of C per year. If, by comparison, emissions had continued to increase at the average annual growth rate of the previous three decades, emissions in 1990 would already have reached nearly 10 Gt of C per year.

The continuing excess of carbon emissions into the atmosphere fills the atmospheric reservoir with ever greater amounts of carbon. Recent analyses suggest that, with no other changes to the atmosphere, if the concentration of CO_2 (now approximately 355 ppmv) were to reach twice the pre-industrial level (approximately 560 ppmv or two times 280 ppmv), the average global surface temperature would increase by 1.5-4.5 °C, relative to the pre-industrial climate (National Research Council, 1983; Bolin , 1986; Houghton, et al., 1990).

This may not seem like a dramatic change, but it would quickly produce a very different world. It is worth noting that a change in average surface temperature of only 1° Celsius is all that separates today's climate from that of the Little Ice Age that occurred from the 14th to the 17th Century. An average warming of only 2 °C would take the planet outside the range of anything that has been experienced during the last 10,000 years - the present inter-glacial period, including the period of written human history. An average warming of 5 °C from today's level would push the planet outside the range of anything experienced in the last million years.

Moreover, the atmospheric concentrations of other greenhouse gases are also increasing over time. At current rates of emissions, carbon dioxide buildup accounts for only about half of the annual increase in radiative forcing on the climate system. Thus, if the share of CO_2 in the total forcing remains the same in the future as at present, to keep the total warming effect on the planet below the warming that would come from doubling the pre-industrial concentration of CO_2 alone will require holding the level of CO_2 to about 150% of the pre-industrial level — about 420 ppmv.

A new carbon cycle model study (Oeschger et al., 1992, in prep.) indicates that to hold the atmospheric concentration of CO_2 at or below 420 ppmv will require stringent reductions in CO_2 emissions. These reductions must first affect not just the rate of growth in emissions, but must go further, causing emissions to decline in an absolute sense. The reduced rate of growth in emissions that has been experienced since 1973 has resulted in average annual emissions for the last decade of about 5.4 Gt of C. From this level, one would need to reduce total emissions by about 50% - to about 2.7 Gt of C - by about 2050, and then continue a slight decrease of emissions level to keep the atmospheric concentration of CO_2 below 420 ppmv.

A reduction of 50% from today's emissions levels will not be easy to achieve given the steady increase in global population, the continuing pattern of excessive consumption in the industrialized countries, and the shift in fuel mix among developing countries from annually cycled biomass supplies to commercial fossil fuels. Achieving the necessary reductions will require a widespread and massive commitment to improving energy efficiency, developing renewable and non-fossil energy technologies, and the rapid transfer of the best available industrial technologies to developing countries.

But it is important to note how much more difficult the challenge would have been if the oil price shocks of the 1970s had not reduced the growth rate in annual emissions from 4.4% to about 2%. Had world fossil fuel use remained on the earlier trajectory, holding the atmospheric concentration of CO_2 at or below 420 ppmv would have required much more stringent efforts and, most likely, more draconian policies. Application of the deconvolution techniques developed for the analysis of paleoclimates suggests that if the earlier rate had persisted and annual emissions thus grown to 10 Gt of C per year in 1990, an 80% reduction in emissions would have been required from the then current level. In today's political environment, given the demographic and industrial momentum of modern societies, an 80% reduction in fossil fuel-derived emissions could only have been achieved through a worldwide economic slowdown or, perhaps, a global economic depression.

6 Conclusions

Some of the evidence from the analysis of air trapped in glacial ice suggests that greenhouse gas buildup has contributed to the onset of global warming in the past. The mathematical technique of deconvolution allows the study of contributions to CO_2 buildup from fossil fuel use, changes in biotic activity, and oceanic uptake. By analysing the observed increase from 1950 to 1990 and comparing the results of the deconvolution to the estimates based on reported use of fossil fuels, the net biotic contribution to recent CO_2 buildup can be identified. This analysis suggests that the fertilization effect of increased CO_2 on the biota was just enough to offset the increased biotic release from deforestation and land use changes during this period.

Application of the deconvolution algorithm has shown the effects of small changes in the rate of growth in fossil fuel emissions on atmospheric CO_2 concentrations. The study concludes that, if the price-induced decline in fossil fuel-derived emissions had not occurred following the oil price shocks of the 1970s, the task of stabilizing future CO_2 concentrations would have been much harder than it is today. The information that will emerge from experiments now under way will improve understanding of the climate system and of the internal feedbacks that amplify or moderate small changes in key parameters. This information may prove critical to improving society's ability to predict future climate changes with confidence.

The analysis of ice core data supports the hypothesis that widespread or regional climate change can occur quite rapidly. During periods of mild climate in the last interglacial, transitions from warm to cold climates have been recorded in the North Atlantic region in episodes as short as 100 years.

Such climate changes have been associated with changes of 50 to 100 ppmv in the atmospheric concentration of CO_2. In some cases, similar changes in concentration have occurred over periods of several millennia. In the last two centuries, human activities have increased the concentration of CO_2 in the atmosphere by approximately 75 ppmv. The evidence is insufficient to say with certainty that this change will induce a shift in climate similar in magnitude to those which have occurred in the past. But it is certain that the damage and disruption which results from climate change will be determined as much by the speed of the change as by the magnitude.

The results of this analysis further emphasize the importance of achieving early reductions in CO_2 emissions in order to minimize the need for draconian and disruptive measures that would otherwise be required to achieve the same ends in the future.

References

Bolin, B., B.R. Döss, J. Jäger and R.A. Warrick, eds. 1986: How much CO_2 will remain in the atmosphere? In *The Greenhouse Effect, Climatic Change and Ecosystems*. SCOPE 29, B. Bolin et al., eds., John Wiley, pp.93-155

Broccoli, A.J. and S. Manabe, 1987: The influence of continental ice, atmospheric CO_2 and land albedo on the climate of the last glacial maximum, *Climate Dynamics 1*, 87-99.

Broecker, W.S. and G.H. Denton, 1989: The role of ocean-atmosphere reorganizations in glacial cycles. *Geochim. Cosmochim. Acta,* **53**, 2465-2501.

Broecker, W.S., D. Peteet and D. Rind, 1985: Does the ocean-atmosphere system have more than one stable mode of operation? *Nature,* **315**, 21-25.

Dansgaard, W., H.B. Clausen, N. Gundestrup, C.U. Hammer, S.J. Johnson, P.M. Kristinsdottir, and N. Reeh, 1982: A new Greenland deep ice core. *Science,* **218**, 1273-1277.

Dansgaard, W., S.J. Johnsen, H.B. Clausen, D. Dahl-Jensen, N. Gundestrup, C.U. Hammer and H. Oeschger, 1984: North American climate oscillations revealed by deep Greenland ice cores. In J.E. Hansen and T. Takahashi, eds., *Geophysical Monograph 29: Climate Processes and Climate Sensitivity.* AGU, Washington, DC, 288-298.

Friedli, H., H. Loetscher, H. Oeschger, U. Siegenthaler and B. Stauffer, 1986: Ice core record of the $^{13}C/^{12}C$ reatio of atmospheric CO_2 in the past two centuries. *Nature* **324**, 237-238.

Houghton, J.T., G.J. Jenkins and J.J. Ephraums, 1990: Greenhouse Gases and Aerosols. In *Climate Change: The IPCC Scientific Assessment.* Cambridge University Press.

Jouzel, J., C. Lorius, J.R. Petit, C. Genthon, N.I. Barkov, V.M. Kotlyakov and V.N. Petrov, 1987: Vostok ice core: a continuous isotope temperature record over the last climatic cycle (160,000 years). *Nature,* **329**, 403-408.

Keeling, C.D., R.B. Bascastow, A.F. Carter, S.C. Piper, T.P. Whorf, M. Heimann, W.G. Mook and H. Roeloffzen, 1989: A three dimensional model of atmospheric CO_2 transport based on observed winds: 1. Analysis of observational data. In D.H. Peterson, (ed.), *Geophysical Monograph 55: Aspects of Climate Variability in the Pacific and the Western Americas.* AGU, Washington, DC, pp.165-236

Langway, C.C. Jr., H. Oeschger and W. Dansgaard, 1984: The Greenland ice sheet in perspective. In C.C. Langway et al., (eds.), *Geophysical Monograph: The Greenland Ice Sheet Program.*, AGU, Washington, DC.

Lorius, C., 1989: Polar ice cores and climate. In A. Berger et al., eds. *Climate and Geo-Sciences.* Kluwer Academic Publ, pp. 77-103.

Marland, G., 1989: CDIAC Communications, Winter 1989, Carbon Dioxide Information Analysis Center, Oak Ridge National Laboratory, Oak Ridge.

National Research Council, (NRC), 1983: *Changing Climate.* Report of the Carbon Dioxide Committee, Board of Atmospheric Sciences and Climate. Washington, DC, National Academy Press.

Neftel, A., E. Moor, H. Oeschger and B. Stauffer, 1985: Evidence from polar ice cores for the increase in atmospheric CO2 in the past two centuries. *Nature,* **315**, 45-47.

Neftel, A., H. Oeschger, T. Staffelbach and B. Stauffer, 1988: CO_2 record in the Byrd ice core 50,000-5,000 years BP. *Nature,* **331**, 609-611.

Oeschger, H., F. Joos and U. Siegenthaler: The CO_2 increase before and after the oil embargo in 1973, (in preparation).

Oeschger, H. 1991: Paleodata, Paleoclimates and the Greenhouse Effect, *Proceedings from the Second World Climate Conference,* 221-224.

Oeschger, H., B. Stauffer, R. Finkel and C.C. Langway, Jr., 1985: Variations of the CO_2 concentration of occluded air and anions and dust in polar ice cores. In E.T. Sundquist and W.S. Broecker, (eds.), *Geophysical Monograph 32: The Carbon Cycle and Atmospheric CO_2: Natural Variations Archean to Present.* AGU, Washington, DC. pp. 132-142.

Revelle, R. and H.E. Suess, 1957: Carbon Dioxide Exchange Between Atmosphere and Ocean and the Question of an Increase of Atmospheric CO_2 during the Past Decades**.** *In Tellus* **IX**, 18-27.

Rotty, R.M. and G. Masters, 1985: In Atmospheric Carbon Dioxide and the Global Carbon Cycle, ed. J. Trabalka, DOE/ ERJ0239, US Dept. of Energy, Washington, DC 20545, pp. 63-80.

Siegenthaler, U. and H. Oeschger, 1987: Biospheric CO_2 emissions during the past 200 years reconstructed by deconvolution of ice core data. *Tellus,* **39B**, 140-154.

Stauffer, B., H. Hofer, H. Oeschger, J. Schwander and U. Siegenthaler, 1984: Atmospheric CO_2 concentration during the last glaciation. Ann Glaciol., **5**, 160-164.

CHAPTER 5

Changes in Climates of the Past: Lessons for the Future

Michael B. McElroy

Editor's Introduction

The record of climate change over the geologic past is one of the most often-cited, but least-understood, pieces of evidence supporting the argument that human activities may have an important effect on future climate. Michael B. McElroy notes that other factors, judging from the record of past changes, may also be important to the future climate. For example, changes in the intensity of solar radiation (caused by the periodic changes in the position of the Earth's orbit relative to the sun) have been a critical driving force in the long-term pattern of climate change over millions of years.

But from the distribution of various atomic isotopes in mud and ice samples that have been laid down over hundreds of thousands of years, McElroy concludes that future human emissions may have a comparable effect on the climates to come as changes in solar intensity and orbital position have had in the past. By studying the natural archives gleaned from mud and ice, he suggests that we can improve our understanding of the dynamics of climate change. With these data, scientists may be able to unravel the puzzling mechanisms that shift ocean currents and atmospheric circulation, possibly triggering the onset of rapid climate change. Emphasizing the complexity of this set of closely coupled, non-linear systems, McElroy observes that human activities are causing changes in atmospheric composition that are on the same scale as those associated with major climate changes (ice ages and sudden warmings) of the geologic past. Despite the potential implications of this analysis, he is careful to avoid even the appearance of sensationalism.

Michael McElroy, Professor and Chairman of the Department of Earth and Planetary Sciences at Harvard University, is widely respected as one of the most creative thinkers in the field of planetary science. He has made important contributions in geology, space science, and paleoclimatology. His great strength is in his ability to see new relationships and patterns among historical data, where others see only a tangle of numbers.

McElroy does not indicate any sense of certainty or predict the climate of the future. He does not assert that human emissions are "shocking" the planet. But given what we now know about past climate change, he demonstrates that such shocks are possible. As one possibility, he suggests that human emissions may have contributed to the end of the Little Ice Age, a cold spell lasting for almost 300 years in Europe, when people could walk across the Baltic and agriculture regularly failed to provide adequate food for the population of continental Europe. Intensely aware of the magnitude of the uncertainties and the scale of unintended effects on global climate from human activity, McElroy urges an acceleration of scientific research coupled with heightened emphasis on international cooperation to reduce the risks of greenhouse gas buildup.

- I. M. M.

1 Introduction

This chapter is concerned with lessons that may be drawn from studies of the geologic past, relevant to our need to predict the climate of the future. Our emphasis is on processes internal to the climate system, and on feedbacks which appear to be significant in understanding climate change.

We believe that some of these processes may be substantially more important than is suggested by their treatment in current climate models.

Climate has varied over a large range in the recent history of the Earth, with extremes represented by warm environments of the Cretaceous and Eocene and the comparatively frigid conditions of the last few million years.

We now have reason to believe that shifts in climate are controlled largely by variations in carbon dioxide (CO_2), modified by fluctuations in ocean circulation and by the interactions of the ocean, the atmosphere, and the biosphere. The concentration of CO_2 reflects, in the short term (100,000 years or less), an equilibrium between the atmosphere and the ocean. That equilibrium, in turn, depends in a complex fashion on the chemical, physical and biological state of the sea, regulated ultimately by climate, modulated in turn by CO_2, defining a system coupled so intimately that it is difficult at any given time to isolate cause from effect.

The behaviour of the climate system is affected by feedbacks from processes involving ice, clouds, water vapour, ocean circulation, terrestrial and marine biota. Taken together, these represent a set of closely coupled, non-linear systems whose interactions are very difficult to simulate in detail using mathematical equations. The best of the mathematical models used to simulate these interactions are as complex as the models used to represent the explosion of nuclear weapons; and the models are run on the world's largest supercomputers. Despite their complexity, these models are unable today to predict with confidence the response of climate to increased concentrations of greenhouse gases. Indeed, most scientists remain pessimistic about the prospects for meaningful near-term predictions of the precise timing (or the regional distribution) of the impacts of future climate change. That is why it is necessary to evaluate the climate record, to arrive at a sense of how much the projections of future global warming may reasonably be trusted.

2 The Past Five Million Years

From our survey of variations in climate over the past several million years, we conclude that changes in CO_2 have had an important influence on climate, at least for the past 160 thousand years (kyrs).

2.1 A climate record preserved as isotopes in shells

The analysis of shells of organisms preserved in the sediments of the deep sea provides a remarkably clear record of the changes in climate over the past five million years. Climate changes are revealed by the changes in the abundance of water stored in continental ice in polar regions. The less water, the warmer the climate. In the shells, the relative abundance of the two primary oxygen isotopes — ^{18}O and ^{16}O — provides a record of the changing abundance of water (Emiliani, 1955; Broecker and Van Donk, 1970; Shackleton, 1967; Shackleton and Opdyke, 1973; Hays et al., 1976; Imbrie et al., 1984). The reasoning for this is based on a simple physical principle: evaporation of H_2O from the ocean favours the light isotope, ^{16}O; growth of continental ice sheets is reflected in a gradual increase in the relative abundance of ^{18}O in the ocean. (This quantity is usually represented by the notation $\delta^{18}O$.) As the ocean becomes isotopically heavy, the change in isotopic composition of ocean water, and inferentially the volume of continental ice, is recorded in the composition of the shells of organisms growing in the sea.

2.2 Matching the climate record to the orbital and axial records

Geologists studying the record of ^{18}O from deep sea cores for the recent past have focused mainly on the role of seasonal changes in insolation (the intensity of solar radiation) in modulating the spatial extent and thickness of continental ice sheets. The central idea behind this research, attributed to Milankovitch (1941), is that expansion of continental ice sheets is likely to occur during periods of low summer insolation. Seasonal variations in the intensity of sunlight are associated with changes in the tilt of the rotation axis with respect to the position of the Earth on its elliptic orbit around the sun, and with changes in the ellipticity of the orbit itself.

As illustrated in Figure 1, the Earth is closest to the sun today during northern hemisphere winter (southern summer). Its distance from the sun is largest in northern summer (southern winter). The position of the Earth on its orbit for a particular season progresses regularly around the orbit, describing a complete cycle every 22,000 years: northern summer coincided with maximum separation of the Earth from the sun 11,000 years ago. 22,000 years ago, the position was similar to today. Specialists refer to this effect as the precession of the equinoxes.

Changes in the intensity of radiation received at a particular location at a particular season arise also as a result of changes in the inclination, or tilt, of the rotation axis with respect to the plane of orbital motion; (see Figure 2). The change in this case also is regular; the axis tilts up and down by 1.5 degrees (a total change of 3°) over a period of 41,000 years. Finally, there are small changes in the intensity of sunlight received at the Earth for a particular season due to slow changes in the ellipticity (i.e., departures from circularity) of the orbit. The repetition period in this case is about 100,000 years.

Figures 1-2 show how the fluctuations in solar radiation correspond, over time, to changes in $\delta^{18}O$ (obtained from a composite of results from deep sea cores). The similarity of the patterns for $\delta^{18}O$ and radiation provides the evidence for a link between the variations in climate and the changes in radiation associated with variations of the Earth's orbit. But the connection is not simple. The largest changes in climate are observed, at least recently, to occur on a schedule that repeats approximately once every 100,000 years. The largest changes in solar radiation are associated with the 22,000 and 41,000 year variations in the Earth's orbit. The amplitude of changes in solar radiation that occur every 100,000 years are

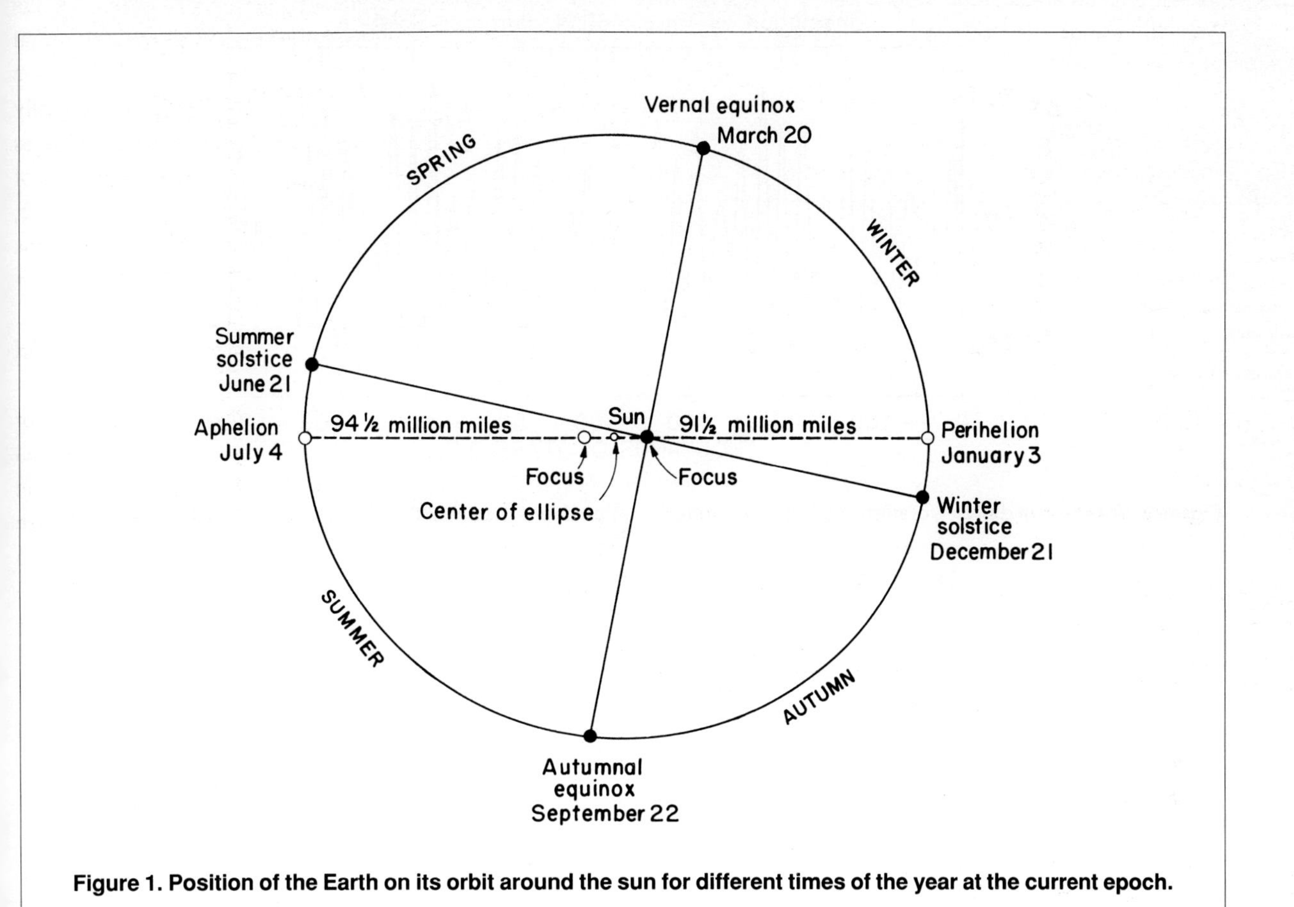

Figure 1. Position of the Earth on its orbit around the sun for different times of the year at the current epoch.

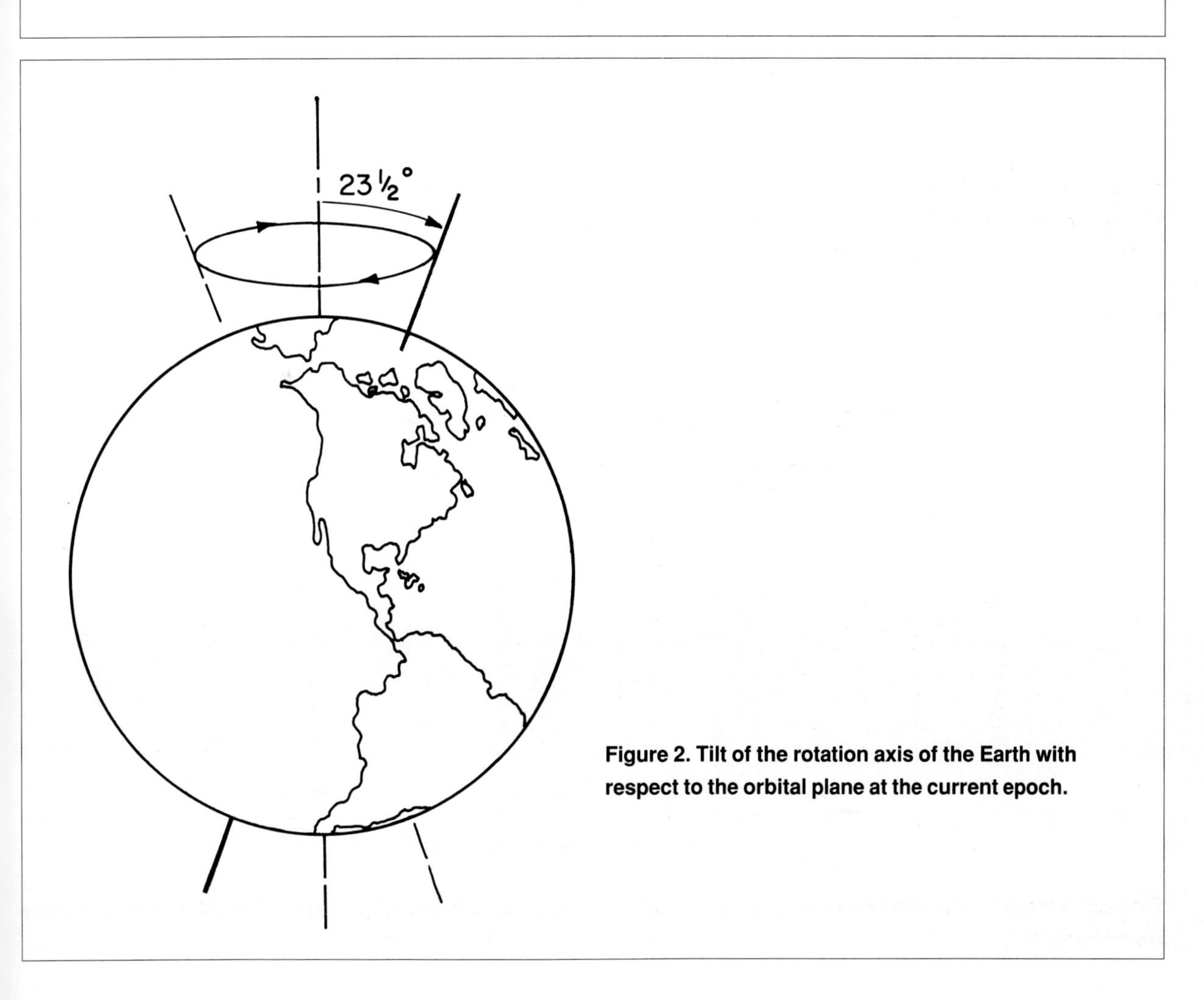

Figure 2. Tilt of the rotation axis of the Earth with respect to the orbital plane at the current epoch.

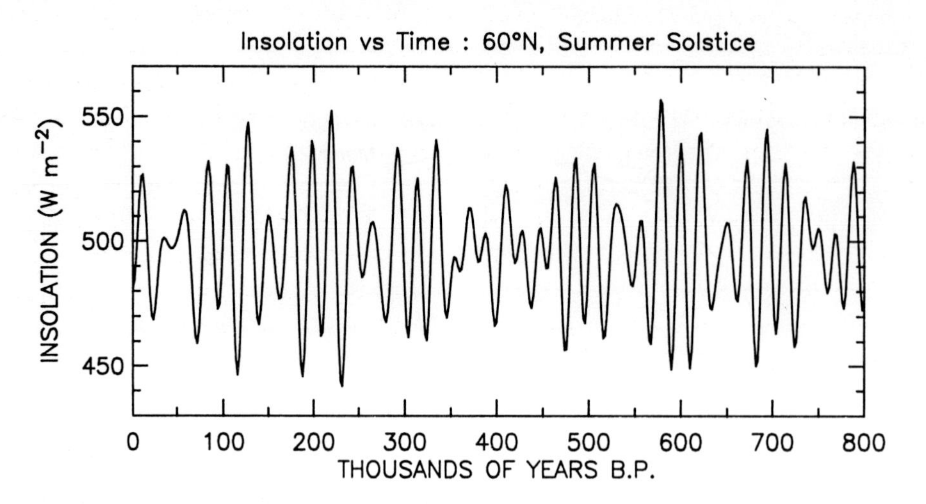

Figure 3. Variation of daily insolation at 60°N for summer solstice over the past 800,000 years. Source: After Berger (1978).

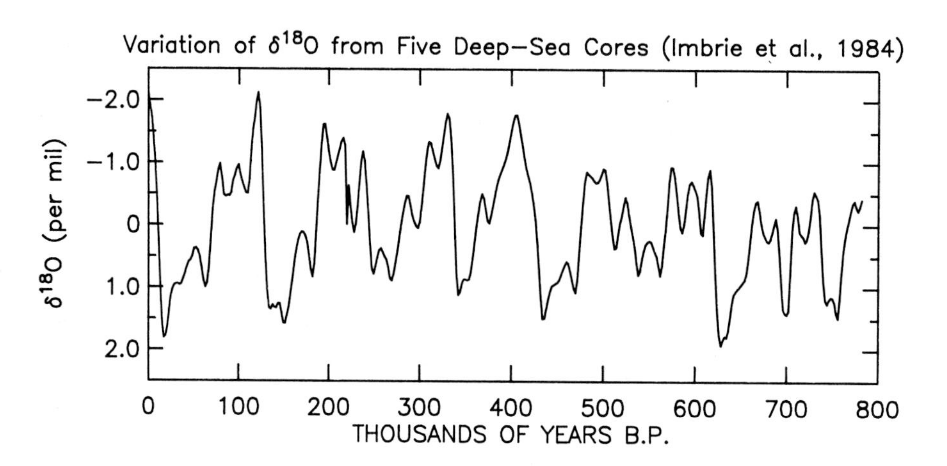

Figure 4. Variation of ^{18}O of shallow-dwelling planktonic foraminifera from five deep sea cores, plotted on the SPECMAP time scale, over the past 800,000 years. Source: After Imbrie et al., 1984, In: A.L. Berger et al., eds., *Milankovitch* and *Climate*, Part1, pp. 269-305. D. Reidel, Dordrecht, Holland.

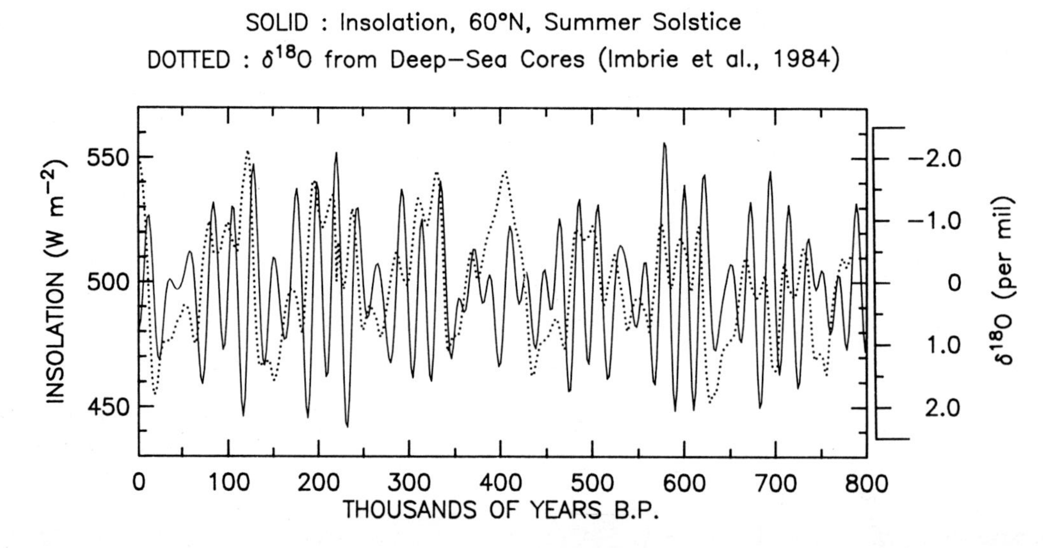

Figure 5. Superposition of insolation at 60°N for summer solstice (solid line, from Figure 3) and the ^{18}O record (dotted line from Figure 4).

relatively small, compared to the amplitude of the observed climate changes for the same period.

3 Climate Change and Carbon: The Long View

Changes in the rhythm of climate about one million years ago may have been due to a decline in the baseline level of CO_2, perhaps associated with an increase in the rate of geologic weathering. The cycles which have been dominant in the pattern of climate change in the last million years are almost undetectable in the geologic record before that time (Joyce et al., 1990; Saltzman, 1990). In the earlier period, the dominant cycle seems to have been one of approximately 40,000 years. In the more recent period, the cycle seems to have shifted, as discussed above, to one with a period closer to 100,000 years.

3.1 Limitations of the non-carbon explanations for climate change

Some analysts have suggested that the climate changes that occur with a period of 100,000 years cannot be fully explained by changes in the Earth's orbit. Thus, a number of alternative explanations have emerged. One set of such explanations focuses on the dynamics of ice sheets (Weertman, 1976; Pollard et al., 1980; Oerlemans, 1980; Birchfield et al., 1981; Pollard, 1982, 1983a, 1983b, 1984; Peltier and Hyde, 1984; Hyde and Peltier, 1987). Another focuses on the effects of variations of CO_2 in the atmosphere (e.g. Pisias and Shackleton, 1984; Genthon et al., 1987). Unfortunately, none of these approaches gives a completely satisfactory explanation of the full set of available data.

The ice dynamic model suggests that the collapse of the ice sheet marking the end of a 100,000 year glacial epoch is triggered by insolation-driven retreat of ice into a depression at the Earth's surface that forms under the thickest part of the ice sheet. This hypothesis gives a good account of the data describing the rapid demise of ice sheets at the end of ice ages for the past few hundred thousand years. But it fails to account for the fact that warming is observed simultaneously in both hemispheres of the globe during such periods (Broecker and Denton, 1989). In addition, this ice dynamic approach fails to explain why the effects of the 100,000 year orbital cycle are not visible in the geologic record for the period prior to about one million years ago.

The CO_2 hypothesis as an explanation of climate variability is more promising.

3.2 Carbon as a key factor in climate change

The concentration of carbon (and consequently of CO_2) in the atmosphere-ocean system can vary over long time scales because of changes in volcanism and tectonic activity. These, in turn, affect the rates at which CO_2 is produced and consumed by the Earth's crust. To some extent, the warm climates of the Cretaceous and Eocene may have been maintained by higher levels of CO_2. The decline in CO_2 since the Cretaceous, for example, may be due to an increase in the rate of consumption of the gas by weathering, relative to the rate at which it is produced by volcanism (Berner et al., 1983).

The geologic evidence suggests that the climate was much warmer when the concentration of CO_2 exceeded its present value by about a factor of two. It may be improper, however, to conclude that a warm, ice-free, planet is the inevitable fate of continued contemporary growth in the concentration of greenhouse gases. Feedbacks involving ice, continental and marine, play an important role in the present climate. Such feedbacks were almost certainly absent, or at least different, during the warmer climates of the early Cenozoic, when the abundance of CO_2 may have been comparable to that expected to arise over the next few decades as a consequence of human activity. Very different climate regimes, each distinguished by different modes of ocean circulation, could exist in equilibrium with a particular level of CO_2.

Changes in climate can proceed rapidly, on time scales as short as centuries or even decades, as indicated by data for the Younger Dryas (a period of globally cold conditions interrupting recovery of the Earth from the last ice age) and the Little Ice Age (a cold snap extending from about 1250 to about 1850 AD). Rapid fluctuations in climate have been attributed to changes in production of deep water in the North Atlantic; they may also be linked to variations in circulation of intermediate waters in the Pacific.

3.3 Isotopic analysis of ice-core data

Additional evidence for the connection between changes in CO_2 concentration and variations in climate comes from the analysis of deuterium (a heavy isotope of hydrogen) and CO_2 trapped in glacial ice.

The isotopic composition of rain and snow reflects the complex history of water from the time it evaporates from the ocean until it returns to the surface in the form of precipitation. Generally, when a core sample of ice is removed from a very cold place, the oldest ice is at the bottom of the core and the most recent ice is at the top. Glaciologists have developed a variety of techniques for dating these ice samples. As discussed earlier in connection with the interpretation of isotopic data from deep sea cores, evaporation favours isotopically light water. Precipitation selectively removes heavy water, leaving lighter water behind in the vapour phase. The isotopic composition of snow falling in Greenland or Antarctica reflects the complex processing that occurs from the time the water enters the atmosphere until the snow is deposited and incorporated in the ice cap. Typically, a mass of air will drop some of its moisture in one place, move on, and drop the rest in other locations. Each precipitation event causes a change in the isotopic composition of the water vapour which remains in the atmosphere.

This process is particularly sensitive to temperature. It has been shown that the ratio of deuterium to hydrogen in snow falling in central Antarctica, from which the Vostok core was extracted, can be interpreted to provide a useful proxy for the annual average temperature of the region.

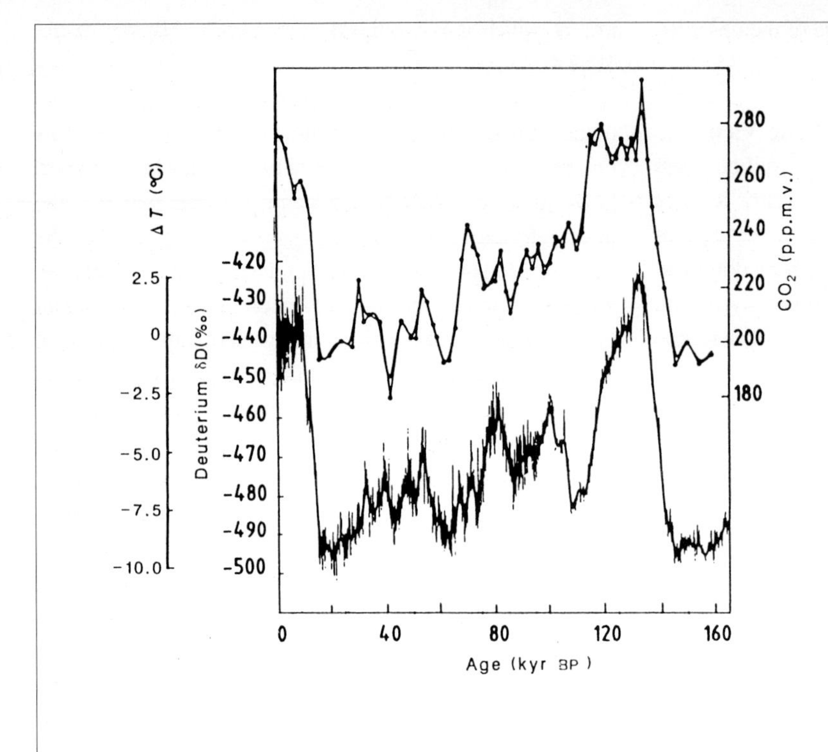

Figure 6. Variation of the concentration of CO_2 (upper curve) and the relative abundance of deuterium (lower curve), over the past 160,000 years, for the Vostok ice core from East Antarctica. Also shown is the temperature change associated with the variation of deuterium (Jouzel et al., 1987). Source: Barnola et al., 1987. Reprinted by permission from *Nature*, vol. 329, pp. 408-414. Copyright (c) 1987, MacMillan Magazines, Ltd.

A close correlation between annual mean local temperatures (indicated by the abundance of deuterium) and CO_2 concentration can be observed in the 160,000 year record extracted from the Vostok ice core (Genthon et al., 1987; Barnola et al., 1987), as indicated in Figure 6. The variations in CO_2 displayed in the ice core data almost certainly reflect complex rearrangements in ocean circulation (Boyle and Keigwin, 1987; Broecker et al., 1988, 1989; Boyle, 1990) and perhaps ocean biology (Broecker, 1982; Knox and McElroy, 1984; Sarmiento and Toggweiler, 1984; Siegenthaler and Wenk, 1984; Martin and Fitzwater, 1988). These studies indicate the complex interaction of factors affecting global climate. Some analysts have suggested that the close correlation between CO_2 and climate in the Vostok ice core may arise as a consequence of controls on CO_2 exercised by climate rather than the reverse.

While variability of CO_2 on a time scale of 100,000 years appears to be controlled mainly by changes in the circulation of the oceans, the ocean-atmosphere system is closely coupled with the circulation of the ocean, which is significantly influenced by the concentration of CO_2 and vice versa.

3.4 Correlation with climate models and remaining discrepancies

Complex general circulation models (GCMs) of the climate suggest that the difference in CO_2 concentration between glacial and inter-glacial periods (approximately 200 ppmv as compared with 280 ppmv) could have an important influence on both regional and global climate (Manabe and Broccoli, 1985; Manabe and Bryan, 1985; Broccoli and Manabe, 1987; Rind et al., 1989; Oglesby and Saltzman, 1990). None the less, despite the overall improvement of the agreement between model and observation when the changes in CO_2 concentration are taken into account, discrepancies remain. A more realistic simulation of clouds, in general, and the hydrological cycle for the tropics, in particular, could resolve some of the remaining discrepancies.

4 Warm Climates of the Cretaceous and Eocene

4.1 Evidence for warmth much greater than in our own era

We turn our attention now from the ice-dominated world of the past five million years to the relatively warm, earlier climates of the Cretaceous — 140 to 65 million years before the present time (myr bp) and Eocene (58 to 37 myr bp). Measurements of the isotopic composition of oxygen in the shells of organisms, dwelling on the ocean floor and preserved in marine sediments, provide a useful record of temporal changes in the temperature of the deep ocean. The data, summarized in Figure 7, indicate a decline in deep water temperatures from about 18 °C in the Cretaceous to near 0 °C today. While interpretation of the isotopic measurements is ambiguous, the pattern exhibited in Figure 7 offers strong support for the view that temperatures at high latitude were much higher during the Cretaceous and Eocene than they are today.[1]

[1] Inferences on temperature depend, for example, on assumptions made with respect to salinity — for which there are no independent data.

The terrestrial flora and fauna tell a similar tale. Frost intolerant vegetation was common on Spitsbergen (paleolatitude 79° N) during the warm Eocene (Schweitzer, 1980). Alligators (Dawson et al., 1976) and flying lemurs (McKenna, 1980) are observed in deposits of comparable age from Ellesmere Island (paleolatitude 78° N), while data from central Asia (paleolatitude 60° N) indicate the presence of palm trees in this region during the Cretaceous (Vakhameev, 1975). As recent as 5 myr bp, the forest-tundra boundary extended to latitudes as high as 82° N, some 2500 km north of its present location, occupying regions of Greenland now perpetually covered in ice (Funder et al., 1985; Carter et al., 1986).

4.2 The once and future Hadley Circulation

The presence of flora incapable of surviving even occasional frost in the interior of continents at high latitude convinced Farrell (1990) that warm climates of the past could indicate that the tropical Hadley circulation had once extended far beyond its present limits (See Figure 7). The Hadley circulation is a pattern of air currents which carry heat away from the tropics, toward higher latitudes. It is observed in both hemispheres. It is an atmospheric system in which heat is transported mainly in the upper troposphere, close to the boundary with the stratosphere (the tropopause). The return flow occurs near the surface.[2]

If there were more friction between the atmosphere and the surface of the planet, the Hadley circulation could once again expand over a larger North-South range. But a number of other factors, besides friction between air and ground, affect the ease with which heat is transmitted from low to high latitudes by the Hadley circulation.[3] These include the height of the tropopausc, the vertical gradient of temperature in the troposphere, the efficiency of radiative relaxation (the ease with which gas molecules release their "trapped" heat), the frictional connection between the atmosphere and the surface, and the horizontal gradient of temperatures at the surface.[4]

5 Tracing the History of Atmospheric CO_2

5.1 Evidence for very high prehistoric levels of CO_2

There are reasons to believe that the abundance of CO_2 was relatively high during the warmer Cretaceous and Eocene epochs. Arthur et al. (1991) — interpreting measurements of the isotopic composition ($\delta\ ^{13}C$) of organic material from phytoplankton preserved in ocean sediments laid down during the Cretaceous — concluded that the level of CO_2 during the Cretaceous was four to 12 times higher than today (see also Rau et al., 1989). More recent data (Freeman and Hayes, 1992) suggest that CO_2 declined more or less steadily from a value near 1000 ppm 100 million years ago, to between 300 and 500 ppm roughly 10 million years ago. It appears from these data that a value between about 600 and 700 ppm may have applied during the most recent warm period of the Eocene. It is tempting to associate the rapid decline in CO_2 since the Miocene with the drop in the temperature of deep water, indicated for about the same time by the data in Figure 8 — and to attribute the eight-fold increase in the parameter γ, that is required to account for these warm geologic periods, to feedbacks associated with a higher level of CO_2.

Recent analyses suggest that warmer temperatures during the Cretaceous and Eocene would have permitted the atmosphere to hold larger quantities of H_2O than is present in today's atmosphere. Release of latent heat, associated with this additional H_2O, would have contributed to a greater level of stability in the atmosphere. Increased convective activity, along with increased burdens of CO_2 and H_2O, could have promoted a longer time span for radiative relaxation. These factors could have combined to provide the necessary change in γ. A warmer climate could have been facilitated further by an increase in the depth of the troposphere as discussed by Farrell (1990).

5.2 New techniques for following the CO_2 record

Recognition that the isotopic composition of the organic fraction of phytoplankton contains information on the abundance of CO_2 dissolved in sea water (Rau et al., 1989; Popp et al., 1989) has led to a promising new technique for the study of past variations of CO_2. Further advances may be

---

[2] In the Hadley Circulation, air parcels moving to higher latitude are deflected to the east by the Coriolis force; this tends to limit the efficiency with which heat is redistributed by meridional motion from low to high latitudes.

[3] The efficiency of the Hadley circulation is measured in terms of a dimensionless number, γ, given by (Farrell, 1990):

$$\gamma = \frac{S\,\tau_R}{\delta_H\,\tau_A}$$

where:

S is a function of the stability of the troposphere;

τ_R denotes the time constant for radiative relaxation (it measures how quickly greenhouse gas molecules give up trapped heat);

δ_H is an indicator of the gradient from hot to cold temperatures that marks the distance from equator to poles; and

τ_A is an index of the friction between the atmosphere and the surface.

Farrell (1990) estimates a value for γ in the present climate of 0.5 and concludes that earlier equable (i.e. very warm) climates would require an increase in γ by about a factor of eight.

[4] A high tropopause provides a deeper troposphere for transport of heat. The more stable the atmosphere, the larger the relative heat content of the air transported in the upper branch of the circulation cell. The longer the time scale for radiative relaxation, the more effective the role of the circulation in transporting heat. The smaller the latitudinal gradient of the temperature of the atmosphere in radiative equilibrium, the lower the demand imposed on the circulation to maintain a specified gradient of temperature.

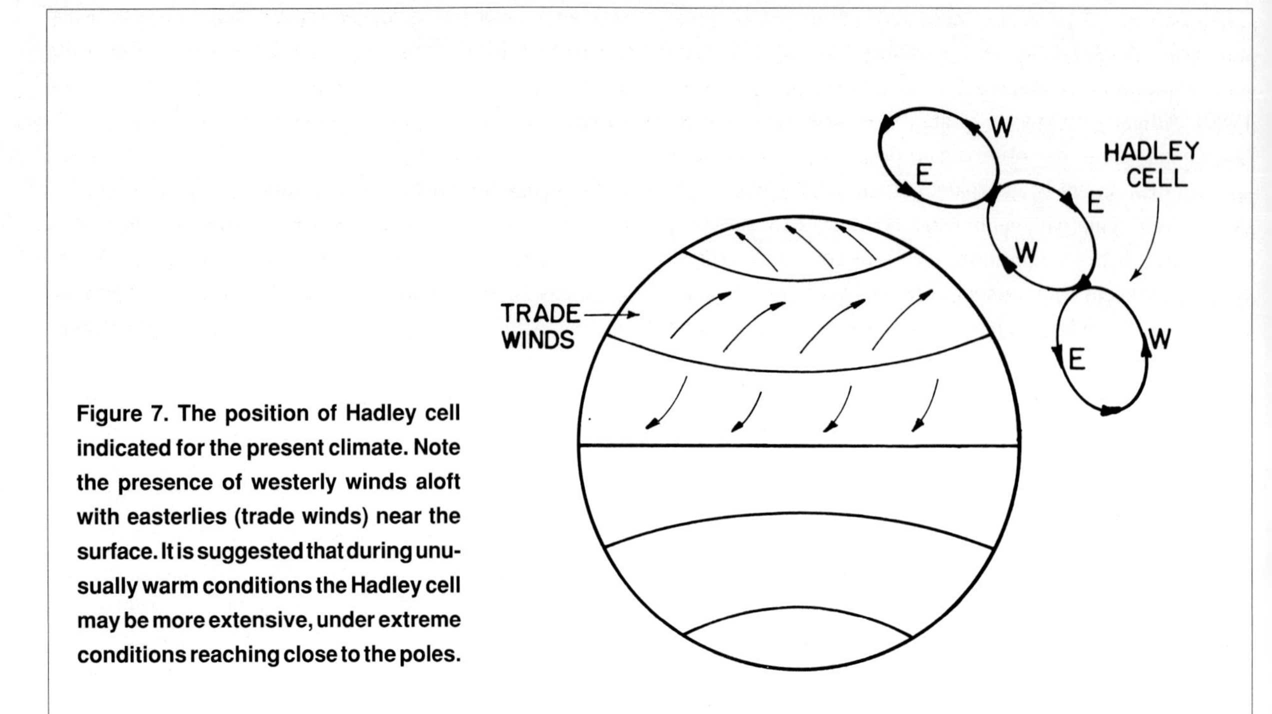

Figure 7. The position of Hadley cell indicated for the present climate. Note the presence of westerly winds aloft with easterlies (trade winds) near the surface. It is suggested that during unusually warm conditions the Hadley cell may be more extensive, under extreme conditions reaching close to the poles.

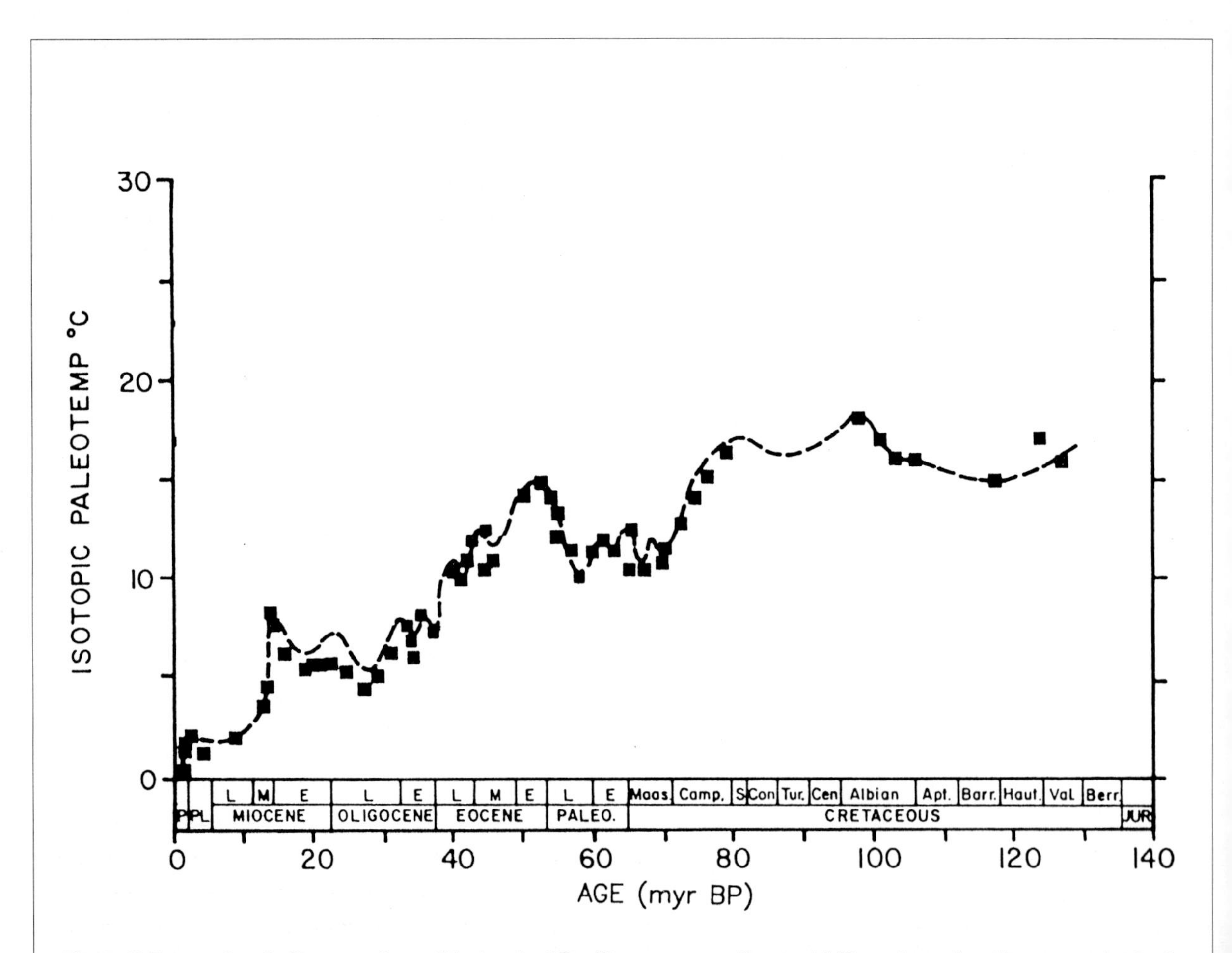

Figure 8. Reconstructed temperature of the tropical Pacific ocean over the past 140 myr based on the oxygen isotopic composition of benthic foraminifera. Source: Douglas and Woodruff, 1981. In: *The Oceanic Lithosphere, The Sea*, vol. 7, pg. 1287, Figure 34.

anticipated from measurements of specific molecular compounds in phytoplankton — the one-celled marine organisms responsible for most of the photosynthetic activity in the oceans (Jasper and Hayes, 1990) and from studies of the isotopic composition of carbonates in paleosols — the prehistoric soils that have been compacted and buried beneath the ground (Cerling, 1991). As discussed above, preliminary results from analyses of phytoplankton (Arthur et al., 1991; Freeman and Hayes, 1992) indicate that the abundance of CO_2 was much higher in the Cretaceous than today, and that it declined steadily through the Cenozoic.

It should be possible with further work to refine this analysis, to identify times in the past when concentrations of CO_2 were comparable to values expected to develop in the future as a consequence of anthropogenic activity. Studies of associated climates could provide valuable insight into conditions anticipated in the future as a response to human activity. Recent work by Budyko and his colleagues may cast some useful light on these complex questions (Budyko, 1990).

6 Climates of the Recent Past

We turn attention now to a discussion of variations in climate observed over the past 10,000 years. Our objective is to establish a framework with which to judge the possible significance of changes taking place today, as they might be attributed to an increase in the concentration of CO_2 and other greenhouse gases. We conclude that shifts in climate can develop rapidly, as illustrated by observations of the Younger Dry (Bard et al., 1987; Heusser and Rabassa, 1989; Kallel et al., 1988; Kennett, 1990; Kennett et al. 1985; Kudrass et al., 1991; Pastouret et al., 1978) as and a number of similar events that punctuate the climate record over the last 10,000 years. The end of the last ice age, for example, was marked by two distinct pulses in the rate at which water was discharged from land to ocean as indicated in Figure 9.

The challenge is to define the long-term trend and natural variability of the climate system in order to isolate the human contribution. It is no simple task. There are indications that temperature has decreased by about 3 °C over the past 6000 years (Nesje and Kvamme, 1991). Imbedded in the long-term trend are short-term fluctuations of as much as 1 °C. It is in the context of these changes that we must assess the significance of the evidence for an increase in globally averaged temperature of about 0.5 °C over the past 150 years (Jones et al., 1986).

6.1 *10,000 years ago: unusual cold*

Figure 10 presents a record of temperature change for the past 10,000 years — inferred for Western Norway in the region of the Jostedalsbreen ice cap, by studying the movements of the ice and the composition of vegetation in the vicinity (Nesje and Kvamme, 1991). The data in Figure 10 document the climatic warming that followed the Younger Dryas, a cold epoch approximately 10,000 years bp. Then, about 9500 years ago, an unusual cold episode occurred (Fairbanks, 1989). This event corresponds with the period of glacial advance known as the Erdalen Event, as indicated in Figure 11. Some scientists attribute this cold snap to a change in ocean circulation, but this has not yet been proven.

6.2 *7000 years ago: peak of a warm period*

Sea level rose rapidly during the warm period between 10,000 and 7000 years BP, reaching within about 15 metres

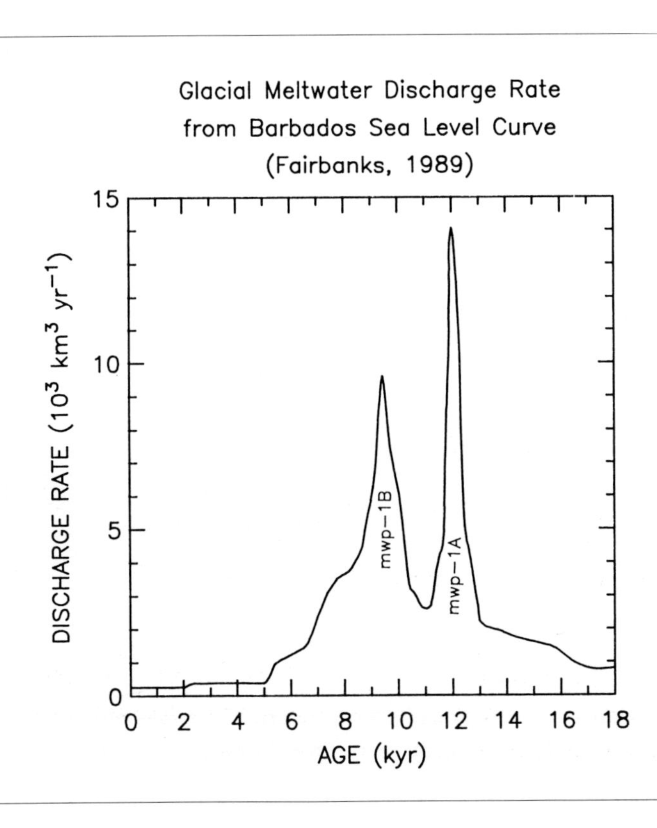

Figure 9. The discharge rate of glacial melt water over the past 18,000 years inferred from a reconstruction of sea level based on studies of coral reefs drilled offshore of Barbados. The time scale corresponds to radiocarbon age. Source: After Fairbanks, R. G., 1989.

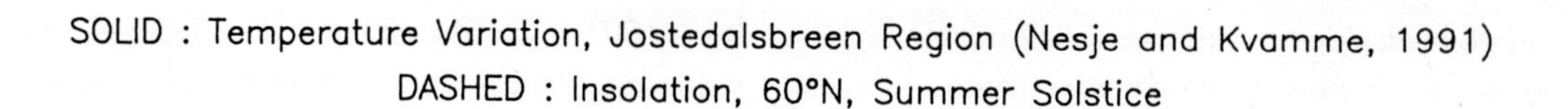

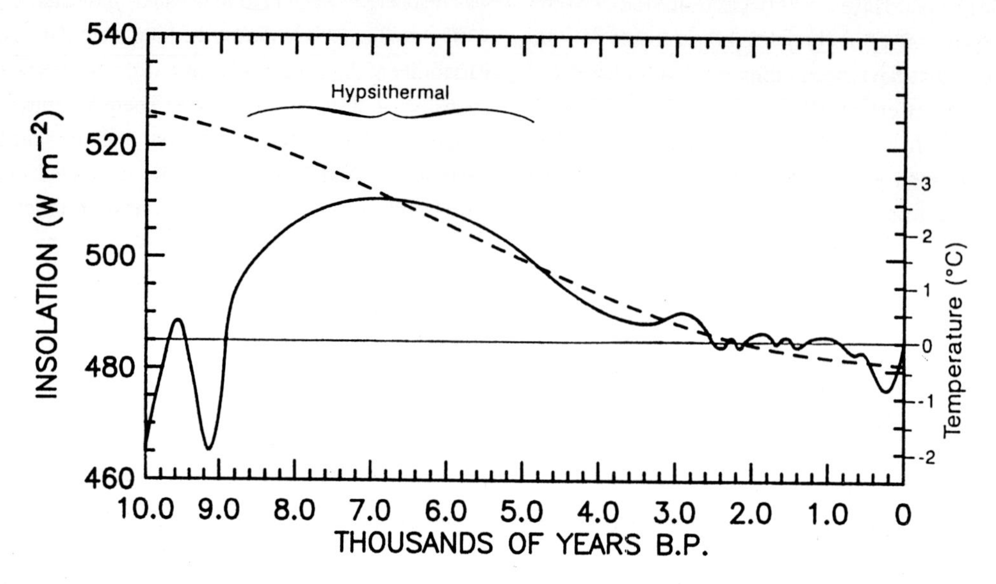

Figure 10. Variation of temperature (solid curve) for the Jostedalsbreen region of Norway (62°N, 7°E), based on a variety of lithostratigraphic and paleobotanical techniques (from Nesje and Kvamme, 1991), and the variation of daily insolation (dashed curve) at 60°N for summer solstice (based on the formulation of Berger, 1978), over the past 10,000 years.

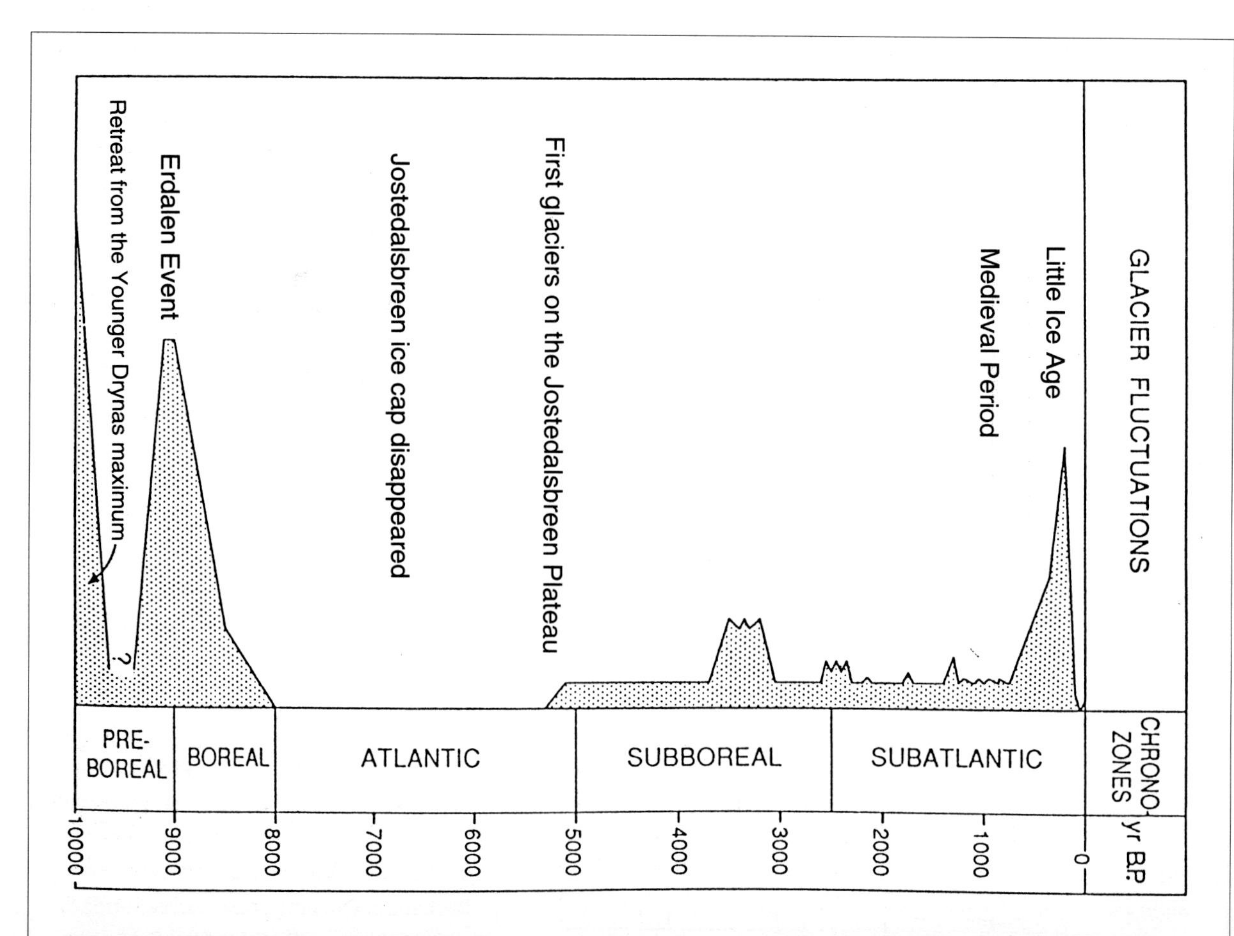

Figure 11. The fluctuation of glaciers for the Jostedalsbreen region of Norway (62°N, 7°E), based on a variety of lithostratigraphic and paleobotanical techniques, over the past 10,000 years. From Nesje and Kvamme, 1991.

of its present value at the end of this period (Fairbanks, 1989). The subsequent rate of increase of sea level was relatively modest, suggesting that the ice sheets had reached close to their present interglacial configuration by 7000 yr bp. The warmest temperatures during the past 10,000 years occurred at this time. The peak temperatures were observed during the period known as the Hypsithermal, from about 8000 to about 5000 radiocarbon years bp. This period was so warm that several long-lived mountain glaciers wholly disappeared. The Jostedalsbreen ice cap in Norway disappeared during the Hypsithermal.

6.3 5000 years ago to 300 years ago: return of cold climate

Temperatures have declined more or less steadily since then, with the Jostedalsbreen ice cap reappearing about 5000 years ago. The ice cap reached its maximum extent since the Erdalen during the period known as the Little Ice Age, which lasted from the 13th to the 17th Century.

6.3.1 The possible influence of solar intensity

Figure 10 includes a curve, following Berger (1978), that shows the variation of insolation over time for the summer solstice at 60° N latitude. It is intriguing to note the similarity in the behaviour of the temperature and insolation curves for the period subsequent to about 6500 yr BP. If the intensity of sunlight is taken as a surrogate for temperature, the Little Ice Age appears as a short cold spell relative to the long-term trend.

Some have suggested that the similarity between the long-term trends of sunlight and temperature after 6500 years bp can account for the major changes in climate that have been observed in this period. In the absence of a clear quantitative model, this hypothesis is hard to prove. The apparent relationship visible in the more recent data seem to contradict the pattern observed in previous eras. The earlier trend could reflect feedbacks associated with the demise of the global ice sheets, associated for example with changes in albedo and/or changes in ocean circulation, relating perhaps to inputs of fresh water to the sea.

6.4 The Little Ice Age

The Little Ice Age was a period of generally cold temperatures, specifically cold winters, lasting from about AD 1550 (or perhaps from as early as AD 1250) to about AD 1850, with extremes observed between about AD 1580 and AD 1700 (Fairbridge, 1987). It was marked by a significant decline in sea surface temperatures, both in the North Atlantic (Fairbridge, 1987) and in the North Pacific (Yoshino and Xie, 1983). The impact appears to have been global, with important changes reported not only for Europe (Lamb, 1984; Weikinn, 1965; Pettersson, 1912; Lindgren and Neumann, 1981), but also for North America (Ludlum, 1966; Baron, 1982; Catchpole and Ball, 1981), China (Wang and Zhao, 1981), Japan (Yamamoto, 1972) and South America (LaMarche, 1975).

There is evidence for worldwide advance of mountain glaciers, with snow lines coming 100-200 metres closer to sea level at mid latitudes and as much as 300 m closer to sea level in the equatorial Andes (Porter, 1975,1981; Hastenrath, 1981; Broecker and Denton, 1989). It appears that the end of the Little Ice Age was abrupt and that it took place globally

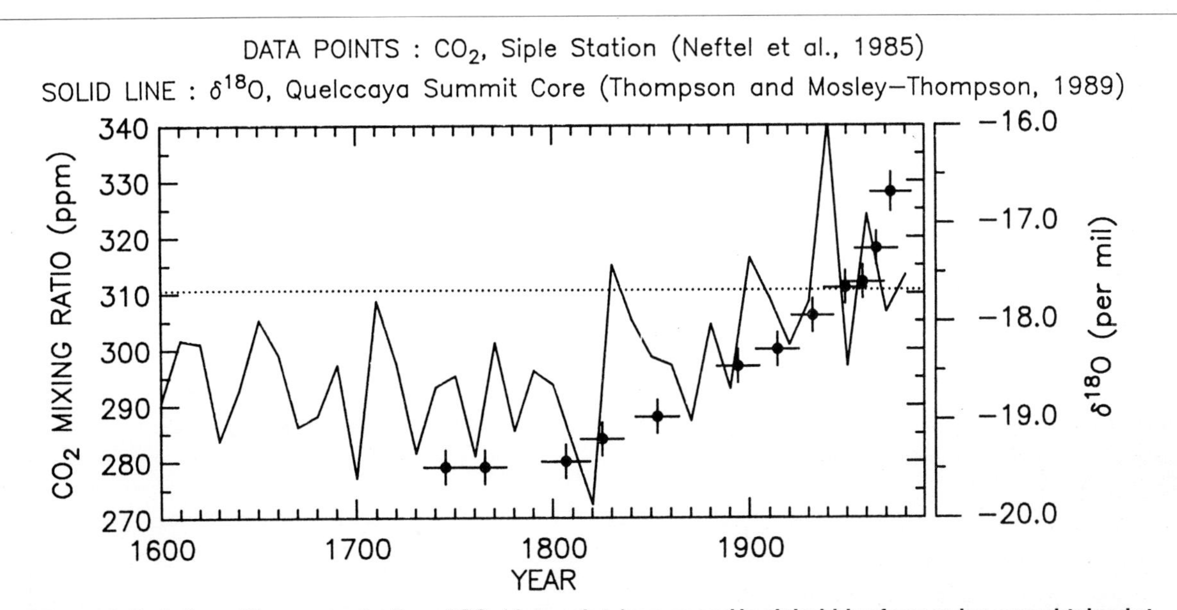

Figure 12. Variation of the concentration of CO_2 (data points) measured in air bubbles from an ice core obtained at Siple Station, Antarctica (76°S, 84°W) (Neftel et al., 1985) and the decadal average value of $\delta^{18}O$ (solid line) from the summit of the Quelccaya ice core. The dotted line represents the mean value of $\delta^{18}O$ between 1880 and 1980. Source: After Thompson L.G. and E. Mosley-Thompson 1989.

and essentially synchronously in the latter half of the 19th Century (Broecker and Denton, 1989).

The recovery from the Little Ice Age may have been a natural response to changes in the cyclic processes of ocean circulation. It may also have been promoted in part by the buildup of greenhouse gases that followed the Industrial Revolution. The changes are essentially synchronous, as illustrated in Figure 12.

6.4.1 *The possible influence of cyclic changes in ocean circulation*

It is tempting to attribute the Little Ice Age to a change in the circulation of the Pacific Ocean, analogous to that suggested earlier for the Younger Dryas and the Erdalen Event. Data on the accumulation of snow in the Quelccaya ice cap in Peru at 14° S latitude (Thompson et al., 1985), reproduced in Figure 13, indicate persistent anomalies in precipitation between about AD 1200 and AD 1900. A general increase in precipitation was observed from about AD 1200 to about AD 1650, followed by a decline from AD 1650 to about AD 1850.

Precipitation at Quelccaya at the present time is sensitive to conditions in the Pacific, as they affect details of the tropical circulation. The region receives most of its precipitation from the east in local summer. Inflow of air from the Atlantic is reduced during an El Niño; indeed annual accumulation rates at Quelccaya provide an excellent proxy for El Niño events, as illustrated in Figure 14 (Thompson et al., 1984).

We suggest that the long-term changes in precipitation at Quelccaya, that can be seen in Figure 13, may reflect cyclic behaviour in the vertical circulation of the Pacific Ocean that occurs every few centuries. The behaviour in the Pacific would be analogous to the oscillatory behaviour postulated for the Atlantic by Broecker et al. (1990b) and by Birchfield and Broecker (1990).[5] Relatively saline waters at the surface

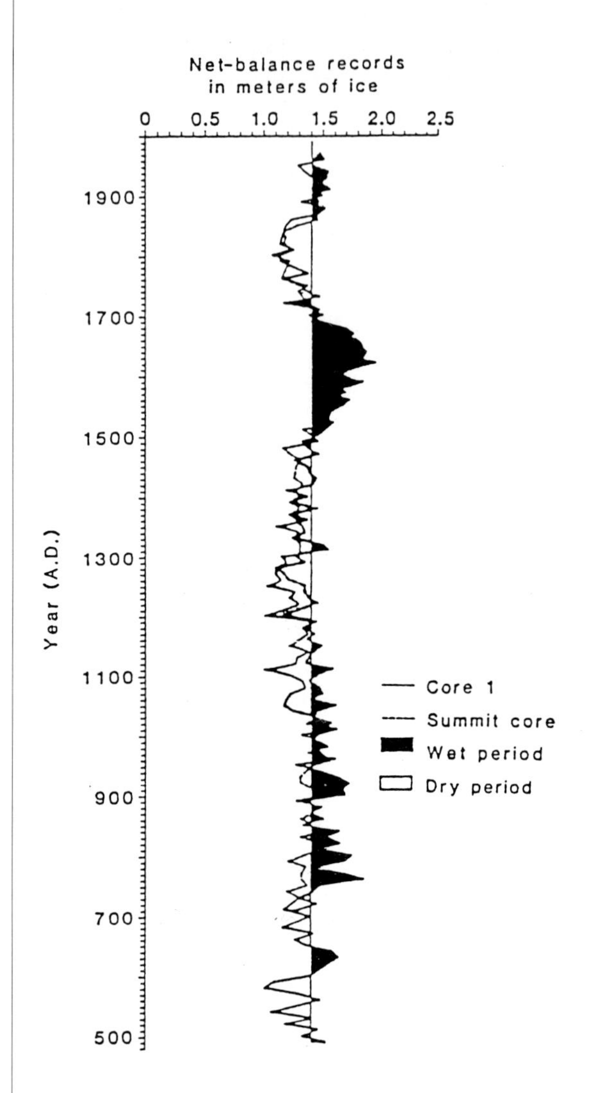

Figure 13. Reconstructed rate of precipitation, based on annual accumulation rates, for two ice cores in the Quelccaya ice cap (14°S, 71°W, 5670 m elevation). Extended periods of aridity and moistness are indicated. Source: Thompson, L. G., et al. 1985. Reprinted by permission from *Science*, vol. 229, pp. 971-973. Copyright (c) 1985 by AAAS.

[5] The Atlantic Conveyor Belt depends in part on a contrast between the salinity of surface and deep waters in the North Atlantic, and in part on a contrast between the salinities of the Atlantic and Pacific. Waters cooling at the surface under the relatively saline conditions of the North Atlantic today are dense enough to sink to the bottom. Deep water flows south, forcing its way into the Pacific and Indian Oceans. The relatively high salinity of the Atlantic is maintained by an excess of evaporation over precipitation: there is a net transfer of water in vapour form from the Atlantic to the Pacific (Broecker et al., 1990a), with a compensating export of salt associated with exchange of waters between ocean basins in the south.

The conveyor belt model for the Younger Dryas proposes that the input of fresh water to the North Atlantic due to glacial melting was sufficient to suppress formation of deep water by altering temporarily the density contrast between the surface and deep. Broecker et al. (1990b) suggest that the Younger Dryas episode may represent an oscillation of the Conveyor Belt, the last of a series of such oscillations that may have affected the climate of the North Atlantic repetitively over the course of the last Ice Age.

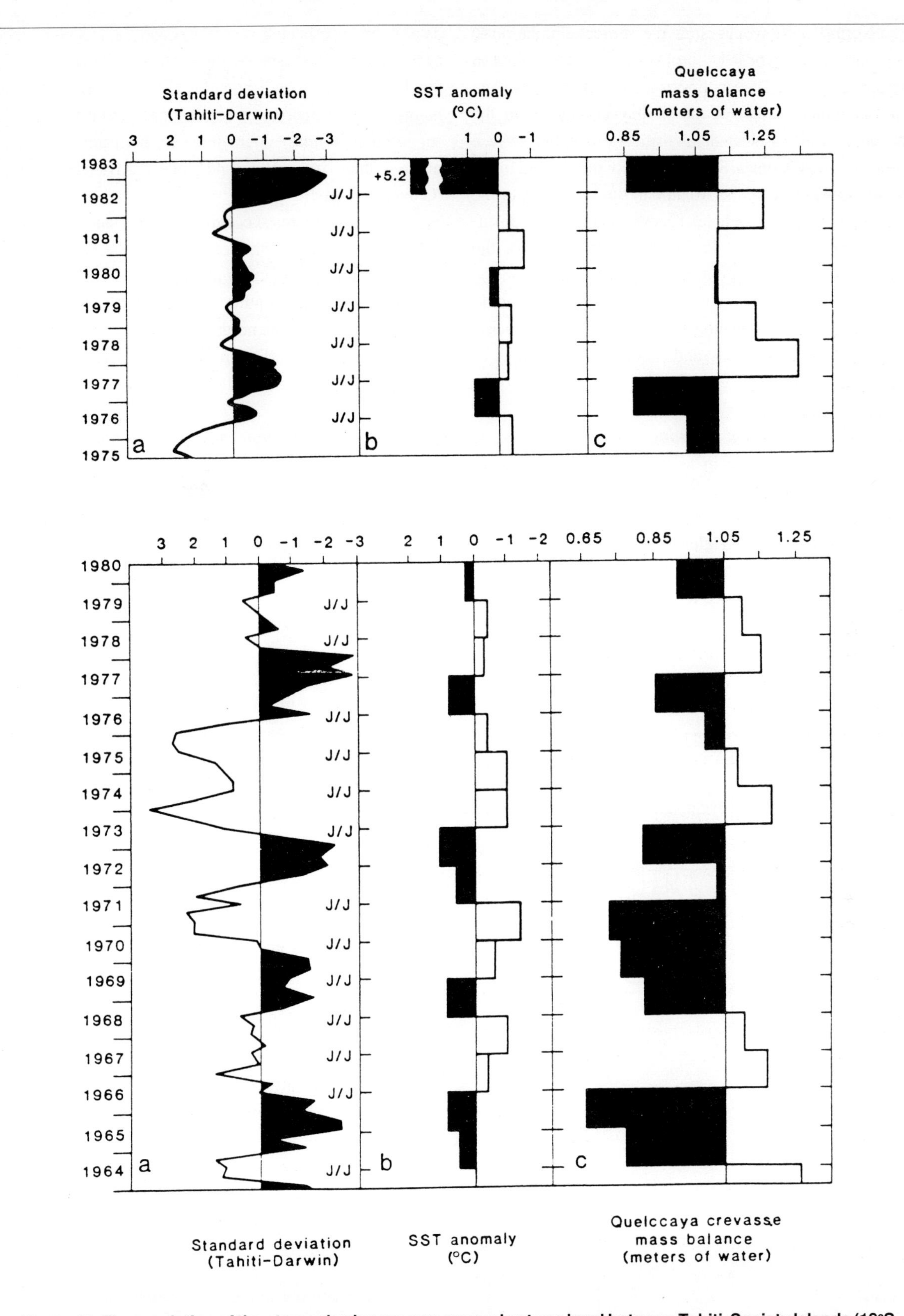

Figure 14. Time variation of the atmospheric pressure anomaly at sea level between Tahiti, Society Islands (18°S, 150°W) and Darwin, Australia (12°S, 131°E), the annual sea surface temperature anomaly at Puerto Chicama, Peru (7°S, 79°W), and the annual accumulation rate of the Quelccaya ice core (in equivalent meters of water). The upper panel is for the summit of Quelccaya; the lower panel is for a crevasse 1 km east of the summit. El Niño events correspond to negative excursions in the Tahiti-Darwin pressure anomaly and positive excursions in the sea surface temperature anomaly. Source: Thompson, L. G. et al. 1984. Reprinted by permission from *Science*, vol 226, pp. 50-53. Copyright (c) 1984, by AAAS.

of the North Pacific would allow the vertical circulation to operate as "a conveyor belt", promoting surface cooling in the tropical Pacific. This cooling could result in a shrinkage of the latitudinal extent of the Hadley circulation in the atmosphere with associated intensification of the surface trade winds in the tropics. Stronger trade winds would move more H_2O vapour from the Atlantic to the Pacific, over the Panama Straights for example. This would offset the postulated increase in salinity, leading to a weakening of the vertical circulation and a return to warm conditions in the tropics. The return of warmer weather in the tropics would expand the Hadley circulation in the atmosphere, reducing the speed of the trade winds, and choking off the supply of excess H_2O vapour from the Atlantic.

6.5 130 years of recent warming

Estimates for the change in temperature of the northern hemisphere, southern hemisphere and globe are presented for the past 130 years in Figure 15 (Jones et al., 1986). In contrast to earlier work (Wigley et al., 1985; Hansen et al., 1981), the study summarized here accounts for data on the time variation of sea surface temperature. Despite gaps in data coverage, especially over the ocean and at high latitudes of the southern hemisphere, the results in Figure 15 provide the best available indication of the trend in global and hemispheric temperatures for the past century. The data suggest that the temperature of the northern hemisphere, and of the southern hemisphere and globe, has risen by about 0.5 °C over the last 100 years.

Much of the increase occurred recently and over the last few decades of the 19th Century, with the earlier change associated almost certainly with the end of the Little Ice Age. The trend observed for the southern hemisphere is more regular than that for the north, where temperatures declined initially during the early part of the 20th Century and then again in the 1930s and 1940s.

6.5.1 The possible influence of cloud reflectivity

Some analysts attribute at least a portion of the difference between the hemispheres to an increase in cloud reflectivity (Charlson et al., 1987). They argue that the change in reflectivity results from an enhanced burden of sulphate-based condensation nuclei. These nuclei are formed from industrial emissions of SO_2 (Schwartz, 1988; Wigley, 1989). Industrial sulphur could also contribute to the lack of a clearly established trend for temperatures over the continental United States (Karl et al., 1988). Indeed there is observational evidence of an increase in cloud cover over North America during the past century, with the largest increase observed between 1930 and 1950 roughly coincident with the period of modest cooling implied for the northern hemisphere as a whole in Figure 15 (Henderson-Sellers, 1989).

6.5.2 The sulphur feedback loop: a fossil fuel dilemma

If the postulated impact of sulphur on climate is real, we are left with a bothersome dilemma: steps taken to reduce emission of industrial SO_2 to mitigate effects of acid rain may have an undesirable consequence: they could result in an acceleration of the warming due to greenhouse gases. This is just one example of the complex feedback loops that link the problems of global warming associated with the greenhouse effect, ozone depletion due to CFCs, and acid

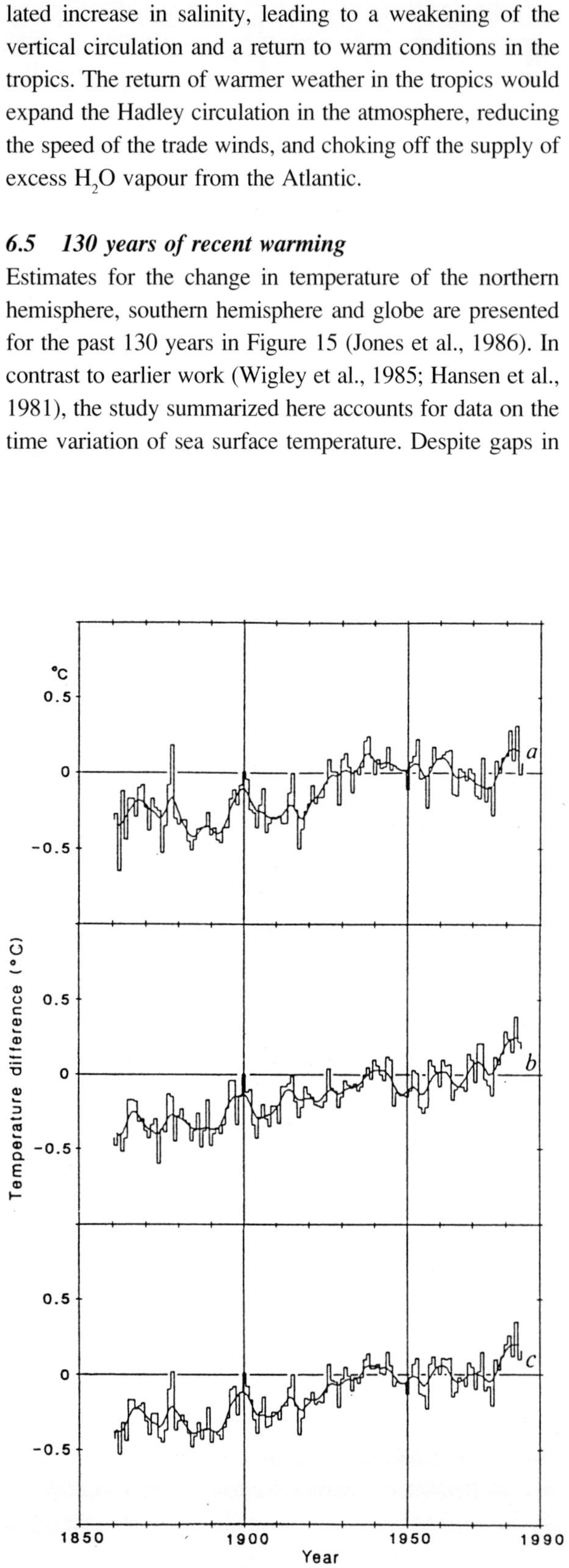

Figure 15. Annual temperature variations since 1861 for the global average (upper panel), the Northern Hemisphere (middle panel), and the Southern Hemisphere (lower panel) based on sea surface and land surface temperature records. The smooth curves represent 10 year gaussian filtered values. Jones, P. G. et al., 1986. Reprinted by permission from *Nature*, vol. 322, pp. 430-434. Copyright (c) 1986, MacMillan Magazines, Ltd.

deposition associated with oxides of sulphur and nitrogen released by burning fossil fuel. (These problems are discussed in greater detail in Crutzen and Golitsyn, this volume.)

7 Concluding Remarks

We conclude that rapid changes in climate may be associated not only with changes in production of deep water in the North Atlantic, but also with variations in the mode of circulation of waters at intermediate depths in the Pacific.

7.1 The long-term influence of carbon concentration changes

Circumstantial evidence suggests that variations in CO_2, combined with shifts in ocean circulation, had a significant influence on climate over much of the past 50 million years of Earth history. Models imply that low levels of CO_2 — about 200 parts per million, by volume (ppmv) — were required to account for the global nature of cooling observed during the two most recent Ice Ages and that changes in CO_2 (increases of about 80 ppmv) may have contributed to the rapidity of warming associated with glacial terminations. It is likely that CO_2 had a similar influence during previous episodes of glaciation. A change in the rhythm of climate variability approximately a million years ago, a shift from a dominant period of about 41,000 years to one of about 100,000 years, may have been due to a long term decline in the level of CO_2.

7.2 Prerequisite for prediction of climate change: better understanding

We view with pessimism, at least for the near term, the prospects for accurate prediction of the response of climate to present and anticipated future levels of greenhouse gases. There are problems with both the models for the atmosphere and the ocean. The resolution of even the most sophisticated models of the atmosphere (the GCMs) is currently inadequate. Many of the most important physical processes, especially those relating to the hydrological cycle, are treated in an *ad hoc* fashion by such models, with mathematical schemes designed to reproduce features of the present climate. Even in this limited application the models are suspect, as shown by Stone and Risbey (1990).

The difficulties get worse when the models are applied to conditions that differ markedly from today's climate, such as the environment of the last Ice Age. We doubt that the problems can be circumvented by resort to faster computers, even though these new machines will allow improved spatial and temporal resolution. The scales of importance are simply too small and the underlying physics too poorly understood. We need to develop a better understanding of relevant processes.

Thus, there is a need for simple models designed to test specific hypotheses, illuminated by observation. With improved understanding, advances can be incorporated in more complex models. Only in this fashion, building first from robust but simple models to more complex and highly refined simulations, can we hope to establish a capability for useful prediction. The acid test of models must involve a demonstration of their ability to simulate climates of the present and past for which useful data exist or for which they may be derived.

Ultimately, we need models for the coupled atmosphere-ocean system. It is likely that small-scale processes are as important for the ocean as for the atmosphere. We need to develop a better understanding of processes regulating the production of deep and intermediate waters. Advances in this area will most likely require major improvements in our capacity to observe the ocean. Studies of the past can be useful in focusing attention on potentially important processes, as discussed above.

7.3 Considering the beneficial possibilities

Some, including Lindzen (1990), have argued that warming over the past century has been comparatively minor, within the natural noise of the climate system, that radiative forcing due to the increase in greenhouse gases is small (about 1% of the incident solar flux), and that models ignore feedbacks that might serve to counteract warming attributable to the direct effect of greenhouse gases. It could be suggested indeed that warming might be advantageous; the world of the latter half of the 20th Century is certainly more pleasant than the harsh conditions of the 17th Century at the depths of the Little Ice Age.

The essence of the argument advanced by Lindzen (1990) is that surface warming, particularly in the tropics, could result in a decrease in the abundance of water vapour in the upper troposphere. The response of the climate system is thought to be particularly sensitive to the abundance of water vapour at higher elevations. A decrease in the concentration of water vapour in the upper troposphere would tend to mitigate surface warming postulated to occur as a consequence of the rise in the concentrations of other greenhouse gases. Surface warming, according to this idea, would tend to drive more vigorous convection (the atmosphere would turn over more rapidly). Air would cool as it rises. Assuming that its water content is controlled by local thermodynamic equilibrium (limited by the equilibrium vapor pressure at the local temperature), and that the warmer surface would cause the air to rise higher, this sequence would result in a reduction in the water content of the upper troposphere, corresponding to a negative feedback. The problem with the model as proposed is that it fails to account for the abundance of water vapour in the contemporary troposphere; the upper troposphere contains much more water vapour than Lindzen's (1990) model would suggest. The resolution of the dilemma lies in the fact that water is transported to the upper troposphere, not simply in vapour form, but also as ice and liquid. Subsequent evaporation of condensed phases provides the dominant source of water vapour in the upper troposphere today, as pointed out for example by Betts (1991).

Lindzen (1991) counters that we are unable to account from first principles for the quantity of water transported as ice and liquid. With a warmer surface, he argues, the rate at which water is delivered as ice and liquid to the upper troposphere could either decrease or increase (the climate feedback could be either negative or positive). The strength of his original proposal is weakened, however. Accepting the uncertainty which he properly highlights, one should contemplate, for consistency, the possibility that the water feedback could exacerbate rather than mitigate warming due to other greenhouse gases. Lindzen (1990) argued that negative feedbacks associated with water should serve as a thermostat for climate. Accepting this view, it would seem difficult to account for large climate shifts in the past, except possibly for the changes driven by fluctuations in continental ice as discussed above. The warm climates of the Eocene and Cretaceous would pose a particular problem; an efficient greenhouse, forced by elevated levels of CO_2, could provide a relatively simple solution to the puzzle.

7.4 *The likelihood of unexpected feedbacks*

McElroy et al. (1992) suggest that there could be additional feedbacks associated with changes in the abundance of stratospheric ozone (O_3). Reductions in the abundance of O_3 in the lower stratosphere could arise as a result of a more vigorous stratospheric circulation, induced perhaps by elevated CO_2. They suggest that reductions in O_3 could cause an increase in the depth of the troposphere, and could be responsible for an expansion of the latitudinal reach of the Hadley cell. The proposal was advanced with appropriate caution, recognizing clearly the limitations of the model on which it was developed. In light of the evidence for large contemporary changes in the abundance of lower stratospheric O_3, the suggestion merits at least further study.

7.5 *Weighing concern according to the record of the past*

The primary justification for contemporary concern relates to the velocity of the change in atmospheric composition now underway. This is underscored by the rapidity with which climate has shifted at particular times in the past, and by the fact that earlier climates were much warmer than today — with levels of CO_2 comparable to values expected in the near future. The potential risks associated with feedbacks not yet identified that could amplify the scale of warming for the next few decades, and the possibility that the introduction of controls on emission of industrial SO_2 might accelerate the pace of warming, suggest the need for caution in managing both the hazards from continued buildup of greenhouse gases and the risks of acid deposition.

The potential significance of social, political and environmental dislocations that could ensue if climate were to undergo a rapid shift argues for policies that would limit the rate of climate change. We are faced with a political choice: to act now to mitigate the rise in the concentration of greenhouse gases when the impact on climate is uncertain; or to wait for an improved understanding when the impact may be already significant. It is clear in either case that the stakes are high.

References

Arthur, M. A., K. R. Hinga, M. E. Q. Pilson, D. Whitaker and D. Allard, 1991: Estimates of CO_2 for the last 120 myr based on the δ^{13} C of marine phytoplankton organic matter. *EOS*, **72**, 166.

Bard, E., M. Arnold, P. Maurice, J. Duprat, J. Moyes and J. C. Duplessy, 1987: Retreat velocity of the North Atlantic polar front during the last deglaciation determined by ^{14}C accelerator mass spectrometry. *Nature*, **328**, 791-794.

Barnola, J. M., D. Raynaud, Y. S. Korotkevich and C. Lorius, 1987: Vostok ice core provides 160,000-year record of atmospheric CO_2. *Nature*, **329**, 408-414.

Baron, W. R., 1982: The reconstruction of eighteenth Century temperature records through the use of content analysis. *Clim. Change*, **4**, 385-389.

Berger, A. L., 1978: Long-term variations of daily insolation and Quaternary climatic changes. *J. Atmos. Sci.*, **35**, 2362-2367.

Berner, R. A., A. C. Lasaga and R. M. Garrels, 1983: The carbonate-silicate geochemical cycle and its effect on atmospheric carbon dioxide over the past 100 million years. *Am. J. Sci.*, **283**, 641-683.

Betts, A.K., 1991: Testimony presented to the Committee on Commerce, Science and Transportation, United States Senate, October 7, 1991.

Birchfield, G. E. and W. S. Broecker, 1990: A salt oscillator in the glacial Atlantic? 2. A "scale analysis" model. *Paleoceanography*, **5**, 835-843.

Birchfield, G. E., J. Weertman and A. T. Lunde, 1981: A paleoclimate model of Northern Hemisphere ice sheets. *Quaternary Res.*, **15**, 126-142.

Boyle, E. A., 1990: Quaternary deepwater paleoceanography, *Science*, **249**, 863-870.

Boyle, E. A. and L. Keigwin, 1987: North Atlantic thermohaline circulation during the past 20,000 years linked to high-latitude surface temperature. *Nature*, **330**, 35-40.

Broccoli, A. J. and S. Manabe, 1987: The influence of continental ice, atmospheric CO_2, and land albedo on the climate of the last glacial maximum. *Climate Dynamics*, **1**, 87-99.

Broecker, W. S., 1982: Glacial to interglacial changes in ocean chemistry. *Prog. Oceanog.*, **11**, 151-197.

Broecker, W. S. and G. H. Denton, 1989: The role of ocean-atmosphere reorganizations in glacial cycles. *Geochim. Cosmochim. Acta*, **53**, 2465-2501.

Broecker, W. S. and J. Van Donk, 1970: Insolation changes, ice volumes and the δ^{18}O record in deep-sea cores. *Rev. Geophys. Space Phys.*, **8**, 169-198.

Broecker, W. S. and T.-H. Peng, 1989: The cause of the glacial to interglacial atmospheric CO_2 change: a polar alkalinity hypothesis. *Global Biochem. Cycles*, **3**, 215-239.

Broecker, W. S., M. Andree, W. Wolfli, H. Oeschger, G. Bonani, J. Kennett and D. Peteet, 1988: *Paleoceanography,* **3**, 1-19.

Broecker, W. S., J. P. Kennett, B. P. Flower, J. T. Teller, S. Trumbore, G. Bonani and W. Wolfli, 1989: Routing of the meltwater from the Laurentide Ice Sheet during the Younger Dryas cold episode. *Nature,* **341**, 318-321.

Broecker, W. S., T.-H. Peng, J. Jouzel and G. Russell, 1990a: The magnitude of global fresh-water transports of importance to ocean circulation. *Climate Dynamics,* **4**, 73-79.

Broecker, W. S., G. Bond and M. Klas,1990b: A salt oscillator in the glacial Atlantic? 1. The concept. *Paleoceanography,* **5**, 469-477.

Carter, L. D., J. Brigham-Grette, L. Marincovich, Jr., V. L. Pease and J. W. Hillhouse, 1986: Late Cenozoic Arctic Ocean sea ice and terrestrial paleoclimate. *Geology,* **14**, 675-678.

Catchpole, A. J. W. and T. F. Ball, 1981: Analysis of historical evidence of climatic change in western and northern Canada. In *Syllogeous 33: Climatic Change in Canada,* 96-148 National Museum of Natural Science, Ottawa, Canada.

Charlson, R. J., J. E. Lovelock, M. O. Andreae and S. G. Warren, 1987: Oceanic phytoplankton, atmospheric sulfur, cloud albedo and climate. *Nature,* **326**, 655-661.

Cerling, T. E., 1991: Stable isotopic constraints on atmospheric CO_2 from paleosols. *EOS,* **72**, 166.

Douglas, R.G. and F. Woodruff, 1981: Deep sea benthic foraminifera. In *The Sea, 7.* C. Emiliani, ed. Wiley-Interscience, New York, pp. 1233-1327.

Dawson, M. R., R. M. West, W. Langston, Jr. and J. H. Hutchison, 1976: Paleogene terrestrial vertebrates: northernmost occurrence, Ellesmere Island, Canada. *Science,* **192**, 781-782.

Duplessy, J. C., A. W. H. Be and P. L. Blanc, 1981: Oxygen and carbon isotopic composition and biogeographic distribution of planktonic foraminifera in the Indian Ocean. *Palaeogeogr., Palaeoclimatol., Palaeoecol.,* **33**, 9-46.

Duplessy, J. C., N. J. Shackleton, R. G. Fairbanks, L. Labeyrie, D. Oppo and N. Kallel, 1988: Deepwater source variations during the last climatic cycle and their impact on the global deepwater circulation. *Paleoceanography,* **3**, 343-360.

Emiliani, C., 1955: Pleistocene temperatures. *J. Geol.,* **63**, 538-578.

Fairbanks, R. G., 1989: A 17,000-year glacio-eustatic sea level record: influence of glacial melting rates on the Younger Dryas event and deep-ocean circulation. *Nature,* **342**, 637-642.

Fairbridge, R. W., 1987: Little Ice Age. In *The Encyclopedia of Climatology,* J. E. Oliver and R. W. Fairbridge, eds., 547-551, New York, Van Nostrand Reinhold Co.

Farrell, B. F., 1990: Equable Climate Dynamics. *J. Atmos. Sci.,* **47**, 2986-2995.

Freeman, K.H. and J.M. Hayes, 1992: Fractionation of carbon isotopes by phytoplankton and estimates of ancient $C0_2$ levels. *Global Bio. Cycles* (in press).

Funder, S., N. Abrahamsen, O. Bennike and R. W. Feyling-Hanssen, 1985: Forested Arctic: evidence from north Greenland. *Geology,* **13**, 542-546.

Genthon, C., J. M. Barnola, D. Raynaud, C. Lorius, J. Jouzel, N.I. Barkov, Y. S. Korotkevich and V. M. Kotlyakov, 1987: Vostok ice core: climatic response to CO_2 and orbital forcing over the last climatic cycle. *Nature,* **329**, 414-418.

Hansen, J., D. Johnson, A. Lacis, S. Lebedeff, P. Lee, D. Rind and G. Russell, 1981: Climate Impact of increasing atmospheric carbon dioxide. *Science,* **213**, 957-966.

Hastenrath, S., 1981: *The Glaciation of the Equadorian Andes,* Balhema, Rotterdam, Holland.

Hays, J. D., J. Imbrie and N. J. Shackleton, 1976: Variations in the Earth's orbit: pacemaker of the ice ages, *Science,* **194**, 1121-1132.

Henderson-Sellers, A., 1989: North American total cloud amount variations this century. *Palaeogeogr., Palaeoclimatol., Palaeoecol.,* **75**, 175-194.

Heusser, C. J. and J. Rabassa, 1989: Cold climatic episode of Younger Dryas age in Tierra del Fuego. *Nature,* **328**, 609-611.

Hyde, W. T. and W. R. Peltier, 1987: Sensitivity experiments with a model of the ice age cycle: the response to Milankovitch forcing. *J. Atmos. Sci.*, **44**, 1351-1374.

Imbrie, J., J. D. Hays, D. G. Martinson, A. McIntyre, A. C. Mix, J. J. Morley, N. G. Pisias, W. L. Prell and N. J. Shackleton, 1984: The orbital theory of Pleistocene climate: support from a revised chronology of the maring $\delta^{18}O$ record. In *Milankovitch and Climate, Part 1,* A. L. Berger et al.,eds., 269-305, Dordrecht, Holland, D. Reidel.

Jasper, J. P. and J. M. Hayes, 1990: A carbon isotope record of CO_2 levels during the late Quaternary. *Nature,* **347**, 462-464.

Jones, P. D., T. M. L. Wigley and P. B. Wright, 1986: Global temperature variations between 1861 and 1984. *Nature,* **322**, 430-434.

Jouzel, J., C. Lorius, J. R. Petit, C. Genthon, N. I. Barkov, V. M. Kotlyakov and V. M. Petrov, 1987: Vostok ice core: a continuous isotope temperature record over the last climatic cycle (160,000 years). *Nature,* **329**, 403-408.

Joyce, J. E., L. R. C. Tjalsma and J. M. Prutzman, 1990: High resolution planktic stable isotope record and spectral analysis for the last 5.35 M.Y.: Ocean Drilling Program site 625 northeast Gulf of Mexico. *Paleoceanography,* **5**, 507-529.

Kallel, N., L. D. Labeyrie, M. Arnold, H. Okada, W. C. Dudley and J. C. Duplessy, 1988: Evidence of cooling during the Younger Dryas in the western North Pacific. *Oceanolog. Acta, 11,* 369-375.

Karl, T. R., R. G. Balwin and M. G. Burgin, 1988: *Time Series of Regional Seasonal Averages of Maximum, Minimum, and Average Temperature, and Diurnal Temperature Range Across the United States: 1901-1984,* Historical Climatology Series 4-5, Asheville, North Carolina, National Climatic Data Center.

Kennett, J. P., 1990: The Younger Dryas cooling event: an introduction. *Paleoceanography,* **5**, 891-895.

Kennett, J. P., K. Elmstrom and N. L. Penrose, 1985: The last deglaciation in Ocra Basin, Gulf of Mexico: high-resolution planktonic foraminifera changes. *Palaeogeogr., Palaeoclimatol., Palaeoecol.,* **50**, 189-216.

Knox, F. and M. B. McElroy, 1984: Changes in atmospheric CO_2: influence of the marine biota at high latitude. *J. Geophys. Res.,* **89**, 4629-4637.

Kudrass, H. R., E. Erlenkeuser, R. Vollbrecht and W. Weiss, 1991: Global nature of the Younger Dryas cooling event inferred from oxygen isotope data from Sulu Sea cores. *Nature,* **349**, 406-409.

LaMarche, V. C., Jr., 1975: Potential of tree rings for reconstruction of past climatic variations in the Southern Hemisphere. In *Proceedings of WMO/IMAP Symposium on Long-Term Climatic Fluctuations,* 21-30 Geneva, Switzerland, World Meteorological Organization.

Lamb, H. H., 1984: Some studies of the Little Ice Age of recent centuries and its great storms. In *Climatic Change on a Yearly and Millenial Basis,* N. A. Mörner and W. Karlén, eds., 309-329, Dordrecht, Holland, Reidel.

Lindgren, S. and J. Neumann, 1981: The cold and wet year 1695: a contemporary German account. *Clim. Change,* **3**, 173-187.

Lindzen, R. S., 1990: Some coolness concerning global warming. *Bull. Am. Met. Soc.,* **71**, 288-299.

Lindzen, R.S., 1991: Testimony presented to the Committee on Commerce, Science and Transportation, United States Senate, October 7, 1991.

Ludlum, D., 1966: *Early American Winters: 1604-1820,* Boston, Mass., American Meteorological Society.

Manabe, S. and A.J. Broccoli, 1985: The influence of continental ice sheets on the climate of an ice age. *J. Geophys. Res.,* **90**, 2167-2190.

Manabe, S. and K. Bryan, Jr., 1985: CO2-induced change in a coupled ocean-atmosphere model and its paleoclimatic implications. *J. Geophys. Res.,* **90**, 11689-11707.

Martin, J. H. and S. E. Fitzwater, 1988: Iron deficiency limits phytoplankton growth in the north-east Pacific subarctic. *Nature,* **331**, 341-343.

McElroy, M.B., R.S. Salawitch and K. Minschwaner, 1992: The changing stratosphere. *Planet. Space Sci.,* in press.

McKenna, M., 1980: Eocene paleolatitude, climate, and mammals of Ellesmere Island. *Palaeogeogr., Palaeoclimatol., Palaeoecol.,* **30**, 349-362.

Milankovitch, M., 1941: *Canon of Insolation and the Ice Age Problem.* Royal Serbian Academy Special Publication 133, Belgrade, translated by Israel Program for Scientific Translation, Jerusalem, 1969.

Neftel, A., E. Moor, H. Oeschger and B. Stauffer, 1985: Evidence from Polar Ice Cores for the Increase in Atmospheric CO2 in the past 2 Centuries. *Nature,* **315**, 45-47.

Nesje, A. and M. Kvamme, 1991: Holocene glacier and climate variations in western Norway: evidence for early Holocene glacier demise and multiple Neoglacial events. *Geology,* **19**, 610-612.

Oerlemans, J., 1980: Model experiments on the 100,000-yr glacial cycle. *Nature,* **287**, 430-432.

Oglesby, R. J. and B. Saltzman, 1990: Sensitivity of the equilibrium surface temperature of a GCM to systematic changes in atmospheric carbon dioxide. *Geophys. Res. Lett.,* **17**, 1089-1092.

Pastouret, L., H. Chamley, G. Delibrias, J. C. Duplessy and J. Thiede, 1978: Late Quaternary climatic changes in Western Tropical Africa deduced from deep-sea sedimentation off the Niger delta. *Oceanol. Acta,* **1**, 217-232.

Peltier, W. R. and W. T. Hyde, 1984: A model of the ice age cycle. In *Milankovitch and Climate, Part 2,* A. L. Berger et al., eds., 565-580, Dordrecht, Holland, D. Reidel.

Pettersson, O., 1912: The connection between hydrographical and meteorological phenomena. *Royal Meteorol. Soc. Quart. Jour.,* **38**, 173-191.

Pisias, N. G. and N. J. Shackleton, 1984: Modeling the global climate response to orbital forcing and atmospheric carbon dioxide changes. *Nature,* **310**, 757-759.

Pollard, D., 1982: A simple ice sheet model yields realistic 100 kyr glacial cycles. *Nature,* **296**, 334-338.

Pollard, D., 1983a: A coupled climate-ice sheet model applied to the Quaternary ice ages. *J. Geophys. Res.,* **88**, 7705-7718.

Pollard, D., 1983b: Ice-age simulations with a calving ice-sheet model. *Quaternary Res.,* **20**, 30-48.

Pollard, D., 1984: Some ice-age aspects of a calving ice-sheet model. In *Milankovitch and Climate, Part 2,* A.L. Berger et al., eds., 541-564. Dordrecht, Holland, D. Reidel.

Pollard, D., A. D. Ingersoll, and J. G. Lockwood, 1980: Response of a zonal climate-ice sheet model to the orbital perturbations during the Quaternary ice ages. *Tellus,* **32**, 301-319.

Popp, B. N., R. Takigiku, J. M. Hayes, J. W. Louda and E. W. Baker, 1989: The post-Paleozoic chronology and mechanism of ^{13}C depletion in primary source organic matter. *Am. J. Sci.,* **289**, 436-454.

Porter, S. C., 1975: Equilibrium-line altitudes of late Quaternary glaciers in the Southern Alps, New Zealand. *Quaternary Res.,* **5**, 27-47.

Porter, S. C., 1981: Glaciological evidence of Holocene climatic change. In *Climate and History: Studies of Past Climates and Their Impact on Man,* T. M. L. Wigley, M. J. Ingram, and G. Farmer, eds., 82-110, Cambridge, England, Cambridge University Press.

Rau, G. H., T. Takahashi and D. J. Des Marais, 1989: Latitudinal variations in plankton δ^{13}C: implications for CO_2 and productivity in past oceans. *Nature,* **341**, 516-518.

Rind, D., D. Peteet and G. Kukla, 1989: Can Milankovitch orbital variations initiate the growth of ice sheets in a general circulation model? *J. Geophys. Res.,* **94**, 12851-12871.

Saltzman, B., 1990: Three basic problems of paleoclimatic modeling: a personal perspective and review. *Climate Dynamics,* **5**, 67-78.

Sarmiento, J. L. and R. Toggweiler, 1984: A new model for the role of the oceans in determining atmospheric CO_2. *Nature,* **308**, 621-624.

Schwartz, S. E., 1988: Are global cloud albedo and climate controlled by marine phytoplankton? *Nature,* **336**, 441-445.

Schweitzer, H. J., 1980: Environment and climate in the early tertiary of Spitsbergen. *Palaeogeogr., Palaeoclimatol., Palaeoecol.,* **30**, 297-311.

Shackleton, N. J., 1967: Oxygen isotope analyses and Pleistocene temperatures re-assessed. *Nature,* **215**, 15-17.

Shackleton, N. J. and N. D. Opdyke, 1973: Oxygen isotope and paleomagnetic stratigraphy of equatorial Pacific core V28-238: oxygen isotope temperatures and ice volumes on a 10^5-year and 10^6-year scale, *Quaternary Res.,* **3**, 39-55.

Siegenthaler, U. and T. Wenk, 1984: Rapid atmospheric CO_2 variations and ocean circulation. *Nature,* **308**, 624-626.

Stone, P. H. and J. S. Risbey, 1990: On the limitations of general circulation models. *Geophys. Res. Lett.,* **17**, 2173-2176.

Thompson, L. G. and E. Mosley-Thompson, 1989: One-half millennia of tropical climate variability as recorded in the stratigraphy of the Quelccaya ice cap, Peru. In *Aspects of Climate Variability in the Pacific and the Western Americas.* D.H. Peterson, ed., American Geophysical Union, Washington, DC, pp. 15-31.

Thompson, L. G., E. Mosley-Thompson and B. M. Arnao, 1984: El Niño-Southern Oscillation events recorded in the stratigraphy of the tropical Quelccaya ice cap, Peru. *Science,* **226**, 50-53.

Thompson, L. G., E. Mosley-Thompson, J. F. Bolzan and B. R.Koci, 1985: A 1500-year record of tropical precipitation in ice cores from the Quelccaya ice cap, Peru. *Science,* **229**, 971-973.

Vakhrameev, V. A., 1975: Main features of phytogeography of the globe in Jurassic and Early Cretaceous time. *Paleontol J.,* **2**, 123-133.

Wang, S. C. and Z. C. Zhao, 1981: Droughts and floods in China, 1470-1979. In *Climate and History: Studies of Past Climates and Their Impact on Man,* T. M. L. Wigley, M. J. Ingram, and G. Farmer, eds., 271-288, Cambridge, England, Cambridge University Press.

Weertman, J., 1976: Milankovitch solar radiation variations and ice-age ice-sheet sizes. *Nature,* **261**, 17-20.

Weikinn, C., 1965: Katastrohale Dürrejahre während des Zeitraums 1500-1850. *Acta Hydrophysica (Berlin),* **10**, 33-54.

Wigley, T. M. L., 1989: Possible climate change due to SO_2-derived cloud condensation nuclei. *Nature,* **339**, 365-367.

Wigley, T. M. L., J. K. Angell and P. D. Jones, 1985: Analysis of the temperature record. In *Detecting the Climatic Effects of*

Increasing Carbon Dioxide, M. C. MacCranken and F. M. Luther, eds., 55-90, Washington, DC, US Dept of Energy.

Yamamoto, T., 1972: On the nature of the climatic change in Japan since the "Little Ice Age" around AD 1800. *Japanese Prog. Climatol,* 97-110.

Yoshino, M. M. and S. Xie, 1983: A preliminary study on climatic anomalies in East Asia and sea surface temperatures in the North Pacific. *Tsukuba Univ. Ins. Geosci. Sci. Repts,* **A4**, 1-23.

CHAPTER 6

Indices and Indicators of Climate Change: Issues of Detection, Validation and Climate Sensitivity

Tom M. L. Wigley, Graeme I. Pearman and P. Michael Kelly

Editor's Introduction

Choosing policy responses to reduce the risks of rapid climate change requires making decisions based on uncertain knowledge of the future. How can we know that the changes observed so far in regional climate records are not just part of the natural background "noise" — i.e., not just reflections of the inevitable stochastic variations of the climate system? In this chapter, three of the world's leading experts on the observation and measurement of atmospheric change address the question, "how can we know when a 'real' change in global climate is taking place?" In the process, Tom Wigley, Graeme Pearman, and Mick Kelly present an overview of the debate on whether signs of a greenhouse warming have been detected with sufficient confidence to prompt policy action by governments.

In the absence of any proof of greenhouse gas-induced warming, how can we validate our models of the global climate system and improve our ability to predict future climate change? One important way, these authors point out, is to compare the predictions from a model with the observations of the Earth's climate. But a number of inherent difficulties hinder our best efforts to validate "realistic" climate models. These are the "dirty window" effects described in this chapter, which include (for example) inherent uncertainties in the observations themselves and differences between the results of different models and types of numerical experiment. Moreover, the omnipresent, apparently meaningless, and clearly inescapable background noise inherent in the climate system obscures the signal that we are trying to see.

In the face of these persistent uncertainties, modellers will probably not be able to "prove" the reality of global warming for several decades. None the less, by carefully selecting a composite set of variables to monitor over time, we may be able to observe what the authors call a "fingerprint" of climate change: indications that give advance warning of what future climate changes are likely to take place. Surface temperature records will remain the foundation of any such composite, but monitoring of biological indicators and changes in the chemical composition of the atmosphere may provide important supplemental information.

Tom Wigley and Mick Kelly, senior scientists in the Climatic Research Unit at the University of East Anglia, Norwich, UK, and Graeme Pearman, Assistant Chief, Division of Atmospheric Research, Commonwealth Scientific and Industrial Research Organization (CSIRO) in Melbourne, Australia, conclude that continuous, systematic monitoring of climate change is essential in order to identify both natural variations and human-induced climate change. Wigley and Kelly are key members of the British scientific team that developed pioneering techniques in the monitoring of global temperature change. They maintain, update, and regularly analyse the best-kept set of global temperature records in the world. Graeme Pearman is responsible for the most extensive atmospheric monitoring network in the Southern Hemisphere. He and the members of his team track the buildup of a broad suite of pollutants and radiatively active trace substances in the atmosphere. All three have made major strides in the development of the "fingerprint" technique of comprehensive analysis.

- I. M. M.

1 Introduction

The purpose of this chapter is to consider the efforts that climatologists are making to monitor and understand the changing climate of the Earth — especially those changes that have arisen, or might arise in the future, from human activities. A significant warming trend has, indeed, been observed over the last 100 years or so — but this does not mean we can claim to have positively identified (or "detected") the enhanced greenhouse effect as the cause.

What, then, *can* we claim? What results can we present to policy makers without full assurance, but with reasonable credibility? To consider these questions, we must consider issues that have not been highlighted in other reports, such as the IPCC report on climate change (Houghton et al., 1990). We will try to dispel some myths; and we will identify which aspects of this inquiry are most significant.

1.1 The value of monitoring

Why do we need to monitor the Earth's climate? Why measure the changes and their rates, and devise indices and indicators of the changing climate?

First, if changes are expected, we must have a well-defined reference state against which to identify and measure these changes. Continual monitoring of the climate system, and comparing it against the past, is the only way to tell objectively whether a single trend is a significant harbinger of more extensive and pervasive climate changes.

Secondly, the observational data of the past form the descriptive basis for our understanding of climatic processes. They thus underlie the theory of climate, which is the basis for formulating climate models. The models, themselves, are predictive descriptions of how climate might react to present and past events. They may be our only tool for providing the reliable early warning necessary for planning — both to adapt to, and to mitigate, climatic change.

Thirdly, to predict future changes, we need first to be able to explain past and current changes. To do this, we need to know what has happened and is happening to the Earth's climate. The data of the past are essential for testing explanations of past changes and for the validation of climate models.

1.2 Climate and climate modelling

Because the climate system is such a complex set of interrelated processes, we must interpret the word climate in its broadest sense. "Climate" means not just "average weather," it includes all the parameters that control or influence what we conventionally think of as the weather. The oceans, the ice masses (from small glaciers to large ice sheets), and the changing chemical composition of the Earth's atmosphere must all be considered part of the climate system.

Observations of the climate system have been made for many decades — in some cases, for centuries — but it is only recently that these data have been harnessed to support climate modelling studies. The spur for this integration has been the realization that human alterations of the environment may be leading us rapidly to climate states that have no precedent in mankind's experience. Already, we have drastically altered the composition of the atmosphere (Pearman, 1988; Watson et al., 1990) — raising carbon dioxide levels by more than 25%, doubling methane concentrations, and introducing new chemicals, the halocarbons (of which the chlorofluorocarbons, CFCs, are best known).

Although these gases exist only at low concentration levels, we know enough to be certain that they have caused changes in the climate system's radiative balance: the net amount of radiation transferred in all directions through the Earth's atmosphere. We know that this radiation balance is one of the primary controls on temperature and rainfall, and (less directly) on sea level. But we do not know enough about the workings of these climatic mechanisms, including the feedback processes described by Crutzen and Golitsyn (1992, this volume), to predict with certainty the extent of ultimate change. Climate models calculate both the changes in the radiative balance and the consequent effects on climate. Our ability to predict future climate hinges, therefore, on our ability to model the climate system. One of the main problems here is assigning a value to the "climate sensitivity" — that is, quantifying how strongly the climate system will respond to any given change in its radiative balance.

1.3 The "detection" issue: isolating significant data

How can we distinguish the temperature and rainfall changes induced by greenhouse gases from those which are just part of the natural variability of the climate system? This is the heart of the so-called "detection" issue. The determination of the climate sensitivity is a closely related problem.

We can, with reasonable accuracy, quantify the radiative forcing changes that have occurred to date, due to greenhouse gas concentration increases. Since pre-industrial times (the late eighteenth century), the change has been about 2.5 watts per square metre. This much more radiation reaches the lower layers of the atmosphere than in 1765 (Shine et al., 1990). This is equivalent to the effect of an increase in the Sun's output of a little more than 1%.

We also know that relatively large changes in climate have occurred since the 18th Century, including an increase in global-mean temperature of around 0.5 °C (Folland et al., 1990; Jones and Wigley, 1990). But are these changes related to the change in radiative forcing? Have we yet detected a human-induced "greenhouse effect," or does the observed climate record merely reflect natural variation?

This is a simple question, but the issue itself is far from simple.

2 How is Detection Important?

What does it mean, then, to say that: "We have detected the greenhouse effect?" If rigorously stated, such a contention should be expressed in terms of probability: "We are so-many-per cent confident that a certain observed change, over such-and-such a period, is attributable to the enhanced greenhouse effect." Quantifying that level of confidence is a difficult task, however, because some of the uncertainties in the greenhouse gas/climate change relationship have yet to be quantified, and some aspects are essentially unquantifiable. Instead, we can (and do) use phrases like, "with high confidence". But the meaning (especially the quantitative equivalent) of "high confidence" varies from person to person.

2.1 *Human activity: Guilty until proven innocent, or innocent until proven guilty?*

Currently (taking into account the evidence of past observed climate trends, our confidence in climate model physics, various forms of model validation, and paleoclimatic evidence), most scientists have considerable confidence that the enhanced greenhouse effect is real, and that climate changes (albeit small to date) are resulting from human activities. This view might seem to contradict published statements that detection has not yet been achieved — for example, as in the IPCC Policymakers' Summary of Houghton et al. (1990), which summarizes Wigley and Barnett (1990). The difference is not one of evidence, or even of interpreting the evidence, but of attitudes towards the issue of proof. Just what does "proof" mean to different people, and what level of proof is required before action should be considered?

Should we use the principle of innocent until proven guilty? That is, should we act only when our current emissions are implicated beyond a shadow of a doubt? Or, should we adopt a precautionary approach, guilty until proven innocent? In statistical analysis, the former approach is standard procedure, and this is the basis of the IPCC's statement that detection has not yet been achieved. From a policy viewpoint, however, the latter approach may be more prudent.

2.2 *Policy making running ahead of detection*

While scientists cannot say that detection has been achieved, many policy makers already seem to feel that there is enough evidence for action on global climate change issues. There are doubtless many reasons for this: the general air of confidence that modellers have in their models (in spite of caveats that appear in the specialist literature); the fact that the world has warmed substantially during this century (even though cause and effect have not been definitively demonstrated); the risk associated with ignoring potential disruptive climate changes; and just plain opportunism in catching on to the coat-tails of a "hot" topic. In some countries, targets for emission control have already been legislated; policy making is running ahead of detection (although not ahead of the science). Many would argue (and do, in other chapters in this book) that it is sensible to act now, given the uncertainties and the realistic possibility of disruptive changes.

So perhaps the detection issue as a guide to *whether* to act is not so important after all. Are there any aspects of detection that may provide a guide for *how* to act? From a policy standpoint, the key question concerns the likely magnitude of future climate effects: "What could take place next?" Policy makers need this information to decide how to respond.

If mitigation (reducing net emissions of greenhouse gases into the atmosphere, or increasing the sinks for these gases by promoting the growth of vegetation) is to be based on some form of cost-benefit assessment, then we must first quantify the amount of climate change which each mitigation measure might help us avoid. To do this, we need to be able to predict the climate's sensitivity. Similarly, if we are to plan adaptive strategies to ameliorate or avoid the consequences of climatic change (such measures as building levees, resettling populations, or reorienting agriculture), we also need to make reliable climate change predictions. Once again, the crucial parameter is the climate sensitivity.[1]

2.3 *Empirical estimates of the climate sensitivity*

In the greenhouse context, climate sensitivity is usually defined as the eventual (i.e. equilibrium) global-mean warming that would occur if the atmospheric CO_2 level were to double (denoted ΔT_{2x}). On the basis of experiments with complex General Circulation Models of the climate system (GCMs), ΔT_{2x} is thought to lie in the range between 1.5 and 4.5 °C. In other words, a doubling of CO_2 would produce a net global warming of between 1.5 and 4.5 °C.

Mitchell et al. (1990), while stressing the uncertainty, give a "best guess" value for ΔT_{2x} of 2.5 °C, based largely on a judgment of model uncertainties and partly on a qualitative comparison of model results and observations.

Several attempts have been made to estimate the climate sensitivity from the observational data on global-mean temperature changes, by comparing these to the changes predicted by models. The best estimate of climate sensitivity is the one that gives the best fit between prediction (models) and observation (data). The task of finding this estimate, however, is complicated by uncertainties about models, observations, possible effects of other climate change mechanisms, and the delays caused by the large thermal inertia of the oceans.

Using this prediction/observation comparison method, and after accounting for uncertainties, Wigley and Raper (1991) find that ΔT_{2x} should lie in range from 0.8 to 4.3 °C. This clearly overlaps with the GCM-based range of 1.5 - 4.5 °C, but it admits a noticeable probability (about 20%) that the

---

[1] As noted by Wigley and Raper (1991), it may well be possible to both identify a signal in the climate record and attribute it to the enhanced greenhouse effect with high confidence, and yet still be unable to improve current estimates of the climate sensitivity. In other words, detection *per se* need not help us in quantifying the climate sensitivity.

GCMs have all overestimated the climate sensitivity.[2] To those who look for reliability in the climate models, this is seemingly a disturbing offset.

However, it is currently thought that the effects of tropospheric sulphate aerosols — which have increased in concentration markedly over the past 50-100 years due to industrial emissions of sulphur dioxide (SO_2)— may explain this discrepancy (Wigley, 1989). Sulphate aerosols cool the Earth both directly (by reflecting away incoming solar radiation under clear sky conditions) and indirectly (by making maritime clouds more reflective) (Charlson et al., 1987, 1991; Wigley, 1989, 1991. When adjusted for the possible effects of aerosol forcing, the "best guess" empirical value of (ΔT_{2x}) rises from 1.4 to 2.3 °C, much closer to the GCM-based best guess.[3]

These results are directly relevant to the detection issue; by demonstrating this relative agreement, they demonstrate that the climate models are probably valid. This is, however, a singularly hollow detection victory. Detection, to be meaningful, must do more than just show that the climate system has a level of sensitivity to human emissions: we can be sure of that *a priori*. As we have said, the point of assigning a number to climate sensitivity (in other words, to detection) should be to help us determine how the enhanced greenhouse effect might alter the climate in the future.

3 The Detection Problem

The most obvious way to test a climate model is to take a prediction derived from the model, and then see whether that prediction "came true," or how close it was to being correct. This process is called model *validation* or *verification*. To practice it, we compare the model's predictions with observations of the real climate (see, e.g., Santer andWigley, 1990). Detection (as described in the last section) can thus be looked at as the ultimate form of model validation: testing the models' views of overall climate sensitivity.

Seen from that context, the human race is currently in the process of performing an experiment on the global environment, by adding a mixture of greenhouse gases to the atmosphere. We have models that allow us to predict the consequence of this experiment so far, and we have observational data with which to test these predictions. Detection is our method of observing and assessing the data early enough that we can adjust the experiment before it gets out of hand, or at least begin earlier to counteract its side-effects.

3.1 *The dirty windows (difficulties in detection)*

In detection we are faced with a series of problems:

- At the outset, we do not know precisely what we are looking for. We know that the models, like all mathematical representations of complex systems, cannot be as comprehensive as the world they are trying to represent. They can only, therefore, give us a rough estimate of the true "signal" of climatic response to greenhouse-gas forcing. In essence, we can consider the model to be a dirty window, through which we are looking to try to isolate the signals that we will seek in our observations.

- Next to this model window is another window, through which we view the real world. But observational data also have well-recognized uncertainties and inadequacies. Thus, the real-world window is also dirty. Our task is to compare these two blurry views. We encounter numerous problems in trying to do so.

- We have more than one model window to look through. Generating the view behind each window is a different climate model, and each model gives a different simulation of the enhanced greenhouse effect. Which of these is the best indicator of the true signal?

- To complicate this issue further, we have two very different types of experiment that have been performed in trying to define the true signal. *Equilibrium* experiments result in a snapshot view, showing the model at a single moment, while *transient response* experiments reveal a signal which evolves with time, like a constantly changing feature of the landscape. Which of these is the best indicator of the true signal?

- Many other factors cause the real world climate to change constantly. Even in the absence of an enhanced greenhouse effect, the view through the dirty observational window would be changing all the time, and in unpredictable ways. This is *noise:* the omnipresent and apparently meaningless background of natural climatic variability. In effect, the signal we are trying to detect through the real-world window is obscured by other irrelevant features in the landscape, which are continually evolving and shifting around. To confuse the issue further, some of these "noisy" features may look like the greenhouse signal, yet be totally unrelated to the greenhouse effect.

- Even if we could minimize all the other problems, the signal we are searching for at present is still, in all probability, a weak one — a minor feature compared to the random variability of the general landscape.

---

[2] Actually, GCMs do not preclude a lower value than 1.5 °C. The range of $1.5 < \Delta T_{2x} < 4.5$ °C must correspond to some confidence interval among the model designers (such as the 90-95% band). But the modellers have not been prepared to quantify their level of confidence.

[3] In the above empirical estimates of ΔT_{2x}, unquantifiable external forcing factors have been ignored. For example, if the Sun's output had increased by, say, 1% over 1890-1990 (there is no direct evidence to suggest this, by the way), then the value of ΔT_{2x} could be as low as 0.5 °C. Alternatively, if there had been a decrease in solar output this century, offsetting the enhanced greenhouse effect, then the ΔT_{2x} value could be substantially higher than 2 °C.

- We must also make the difficult choice of deciding which part of the window to look at. Do we scan the whole range of signals, or do we narrow the focus on part of the observational record? Should we consider some overall characteristic, or do we try to look for some particular pattern?

- Finally, how do we quantify the similarity between the views we see through the observational and model windows — and how do we determine whether the similarity is simply due to chance?

3.2 How to peer through dirty windows

None of these aspects of the detection problem can be answered thoroughly, but they are not insurmountable. Remember, we are looking not for a perfect model, but to learn as much as possible about the link between human activity and climate change. Ultimately, we are looking for effective leverage, in the real world, for dealing with climate change problems. Our aim is to develop models to the point where they can be used with confidence in testing our ideas for ways to reduce or ameliorate the effects. Detection is one way to build confidence. The technical issues in the detection process are:

(1) How do we cope with uncertainties in the signal?

(2) Which signal should we look for, given the choice of different models and types of experiment?

These two questions are related. Differences between, and uncertainties in, models or experiments can be partly overcome by trying to identify a signal common to (or similar in) all models and experiments. It is not enough to look at all of the possible signals, since this merely increases the chance of seeing one of them in the observations by chance. This would be a meaningless result. Residual uncertainties must still remain. If we could eliminate these completely (i.e. if we knew the true signal), then we wouldn't have to bother with detection at all. We could simply run our all-certain global climate model, enter in our proposed policies, and judge them according to the results.

(3) How do we cope with uncertainties in the observational data?

These can be minimized by careful quality control, and by estimating the magnitudes of these uncertainties so that their effect can be entered into the calculations. For example, despite claims made by various critics, we know enough to eliminate the effects of urban warming (the heat and activity of cities) as a factor in global temperature change (Jones et al., 1990). Similarly, the consequences of the uneven geographic coverage of temperature measurements have been assessed (much more data exists, for example, for North America than for Africa), and error margins have been developed to account for these statistically (Jones et al., 1986a, b; Folland et al., 1990).

(4) How do we account for the "noise" of natural climatic variability — especially the low-frequency variability that occurs on time scales most relevant to the greenhouse problem?

This is a vexing problem. The central problem of detection is precisely here: to separate the signal (meaningful variations) from the noise (variations with no significance to the changes we are trying to understand). We can try to estimate variability from the existing data by applying suitable statistical techniques to as long a data record as possible. This is what has been done so far. But, because temperature measurement records begin about the same time that human-made greenhouse gases began to accumulate in the atmosphere, these data are partly "contaminated." We do not have a "pure" record of natural variability from, say, the time of Aristotle to compare with the record from the time of Fourier.

Furthermore, even if we could discount the "purity" issue, we still only have instrumental data for the past 100 years or so, and this is not enough to reliably quantify the full range of relevant natural climate variability. Many centuries of data would be needed for this.

Alternatively, we can try to use models to determine the noise level, although our problem then becomes circular: in order to build a model of the noise (with which to develop an understanding of the mechanisms of climate change) we must already understand much of the mechanisms of climate change. On the positive side, we are not limited by data duration with this approach. Models can, in principle, be run for as long as is required to define their noise characteristics.

(5) How do we cope with the fact that the signal to date is probably of similar magnitude to the noise?

(6) What type of statistical procedure do we use to identify the signal?

These two questions concern the statistical methods for differentiating between signal and noise in the long-term record of observations. Numerous methods exist that are able to detect a weak signal in noisy data. The task is to select the most appropriate methods for the task.

(7) What aspect of the signal should we look for — an effect which integrates overall data (such as a change in global-mean temperature), or a multivariate "pattern" of change?

Perhaps we can optimize our chances of detection by choosing which parts of the windows to look through — looking for patterns involving several indicators. This approach will be described later in this chapter as the "fingerprint method."

Or we could seek out a single variable, wherever it may appear in our "window." This, in effect, would be like wearing sunglasses that filter out glare but allow us to see some parts of the visible spectrum more clearly. (To carry the analogy further, an example might be if the signal were largely in blue light, while the noise covered the whole spectrum. Wearing blue glasses might then allow us to see a signal invisible to the naked eye.)

4 Using Detection in the Real World: Global-Mean Temperature

4.1 Advantages of the global-mean record

With all of these limitations in mind, it appears that surface temperature is the best available parameter for detection of the enhanced greenhouse effect. This indicator has many advantages. First, we know that, almost by definition, it is a signal we seek. Among possible temperature signals (from different regions, for example), the most obvious to look for is changes in the average temperature of the Earth. (Changes in global-mean temperature to date are shown in Figure 1.) After all, "global warming" is the most often cited symptom of the enhanced greenhouse effect. All models agree that the enhanced greenhouse effect will cause a global-mean warming. The issue is... by how much?

Secondly, the record of global-mean temperature change is long enough (it goes back more than 100 years) to maximize the signal and reduce the chance of confusing signal and noise. (Unfortunately, it is not long enough to reduce this chance as much as we would like.)

Thirdly, as noted earlier, much effort has gone into producing a homogeneous record and minimizing (and quantifying) data uncertainties.

Fourthly, statistical techniques are well developed for examining series of one variable's changes over time (although there are still some tricky choices to be made in the application of these techniques).

What we find is that the observed warming trend from around 1900 (when records first have satisfactory reliability) to the present is highly statistically significant. In other words, such a warming is very unlikely to have occurred by chance. This has been demonstrated statistically by Wigley et al. (1989) and Tsonis and Elsner (1989). It has also been demonstrated using a different, climate-model-based approach (Wigley and Raper, 1990).

4.2 Difficulties in interpreting the global-mean temperature record

The existence of a significant global warming trend over the past 100 years or so is clearly an interesting and important result — but what does it mean in the context of the greenhouse effect? It does not prove that the enhanced greenhouse effect is real. To make that demonstration, we need to prove not only the existence of a "real" (i.e. statistically significant) trend, but also that greenhouse gases are the cause.

Furthermore, there are aspects of the global-mean temperature record that, at first, seem inconsistent with the greenhouse effect hypothesis. Between the 1940s and 1970s, for example, the global mean temperature decreased slightly. On shorter time scales, decades, there were many such cooling trends. Such deviations from the long-term warming that is expected under the greenhouse hypothesis are not really evidence against the hypothesis, but they do introduce an element of uncertainty that must be resolved.

There are many possible explanations for these departures from expectation. The lack of warming between the

Figure 1. Observed global-mean temperature changes and extreme predictions of future change. The extremes are based on the lowest and highest emissions scenarios produced for the 1992 Supplementary Report to the IPCC Scientific Assessment, with a low climate sensitivity (ΔT_{2x} = 1.5 °C) applied to the low emissions case and a high climate sensitivity (ΔT_{2x} = 4.5 °C) applied to the high emissions case. The observed temperature data have been smoothed using a 10-year Gaussian filter. If a further 0.5 °C warming were chosen as a threshold for detection of the enhanced greenhouse effect on the basis of an "unprecedented change" argument (upper horizontal line), then this would be reached some time between 2009 and 2033. In practice, detection should be based on more sophisticated methods which would bring these dates closer to the present. (This Figure is an update of Figure 8.5 in Wigley and Barnett, 1990.)

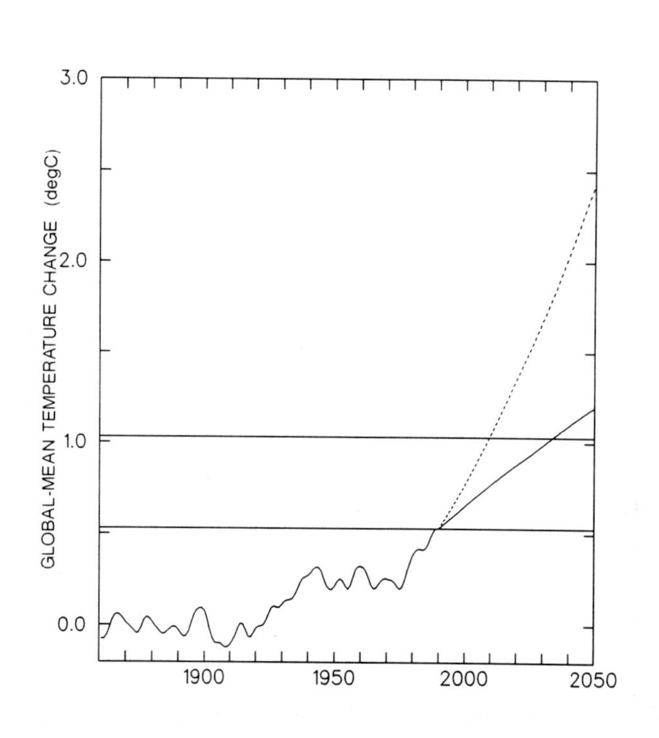

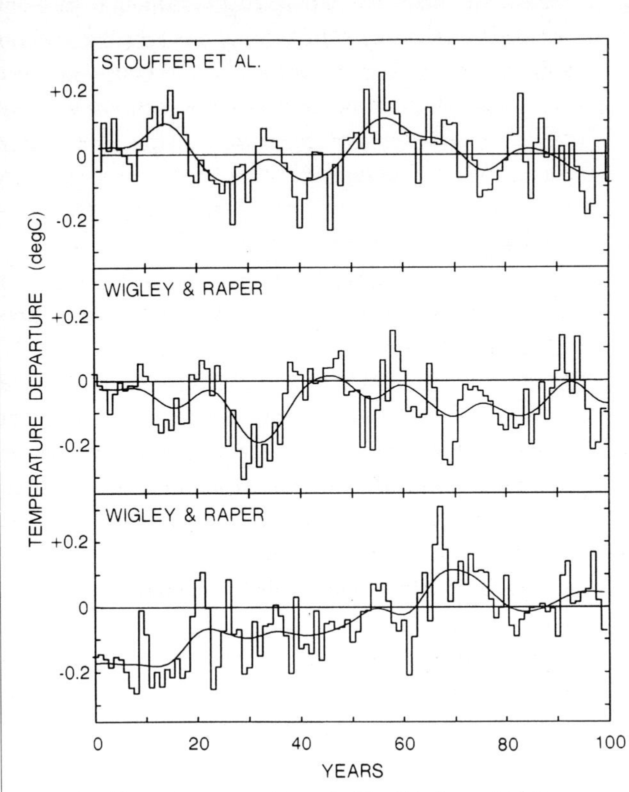

Figure 2. Simulated natural variability of global-mean termperature in the absence of any external forcing influences. The top panel shows results from the 100-year control run with the coupled ocean/atmosphere GCM of Stouffer et al. (1989). The lower two panels are 100-year sections selected from a 100,000 year simulation using the simpler climate model of Wigley and Raper (1990). In both cases, the climate sensitivity was ΔT_{2x} = 4 °C. The simpler model is forced with random inter-annual radiative changes whose statistical properties are chosen to match observed inter-annual variations in global-mean temperature. The consequent low-frequency variability arises due to the modulating effect of oceanic thermal inertia. Most 100-year sections are similar to the middle panel and are qualitatively indistinguishable from the single coupled ocean/atmosphere GCM results. However, a significant fraction shows century time-scale trends as large or larger than that in the bottom panel. Longer GCM simulations may therefore reveal similar century time-scale trends. The existence of trends like this is part of the reason why detecting the enhanced greenhouse effect is so difficult. A significant fraction of the observed warming could be due to natural variability, implying only a small greenhouse effect. Alternatively, a large greenhouse effect could have been partly offset by such natural variability. (From Wigley and Barnett, 1990.)

1940s and 1970s could, for example, simply be a natural, internal fluctuation of the climate system offsetting greenhouse-gas induced warming, or it could be the result of some other external forcing effect, such as a change in solar radiation. We know that the Earth's climate has varied markedly in the past, long before human influences began, so fluctuations like this are just what one would expect: smaller peaks and valleys that may last several decades, or longer.

As alluded to above, natural climate variations may occur as a result of purely internal fluctuations in the system — "free oscillations" — or as a result of external forcing factors. Examples of the former are shown in Figure 2.

Figure 2 shows that variations in global-mean temperature of up to 0.5 °C may occur from one decade to the next, even in the absence of external forcing. From one century to the next, however, these internal fluctuations must be much smaller, almost certainly less than 0.2-0.3 °C — as shown in the bottom panel of the Figure (Wigley and Raper, 1990).

On the century time scale, however, *external* forcing effects may cause somewhat larger trends. An example is the cold period that occurred between (roughly) 1400 and 1800, known as the "Little Ice Age" (Grove, 1988). This was certainly a natural event, one of a number of similar cold periods during the past 10,000 years that were probably due to century time-scale fluctuations in the Sun's output by a few tenths to 0.5% (Baliunas and Jastrow, 1990; Wigley and Kelly, 1990).

It is possbile, therefore, that the observed global warming of 0.5 °C over the past century could be largely a natural event, part internally generated and part externally forced. This is what makes detection difficult. While it is very unlikely that the warming is largely natural, the possibility that it is is quite difficult to rule out.

We can, however, be sure that the 20th Century warming is not due solely, or even largely, to natural *external* forcing effects. This is because greenhouse-gas forcing since the late 18th Century is so large, about 2.5Wm^{-2} (10%) (Shine et al., 1990)[4]. Of the natural external forcing agents that might possibly have contributed to the 20th Century forcing, changing solar output is the one which has the largest potential effect. But, solar forcing changes over the past few centuries were certainly less than 2.5Wm^{-2} (i.e. about 1% of total solar

[4] This leads us to an important and little appreciated aspect of detection. The detection issue is not one of proving that the enhanced greenhouse gas *forcing* is real, but of proving that this forcing has had a substantial impact on the Earth's climate. The major uncertainty is the link between forcing changes and climate changes, a link which is quantified by the climate sensitivity (see Section 2.3).

output). We know this because, even though early observations of the Sun's output were very inaccurate, a 1% change would have been observed. We also have other evidence that indicates that the Sun's output has not varied by more than 0.5% over the past 100 years (Baliunas and Jastrow, 1990; Wigley and Kelly, 1990).

While greenhouse-gas forcing is almost certainly the main external factor that has operated over the past few centuries, this still leaves us with many ways to explain the recent global warming. The contributors are, with their associated directions; the enhanced greenhouse effect, warming; the effect of man-made emissions of sulphur dioxide, cooling; other natural external forcing effects, warming *or* cooling; internally generated variability, warming *or* cooling. None of these factors can be reliably quantified, and their relative importance is uncertain. Because of this, all we can say about the observed global warming record is that it is qualitatively consistent with the greenhouse hypothesis. On the basis of the observed global warming alone, we are unable to either prove or disprove the hypothesis.

There are three ways to proceed from this lack of proof.

- *Eliminating alternative causes:* First, we could monitor all of the possible alternative factors so that, at some time in the future, specific causes of possible warming could be eliminated. The reason why solar irradiance — the intensity of solar radiation impinging on the Earth — cannot be completely eliminated as a cause of the century time-scale warming is because early, ground-based measurements of the Sun's output are so inaccurate. Irradiance has only been accurately monitored for 10 years, since suitable instruments were sent above the atmosphere on satellites. It is essential that it continue to be monitored, since it is a primary alternative candidate to explain global warming.

  Other possibly important casual factors, such as tropospheric and stratospheric aerosols, are more difficult to monitor comprehensively — but they may be no less important.

- *Waiting for proof:* Secondly, we could simply wait until the global-mean warming reached such proportions that the enhanced greenhouse effect was the only possible explanation. This "unprecedented change" approach has been considered by Wigley and Barnett (1990), who conclude that we would have to wait 10-50 years before it was convincing , see Figure 1. (The waiting time depends on the climate sensitivity, and on how rapidly greenhouse gases accumulate in the atmosphere.) By that time, because of the delays built in to the system, it could be much more difficult and expensive, if not impossible, to take effective corrective steps.

- *The fingerprint method:* The third way is to abandon global-mean quantities as indicators of climatic change (at least in the detection context) and to use more sophisticated indicators. This approach is generally referred to as the "fingerprint method". In practice, the method is severely limited by available data, and by the fact that certain patterns of change are not unique to greenhouse-gas forcing. Nevertheless, it appears to be the best chance we have of detecting the enhanced greenhouse effect in the near future (Wigley and Barnett, 1990).

5 Using the Fingerprint Method

In the fingerprint method, the idea is to monitor a number of different variables simultaneously, and to look for a specific pattern of climate change. An example could be a three-region pattern: large warming in high latitudes, less warming in the tropics and cooling in the stratosphere. The terms "variable" and "pattern," however, are used in a more general sense than in this example. The variables may be the same type (such as temperature) measured at different locations on or above the Earth's surface or in the oceans, or different types (such as precipitation, temperature, atmospheric moisture content, etc.) measured at the same or different locations; while pattern refers to the way the measurements are distributed over space, including vertically (at different heights in the atmosphere).

It is this multivariate pattern, tracked over time, that is the fingerprint. There are virtually an infinite number of possible fingerprints that could be chosen, but there are some obvious and important characteristics that an efficient fingerprint should have.

5.1 *Ideal characteristics of the efficient fingerprint*

- *Uniqueness of the fingerprint:* The fingerprint pattern should be unique to the enhanced greenhouse effect. In other words, the specific pattern of change we seek should be distinctive, unlike the patterns associated with any alternative causes (either external or internal), and unlike natural variability.

  Defining a greenhouse-unique fingerprint is far from a simple task, in part because similar "signals" may arise from a variety of different causes. For example, surface temperature change patterns stemming from the enhanced greenhouse effect appear similar to the patterns of change that would result from a change in solar irradiance (Hansen et al., 1984). In this case, the two causes (greenhouse and Sun) could be distinguished by choosing a different fingerprint. The enhanced greenhouse effect causes warming near the surface of the Earth, and cooling higher in the stratosphere; while solar irradiance should cause warming at every level of the atmosphere. Thus, a fingerprint incorporating vertical information may help to distinguish between the two causes.[5]

[5] Unfortunately, these vertical differences cannot easily be capitalized on because the contrasting directions of change are also characteristic of natural variability (Liu and Schuurmans, 1990). Thus, we may be able to eliminate the solar cause in this way, but we cannot eliminate all other possible causes.

• *Strength of the signal:* The individual components of the fingerprint should, themselves, have high "signal-to-noise ratios" (SNR). In other words, their changes under an enhanced greenhouse effect should be large relative to their natural variability (Barnett and Schlesinger, 1987; Santer et al., 1991). Modelling studies have shown that variables such as precipitation and mean-sea-level pressure have low SNR; their predicted changes are either too small (pressure), or — as in the case of precipitation — they are too variable (i.e. noisy) in both space and time today.

Variables with high SNR include temperature and atmospheric moisture content. These two variables, however, are highly correlated, so including both in a fingerprint detection variable is little better than using just one.[6]

• *Availability of suitable data:* Thirdly, there must be sufficient high-quality data available to be able to identify the selected fingerprint in the past observational record. Since we know (or suspect) that all climate variables vary considerably in nature from decade to decade, and since the enhanced greenhouse effect operates on a 100-year time scale, a few decades of observational data is not enough to use in a search for the signal.

This strongly detracts from the usefulness of all satellite data (unless these are part of a longer, composite "ground-truth" and satellite record); satellite data go back only to the 1960s. Upper atmosphere data are only of suitable quality after about 1950 — insufficient time to rule out the possibility that the warming trend is merely natural variability.

The only variables with long enough records are near-surface air temperatures (over land and ocean areas), sea surface temperatures, precipitation (only over the land), mean sea level pressure and (at relatively few points and sometimes difficult to interpret) sea level. There are also so-called "proxy" climate variables (generally biological or geological quantities, such as tree rings and glacier movements) which have longer records, but, although these variables are clearly climate-dependent, their links to climate are generally complex and imperfectly understood.

5.2 *Selecting fingerprint variables*

On the basis of these criteria, it appears that near-surface air and/or marine temperatures are the best candidate for a fingerprint. The problem with these data is that the spatial pattern of greenhouse-gas-induced temperature change is not unique. As noted above, a similar pattern of change occurs with solar forcing changes — at least where these changes relate to solar emission, rather than to the orbital characteristics (e.g. tilting) of the Earth. Moreover, the pattern of low-frequency natural variability seems similar to the greenhouse-gas signal pattern.

Nevertheless, pursuing a near-surface, temperature-based fingerprint is a worthwhile exercise. Our best hopes for early success probably lie in the possibility of finding a convincing upward trend in an appropriate detection statistic (i.e. indicator of signal strength), clearly distinguished from background variability[7], becoming more and more obvious in the later measurements as the radiative forcing from the enhanced greenhouse effect increases.

6 A Broader Perspective on Detection

Let us now consider the broader issues behind detection: the need to understand the whole climate system, and to predict or anticipate future changes. So far, the process of monitoring and evaluating data has been considered for its use in detection: as an aspect of validating climate models. But there are (at least) two more general aspects: anticipating surprises, and identifying symptoms.

6.1 *Preparing for surprises*

Surprises, like the Antarctic ozone hole, come in two forms — those that are entirely unexpected, and those that have been anticipated but given very low probability. We know of no important climatological surprise in the first category. For example, although the spatial character of the Antarctic ozone hole was a surprise, the idea of stratospheric ozone depletion, from halocarbons, had been suggested as a possibility in the 1970s. It was the more general threat of ozone depletion that was driving the negotiation of international control measures for halocarbons emissions, before the ozone hole was discovered. The ozone hole added a new perspective and increased the urgency of the negotiations. It should be noted that scientists were correct in their concern that the halocarbons may be environmentally damaging, but that they had underestimated the scale of the problem.

Many low-probability climate "surprises" have been proposed as potential consequences of the enhanced greenhouse effect. They include changes in the frequency of extreme events like tropical cyclones; rapid melting of the West Antarctic ice sheet (and the consequent accelerated rise in sea level); rapid changes in the ocean's thermohaline circulation (i.e. the vertical component of the circulation); and feedback effects associated with the emissions of CO_2 or CH_4

---

[6] Relatively little work has been done on selecting optimum variables using SNR estimates, and what has been done has used SNR values based on model estimates of the noise levels. Observed noise levels may be preferable. There is scope for further work here.

[7] The main statistical problem with the fingerprint method is to define a "statistic" which quantifies the similarity between an observed pattern of change and a given signal pattern, and to determine that the similarity is not just a chance phenomenon. The only method explored in any detail to date is a correlation approach (Barnett, 1986; Barnett and Schlesinger, 1987). Here, observed patterns of climate change relative to an initial reference period are constructed, and each pattern compared (using some type of spatial correlation coefficient as the detection statistic) with a model-based signal. The results produced are in the form of a time series of correlation coefficients.

from terrestrial sources, or dimethylsulphide (DMS) from the oceans (the latter leading to a large increase in sulphate aerosol concentrations).

6.2 Identifying symptoms

In many cases, suitable observational networks are in place to ensure that we are able to identify symptoms of these processes soon after they begin. For example:

- Tropical storms are well monitored as part of the World Weather Watch system for weather forecasting. However, a radical, long-term change in tropical cyclone frequencies would require at least a few decades to be sure that the change was more than just an ephemeral manifestation of natural variability. Surprises are also possible in mid-latitude cyclones. The fact that models predict only modest changes — while the real world exhibits considerable variability — may point to weaknesses in the models, perhaps associated with their poor spatial resolution.

- Melting (or, more precisely, surging) of the West Antarctic ice sheet could be detected by satellite altimetry, using systems soon to be in operation. Ice sheet melting would, however, be preceded by a rapid decline in the extent of the floating ice shelves which buttress the continental ice sheet, and this is something that present satellite monitoring would identify "on line". Once again, some decades of data would probably be required to eliminate natural variability as an explanation.

- Changes in the ocean's thermohaline circulation may be detectable directly, or indirectly, through "knock-on" effects on the climate system. For example, a slow down in the circulation should lead to a reduced atmospheric warming rate. Direct ocean observations will be expanded in the coming decade as part of the World Ocean Circulation Experiment (WOCE), and some suitable observations already exist. These suggest that the thermohaline circulation varies naturally on a decadal time scale; thus, as with other "surprises", even if a substantial shift is observed, we will still be faced with the difficult problem of associating cause and effect.

6.3 Difficulties in associating small-scale events with the greenhouse effect

In the context of detecting the enhanced greenhouse effect, unexpected events or changes are only important if they can be shown to be more than just manifestations of natural variability.[8] It is already common for the media to "blame" events on the greenhouse effect — or, in the scientific community, to say: "This is the sort of event that could occur more frequently as a result of the greenhouse effect." Unwarranted attributions and deceptive statements are sure to continue to be popular as time goes by. While it is likely that an increasing number of anomalous events or short-term trends will, in fact, be at least partly due to the greenhouse effect, demonstrating cause and effect will always be difficult. Part of the reason for this is because these events will take place on a regional, or smaller spatial scale. Usually, the smaller the scale, the lower the signal-to-noise ratio, and thus the more difficult it is to detect the influence of the greenhouse effect.

Two examples will illustrate the point. A study of tree-rings in northern Scandinavia (Briffa et al., 1990) has recently made use of the long dendrochronological records available to get a better idea of the level of low-frequency natural variability. On this basis, the authors concluded that the expected greenhouse-related increase in summer temperatures in this region would not rise above the noise of natural variability for many decades — until long after global-scale temperature effects were detectable.

A second example concerns a biological indicator, coral reefs. Here the situation is more complex, but it involves the same issues. There have been a number of examples of recent bleaching (or whitening) of coral reefs, attributed by some to an increase in marine temperatures (although there are other explanations). Do these events have any significance in the context of the greenhouse effect? It would be pointless to use this sort of proxy information to infer temperature changes, when we have direct measurements of marine temperatures which are clearly superior to any indirect indicator. At first glance, it would be even more foolish to try to attribute these events to the enhanced greenhouse effect. Even if ocean warming were the cause of reef bleaching, we cannot yet attribute even the global-mean observed ocean warming to the greenhouse effect, so to attribute a regional change to this specific cause is impossible. But what makes this case interesting is that it may relate to the "unprecedented change" method mentioned in Section 4.2. Are these bleaching events unprecedented in the record of, say, the Holocene, or even the last 1000 years? If this were undeniably the case, then the bleaching may point to unprecedentedly high temperature levels which may, in turn, be a consequence of the enhanced greenhouse effect. Even that would be only a first step towards convincingly demonstrating a cause-effect relationship.

6.4 Can biological "canaries" for climate change exist?

Coral reef bleaching has been seen as a "biological canary" — a life form which integrates different effects to give an early indicator of larger and more extensive changes, like the famous canary carried into mines so that it will faint first when air becomes short. Although the concept may not apply to coral reef bleaching, it is clearly valuable; and other types of biological canary may be identified in the future. These ideas, while speculative and lacking concrete examples, are worth pursuing.

---

[8] Even if not associated with the greenhouse effect, such changes could have devastating effects on society. An example is the Sahel drought, which began some 30 years ago and continues today.

6.5 Atmospheric chemistry surprises

The list of possible surprises contains items that are, strictly speaking, not an issue of climate *per se*. In addition to changing the climate, greenhouse gases affect the balance of chemical composition in the atmosphere. Thus, the greenhouse issue applies not only to climate models, but also to other types of models, whose validation is equally important. It is here where there is considerable scope for improvements in monitoring, and in the use of data analysis tools.

Our current understanding of atmospheric chemistry, for example, is still rudimentary; atmospheric chemistry is, in many ways, an even newer science than climatology (Pearman, 1991). But it is central to making predictions of changes in the concentrations of greenhouse gases (other than carbon dioxide and water vapour). It is virtually certain that surprises are in store here, albeit perhaps not as dramatic as the ozone hole. For example, the level of the atmosphere's primary oxidizing agent, the hydroxyl radical (OH), is poorly known, and the variability of its atmospheric concentration is virtually unknown. Since individual OH radicals exist only temporarily, during the intermediate stages of chemical reactions, they cannot be measured directly. All we know about OH must be inferred from other species and/or from models. Yet OH is the main sink for methane, and for all of the hydrogenated halocarbons; and it is an important aspect of the chemistry of tropospheric ozone. It is therefore crucial for determining future changes in these substances.

To compound these atmospheric chemistry uncertainties, the sources of gases such as methane and nitrous oxide are highly uncertain, both in magnitude and geographical location (Watson et al., 1990). Consider, for example, the possibility of an enhanced high-latitude methane source (from reservoirs of methane possibly locked in permafrost, or from the chemical decomposition of organic matter) resulting from greenhouse-gas-induced warming. This would be of some importance (although, even if the most pessimistic scenario is assumed, the added global warming is still only a few tenths of a degree Celsius). How would we know if such a process were occurring? A recent study by Fung et al. (1991) provides one possible answer. By using a General Circulation Model to specify the transport of gases by the atmospheric circulation, they were able to determine the spatial pattern of methane concentrations that would result from different spatial configurations of sources. They then compared these with observations. In this way, one may be able to choose the most likely source configuration. If a high-latitude source were suddenly increased, it could, in principle, be identified in this way.

Analyses like this could not only allow us to detect changes in high latitude sources but also to decide on optimum locations for monitoring methane concentrations, to facilitate further detection. Fung et al. showed that a small enhancement in the observation network would considerably improve our ability to pin down the sources.

7 Conclusions

Monitoring climatic change is essential for several reasons: to identify natural and (potential) man-made climate trends objectively; to provide a fundamental basis for the understanding of climate processes, in part so that we may develop better predictive climate models; and to provide appropriate data for testing and validating these models. We have seen that (despite limitations) detection studies and the fingerprint approach in particular offer the most hope for definitively establishing — one way or the other — that emissions from human activities have, through the enhanced greenhouse effect, caused a change in the Earth's climate.

Given the importance of climate monitoring, and the relative sparseness of some observational networks (which have, so far, been primarily based on weather forecasting needs, not on climate change monitoring needs), it is not surprising that there is currently a call for enhanced observational efforts. Data needs depend on the task, and, for some aspects of detection, existing networks may be adequate. For most aspects, however, present networks are far from adequate. Improved monitoring should:

- reflect the primary objectives of describing the climate system and its changes;
- provide data for the understanding of climate processes;
- provide data for detection studies and, more generally, for the validation of models and the elucidation of the causes of climatic variability and change.

For detection, priority should be given to the maintenance of data sets that already have long records (such as temperature records), to the infilling of gaps in coverage (such as the high-latitude southern oceans), and to data that define other external forcing factors (solar output and sulphate aerosol concentrations in particular).

It should be noted, however, that detection itself is not as important as narrowing uncertainties in our knowledge of the climate sensitivity, since this is the primary factor in determining how severely our climate could change. The key issue, both from a scientific and policy point of view, is not whether we have identified the greenhouse-gas warming signal, but determining the strength of the signal.

Fortunately, our understanding of climate — and the incorporation of that understanding into more and more realistic global and regional models — is advancing at a rapid rate. It is therefore essential that the extensive scientific basis for both the confidence and uncertainties in the models be adequately communicated to the policy-making communities of the world — the people who need to weigh uncertainty against the costs of either action or inaction. This lack of information transfer is a major (albeit not the only) contributor to polarization of opinion in the policy debate.

Acknowledgements

TMLW and PMK wish to acknowledge support from the US Department of Energy towards research into the detection issue.

References

Baliunas, S. and R. Jastrow, 1990: Evidence for long-term brightness changes of solar-type stars, *Nature*, **348**, 520-523.

Barnett, T.P., 1986: Detection of changes in global tropospheric temperature field induced by greenhouse gases, *Journal of Geophysical Research,* **91**, 6659-6667.

Barnett, T.P. and M.E. Schlesinger, 1987: Detecting changes in global climate induced by greenhouse gases, *Journal of Geophysical Research*, **92**, 14772-14780.

Briffa, K.R., T.S. Bartholin, D. Eckstein, P.D. Jones, W. Karlén, F.H. Schweingruber and P. Zetterberg, 1990: A 1400-year tree ring record of summer temperatures in Fennoscandia. *Nature*, **346**, 434-439.

Charlson, R.J., J.E. Lovelock, M.O. Andreae and S.G. Warren, 1987: Oceanic phytoplankton, atmospheric sulphur, cloud albedo and climate. *Nature*, **326**, 655-661.

Charlson, R.J., J. Langner, H. Rodhe, C.B. Leovy and S.G. Warren, 1991: Perturbation of the Northern Hemisphere radiative balance by backscattering of anthropogenic sulfate aerosols. *Tellus,* **43AB**, 152-163..

Folland, C.K., T.R. Karl and K.Y. Vinnikov, 1990: Observed climate variations and change. In *Climate Change: The IPCC Scientific Assessment*, eds. J.T. Houghton, G.J. Jenkins, and J.J. Ephraum, pp 195-238. Cambridge, Cambridge University Press.

Fung, I., J. John, J. Lerner, E. Matthews, M. Prather, L.O. Steele and P.J. Fraser, 1991: Three-dimensional model synthesis of the global methane cycle. *Journal of Geophysical Research,* **96**, 13033-13065.

Grove, J.M., 1988: *The Little Ice Age*. London, Methuen, 498 pp.

Hansen, J., A. Lacis, D. Rind, G. Russell, P. Stone, I. Fung, R. Reedy and J. Lerner, 1984: In *Climate Processes and Climate Sensitivity*, ed. J. Hansen and T. Takahashi, *Geophys. Mono.,* **29**, pp 130-163, American Geophysical Union.

Houghton, J.T., G.J. Jenkins and J.J. Ephraum (Eds.), 1990: *Climate Change: The IPCC Scientific Assessment*, Cambridge, Cambridge University Press, 365 pp.

Jones, P.D. and T.M.L. Wigley, 1990: Global warming trends. *Scientific American.* **263**(2), 84-91.

Jones, P.D., S.C.B. Raper, R.S. Bradley, H.F. Diaz, P.M. Kelly and T.M.L. Wigley, 1986a: Nothern Hemisphere surface air temperature variations: 1851-1984. *Journal of Climate and Applied Meteorology*, **25**, 161-179.

Jones, P.D., S.C.B. Raper and T.M.L. Wigley, 1986b: Southern Hemisphere surface air temperature variations: 1851-1984. *Journal of Climate and Applied Meteorology*, **25**, 1213-1230.

Jones, P.D., P.Y. Groisman, M. Coughlan, N. Plummer, W.-C. Wang and T.R. Karl, 1990: Assessment of urbanization effects in time series of surface air temperature over land *Nature*, **347**, 169-172.

Liu, Q. and C.J.E. Schuurmans, 1990: The correlation of tropospheric and stratospheric temperatures and its effect on the detection of climate changes. *Geophysical Research Letters,* **17**, 1085-1088.

Mitchell, J.F.B., S. Manabe, T. Tokioka and V. Meleshko, 1990: Equilibrium climate change. In *Climate Change: The IPCC Scientific Assessment*, ed. J.T. Houghton, G.J. Jenkins and J.J. Ephraum, pp 131-172. Cambridge, Cambridge University Press.

Pearman, G.I., 1988: Greenhouse gases: evidence for atmospheric changes and anthropogenic causes. In *Greenhouse: Planning for Climate Change*, ed. G.I. Pearman, pp 3-21. Melbourne, CSIRO.

Pearman, G.I., 1991: Changes in atmospheric chemistry and the greenhouse effect: A Southern hemisphere perspective. *Climatic Change* **18**, 131-146.

Santer, B.D. and T.M.L. Wigley, 1990: Regional validation of means, variances and spatial patterns in GCM control runs. *Journal of Geophysical Research*, **95**, 829-850.

Santer, B.D., T.M.L. Wigley, P.D. Jones and M.E. Schlesinger, 1991: Multivariate methods for the detection of greenhouse-gas-induced climate change. In *Greenhouse-Gas-Induced Climatic Change: A Critical Appraisal of Simulations and Observations*, ed. M.E. Schlesinger, pp 511-536. Elsevier, Amsterdam.

Shine, K.P., R.G. Derwent, D.J. Wuebbles and J.J. Morcrette, 1990: Radiative forcing of climate. In *Climate Change: The IPCC Scientific Assessment*, ed. J.T. Houghton, G.J. Jenkins, and J.J. Ephraum, pp 41-68. Cambridge, Cambridge University Press.

Stouffer, R.J., S. Manabe and K. Bryan, 1989: Interhemispheric asymmetry in climate response to a gradual increase of atmospheric carbon dioxide. *Nature*, **342**, 660-662.

Tsonis, A.A. and J.B. Elsner, 1989: Testing the global warming hypothesis. *Geophysical Research Letters*, **16**, 795-797.

Watson, R.T., H. Rodhe, H. Oeschger and U. Siegenthaler, 1990: Greenhouse gases and aerosols. In *Climate Change: The IPCC Scientific Assessment*, ed. J.T. Houghton, G.J. Jenkins, and J.J. Ephraums. pp 1-40. Cambridge, Cambridge University Press.

Wigley, T.M.L., 1989: Climatic change due to SO_2-derived cloud condensation nuclei. *Nature*, **339**, 365-367.

Wigley, T.M.L., 1991: Could reducing fossil-fuel emissions cause global warming? *Nature*, **349**, 503-506.

Wigley, T.M.L. and T.P. Barnett, 1990: Detection of the greenhouse effect in the observations. In *Climate Change: The IPCC Scientific Assessment*, ed. J.T. Houghton, G.J. Jenkins, and J.J. Ephraums. pp 239-255. Cambridge, Cambridge University Press.

Wigley, T.M.L. and P.M. Kelly, 1990: Holocene climatic change, ^{14}C wiggles and variations in solar irradiance. *Philosophical Transactions of the Royal Society*, London, **A330**, 547-560.

Wigley, T.M.L. and S.C.B. Raper, 1990: Natural variability of the climate system and detection of the greenhouse effect. *Nature*, **344**, 324-327.

Wigley, T.M.L. and S.C.B. Raper, 1991: Detection of the enhanced greenhouse effect on climate. In *Climate Change: Science, Impacts and Policy,* ed. J. Jäger and H.L. Ferguson, pp 231-242. Cambridge, Cambridge University Press.

Wigley, T.M.L., P.D. Jones, P.M. Kelly and S.C.B. Raper, 1989: Statistical significance of global warming. In *Proceedings of the Thirteenth Annual Climate Diagnostics Workshop*, ed. D.R. Rodenhuis, pp A1-A8. U.S. Department of Commerce, NOAA.

CHAPTER 7

Future Sea Level Rise: Environmental and Socio-Political Considerations

Richard A. Warrick and Atiq A. Rahman

Editor's Introduction

Many people assume that sea level and climate have been stable over the eons of the geologic past. In both cases, this assumption is far from the truth. The average level of the seas has risen and fallen substantially over the last 100,000 years — principally due to the advance and retreat of the large continental ice sheets associated with the last ice age. Changes in sea levels have also been observed over shorter time horizons. These changes have reflected both oceanographic fluctuations (e.g. shifts in ocean currents) and atmospheric changes (e.g. atmospheric pressure changes).

In this chapter, Richard Warrick and Atiq Rahman point out that global warming is expected to cause the Earth's mean sea level to rise. The extent of that rise is uncertain, but it is likely to be caused by two factors: the melting of land ice and the thermal expansion of the ocean. By contrast, the melting of sea ice floating in a warmer ocean will not raise the level of the seas, just as the melting of an ice cube in a glass of water does not raise the level of liquid in the glass.

Journalistic sensationalism has frightened much of the public with visions of a catastrophic 5-metre sea level rise that will inundate huge coastal areas. But these authors suggest (and most responsible scientists agree) that this is not very likely during the next century. Instead, today's best scientific estimate of average global sea level rise is about 60 cm, with a large uncertainty range of ±45 cm, if we continue on a "business-as-usual" path of greenhouse gas emissions.

But, just as no one lives in the average global climate, no coastline will necessarily be flooded by the average global amount. As we shall see in this chapter, the important impacts — both social and ecological — will be determined by the diverse interactions of sea level rise with the natural and human systems of each locale. In Bangladesh, for instance, large riverine sediment flows in the Ganges Brahmaputra-Meghna system may change the shape and topography of the delta, perhaps resulting not in general inundation of the coast, but rather in significant inland flooding. In the Mississippi valley, the balance of controlling factors is different: loss of wetlands may have more to do with river channelisation and dyke construction than with global sea level changes. A collection of examples from both developed and developing countries add authenticity and urgency to this chapter's argument.

Richard Warrick, a senior member of the Climatic Research Unit, University of East Anglia, UK, and Atiq Rahman, Director of the Bangladesh Centre for Advanced Studies, join forces to provide an interdisciplinary perspective on issues of sea level rise. Warrick has been an influential member of the IPCC Scientific Working Group (WG-1). Rahman is an outstanding chemist as well as a powerful and articulate advocate of balanced and sustainable development. Their careful consideration of the technical aspects of sea level rise lead them to conclude that the most efficacious strategy for many developing countries at risk from sea level rise, like Bangladesh, is to modify the human systems that must respond to these changes — in particular, to accelerate social and economic development. Poverty restricts choice in responding to environmental threat. Prosperity provides opportunities — for educating people to the risks, building more resilient structures in coastal regions, developing better disaster management schemes, and encouraging settlement in less vulnerable areas. All such actions will help to reduce future damages from sea level rise.

The problems — political as well as technical — of reducing greenhouse gas emissions sufficiently to slow sea level rise are formidable. We may already be committed to a substantial rise (about 10 cm over the next forty years) from increases in greenhouse gas concentrations that have already *occurred. Most of this rise can be attributed to emissions from the industrialized countries of the world. But as Warrick and Rahman show, preventing any additional future rise cannot be accomplished effectively by these same countries acting in isolation from the rest of the world; even the strongest emission policies on their part would be largely ineffectual in stemming the tide of sea level rise.*

Thus, for both adaptation and prevention, strong international cooperation is urgently needed. Achieving such goals will not, however, necessarily mean an altruistic commitment to the fantasy of "one-worldism", but it will require closer international cooperation in monitoring, in research, and in technological and economic development. Warrick and Rahman conclude that a combination of such measures can reduce future damage and disruption from sea level rise, and, as a consequence, decrease the likelihood of mounting inter-regional and intra-regional tensions.

- I. M. M.

1 Introduction: Sea Level Variations and Impact Issues

Discussions of future sea level rise often give the impression that sea level is, and has been, stable. Nothing could be further from the truth. The level of the sea surface varies across a broad range of time and space scales for many reasons. These include very long-term changes in ocean basin volume (from, for example, sea floor spreading or sedimentation) and medium-term changes in ocean mass from variations in groundwater, surface water or land-based ice. There are also shorter-term dynamic changes due to oceanographic (e.g. ocean current) or meteorological (e.g. atmospheric pressure) factors on the local or regional scale. The schema in Figure 1 is suggestive of a number of such factors.

As a result of global warming, mean sea level (MSL) may change for two reasons: expansion of the oceans due to higher sea temperatures, and changes in land-based ice. The latter would include changes in the mass of the large Greenland and Antarctic ice sheets, and of smaller ice caps and mountain glaciers.

Various projections of future sea level change are summarized in Table 1. In general, all agree that as a result of global warming, MSL should rise. The most recent projections depict a rise of less than one metre before the end of the next century — a change to which many communities might

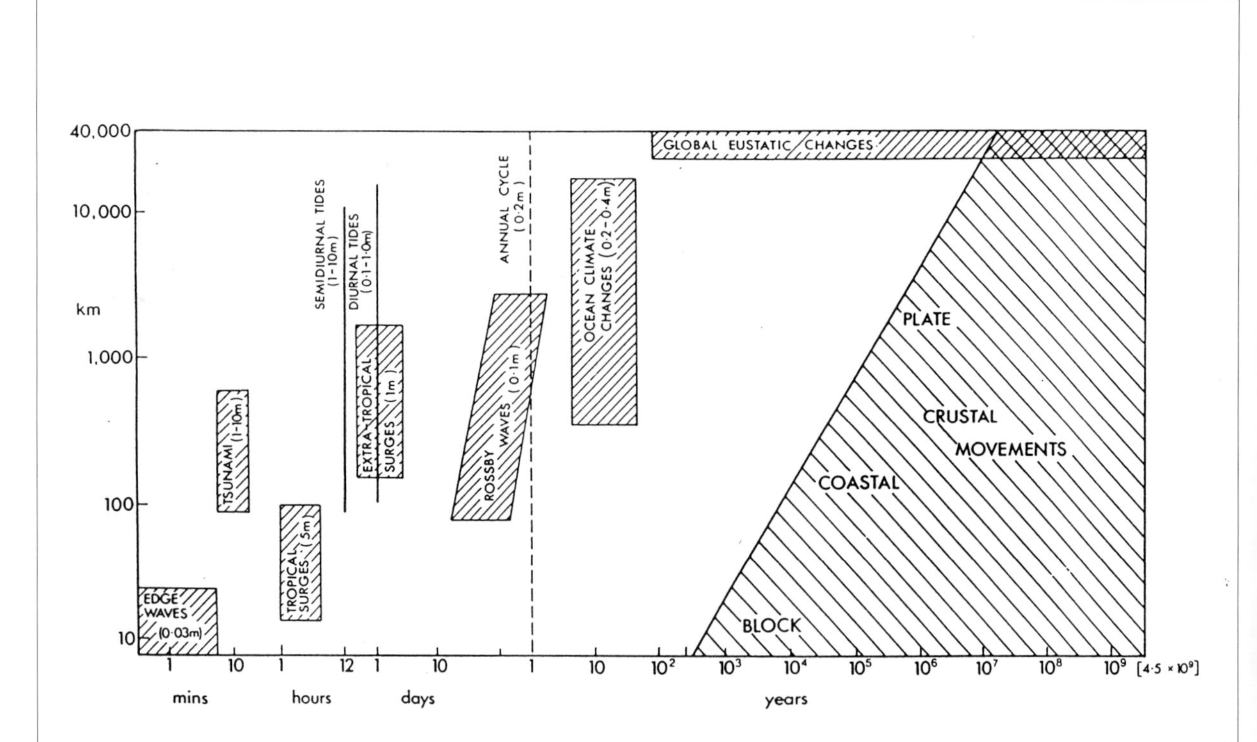

Figure 1. The level of the sea surface varies across a broad range of time and space scales. This figure shows some of the physical and geological processes responsible for sea-level changes, with indications of the length (vertical) and time-scales (horizontal) for each. Global processes have a length scale of 40,000 km. Source: Pugh, 1992.

Table 1. Estimates of future global sea level change (in centimetres). Source: Warrick, 1992; as modified from Warrick and Oerlemans, 1990 and Raper et al., 1991.

	Contributing factors				Total rise[a]		
	Thermal expansion	Alpine	Greenland	Antarctica	Best estimate	Range[f]	To (year)
Gornitz (1982) 2050	20		20 (combined)		40		
Revelle (1983)	30	12	13		71[b]		2080
Hoffman et al. (1983)	28 to 115	28 to 230 (combined)				56 to 345 26 to 39	2100 2025
PRB (1985)	[c]	10 to 30	10 to 30	-10 to 100		10 to 160	2100
Hoffman et al. (1986)	28 to 83	12 to 37	6 to 27	12 to 220		58 to 367 10 to 21	2100 2025
Robin (1986)[d]	30 to 60[d]	20±12[d]	to +10[d]	to -10[d]	80[i]	25 to 165[i]	2080
Thomas (1986)	28 to 83	14 to 35	9 to 45	13 to 80	100	60 to 230	2100
Villach (1987) Jaeger, 1988)[d]					30	-2 to 51	2025
Raper et al. (1991)	4 to 18	2 to 19	1 to 4	-2 to 3	21[g]	5 to 44[g]	2030
Oerlemans (1989)					20	0 to 40	2025
Van der Veen (1988)[h]	8 to 16	10 to 25	0 to 10	-5 to 0		28 to 66	2085
Warrick and Oerlemans (IPCC; 1990)[j]	28 to 66	8 to 20	3 to 23	-7 to 0	66	31 to 110	2100
Wigley & Raper (1992)[k]					46[k]	3 to 124[l]	2100

[a] from the 1980s
[b] total includes additional 17cm for trend extrapolation
[c] not considered
[d] for global warming of 3.5° C
[f] extreme ranges, not always directly comparable
[g] internally consistent synthesis of components
[h] for a global warming of 2-4° C
[i] estimated from global sea level and temperature change from 1880-1980 and global warming of 3.5±2.0° C for 1980-2080
[j] for IPCC "Business-as-usual" forcing scenario only
[k] for IPCC Policy Scenario B
[l] for all IPCC forcing scenarios

easily adapt. For instance, the best estimate given by the report of the Intergovernmental Panel on Climate Change (IPCC) is an average rate of rise of 6 millimetres per year (mm/yr) over the next century, if no actions are taken to stem the emissions of greenhouse gases (Warrick and Oerlemans, 1990).

None the less, the large uncertainties around most projections suggest that the possibility of a much larger rise cannot be ruled out. Whether slow or fast, however, the crucial factor in many situations will not simply be the rate of global sea level rise, but the circumstances at the water's edge — the underlying rate of vertical land movement, the response of natural systems, and the way in which each community chooses to meet the challenge.

At coastal locations, relative sea level (RSL; sea level relative to land) is affected by vertical movements of the land as well as changes in the actual sea surface elevation. Large-scale crustal movements can be regionally important. In parts of Scandinavia and Hudson's Bay, for example, relative sea level is falling one metre per century as a result of "isostatic rebound", i.e. crustal uplift due to the disappearance of the large continental ice sheets following the last glaciation (Peltier, 1986). In some areas like the East Coast of North America and Southeast Britain, sea level is still rising from land subsidence due to collapse of the "forebulge" at the margins of the last glacial advance (Emery and Aubrey, 1991). In other areas, land accretion or degradation processes interact with changes in the sea surface and with crustal uplift/subsidence to affect a dynamic sea level balance. For instance, in the Mississippi Delta, RSL change is determined by the difference between deltaic subsidence and global MSL rise on the one hand, and land accretion from sedimentation on the other (Day et al., 1992). In short, the meaning of global MSL rise for specific locations must be interpreted in relation to the mix of other factors which can affect relative sea level in complex ways (for recent reviews see NRC, 1990; Emery and Aubrey, 1991; Warrick et al., 1992).

The risks of sea level rise affect many countries, with potentially important international complications. With their emissions of CO_2 and other greenhouse gases, the industrialised nations of the world cause the bulk of the global warming problem. Warming, in turn, affects ice sheets at the extreme high latitudes, glaciers in remote mountain areas, and ocean temperatures worldwide, raising global sea level. The threat of environmental and socio-economic impacts is worrisome to many small island states and low-lying river deltas in developing countries, many of which have so far contributed negligibly to greenhouse gas emissions. Thus, the political issue of attributing responsibility — who is to blame? — arises and lies at the heart of difficult decisions regarding who should bear the burden of greenhouse gas emission reductions, and whether compensation to the "victims" should be forthcoming.

Few countries serve to illustrate these issues better than Bangladesh. Bangladesh is one of the poorest of the world's Least Developed Countries (LDCs), with an average per capita GNP of US$170 and a low per capita use of fossil fuels. Bangladesh rates poorly on indicators of development: less than one-third of the population is literate, and the population of approximately 115 million people (in 1990) is growing at an annual rate of about 2.2%. The densely populated, low-lying country of about 144,000 km^2 consists largely of the delta of three major rivers of the world, namely the Ganges, the Brahmaputra and the Meghna, and is the victim of recurrent, climate-related natural disasters: the devastating floods of 1987 and 1988 and the unprecedented cyclone of April 1991 are examples. For these reasons, Bangladesh may be one of the countries least responsible for the causes of climate change and most vulnerable to the effects, particularly sea level rise (Rahman and Huq, 1989).

In this chapter, the case of Bangladesh illustrates the issues we wish to examine. The chapter is divided into two parts. In the first, we outline some considerations for local environmental and socio-economic impact assessments, in light of the dynamic responses of natural and human systems. In the second part, we consider the global geo-political dimensions, particularly with regard to the issue of attribution of responsibility for the enhanced greenhouse effect and future sea level rise.

2 Environmental And Socio-economic Impacts: Some Considerations[1]

Too often, in the interests of expediency, projections of global mean sea level are applied statically for purposes of local impact assessment. By this, we mean that a projected rise — say, one or two metres — is simply added to mean sea level and the shape of the "new" coastline laid out. The adverse environmental effects of salt intrusion (into fresh groundwater, surface waters, wetlands or soils), erosion (of beaches and cliffs) or inundation (permanent or temporary storm surge flooding) are estimated and evaluated separately, if at all.

For example, a number of such studies have been carried out for Bangladesh. Milliman et al. (1989) and Broadus (1992), for instance, postulated an inundation of up to one-third of the country due to 1 m and 3 m rises in relative sea level by estimating the area below the 1 m and 3 m contour lines. The population, land, agricultural production, etc., within the affected area were then calculated as a basis for estimating impacts. A recent study by the Bangladesh Centre of Advanced Studies in association with the Centre for Global Change, University of Maryland (Ali and Huq, 1991) followed a similar approach for a 1m rise, albeit in a more detailed manner. Such studies provide a first rough approximation of potential impacts. However, it is extremely unlikely that the effect will be so straightforward. The real world works in vastly different, more complex and dynamic ways.

2.1 Dynamic influence of local natural systems

The actual physical impacts on Bangladesh are difficult to predict, because the coastal system will respond dynamically as sea level rises. Huge quantities of sediment (1.5-2.5 billion tonnes per year) are carried by the rivers — whose combined flood level can exceed 140,000 m^{3}s^{-1} — into Bangladesh from the whole of the Himalayan drainage system, including the countries of Nepal, China and India. About two-thirds of this sediment goes into the Bay of Bengal and, over the long term, causes land subsidence within the Delta region. In some coastal areas, the sediment rates are sufficient to more than compensate for deltaic

[1] In this section we choose not to reiterate the range of possible impacts of sea level rise. These have summarised elsewhere (e.g. Titus, 1992; Teggart et al., 1990) and exemplified in numerous case studies (e.g. Frascetto, 1991; Titus et al., 1990).

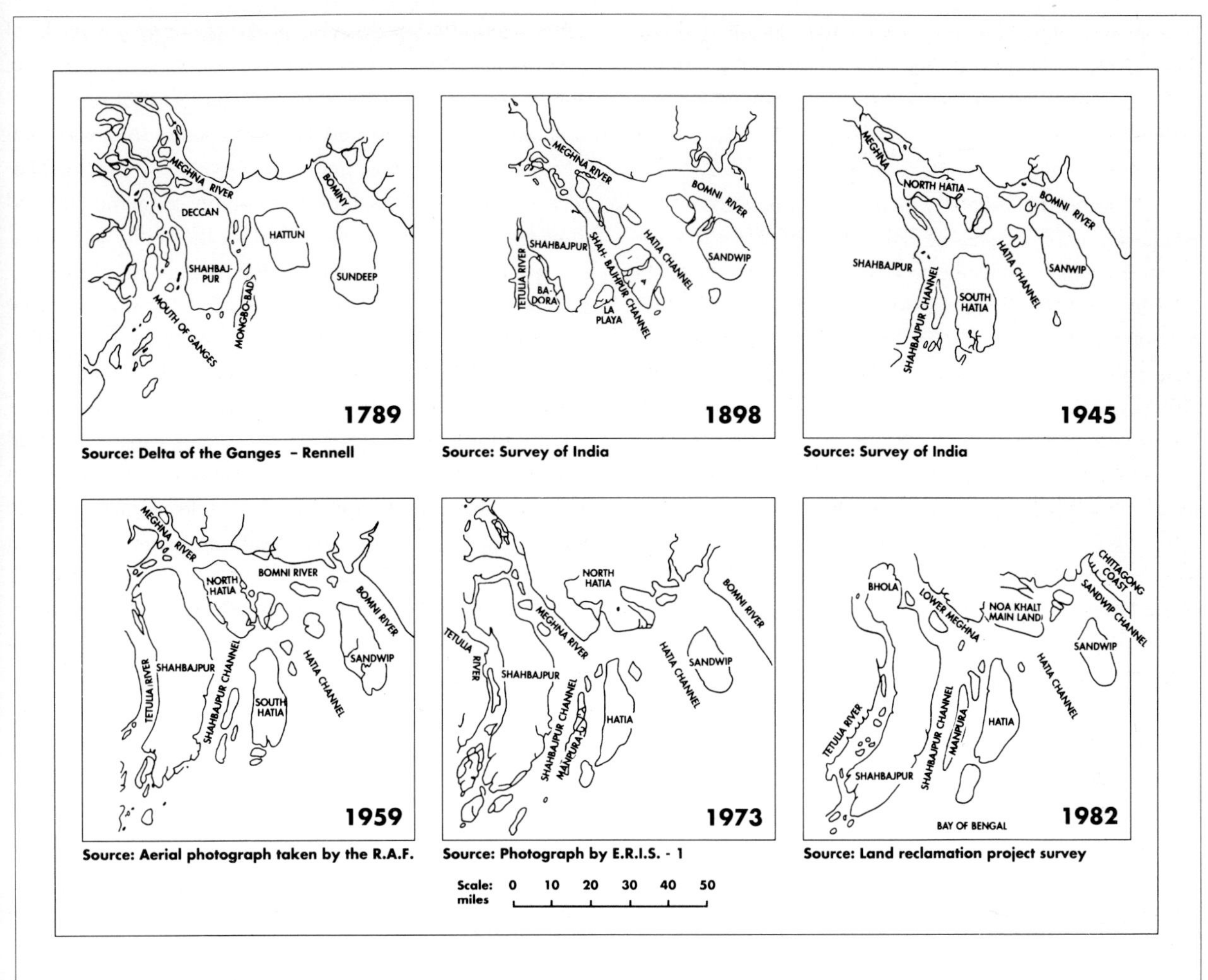

Figure 2. The historical development of the lower Meghna estuary, on the Bay of Bengal, shown over two centuries. Against this large natural variation in coastal configuration at the regional level, it is unclear whether a global sea level rise of six mm per year would be discernible.

subsidence and the land accretes; elsewhere, insufficient sediment results in erosion.

The result is that, naturally, the coastal configuration undergoes large changes, as shown in Figure 2. Against this large natural variation, it is unclear whether a global sea level rise of, say, 6 mm/year would, in fact, even be discernible. Furthermore, Brammer (1989; 1992) believes that, due to changes in river gradients, sedimentation and drainage, the primary impact of global MSL on Bangladesh will be an increase in flooding in the depressed basins *upstream* rather than only at the coast or within the tidal limits. Thus, the fate of Bangladesh is unlikely to follow simple predictions of an inexorably retreating coastline. Predicting the effects of global sea level rise on Bangladesh depends critically on an understanding of accretion and erosion rates and how they interact with sea level variations.

A similar set of relationships exists in the marshlands of the East Coast of the United States. In many locations, sedimentation rates have kept up with sea level rise, aided to a large extent by the marshland vegetation (Stevenson et al., 1986). If the rate of relative sea level rise becomes too high, however, salt stress and other factors would adversely affect the health of plants, thereby reducing sedimentation rates and exacerbating the changes. In this case, the picture that emerges is one of a dynamic, non-linear process, a central element of which is biological and thus dependent on the health of local life forms.

Biological health is also important for many low-lying, coral-fringed island-states, like the Republic of Maldives. The Maldives would not necessarily simply be "drowned" by rising sea level — provided that the rise is not too fast. Healthy coral can grow and keep up, maintaining the coral platform necessary for the dissipation of wave energy and the accretion of sediments, in effect compensating for the change in sea level. For the Maldives, the coral growth rate is not known precisely, but may range between 3 mm/yr and 10 mm/yr (Woodroffe, 1989; Edwards, 1989). If the threshold speed is exceeded, or if ocean temperatures become too warm, the coral would die, with the possibility of accelerating erosion and land loss. Again, it appears that dynamic, non-linear processes are at work.

In these cases and others, accurately projecting the rate and effects of RSL rise requires understanding of the interrelated roles of a number of complex environmental feedbacks.

2.2 *Dynamic influence of local human activity*

A related point to emphasize is that, in many cases, human interference in the dynamic processes can change local sea level by amounts comparable to those projected as a result of future global warming. The Mississippi River Delta is once again a good example. As described by Day et al. (1992), the Delta was roughly in a state of dynamic balance (or slowly growing) for thousands of years before the present century. Since then, levee construction along the banks of the Mississippi River, the construction of smaller-scale dykes, the damming of distributaries, and the development of numerous canals and diversions have effectively starved the wetlands of needed freshwater and sediments. Now, RSL is rising at a rate of 1 metre per century and up to 100 km^2 of wetlands are lost each year (Gagliano et al., 1981).

In the Republic of Maldives, pollution and mining of coral for construction purposes are threatening to disrupt the islands' natural capacity to respond to rising sea level (Pernetta and Sestini, 1989). The coral fringing the capital island of Male, where most Maldivians currently reside, is effectively dead, leaving the island vulnerable to rising sea level. Mining of coral elsewhere in the Maldives continues in response to high demand for scarce building material, largely for breakwaters for docks to support a growing tourism industry and for construction of Western-style houses in preference to traditional styles.

In Bangladesh, it has long been purported, but not proved, that deforestation in the headwaters of the Ganges-Bramaputra-Meghna river system has affected runoff, sediment flow and deposition rates, with consequent changes in coastlines and inland flooding. The possible effects of this deforestation, however, may be minor compared to effects that could accrue from a number of interventions being considered under the Flood Action Plan (FPCO, 1991). The comprehensive plan includes major structures and extensive embankments. Fears have been expressed that such interventions in the major river systems could have unexpected adverse effects on sedimentation and hydrological processes that could exacerbate the effects of future sea level rise (Huq and Rahman, 1990).

In these cases, the effects of local human interference — as a result of settlement in the area, and not in response to any threat of global warming or sea level rise — have to be understood. Every realm of human activity takes place within a cultural-historical context, with its own shared perception of both the natural resources and the alternative technologies available. Understanding of the "human ecology" of coastal environments is a fundamental element in predicting changes in, and seeking adequate solutions to, local sea level rise.

2.3 *Manipulating dynamic processes to favourable advantage*

The final point is that the complex dynamics of this human-environmental system can be manipulated to favourable advantage. The threat of future global MSL rise can provoke actions and policies that nurture the ability of natural systems to respond (by reducing RSL rise), or the ability of human systems to respond (by reducing the hazards posed by RSL rise).

Often, this can be accomplished indirectly by altering the management contexts in which decisions are made or by broadening the range of perceived choices in resource use. For example, eliminating the dikes and canals that are obstacles to freshwater and sediment flow could help re-establish land accretion processes in selected areas of the Mississippi Delta region. This could be promoted through incentives for less intensive land uses (like wetland and wildlife reserves) that preclude the need for such obstacles (Day et al., 1992). For the Maldives, it has been suggested that the best sea defence may be achieved by putting a stop to coral mining and encouraging the development of alternative building materials, thus preserving the health of coral to respond to slowly rising seas (Commonwealth Group of Experts, 1989).

The potential options for manipulating the natural systems in Bangladesh are less clear. It has been suggested that *controlled* flooding, rather than strict reliance on flood prevention through embankments and other engineering structures, might be one way of directing sediment and ensuring land accretion at critical coastal areas in the face of sea level rise (Mahtab, 1989). However, it must be admitted that, because of the complex dynamic processes involved, the uncertainties regarding the consequences of this and other more conventional interventions in the hydrological system are currently immense.

So, too, are the uncertainties regarding the environmental and socio-political effects of sea level rise on the coastal region of Bangladesh. Certainly, the spectre, often put forth, of millions of migrating homeless fleeing a retreating coastline and spilling over international borders — "eco-migration" — is not a foregone conclusion, maybe not even a realistic one. It is becoming clear that climate and sea level change must be viewed as one set of components amongst many interrelated factors — poverty, indebtedness, high population growth and meagre resources — that define vulnerability. Whether the future brings an inflammation of festering international tensions and regional conflicts may depend as much on the socio-economic conditions of the population-at-risk as on the rate of sea level rise. The more desperate the conditions, the larger the impacts and subsequent ramifications are likely to be.

For this reason, the most efficacious strategy for countries like Bangladesh — for reducing possible adverse impacts in the face of large uncertainty — lies in modifying the *human* systems, e.g. by accelerating social and economic development. It is unlikely that the main threat to Bangladesh stems directly from sea level creeping slowly higher. Rather, the cause for alarm arises principally from the prospect of more frequent, severe extreme events (see Mitchell and Ericksen, this volume) causing storm surge inundation as a result of higher sea level and/or increased severe tropical storm activity. As during the cyclone of April 1991, many thousands lost their lives due to a lack of opportunities for alternative livelihoods in less hazardous locations, inadequate communication and transportation systems, inadequate or too few shelters, and a reluctance to leave their few worldly goods unattended, even at the risk of death. Development creates *options* for vulnerable inhabitants who currently occupy lands susceptible to the long- and short-term environmental changes, such as sea level rise and storm surges. In short, development reduces poverty — which reduces vulnerability.

2.4 The need for integrated assessments

Effectively changing the pattern of human activities requires, foremost, a thorough interdisciplinary knowledge of the interplay of human activities and nature, especially as they pertain to the specific coastal situations at hand. Unfortunately, in many parts of the world which are vulnerable to sea level rise, such knowledge is meagre, and the uncertainties about future RSL rise and its effects are large. The next step forward is to move away from the simplistic, static approaches to impact assessments — rife in the literature — and to develop methods of integrated assessment that take account of the dynamic influences of both human and natural systems on relative sea level and its effects on the coastal environment.

3 Socio-Political Issues In International Response

The prospect of rising sea level is certain to motivate certain countries, like Bangladesh, to endorse (at least in principle) proposals for reducing greenhouse gas emissions and slowing climatic change. But although Bangladesh contains over 2% of the world's population and is one of the world's most densely populated countries, the country has one of the lowest fossil fuel consumption rates per capita. Most of the people consume non-commercial energy such as fuelwood, which is mostly used for cooking. Lighting in rural areas is confined to only 2 hours per night and is derived mostly from kerosene. Rural electrification exists but is available to less than 10% of the households. Consequently, Bangladesh contributes a miniscule 0.06% of the world's annual emissions of carbon into the atmosphere. Surely, the burden of CO_2 emissions reduction should fall elsewhere.

In the global context of environment and development the case of Bangladesh thus highlights delicate international issues. Who is responsible for the enhanced greenhouse effects and the prospective rise in sea level? Who, therefore, should accept responsibility for the costs of reducing greenhouse gas emissions? By extended implication, who should pay for damages accruing from any sea level rise? Such issues are clouded by the uncertainties regarding the rates of rise and the effects of emission policies in slowing the changes. To clarify the issues, we must first address the question of accountability with a clear understanding of the scientific and political realities.

3.1 Accounting for the enhanced greenhouse effect: procedural difficulties

Lately, the question of the attribution of responsibility for the enhanced greenhouse effect has been hotly debated. The debate has been sparked by the somewhat contentious publication by the World Resources Institute (WRI, 1990) of country-by-country estimates of greenhouse gas emissions and their combined radiative forcing effects, and the equally contentious critique by the Centre for Science and Environment in New Delhi (CSE; Agarwal and Narain, 1991; also see McCully, 1991). We shall not review the relevant literature since this has been done elsewhere (Hammond et al., 1991; *Environment*, 1991; *Nature* 349, 1991). Rather, let us first elucidate some of the difficulties involved in resolving the question of responsibility. Then, we shall offer a range of estimates that we feel is reasonable.

3.1.1 Inadequacy of scientific knowledge

First, in some critical areas the current scientific knowledge concerning greenhouse gas sources, sinks and fluxes is meagre. For example, most carbon cycle models fail to "balance," predicting current atmospheric concentrations that are too high compared to observations — the "missing sink" problem (see further Hoffert, 1991, this volume). Estimates of current carbon emissions from changing land use (including deforestation) differ by as much as a factor of ten. With respect to methane, there is a slim base of knowledge upon which to estimate local emissions. Together, such problems create large uncertainties in attempts to disaggregate GHG forcing changes on a country-by-country basis.

3.1.2 The problem of acceptable accounting procedures

There is also a lack of general agreement regarding the accounting procedures for allocating greenhouse gas emissions, concentrations and consequent changes in radiative forcing on a country-by-country basis. Moreover, the choice of procedures can be strongly influenced by socio-political judgements.

For example, should each country's annual greenhouse gas emissions be calculated net or gross — that is, with or without taking into account the capacity of the ocean, atmosphere and land to "absorb" emissions (the "sinks")? If on a net basis, how should the absorptive capacity be allocated — in proportion to the country's emission rate (e.g. by WRI, 1990) or weighted on a per capita basis (e.g. by Agarwal and Narain, 1991)? A per capita allocation scheme reduces the alleged Third World responsibility, particularly for relatively high greenhouse gas-emitting, highly populated countries (e.g. China and India). (See further Grubb et al., 1992; this volume.)

Furthermore, there is the issue of entitlement to any *growth* in the absorptive capacity. For instance, if the CO_2 "fertilization" effect increases the annual biospheric uptake of atmospheric CO_2, who is entitled to claim it — those countries with the highest emission rates, those with the positively affected vegetation, or every person equally?

There is also the question of whether accounting should be based on current emissions, or cumulative (e.g. from pre-industrial times). Some would argue (Smith, 1991) that the cumulative approach is technically more correct, because of the long residence times of most greenhouse gases in the atmosphere, and a fairer way to estimate responsibility for the enhanced greenhouse effect. But one conceptual difficulty is that actual greenhouse gas emissions from earlier periods have been largely "washed out" of the atmosphere; a molecule of CO_2 emitted in 1800 is unlikely to be in the atmosphere today. Another difficulty with the cumulative approach is that most emission data for earlier periods are scanty; for example, Third World fossil fuel carbon emissions can be estimated fairly accurately back only to 1950. Data on carbon emissions from land use change (deforestation) are poor for any period, even today. Notwithstanding these difficulties, the cumulative approach would attribute less responsibility for the enhanced greenhouse effect to the Third World.

3.1.3 Ethical considerations

Finally, there are some purely moral/ethical aspects that should be considered. For example, to paraphrase Agarwal and Narain (1991), does it serve the cause of global justice to assign equal weight to units of "climate warming potential" from greenhouse gases, regardless of origin? That would make a unit of methane from a Third World farm equal to a unit of carbon dioxide from a gas-guzzler in the industrialized West. Another such issue relates to production versus comsumption: To whom should emissions be allocated, the country of origin or the country that eventually consumes the product for which the greenhouse gas was emitted? This issue becomes increasingly important as more energy-intensive "metal-bashing" activities (such as manufacture of car components) move to developing nations, with the productive output designated for export to the West.

One thing is clear: there are no objective answers to these questions. But there may be cooperative political agreements that can address these issues.

3.2 How much of the enhanced greenhouse effect can be blamed on the Third World?

Bearing in mind the above issues, we now suggest a method for allocating the shares of responsibility. We do so by apportioning the radiative forcing changes (from IPCC; Shine et al., 1990) from the increased concentrations of the major greenhouse gases according to the fraction of total emissions contributed by countries of the developing world[2]. Our data come largely from the World Resources Institute (1990) and *Trends '90* (Boden et al., 1990).

Two approaches were followed. In the first, calculations were based on the relative emissions and radiative forcing changes which occurred over the decade of the 1980s — a time of high growth in energy use rates for the developing nations. But contemporary trends do not tell the whole story: less than one-quarter of the full radiative forcing change (since pre-industrial time) occurred during the 1980s. Accordingly, we also considered the cumulative changes in greenhouse gas emissions and radiative forcing since 1765, at the dawn of the industrial era.

As mentioned above, for the cumulative approach the lack of historical emission data is a problem. For fossil fuel carbon emissions from the developing world prior to 1950, we linearly interpolated to zero in 1900 (Figure 3). To infer the past deforestation contribution, we used a balanced *inverse* carbon cycle model (Wigley, 1992), that is, a model that calculates the CO_2 emission history required to produce a given history of atmospheric CO_2 concentrations. Assuming a current biospheric carbon emission rate of 1.6 gigatonnes of carbon (GtC) per year (Bolin et al., 1986; Houghton et al, 1990), the model results suggest that, on a cumulative basis, the total carbon emission from deforestation nearly matches that from fossil fuel burning (Figure 4). Interestingly, developing versus developed world partitioning (albeit highly judgemental) suggests that the industrialized nations have been responsible for at least three-quarters of cumulative deforestation carbon emissions, presumably from large-scale deforestation in temperate latitudes during the last century[3]. For methane and CFC emissions, we adopt the WRI estimates (for 1987) and assume that the ratio between developed and developing nations applies to earlier periods.

Tables 2 and 3 show the relative contributions of the two groups of nations for, respectively, the current and cumulative approaches. These estimates suggest that the developing world is responsible for 27-35% of the enhanced greenhouse effect[4].

---

[2] We chose not to make allocations on a per capita basis, since we are interested in investigating the world-regional differences. Nor did we consider allocating the absorptive capacity (sinks) according to region or population, because of the large uncertainties in greenhouse gas sinks and, importantly, because of the lack of procedural guidance for doing so, as discussed above. Thus we avoid the "minefield" and keep the approach simple. We consider the "developing world" to be all countries with the exception of Western and Eastern Europe, the former Soviet Union, the United States, Canada, Japan, Australia and New Zealand.

[3] We assume that net deforestation in the developing world regions was zero prior to 1900.

[4] In comparison, the World Resources Institute (WRI, 1990) obtains 44% (assuming a high Third World contribution from deforestation) and Agarwal and Narain (1991) obtain about 20% (with a lower deforestation estimate and absorptive capacity allocated on a per capita basis).

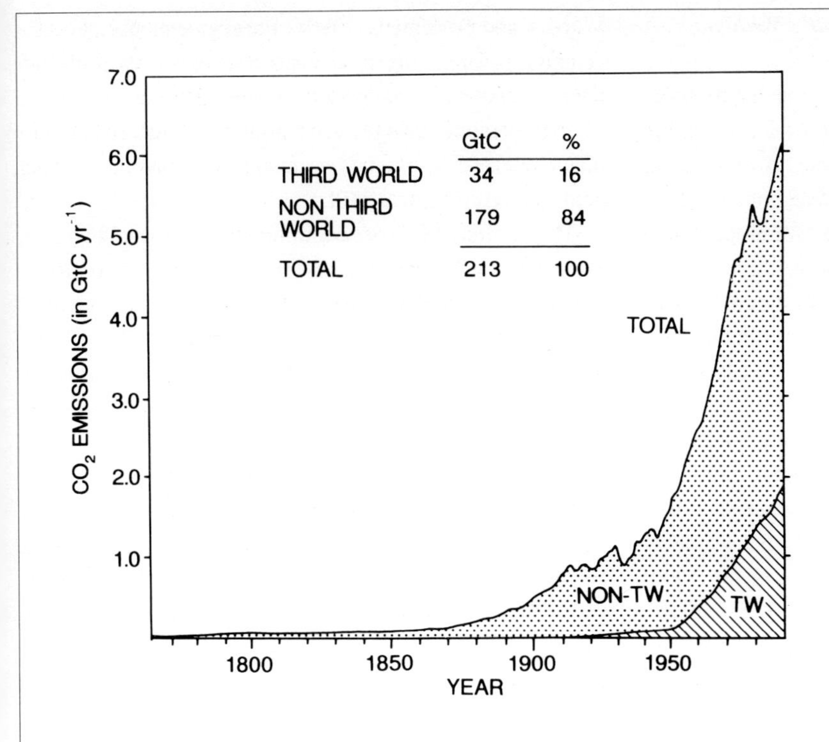

Figure 3. Comparison of fossil fuel-derived carbon dioxide emissions from Third World ("TW") and other ("non-TW") countries since 1765. On a cumulative basis, the non-TW countries have accounted for more than four-fifths of total emissons.

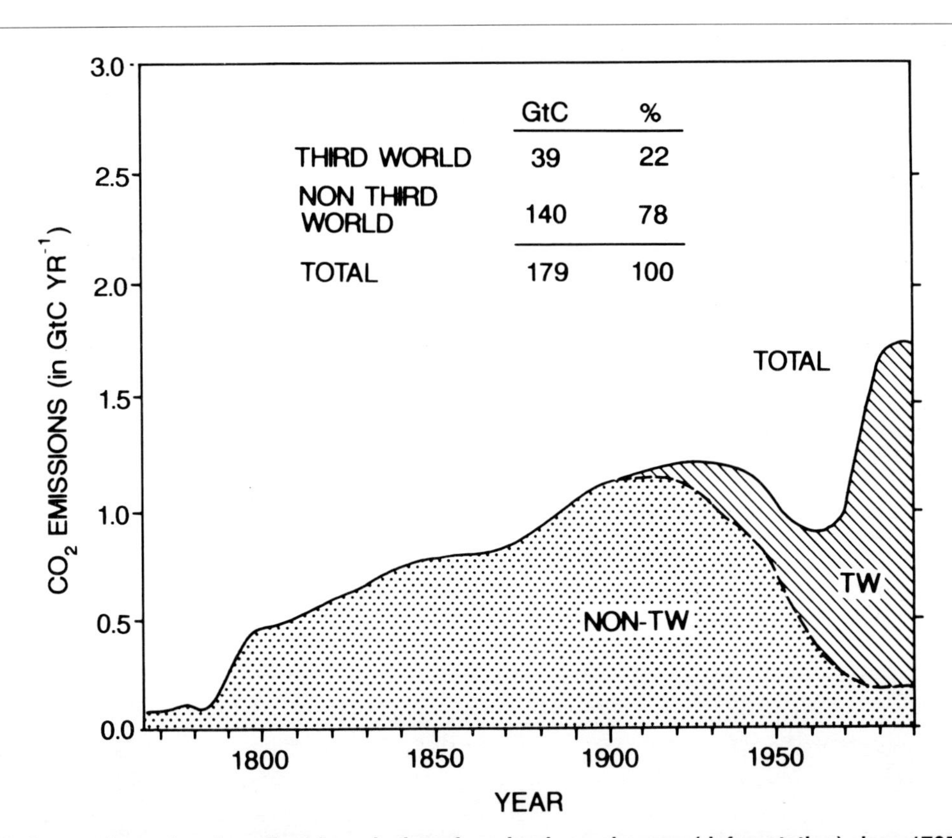

Figure 4. Comparison of carbon dioxide emissions from land use changes (deforestation) since 1765, as derived from an inverse carbon cycle model. Emissions are divided between Third World ("TW") and other ("non-TW") countries. On a cumulative basis, the total carbon emission from deforestation nearly matches that from fossil fuel burning. These results suggest that the non-TW countries have been responsible for about three-quarters of the cumulative deforestation-related carbon emissions.

3.3 Future sea level rise "commitment": Who's responsible?

These estimates allow us to make some judgements about responsibility for possible sea level rise. We start by asking, how much will *future* sea level rise due to changes in greenhouse gas concentrations that have *already* taken place? We shall call this the mean sea level rise "commitment" (a variation on the term introduced by Mintzer, 1987)[5]. The commitment results from lags in the climate system, including the response of the oceans and land-ice; one could think of it as the MSL rise that is "in the pipeline" and essentially unavoidable[6].

In order to estimate the magnitude of the commitment, we used the same model (with minor modifications), developed by Wigley and Raper (1987; 1992; also Wigley et al., 1991), that provided the time-dependent climate and sea level changes for the IPCC scenarios (Bretherton et al., 1990; Warrick and Oerlemans, 1990)[7]. From greenhouse gas concentration changes, the model generates estimates of global-mean temperature change and oceanic thermal expansion. The temperature changes drive a global alpine glacier ice-melt model and determine Greenland and Antarctic ice sheet contributions to global MSL.

The resulting MSL rise commitment is shown in Figure 5. To isolate the "committed" rise from any rise due to further increases in greenhouse gas concentrations, we "shut off" additional changes in radiative forcing in the model in 1990. Even so, mean sea level continues to rise sharply during the first few decades, at a (declining) rate of about 3 mm/yr, which is approximately 1.5-3.0 times faster than the average rate of MSL rise observed over the past 100 years.

During these decades, heat continues to penetrate to, and be transported within, the deeper layers of the modelled ocean. This retards surface warming and causes further

[5] For forecasting purposes, we define the term here as the difference between model-predicted global MSL rise by 1990 (our cut-off date for current observational data) and a specified future date, with the model being forced by observed increases in greenhouse gas concentrations to 1990, with no additional increases thereafter.

[6] This interpretation is given some credence by the fact that global CO_2 emissions would have to be reduced immediately by 60-80% just to achieve a stabilisation of CO_2 concentration (Watson et al., 1990).

[7] The core model is a box-upwelling-diffusion energy-balance model is forced by greenhouse gas concentrations changes (from 1765 to 2100) with model parameters that can be "tuned" to reflect uncertainties in the climate sensitivity to greenhouse forcing, the penetration of heat from the surface to deeper layers of the ocean, upwelling rates and other processes. The model assumptions were the same as the "best-estimate" values used by IPCC.

Table 2. Estimated contributions to radiative forcing change (in WM^{-2}) during the 1980s

Greenhouse gas	Third	Non-third	Total
CO_2	0.123 41%	0.177 59%	0.30
(fossil fuel)	(0.063) (21%)	(0.169) (56%)	
(deforestation)	(0.060) (20%)	(0.008) (03%)	
CH_4	0.046 57%	0.034 43%	0.08
CFCs	0.013 10%	0.117 90%	0.13
N_2O	0.006 20%	0.024 80%	0.03
Total forcing change	0.188 35%	0.352 65%	0.54 100%

Source: Shine et al., 1990; Boden et al., 1990;WRI, 1990.

Table 3. Estimated contributions to radiative forcing change (in WM^{-2}) from 1765 to 1990

Greenhouse Gas	Third	Non-Third	Total
CO_2	0.28 19%	1.22 81%	1.50
(fossil fuel)	(0.13) (9%)	(0.68) (45%)	
(deforestation)	(0.15) (10%)	(0.54) (36%)	
CH_4	0.32 57%	0.24 43%	0.56
CFCs	0.03 10%	0.26 90%	0.29
N_2O	0.02 20%	0.08 80%	0.10
Total forcing change	0.65 27%	1.8 73%	2.45 100%

Source: Shine et al., 1990; Boden et al., 1990; WRI, 1990.

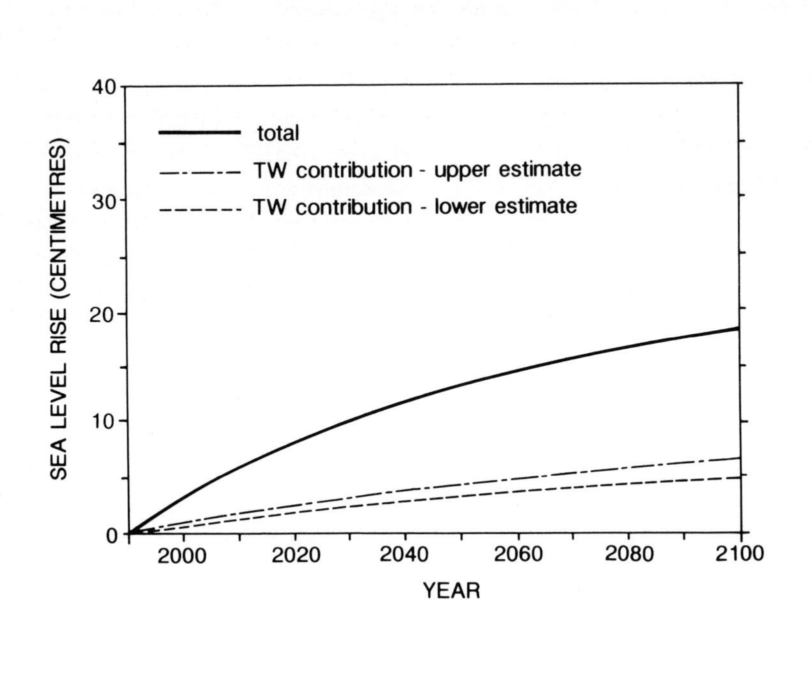

Figure 5. The present commitment to future sea level rise. Even if all new additions to greenhouse gas concentrations were shut off tomorrow, the sea level would continue to rise. Because of lags in the climate system, a certain "commitment" to future sea level rise is unavoidable as a consequence of concentration changes that have already taken place. This chart shows the unavoidable sea level rise to which we were "committed" by 1990, and the proportion of that rise attributes to Third World ("TW") countries. Greenhouse gas emissions by the developing world have committed us to a mean sea level rise of only 2.7-3.5 cm by 2030, and 5.0-6.5 cm by 2100.

thermal expansion of the oceans. Land ice changes continue to take place, both from the model-predicted warming attained by 1990 and from additional warming thereafter (still not including any new increases in greenhouse gas concentrations, however). The 1990-2030 sea level rise is about 10 cm from already-existing commitments — more than half that expected under the "business-as-usual" reference scenario (see Section 3.4 below). Near the end of the 21st Century, the curve has flattened considerably, returning to a pre-1990 rate of MSL rise (about 1.5 mm/yr). The 1990-2100 sea level rise commitment is 18.5 cm.

These figures, if accurate, are the bottom limit; they assume that emissions are reduced sufficiently to stabilize atmospheric concentrations of greenhouse gases at today's levels — an unlikely prospect. But we make the distinction anyway, because the "committed" rise is one way of analysing past responsibility. As shown in Figure 5, greenhouse gas emissions by the developing world have committed us to a MSL rise of only 2.7-3.5 cm by 2030 and 5.0-6.5 cm by 2100. This rate is well below the overall rate of global sea level rise experienced during the last 100 years. Clearly, the developing world, which contains about three-quarters of the world's population and which may be most vulnerable, has contributed to future sea level rise in only a minor way so far.

3.4 Future "un-committed" sea level rise: how much?

We now consider future sea level rise to which we are *not* committed — that is, the portion of a projected future rise which could conceivably be prevented. Let us define this "un-committed" sea level rise as the difference, at any future date, between the committed rise (as above) and any projection resulting from a baseline emissions scenario that gives positive changes in radiative forcing.

For the baseline scenario, we have elected to use a modified version of the IPCC Business-as-Usual scenario (BaU) (Swart, 1991). In comparison to the previous BaU case, this NEWBaU scenario, as we shall call it, has slightly higher CO_2 emissions than its predecessor (but this is partly offset by our use of a carbon cycle model with biospheric feedbacks that gives lower CO_2 concentration changes) and includes larger CFC production cuts as per the London amendment to the Montreal Protocol. As shown in Figure 6, this results in a 1990-2100 mean sea level rise "best estimate" of 58 cm, with a range of uncertainty from 21 cm to 105 cm. The range of uncertainty is large because the uncertainties in the land-ice response are compounded by the uncertainties in the global-mean temperature change.

Figure 7 shows both the "un-committed" and "committed" components of the NEWBaU best-estimate projection. It is evident that, even with the strongest political will, there is little hope of immediately slowing the rate of MSL rise; most of this rise has already been determined by past changes in greenhouse gas forcing. There is comparatively little un-committed sea level rise during the first part of the 21st Century; for example, by the year 2020 more than half of the projected rise is still un-preventable. It is not until the year

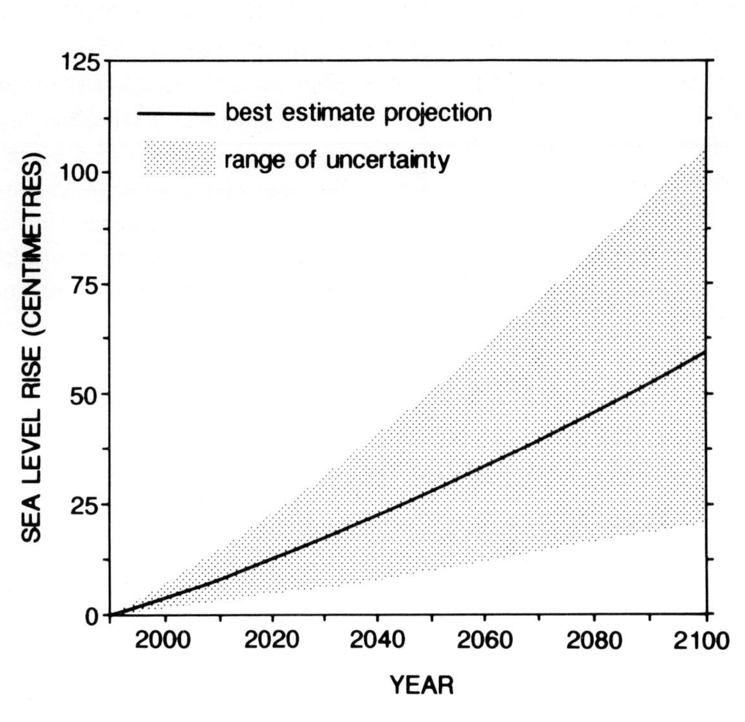

Figure 6. Estimates of sea level rise between 1990 and 2100, for a modified version of the IPCC Business-as-Usual emissions scenario. This produces a best estimate (solid line) of a 58 cm rise by the year 2100, with a range of uncertainty (grey area) from 21 cm to 105 cm.

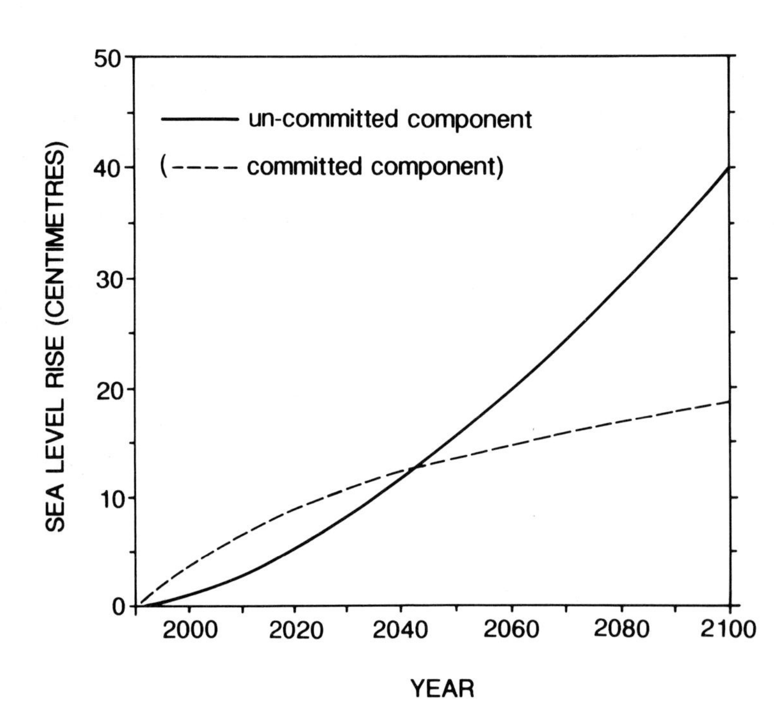

Figure 7. Committed and uncommited (preventable) components of the sea level rise estimates shown in Figure 6. By the year 2020, more than half of the projected rise is still un-preventable. It is not until the year 2040 that the un-committed component overtakes and surpasses the committed component; the reductions in sea level rise due to cutting emissions today will accrue only after many years into the future.

2040 that the uncommitted component overtakes and surpasses the committed sea level rise. Decision makers thus face a largely unprecedented situation in the field of environmental issues: they must consider actions with potentially large short-term costs, but with benefits which are uncertain and which would accrue only after the passage of many decades into the future.

3.5 *Unilaterally setting one's house in order: to what avail?*

We have avoided making any direct attribution of responsibility for future un-committed MSL rise. We do not think it entirely proper to be pointing fingers at groups of countries which are guilty of greenhouse gas emissions only in hypothetical scenarios. Nevertheless, we would like to examine and evaluate some prescriptions that can be found in the literature that do imply such responsibility.

3.5.1 *CO_2 emission reduction: the OECD nations acting alone*

First, we take the assertion by Agarwal and Narain (1991) that it is time for the West to stop "preaching environmental constraints and conditionalities to the developing countries...[and to] first set their own house in order" (p. 23). Aside from the moral and ethical imperatives of this statement (with which we agree), there is the practical issue: what difference would it make to projections of global MSL rise?

For the purpose of a sensitivity test, we apply a scenario in which the OECD countries reduce all fossil fuel-related CO_2 emissions to zero by the year 2083. For other greenhouse gases and all other countries including developing nations, the assumptions remain the same as NEWBaU. The results are shown in Figure 8 (top panel). As compared to the NEWBaU case, such a Draconian policy would make little difference to the projected MSL rise — about 6 cm prevented by 2100 — if the rest of the world carried on its merry way.

Of course, in the NEWBaU reference case, CO_2 emissions were assumed to grow much more slowly in the OECD than in the rest of the world, particularly the developing countries. None the less, in general, it might be concluded that a policy of the West "setting its own house in order" cannot, alone, do the trick.

3.5.2 *Stopping deforestation in the developing world*

Second, let us examine an assertion frequently made in the West: that the developing world must cease deforestation in order to slow global warming and sea level rise. Currently, action by developing nations depends largely on their ability to halt deforestation. In the NEWBaU reference scenario, deforestation carries on until tropical forests are depleted by 2100. What difference would it make if, instead, deforestation ceased altogether by the year 2010, and the biosphere, through reafforestation, became a sink for 1GtC per year by 2025 and for the remainder of the 21st Century?

The results are shown in Figure 8 (middle panel). In terms of future MSL rise, deforestation — or not — makes little difference, about 1 cm by 2100. In the context of future climate change, the deforestation issue is largely a "red herring" — notwithstanding the fact that "every little bit counts," and that there are indisputably sound environmental, social and economic reasons for putting forest resource management on a sustainable basis. It would be a tragedy to lose most of the world's rainforests, but it would not elevate the oceans much.

4 The Need for International Cooperation

In the lexicon of the global change issue, "one-worldism" has come under some heavy criticism lately by some who see the concept as a means of unduly shifting blame and responsibility for the enhanced greenhouse effect to undeserving countries, like Bangladesh. There may be more than a grain of truth in the assertion. But, on the other hand, from a practical standpoint "unilateralism" would appear to be an ineffectual basis for slowing sea level rise. As we showed above, relying solely on a region or sub-set of nations, like the industrialized West, to solve the problem is tantamount to postponing action. Can developing countries like Bangladesh afford the delay?

Let us speculate on some possible geopolitical consequences. The combination of sea-level rise and changing climate, superimposed on increasing population density and poverty in Bangladesh, could increase the political tensions that already exist in the region. While global climate change in general might render life difficult, sea level rise could render life in some areas impossible. This could apply not just to Bangladesh, but to island-states like the Maldives or to the low-lying coasts of Southeast Asia where mangrove swamps that have supported coastal communities for centuries could cease to be productive. Aside from increased cries for financial aid, there could be growing migration pressure. Even where people remain, sea level rise could change the productivity of coastal environments that have traditionally supported food production, recreation, transportation and other important activities.

Of course, it would not be right to assume that global sea level rise would automatically lead to local sea level rise at the same rate, as discussed earlier. Nor would it automatically lead to increased political conflict or major migration. But it certainly has the potential for exacerbating existing tensions that have, in the past, contributed to such regional unrest and disruption. These political stresses, distracting any opportunity for mutual aid, could make dealing with the underlying problem of sea level rise more difficult. Thus, while the uncertainties regarding future sea level rise are very large, so, too, are the potential geopolitical consequences of doing nothing to slow it.

Because of the large "commitment", the opportunities for slowing the sea level rise predicted by current models are somewhat limited, particularly during the next several decades. In the longer term, a larger portion of the rise is "preventable," but will require a major deviation from the "business-as-usual" course. For instance, in order to keep the

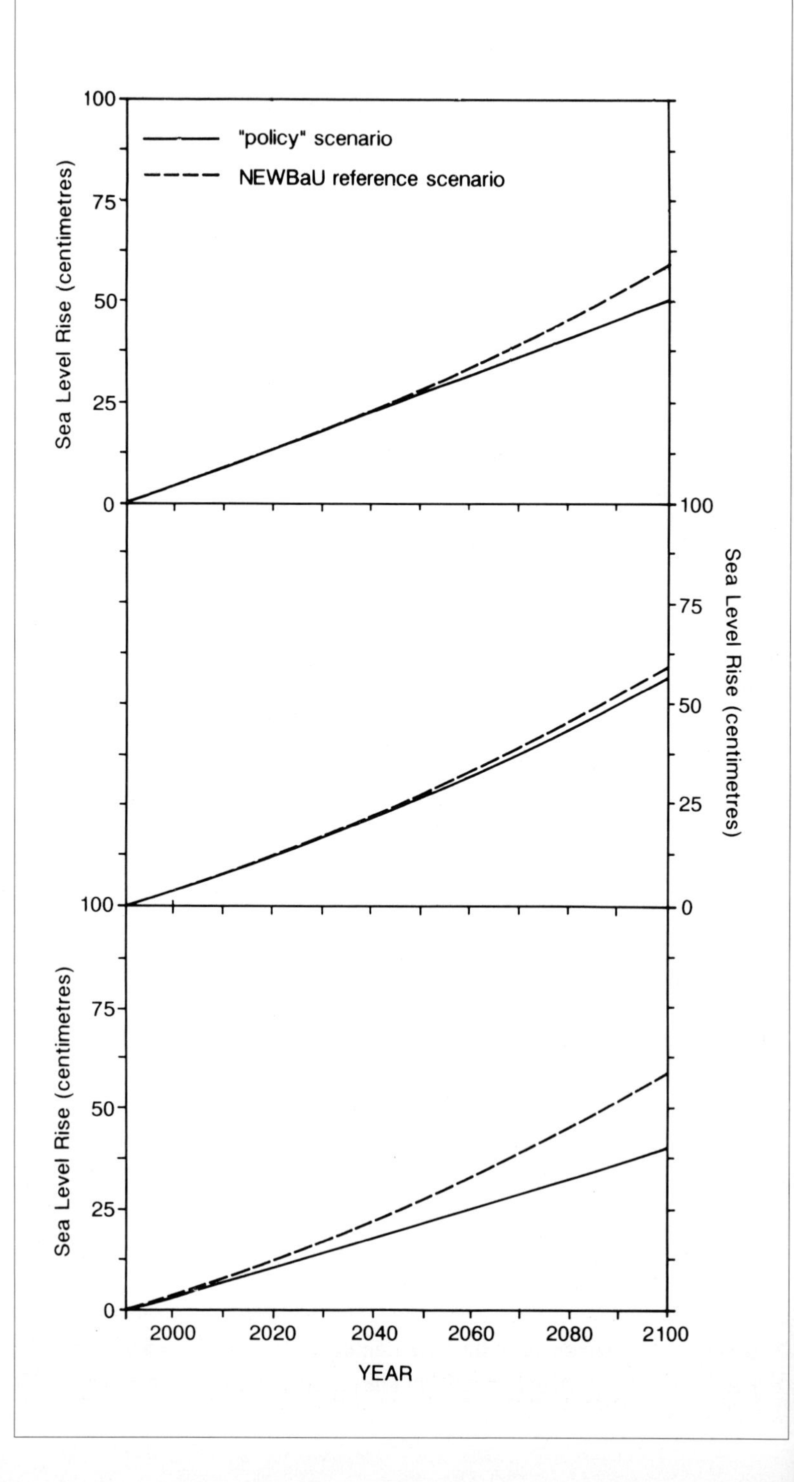

Figure 8. The effects of three hypothetical emission "policies" in reducing the rate of sea level rise: reducing OECD carbon emissions to zero by 2083 (top panel); stopping deforestation by 2010 and re-afforestion thereafter (middle panel); and halving the growth of global emissions of carbon dioxide, methane and nitrous oxide (bottom panel). All projections are compared to the modified IPCC Business-as-Usual scenario (dashed line). Neither strong OECD action nor strong deforestation controls would, alone, make much difference from the Business-as-Usual case. The lower panel suggests that concerted global cooperation would be required to make a significant reduction in the rate of sea level rise.

model-predicted sea level rise to about 40 cm by 2100 (a reduction of only about half of the "un-committed" sea level rise component) it is necessary to cut by half the projected growth of global CO_2, CH_4 and N_2O emissions, as compared to the NEWBaU scenario, (Figure 8, bottom panel). Clearly, such cuts are not possible without suppressing the high rate of growth of fossil fuels in the developing world exhibited over the last several decades (around 4.5% per annum). The crux of the problem is how to do this without also suppressing social, economic and technological development.

In short, international cooperation and participation in reducing emissions of greenhouse gases will be required in order to alter substantially a "business-as-usual" path of sea level rise in the future. Although most of the responsibility for the future sea level rise *commitment* falls directly into the lap of the most industrialised countries of the world, concerted action by those countries alone will do little to slow the overall rate of future rise. This means that, if the most dangerous outcomes are to be avoided, all countries, including developing countries, must cooperate in an international effort to reduce the rate of growth in greenhouse gas emissions. This fact must be grasped clearly by those in the Third World. At the same time, the industrialized nations must take greater responsibility for fostering "trajectories" of international development that can achieve dual objectives: improving the economic conditions of those in the South and ensuring future energy paths that are compatible with international goals of greenhouse gas emission reductions for the world as a whole.

Acknowledgements

Research support provided by the EPOCH Programme of the Commission of the European Communities is gratefully acknowledged. The authors wish to thank T.M.L. Wigley and S.C.B. Raper for use of their models. P.M. Kelly provided helpful comments on early drafts of this chapter.

References

Agarwal, A. and S. Narain, 1991: *Global Warming in an Unequal World: A Case of Environmental Colonialism.* Centre for Science and Environment, New Delhi.

Ali, S.I. and S. Huq, 1991: International Sea Level Rise: A National Assessment of Effects and Possible Responses for Bangladesh, prepared for Centre for Climate Change, University of Maryland, Maryland, USA.

Boden, T.A., P. Kanciruk and M.P. Farrell (eds), 1990: *Trends '90: A Compendium of Data on Global Change.* Carbon Dioxide Information Analysis Center, Oak Ridge National Laboratory, Oak Ridge, Tennessee.

Bolin, B., B.R. Doos, J. Jäger and R.A. Warrick (eds), 1986: *The Greenhouse Effect, Climatic Change and Ecosystems.* John Wiley and Sons, Chichester.

Brammer, H., 1989: Monitoring the evidence of the greenhouse effects and its impact on Bangladesh. In *Proceedings of the Conference on the Greeenhouse Effect and Coastal Area of Bangladesh*, ed. H. Moudud, A. Rahman et al., Dhaka, Bangladesh.

Brammer, H., 1992: The complexities of detailed impact assessment for the Ganges-Bramaputra-Meghna delta of Bangladesh. In *Climate and Sea Level Change: Observations, Projections and Implications,* ed. R.A. Warrick, E.M. Barrow and T.M.L. Wigley. Cambridge University Press, Cambridge.

Bretherton, F.P., K. Bryan and J.D. Woods, 1990: Time-dependent greenhouse-gas-induced climate change. In *Climate Change: The IPCC Scientific Assessment*, ed. J.T. Houghton, G.J. Jenkins and J.J. Ephraums. Cambridge University Press, Cambridge.

Broadus, J.M., 1992: Possible impacts of and adjustments to sea level rise: the cases of Bangladesh and Egypt. In *Climate and Sea Level Change: Observations, Projections and Implications,* ed. R.A. Warrick, E.M. Barrow and T.M.L. Wigley. Cambridge University Press, Cambridge.

Commonwealth Group of Experts, 1989: *Climate Change: Meeting the Challenge.* Commonwealth Secretariat, London.

Day, J.W., W.H. Conner, R. Costanza, G.P. Kemp and I.A. Mendelssohn, 1992: Impacts of sea level rise on coastal ecosystems. In *Climate and Sea Level Change: Observations, Projections and Implications,* ed. R.A. Warrick, E.M. Barrow and T.M.L. Wigley. Cambridge University Press, Cambridge.

Edwards, A.J., 1989: The implications of sea-level rise for the Republic of Maldives. *Report to the Commonwealth Expert Group on Climate Change and Sea Level Rise.* Centre for Tropical Coastal Management Studies, University of Newcastle-upon-Tyne.

Emery, K.O. and D.G. Aubrey, 1991: *Sea Levels, Land Levels and Tide-Gauges.* Springer Verlag, New York.

FPCO (Flood Plain Coordinating Organization), 1989: *Flood Action Plan.*

Frascetto, R. (ed), 1991: *Impact of Sea Level Rise on Cities and Regions.* Marsilio Editori, Venice.

Gagliano, S.M., K.J. Meyer-Arendt and K.M. Wicker, 1981: Land loss in the Mississippi River deltaic plain, *Trans. Gulf Coast Assoc. Geol. Soc.*, **31**, 295-300.

Hammond, A.L., E. Rodenburg and W.R. Moomaw, 1991: Calculating national accountability for climate change. *Environment,* **33**(1), 11.

Houghton, J.T., G.J. Jenkins and J.J. Ephraums (eds.), 1990: *Climate Change: The IPCC Scientific Assessment.* Cambridge University Press, Cambridge.

Huq, S & A.A. Rahman, 1990: Global Warming and Bangladesh: Implications and Response. In *Responding to the Threat of Global Warming: Options for the Pacific,* eds., D. G. Street and T.A. Siddiqi, Argonne National Laboratory, Argonne, USA, 1990.

Mahtab, F.U., 1989: *Effect of Climate Change and Sea Level Rise on Bangladesh.* Report for the Commonwealth Secretariat.

McCully, P., 1991: Discord in the greenhouse: how WRI is attempting to shift the blame for global warming. *The Ecologist*, **21**(4)157-165.

Milliman, J.D., J.M. Broadus and F. Gable, 1989: Environmental and economic impacts of rising sea level and subsiding deltas: the Nile and Bengal examples. *Ambio*, **18**(6), 340-345.

Mintzer, I., 1987: *A Matter of Degrees: Potential for Controlling the Greenhouse Effect.* World Resources Institute Research Report No. 5. World Resources Institute, Washington, D.C.

NRC (National Research Council), 1990: *Sea-Level Change*. National Academy Press, Washington, DC.

Peltier, W.R., 1986: Deglaciation-induced vertical motion of the North American continent and transient lower mantle rheology, *J. Geophys. Res.*, **91**, 9099-9123.

Pernetta, J.C. and G. Sestini, 1989: The Maldives and the impact of expected climatic changes, UNEP Regional Seas Reports and Studies No. 104, UNEP, Nairobi.

Pugh, D., 1992: Improving sea level data. In *Climate and Sea Level Change: Observations, Projections and Implications,* ed. R.A. Warrick, E.M. Barrow and T.M.L. Wigley, Cambridge University Press, Cambridge.

Rahman, A.A. and S. Huq, 1989: Greenhouse effect and Bangladesh: a conceptual framework. In *Proceedings of the Conference on the Greenhouse Effect and Coastal Area of Bangladesh,* eds. Moudud, H. et al., Dhaka.

Shine, K.P., R.G. Derwent, D.J. Wuebbles and J.J. Morcrette, 1990: Radiative forcing of climate. In *Climate Change: The IPCC Scientific Assessment*, ed. J.T. Houghton, G.J. Jenkins and J.J. Ephraums. Cambridge University Press, Cambridge.

Smith, K.R., 1991: Allocating responsibility for global warming: the Natural Debt Index. *Ambio*, **20**.

Stevenson, J.C., L.G. Ward and M.S. Kearney, 1986: Vertical accretion in marshes with varying rates of sea level rise. In *Estuarine Variability*, ed. D.A. Wolfe, Academic Press, New York.

Swart, R., 1991: Personal communication.

Titus, J.G., 1992: Regional impacts and responses. In *Climate and Sea Level Change: Observations, Projections and Implications,* ed. R.A. Warrick, E.M. Barrow and T.M.L. Wigley. Cambridge University Press, Cambridge.

Titus, J.G., R. Wedge, N. Psuty and J. Fancher (eds). 1990: *Changing Climate and the Coast,* Vol. 1 and 2. Report to the Intergovernmental Panel on Climate Change from the Miami Conference on Adaptive Responses to Sea Level Rise and Other Impacts of Global Climate Change. US Environmental Protection Agency, Washington, DC.

Warrick, R.A. and J. Oerlemans, 1990: Sea level rise. In *Climate Change: The IPCC Scientific Assessment*, ed. J.T. Houghton, G.J. Jenkins and J.J. Ephraums. Cambridge University Press, Cambridge.

Warrick, R.A., E.M. Barrow and T.M.L. Wigley (eds.), 1992: *Climate and Sea Level Change: Observations, Projections and Implications*. Cambridge University Press, Cambridge.

Watson, R.T., H. Rodhe, H. Oeschger and U. Siegenthaler, 1990: Greenhouse gases and aerosols. In *Climate Change: The IPCC Scientific Assessment*, ed. J.T. Houghton, G.J. Jenkins and J.J. Ephraums. Cambridge University Press, Cambridge.

Wigley, T.M.L., 1992: A simple inverse carbon cycle model. *Global Biogeochemical Cycles* (in press).

Wigley, T.M.L. and S.C.B. Raper, 1987: Thermal expansion of sea water associated with global warming. *Nature*, **330**, 127-131.

Wigley, T.M.L. and S.C.B. Raper, 1992: Future changes in global-mean temperature and sea level. In *Climate and Sea Level Change: Observations, Projections and Implications,* ed. R.A. Warrick, E.M. Barrow and T.M.L. Wigley. Cambridge University Press, Cambridge.

Wigley, T.M.L., T. Holt and S.C.B. Raper, 1991: *STUGE (an Interactive Greenhouse Model): User's Manual.* Climatic Research Unit, University of East Anglia, Norwich.

Woodroffe, C., 1989: Maldives and sea-level rise: an environmental perspective. Department of Geography, University of Wollongong (unpublished).

WRI, (World Resources Institute), 1990: *World Resources 1990-1991 — A Guide to the Global Environment.* Oxford University Press, Oxford.

CHAPTER 8

Effects of Climate Change on Food Production

Martin L. Parry and M.S. Swaminathan

Editor's Introduction

Like hunger, the stresses that arise from rapid climate change will fall most heavily on the poorest, the most vulnerable, and those least able to adopt new technology. Food supply is perhaps the arena in which the potential crisis could prove most severe. As Martin Parry and M.S. Swaminathan show in this chapter, global climate change will increase the stress on agricultural systems, potentially decreasing yields at the very time when demand for food is growing dramatically.

This chapter, an overview of the most likely impacts of climate change on food production, emphasizes that current scientific knowledge is insufficient to allow us to predict the regional impacts of climate change on food production with much confidence. The news is not all bad, however. Increased concentrations of carbon dioxide have had a fertilization effect on some crops in the laboratory, when all other environmental conditions have been tightly controlled. Unfortunately, field conditions cannot always be so carefully controlled. And not all plants benefit equally from this effect. Wheat and barley have shown strongly positive effects; corn, a less positive response.

Parry and Swaminathan also note some strong and troubling trends in recent research results. The more pronounced warming in high latitude areas that is predicted by climate models will extend the Northern range of agricultural potential. But it may also decrease grain yields in the core of traditional agricultural areas. In the low-latitude countries, changes in temperature regimes will require farmers to alter crop-timing patterns that have been fine-tuned over centuries of experience with traditional regional climates.

Perhaps more important for food production, climate model results also suggest that soil moisture will decrease in some already stressed areas as global temperatures rise. The areas most likely to see decreases in yield include North and West Africa, parts of Eastern Europe, North and Central China, Western Australia, and Eastern Brazil. In most cases, these are areas already strained by the pace of economic (and in some cases political) changes that make rapid, well-coordinated institutional responses unlikely.

But the most important impacts may be due to the least predictable aspects of the climate system — changes in the frequency, distribution, and severity of extreme weather events. (See further Mitchell and Ericksen (1992), this volume.) In a world warmed by the buildup of greenhouse gases, the number of days with temperatures above 30 °C is likely to increase significantly while the number of days below 0 °C is likely to decline. Such changes, along with changes in the frequency of floods, droughts, and wind storms, will necessitate major adjustments in the agricultural sector to prevent significant decreases in yield.

Martin Parry, Director of the Environmental Change Unit at the University of Oxford University, has been studying the effects of climate change on agriculture for more than a decade. His has been one of the few clear and stable voices urging caution, systematic analysis and careful preparation of adaptive response strategies to minimize the risks of rapid climate change. M.S. Swaminathan, former President of IUCN, is an eloquent spokesman for the shared requirements of environment and development. Swaminathan, as former director of the International Institute for Rice Research in Manila, has made major scientific contributions to the development of practical techniques to enhance agricultural yields. Between them, these two leading scientists bring an unmatched depth of vision and compassion to this issue.

Parry and Swaminathan conclude that global food production can be sustained at levels that might occur in the absence of climate change, but only at a high cost. Vast new areas may have to be brought under cultivation. They caution that the gains in production in some traditional agricultural areas would be unlikely to balance possible large-scale reductions in these areas. Adequate response strategies may require changes in land use and tenure patterns, changes in management approach, and changes in crop and livestock husbandry. Successfully integrating these responses in ways which do not increase cross-border tensions within and between regions may produce challenges to political leadership that are every bit as difficult as the problem of breeding better crops for the new world ahead.

If we do not solve these food production problems, the results may well be devastating—for reasons that link this chapter with Nathan Keyfitz' work on population (1992; this volume). By the end of the next century, the human population of our planet is likely to be at least twice as large as it is today. But even at today's level of about 5.2 billion, the human race is hard pressed to keep itself well-fed. Current annual rates of food production (approximately 5.2×10^{15} Calories per year), could provide enough food for approximately 6 billion people, if all the food were shared equally and nearly all the caloric input came from grains. (Each person would receive about 2350 Calories per day, the UN recommended daily dietary input.) If the world's current food output were used to provide every person with a diet like that typical in South America—containing about 10% of the caloric input from animal products—only 4 billion people could be fed at today's level of production. If everyone were to eat on a European-style diet, with about 30% of the calories from animal products, only 2.5 billion people could be fed from the food produced worldwide today (World Resources Institute, 1988).

Of course, the world's food output is not distributed equally—neither within countries nor between countries. Kates et al. (1988) note that among 46 countries where food consumption levels were below those needed to support a healthy, productive life, 43 were in sub-Saharan Africa, Asia, Latin America or the Caribbean. A World Bank study (1986) of 87 developing countries, indicated that in 1980, 34% of the population or nearly 730 million people, got less than 90% of the FAO/WHO minimum daily requirements—not enough calories to support an active working life. What's more distressing, 16% (nearly 340 million people) received less than the amount of food required each day to prevent stunted growth and major health problems.

Even in poor countries, hunger is unevenly distributed. The Fifth World Food Survey (FAO, 1985) noted that in Bangladesh, for example, urban dwellers are better fed than their rural cousins. In rural areas, landless peasants averaged about 1900 Calories per day while members of household owning more than 7 hectares received an average of almost 2400 Calories per day.

In 1980, nearly 70% of the world's arable land was under the plough. Dramatic increases in productivity would be necessary in both food production and in the management of food distribution systems, just to keep pace with expected population growth over the next century. In other words, the situation is worthy of concern even without the threat of climate change; Parry and Swaminathan, in this chapter, put the challenges of maintaining food production and the risks of rapid climate change, finally, in perspective to each other.

- I. M. M.

1 Introduction

The purpose of this chapter is to consider the likely effects of changes of climate on the ability of the world to feed its population. We consider, first, aspects of climate change that will be most important in their effect on agricultural potential. Secondly, we discuss the "direct" or fertilization effect of CO_2 on crop production. This is followed by a discussion of the possible shifts in potential production areas that could result from geographical shifts of climate, and the changes in yield levels that may occur in these regions. The chapter concludes with an overall assessment of the consequences of climate change for food production. We outline the responses that should be taken now, to minimize the negative effects and maximize the potentially positive benefits of climate change.

We conclude that changes of climate could provoke serious reductions in food production potential, particularly in regions that are at present vulnerable to climate variability. The most vulnerable areas include traditional agricultural zones in the semi-arid and humid tropics. Many of these regions are characterized by low incomes, and have a limited ability to adapt through technological change. Sharing of knowledge and technological assistance with these regions by those in the advanced industrialized societies will be critically important. The co-development of agricultural technologies specifically adapted to local conditions will strengthen each nation's capabilities to respond and adapt to food shortages and new agricultural needs.

We note that the rate of future climate change will be critical in determining the sign and magnitude of climate change impacts. The costs will be made high if the changes are rapid, or if the perturbations in local weather are unstable and erratic. And, if the frequency, severity, and duration of extreme weather events increases, the risks to farmers and farm economies will be amplified.

Farmers world wide are generally conservative people. They prefer to continue what has worked well in the past, rather than to embrace radical change. If the patterns of regional climate change gradually, farmers and the institutions that support them will respond positively and adapt successfully. Some will diversify their holdings and their behaviour. Others will switch to new crops and alternative cultivars. Thus, we expect that if changes occur slowly, and with an easily predictable pattern, losses will be minimized.

2 The Most Threatening Types of Climate Change

Although increases in average temperature world-wide are the most predictable consequence of greenhouse gas emissions, many other aspects of global and regional climate will also change. Some of these changes will be much more significant than others, with respect to agricultural output.

2.1 Climatic extremes

It is not clear whether changes in the variability of temperature will occur as a result of climate change. However, even if variability remains unaltered, an increase in average temperatures would result in the increased frequency of temperatures above particular thresholds. Changes in the frequency and distribution of precipitation are less predictable, but the Intergovernmental Panel on Climate Change (IPCC) concluded that the combination of elevated temperatures, and drought or flood, probably constitutes the greatest risk to agriculture (Parry and Duinker, 1990).

2.2 Warming in high latitudes

There is relatively strong agreement among general circulation model (GCM) predictions that greenhouse-gas induced warming will be greater at high latitudes (IPCC, 1990). This will reduce temperature constraints on high latitude agriculture, and increase the competition for land here, by bringing previously uncultivable land within the range that can support traditional cultivars (Parry and Duinker, 1990). Warming at low latitudes, although less pronounced, is also likely to have a significant impact on agriculture — to the extent that it alters availability of soil moisture or raises temperatures above the threshold of tolerance for key crop cultivars.

2.3 Poleward advance of monsoon rainfall

In a warmer world, the intertropical convergence zones and polar frontal zones may advance further poleward, as a result of an enhanced ocean-continent pressure gradient. Because air over the continental land masses will warm more rapidly than air over the oceans, the key zones of interaction between weather fronts are likely to advance poleward. If this were to occur, then total rainfall could increase in some areas of Asia, Africa and Australia that are now exposed to monsoon rainfall, but there is currently little agreement on which regions might be most affected (IPCC, 1990). Rainfall episodes could also be more intense, and thus flooding and erosion could increase.

2.4 Reduced soil water availability

Probably the most important consequences for agriculture could stem from higher potential evapotranspiration (the process by which water returns from plants and soils to the atmosphere). Enhanced evapotranspiration would be primarily a consequence of higher air and land surface temperatures. Even in the tropics, where temperature increases are expected to be smaller than elsewhere, and where precipitation might increase, the increased rate of loss of moisture from plants and soil could be considerable (Parry, 1990; Rind et al., 1989). Such losses may be somewhat reduced by greater air humidity and increased cloudiness during the rainy seasons, but could be pronounced in the dry seasons.

3 Potential Direct Effects of Greenhouse Gases on Crops and Livestock

3.1 Effects of CO_2 increases on plant growth

The impacts of elevated concentrations of carbon dioxide (CO_2) on crop growth and yield have been studied for nearly 100 years (Wittwer, 1985). The effects of other greenhouse gases — methane, dinitrogen oxides and chlorofluorocarbons — have yet to receive similar attention. Early experiments that increased the CO_2 concentration in greenhouses resulted in yield enhancement in lettuce, tomatoes, cucumbers and several ornamental plants. Kimball (1985) assembled 770 observations on the impact of higher levels of atmospheric CO_2 in crop yield, and found that the yield increase overall was of the order of 32% (if all other growth-related factors were held constant).

Why does this take place? The most direct effect of high levels of CO_2 that has been verified in plants is an increase in leaf and canopy photosynthetic rates. It has been observed that an increase in photosynthesis may continue at concentrations up to four times the pre-industrial level of CO_2, or 1000 parts per million (Wittwer, 1985). Increased metabolic activity in plants may lead to conversion of up to 51% of the CO_2 that was transformed into carbohydrates being oxidized through photorespiration, and released again as CO_2. In the long term, photorespiration rates are likely to decrease with a rise in atmospheric CO_2 levels.

An increase in biomass production invariably occurs with CO_2 enrichment. This may not always come from an increase in net photosynthesis. Increasing CO_2 concentration leads to a partial closure of leaf stomates. This reduces leaf transpiration rates, and promotes water use efficiency. Increased water use efficiency is an important outcome of higher levels of CO_2. Some studies have shown that the yields of wheat varieties subjected to water stress were as high or even higher at high CO_2 levels, as compared to wheat grown with adequate irrigation but at normal CO_2 levels (Gifford, 1979, Havelka et al., 1984).

3.1.1 C3 and C4 plants

The metabolic processes of several hundred thousands of plant species have been described and analysed by scientists. These plants are categorized by the particular pattern of photosynthetic processes that each plant uses to convert CO_2 and H_2O to carbohydrates. The two most important groups are called C3 and C4 plants. These terms reflect the number of carbon atoms characteristically involved in photosynthesis.

Of the many thousands of plants on the planet, global food security depends only on about 20 plant species. Of these, 16 have a C3 photosynthetic pathway. C3 plants include: rice, wheat, barley, oats, mung bean, cowpea, chickpea, pigeon pea, potato, sweet potato, cassava, sugar beet, banana and coconut. Most fruits and vegetables are C3 plants. In general, C3 crops exhibit a positive response to higher levels of CO_2. Cure (1985) has reported a yield increase of nearly 40% with CO_2 enrichment in wheat and barley. In C3 species, increases in leaf area, weight per unit area, leaf thickness, stem height, seed and fruit numbers and weight have been observed to increase at higher CO_2 levels. The carbon-nitrogen (C:N) ratio has been found to increase and the harvest index has generally been altered in favour of the economically important part of the plant.

Maize, by contrast, is a C4 plant. In maize, higher than normal levels of atmospheric CO_2 have been reported to induce greater water use efficiency (Rogers et al., 1983).

Though stomatal conductance declines with increasing CO_2 concentration in both C3 and C4 plants, C3 plants seem to benefit in dry matter production more from higher CO_2 level due to the following reasons:

a. higher leaf expansion
b. increase in the photosynthetic rate per unit leaf area and
c. increase in water use efficiency.

Photorespiration rates may decrease with an increase in atmospheric CO_2.

3.1.2 The limits of knowledge about CO_2 effects

These results would tend to suggest that a higher CO_2 content in the atmosphere may be an advantage, particularly in areas where water may be a limiting factor. Unfortunately, most studies on direct CO_2 effects have involved a CO_2 concentration of 600 ppm. We need studies using levels ranging from 450 to 500 ppm.

More important, under field conditions, the anticipated gains may not be realised due to the complex nature of the linkages involved between the biological effects of photosynthetic efficiency and temperature, sunlight and moisture availability. Furthermore, the optimistic results of the laboratory studies presume that rainfall conditions and soil moisture availability will remain unchanged in a CO_2-enhanced world. This is not necessarily a good assumption. The combination of increased atmospheric CO_2 with, for example, reduced rainfall (or reduced soil moisture availability at critical periods in the growth cycle) could result in no change or in significant yield reductions. In some cases, the integrated impact of a rise in temperature and CO_2 concentration on the yield of wheat and rice may be negative (Sinha and Swaminathan, 1991). Under conditions in North India, every 0.5 °C increase in temperature resulted in a drop in the duration of wheat crop by 7 days, which in turn reduced yield by 0.45 tonnes per hectare (t/ha). If mean temperatures rise by 1-2 °C, the adverse impact on yield will be even greater.

4 Potential Effects of Climate Change

There are two broad types of effect that changes of climate may have on agriculture: first, effects on the geographical limits to the regions where different types of crops and livestock can be produced; and secondly, effects on the potential yields of crops and livestock in these regions.

4.1 Effects on geographical limits to agriculture

4.1.1 Changes in thermal limits to agriculture

Increases in temperature can be expected to lengthen the growing season in areas where agricultural potential is currently limited by insufficient warmth — resulting in a poleward shift of thermal limits of agriculture. The consequent extension of potential will be most pronounced in the northern hemisphere, because of the greater extent here of temperate agriculture at higher latitudes.

There may, however, be important regional variations in our ability to exploit this shift. For example, the greater potential for exploitation of northern soils in Siberia than on the Canadian Shield may mean relatively greater increases in potential in northern Asia than in northern N. America (Parry, 1990).

A number of estimations have been made concerning the northward shift in productive potential in mid-latitude northern hemisphere countries. These relate to changes in the climatic limits for specific crops under a variety of climatic scenarios, and are therefore not readily compatible (Newman, 1980; Blasing and Solomon, 1983; Rosenzweig, 1985; Williams and Oakes, 1978; Parry and Carter, 1988; Parry, et al., 1989). They suggest, however, that a 1 °C increase in mean annual temperature would tend to advance the thermal limit of cereal cropping in the mid-latitude northern hemisphere by about 150-200 km, and to raise the altitudinal limit to arable agriculture by about 150-200 m.

While warming may extend the margin of potential cropping and grazing in mid-latitude regions, it may reduce yield potential in the core areas of current production. This is because higher temperatures encourage more rapid maturation of plants and shorten the period of grain filling (Parry and Duinker, 1990). An important additional effect, especially in temperate mid-latitudes, is likely to be the reduction of winter chilling (vernalization). Many temperate crops (such as barley and oats) require a period of low temperatures in winter, to either initiate or accelerate the flowering process. Reduced vernalization results in low flower bud initiation and, ultimately, reduced yields. It has been estimated that a 1 °C warming would reduce effective winter chilling by between 10 and 30%, thus contributing to a poleward shift of temperate crops (Salinger, 1989).

Increases in temperature are also likely to affect the crop calendar in low latitude regions, particularly where more than one crop is harvested each year. For example, in Sri Lanka and Thailand a 1 °C warming would probably require a substantial re-arrangement of the current crop calendar, which is finely tuned to present climatic conditions (Kaida and Surarerks, 1984; Yoshino, 1984).

4.1.2 Shifts of moisture limits to agriculture

There is much less agreement between GCM-based projections concerning greenhouse gas-induced changes in precipitation, than there is about temperature - not only concerning changes of magnitude, but also of spatial pattern and distribution through the year. For this reason it is difficult to identify potential shifts in the moisture limits to agriculture. This is particularly so because relatively small changes in the seasonal distribution of rainfall can have disproportionately large effects on soil moisture, and thus on the viability of agriculture in tropical areas. This influence takes place largely through changes in growing period when moisture is sufficient, and thus through the timing of critical episodes such as planting.

However, recent surveys for the IPCC have made a preliminary identification of those regions where there is some agreement among the GCM experiments, concerning the regional implications for soil water availability if a doubled level of atmospheric CO_2 comes to pass (Parry, 1990; Parry and Duinker, 1990). It should be emphasized that coincidence of results for these regions is not statistically significant. The regions where soil water may be reduced are shown in Table 1.

4.1.3 Regions affected by drought, heat stress and other extremes

Probably among the most important impacts of global warming for agriculture, but about which least is known, are the possible changes in climatic extremes — such as the magnitude and frequency of drought, storms, heat waves and severe frosts (Rind et al., 1989; Mitchell and Ericksen, 1992, this volume). Some modelling evidence suggests that hurricane intensities will increase with climatic warming (Emanuel, 1987). This has important implications for agriculture in low latitudes, particularly in coastal regions.

Table 1. Regions where soil water may be reduced by global climate change.

	Decreases of soil water in Dec, Jan, Feb	Decreases of soil water in June, July, Aug
Africa	North-east Africa Southern Africa	North Africa West Africa
Europe		parts of Eastern Europe
Asia	Western Arabian Peninsula Southeast Asia	North and central China Parts of Soviet central Asia and Siberia
Australasia	Eastern Australia	Western Australia
N. America	Southern USA	Southern USA and Central America
S. America	Argentine Pampas	Eastern Brazil

The effects of heat or cold stress on crops are often disproportionate to the magnitude of the stress itself, and crop yields often exhibit a nonlinear response to heat or cold stress. Thus, changes in the probability of extreme temperature events can be significant (Mearns et al., 1984; Parry, 1976). In addition, even assuming no change in the standard deviation of temperature maxima and minima, we should note that the frequency of hot and cold days can be markedly altered by changes in mean monthly temperature. To illustrate, in a climate where CO_2 levels had doubled (a 2 x CO_2 climate), the number of days in which temperatures would fall below freezing would decrease from a current average of 39 to 20 in Atlanta, Georgia (USA), while the number of days above 90 °F would increase from 17 to 53 (EPA, 1989).

The frequency and extent of area over which losses of agricultural output could result from heat stress, particularly in tropical regions, is therefore likely to increase significantly. Unfortunately, no studies have yet been made of this. However, the apparently small increases in mean annual temperatures in tropical regions (about 1 to 2 °C under a 2 x CO_2 climate) could sufficiently increase heat stress on temperate crops such as wheat, so that these are no longer suited to such areas. Important wheat-producing areas such as North India could be affected in this way (Parry and Duinker, 1990).

There is a distinct possibility that, as a result of high rates of evapotranspiration, some regions in the tropics and subtropics could be characterized by a higher frequency of drought, or a similar frequency of more intense drought, than at present. Current uncertainties about how regional patterns of rainfall will alter mean that no useful prediction of this can at present be made. However, it is clear in some regions that relatively small decreases in water availability can readily produce drought conditions. In India, for example, lower-than-average rainfall in 1987 reduced food grains production from 152 to 134 million tonnes (Mt), lowering food buffer stocks from 23 to 9 mt. Changes in the risk and intensity of drought, especially in currently drought-prone regions, represent potentially the most serious impact of climatic change on agriculture both at the global and the regional level.

4.1.4 Effects on the distribution of agricultural pests and diseases

Studies suggest that temperature increases may extend the geographic range of some insect pests currently limited by temperature (EPA, 1989; Hill and Dymock, 1989; Porter et al., 1992). As with crops, such effects would probably be greatest at higher latitudes. The number of generations per year produced by multivoltine (i.e. multigenerational) pests would increase, with earlier establishment of pest populations in the growing season and increased abundance during more susceptible stages of growth.

An important unknown, however, is the effect that changes in precipitation amount and air humidity may have on the insect pests themselves — and on their predators, parasites and diseases. Climate change may significantly influence interspecies interactions between pests and their predators and parasites.

Under a warmer climate at mid-latitudes there would be an increase in the over-wintering range and population density of a number of important agricultural pests, such as the potato leafhopper — which is a serious pest of soybeans and other crops in the USA (EPA, 1989). Assuming planting dates did not change, warmer temperatures would lead to invasions earlier in the growing season, and probably to greater damage to crops. In the US Corn Belt, increased damage to soybeans is also expected, due to earlier infestation by the corn earworm.

Examination of the effect of climatic warming on the distribution of livestock diseases suggests that those at present limited to tropical countries, such as Rift Valley fever and African Swine fever, may spread into the mid-latitudes. For example, the horn fly, which currently causes losses of $730.3 million in the US beef and dairy cattle industries, might extend its range under a warmer climate — leading to reduced weight gain in beef cattle and a significant reduction in milk production (Drummond, 1987; EPA, 1989).

In cool temperate regions, where insect pests and diseases are not generally serious at present, damage is likely to increase under warmer conditions. In Iceland, for example, potato blight currently does little damage to potato crops, because it is limited by the low summer temperatures. However, under a 2 x CO_2 climate, those summer temperatures may be 4 °C warmer than at present. As a result, crop losses to disease may increase by up to 15% (Bergthorsson et al., 1988).

Most agricultural diseases have greater potential to reach severe levels under warmer and more humid conditions (Beresford and Fullerton, 1989). Cereals would be more prone to diseases such as Septoria. In addition, increases in population levels of disease vectors may well lead to increased epidemics of the diseases they carry. To illustrate, increases in infestations of the Bird Cherry aphid (*Rhopalosiphum padi*) or Grain aphid (*Sitobian avenae*) could lead to increased incidence of Barley Yellow Dwarf Virus in cereals.

4.2 *Effects on crop yields*

4.2.1 In humid and semi-arid low latitude regions

An analysis of more than 100 cooperative trials — conducted by the International Rice Research Institute (IRRI) under irrigated conditions in 22 countries over eight years — indicates that rice yields are significantly affected by temperature and solar radiation. When the minimum temperature increases from 18 to 19 °C, there is a decrease in yield of about 0.7 tonnes per hectare (t/ha). At a range of sites, varying in latitude from 31° N to 6° N, yields drop by 10-20% with a rise in temperature of 1-2 °C during the November-December harvest period. Yields are most severely affected at higher latitudes.

Re-adjustment of the harvest period prior to the onset of higher temperatures would help, but this would be difficult where the crop season has to be tailored to the monsoon rainfall period. Flowering response in crops to integrated factors needs study, since this is a major determinant of yields. Because yields are largely determined by flowering responses, it would be useful to study the integrated effects of increased temperatures, elevated CO_2 levels, and altered rainfall patterns on overall yields.

Considerable genetic variability occurs among genotypes of rice, wheat and other plants in relation to tolerance to several biotic and abiotic stresses. More studies at the micro-level are required, in order to understand the interrelationships between CO_2 concentration, temperature and precipitation. We also need integrated studies to understand the impact of other greenhouse gases on yield. It is possible to assemble genes from a wide range of land races through a properly planned breeding programme.

Another aspect of this situation which needs careful study is the impact of changes in CO_2 levels, temperature, precipitation, ultraviolet-B radiation on the incidence of pests. Most of the studies on the direct impact of CO_2 have been carried out under conditions of optimum pest management. Pest and disease problems are more severe in tropical regions, due to multiple cropping and the availability of alternate hosts throughout the year (Table 2). Temperature influences the multiplication patterns of insect pests. For example, the seed potato industry in northern India is based on the fact that aphid transmitted virus diseases do not occur during November-December. Hence, the potato crop raised during this aphid-free season could be kept for seed purposes.

Weeds take a heavy toll on crop yield. The widely prevalent weeds that most severely affect C3 crop plants tend to incorporate the C4 metabolic pathway. Conversely, for a C4 plant such as maize, 19 of the 38 major weeds reported in the United States are C3 plants. Some of the serious weeds, insect pests and pathogens of today may become less impor-

Table 2. Crop disease in temperate and tropical regions

	Number of diseases reported	
Crop	**Temperate**	**Tropical**
Rice	54	500-600
Maize	85	125
Citrus	50	248
Tomato	32	278
Beans	52	250-280

Source: Swaminathan, 1986.

tant under altered climate conditions, and some of the less important ones may become more serious. Experience with high yielding varieties of rice in Asia shows that, when the micro-environment is altered through irrigation and fertilizer application, pests which were not important before, such as the brown plant hopper, become a serious concern.

In the semi-arid tropics, farmers face enormous uncertainty about future yields. That uncertainty stems from our ignorance of future changes in precipitation - both its quantity and its distribution over time and space. For example, a series of studies conducted at the International Crops Research Institute for the Semi-Arid Tropics (ICRSAT) modelled the impact of climate on the yield of sorghum in two semi-arid, tropical regions of India. The two regions, otherwise, had contrasting climate: Anantapur is dry with a highly variable, erratic rainfall, while Hyderabad experiences a relatively dependable, 6-month rainy season. The study, using a sorghum growth model developed at Texas A&M University, found that yields increased when both rainfall and temperature increased. (In Antanapur, under those conditions, yields would more than triple.) On the other hand, if temperature increase alone, without any related changes in rainfall, would lead to more than 10% losses of grain and dry matter in both locales. This is due to increased transpiration and subsequent soil moisture stress.

In general, much depends upon whether increases in temperature are accompanied by increases in available moisture. Increases in temperature, — even the relatively small increases of about 1.5 °C projected in lower latitudes for a doubling of greenhouse gases — would increase rates of evapotranspiration by 5 to 15%. If there were not compensating increases in rainfall, this would tend to reduce yields. In northern India, for example, temperature increases of even 0.5 °C probably would reduce wheat yields due to heat stress, by about 10% if rainfall did not increase (Van Diepen et al., 1987). The effects on maize in semi-arid areas may be similar. In central China, for example, maize yields are estimated to decrease on average by 3% per 1 °C (Terjung et al., 1989).

No impact studies based on climatic scenarios from global climate model (GCM) experiments have, at the time of writing, been made in Africa, primarily because there is little agreement between GCMs concerning changes in rainfall — in this region as in others, rainfall is the climatic variable that most determines variations in agricultural yield. Some preliminary work has been completed in South Africa, indicating that an increase in mean annual temperatures of 2 to 4 °C in Natal Province (consistent with the range of GCM projections for 2 x CO_2 climates) would increase rates of evapotranspiration by 5% to 15% and reduce plant productivity by about the same amount (Schulze, personal communication.).

In Kenya, a recent case study considered the effects of the driest 10% of years from the recent past. These suggest reductions of maize yields by 30%-70% (Akong'a et al., 1988). Any changes in the frequency of such years would substantially affect the average output of agriculture in the region, but we cannot at present estimate how these might change.

4.2.2 In the midlatitude grainbelts

One of the most serious threats to world food supply could result from decreased productive potential in the mid-latitude "grainbelt" (i.e. the grain-exporting regions of North America, Australia, Southern Africa, South America, and the southern regions of what was the USSR. In the USA increased temperatures and reduced crop water availability are estimated to lead to a decrease of yields of all the major unirrigated crops (EPA, 1989). The largest reductions are projected for the south and south-east. In the most northern areas of the US, however, where temperature is currently a constraint on growth, yields of unirrigated maize and soybeans could increase as higher temperatures increase the length of the available growing season.

When the direct effects of increased CO_2 are considered, yields may increase more generally in northern areas; but the production of most crops in the US will probably be reduced because of heat stress and diminished rainfall. The largest reductions are expected to be in sorghum (-20%), corn (-13%) and rice (-11%), with an estimated fall in net value of agricultural output of $33 billion. If this occurred, American consumers would face slightly higher prices, although supplies are estimated to meet current and projected demand. However, exports of agricultural commodities could decline by up to 70%, and this could have a substantial effect on the pattern of world food trade (EPA, 1989).

On the Canadian prairies, where growing season temperatures under a 2 x CO_2 equilibrium climate would be about 3.5 °C higher than today, uncertain changes in rainfall could decrease average potential yields by 10 to 30%. Spring wheat yields in Saskatchewan are estimated to fall by about 30% (Williams et al., 1988). Since Saskatchewan at present produces 18% of all the world's traded wheat, such a reduction could have global implications.

In Ontario, even sizeable increases in precipitation would be offset by increases in evapotranspiration — with consequent increased moisture stress on crops. Maize and soybean would thus become very risky in the southern part of the province. In the north, where maize and soybean cannot currently be grown commercially because of inadequate warmth, cultivation may become profitable but this is not expected to compensate for reduced potential further south.

In Australia, wheat production could increase, assuming increased summer rainfall, decreased winter rainfall, and a general warming of 3 °C. Increases are expected in all states except Western Australia, where more aridity might cause a significant reduction in output (Pittock, 1989). The major impact would probably be on the drier frontiers of arable cropping. For example, increases in rainfall in subtropical northern Australia could result in increased sorghum production at the expense of wheat. Increased heat stress might shift livestock farming and wool production southward with sheep possibly replacing arable farming in some southern regions. Many areas currently under fruit production would no longer be suitable under a 3 °C warming, and would need to shift southwards or to higher elevations in order to maintain present levels of production.

All of these changes would also be affected by changes in the distribution of diseases and pests.

4.2.3 In northern marginal regions

Some of the most pronounced effects on agriculture would be likely to occur in high latitude regions, because GHG-induced warming is projected to be greatest here and because this warming could remove current thermal constraints on farming. Inappropriate terrain and soils are, however, likely to limit the increase in extent of the farmed area. Thus, in global terms, production increases would probably be small (Parry, 1990). A summary of available information is given below.

In Iceland, with mean annual temperatures increased by perhaps 4.0 °C, hay yields on improved pastures would increase by about two-thirds and herbage on unimproved rangelands by about a half (Bergthorsson, et al., 1988). The numbers of sheep that could be carried on the pastures would be raised by about 250%, and on the rangelands by two-thirds, if the average carcass weight of sheep and lambs is maintained as at present. At a guess, output of Icelandic agriculture could as much as double with a warming of 4 °C.

In Finland, an increase in summer warmth by about a third and precipitation by about half (consistent with a 2 x CO_2 climate) could increase barley and spring wheat yields by about 10% in the south of the country, but slightly more in the north — due to relatively greater warming and lower present-day yields (Kettunen et al., 1988).

The only other northern region for which an integrated impact assessment has been completed is in north European Russia. Here a 2.5 °C warming, (consistent with a 2 x CO_2 climate), is estimated to decrease winter rye yields by about a quarter, due to faster growth and increased heat stress under the higher temperatures (Pitovranov et al., 1988). However, crops such as winter wheat and maize, which are currently low yielding because of the relatively short growing season in these regions, are better able to exploit the higher temperatures and exhibit yield increases.

Production could increase in regions currently near the low-temperature limit of grain growing. In the northern hemisphere, those regions would include the northern Prairies, Scandinavia, and north European Russia; in the southern hemisphere, southern New Zealand, and southern parts of Argentina and Chile could be affected. But it is reasonably clear that, because of the limited area unconstrained by inappropriate soils and terrain, increased high-latitude output will probably not compensate for reduced output at mid-latitudes. The implications of this for global food supply and food security are considered later in this chapter.

4.3 Effects on livestock carrying capacity

Here, the information is extremely scant. The only model-based study using GCM-derived scenarios of future climate has been completed in Iceland. There, increases in mean annual temperature of 4 °C and in rainfall of 15% are estimated to increase the carrying capacity of sheep on improved grassland by about two-and-a-half times, and on rough pasture by more than a half (Bergthorsson et al., 1988).

Few other regions, however, are currently as constrained by inadequate warmth as Iceland. One other beneficiary might be Patagonia, in the southern part of Argentina and Chile, where grass production and cattle grazing are limited by temperature rather than rainfall. Further north, in the southern pampas of Argentina, increases in rainfall would be needed to compensate for the higher rates of evapotranspiration (about 10%) that would stem from higher mean temperatures (2 °C to 4 °C under the GCM 2 x CO_2 climates).

The productivity of the rangelands of Africa depends almost wholly on the amount and timing of rainfall. In Kenya, for example, forage yield in the driest 10% of years is reduced by 15% — to 60% — from its average (Akong'a, et al., 1986). In these dry years, the carrying capacity of livestock can fall by 10 to 40% and milk yields to zero.

Projected increases in summer rainfall in eastern Australia are expected to increase grass growth, but this is likely to be offset by the poorer nutritive value of tropical compared with temperate species. Loss of the Mediterranean-type climatic zones of Victoria and Western Australia — which are the current principal lamb and wool producing areas — together with increased heat stress of both cattle and sheep, could mean that livestock productivity would decrease (Pearman, 1988).

5 Likely Effects on World Food Supply and Food Security

Although, on average, global food supply currently exceeds demand by about 10 to 20%, its year-to-year variation (which is about ±10%) can reduce supply in certain years to levels where it is barely sufficient to meet requirements. In addition, there are major regional variations in the balance between supply and demand, with perhaps a billion people (about 15% of the world's population) not having secure access to sufficient quantity or quality of food to lead fully productive lives.

For this reason, the working group on food security at the 1988 Toronto Conference on The Changing Atmosphere concluded that: "While averaged global food supplies may not be seriously threatened, unless appropriate action is taken to anticipate climate change and adapt to it, serious regional and year-to-year food shortages may result, with particular impact on vulnerable groups" (Parry and Sinha, 1988). Statements such as this are, however, based more on intuition than on knowledge derived from specific study of the possible impact of climate change on the food supply. No such study has yet been completed, although one is currently being conducted by the US Environmental Protection Agency and is due to report in 1992.

The information available at present is extremely limited. It has for example, been estimated that increased costs of food production due to climate change could reduce per capita global GNP by a few percentage points (Schelling, 1983). Others have argued that technological changes in agriculture will override any negative effects of climate changes and, at the global level, there is no compelling evidence that food supplies will be radically diminished (Crosson, 1989). Recent reviews have tended to conclude, however, that at a regional level, food security could be seriously threatened by climate change — particularly in less developed countries in the semi-arid and humid tropics (Sinha et al., 1988; Parry, 1990; Parry and Duinker, 1990).

Analyses conducted for the IPCC, designed to test the sensitivity of the world food system to changes of climate, indicate what magnitudes and rates of climatic change could possibly be absorbed without severe impact and, alternatively, what magnitudes and rates could seriously perturb the system (Parry, 1990; Parry and Duinker, 1990). These suggest that yield reductions of up to 20% in the major mid-latitude grain exporting regions could be tolerated without a major interruption of global food supplies. However, the increase in food prices (perhaps, as much as 7% under a 10% yield reduction) could seriously influence the ability of food-deficit countries to pay for food imports, eroding the amount of foreign currency available for promoting development of their non-agricultural sectors. It should be emphasized that these analyses are preliminary; more work is necessary before we have an adequate picture of the resilience of the world food system to climatic change.

On balance, the evidence is that food production at the global level can, in the face of estimated changes of climate, be sustained at levels that would occur without a change of climate, but the cost of achieving this is unclear. It could be very large. As we have seen, increases in productive potential at high mid-latitudes and high latitudes, while being of regional importance, are not likely to open up large new areas for production. The gains in productive potential would be unlikely to balance possible large-scale reductions in some major grain-exporting regions at mid-latitude. Moreover, there may well occur severe negative impacts of climate change on food supply at the regional level, particularly in regions of high present-day vulnerability, least able to adjust technically to such effects.

The average global increase in overall production costs could thus be small (perhaps a few per cent of world agricultural GDP). Much depends, however, on how beneficial are the so-called 'direct' effects of increased CO_2 on crop yield. If plant productivity is substantially enhanced, and more moisture is available in some major production areas, then world productive potential of staple cereals could increase relative to demand, with food prices reduced as a result. If, on the contrary, there is little beneficial direct CO_2 effect, and climate changes are negative for agricultural potential in all or most of the major food-exporting areas, then the average costs of world agricultural production could increase significantly. These increased costs could amount to more than 10% of world agricultural GDP.

6 Response Strategies

Three broad types of adjustment strategies may be anticipated: changes in land use, changes in farm management, and changes in crop and livestock husbandry (Parry, 1990; Parry and Duinker, 1990).

6.1 Changes in land use

6.1.1 Changes in farmed area

Farmed area can be extended where environmental factors and economic incentives permit. Expansion may be most marked in Russia and northern Europe, where terrain and soils will permit further reclamation (Pitovranov, et al., 1988; Squire and Unsworth, 1988). But it may be more limited by inappropriate soils in much of Canada, with the exception of the Peace River region in northern Alberta, and parts of Ontario (Smit, 1987, 1989). There may also be potential for high-latitude reclamation of farmland in some of the valleys of central Alaska, in northern Japan, and in southern Argentina and New Zealand (Jäger, 1988; Salinger et al., 1990).

Warming may also tend to induce an upward extension of farmed area in highland regions. For example, in the European Alps a 1 °C warming could raise climatic limits of cultivation by about 150 meters (Balteanu et al., 1987). However, these incusions on the semi-natural environment may confront the extensive existing rangeland economies of mountain regions, which will come under pressure both from the upward advance of more intensive agriculture, and from afforestation.

The broadscale changes in crop location imply a general poleward shift of present-day agricultural zones. This is likely to be most pronounced in mid- and high latitudes, partly because warming will be most marked here, but largely because latitudinal zoning is most evident in these regions, as a result of differences in available warmth for crop maturation.

The northern limit of maize production in the UK, for example, is at present located in the extreme north of the country. It would shift about 300 km northwards for each °C rise in mean annual temperature (Parry et al., 1989). In southern Europe, higher temperatures imply a more northerly location of present limits of citrus, olives, and vines (Imeson et al., 1987). In central North America, zones of farming types are estimated to shift about 175 km northwards for each 1 °C of warming, resulting in reduced intensity of use in the south, but increased intensity in the north where soils and terrain permit. A sizeable area of output, currently located in the northern Great Plains of the USA, would relocate in the sourthern Canadian prairies. Other losses of Great Plains cropped acreage may be partially compensated for by increases in cultivated areas in the Great Lakes region (EPA, 1989).

Similar southward shifts of land have been suggested for the southern hemisphere, perhaps up to six degrees of latitude (670 km) under a 3-4 °C warming (Salinger et al., 1990). It should be emphasized, however, that these broadscale effects will be much influenced, at local levels, by regional variations in soils, by the competitiveness of different crops, and most importantly, by regional patterns of rainfall — none of which can be predicted at present.

6.1.2 Changes in crop type

In regions where there are substantial increases in the warmth of the growing season (and where output is currently limited by temperature, rather than rainfall), substitution of crops with higher thermal requirements — that would make fuller use of the extended and more intense growing season — should allow higher yields. In Hokkaido (in Northern Japan), for example, the adoption of a late-maturing rice variety (at present grown in Central Japan) would enable greater advantage to be taken of the now-warmer climate, with yields increased by about 25% (Yoshino, et al., 1988).

Where moisture, rather than temperature, is the climatic constraint on output, or where increases in temperature could lead to higher rates of evapotranspiration — and thus to reduced levels of available moisture — there may occur a switch to crops with lower moisture requirements. Once again, the lack of information on likely changes in rainfall makes further speculation on this unprofitable, particularly at lower latitudes. However, there is some evidence that, at high mid-latitudes, a switch from spring to winter varieties of cereals would be one strategy for avoiding losses resulting from more frequent dry spells in the early summer. This might be the case in Scandinavia and on the Canadian prairies (Koster et al., 1987; Williams et al., 1988).

6.2 Changes in management

6.2.1 Irrigation

Substantial increases in the need for irrigation (and its costs) are likely, as it substitutes for moisture losses. The most detailed estimates yet available are for the USA, where irrigation requirements may increase by about 25% in the southern and 10% in the northern Great Plains under a 2 x CO_2 climate. Given the likely increased rate of groundwater depletion, this will probably lead to significantly higher costs of production, with consequent shifts to less water-demanding uses in the most affected areas (EPA, 1989). Substantially increased irrigation needs are also projected for most of western and southern Europe (Imeson et al., 1987). In other places where available water will be reduced, although data are not available, it is probable that effects will be similar. Tighter water-management practices (Gleick, 1992, this volume) will be necessary to yield higher irrigation efficiency.

6.2.2 Fertilizer use

More use of fertilizers may be needed to maintain soil fertility where increases in leaching result from increased rainfall. In other regions, warming may increase productive potential to the extent that current levels of output can be achieved with substantially lower amounts of fertilizers. In Ireland, for example, fertilizer use could possibly be halved under a warming of 4 °C, while maintaining present-day output (Bergthorsson et al., 1988). Much will depend on other factors: the extent that higher CO_2 levels make nutrients more limiting (thus requiring more use of fertilizers), or changes in future energy prices (and their effect on the cost of fertilizers).

6.2.3 Control of pests and diseases

The costs of these are likely to alter substantially, although it is quite impossible to specify them with any degree of detail. Possibly most important for global cereal production may be the costs of controlling the spread of subtropical weed species into current major cereal-producing regions (EPA, 1989).

6.2.4 Soil drainage and control of erosion

Adjustments in management are likely to be needed in tropical regions, particularly those characterized by monsoon rainfall, where there may be an overall increase in rainfall receipt and intensity. Recent assessments have indicated that, over the longer term, reduced soil fertility, increased salinity, and the costs of erosion control may more than offset the beneficial effects of a warmer climate, leading ultimately to reduced yields and higher production costs (Pitovranov et al., 1988).

6.2.5 Changes in farm infrastructure

Regional shifts of farming types and altered irrigation requirements imply major changes in capital equipment for farm layout, and in agricultural support services (credit, marketing, etc.). In the USA, it has been estimated that these will be substantial (EPA, 1989). Because of the large costs involved, only small, incremental adjustments may occur without changes in government policies.

6.3 Changes in crop and livestock husbandry

In particular, there are likely to occur many alterations in the timing of farm operations such as tillage (ploughing, sowing, harvesting), fertilizing, and pest and weed control (spraying). The timing of these in the present farming calendar is frequently affected by climate. Particular aspects of husbandry are also likely to be affected — density of planting, the use of fallowing and mulching, and the extent of intercropping. A change of climate implies a retuning of these strategies to harmonize with the new set of climatic conditions.

6.4 Integrating the responses

The overall response to the predicted changes in climate has to be in the form of an integrated package of long-term and short-term measures. The longer-term measures which can help to promote a balance between carbon emission and carbon absorption, including "green" energy policies and afforestation, are well known. In the short term, measures will have to be developed for each agro-ecological region, to help insulate the human and animal populations from the adverse impact of drought or floods, and impart greater stability to agricultural production.

Swaminathan (1982, 1986) has described in detail the ways in which these twin goals can be achieved. Contingency plans for planting alternate crops will have to be developed for drought and flood-prone areas. This will call for building seed reserves of alternate crops as well as alternate photo-insensitive short-duration varieties. In many areas in the tropics, seed reserves are essential for crop security, just as food grain reserves are important for food security.

The contingency plans should include optimizing production in the Most Favourable Areas (MFA) through improved input delivery systems and remunerative output pricing policies. MFAs are generally those with assured irrigation facilities and with soil conditions which permit sustainable intensification. In the areas which are liable to be affected most seriously (MSA), effective public distribution systems and fodder and feed banks for animals will have to be developed. The idea is similar to that of building storm shelters along the coastal areas prone to cyclonic damage. Drinking water security for the human and animal populations is another area which needs location specific planning.

Ultimately, the most effective method of both avoidance and adaptation is public education and awareness. The Crop-Weather-Watch Groups recommended by Swaminathan (1982) at the village level could be a method of awareness generation as well as of micro-level monitoring of the interaction between climate factors and crop productivity.

The response strategies should include steps to diversify employment opportunities. Famines in India in the 19th Century were not famines of food, but of work. Where there is work, there is purchasing power; and where there is purchasing power, there is invariably food, except when communication is disrupted due to war or civil strife. Crop failures affect most severely societies which rely heavily on the primary sector for the security of their livelihoods. For example, a study by Resources for the Future found the greatest economic threat in the MINK (Missouri, Iowa, Nebraska and Kansas) region of the US, (where the percentage of farm income to total income is 3.4 times the percentage of farm income to total income found in the USA as a whole). Even this region may not suffer seriously due to changes in precipitation and temperature, since the share of the farm sector in the income of this region is only 3.87%.

A diversification of employment opportunities can be brought about only through appropriate technologies and trade opportunities. Unless developing countries are assisted technologically and economically to put in place appropriate response mechanisms, the sufferings of their people may increase as a result of adverse changes in climate caused by factors beyond their control.

Adaptive and anticipatory research strategies are important. The Centre for Research on Sustainable Agricultural and Rural Development, Madras (CSARD) has, for example, established a Genetic Resources Centre for Adaptation to Sea Level Rise in Tamil Nadu (India). At this centre, genotypes are being assembled to process "candidate genes" for conferring resistance to sea water intrusion and flooding.

7 Conclusions

Certain steps that are necessary to avoid and/or to adapt to new climatic situations may also be important even if there is no global warming. These include improving energy efficiency, planting trees and the adoption of sustainable lifestyles. Without changes in current high-consumption lifestyles, it will be difficult to promote sustainable development.

While development action must continue, there is need for more inter-disciplinary and inter-institutional collaborative research on all aspects of climate change. The organization of an international research network on climate change and food security with the following aims would be useful in this context:

- Estimation and monitoring of the emission of greenhouse gases as a result of farm operations under different agro-ecological and technological conditions.

- Relative, as well as integrated, analysis of the agricultural impact of higher concentrations of CO_2 methane, dinitrogen oxides, and CFCs.

- Standardization of techniques — both for minimizing agriculture's contributions to the accumulation of greenhouse gases, and for withstanding additional biotic and abiotic stresses on crop and farm animals arising from climate change.

- Promotion of research and training in the field of restoration ecology, in the afforestation of degraded forests and in the sustainable management of tropical rain forests.

- Anticipatory research and development measures in coastal areas to avoid or minimize the adverse impact of potential changes in sea levels.

- Standardization of methods for deriving benefit from higher atmospheric CO_2 concentration.

- Standardization of post-harvest technologies for perishable agricultural commodities, based on non-CFC dependent refrigeration methods.

- Stimulating policy research designed to strengthen the public policy back-up both for avoiding adverse changes in climate and for adapting to new growing conditions.

To date, less than a dozen detailed regional studies have been completed that serve to assess the potential impact of climatic changes on agriculture. It should be a cause for concern that we do not, at present, know whether changes of climate are likely to increase the overall productive potential for global agriculture, or to decrease it. Neither do we know which regions are likely to gain or lose in terms of productive potential, although we have indicated what the regional-level impacts might be.

Huge uncertainties remain, and the risks attached to such levels of ignorance are great. A comprehensive, international research effort is required, now, to redeem the situation. In the meantime, actions should be taken to increase the resilience of the food production system world-wide, but especially in the vulnerable areas of developing countries. Additional actions to reduce post-harvest losses, and to increase the efficiency of food processing and distribution systems makes sense today — whether or not the world is on the edge of a major global climate change.

References

Akong'a, J., T.E. Downing, N.T. Konijn, D.N. Mungai, H.R. Muturi and H.L. Potter, 1988: The effects of climatic variations on agriculture in Central and Eastern Kenya. In *The Impact of Climatic Variations on Agriculture, Volume 2, Assessments in Semi-Arid Regions,* Parry, M.L., Carter, T.R., and Konijn, N.T., eds., Kluwer, Dordrecht, The Netherlands.

Balteanu, D., P. Ozenda, M. Huhn, H. Kerschner, W. Tranquillini and S. Bortenschlager, 1987: Impact analysis of climatic change in the central European mountain ranges, **G**, *European Workshop on Interrelated Bioclimatic and Land Use Changes*, Noordwijkerhout, The Netherlands, October, 1987.

Beresford, R.M. and R.A. Fullerton, 1989: Effects of climate change on plant diseases. Submission to Climate Impacts Working Group, May 1989, 6pp.

Bergthorsson, P., H. Bjornsson, O. Dyrmundsson, B. Gudmundsson, A. Helgadottir and J.V. Jonmundsson, 1988: The effects of climatic variations on agriculture in Iceland. In *The Impact of Climatic Variations on Agriculture, **1** Cool Temperate and Cold Regions*, M.L. Parry, T.R. Carter and N.T. Konijn, eds., Kluwer, Dordrecht, The Netherlands.

Blasing, T.J. and A.M. Solomon, 1983: *Response of North American Corn Belt to Climatic Warming*, prepared for the US Department of Energy, Office of Energy Research, Carbon Dioxide Research Division, Washington, DC, DOE/N88-004.

Crosson, P., 1989: Greenhouse warming and climate change: why should we care? *Food Policy*, **14** (2), 107-18.

Cure, J.D., 1985: Carbon dioxide doubling responses: a crop survey. In *Direct Effects of Increasing Carbon Dioxide on Vegetation*, B.R. Strain and J.D. Cure, eds., US DOE/ER-0238, Washington, USA, pp.100-116.

Drummond, R.O., 1987: Economic aspects of ectoparasites of cattle in North America. In *Symposium, The Economic Impact of Parasitism in Cattle*, XXIII World Veterinary Congress, Montreal, pp.9-24.

Emanuel, K.A., 1987: The dependence of hurricane intensity on climate: mathematical simulation of the effects of tropical sea surface temperatures. *Nature*, **326**, 483-485.

Environmental Protection Agency (EPA), 1989: *The Potential Effects of Global Climate Change on the United States*, Report to Congress, Washington, DC.

Gifford, R.M., 1979: *Austral. J. Plant Physiol.*, **6**, 367-378.

Havelka, U.D., V.A. Wittenbach and M.G. Boyle, 1984: *Crop Sci.* **24**, 1146-1150.

Hill, M.G. and J.J. Dymock, 1989: Impact of Climate Change: Agricultural/Horticultural Systems. DSIR Entomology Division, submission to New Zealand Climate Change Programme, Department of Scientific and Industrial Research, New Zealand, 16pp.

Imeson, A., H. Dumont and S. Sekliziotis, 1987: *Impact analysis of climate change in the Mediterranean region*, **F**, European Workshop on Interrelated Bioclimatic and Land Use Changes, Noordwijkerhout, The Netherlands.

Intergovernmental Panel on Climate Change (IPCC), 1990: *The IPCC Scientific Assessment of Climate Change: Policymakers Summary*, Geneva and Nairobi: WMO and UNEP.

Jäger, J., 1988: *Developing Policies for Responding to Climatic Change*, Report of the Workshops in Villedi, Austria and Bellaggio, Hal, Stockholm Environment Institute, Stockholm, Sweden.

Kaida, Y. and V. Surarerks, 1984: Climate and agricultural land use in Thailand. In *Climate and Agricultural Land Use in Monsoon Asia,* M.M. Yoshino, ed., University of Tokyo Press, Tokyo, 231-253.

Kettunen, L., J. Mukula, V. Pohjonen, O. Rantanen and U. Varjo, 1988: The effects of climatic variations on agriculture in Finland. In *The Impact of Climatic Variations on Agriculture: **1**, Assessments in Cool Temperate and Cold Regions,* M.L. Parry, T.R. Carter and N.T. Konijn, eds. Kluwer, Dordrecht, The Netherlands, pp.511-614.

Kimball, B.A., 1985: In *Direct Effects of Increasing Carbon-Dioxide on Vegetation*, B.A. Strain and J.D. Cure, eds., National Technical Information Service, US Dept. Commerce, Springfield, Virginia, pp.185-204.

Koster, E.A., E. Dahl, H. Lundberg, R. Heinonen, L. Koutaniemi, B.W.R. Torssell, R. Heino, J.E. Lundmark, L. Stronquist, T. Ingelog, L. Jonasson and E. Eriksson, 1987: *Impact analysis of climatic change in the Fennoscandian part of the boreal and sub-arctic zone*, **D**, European Workshop on International Bioclimate and Land Use Changes, Noordwijkerhout, the Netherlands, October 1987.

Mearns, L.O., R.W. Katz and S.H. Schneider, 1984: Extreme high temperature events: changes in their probabilities with changes in mean temperatures. *Journal of Climatic and Applied Meteorology*, **23**, 1601-13.

Newman, J.E., 1980: Climate change impacts on the growing season of the North American Corn Belt. *Biometeorology*. **7(2)**, 128-142.

Parry, M.L., 1976: *Climatic Change, Agriculture and Settlement.* Dawson, Folkestone, England.

Parry, M.L., 1990: *Climate Change and World Agriculture.* Earthscan, London, 165pp.

Parry, M.L. and T.R. Carter, 1988: The assessments of the effects of climatic variations on agriculture: aims, methods and summary of results. In *The Impact of Climatic Variations on Agriculture, **1**, Assessments in Cool Temperate and Cold Regions*, M.L. Parry, T.R. Carter and N.T. Konijn, eds., Kluwer, Dordrecht, The Netherlands.

Parry, M.L. and P.N. Duinker, 1990: *The Potential Effects of Climatic Change on Agricultur*. In Intergovernmental Panel on Climate Change (1990) the IPCC Impacts Assessment. WMO and UNEP, Geneva, Switzerland, pp.2-1--2-45

Parry, M.L. and S.K. Sinha, 1988: Food Security, Working Group Report. In *Conference Proceedings, The Changing Atmosphere: Implications for Global Security*. Toronto, Canada, 27-30 June, WMO-No, 170, World Meteorological Organisation, pp.321-3.

Parry, M.L., T.R. Carter and J.H. Porter, 1989: The greenhouse effect and the future of UK agriculture. *Journal of the Royal Agricultural Society of England*, pp.120-131.

Pearman, G.I., 1988: *Greenhouse: Planning for Climate Change*. Melbourne, Australia: CSIRO.

Pitovranov, S.E., V. Iakimets, V.E. Kislev and O.D. Sirotenko, 1988: The effects of climatic variations on agriculture in the subarctic zone of the USSR. In *The Impact of Climatic Variations on Agriculture: Volume 1, Assessments in Cool, Temperate and Cold Regions*, M.L. Parry, T.R. Carter and N.T. Konijn, eds., Kluwer, Dordrecht, The Netherlands.

Pittock, A.B., 1989: Potential impacts of climatic change on agriculture, forestry and land use, personal communication, pp.5.

Porter, J.H., M.L. Parry and T.R. Carter, 1992: The potential effects of climatic change on agricultural insect pests. *Agricultural and Forest Meteorology*, in press.

Rind, D., R. Goldberg and R. Ruedy, 1989: Change in climate variability in the 21st Century. *Climatic Change*, **14**,5-37.

Rogers, H.H., J.F. Thomas and G.E. Bingham, 1983: Response of Agronomic and Forest Species to Elevated Atmospheric Carbon-Dioxide. *Science*, **220**,428-429.

Rosenzweig, C., 1985: Potential CO_2-induced climate effects on North American wheat-producing regions. *Climatic Change*, **7**:367-389.

Salinger, M.J., 1989: *The Effects of Greenhouse Gas Warming on Forestry and Agriculture*. Draft report for WMO Commission of Agrometeorology, 20pp.

Salinger, M.J., W.M. Williams, J.M. Williams and R.J. Martin eds., 1987, 1988, 1990: *Carbon Dioxide and Climate Change: Impacts on Agriculture*. New Zealand: New Zealand Meteorological Service; DSIR Grasslands Division; MAFTech, 1990.

Schelling, T., 1983: Climate change: implications for welfare and policy. In *Changing Climate: Report of the Carbon Dioxide Assessment Committee*, Washington, DC: National Academy of Sciences.

Schulze, R.E., 1989: *Hydrological responses to long-term climatic change*, Personal Communication, 9pp.

Sinha, S.K. and M.S. Swaminathan, 1991: Deforestation, Climate Change and Sustainable Nutrition Security. In *Climatic Change*.

Sinha, S.K., N.H. Rao and M.S. Swaminathan, 1988: Food security in the changing global climate. In *Conference Proceedings, The Changing Atmosphere: Implications for Global Security*, Toronto, Canada, 27-30 June, 1988, WMO-No. 170, World Meteorological Organisation, 1988, pp.167-92.

Smit, B., 1987: Implications of climatic change for agriculture in Ontario. In *Climate Change Digest*, CCD 87-02, Downsview, Ontario: Environment Canada.

Smit, B., 1989: Climate warming and Canada's comparative position in agriculture. In *Climate Change Digest*, CCD89-01. Downsview, Ontario: Environment Canada.

Smit, B., M. Brklacich, R.B. Stewart, R. McBride, M. Brown and D. Bond, 1989: Sensitivity of crop yields and land resource potential to climate change in Ontario. In *Climatic Change*, **14**, pp.153-74.

Squire, G.R. and M.H. Unsworth, 1988: Effects of CO_2 and climatic change on agriculture, *Contract Report to the Department of the Environment*. Sutton Bonnington, UK: Department of Physiology and Environmental Science, University of Nottingham.

Swaminathan, M.S., 1982: *Science and Integrated Rural Development*. Concept Publishing Company, New Delhi. 354pp.

Swaminathan, M.S., 1986: Building national and global food security systems. In *Global Aspects of Food Production*, M.S. Swaminathan and S.K. Sinha, eds., Tycooly Press, pp.417-449.

Terjung, W.H., H-Y Ji, J.T. Hayes, P.A. O'Rourke and P.E. Todhunter, 1989: Actual and potetial yield for rainfed and irrigated maize in China. *International Journal of Biometeorology*, **28**,115-135.

Van Diepen, C.A., H. Van Keulen, F.W.T. Penning de Vries, I.G.A.M. Noy and J. Goudriaan, 1987: Simulated variability of wheat and rice yields in current weather conditions and in future weather when ambient CO_2 has doubled. In *Simulation Reports CABO-TT*, **14**., Wageningen, The Netherlands: University of Wageningen.

Williams, G.D.V. and W.T. Oakes, 1978: Climatic resources for maturing barley and wheat in Canada. In *Essays on Meteorology and Climatology*: in Honour of Richard W. Longley, Studies in Geography Mono.3., 367-385, Haye, K.D. and E.R. Reinelt, eds., University of Alberta, Edmonton.

Williams, G.D.V., R.A. Fautley, K.H. Jones, R.B. Steward and E.E. Wheaton, 1988: Estimating effects of climatic change on agriculture in Sasketchewan, Canada. In *The Impact of Climatic Variations on Agriculture: **1**, Assessments in Cool Temperate and Cold Regions*, M.L. Parry, T.R. Carter and N.T. Konijn, eds., pp.219-379, Kluwer, Dordrecht, The Netherlands.

Wittwer, S.H., 1985: Carbon Dioxide Levels in the Biosphere Effects on Plant Productivity. *Crit. Rev. Plant. Sci.*, **2**(3), 171-198.

Yoshino, M.M., 1984: Ecoclimatic systems and agricultural land use in Monsoon Asia. In *Climate and Agricultural Land Use in Monsoon Asia*, M.M. Yoshino, ed., University of Tokyo Press, Tokyo, pp.81-108.

Yoshino, M., T. Horie, H. Seino, H. Tsujii, T. Uchijima and Z. Uchijima, 1988: The effects of climatic variations on agriculture in Japan. In *The Impact of Climatic Variations on Agriculture, **1**, Assessments in Cool Temperate and Cold Regions*. Dordrect, The Netherlands: Kluwer, pp.725-868.

CHAPTER 9

Effects of Climate Change on Shared Fresh Water Resources

Peter H. Gleick

Editor's Introduction

Climate change will not only affect the level and location of the seas. It will also alter the timing, extent, and distribution of precipitation and runoff—the renewable sources of fresh water on which human societies and natural ecosystems depend. Peter Gleick analyses the implications of general circulation modelling experiments for rainfall, soil moisture, and streamflows, and notes that the potential impacts in some regions may be severe. Growing populations may add more demand for this water. These effects may be particularly important where two or more nations depend heavily on shared rivers or lakes. Where water resources are already tightly stretched, fresh water availability could become a military security concern.

Will nations go to war over water? Probably not, but in some river valleys — including the Jordan/Litani and the Tigris/Euphrates systems — heightened tensions over disrupted or shrinking water supplies, exacerbated by climate change, can only add to what is already a highly volatile mix of political tensions. As Gleick points out, we have seen early evidence of "water warfare" in the last several years. Water supplies may become strategic targets; control over access to shared water resources may be used as an economic and political weapon. Conflicts over water rights may become a brake on development for some struggling nations. Even if climate change does not take place, concerns over shared water resources will probably become an increasingly important part of international relations in the future.

Peter Gleick, the Director of the Global Environment Program at the Pacific Institute for Studies in Development, Environment and Security, is a world-renowned expert on the linkages between environment and security issues. The principal focus of his research has been on the implications of climate change for regional hydrology. Gleick takes a solution-oriented approach to the management of shared fresh water resources. He is realistic, but remarkably optimistic, about the prospects for meeting these complex and difficult challenges. Minimizing potential damage and conflict will require changes in water management strategies, improved water efficiencies (in supply and use), and a systematic evolution of international strictures on military/environmental aggression. Ultimately, sharing water resources will require unprecedented levels of regional cooperation. The magic of Gleick's approach is the essence of the "no regrets" strategy—it will pay big dividends in both the Northern and Southern nations, whether or not the world sits on the edge of a major climate change chasm.

- I. M. M.

1 Introduction

The links among climate change, economic development, and international security are increasingly evident. Economic development can either improve or worsen global environmental problems. Global environmental problems can hinder economic development. Inequities in resource availability and use can worsen international tensions. And disputes over international environmental problems can spill over into the geopolitical realm. Understanding these links is vital. Failing to perceive them and respond appropriately can worsen environmental risks and increase the likelihood of resource-related disputes and conflict.[1]

The links among these problems are especially strong in the area of fresh water resources. Many rivers and sources of fresh water are shared by two or more nations. This geographical fact has led to the geopolitical reality of disputes over shared water resources — including the Nile, Jordan, and Euphrates in the Middle East, the Indus, Ganges, and Brahmaputra in southern Asia, and the Colorado and Rio Grande in North America. As growing populations demand more water for agriculture and economic development, strains on limited water resources will grow, and international disputes in water-short regions will worsen.

While various regional and international legal mechanisms exist for reducing water-related tensions, these approaches have never received the support or attention necessary to resolve all conflicts over water. Indeed, there is growing evidence that existing international water law may be unable to handle the strains of future problems.

In the past, water planners struggled with the problem of estimating future demand. It now appears that our ability to estimate future water *supply* is equally uncertain. In particular, the likelihood of global climatic change greatly complicates our ability to determine how much water will be available for human use at any given place or time, and what its quality will be.

Unless new approaches to water management are developed that take into account these new uncertainties, future conflicts over water resources are certain to increase. This chapter identifies the greatest climatic threats to water supplies, evaluates the types of disputes and tensions that might be engendered, and suggests approaches for reducing the risk of climate-induced water conflicts.

2 Climate Changes and the Hydrologic Cycle

Most attention given recently to the impacts of global climatic change has focused on a very limited aspect — global warming, the question of increases in annual average temperature. But in fact, some of the most severe impacts to society and natural ecosystems are likely to result not from changes in temperature, but from changes in precipitation, evapotranspiration, runoff, and soil moisture. Furthermore, global and annual averages are far less important than changes on much smaller spatial and temporal scales (Gleick, 1989).

This partially misdirected focus of public attention results primarily from limitations in the ability of global climate models — including the most complex representations, the general circulation models or GCMs — to incorporate and reproduce important aspects of the hydrologic cycle (see Wigley et al., 1992, this volume). Many important hydrologic processes, such as cloud formation and rain-generating storms, occur on spatial scales far smaller than can be modelled in present-day GCMs. At the same time, the hydrologic processes included in the models are far simpler than the real world processes.[2] We thus know much less about how the global water cycle is likely to change than we would like to know in order to develop intelligent social policies.

Despite these limitations, we do know some things about how hydrology and water supplies will be affected by climatic changes. In the next section we review this information and try to place what we know in the context of what policy makers need to know to make rational decisions about large-scale water systems. Much of this information is summarized in Tables 1 and 2.

3 Global and Regional Changes in Precipitation and Temperature

3.1 An increase in global precipitation of 3-15%

As global average temperatures rise, the evaporation of water from land and water surfaces will also increase. While the relationship between evaporation and temperature depends on many factors, large-scale (regional to continental) changes are often described fairly simplistically in global climate models. The analysis of Budyko (1982), for example, suggests a four per cent increase in potential evapotranspiration for each degree Celsius rise in temperature. (Evapotranspiration is the process in which water is drawn from the soil into plant roots, and then is evaporated or transpired from leaves and other plant surfaces into the air.) Similarly, Wetherald and Manabe (1981) found that global evaporation changes by 3% when temperature changes by 1 °Celsius. This simple level of analysis suggests that global warming of about 3 °C would lead to an increase in evaporation of about 10% and thus an increase in global average precipitation of about 10%.

State-of-the-art GCMs incorporate more complex climatic feedbacks. Most recent reviews of experiments with these models suggest that global average precipitation may increase (as does global average evaporation) by 3-15% (IPCC, 1990) for a buildup of greenhouse gases equivalent to a doubling of the pre-industrial concentration of atmospheric carbon dioxide. The greater the warming, the larger the expected precipitation increases.

---

[1] There is some contention over the role of resources and environmental problems in affecting international "security", but much of the argument rests over definitions of the term "security" and the applicability of methods of analysis and conflict resolution to environmental problems. These issues are reviewed in more depth by Gleick (1990a, 1991a) and Deudney (1991). There is little doubt, however, that resources and environmental concerns are playing an increasingly important role in international politics.

[2] Almost all work on the large-scale hydrologic implications of climatic changes comes from equilibrium climate models. The few analyses using time-varying studies (the transient runs) do not show significant differences in major hydrologic parameters from the equilibrium results.

Table 1. Summary of expected climate and hydrology changes

Phenomenon	Projection of probable global annual average change[a]	Distribution of Change: regional average	change in seasonality	interannual variability	significant transients	Confidence of projection: global average	regional average
Temperature	+2 to +5 °C	-3 to +10 °C	Yes	Down?	Yes	High	Medium
Sea level	+10 to +100 cm		No	?	Unlikely	High	Medium
Precipitation	+7 to +15%	-20 to +20%	Yes	Up	Yes	High	Low
Direct solar radiation	-10 to +10%	-30 to +30%	Yes	?	Possible	Low	Low
Evapotranspiration	+5 to +10%	-10 to +10%	Yes	?	Possible	High	Low
Soil moisture	?	-50 to +50%	Yes	?	Yes	?	Medium
Runoff	Increase	-50 to +50%	Yes	?	Yes	Medium	Low
Severe storms	?	?	?	?	Yes	?	?

Notes:[a] For an "equivalent doubling" of atmospheric CO_2 from the pre-industrial level. These are equilibrium values, neglecting transient delays and adjustments.
Source: Schneider et al., 1990.

Table 2 . Additional detail on GCM estimates of regional hydrologic changes

Global average precipitation increase:3 to 15%

Increases in precipitation are expected to occur more consistently and intensely at high latitudes throughout the year.

All models produce greater precipitation in high latitudes and tropics throughout the year and in mid-latitudes in winter.

In certain zones (such as 35-55° N), overall precipitation may increase by as much as 10 to 20%.

In many of the models summer rainfall decreases slightly over much of the northern mid-latitude continents.

In many models the Southwest Asian monsoon strengthens.

All GCMs simulate a general increase in soil moisture of the high northern latitudes in winter.

Most models simulate large-scale drying of the Earth's surface over northern mid-latitude continents in northern summer.

3.2 *Moister regional climates, particularly at high latitudes and in the tropics*

More valuable for policy makers, and for forecasts of social impact, would be detailed information on the expected changes in specific regions. Such information is only slowly becoming available from climate models. Over the last two decades, improvements in hydrologic and atmospheric modelling have led to the incorporation of far more detailed evaporation and soil processes, which permit more realistic estimates of regional evaporation and precipitation patterns. Despite remaining limitations, certain trends are becoming evident.

All models produce a moister atmosphere (climatologists refer to this as "increased specific humidity"), and greater precipitation in high latitudes and tropics throughout the year and in mid-latitudes in winter. In certain zones (such as 35-55° N), overall precipitation may increase by as much as 10-20% (IPCC, 1990). In many of the models, summer rainfall decreases slightly over much of the northern mid-latitude continents, and the Southwest Asian monsoon strengthens. Other changes in mid-latitudes remain highly variable and ambiguous.

3.3 *Little change in arid zones*

There are few consistent and large-scale changes in precipitation in subtropical arid regions (IPCC, 1990). It must be noted here that even small changes in these arid zones can have significant implications for ecological and human systems.

3.4 *Net summer drying of soil in northern latitudes*

An important variable, often overlooked in climate impact studies, is soil moisture.[3] It is vitally important to remember that an increase in precipitation does not necessarily mean a wetter land surface or more soil moisture (IPCC, 1990, p.139). In regions where precipitation increases, increases in

[3] For our purposes here, we are simply interested in the water held in soils and groundwater that does not run off or return to the atmosphere.

evaporation owing to higher temperatures may be even greater, leading to a net drying of the land surface. Indeed, one important recent finding was that the incidence of droughts in the United States, measured by an index that looks at soil wetness conditions, is likely to dramatically increase as temperatures go up — despite an accompanying increase in precipitation — because of the increased evaporative losses (Rind et al., 1990). This finding is also evident in some of the detailed hydrologic modeling of river basins (Gleick, 1987; Nash and Gleick, 1991), where large increases in precipitation may be necessary to simply maintain river runoff at present historical levels as temperatures and evaporative losses rise.

An exception to the general drying trend is winter in the high northern latitudes, where all GCM results simulate a general increase in soil moisture due to large increases in rain and snow. But in summer, most models also simulate large-scale drying of the Earth's surface over northern mid-latitude continents due to higher temperatures and either insufficient precipitation increases or actual reductions in rainfall. Since these latitudes include many productive grain-growing regions, drying could have significant impacts, particularly on agricultural production and water demand. In a review of the soil moisture results from five different GCMs, Kellogg and Zhao (1988) showed that all five models agreed upon decreased soil moisture in the central part of the United States.[4]

A limitation of GCMs in assessing soil moisture changes is the exclusion of the direct effect of CO_2 concentrations on vegetation. The question of how climate change may alter evapotranspiration is complicated, for many reasons. Greenhouse warming may alter temperature, cloudiness, wind conditions, humidity, plant growth rates, rooting, leaf area, or other factors. Higher CO_2 levels have been shown in laboratory studies to alter the efficiency with which water is used by certain plants. But the net effect on the water balance is still unclear, because of many complicating factors and feedbacks.

According to Rosenberg et al. (1990), evapotranspiration in certain plant communities is more responsive to changes in air temperature, the resistance of plant stomates, and net radiation, and less responsive to leaf area index, vapour pressure, and windspeed. They urge more careful study of the many complicating factors. Because of these complexities, most analyses simply assume zero (or limited) net change in plant-water interactions. As our understanding of these phenomena improves, more accurate estimates of plant-water interactions can be incorporated into the analysis.

[4] While there is a 1-in-32 chance that any point would have decreased soil moisture in all five models if the results were purely random and uncorrelated, there is also a plausible physical explanation for this result: substantial warming in the summer months and changes in the timing of snowmelt and soil drying (see the next section).

3.5 *Shorter seasons of less abundant snow cover*

One of the most important hydrologic impacts of climatic change is a regional effect not well represented in large-scale models — snowfall and snowmelt changes in high-altitude watersheds, or in areas with strong snowmelt runoff dynamics. In these watersheds, changes in temperature alone will lead to important changes in water availability and quality.

In basins with substantial snowfall and snowmelt, temperature increases have three effects. They increase the ratio of rain to snow in cold months; they shorten the overall duration of the snow season; and they increase the rate and intensity of spring snowmelt. As a result of these three effects, winter average runoff increases; peak runoff increases and occurs earlier in the year; and in warm seasons, the soil dries more quickly and extensively. Manabe and Wetherald (1986) and Wilson and Mitchell (1987) identified this mechanism, in models of general circulation, as a leading cause of summer-soil moisture drying. This effect was first observed in regional hydrologic modeling by Gleick (1987), and has since been observed for other basins with substantial snowfall and snowmelt runoff (Lettenmaier et al. 1988, Bultot et al. 1988, Mimikou and Kouvopoulos 1991, Nash and Gleick 1992, Shiklomanov 1987, Rango and van Katwijk, 1990).

3.6 *Increased risk of storms and floods*

Because of the increase in runoff from snowmelt, far more attention needs to be paid to the risk of floods in some regions. In many mid- and high-latitude river basins, the worst floods occur during snowmelt runoff periods. One of the greatest concerns about the effect of higher temperatures is therefore the increased probability and intensity of flood flows. When combined with possible precipitation increases in many regions, flooding becomes a critical concern.

Another vitally important question is what happens to the variability of climatic conditions — that is, the frequency and intensity of extremes. Although little work has been done on this issue, there are some indications that the variability (interannual standard deviation) of the hydrologic cycle increases wherever mean precipitation increases, and vice-versa. In one model study, the total area over which precipitation fell over the earth decreased, even though global mean precipitation increased (Noda and Tokioka, 1989). This implies more intense local storms and hence runoff.

3.7 *Unpredictable storm effects*

Other changes in variability are also likely, though at this point we have little confidence that we can predict what they will be. There is some indication of a general reduction in day-to-day and interannual variability of storms in the mid-latitudes. At the same time, there is evidence from both model simulations and empirical considerations that the frequency, intensity, and area of tropical disturbances may increase (Emanuel 1987, see also IPCC 1990, pages 153-155 for summary). Far more modeling and analytical efforts are needed in this area.

3.8 Changes in runoff

Estimates of surface runoff in global climate models are not generated directly, but are typically derived from the difference between precipitation and evaporation at the land surface. As a result, these estimates do not include direction of flow, discrete river basin results, or realistic surface runoff processes. Thus, the global model runoff estimates do not always agree with more detailed regional model results, even in the broadest expectations of possible changes.[5]

The regional runoff models are more realistic, in part, because they have been based upon detailed regional hydrologic models — that is, models of the interrelationships between geography, soil, and water in specific river basins. These models suggest that some significant changes in the timing and magnitude of runoff are likely to result from quite plausible changes in climatic variables.

For example, in a series of studies of basins in the western United States and in Greece, temperature increases of 2 to 4 °C result in decreases in runoff of up to 20%, with no changes in precipitation, depending on the characteristics of the basin (Gleick, 1987; Lettenmaier et al., 1988; Mimikou and Kouvopoulos, 1991; Nash and Gleick, 1991). Increases or decreases of precipitation of 10 and 20% increase or decrease runoff by about 10 and 20%, respectively. Other basin studies show similar results (see, for example, Bultot et al., 1988 for results from Belgium).

In basins where demands for water are close to the limit of reliable supplies, such changes will have enormous policy implications. Where the river basins are shared by two or more nations, such changes will have political, and perhaps security, implications.

4 The Geopolitics of Shared Water Resources

Before discussing how climatic changes may affect the politics of shared water resources, it will be worthwhile to look at the different roles that fresh water resources already play in the international arena. There is a long history of water-related disputes, from conflicts over water supply to intentional attacks on water systems during wars (Gleick, 1991a). Even in the absence of future climatic changes, these water wars will continue and grow more intense in some places, as increased development and growing populations compete for limited supplies.

Conflicts between nations are caused by many factors — religious animosities, ideological disputes, arguments over borders, and economic competition. Resources and environmental factors are playing an increasing role in such disputes, but it is difficult to disentangle the many intertwined causes of conflict (Gleick, 1990a, 1991a). This section identifies four classes of water-related disputes and presents brief historical case studies of each: water resources as strategic goals, water systems as targets during war, water resources as tools of conflict, and resource inequities as roots to conflict.

4.1 Water resources as strategic goals

Conflicts over resources have occurred for many thousands of years. Thucydides, writing in ancient Greece, describes a conflict between the Athenians and Thasians over control of mineral resources during the Peloponnesian war, nearly 2500 years ago. Conflict over water resources has a more recent, but equally important history, most notably in the Middle East. This region, with its many conflicting ideological, religious, and geographical disputes, is also extremely water-poor. Even those parts of the Middle East with relatively extensive water resources, such as the Nile, Tigris, and Euphrates River valleys, are coming under increasing population, irrigation, and energy pressures. And every major river in the region crosses borders.

The characteristics that make water likely to be a source of political rivalry are:

(1) scarcity;
(2) the extent to which a source of supply is shared;
(3) the relative power of the basin states; and
(4) the extent to which each nation depends on the shared supply.

Perhaps the best example of all four is the Jordan River. This region has been the locus of intense interstate conflict since the establishment of Israel, and the riparian dispute over the Jordan River is an integral part of the conflict (Lowi, 1990). One outcome of the 1967 Arab-Israeli war was the occupation of significant portions of the headwaters of the Jordan by Israel, and the loss to Jordan of a significant fraction of its available water. Approximately 40% of the groundwater upon which Israel now depends — and more than 33% of its sustainable annual water yield — originates in the disputed territories. Indeed, almost the entire increase in Israeli water use since 1967 derives from the waters of the West Bank and the Upper Jordan River, (Lowi, 1990).

The Nile River is also a river of tremendous regional importance. It flows through some of the most arid regions of northern Africa, and is vital for agricultural production in Egypt and the Sudan. Ninety-seven percent of Egypt's water comes from the Nile River, and more than 95% of the runoff which flows into the Nile originates outside of Egypt, in eight other nations. A treaty signed in 1959 resolves a number of important issues, but was negotiated and ratified by only two nations, Egypt and the Sudan. Additional water develop-

[5] In general, GCMs underestimate increases in basin-scale runoff and overestimate increases compared to corresponding outputs from hydrologic models (Nash and Gleick, 1991). In an analysis of how one GCM reproduces the major hydrologic processes at a continental scale, significant differences in observed runoff and estimated runoff were seen (Thomas and Henderson-Sellers, in press, 1992).

ment in other upstream nations, particularly Ethiopia, could greatly increase tensions over water in this arid region. Indeed, the Egyptian Foreign Minister was quoted in 1985 as saying that "The next war in our region will be over the waters of the Nile, not politics."

4.2 Water resource systems as strategic targets

In addition to provoking conflict, water resource systems are regularly among the targets of war. Hydroelectric dams were routinely bombed during World War II and the Korean War. Irrigation systems in North Vietnam were bombed by the United States in the late 1960s. Dams, desalination plants, and water conveyance systems were targeted by both sides during the recent Persian Gulf War. As water supplies become increasingly valuable in water-scarce regions, their value as military targets increases.

4.3 Water supply as a weapon

In the last few years, non-military tools — including resource "weapons" and embargoes — have increasingly been used to achieve military ends. In some instances, the resource "weapon" manipulated was water. While fresh water resources are renewable, in practice they are finite, poorly distributed, and often subject to substantial control by one nation. In such circumstances, the temptation to restrict water flow for political purposes may prove irresistible. Even the perception that access to fresh water could be used as a political tool by another nation may lead to violence.

A strange twist on this problem surfaced in 1986, when North Korea announced plans to build a dam on the Han River, upstream of South Korea's capital, Seoul. This raised fears in South Korea that the dam could be used as a tool to disrupt their water supply, or to upset the ecological balance of the area. South Korean military analysts even predicted that the intentional destruction of the dam, and resultant flooding, could be used as a military weapon of attack. A formal request was made to halt construction, and South Korea built a series of levees and check dams above Seoul to try to mitigate possible impacts.

Similarly, in the early days of the Persian Gulf War, there were behind-the-scenes discussions at the United Nations about the Euphrates River in Turkey to shut off the flow of water to Iraq. While no such action was taken, the threat of the "water weapon" was made clear earlier in 1990 by the Turkish President, who threatened to restrict water flow to Syria to force it to withdraw support for Kurdish rebels operating in southern Turkey. While this threat was later disavowed, Syrian officials argue that Turkey has already used its power over the headwaters of the Euphrates for political goals, and could do so again (Cowell, 1990).

4.4 Water resource inequities as roots of conflict

Of all of the issues discussed so far, this one has the most direct ties to the question of economic and sustainable development. Like certain mineral resources, some renewable resources (such as water) suffer great maldistribution.

Unlike minerals, however, water is quite difficult to redistribute economically. In some regions, water availability is coming up against the limits of minimum water requirements — the so-called water barrier described by Falkenmark (1986). Although there is no doubt that great improvements in the efficiency of water use can be made, as can trade-offs between water-consumptive and water- efficient sectors, these actions only push back the barrier, they do not eliminate it. Some limits to supply will eventually be reached in fast-growing semi-arid nations and regions, despite efforts to reduce wasteful use and to redirect priorities.

The importance of clean water cannot be overstated. Enormous human suffering occurs because of the lack of satisfactory water for health and sanitation, despite great efforts in the 1980s — the International Drinking Water Supply and Sanitation Decade. In fact, there are as many people (or more) without access to safe drinking water or sanitation services today as there were a decade ago. Over one billion people — 30% of the population in developing countries — have no access to safe, clean water, and 1.8 billion — 43% of the developing world population — have no access to appropriate sanitation facilities (UNEP, 1991).

Lack of progress in this area during the 1980s was due in large part to population growth, the enormous (and growing) debt burden carried by developing countries, and the lack of industrial infrastructure for building and maintaining sanitation and water supply projects. Unless there is a continuing effort on the part of the richer nations to fill this gap, and an effort by the poorer nations to curb population growth and institute other changes, the world's water-related health burden will rise.

Similar inequities exist in the use of water for irrigation and energy. Only eight countries in Africa irrigate more than 10% of their cropland; over 60 countries worldwide fall in this category. In fact, nearly 20 nations in Africa have effectively *no* irrigation supply systems. Less than 4% of global hydroelectricity comes from Africa, with 12% of the world's population; in contrast nearly 30% comes from North America, with 6% of the world's population.

In most cases, these resource inequities will lead to more poverty, shortened lives, and misery rather than to direct violent conflict. But in some cases, these resource gaps will increase the likelihood of international disputes, the number of refugees who cross borders, and the inability to resist economic and military aggression. Although these ties to security issues are often indistinct and indirect, they remain real.

5 Determining a Nation's Water-Resource Vulnerability

As described above, four criteria are useful measures of a national or regional vulnerability to water resource conflicts:

(1) scarcity;
(2) the extent to which a source of supply is shared;

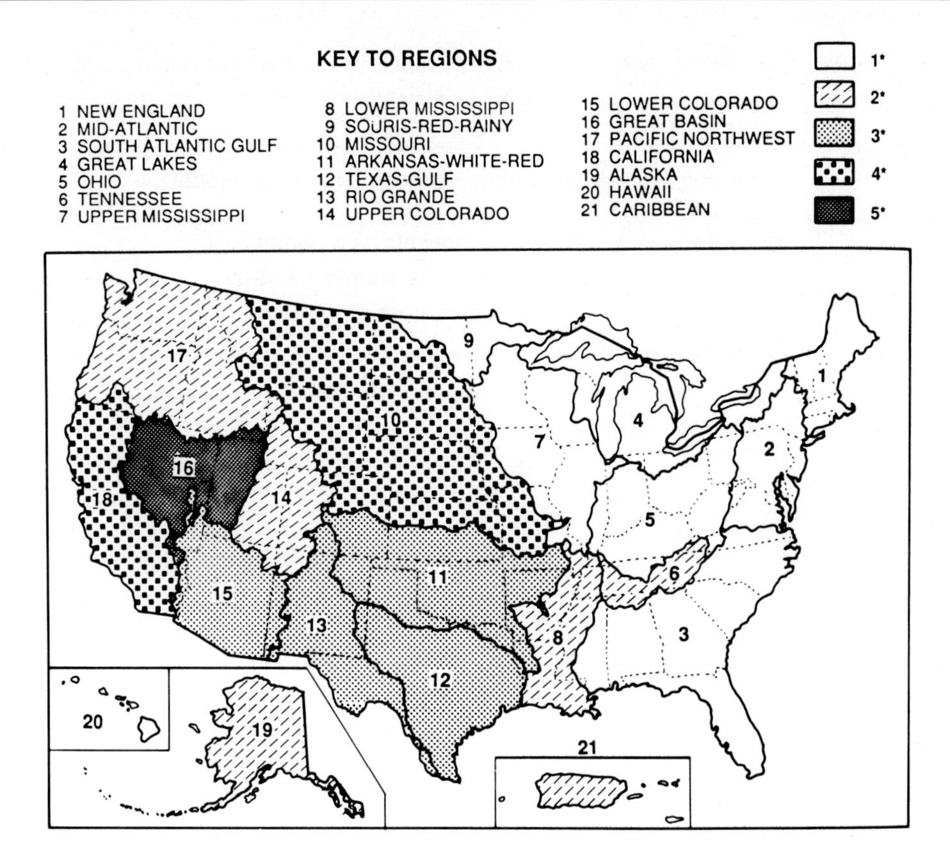

Figure 1. The 21 hydrologic regions of the United States, rated according to five important regional indices of water resources vulnerability: supply, demand, dependence on hydroelectricity, overpumping of groundwater, and hydrologic variability (Gleick, 1990b). White regions are vulnerable to one of these measures. The darkest region, the Great Basin, is vulnerable to all five. Much of the Western US is vulnerable to three or more measures. Source: Gleick, 1990, in P. Waggoner, ed., *Climate Changes and US Water Resources*, Copyright (c) 1990 by John wiley and Sons, Inc. Reprinted by permission of John Wiley and Sons, Inc.

(3) the relative power of the basin states; and
(4) the dependence on the shared supply.

Using these measures, rough indices of vulnerability can be developed and, to some extent, quantified. One series of regional indices of water resources vulnerability was developed in 1990, and applied first to the United States, though it could be applied to any region. It embodied measures of supply, demand, dependence on hydroelectricity, overpumping of groundwater, and hydrologic variability (Gleick, 1990b) These indices allowed the delineation of "regions at risk" in the US, which are at risk even in the absence of future climatic changes. See Figure 1.

Such indices cannot be complete. In many regions, water resource data are limited and unreliable, making the quantification of these indices uncertain. For some issues, more detailed regional data, or data on a seasonal basis rather than an annual average basis, are needed (Gleick, 1990b). Nevertheless, such measures can be extremely useful in highlighting regions and problems that may be of interest now or in the future. Several are presented here and more work should be done on refining these and developing new ones.

5.1 The shortage index

One way to evaluate water scarcity is the ratio of water demand (withdrawals) to available renewable water supply — a "shortage index" (Gleick 1990b). Countries whose present water withdrawals exceed one-third of their total renewable supply are listed in Table 3. In these countries, either water supply could be low or demands could be high. Nine nations are already forced to import freshwater, pump groundwater at a non-renewable rate, or desalinate non-potable sources at great expense in order to meet demand. All nine are in the Middle East.

5.2 Water availability per capita

Another measure of scarcity that takes into account changing populations is shown in Table 4. This Table lists countries where annual water availability (either in 1990 or by 2025) will fall below 1000 cubic metres per person.[6] For many of these countries, annual availability falls below 250 cubic

[6] Or where more than 1000 people will need to be supplied per million cubic metres of water — an equivalent measure proposed by Falkenmark (1986) and others.

Table 3. Nine countries use more than 100% of renewably available supply. This means they partly depend for their water supplies on imports of fresh water, groundwater, or desalination of brackish or salt water.

Water resources vulnerability: ratio of demand to supply

Country	Water withdrawals as a percentage of internal renewable supplies and river flows from other countries
Libya	374
Qatar	174
United Arab Emirates	140
Yemen	135
Israel	110
Jordan	110
Saudi Arabia	106
Bahrain	>100
Kuwait	>100
Egypt	97
Malta	92
Belgium	72
Cyprus	60
Tunisia	53
Afghanistan	52
Barbados	51
Pakistan	51
Iraq	43
Madagascar	41
Iran	39
Morocco	37

Note: Based on data from World Resources Institute 1990; Thomas Naff, personal communication.

metres per person. No developed country uses this little water. Even Israel, which has done a great deal to increase its water-use efficiency and minimize water-intensive development, uses over 400 cubic metres/year/person.

Note the substantial numbers of water-limited nations in Africa and Asia, and the few nations in Europe, the Pacific, or the Americas with these constraints. It is true that some developing nations (Argentina, Pakistan, Iraq, Egypt) exist in water-rich regions or regions highly dependent on irrigation. But major improvements in water-use efficiency are possible in all of those, should the demand for water begin to approach the supply, as it has in Egypt, Iraq, the Sudan, and regions of Europe and the United States.

5.3 Dependence on shared water resources

An index that measures the extent to which water supplies are shared between nations — and are hence vulnerable to competing interests — is shown in Table 5, which lists those nations with a high dependence on sources of water originating outside of their borders, or under the control of other nations. Besides the example already given of Egypt, 30 other nations receive more than one-third of their surface water across national borders. An alternative way to measure this vulnerability, also shown in Table 5, is the ratio of external supply to internal supply. Where this ratio is greater than 1, more than half of a nation's water supply could be subject to external political pressures and constraints.

6 The Role of Future Climate Changes in Altering the Geopolitics of Water Resources

At different times and different places, global climatic changes have the potential to both increase and decrease the likelihood of international frictions and tensions over water resources. Our challenge is to identify those cases in which conflicts are likely to be exacerbated, and to work to reduce the probability and consequences of those conflicts. Based on the changes in the water cycle that we expect from climate changes, each of the classes of water disputes described above is explored for its potential to alter international disputes and tensions over water.

Regional vulnerabilities to climate change, as well as to water-induced conflicts, are highly variable. Regions subject to droughts and water competition will benefit from increases in rainfall and suffer from decreases in rainfall. Areas vulnerable to periodic floods will suffer from certain climate-induced increases in runoff, and benefit from reductions in peak flows. Regions dependent on hydroelectricity for a substantial fraction of their energy production will be

Eighteen countries have a water availability per capita, in 1990, of 1000 cubic metres or less. This includes internally renewable flow, and flow from other nations. Falkenmark (1986) suggests that 500 cubic metres per person is a tentative minimum for a modern society in a semi-arid region, using extremely sophisticated water management techniques. In Oceania, no countries fall into this category, although some Pacific island nations may. Note the large numbers of countries in Africa and Asia with these water constraints. By 2025, 33 countries will fall in this category.

Table 4. Water resources vulnerability: per capita water availability today and in 2025

Country	Water availability per capita in 1990 cubic metres/person/year	Projected water availability per capita in 2025 cubic metres/person/year
AFRICA		
Algeria	750	380
Burundi	660	280
Cape Verde	500	220
Comoros	2040	790
Djibouti	750	270
Egypt	1070	620
Ethiopia	2360	980
Kenya	590	190
Lesotho	2220	930
Libya	160	60
Morocco	1200	680
Nigeria	2660	1000
Rwanda	880	350
Somalia	1510	610
South Africa	1420	790
Tanzania	2780	900
Tunisia	530	330
NORTH AND CENTRAL AMERICA		
Barbados	170	170
Haiti	1690	960
SOUTH AMERICA		
Peru	1790	980
ASIA/MIDDLE EAST		
Cyprus	1290	1000
Iran	2080	960
Israel	470	310
Jordan	260	80
Kuwait	< 10	<10
Lebanon	1600	960
Oman	1330	470
Qatar	50	20
Saudi Arabia	160	50
Singapore	220	190
United Arab Emirates	190	110
Yemen (both)	240	80
EUROPE		
Malta	80	80

Sources: Computed from United Nations population data and estimates; water availability data from World Resources Institute, 1990.

strongly affected by reductions in reservoir levels that result from prolonged shortages. Because of the weak regional predictive ability of climate models, this section evaluates qualitatively the greatest regional risks of climate-induced water conflicts, the uncertainties added by climatic changes, and mechanisms for reducing water-related tensions. Brief case studies of critical areas and present-day analogies are included where illustrative.

6.1 Rising temperature in arid lands

Water will continue to be a vital resource in arid and semi-arid regions of the world, and conflicts over access and possession are likely to worsen in regions that are particularly water-poor, such as the Middle East and Saharan Africa. The greatest certain threat from climatic change is the increase in evaporative losses and water demands caused by higher average temperatures. Even without changes in precipitation, as noted above, water availability can decrease by 10% or more simply owing to temperature increases of 2 °C — well within the range of expected changes. And these effects are independent of the increased demands from both human users and natural ecosystems that will occur at the same time.

6.2 Water shortages from unstable precipitation

When combined with temperature changes, plausible precipitation changes of ±10 to 20% can alter average annual runoff by 10 to 25%. Even greater percentage changes may

Table 5. Imported water

Country	Countries with more than 33% of total flow originating outside of border	Ratio of external water supply to internal supply[a]
Egypt	97	31.0
Hungary	95	18.2
Mauritania	95	17.5
Botswana	94	17.5
Bulgaria	91	10.4
Netherlands	89	8.0
Gambia	86	6.3
Kampuchea	82	4.7
Romania	82	4.6
Luxembourg	80	4.0
Syria	79	3.7
Sudan	77	3.3
Paraguay	70	2.3
Czechoslovakia	69	2.2
Niger	68	2.1
Iraq	66	1.9
Albania	53	1.1
Uruguay	52	1.1
Germany	50	1.0
Portugal	48	0.9
Congo	44	3.4
Yugoslavia	43	0.8
Bangladesh	42	0.7
Thailand	39	0.6
Austria	38	0.6
Jordan	36	0.6
Pakistan	36	6.6
Venezuela	35	0.5
Senegal	34	0.5
Belgium	33	0.5
Israel[b]	21	0.3

Notes

[a] Using national average annual flows. "External" represents river runoff originating outside national borders. "Internal" includes average flow of rivers and aquifers from precipitation within the country. Original data come from many different sources (World Resources Institute, 1990). Work is needed to improve the quality and consistency of water supply data.

[b] Although only 21% of Israel's water comes from outside current borders, a significant fraction of Israel's fresh water supply comes from disputed lands, complicating the calculation of the origin of Israel's surface water supplies. This would be affected by a political settlement of the Middle East conflict.

occur on a monthly basis. This is more than enough to cause serious problems in certain regions. By combining this information with data on per capita water availability and supply, highly vulnerable regions become apparent. One is the Middle East (particularly Jordan, Israel, Egypt, and the Arabian Peninsula). A recent review of the GCM-generated climate changes in the Middle East (Lonergan, 1991) shows both the uncertain nature of the changes and the possibility that the changes will be severe. For the region of the Jordan and Litani Rivers, three different GCMs estimate that precipitation could change by between -14 and +48%. For the region of the Nile, even larger changes in runoff in both directions are possible (Gleick, 1991b).

6.3 *Floods*

Floodplains and mountainous areas are vulnerable to increases in flow. The risks of flooding depend on the intensity of storms, the level of floodplain development, geomorphology, and the extent of physical protection, such as levees and dams. If estimates of increased intensity of monsoons are correct, southern Asia (Bangladesh, Bhutan, Kampuchea, Laos, and others) would be particularly vulnerable. Other regions of concern include central Sudan, Turkey, Zaire, eastern India, and Guyana. Fewer "security" risks arise from floods than from shortages — with the exception of regions where there are disputes over water management or the construction and operation of dams (such as south Asia), and regions where large numbers of displaced refugees may cross national borders, such as between Bangladesh and India.

6.4 *Water supplies as targets*

The increasing use of water and water supply systems as targets during conflict reflects their rising strategic value. As water becomes more important, water resources systems will become increasingly attractive as targets during wars. The Persian Gulf war provides the most disturbing example of this, with water-supply facilities, dams, desalination plants, and other water-delivery systems attacked by both sides. Nevertheless, this problem is secondary to more direct changes in water supplies or demand. It is an outcome of conflict, rather than a source of conflict.

To the extent that the likelihood of conflicts, in general, can be lessened by either changes in climate or by policy responses to the greenhouse effect, the targeting of water systems will also be lessened. There is also a question about the legality of attacks on environmental services and water resources systems, discussed in the concluding section of this chapter.

6.5 *Water supply as a weapon*

The more vulnerable a nation is to climate-induced water disruptions, or the weaker it is militarily, the more tempting the water weapon. Egypt, for example, is extremely vulnerable to intentional reductions in the flow of the Nile, and periodically offers rhetorical threats to its upstream neighbors. Yet Egypt has by far the stronger position militarily, and is therefore less vulnerable to actual disruption because of its ability (and implied willingness) to intervene with force. Of greater concern to Egypt is the potential for increased development in neighbouring states that would incrementally withdraw water from Egypt's current supply. This is discussed more fully below.

At the other extreme, Bangladesh, Iraq, and Mexico represent weaker, downstream nations. These nations would suffer from a climatic change that reduced flows, since the stronger, upstream nations are in a position to dominate the use of remaining water and require that shortages be borne by the downstream parties. In the case of Mexico, a treaty adopted by the United States and Mexico allocates water and, in theory, shortages. In practice, a long-term shortfall caused by climatic change would be likely to lead to disputes and renegotiations over deliveries (Gleick, 1988).

6.6 *Water as a brake on development*

A more subtle problem is the way in which climatic changes will worsen the disparities in water use for irrigation, development, and sanitation. Growing levels of economic development will increase the pressure on water resources. In water-limited regions, growing demands for fixed supplies are likely to increase the tensions among nations, and may play a direct role in increasing or decreasing economic disparities.

In Northern Africa, the Sudan is considered to be one of the few nations with great potential for increased irrigation: there is sufficient arable land, and there is, in theory, sufficient water in the Nile. In reality, however, a significant fraction of the water in the Nile belongs to Egypt — according to a treaty signed in 1959, which allocates the flow of the Nile between the two countries (Gleick, 1990c, 1991b). In order for the Sudan to increase its agricultural production, it would have to increase its use of Nile River water, or the greenhouse effect would have to increase water availability substantially. Similarly, Ethiopia has periodically talked about using more water from the Blue Nile to meet growing demands for food, but Egypt always objects on grounds that this would reduce its own total water supply.

7 The Role of Uncertainty

Perhaps the most important effect of climatic change for water resources will be to greatly increase the overall uncertainty associated with water management and supply. Rainfall, runoff, and storms are all (in the language of hydrologists) stochastic — they are all natural events with a substantial random component to them. In many ways, therefore, the science of hydrology is the science of estimating the probabilities of certain types of events. Dams are sized based on the probability of floods of a certain size occurring and the expected frequency of droughts. These "expectations" come from the study of the recent past, in some cases no more than 30 or 40 years of records.

Already, we often find ourselves in trouble when we try to estimate climatic probabilities. Two examples from the long history of water in the western United States highlight this problem. The Folsom Dam in California was designed in the 1950s for what was thought to be the 100-year flood — that size of flood that comes only once in 100 years. But owing to inadequate records, the flood for which the dam was designed was actually more common than had been expected, and the flood that comes only once in 100 years is substantially larger than expected. As a result, the Folsom Dam will be subjected to unanticipated, severe flooding. This may not have been caused by "climate change," but it is the sort of problem which climate change will make much more common.

Similarly, in 1922 the Colorado River was apportioned among the seven states that share the water. At the time, planners used the best — but quite limited — information on total flows in the river. In fact, the estimates of available flow in the river were based on what we now know to have been an unusually wet period. As a result, there is considerably less water available on average to users in some states than initially anticipated. If we had known then what we know now about the hydrology of the river, the allocations and the methods used to allocate the water would have been very different.

Recent studies, in fact, suggest that the present methods of allocation and operation will leave the Colorado and its users open to significant water supply and quality problems (Nash and Gleick, 1991, 1992). But none of the agencies responsible for Colorado water supply has yet indicated a willingness to even consider the effectiveness of changing their operating rules in order to improve the system's ability to handle possible changes.

Part of the reluctance of resource managers to begin planning for such changes stems from the difficulty in identifying long-term permanent changes as opposed to short-term climatic variability. The kinds of institutional changes necessary to deal with each of these are, in many instances, quite different. For example, drawing down a reservoir to meet short-term shortfalls is very different from changing demand structures and supply commitments. Adding to this problem is the fact that much water data is still classified as secret by national governments. Changes in flow could therefore be perceived and misinterpreted by downstream nations as intentional manipulations rather than geophysical events.

8 The Future of Water Resources

Future climatic changes effectively make obsolete all our old assumptions about the behaviour of water supply. Perhaps the greatest certainty about future climatic changes is that the future will not look like the past. We may not know precisely what it will look like, but changes are coming.

When are the worst problems likely to appear? We cannot say. Some regions may already be experiencing changes in extremes, but this is uncertain and likely to be uncertain for some time. Climatic changes will occur within the context of an already highly variable climate. Indeed, much of the current dispute over the greenhouse effect centres on the question of "detection" and whether the greenhouse effect can yet be seen. A better question is: How should we react given the different levels of certainty and uncertainty? How can we minimize water-related disputes regardless of the direct cause?

8.1 Changes in water use patterns

Whatever happens to the climate, human needs for water will continue to grow, because of increasing development and rapidly growing populations. In many regions of the world, water is already a scarce resource, to be used and reused many times, and occasionally to be fought over. Many countries in the Middle East and elsewhere already use water at a rate faster than natural processes can replenish it, leading to falling ground water levels, expensive desalination projects, and imports of water across borders. Oddball schemes that would have been laughed at a few decades ago are now being implemented or seriously considered, including the importation of water in tankers, pipelines thousands of kilometres long, or the diversion of icebergs from the poles.

We have already noted Israel's strategic emphasis on water conservation. But all countries, including Israel, can do more to reduce the wasteful use of water, improve water efficiency, and thus extend the amount of water available for other use. Any nation which accomplishes this could, in the geopolitical future, be in a stronger military position. The task is hard, however, in part because most data on water use do not differentiate between water withdrawn from the system and water consumed. In other words, better data on water consumption, along with better economic analyses of the true cost of water consumption, are needed.

8.2 International strictures on military/environmental aggression

International law and intergovernmental institutions must play a leading role in reducing environment and resource-related conflicts. There have already been some attempts to develop acceptable international law protecting environmental resources. For example, the Environmental Modification Convention of 1977, negotiated under the auspices of the United Nations states:

> *"Each State Party to this Convention undertakes not to engage in military or any other hostile use of environmental modification techniques having widespread, long-lasting or severe effects as the means of destruction, damage or injury to any other State Party."* (Article I.1)

In 1982 the United Nations General Assembly promulgated the World Charter for Nature, supported by over 110 nations, which states:

"Nature shall be secured against degradation caused by warfare or other hostile activities" (Article V) and "Military activities damaging to nature shall be avoided" (Article XX).

Other relevant international agreements include the 1977 Bern Protocol on the Protection of Victims of International Armed Conflicts (additional to the Geneva Conventions of 1949), which states:

"It is prohibited to employ methods or means of warfare which are intended, or may be expected, to cause widespread, long-term and severe damage to the natural environment." (Article XXXV.3)

and

"Care shall be taken in warfare to protect the natural environment against widespread, long-term and severe damage. This protection includes a prohibition of the use of methods or means of warfare which are intended or may be expected to cause such damage to the natural environment and thereby to prejudice the health or survival of the population." (Article LV.1)

Unfortunately, these agreements carry less weight than might be desired in the international arena when politics, economics, and other factors in international competition are considered more important. Until such high-sounding agreements have some enforcement teeth and are considered true facets of international behavior, they will remain ineffective.

8.3 Water rights dispute resolution

There are also important roles to be played by international water law and relevant intergovernmental institutions. Unfortunately, the history of international water law is also incomplete and offers only limited guidance. No satisfactory water law has been developed that is widely acceptable among nations, despite years of effort by the International Law Commission (ILC of the United Nations). Indeed, in 1991, the ILC completed the "first reading" or provisional adoption of the set of 32 draft articles on the Law of the Non-Navigational Users of International Watercourses. Among the general principles set forth are those of equitable utilization, the obligation not to cause harm to other riparian states, and the obligation to exchange hydrologic data and other relevant information on a regular basis.

Most effective have been a wide range of water treaties negotiated and signed by a small number of parties in an affected region. There are dozens of major international river treaties covering everything from navigation to water quality to water rights allocations. These treaties have helped reduce the risks of water conflicts in many areas, but none of them explicitly incorporates the issue of climate change, and not all of them are likely to prove effective in holding up to significant, permanent changes in the hydrologic cycle, such as those that may be produced by the greenhouse effect. In addition, some of them are beginning to fail on their own accord, as changing levels of development alter the water needs of regions and nations.

To make both regional treaties and broader international agreements over water flexible, detailed mechanisms for conflict resolution and negotiations need to be developed, basic hydrologic data need to be acquired and completely shared with all parties, flexible proportional water allocations are needed, rather than fixed allocations, and strategies for sharing shortages and apportioning responsibilities for floods need to be developed before climatic changes become an important factor.

To do less is, in essence, to do nothing.

Water is life.
The Greek poet, Pindar

References

Budyko, M.I., 1982: The Earth's climate: past and future. *International Geophysics Series,* **29**, 307, New York, Academic Press.

Bultot, F., A. Coppens, G.L. Dupriez, D. Gellens and F. Meulenberghs, 1988: Repercussions of a CO_2 doubling on the water cycle and on the water balance: a case study for Belgium. *J. Hydrol,* **99**, 319-347.

Cowell, A., 1990: Water rights: plenty of mud to sling. *The New York Times,* February 7, A4.

Deudney, D., 1991: Environment and security: muddled thinking. *The Bulletin of the Atomic Scientists,* **47**, 22-28.

Emanuel, K.A., 1987: The dependence of hurricane intensity on climate. *Nature* **326**, 483-485.

Falkenmark, M., 1986: Fresh water — time for a modified approach. *Ambio,* **15**, 19.

Gleick, P.H., 1987: Regional hydrologic consequences of increases in stmospheric CO_2 and other trace gases. *Climatic Change,* **10**, 137-161.

Gleick, P.H., 1988: The effects of future climatic changes on international water resources: The Colorado River, the United States, and Mexico. *Policy Sciences,* **21**, 23-39.

Gleick, P.H., 1989: Climate change, hydrology, and water resources. *Reviews of Geophysics,* **27**(3), 329-344.

Gleick, P.H., 1990a: Environment, resources, and international security and politics. In *Science and International Security: Responding to a Changing World,* E. Arnett, ed., American Association for the Advancement of Science AAAS, Washington, DC, pp. 501-523.

Gleick, P.H., 1990b: Vulnerability of water systems. In *Climate Change and US Water Resources,* P.E. Waggoner, ed., New York, John Wiley and Sons, pp. 223-240.

Gleick, P.H., 1990c: Climate changes, international rivers, and international security: the Nile and the Colorado. In *Greenhouse Glasnost*, R. Redford, T.J. Minger, eds., New York, The Ecco Press, pp. 147-165.

Gleick, P.H., 1991a: Environment and security: clear connections. *The Bulletin of the Atomic Scientists,* **47**(3), 17-21.

Gleick, P.H., 1991b: The vulnerability of runoff in the Nile Basin to climatic changes. *The Environment Professional,* **13**, 66-73.

IPCC (Intergovernmental Panel on Climate Change), 1990: *Climate Change: The IPCC Scientific Assessment,* Cambridge, Cambridge University Press, pp. 365.

Kellogg, W. and Zong-Ci Zhao, 1988: Sensitivity of soil moisture to doubling of carbon dioxide in climate model experiments. Part I: North America. *J. Climate,* **1**, 348-366.

Lettenmaier, D.P., T.Y. Gan and D.R. Dawdy, 1988: Interpretation of hydrologic effects of climate change in the Sacramento-San Joaquin River Basin, California. *Water Resources Technical Report,* **110**, Seattle, University of Washington.

Lonergan, S., 1991: Climate warming, water resources, and geopolitical conflict: a study of nations dependent on the Nile, Litani and Jordan river systems, *ORAE* Paper No. 55, Ottawa Canada, National Defence.

Lowi, M. R., 1990: The politics of water under conditions of scarcity and conflict: The Jordan River and Riparian States, Department of Politics, Princeton, New Jersey, Princeton University.

Manabe, S. and R.T. Wetherald, 1986: Reduction in summer soil wetness induced by an increase in atmospheric carbon dioxide. *Science,* **232**, 626-628.

Mimikou, M.A. and Y.S. Kouvopoulos, 1991: Regional climate impacts: I. impacts on water resources. *Bulletin of Intern. Assoc. Hydrol. Sciences* (In press.)

Nash, L.L and P.H. Gleick, 1991: The sensitivity of streamflow in the Colorado Basin to climatic changes, *Journal of Hydrology,* **125**, 221-241.

Nash, L.L. and P.H. Gleick, 1992: The sensitivity of streamflow and water supply in the Colorado Basin to climatic changes. US Environmental Protection Agency, Washington, DC.

Noda, A. and T. Tokioka, 1989: The effect of doubling the CO_2 concentration on convective and non-convective precipitation in a general circulation model with a simple mixed layer ocean. *J. Met. Soc. Japan* **67**, 1055-1067.

Rango, A. and V. van Katwijk, 1990: Developing and testing of a snowmelt-runoff forecasting technique. *Water Resources Bulletin* **26**(1), 135-144.

Rind, D., R. Goldberg, J. Hansen, C. Rosenzweig and R. Ruedy, 1990: Potential evapotranspiration and the likelihood of future drought. *J. Geophys. Rev.* **95**, 9983-10004.

Rosenberg, N.J., B.A. Kimball, P. Martin and C.F. Cooper, 1990: From Climate and CO_2 enrichment to evapotranspiration. In *Climate Change and US Water Resources.* P.E. Waggoner, ed., New York, John Wiley and Sons, pp. 151-176.

Schneider, S.H., P.H. Gleick and L.O. Mearns 1990: Prospects for climate change. In *Climate Change and US Water Resources.* P.E. Waggoner, ed., New York, John Wiley and Sons, pp. 41-74.

Shiklomanov, I., 1987: Effects of climatic changes on Soviet rivers. *International Symposium of the XIXth General Assembly, International Union of Geodesy and Geophysics (IUGG),* Vancouver, BC. Canada, August 9-22, 1987.

Thomas, G. and A. Henderson-Sellers, 1992: Global and Continental water Balance in a GCM. *Climatic Change* (in press, April 1992).

UNEP, United Nations Environment Programme. 1991: The state of the world environment in 1991. *Climate Change: Need for Global Partnership,* Nairobi, Kenya.

Waggoner, P.E., ed., 1990: *Climate Change and US Water Resources,* New York, John Wiley and Sons.

Wetherald, R.T. and S. Manabe, 1981: Influence of seasonal variation upon the sensitivity of a model climate. *J. Geophys. Res.,* **86**(C2), 1194-1204.

Wilson, C.A. and J.F.B.Mitchell, 1987: A doubled CO_2 climate sensitivity experiment with a global climate model including a simple ocean. *J. Geophys. Res.,* **92**, 13, 315-13, 343.

WRI (World Resources Institute), 1990: *World Resources 1990-91: A Guide to the Environment.* New York, Oxford University Press.

CHAPTER 10

Effects of Climate Change on Weather-Related Disasters

James K. Mitchell and Neil J. Ericksen

Editor's Introduction

From the beginnings of history, human societies have been rocked by the shocks of weather-related disasters. Hot spells, cold snaps, hurricanes, typhoons, wind storms, droughts, and floods have killed people and devastated communities. The brief but painful moments of disaster — and their lengthy aftermaths — have long been the subjects of poems, folktales, novels, and historical writing. Among non- weather-related hazards, only earthquakes have killed or displaced so many people in such brief and bitter episodes.

With or without climate change, the precise timing, frequency and location of weather-related disasters are among the most difficult things to predict. We do not now (and perhaps can not ever) know when the next major disasters will occur. Only one thing is certain: extreme weather events will continue to menace civilization in the future, as they have in the past.

The situation may become more complicated in the future, however. Climate model studies indicate that a global warming resulting from the buildup of greenhouse gases will not be evenly distributed around the planet. The warming will be amplified in high-latitude regions to two or three times the global average effect. The temperature increase in the tropics is expected to be only 50-75% of the global average warming. This differential warming effect will reduce the thermal gradient from equator to poles, shifting the traditional patterns of winds and ocean currents. Some analysts believe that this change in air and ocean currents will increase the frequency of extreme weather events, and shift them, in some cases, to new locales.

Ken Mitchell, Professor of Geography at Rutgers University in the US, and Neil Ericksen, Professor of Geography at the University of Waikato in New Zealand, have spent most of their adult lives studying natural hazards and their effects on human societies. Mitchell has been one of the leading voices in the organization of the UN's International Decade for Natural Disaster Reduction. Mitchell and Ericksen observe that extreme weather events do not necessarily result in disasters. Disasters, events with extensive loss of life and damage to property, result from the interaction of natural hazards and human societies. While extreme weather events are inevitable, most disasters are probably avoidable.

Mitchell and Ericksen have analyzed the recent record of damages due to weather-related disasters. They note that, during a recent 20-year period, natural disasters resulted in about 2.8 million deaths and adversely affected over 820 million people. Costs were high in both industrialized and developing countries, but they were not evenly distributed or equally important to the effected economies. In 1989, approximately US $7.6 of losses from natural disasters were recorded in the US, with more than half due to one event, Hurricane Hugo. For comparison, three countries in Latin America — Bolivia, Ecuador, and Peru — suffered losses of almost US$4 billion as a consequence of the El Niño weather event of 1982-83. But while the 1989 losses in the US from Hurricane Hugo represented less than 0.1% of US GNP, the losses from El Niño represented almost 10% of GNP (a fraction almost 100 times larger) for the three Latin American countries. The disaster cost them the equivalent of nearly 50% of their annual tax revenues for those years.

The disproportionate impact of these natural disasters on the poorer and more vulnerable countries of the South is typical of recent decades. The heavy social and economic costs from these events, imposed on societies that are already under heavy stress from other factors, may contrib-

ute, according to Mitchell and Ericksen, to the evolution of waves of environmental refugees — mostly poor and now homeless people who feel forced to leave their homelands in search of more hospitable economic and physical climates. Recent experiences with the relatively small numbers of Vietnamese "boat people" and Cuban Marielitos indicate that most governments are ill-prepared to provide emergency shelters for such people and disinclined to accept them as permanent residents of their adopted countries.

But as a result of their careful study of recent experiences with natural hazards, Mitchell and Ericksen are optimistic that much can be done to reduce the expected damages from natural disasters in general, and from extreme weather events in particular. The key in both cases is to shift the emphasis from post hoc damage control to anticipatory strategies that strengthen local capacity to deal with the stress of natural disasters and include disaster reduction as an integral component of investment and development activities. The programme, already embodied in the International Decade for Natural Disaster Reduction, emphasizes three things: (1) assessments of damage potential, before the event; (2) development of institutional capability to provide warning of impending events and to prepare local communities; and (3) the development of disaster-resistant infrastructures and institutions in the context of disaster-sensitive investment programmes.

These measures will provide significant economic benefits, especially in developing countries, irrespective of whether the buildup of greenhouse gases causes a significant increase in the frequency and severity of extreme weather events. If broadly implemented, they will "pay off," whether or not we improve the ability of the scientific community to predict the timing and duration of these events. Of course, scientific improvement would further reduce the damages of sudden-onset disasters and potentially help nullify the losses from rapid or gradual climate change. A joint strategy that simultaneously addresses the risks of natural hazards and the potential damages from weather-related disasters will capture important synergisms. The most valuable of these will be the strengthening of institutional competence and coordination. By linking the interests of members of the disaster reduction community with those in the science and policy communities, especially scientists concerned with climate change, strategies such as these will build a larger, stronger, and more effective community to support sustainable development.

- I. M. M.

1 Weather Extremes in a Warming World

The Bangladesh cyclone of April 1991 killed at least 125,000 people, devastated hundreds of settlements, laid waste vast areas of land, and inflicted a major setback to economic and social development in one of the world's poorest countries (World Bank, 1991). The cyclone was a grim reminder that much of the global population maintains an uncertain existence in the face of extreme natural events. It illustrated both the continuing vulnerability of society to such events and it underscored the central role of humans as agents of disaster. Humans are not simply pawns of nature; we modify natural processes - sometimes deliberately, sometimes inadvertently. Human interactions with extreme events determine whether floods, droughts and other phenomena merely pose hazards or lead to major losses. We have the capacity to avoid, prevent or reduce losses by engaging in protective actions.[1]

If global temperatures continue to rise in accordance with current predictions (Kerr, 1991), increases in the number and severity of storms, floods, droughts, and other short-term weather extremes may be one of the earliest observed and most dramatic effects (see Figure 1). Societies that are unable to increase capacities for dealing with extreme events will experience more disasters. In short, it is feared that the road to a warmer world may be paved with weather-related disasters.[2]

The human implications of increased disasters are large, partly because disasters inflict heavy losses on society. They are also important because disaster prevention and protection institutions play important roles in contemporary public life. Planning for disaster — and coping with it — is a widely acknowledged responsibility of governments, intergovernmental agencies, voluntary organizations, insurance companies, and many other groups. In recent years the policies, programmes and measures developed to cope with natural

[1] A natural hazard is an extreme event that threatens humans or what they value. A "natural disaster" is a natural hazard that overwhelms societal coping capabilities and inflicts heavy losses. Depending on human actions, all extreme events have the potential to become natural disasters. Likewise, the success of disaster reduction efforts is also a function of human actions. These relationships can best be appreciated by way of example. On average about 45 fully developed tropical cyclones occur each year somewhere on Earth. They comprise one set of extreme events. Approximately 15 of these storms approach or cross land, thereby constituting natural hazards. Perhaps two or three of the latter inflict substantial losses and are recorded as natural disasters.

[2] As used here the terms "weather" and "climate" serve to distinguish between short-term and long-term atmospheric conditions. "Weather" refers to atmospheric conditions that are experienced by humans on time scales of seconds to weeks or months. Most extreme natural events that threaten society fall into this class: lightning strikes, tornadoes, tropical cyclones, etc. "Climate" refers to the mean state of the atmosphere over periods of years to decades and perhaps longer. Long-lasting droughts and some other atmospheric hazards might be considered as features of climate.

Figure 1. Global warming may trigger a cascade of hazard effects - both directly through the mechanism of climate change and indirectly via sea level rise. Source: *Climate Change and Natural Disasters* (adapted from Fig. 3.11 in Campbell and Ericksen, 1990).

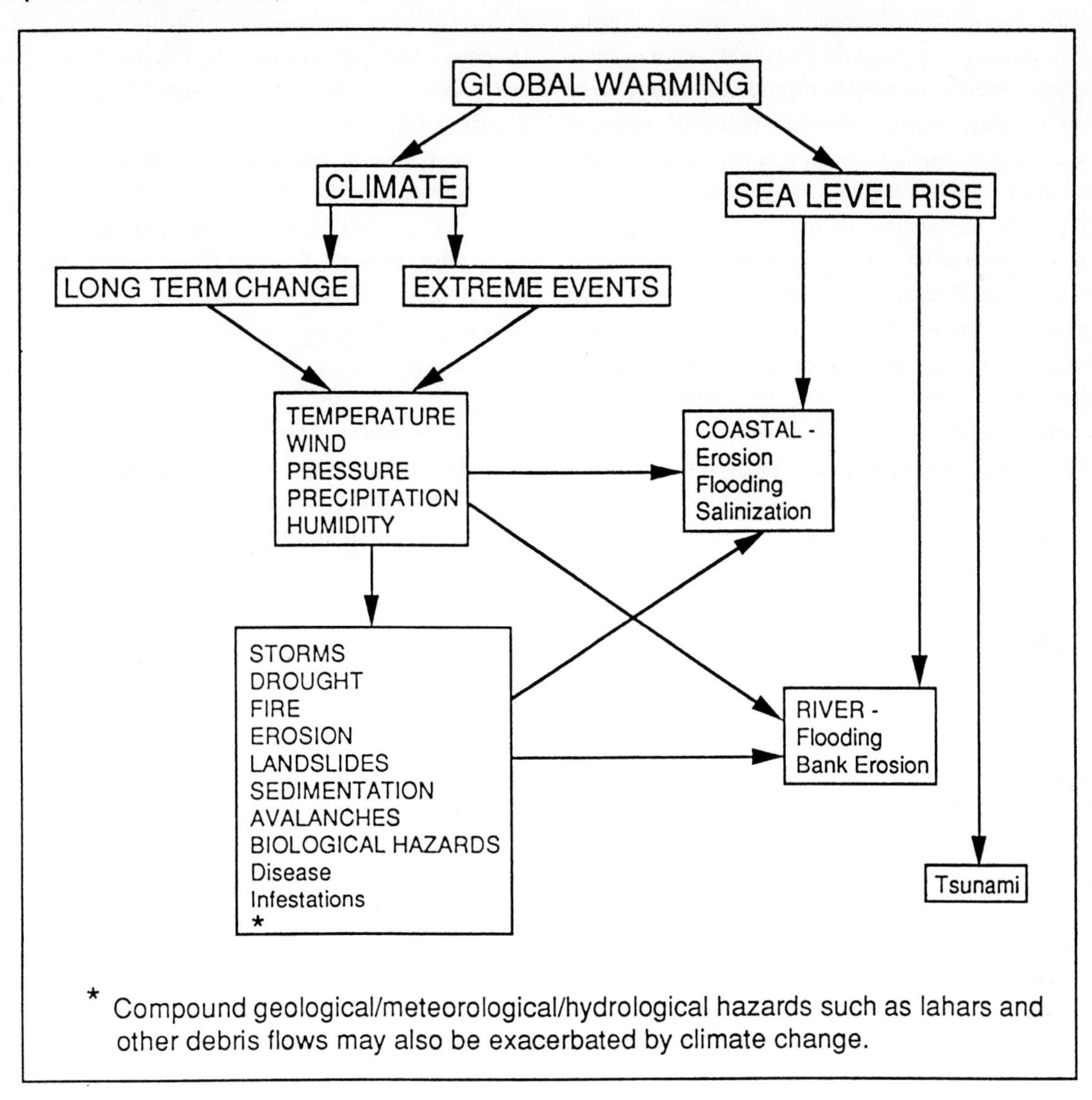

disasters have become well known and widespread. In this chapter, we suggest that publicly-acceptable mechanisms for managing one type of environmental risk (i.e. natural disasters), may prove to be valuable vehicles for reducing the risks and impacts of another type of environmental risk - climate change.

The chapter addresses a variety of questions concerned with managing the impacts of rapid climate change. How valid is the concern that there will be more extreme events? Will worsening weather extremes lead to larger disasters? What are the trends in disaster losses? Why are losses not being reduced? What should be done to prepare societies for increasing disasters in a warmer world? Our answers to these questions suggest that a major effort to apply existing knowledge about the reduction of disasters will yield large benefits if climate change occurs as now anticipated — and will remain amply worthwhile even if existing patterns of regional climate persist unchanged into the foreseeable future.

2 The Potential for Future Extreme Weather

2.1 Will climate change produce more extreme weather?

One question often asked by policy makers is: "Will global climate change produce more extreme weather?" Some atmospheric scientists are convinced that climate change will exacerbate weather extremes, but many are uncertain about the outcomes. Theoretical and empirical evidence in support of more extreme weather is not difficult to find, though it is rarely definitive. As Parry points out, "...if there is a change in the central tendency of the climate, an important medium through which (it) is felt may be...a change in the frequency of the extreme anomalies"(Parry, 1986). The implications of such a change are significant in cases where existing extremes already pose risks for human activities. Moreover, it is entirely possible that anticipated climate warming might increase the probability of extreme events disproportionately. According to Wigley, a shift of one standard deviation in the mean annual surface temperature of the planet might cause the probability of an extreme (temperature) event to increase from once in 20 years to once in 4 years (Wigley, 1988). An event that previously occurred once in a 100 years would occur once in 11 years. Precipitation and other weather variables may be affected similarly.

The societal implications of such a "shift to risk" are considerable because the probability of catastrophic events greatly increases. Events such as the 100 year flood, that were unlikely to happen within an average lifetime, become events that almost everyone can expect to experience.

Some analysts point to evidence that sub global atmospheric and oceanic processes are sensitive to changes in energy fluxes of a magnitude that might accompany anticipated climate warming. For example, warming may increase the average global sea level or the mean oceanic surface temperatures. Studies of El Niño and La Niña phenomena associated with the Southern Oscillation clearly illustrate that annual changes of sea-surface temperatures in the Pacific Ocean are followed by shifts in the world-wide occurrence of extreme weather (Nicholls, 1989; Swetnam and Betancourt, 1990; Leetmaa, 1989). Processes of similar scale affect much of South Asia where increased flood and drought losses are often directly connected with the occurrence of abnormal region-wide monsoons (Swaminathan, 1987). Gray's recent research on relationships between rainfall over West Africa and hurricanes in the United States also suggests that changes in the atmospheric energy regime of one region can have repercussions for hazards in distant locations (Gray, 1990). All of this circumstantial evidence points towards increased weather extremes as a consequence of global warming.

Some of the most convincing evidence has been assembled by Pittock and his colleagues at Australia's Commonwealth Scientific and Industrial Research Organization (CSIRO). This work examines the behaviour of monsoons, tropical cyclones and similar phenomena in a warming world using information drawn from General Circulation Models (GCMs), paleo-climatic and historical data, and high-resolution limited area models (Whetton and Pittock, 1991a, b; Pittock, Fowler and Whetton, 1991; Whetton, Fowler and Haylock, 1992). Among other findings, Pittock concludes that flood frequency will increase because rainfall is likely to become more intense, and that dry spells will also be more common because rain will fall on fewer days (Whetton and Pittock, 1991a,b). Similar conclusions have been reached by New Zealand analysts (Thompson, 1990).

For countries like Australia and India, the consequences of such changes are portentous. Climate change may redraw the map of extreme natural events. In the words of the late Walter Orr Roberts (President Emeritus of the University Corporation for Atmospheric Research): "Hazards such as storm surges, riverine flooding, and drought will change in frequency, magnitude and location. There will be winners as well as losers; hazards may be reduced in some areas, but most areas will lose because the rate of change threatens to throw resource management systems out of balance." (Roberts, 1989) Other disaster experts draw similar conclusions about the risks of rapid climate change (Henderson-Sellars and Blong, 1989; Walker, 1989).

Though most climate experts recognize the possibility of links between climate change and weather extremes, interpretations are often circumspect about the timing and severity of future impacts. Working Group I of the Intergovernmental Panel on Climate Change (IPCC) offers a guarded assessment of the risks of climate change during the next century:

> *"If large scale weather regimes, for instance depression tracks or anticyclones, shift their position, this would affect the variability and extremes of weather at a particular location, and could have a major effect. However, we do not know if, or in what way, this will happen. ...Although the theoretical maximum intensity is expected to increase with temperature, climate models give no consistent indication whether tropical storms will increase or decrease in frequency or intensity as climate changes; neither is there any evidence that this has occurred over the past few decades"* (World Meteorological Organization, 1990).

A recently completed US report on policy implications of global warming comes to similar conclusions about mid-latitude storms as well as tropical storms (National Research Council, 1991).

In summary, there is cautious general agreement that worsening weather extremes are possible and portentous; but few are willing to predict the magnitude and timing of future extremes, much less the regional pattern of events.

2.2 *Will worsening weather extremes lead to larger disasters?*

Climate change experts are reluctant to predict how weather extremes may change, but are quick to recognize the possible implications if changes occur. Increased impacts of drought on agriculture and water supplies are feared. Some analysts are concerned that persistent, multi-year droughts may increase the likelihood of large brush and timber fires. The 1991 fire near Oakland, California (which caused more than $3 billion in damages) is indicative of the kinds of events that could occur with increasing frequency as a result of accelerated greenhouse warming.

Another concern often raised is that flooding will become worse, particularly in burgeoning urban areas of developing countries — such as Calcutta, Bombay, Shanghai, Rio de Janeiro, Buenos Aires, Jakarta, Karachi, Dhaka, Manila, and Bangkok. Some of these cities — and many heavily settled maritime plains — are also exposed to increased risks of tropical cyclones and rising sea levels (Silver and DeFries, 1990). Expansion of warm surface waters in the tropical oceans might shift hurricane tracks and extend their poleward range. A warmer ocean might also mean a longer hurricane season and a more intense one. Areas that are presently exempt from such storms might receive many. Populations that have not experienced the hurricane's combination of storm surges, high winds and heavy rainfall might be placed in jeopardy. Existing storm tracking and warning systems might have to be extended and new forecasting procedures might become necessary. Comparable changes associated with other types of hazards might occur elsewhere along the fringes of the ecumene (the habitable Earth) as semi-arid areas expand, permafrost acreage shrinks, ocean waters penetrate further inland, and agriculture invades higher altitude regions.

2.3 Will human activities mitigate or exacerbate disasters?

But it is important to realize that, even if weather extremes worsen, disasters are not inevitable. Physical risks are only one element — and sometimes not even the most important element — in the calculus of disaster (Mitchell, Devine and Jagger). Human-related factors such as exposure, vulnerability, response, and thresholds of acceptable loss, as well as contextual variables, also help to determine the scale of damage that results from weather events. The major components of natural hazard systems and feedback relationships are shown in Figure 2.

In recent decades researchers have highlighted human contributions to disaster. People are now viewed as major contributors to disasters — not just their victims. By altering the paths of major rivers, building new reservoirs where lakes never previously existed, and constructing urban "heat islands", engineered systems can modify geological, hydrological and meteorological processes — both deliberately and inadvertently. Moreover, the scale of human intervention is judged to be growing rapidly as pollution endangers coral reefs and governments consider projects that would reverse the flow of water in major river systems. The rate of land conversion from unmanaged forest, wetland, or range to agriculture and other managed systems is burgeoning almost everywhere except North America and Europe (Richards, 1990). Coasts that once were relatively free of artificial structures are now festooned with jetties, seawalls, bulkheads, and similar devices (Walker, 1990). In nearly every country, expanding urban settlements are pushing onto steep slopes, wetlands, or other hazardous areas.

Humans also affect the potential for disasters in more subtle ways. For example, we establish the criteria for designating disasters, and for defining acceptable thresholds of loss. Disaster criteria often differ among national governments, levels of government, and types of designating institutions. In the United States, an average of 20 to 30 Presidential disaster declarations are issued every year, while the American Red Cross typically records 30,000-40,000 (lesser) disasters during the same period. Furthermore, losses that are regarded as exceptional in one country may be defined as routine in another. Perhaps, most important, vulnerability to loss and resilience in the wake of disaster are mostly determined by sociopolitical and economic considerations - not natural phenomena or characteristics (Smith, 1991; Liverman, 1990). Groups that are usually less able to protect themselves against disasters include the poor (particularly women), young children, the elderly, the infirm, and the handicapped (Pankhurst, 1984). Those who live in flimsy or poorly engineered houses are also disproportionately likely to be victims.

Geographic isolation, political dependency, economic vulnerability, and ethnic tensions are important contributors to many disasters, from typhoons in the Pacific to drought in Africa. For example, the extended drought that struck the Horn of Africa in the mid-1980s became a major disaster

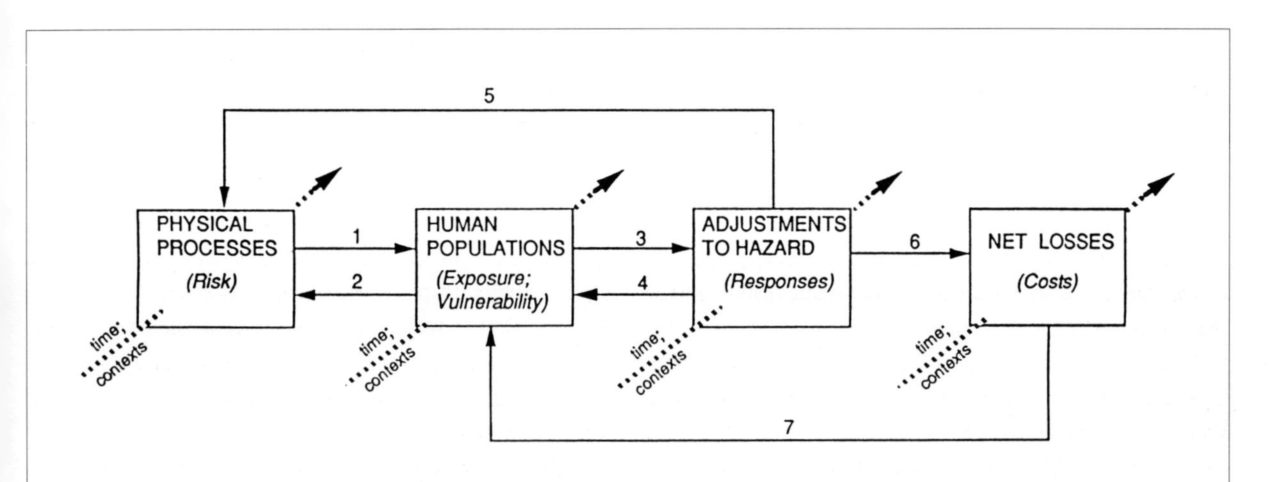

Figure 2. The major components of natural hazard systems and feedback relationships. At any one time a natural hazard represents the outcome of direct interaction among several different physical and human components, plus the cumulative effects of long-term external influences. Source: Mitchell, Devine and Jagger, 1989.

because local civil strife effectively prohibited relief workers from delivering supplies to the isolated areas that were experiencing the most severe impacts. Political rivalries within the country created institutional obstacles that kept disaster relief teams from rebuilding the local infrastructure and made it impossible for these workers to assist local farmers in adjusting to the new weather conditions.

Small, geographically remote, isolated, economically and politically dependent communities and institutions are more vulnerable than large, spatially contiguous, interdependent communities that can tap into extended assistance networks. There is good evidence to show that the occurrence of extreme events simply exacerbates the plight of those who are already marginalized by social, economic, and political forces (Allan, 1988).

Finally, the human factors involved in disasters are highly changeable. For example, the present era of societal restructuring — in Eastern Europe and Western Europe; the former Soviet Union, South Africa, the Middle East, China and the countries of the Pacific Rim — may produce fundamentally different patterns of disaster vulnerability in the years ahead. Such rapid changes may temporarily paralyse existing institutions of disaster management. They may also introduce sufficient local instability that it will take years for the indigenous institutions to recover their historic levels of efficiency. In short, the modern interpretation of disasters views natural extremes as potentially dangerous phenomena that can be prevented, avoided, or mitigated. Catastrophic losses can be avoided or prevented if societies undertake appropriate measures. Among others these include: becoming aware of the risks; practising careful environmental management; reducing levels of vulnerability; and increasing the resilience of human and technological systems.

3 The Costs of Disasters

According to the US National Research Council, in a recent 20-year period natural disasters claimed about 2.8 million lives and adversely affected 820 million people (National Research Council, 1987). In other words, natural disasters burden the lives of about one-fifth of the world's population, at least once each generation. However, because of data gaps and inconsistencies, it is impossible to be precise about the global toll of natural disasters. Most international records are believed to underestimate the losses (Mitchell, 1989). Because many of the worst episodes occur in remote rural areas and because no international relief is provided, there is neither the means nor the incentive for accurate tabulation of the losses.

Disaster death rates are declining in most rich countries and in some poor countries but elsewhere they are rising (e.g. Phillipines) (Mitchell, 1989). Estimates of global economic losses due to natural disasters range from US $25-100 billion (Kates, 1979). Those figures may be about 40% larger if investments in prevention and mitigation are included. Nearly everywhere the economic bill for weather-related disasters is increasing, often at rates that exceed growth of population and national wealth. These loss estimates are increasing because people continue to establish new developments in areas of known vulnerability to extreme weather events — areas such as pristine barrier islands and alluvial flood plains. Because investments in these areas are increasingly valuable, the damage done by the inevitable occurrence of extreme weather events is increasingly costly.

Four types of phenomena account for about 90% of known disaster deaths and a disproportionate share of damage. With the exception of earthquakes, they are all weather-related risks (i.e. tropical cyclones, floods, droughts), that may be exacerbated by climate change. Three quarters of all disaster deaths and economic losses are inflicted by weather-related risks (Kates, 1979).

But the global pattern of losses is uneven. Up to 95% of recent disaster deaths have occurred in poor countries (United Nations Conference on Trade and Development, 1983). Within this group it is believed that societies in transition to a different stage of development are particularly prone to losses (White, 1987). Absolute economic losses are heavier in rich countries but poor countries are disproportionately affected; the impact of weather-related disasters can be 20-30 times larger, relative to the size of GNPs, than for their wealthy counterparts (Table 1). A single disaster can completely negate any real economic growth for several years in a poor country, by diverting scarce domestic resources into relief and recovery programs and by increasing the burden of international debt. When severe typhoons strike islands in the southwest Pacific, it is not unusual for the entire annual agricultural output to be lost. And even in the United States, where in 1989 there was an unprecedented $7.6 billion spent on losses, more than half of the loss (55%) was incurred when Hurricane Hugo struck the comparatively undeveloped Virgin Islands, Puerto Rico, and southeast coast of the United States (Insurance Information Institute, 1988 and 1991).

Indeed, it is instructive to consider disaster-related expenses as a percentage of GNP. The United States spent at least 0.15% of its GNP on disaster-related expenses during 1989. During the drought years of 1988 and the mid-1970s losses may have ranged from 1% to 4% of US GNP (National Research Council, 1991). In Japan, where annual death tolls are in the range of 100-500, damage averages around 0.25% of GNP, and about 10% of the national budget is allocated to disaster protection. In Latin America regional economic losses average $1.5 billion per year — amounting to 2% to 3% reduction of GNPs over periods of several decades (Jovel, 1989). In particular years, losses have been much heavier. During floods and drought associated with the El Niño of 1982-83, total losses of $3.97 billion occurred in Bolivia, Chile, Ecuador and Peru. This represented about 10% of the affected countries' Gross National Products or 50% of their annual public revenues (Jovel, 1989). Thus, relative to their effects on industrialized countries, weather-related disasters place proportionately a much heavier human and economic burden on developing countries, which typically have less financial flexibility to begin with.

4 The Geopolitical Implications of Weather-Related Disasters

When the impacts of severe weather-related disasters are tallied, the focus is on dead bodies and economic losses. But many other types of deep scars are not tabulated. For example, in developing countries, thousands of poor citizens are typically left homeless and without access to traditional means of employment. Weather-related disasters of the future may create waves of environmental refugees who will face great difficulties finding new homes in a world that possesses few open settlement frontiers and where governments are inhospitable to the powerless and destitute from neighbouring states (El-Hinnawi, 1985). If the experiences of Vietnamese boat people, "Marielitos" and Haitians are any indication, such governments are both ill-prepared to provide temporary care for abandoned citizens of adjacent countries and are even more reluctant to accept them as permanent residents.

This situation presents a complex puzzle. Where should such refugees go? Who will welcome and nurture them? If, for example, one million victims of the next disastrous cyclone in Bangladesh were to cross the border to higher ground in India, would they be welcomed? Or would they be seen as an irritant, upsetting the already fragile balance of political and economic forces at work on the Indian subcontinent? Under certain circumstances, it seems that refugee flows could increase both intraregional and interregional tensions. And, as a consequence, such flows could increase the risks that conflicts, which had been slowly simmering over other issues, might come to a violent conclusion.

5 The Reduction of Losses due to Weather Extremes

5.1 Progress in reducing loss

Fortunately, people do not have to remain passive in the face of natural disasters. It is possible to significantly reduce losses, if governments and populations at risk employ existing knowledge and known adjustment systems. A wide variety of tools are available, as listed in Table 2. Indeed, many countries have already made impressive progress toward the reduction of losses due to weather-related disasters.

Reductions of death rates have been achieved through the use of monitoring, tracking, prediction and warning systems, especially when coupled with improved evacuation procedures, and community information programmes. Residents of highly developed countries have been the chief beneficiaries. Because of clear advantages in their communications and early warning systems, rich industrialized countries have often been able to move people and their most valuable possessions out of the path of on-coming danger.

Adoption of sophisticated prediction, warning and evacuation measures has been slower in less developed countries; but progress is occurring in some places (Smith, 1991). Table 3 illustrates trends in death rates due to atmospheric and hydrological hazards, standardized to take account of population growth in four sample Asian countries. In India and the Philippines, per capita death tolls from most types of weather extremes are increasing at rates faster than population growth. On the other hand, Japan and Korea have succeeded in reducing heavy losses.

Advances have also been made on other fronts. One of the most spectacular campaigns against natural disaster losses was mounted by the government of Japan beginning in 1959. During that year, the disastrous Ise Bay typhoon devastated

Table 1. Typical weather-related disaster costs in more-developed and less-developed countries

	USA	Bolivia Ecuador Peru
Population (millions)	248	31
GNP (billions $US)	5200	37
Losses in largest recent disaster (millions $US)	4195[a]	3970[b]
Losses in largest recent disaster (% of GNP)	0.08	9.3

[a] Hurricane Hugo, September 17-22, 1989
[b] El Niño floods and droughts, 1982-83

Table 2. Main adjustments to weather-related disasters

Long-term adjustments

1. Controls on investment, planning, development and management of land, resources, infrastructure, and facilities
2. Education and training
3. Public information
4. Hazard insurance
5. Weather modification
6. Structural engineering works
7. Hazard-resistant construction techniques

Immediate pre-impact adjustments

8. Monitoring/tracking
9. Prediction/forecasting
10. Warning
11. Evacuation
12. Shelters

Impact and immediate post-impact adjustments

13. Emergency management
14. Hazard-fighting
15. Search, rescue, and salvage
16. Disaster relief
17. Loss bearing

much of the Nagoya region and killed over 5000 people. In response, a determined programme of structural engineering works, emergency preparedness schemes, meteorological warning networks, and hazard-sensitive land use management was implemented with major government assistance. As a result, annual deaths tolls quickly fell to below 1000; they have rarely exceeded 300 during the last 20 years.

These examples involve sophisticated science and technology. But striking success has also been achieved with low-technology measures such as inexpensive building modifications, floodplain management, and wetlands preservation and restoration programmes. These are alternatives that are often more appropriate for less developed countries. For example, training of local carpenters and masons in hazard-resistant construction practices has returned significant gains for limited investment in India and the Caribbean (Cuny, 1983). India and Botswana are among developing countries with effective drought relief programmes that rely on low-technology food stockpiling and market intervention mechanisms (Walker, 1989). Crop insurance schemes, developed by indigenous farmers' cooperatives in the Caribbean and Fiji have helped to cushion losses due to severe windstorms (Jodha and Mascarenhas, 1985). Low cost hazard zoning techniques have been developed in Peru, and improved hazard information, preparedness and disaster prevention arrangements have been stimulated by regional cooperative bodies like ESCAP's Typhoon Committee and the Asian Disaster Preparedness Center.

The People´s Republic of China has pioneered labour-intensive flood defences and flood-spreading techniques to cope with catastrophic events that exceed the capacities of levees and reservoirs (Kates, 1979.) Major sections of floodplains that are often intensively farmed, are designed to act as retention pools in the event of major floods. Settlements located within these areas are provided with higher levees, and trees are planted to provide emergency refuges. Even perennially threatened states like Bangladesh have recorded some progress. Over the last 20 years an innovative cyclone preparedness programme has been developed. It uses more than 20,000 locally organized volunteers to disseminate cyclone warnings to local populations, and to guide evacuees from their homes to 175 earthen mounds and over 200 cyclone shelters. During the non-cyclone periods shelters are used as schools, credit unions, health care centers, and agricultural extension facilities (Quareshi, 1990). This programme is credited with holding down cyclone death rates throughout the 1980s, especially on exposed islands at the mouth of the Ganges delta along the western and central sectors of the coast. (However, its limitations were clearly exposed by the cyclone which struck the eastern coast in April 1991.)

Table 3. Trends in weather disaster death rates for Asian countries (Mitchell, 1989).

	Japan	Rep. of Korea	India	Philippines
1965 pop. (million)	99	26	488	32
1985 pop. (million)	120	40	761	55
1960s mean annual deaths due to weather disasters	416	285	404	168
1980s mean annual deaths due to weather disasters	122	262	1819	982
1960s weather deaths/million	4.2	10.0	0.8	5.2
1980s weather deaths/million	1.0	6.6	2.4	17.9

There are many additional examples of successful disaster reduction initiatives but the implications are already clear. No country must accept heavy loss of life due to natural disasters; such death tolls can be curbed. Moreover, technologically sophisticated, imported and costly methods are not the only effective alternatives. The growth of economic losses can also be slowed — and perhaps stopped — by the use of known and available measures. Although the prospects for reversing economic loss trends are less bright, that too is not an impossibility. Yet the reality falls well short of the potential.

5.2 *Why are losses not being reduced?*

When so much might be done to reduce disaster losses, why have death tolls been slow to decline in many countries and why are material losses spiralling upwards almost everywhere? The central dilemma of global disaster management is that the knowledge and the means to reduce losses exist but they are not used appropriately. Many factors contribute to the problem. Lack of awareness of hazards or adjustments to hazards is sometimes a barrier to reduction of losses; sometimes the problem is lack of choice, lack of expertise, or lack of resources; sometimes it is misperception of risks; sometimes it is conflicts between the goals of hazard reduction and the goals of economic investment interests, sometimes it is lack of coordination among agencies and institutions that should be working in concert.

If disaster losses are to be effectively reduced throughout the world it will be necessary to overcome these barriers so that available knowledge and expertise can be put to work. Increasingly, it is also evident that disaster-reduction cannot be accomplished alone. It must also take account of — and be reflected in — programs for economic development; environmental management; population, land use and facilities planning; and community health, among others. Indeed, a high level of reciprocity is needed among these programs if any or all are to be successful (Mitchell, 1988).

6 What Should be Done?

The preceding critique has been accepted by the international disaster research and management community. To a greater or lesser extent its precepts are slowly and unevenly beginning to change the operations of international agencies, non-governmental organizations and the governments of some countries. These ideas also provide an underlying rationale for a new world wide programme of disaster reduction that was initiated by the United Nations in 1990.

6.1 The international decade for natural disaster reduction 1990-2000

The International Decade for Natural Disaster Reduction (IDNDR) is intended to mobilize existing, but underused knowledge, about natural disaster reduction measures. Its purpose also includes developing the scientific basis for confronting new types of environmental hazards that are expected to pose major societal problems in the 21st Century. The immediate focus is on tropical cyclones, floods, drought, earthquakes, and other acute risks of meteorological, geological and hydrological nature.

From the standpoint of policy, the chief distinguishing characteristic of the Decade is its attempt to shift the focus of disaster management away from reacting to disasters that have already occurred, and towards anticipating those that have yet to happen (i.e. moving from a relief-oriented policy to a mitigation-driven one). UNDRO (Office of the United Nations Disaster Relief Coordinator) and other international disaster relief organizations have endorsed similar policies in the past, but this is the first time that a comprehensive range of disaster interest groups — including scientists and engineers, emergency managers, relief officials and development planners — have the opportunity to join in a global effort to realize that end.

The programme emphasizes three disaster reduction tools:

(1) assessments of damage potential;
(2) warning systems; and
(3) disaster-resistant structures.

Existing knowledge is intended to provide the basis for action, but provisions have also been made to acquire and disseminate new scientific and technical information. The means of implementation include technical assistance programmes, technology transfer, education and training, and demonstration projects.

Although the UN supplies administrative oversight from a Geneva-based Secretariat and assists in fund-raising, individual countries provide the heart of the programme. Governments are encouraged by the Secretariat to formulate national disaster mitigation programs and to integrate them into economic development programs. They also are expected to mobilize public and private support, to increase awareness of disaster reduction, to stockpile emergency supplies, and to pay particular attention to the impacts of disasters on health care, food storage facilities, housing and infrastructure.

Participation in the programme has different manifestations in different countries. In Germany the emphasis is on management of hazards in the context of economic development and environmental protection; and in the United States the programme favors scientific research and the sharing of proprietary technical information among participating countries.

If properly targeted and supported, the International Decade for Natural Disaster Reduction (IDNDR) has the potential to significantly increase societal security against a wide range of weather-related disasters. It could also make a major contribution to saving money and other resources that are presently spent on disaster relief, in stop-gap measures that do nothing to address the basic dimensions of natural disaster.

Unfortunately, this promise is far from being realized and, without some course corrections, it may be unobtainable. Part of the problem is that the IDNDR is seriously underfunded, and relatively unknown outside the global community of disaster professionals. It has also been hampered by bureaucratic wrangling both inside and outside the UN system. But more important, the decade has not effectively engaged key issues that lie at the heart of disaster reduction. Chief among these is the process of investment decision-making by private and public bodies. Many - perhaps most - ostensibly "natural" disasters are triggered or gravely exacerbated by human mismanagement of the biosphere and narrowly conceived disaster response programs (Ives, 1988). Investment decision-making plays a crucial role in this process (Blackwell et al., 1991). Thus, sensitive use of investment incentives and related tools can make important contributions to reducing both the likelihood of disasters and their impacts (Silver and DeFries, 1990). Disaster-reduction programs are likely to fail if the linkage between disasters and development is ignored, or if measures that influence human actions are neglected in favor of technological fixes (Maskrey, 1989).

Whether the IDNDR is successful or not, the path to a less-disaster susceptible world is clear and some countries — mostly the more developed — will continue to move along it. But growth in disaster potential threatens to swamp the gains that are occurring in those countries and to prevent even temporary gains in a majority of less developed countries.

Increasing exposure and vulnerability of human populations, and lack of investment in improved disaster management are the chief culprits. These factors will continue to drive up the potential for disasters, even if climate change does not occur. The prospect of a warmer world with more extreme weather simply compounds existing difficulties.

7 The Value of Disaster-Reduction for Coping with Climate Change

It is time for the disaster community and the climate change community to make common cause. Strategies for reducing disasters can be "tied in" to strategies for adjusting to a warmer world, and vice versa (Silver and DeFries, 1990).

Sudden-onset disasters of the type that are addressed in the IDNDR and slow-developing disasters of the kind envisaged in global change scenarios are, in effect, linked problems that can benefit from coordinated responses. Indeed, modern interpretations of natural hazards frequently recognize the connections. Books about drought now customarily include sections on desertification; writings about coastal hazards make reference to the forcing mechanism of sea level rise; and analyses of hurricanes and coastal storms recognize the changing role of ocean temperatures.

The benefits of a joint strategy are considerable. The principal link between the components lies in building institutional competence and coordinating institutional responses. First, a joint strategy promises to reduce losses from sudden-onset disasters, thereby offsetting — and perhaps nullifying — losses induced by climate change. If climate does not change, the gains are still large. Second, increased consideration of climate change hazards reinforces commitments to long-term planning horizons among disaster professionals. It also consolidates preferences for cost-effective anticipatory mitigation strategies instead of expensive post-disaster responses. Third, climate interest groups benefit because the range of alternative responses to climate change is broadened to include a programme of presently available and effective adjustments. Disaster reduction measures provide known and immediate responses to climate change that can buy time for more fundamental changes in human behaviour and resource use to take effect. In other words, they are part of a climate change "insurance" package. Moreover, by making common cause with disaster managers, those who are concerned about climate change gain the support of a significant group of "hands-on" professionals who can translate broad policies into practical action and interact directly with endangered publics at the grassroots level.

But the potential for partnerships among disaster interest groups and climate change interest groups is presently unrealized. Though there are budding intellectual linkages between the two they have not yet carried over into joint lobbying or other cooperative activities. The respective constituencies have not yet mobilized for mutual effectiveness. Advocates of adaptations to climate change need to support programs for reduction of sudden-onset events (e.g. the IDNDR) and members of the disaster-reduction community need to lengthen their horizons to take account of the acute and cumulative long-term risks of climate change. Such actions will build a larger, more versatile, and ultimately more effective constituency for sustainable planetary management.

References

Allan, B.J., 1988: Adaptation to frost and recent political change in highland Papua New Guinea. In *Human Impact on Mountains*, ed. N.J.R. Allan, G.W. Knapp, and C.Stadel, Totowa, NJ, Rowman and Littlefield, pp. 255-264.

Blackwell, J.M., R.N. Goodwillie and R. Webb, 1991: *Environment and development in Africa: selected case studies*, EDI Development Policy Case Series, Analytical Case Studies No, 6 Washington, DC, Economic Development Instuitute of the World Bank, pp. 77-93.

Campbell, John R. and Neil J. Ericksen, 1990: Change, extreme events, and natural hazards. In *Climate change impacts on New Zealand*. Wellington: Ministry for the Environment, 1990, pp. 19-28.

Cuny, F.C., 1983: *Disasters and Development*, New York, Oxford University Press, pp. 164-193.

El-Hinnawi, E., 1985: *Environmental refugees*, Nairobi: United Nations Environment Programme.

Gray, W.M., 1990: Strong association between West African rainfall and US landfall of intense hurricanes, *Science*, **249**.

Henderson-Sellars, A. and R. Blong, 1989: *The Greenhouse Effect: Living in a warmer Australia*, Kensington, New South Wales University Press.

Insurance Information Institute, 1988: *Insurance Facts: 1988-89 Property/Casualty Fact Book*, New York, Insurance Information Institute, p. 77.

Insurance Information Institute, 1991: *Insurance Facts: 1990-91 Property/Casualty Fact Book*, New York: Insurance Information Institute, p. 68.

Ives, J.D., 1988: Mapping of mountain hazards in Nepal. In *Human Impact on Mountains*, ed. N.J., R. Allan, G.W. Knapp, and C. Stadel, Totowa, NJ Rowman and Littlefield, pp. 154-164.

Jodha, N.S. and A.C. Mascarenhas, 1985: Adjustment in self-provisioning societies. In *Climate Impact Assessment*, ed. R.W. Kates, J.H. Ausubel, and M. Berberian, Wiley and Sons, Ltd., pp. 437-464.

Jovel, J.R., 1989: Natural disasters and their economic and social impact, *CEPAL Review*, **38**, Santiago, Chile: Economic Commission for Latin America and the Caribbean, pp. 133-145.

Kates, R.W., 1979: Climate and society: lessons from recent events, World Climate Conference, World Meteorological Organization, Geneva. Updated in W.J. Maunder, *The Uncertainty Business: Risks and Opportunities of Weather and Climate*, London and New York, Routledge.

Kates, R.W., 1980: Disaster reduction: links between disaster and development. In *Making the most of the least: Alternative ways to development*, ed. L. Berry and R.W. Kates, New York, Holmes and Meier Publishers, Inc., pp. 135-170.

Kerr, R.A., 1991: Global temperature hits record again, *Science*, **251**, p. 274. See also: National Research Council, *Changing Climate: Report of the Carbon Dioxide Assessment Committee*, (Washington, DC: Board on Atmospheric Sciences and Climate, Commission on Physical Sciences, Mathematics and Resources, 1983); World Meteorological Organization, United Nations Environment Programme, *IPCC First Assessment Report*, 2 vols. (Geneva: Intergovernmental Panel on Climate Change, 1990).

Leetmaa, A., 1989: The interplay of El Niño and La Niña. *Oceanus*, **32**(2), 30-34.

Liverman, D.M., 1990: Vulnerability to global environmental change. In *Understanding Global Environmental Change: The Contributions of Risk Analysis and Management*, ed. R.E.

Kasperson, K. Dow, D. Golding, and J.X. Kasperson. A Report on an international workshop, Clark University, October 11-13, 1989. (Worcester, Massachusetts: Center for Technology, Environment and Development, 1990), pp. 27-44.

Maskrey, A., 1989: *Disaster mitigation: A community based approach*, Oxford, Oxfam.

Mitchell, J.K., 1988: Confronting natural disasters: an international decade for natural hazard reduction, *Environment*, **30**,25-29.

Mitchell, J.K., 1989: *Where should the international decade for natural disaster Reduction concentrate its efforts? A comparative analysis of international data on natural disasters*, Decade for Natural Disaster Reduction Working Paper No 2, Boulder, Natural Hazards Research and Applications Information Center, 14 pp.

Mitchell, J.K., N. Devine and K. Jagger, 1989: A contextual model of natural hazard, *Geographical Review*, **79**, 391-409.

National Research Council, 1987: *Confronting Natural Disasters: An International Decade for Natural Hazard Reduction*, Washington, DC, National Academy Press, p. 1.

National Research Council, 1991: *Policy implications of greenhouse warming*, Synthesis Panel on policy implications of Greenhouse Warming; Committee on Science, Engineering, and Public Policy, Washington, DC, National Academy Press, p. 95.

Nicholls, N., 1989: Global warming, tropical cyclones and ENSO. In *Responding To the Threat of Global Warming: Options for the Pacific and Asia*, ed. D. G. Streets and T. A. Siddiqi, pp. 2-19 to 2-36. Proceedings of a Workshop sponsored by Argonne National Laboratory and the Environment and Policy Institute, East-West Center, June 21-27, 1989, Honolulu, Hawaii, ANL/EAIS/TM-17 (Argonne National Laboratory, 1990)

Pankhurst, A., 1984: Vulnerable groups, *Disasters*, **8**, 206-213.

Parry, M.L., 1986: Some implications of climatic change for human development. In *Sustainable Development of the Biosphere*, ed. W. C. Clark and R. E. Munn, Cambridge, Cambridge University Press, pp. 378-407.

Pittock, A.B., A. M. Fowler and P. H. Whetton, 1991: Probable changes in rainfall regimes due to the enhanced greenhouse effect, Proceedings of the International Hydrology and Water Resources Symposium, Perth 2-4 October 1991, pp. 182-186.

Quareshi, A.H., 1990: Cyclone preparedness programme of Bangladesh. In *Towards a less hazardous world in the 21st Century*, pp 395-399. Proceedings of the IDNDR International Conference September 27-October 3, 1990, Yokohama and Kagoshima.

Richards, J.F., 1990: Land transformation. In *The Earth as Transformed by Human Action: Global and Regional Changes in the Biosphere over the Past 300 Years*, ed. B.L. Turner II, W.C. Clark, Robert W. Kates, John F. Richards, Jessica T. Mathews, and William B. Meyer, Cambridge and New York, Cambridge University Press, pp. 163-178.

Roberts, W.O., 1989: Global climate change as a hazard, *Natural Hazards Reporter*, **13**,1-2.

Silver, C.S. and R.S. DeFries, 1990: *One Earth, One Future: Our Changing Global Environment*, Washington, DC, National Academy Press, p. 95.

Smith, K., 1991: *Environmental Hazards: Assessing Risk and Reducing Disaster*, London, Routledge.

Swaminathan, M.S., 1987: Prediction and warning systems and International, Government, and Public Response: A Problem for the Future. In *Monsoons*, ed. J.S. Fein and P.L. Stephens, New York, John Wiley and Sons, pp. 607-620.

Swetnam, T.W. and J.L. Betancourt, 1990: Fire-Southern Oscillation relations in the south western United States, *Science*, **249**, 1017-1020.

Thompson, C.S., 1990: Extreme rainfalls under greenhouse scenarios, unpublished manuscript, Wellington, New Zealand Meteorological Service, 1989. Cited in: J.R. Campbell and N.J. Ericksen: Change, extreme events, and natural hazards, *Climate Change: Impacts on New Zealand*, Wellington, Ministry for the Environment, pp. 19-28.

United Nations Conference on Trade and Development, 1983: *The incidence of natural disasters in island developing countries*, New York, United Nations, TD/B/961, p 2.

Walker, P., 1989: *Famine Early Warning Systems: Victims and Destitution*, London, Earthscan Publications, pp. 101-109; pp. 112-116; pp. 170-172.

Walker, H.J., 1990: The coastal zone. In *The Earth as Transformed by Human Action: Global and Regional Changes in the Biosphere over the Past 300 years*, ed. B.L. Turner II, W.C. Clark, R.W. Kates, J.F. Richards, J.T. Mathews, and W.B. Meyer, Cambridge and New York, Cambridge University Press, pp. 271-294.

Whetton, P. and A.B. Pittock, 1991a: Australian region intercomparison of the results of some general circulation models used in enhanced greenhouse experiments, *Technical Paper No 21*, (Mordialloc: Commonwealth Scientific and Industrial Research Organization (Australia), Division of Atmospheric Research, 1991), 73pp.

Whetton, P. and A.B. Pittock, 1991b: Toward Regional Climate Change Scenarios: How Far Can We Go? Paper presented at the conference on Global Climate Change: Its Mitigation through Improved Production and Use of Energy, Los Alamos, New Mexico, 21-24 October 1991, 13pp.

Whetton, P.H., A.M. Fowler, M. R. Haylock and A. B. Pittock, 1992: *Simulated changes over Australia in daily rainfall intensity due to the enchanced greenhouse effect: Implications for floods and droughts*, (in preparation).

White, G.F., 1987: Foreword. In *Lands at Risk in the Third World: Local-level Perspectives*, ed. P.D. Little and M.M. Horowitz, with A.E. Nyerges Boulder, Westview Press, pp. x-xi.

Wigley, T.M.L., 1988: The effect of changing climate on the frequency of absolute extreme events. In *Climate Monitor*, **17**, 44-55.

World Bank, 1991: *Annual Report 1991* (Washington, DC, 1991), p. 121.

World Meteorological Organization, 1990: United Nations Environment Programme, *IPCC First Assessment Report*, **Vol 1**, *Working GroupI - Policymakers Summary* (Geneva: Intergovernmental Panel on Climate Change), pp. 17-18.

CHAPTER 11

The Effect of Changing Climate on Population

Nathan Keyfitz

Editor's Introduction

The world's population is growing and will continue to grow in the decades ahead. It took human beings almost 2 billion years of evolution to achieve a population of 5 billion people. If current trends continue, global population will reach at least twice that level within the next century.

Nathan Keyfitz notes that in the past, most population growth has occurred in regions where climate and ecology could support the increased needs of a growing human community. In many cases, especially in societies where agriculture was the main industry, population growth was limited by soil fertility. As manufacturing and trade have become more prominent, the connection between soil quality and population size has become less important. Still, for countries in which more than half the population is engaged in agriculture, rapid climate change could significantly affect the ability to support a growing population. For these countries, a shift in climate, even without a diminution of rainfall, could cause great hardship. But for countries, such as Bangladesh, with high rates of population growth, the difficulties that arise from climate change may pale in comparison to the challenges of managing economic development for a growing population.

As we have seen in previous chapters, the effects of climate change will not be evenly distributed. The Keyfitz analysis suggests that the poor, both among countries and within countries, will be most severely affected by climate change. For example, to the extent that global warming causes agricultural zones to shift, it will increase migratory pressures on the rural poor. Poor countries (and the poor within all countries) are often the very ones that can least afford to participate in or take advantage of international measures of protection against the risks of rapid climate change. Relatively affluent, mobile, urban cultures will be less affected than geographically stable, rural societies.

Keyfitz concludes that the challenges of responding to rapid climate change — and to many other environmental problems — will be less severe if population growth is slow than if population grows rapidly. If, for example, Bangladesh is partially inundated by a 1-metre sea level rise, it would be easier to relocate the affected fraction of a Bangladeshi population equivalent to its 1950 level of 42 million, than it would be to relocate the same fraction of the expected 2020 population of 206 million.

Nathan Keyfitz, leader of the population programme at the International Institute for Applied Systems Analysis in Laxenburg, Austria, is an environmental demographer. For three decades, he has published extensively on technical issues in demography. In the last ten years, his work has focused on the relationships between population growth, energy use, economic development, and international environmental problems. Aware that uncertainties about the impacts of climate change complement and complicate the uncertain impacts of future population growth, Keyfitz concludes that rational actions taken in the next several decades can reduce the risks of rapid climate change. In a rational world, free movement of food and other commodities could buffer the impacts of future climate change on food production. Systematic shifts in cereal cropping patterns and agricultural management practices could turn the effects of future climate change from problems into potential benefits.

Many have noted that ours is not a wholly rational world. In our world, where we cannot assume that genuinely utopian policies or world-wide cooperation will prevail, several simple steps can help human societies to cope with growing populations amidst environmental stress. Keyfitz suggests that free passage on the seas (and, as far as possible over land), increasingly free and equitable trade regimes, and international agreements to overcome economic externalities would move human societies in the right direction. Other pressures and disruptions — including wars, ethnic clashes and civil disturbances, economic fluctuations, and political upheaval — will divert public concern and policy attention from the need for systematic response strategies to the risks of rapid climate change. This chapter makes a good case for less discussion of long-term plans of the path to paradise, and more attention on short-term, incremental steps — to assure equitable economic development, to reduce the rate of population growth, and to increase the resilience of existing institutions through international cooperation. Measures such as these are the essential hedges needed by a growing population in a period of high stress and substantial uncertainty.

- I. M. M.

1 Population Distribution in an Industrializing World: No Longer Configured to Climate

Population has generally grown in those parts of the world where the ecology could support it. The rich soil of Java, nourished by lavish rainfall, is the base on which its 100 million population grew, and similarly with the delta of the Ganges, now West Bengal and Bangladesh. Density in the mountains of China is trifling compared with the rich coastal plains.

This perception of population density as a function of soil fertility carries conviction where agriculture is the main industry, but beyond that stage in economic development, population distribution is governed by quite other considerations. The correspondence between population on the one side and soil on the other diminishes in importance as manufacturing industry and trade grow.

For countries of dense population, with half or more of the people engaged in agriculture, climate change could be especially troublesome. But in considering the prospective effects of climate on the population of any country one must take into account not only climate change, but the way the impact of that change is intertwined with other changes in its economy and society.

2 Population and Climate: Linked as Problems

Problems are diverse. Algeria's population is expected to nearly double in the 30 years ahead. If that happens, and Algeria runs out of oil, and also suffers warming and drying out, its position will be unhappy indeed. Afghanistan is to multiply its population by 2 . Egypt is to add 35 million over the 30 years, and the waters of the Nile may, or more likely may not, be adequate.

Bangladesh adds more than two million to its population every year; one can visualize the intensity of its agrarian problem, superimposed on difficulties of caring for ecological refugees from its coastal area. China is the most populous country, but it seems to be making progress with the freeing of its economy, and certainly it is controlling its population. India has been less successful in controlling population growth, but on the other hand its soil is richer. It depends so crucially on the monsoon, that if the warming moves the seasonal winds even a little out of their usual paths the effect could be serious. Somewhat the same, though much less acutely, applies to Indonesia — whose population is closer to control, but even now has the inconvenience of pressure of the ocean on Java's densely settled north shore.

With Iran and Iraq, we have two states that will at least double their populations in the next 30 years; dependence on oil is a temporary strength and a long-term weakness. Japan could have some difficulties with ocean levels, but its slow rate of population increase, together with its great wealth, provide assurance that it will come through. Despite the recent slowdown, it is and is likely to remain the world's largest creditor. There it contrasts with Mexico, also an oil producer, that has borrowed on the security of its oil, is now deeply indebted, and has its own severe population problems.

Observers have often noted the many ways in which population growth makes trouble by itself, and they have also noted the dangers of climate change. But what about the two — population and climate change — acting together? Do they have the same effect as if they were acting separately, being simply additive — or is there synergy? Do they accentuate one another, or do they offset one another?

It is not hard to show that, in fact, together they give more trouble than the sum of what they would give acting independently. If an area is hit with drought or with flood, and its population has to be assisted in resettling elsewhere, then the greater the population the greater the arduousness of the task. The difficulty is more than in proportion; there is not only the work of relocating the individuals, but the even bigger task of finding land for them.

3 The Poor Will Feel it Worse

The distribution of income in the world, and especially between the Developing Countries (DCs) and the Industrialized Countries (ICs), can only be made less equal as a result of any likely change of climate. The poorer they are, the more vulnerable to the risks of rapid climate change. For example, if countries whose income is unequal are all put to a given amount of expense, then the ratio of the net incomes of the richer to those of the poorer will be increased. But we can say more than this.

Because they are poorer in incomes, skills, and capital, DCs will be less capable of taking action to protect themselves — building dykes against the sea, altering agricultural practices. That the DCs are less mobile will make it harder for them to relocate any ecological refugees that the warming produces. That they are mostly in warm climates anyhow means that they will not be benefited in the way northern countries will be benefited, by having to use less fuel for home heating. Instead, they will need more fuel if their middle-class citizens choose to use air conditioning.

But the main impact of a warming that puts some land under water is bound to be increased population pressure on the agricultural land that is left. When all land is occupied and used, the loss of any land, or equivalently, the loss of rainfall essential to its fertility, lowers its population-supporting capacity. In the face of the rising oceans and other changes that come with warming, it helps to be a sparse population on high ground that has plenty of options on where to live within its national territory, and it helps most to be an industrial country producing for a world market, more than capable of meeting any price rise in the agricultural products that it imports.

Mustafa K. Tolba, executive director of the United Nations Environmental Programme (UNEP), makes the point that it is the DCs that will be hardest hit by any likely changes in climate:

> *"The most devastating effects of climate change will be felt in low-lying countries, including Bangladesh, Indonesia, the Maldives and Egypt...."* (Tolba, 1990)

Perhaps by coincidence, some of the countries that have had the greatest misfortunes in the recent past will be those most adversely affected. As the UNEP Executive Director noted:

> *"Half the world's population dwells in coastal regions which are already under great demographic pressure, and exposed to pollution, flooding, land subsidence and compaction, and to the effects of upland water diversion..... In developed countries, protection for some regions will be possible, whereas in developing countries without adequate technical and capital resources, it may not be."* (Tolba, 1990; GESAMP, 1990).

There is a northern limit of sufficiently reliable rainfall for cereal farming in the Sahel, and a southern limit of sufficiently high temperatures for coffee production in southern Brazil (Parry, 1990). These limits could be drastically affected by climate change.

A poor country may need much more than a rich one to preserve the existing climate, but the outlays required today to capture these anticipated benefits tomorrow just cannot be afforded now. For one thing capital is more expensive, as indicated by much higher interest rates. Worse than that, the DCs rightly fear any drift of the debate in the direction of freezing development at wherever countries are now.

Thus, although the most devastating effects will be experienced by the densely populated DCs, their governments are less disposed to support measures that will control the warming, and oppose measures put forward by the ICs that would distract from their problems of debt and development in general. Again Tolba:

> *"[The] climate change issue is creating a tension between developing countries who need to develop and use more energy and the north trying to curb the emission of greenhouse gases inducing the climate change."* (Tolba, 1990)

For the developed countries, industrial and other changes are far more rapid, so warming will add less to the inconvenience they cause. Industry in the ICs is footloose, its population growing slowly if at all. The DCs on the other hand are agricultural, tied to place, their populations growing rapidly; they will find it much harder to adapt to warming. They will be only partially protected by the lesser warming in the tropics. Yet the DCs have given the issue low priority. Paradoxically, the countries that have most to lose by the warming are the least anxious for something to be done now to anticipate it, while those that could actually gain (for instance Sweden and Canada) are strongly promoting measures to control the carbon and other emissions that are causing the trouble.

4 Mobile, Urban Cultures Will Be Less Affected

When automobile manufacturing shifted away from Detroit in the 1970s and 1980s, mostly occasioned by changing labour markets and international competition, the change took place fast enough that still-useful buildings had to be abandoned or converted to lesser uses. Abandoned factories were seen in New England for a long time after textile, leather, lumber and other manufacturing shifted to the South. These changes were indeed damaging, but the cities have nonetheless survived.

Within the United States we can imagine that warming will make the southern part of the country somewhat less attractive, and, all other things being equal, will draw population to the north. To judge whether these are serious matters, we have to set them against the changes that take place anyhow, not for geophysical but for economic reasons. Experience shows that, because of its wealth, the capacity of the United States to respond to change is incomparably greater than that of a developing country. And because of throwaway habits that apply to clothing and automobiles but carry over to houses and soils, the United States loses less by a shift of population than does Europe. Wood-frame houses have a short life in any case; like such infrastructure as roads, they can be built in the places to which people migrate and abandoned in the places they leave, at less cost than in Europe where housing construction is more permanent.

If the impact of global climate change in industrialized countries is gradual (as seems likely), and its effects can be foreseen, then its costs could be slight. This applies especially to cities. After all, houses last less than 100 years before they deteriorate or become obsolete. The stoutest public buildings do not retain their usefulness for much longer. Thus cities would gradually change their location as individual builders sought higher ground.

But though there may not be overall losses of human habitat, one must anticipate large commercial gains and losses through changing land values. Shifting real estate prices need not affect the wealth of a nation, but it would affect regions. Suppose that water overran one-third of Manhattan. New York would diminish in importance, and other cities, perhaps thousands of miles away, would take over its functions. (None the less, judging from current press reports, New York is in far greater danger from its financial condition than from incursions of the oceans.)

American society is accustomed to movement and change. During the 1970s, the shock of oil price change made Texas richer and New England poorer, and induced rapid migration southward. It is hard to imagine that a degree or two of warming (say over half a century) would have as much economic effect as fluctuations in the price of oil (that take place over months). Climate change would similarly have a minuscule economic effect in comparison with the closing of an air force base in a locality, or the exhaustion of supplies of fossil water in the western states (which is going to occur in any case).

Part of the reason a modern economy can handle shocks much greater than those predicted for climate is that modern industry is not strongly tied to place. In contrast to agriculture and mining, even to steel and shipbuilding, modern industry is movable; one can set up aircraft or computer manufacture anywhere. And service industries that use relatively little fixed capital can follow population movements even more readily than can light manufacturing.

Thus we have to regard the movement of population (in general towards the north) occasioned by global warming against the background of man-made population changes that are taking place anyway. On most forecasts the effects of climatic change will be slow and small compared with the customary mobility of Americans.

Such considerations apply somewhat less to Europe and to the former Soviet Union. Their populations are less mobile, and for some individual countries economies may be more tied to climate. A succession of winters as warm as 1989-90 would be disastrous for Austria, one component of whose economy depends on attracting skiers.

In so far as international migration is more difficult than intranational, small countries will find it harder to adapt than will large ones. On the other hand, small countries are more likely to face up effectively to their physical problems, to have the flexible organizations that will cope with them. This applies especially to industrial countries that are culturally homogeneous.

5 The Moving Food Supply: More Pressure to Migrate

It is possible there will be more rainfall in high latitudes and in the coastal regions of the temperate zones, along with drier continental interiors — changes that might offset one another, for example, for Australia and the United States. Once again, large countries will have the advantage of averaging the losses and benefits of warming. Smaller countries will find their food and forest supplies enhanced or damaged, without appeal, in a sort of planetary lottery.

Scott et al., in their review of the several directions of impact (1990), conclude that for the next 50 to 75 years with an expected doubling of CO_2, the rate of growth in the worldwide demand for food and fiber can be met even with the projected climate changes. They estimate that effect on tropical forests would be small; boreal forests would migrate northward, both to higher ground and to more northerly latitudes. This natural process, in which forests that have to replant themselves continuously do so in slightly different areas, has some analogy with the moving of cities. The difference is that the forests must do their replanting at their own speed, and there have been warnings that the speed of warming may be too rapid for them.

Effects on yield potential will tend to be positive in northern areas. "Agriculture in Scandinavia stands to gain more from global warming than perhaps any other region of the world." In southern Finland spring wheat yields could increase by about 10 percent, in the center by up to 20 percent, and by more than that in the north. In Finland, yields of barley and oats are raised by 9 to 18 percent, depending on the region. "Iceland's climate under the GISS scenario is similar to that of Northern Britain today," which is to say much warmer than at present (Parry, 1990).

Many doubt, however, whether the rate of migration could be rapid enough to keep up with climate change. There could be a diminution of the forested area. As for the fisheries, the distributions of most fish species are expected to move polewards. With global warming, some freshwater fisheries would be advantaged, some would diminish; the net could well be positive. In all these cases, the pressure will be strongly felt by people in both DCs and ICs, to move to keep up with the availability of changing agriculture — or to expend even more fossil fuels in having food transported to them.

6 The Advantage of a Smaller Population

We cannot generalize overmuch about developing countries versus industrialized countries in regard to these conditions, since each of those two groups includes a wide range of densities and altitudes among their members. But one thing can be said with certainty: that for any given country that faces a threat, the less its population increases from this point

Table 1. Eleven countries expected to exceed a population of 100 million by 2020, as estimated and forecast by the United Nations Population Branch, 1950-2020. Millions of persons.

Year	Japan	USA	USSR	Brazil	Mexico	Nigeria	Bangladesh	China	India	Indonesia	Pakistan
1950	84	152	180	53	27	33	42	555	358	80	40
1980	117	228	265	121	69	81	88	996	689	151	86
2000	130	268	315	179	109	162	146	1256	964	211	141
2020	133	304	358	234	146	302	206	1436	1186	262	198

on, the easier will be the protection of its citizens against the threat. That applies not just to climate change, but to most contingent difficulties. In a country with occasionally active volcanoes, a larger population requires more citizens to take chances by inhabiting and tilling the land up the sides of the volcanoes.

Of the 11 countries expected to exceed 100 million by 2020, fully 8 will be DCs (Table 1).

While much of what is said above for individual countries is speculation, there is one thing that we know for sure: every one of the adjustments required by climatic change will be made more easily with fewer than with more people. If sea level rises and parts of Bangladesh are under water, the 1950 population of 42 million could be redistributed through the territory and otherwise protected more easily than the population of 206 million projected for Bangladesh in 2020 (as shown in Table 1). The statement is true whatever the cause of the rise of sea level, and whatever the institutions responsible for defence against it.

The population of 42 million is that of more than 40 years ago, and Bangladesh could get back to it by one-child families continuing for two generations; the 206 million represents the (much more likely) continuation of present trends, including what success or lack of success has been shown in population control to date.

Table 1 shows the DC population increasing well over three-fold between 1960 and 2020, while the ICs grow by only 42%. Table 2 shows the progression expected decade by decade. Dense populations growing rapidly are a serious problem for many reasons — not the least of which is that, in themselves, they make adaptation to global warming much more difficult.

Table 2. Populations of the developing and industrialized countries decade by decade, as estimated and forecast by the United Nations Population Branch, 1960-2020. Millions of persons.

	1960	1970	1980	1990	2000	2010	2020
World	3019	3697	4450	5292	6251	7190	8062
DC	2074	2648	3313	4086	4988	5883	6722
IC	944	1049	1136	1205	1262	1307	1340
Europe	425	460	484	497	508	513	513

7 Population and Water Resources

We talk much of warming, but warming as such is not likely to be a serious problem in most places. In a few locations the temperature might pass from just tolerable to not livable, but in most it would mean a little more running of air conditioners, offset by less fuel for winter heating in other places. However, climate change would have very severe effects through the alteration of rainfall patterns. Peter Rogers (1985) tells us

> *"...[small] changes of temperature could themselves lead to marked changes in precipitation and runoff... Furthermore, because of increasing temperatures, large increases in precipitation in the high northern latitudes and decreases in the lower latitudes are predicted. For the Colorado Basin at about 40 degree latitude, there would probably be a 115 percent decline in precipitation. Thus the temperature effect could lead to a 50 percent reduction in the runoff of the Colorado River.*
>
> *Of course, there would be some big winners if this happened. Both Canada and the USSR would have much better temperatures and precipitation for grain harvesting, and other northern regions could be similarly helped. In northern Africa, the average flow of the Niger, Chari, Senegal, Volta and Blue Nile rivers would increase substantially; they would receive a 10-20 percent increase in precipitation and only a small increase in temperature. Other areas could be severely affected by greatly decreased flows, for example the Hwang Ho in China, the Amu Darya and Syr Darya in the prime agricultural areas of the USSR, the Tigris-Euphrates system, the Zambesi, and the Sao Francisco in Brazil."*

In so far as population has settled where the resources, and especially rainfall, have been reliably available, these changes could have a major effect. Roger Revelle (1982) suggests some major changes in the availability of the world's fresh water are likely to occur:

> *"The average flow of the Colorado river could diminish by 50 percent. And even today's flow, backed up by fossil water, is barely enough to meet the demands of irrigated agriculture. In northern Africa*

> *the average flow of the Niger, Chari, Senegal, Volta and Blue Nile rivers could increase substantially... In many other rivers the flow could greatly decrease: the Hwang Ho in China, the Amu Darya and Syr Darya in one of the prime agriculture areas of the USSR, the Tigris-Euphrates system in Turkey, Syria and Iraq, the Zambezi in Zimbabwe and Zambia and the Sao Francisco in Brazil...Also somewhat smaller runoff [could be expected] in the Congo River in Africa, the Rhone and the Po in western Europe, the Danube, the Yangtze and the Rio Grande."*

Where large populations are directly dependent on irrigation from these rivers the effects can be drastic.

> *"In the breadbasket regions of the world, (the US Great Plains, Canadian prairies, North European lowlands, the Soviet Ukraine, the Australian wheat belt and the Argentine pampas) much depends on future changes in precipitation, about which we know little."* (Parry, 1990)

Parry summarizes available studies by showing the effect under three scenarios in the alteration of yields with a doubling of CO_2. Supposing that the impact on climate is moderate, with small rise in temperature and not much change in precipitation, then all regions but the United States will either show no change for most crops, or will show an increase in yield. If on the other hand the precipitation increases were insufficient to compensate for the increased rates of evapotranspiration, then only Northern Europe would gain and all other regions would have lower yields, running as high as 20% decline in Canadian wheat yields and 30% decline in American maize and soybeans.

But the risk is not only that precipitation and runoff will be insufficient; there is also the risk of having too much water at one moment. Increased runoff from certain rivers would mean frequent, perhaps annual, flooding of settlements created in places that in the past were secure. For example, if there were simultaneously large increases in the flow of the Mekong and the Bramaputra rivers, the resulting streamflow could lead to frequent and quite destructive floods.

Revelle refers to the disintegration of the West Antarctic ice sheet, thought to be somewhat unstable in the face of climate change. If its 2 million cubic kilometres above sea level were to be carried into the ocean, the sea would rise by 5 to 6 meters. Populated river deltas everywhere would be flooded. Half of the state of Florida would be covered by seawater. This is unlikely to occur in the next century. However, the possibility suggests that thought should be given to the contingency of it taking place, and particularly to the task of moving large populations from such thickly settled, low-lying areas.

8 The Relative Importance of Population Control

Martin Parry (1990; and 1992, in this volume) tells us what would happen with a 1.5 meter sea level rise. In Bangladesh, 15% of all land (and about one-fifth of all farmland) would be inundated, and a further 6% would become more prone to frequent flooding. Altogether 20% of agricultural production would be lost. About the same applies to Egypt.

But to put this change in perspective, consider the time frame of the sea level rise. It might take place over the course of 75 years. The loss of agricultural production would amount to approximately 0.3% per year. The warming would take place, of course, along with other changes. As the oceans rose, the Bangladeshi population would increase at its continuous rate. Neither would seem very different from one year to the next, but over the course of a century one-third of the population might find themselves on land that came to be awash year by year. They would have no choice but to flee to high ground that is already fully exploited.

Suppose that the mean sea level rose annually by ten millimetres (the highest estimate given). Every year the equivalent of a strip of land would be lost 10 metres wide running down a 500-kilometre coast. The annual loss would be about five million square meters, or about 500 hectares, from which the population will have fled to higher ground. Five hundred hectares is the land on which, at most, 10,000 people might be sustained. Meanwhile, about two million people would have been added to the country's population.

Clearly, over 1 year at current rates, as far as we can guess them, the effect of population growth would overshadow the loss-of-land effect. Despite its crudity, such a calculation should suffice to make the point that population growth is two or three orders of magnitude more serious, from the point of view of sustenance of population than is the prospective rise of sea levels, even for the country that is frequently cited as the most vulnerable. Unfortunately, the sea level rise does not come on gradually, year by year — but in the form of intermittent severe floods of which Bangladesh has had examples in the last few years. This is not to argue that prospective climate change is unimportant for a poor country, but only that population control is even more important. Loss of Bangladeshi land to the ocean is of course serious, but its effect on the per capita food supply, and on per capita incomes, is less than the effect of the population increase now taking place.

Our arithmetical exercise is too conservative, and requires much elaboration to represent a real country, like Bangladesh. The refinements would be in the loss of land, the increase of population, and the nature of the economic activities supporting the population (See Warrick and Rahman, 1992; this volume).

9 Responsibility for Causing the Problem

Modification of the atmosphere is the most international effect of all, and institutions for its adequate control have yet to be put in place. One measure of local responsibility for the cause is emissions per head of population. Unfortunately, far from being able to say what the national sources of emission will be in the future, we cannot even say what present emissions are. Thus the IPCC speaks of

> *"poor or inconsistent measurements of emissions, as well as a lack of information on activities that cause emissions, such as the area of tropical forest cleared, and the amount of biomass burned as a result of the clearing."* (IPCC, 1991)

Estimates by the World Resources Institute show Canada, the United States and Australia as the leaders among the OECD countries, while Japan comes near the bottom among developed countries.

The most potentially disruptive element that interferes with attempts to secure international agreement is who is responsible for the trouble as between the ICs and the DCs. The Stockholm Environment Institute Boston Center has provided some figures (Table 3).

As a result of these figures, allocating blame for the warming that has taken place so far is relatively straightforward. Published estimates range up to 80% for the presently developed countries, and 20% for the less developed. Since most of the population is in the DCs, this has been said to prove that population does not make pollution, but rather development of the style now current in the ICs. Hence above all, it is said, the ICs ought to be controlling both their populations and their economic growth.

That would be perfectly sound if the DCs were not themselves intending to go ahead with development. Far from having dropped out, they desire more than ever to proceed along the road already travelled by the ICs. Thus looking back at the past tells us nothing about the need for population control; when we agree that a marginal addition to India's 850 million population does less damage to the environment than is done by one more person added to the population of Denmark or the United States, we had better add that the present situation of relative poverty of India is not the one that we want to see continuing into the future.

Table 3. Indicators of global change: ratio of CO_2 emissions of DCs and ICs to world average; current 1986 and cumulative 1860-1986.

	DCs	ICs
Current		
Nation	0.54	2.39
Per capita	0.52	3.68
Cumulative		
Nation	0.36	2.92
Per capita	0.35	3.18

(Cumulative data refer to the period from 1860 to 1986; per capita data have been calculated on the basis of current population as denominator; all figures are ratios to the world as a whole; I am grateful to N. Nakicenovic for drawing this material to my attention.)

The DCs are bound to reflect that it was the ICs that, at least up to now, have been the chief culprits in releasing carbon dioxide to the atmosphere. The concern of the ICs, as the DCs come to resemble them in patterns of consumption, is the future release of carbon dioxide in prospect, and in consequence the acceleration in the rate of warming. Potential conflict is built into such a situation. The confrontation need not be in the form of rational argument; nor need it come to all-out wars. But it could show itself in religious fundamentalism, kidnappings and terrorism generally, and in other anti-western cultural manifestations.

The fact seems to be that the capacity of the atmosphere to absorb carbon with only tolerable heating can be thought of as a resource, and indeed one of the most valuable of resources, even though it is not expressible in money terms, cannot be bought and sold, and will gradually increase as carbon is absorbed into the oceans. Once it is seen that way, and recognized that the atmosphere can absorb 600 gigatonnes in all, and that so far the ICs have put 200 gigatonnes into it, then there are 400 gigatonnes to go, and the question is who has the right to make further deposits. Formulated this way, the atmosphere is a storage cupboard with only so much room, and putting carbon into it being essential for development, there is to be only so much development.

If most of the carbon so far released into the atmosphere is by industry in ICs, then there is no question where the moral responsibility for doing something about it lies. But if the heaviest incidence of damage will be on the DCs, they have strongest practical reasons for doing something about it. No quick cleanup is possible but steps can certainly be taken to slow down the increase of carbon. Most of the DCs are introducing family planning programmes and that is an important part. Beyond that an obvious division of labour is for the DCs to moderate the extraction of wood from their forests, but since much of the woodcutting is to raise the foreign exchange to service debts, at least certain of the ICs could trade debt for forests — an exchange that has started, but on much too small a scale.

10 Rationality in Social Behaviour: Encouraging Institutions to Confront Physiographic Changes Effectively

Global warming has to be seen against the background of social and economic interrelationships between nations.

Even if the worst is actually occurring, and even if we cannot count on disregarding the positive effects on crops of more CO_2 and higher temperatures, warming need not greatly affect human life on the planet, if the right political and institutional decisions are made.

Yet not only do we not know the future, but we lack any history that provides probabilities for such decisions. It is exactly this lack in other fields that makes the economy fall back on personal intuition; those whose intuition is proven good survive in the market, while the others go bankrupt and are heard from no more.

10.1 Examples from individual actors

Robert E. Munn (1990) tells us the implicit evaluation of the probability of ocean warming contained in a decision of Shell Oil. It is about to build the world's largest offshore natural gas platform, with an expected lifetime of 70 years and it is building in an allowance for sea level rise:

The cost, between 10 and 20 million sterling, is to be added to the estimated 2 billion cost of the entire development (McCarthy, 1989).

Apparently Shell Oil is prepared to wager that predictions of sea level rise may be right. For collective decisions, where there is no market, Clark (1985) urges the recasting of the policy aspects of global change in a risk management context, whose intellectual framework still remains to be elaborated. What individual decision makers do, when they have a stake in the same problem, can be a guide to the collective decision maker.

10.2 The rational "solution" — How a rational world might cope with migration problems

We have seen how climate change will, inevitably, drive people to migrate; but we have not yet examined the political realities. Consider one conceivable example of climate change effects on a nation's people. A small shift in the precipitation pattern might move Java's abundant winter rains a few hundred kilometres to the east. That could make the presently dry islands to the east wet, while Java becomes dry. The national authorities would be forced to move either people or foodstuffs in large quantities. That would be difficult indeed, as transmigration programmes have shown. But even worse, a shift of the monsoon to the south could make the centre of Australia wet and fertile. In that case, shifting population between Java and central Australia would be many times as difficult as shifting within Indonesia.

To continue along this line, suppose that Australia became wet and tropical — so that its interior came to be covered with rain forest — and India became as dry as Australia is now. By that time, there would most likely be 1.5 billion people in India. If they were as productive as the population of Japan in industries that did not depend on soils, then the difficulties might be surmounted. They would have plenty of artifacts to exchange, through international markets, for food grown in Australia and elsewhere. Free movement of commodities would certainly aid adaptation, given the presumed ability and willingness of the food-surplus countries to absorb indefinite numbers of video recorders and similar secondary goods. The condition of many African countries is that they are unable to grow enough food or to make the things that would exchange for food. In a rational world, under the impetus of the free markets, they would quickly learn to make them.

Among many other changes would be a shift from one cereal to another. Parry gives as an example the Moscow region of the former USSR, that is near the northern limit for wheat, and where yields of barley are higher and less variable. But beyond a certain level of warming, barley becomes heat-stressed and wheat yields increase. A small increase in average temperature could result in a major shift from barley to wheat. Such shifts are far from suggesting that calorie supplies would diminish, though in local markets there could be considerable price changes for particular cereals.

If we were rational beings united under a perfect world government, the expected warming could be handled in an orderly fashion and certainly without disastrous effects. Total food production need not be drastically reduced; some coastal areas would be under water and it might or might not be worthwhile to dyke and drain them. In other cases, it would be easier to move the population. There would be some migration of plant life, and some corresponding migration of people. The monsoons would change both in location and in timing; deserts would move to different places. Adaptation would require some expenditure, and it would be routinely charged each year to the taxpayers of the world. But adaptation would occur with reasonably great overall effectiveness, and reasonably low cost.

Unfortunately, this sort of rationality — in which a sufficient number of people and governments manage their affairs for the greatest long-term good of the world's population as a whole — can be dismissed along with other unrealizable Utopias. Utopias have been found too unrealistic for fiction, and now find a lodging in social and political science — especially among rationalistic rulers who believe they can control any irrational impulses in their populations, or in themselves. Instead, any workable solution to population problems, in the face of global climate change, must take into account — and prepare for — the deep irrationality inherent in both personal and institutional behaviour. It must take account of spreading fundamentalism, of ideological extremism, and of war-making nationalism.

10.3 Planning for climate change in our irrational world

In that context, a crucial factor will be the interaction of global climate change with the institutional system of our time and of the future. To prepare effectively, we need to know the politically unknowable: what will be the effect of warming on the 12 billion or so inhabitants of the latter half of the 21st Century, distributed into unfamiliar national states, with their very different regimes, alliances and enmities. Without making a political-economic forecast of the arrangement of the world by the year 2050, we could not say anything useful about the consequences of warming, even if we knew exactly the distribution of population, how much climatic change there would be in each place, and its effect on rainfall and on crops. And the social scientists who know most are, on their side, the least inclined to say what sorts of institutions will be important in the next century.

Any serious thinking about policy to avert the causes or mitigate the consequences of global warming must take account of how the problem stands amid the array of other problems. Countries engaged in a life and death struggle — like that of Somalia against Ethiopia, or Chad against Libya

— will not declare peace and begin to cooperate in dealing with the greenhouse problem the moment a western scholar tells them that warming is going on. And they will pay even less attention if the scholar also admits that the change is not certain, that at the worst it will be gradual, that it is not known what overall effect the change is going to have, that it will take decades of research before one can say what difference (if any) it will make to the physical geography of the countries concerned, and that it will take even longer to find the effects on their economies. In fact, they will think anyone is out of their mind who asks these countries to stop their fighting for so trivial a reason.

World government being out of the question, we have to think what kind of arrangements among sovereign states will secure some at least of the benefits of rational response to the changes induced by warming. Free passage on the seas and, as far as possible, on land would contribute to rationality. So would free markets in foodstuffs and in manufactured goods (that can be exchanged for food). Free trade would help adaptation to changes in the environment that made one part more fertile, another part less. Free exchange among sovereign states would go some considerable way to providing the mutual assistance that some hope (in vain) would be obtainable by a world government; rules and agreements to overcome externalities would take us the rest of the way. And among externalities, those impinging on the generations following us are of special importance.

The importance of trade is suggested in scenarios that have been set up, in which increases in productive potential in Australia, China and the former USSR broadly match losses in North America and Europe. If there is a single world market the supplies would not be much altered, and prices would remain what they are. With restrictions in trade, prices would surge in some regions, and plummet in others, creating considerable instabilities in the interwoven world economy.

A part of what makes it difficult for scholars to investigate the linkages between climate and population, and for the public to attend to them, is the other things that are happening. In relation to the population-climate interaction, such matters as wars, civil disturbance, and economic fluctuations are mere noise; they have nothing directly to do with the matter, being mere short-term effects that distract attention. They will be forgotten soon, while population growth marches inexorably on. Yet these short-term disturbances do enormous damage, both directly and by exacerbating the long-term problems. Dealing with the climate-population interaction is hard enough; to deal with it when those involved are distracted by war is to encounter one further layer of difficulty.

Long-term plans for protection make little sense. We need incremental, short-term measures, and we should be prepared to revise them as new knowledge comes into view. We need both better estimates of what will happen in the future, and better estimates of the errors of those estimates. The lack of certainty is not the decisive factor — need is. The amount that should be done in the way of protection against any specific danger is directly proportional to the estimated need — that is, the extent of damage threatened — and to the certainty of knowledge of the need; and it is inversely proportional to the cost.

References

Clark, W.C., 1985: *On the Practical Implications of the Carbon Dioxide Question.* WP-85-43, Laxenburg, Austria: International Institute for Applied Systems Analysis.

GESAMP, 1990: The State of the Marine Environment: UNEP Regional Seas Reports and Studies No. 115, Nairobi, Kenya, p. 81.

IPCC, 1991:*Climate Change: The Response Strategies,* Working Group III, 1991, p. 19

McCarthy, M., 1989: Greenhouse effect gives a life to gas platform. *The Times,* September 7, 1989.

Munn, R.E., 1990: Towards sustainable development, *Managing Environmental Stress.* University of New South Wales. Proceedings of a Symposium.

Parry, M., 1990: *Climate Change and World Agriculture*, London: Earthscan Publications Limited, p. 7, 60, 72, 78, 80.

Revelle, R., 1982: Carbon dioxide and world climate. *Scientific American,* **247**, 35-43.

Rogers, P., 1985: Fresh Water. In *The Global Possible: Resources, Development, and the New Century*, R. Repetto, ed., New Haven: Yale University Press.

Scott, M.J., N.J. Rosenberg, J.A. Edmonds, R.M. Cushman, R.F. Darwin, G.W. Yohe, A.M. Liebetrau, C.T. Hunsaker, D.A.Bruns, D.L. DeAngelis and J.M. Hales, 1990: Consequences of climatic change for the human environment. *Climate Research*, **1**, 63-79.

UNEP, 1989: Criteria for Assessing Vulnerability to Sea Level Rise: A Global Inventory to High Risk Areas. United nations Environment Programme and the Goverment of the Netherlands.

Tolba, M., 1990: *World Development Forum*, **8**, 3.

CHAPTER 12

The Energy Predicament in Perspective

John P. Holdren

Editor's Introduction

Paradoxically, the use of energy contributes considerably to human well-being — and to the environmental risks that threaten the quality of human life. Energy is the essential engine of economic development, but the extraction, mobilization, and supply of energy generate risks to human health and may also endanger natural ecosystems. The consumption of energy releases dangerous pollutants to air, water, and soil.

In this chapter, John Holdren shows us that the world is not running out of energy — not in any absolute sense. Instead, the most important and problematic costs of energy supply are the environmental and sociopolitical ones. Holdren's list of what we are running out of—from political will to patience to atmospheric room for the pollutants from fossil fuels—is one of the most compelling arguments made about energy policies today. He particularly focuses on the imbalances and inequities in the current pattern of global energy use: fully three-quarters of global primary energy is supplied by fossil fuels, and two-thirds of the total supply is consumed by the 20% of today's population which lives in the industrialized world. As with food supply and disaster relief, the poor bear the brunt of the energy problem. They often have too little energy to meet basic human needs and pay too high a cost for what they can get.

Among the environmental problems caused by energy, Holdren suggests that the most troublesome are not local impacts — which can certainly be painful and disruptive — but the transboundary effects that stretch across continents and, indeed, span the globe. These include radioactive plumes, acidic precipitation, and the atmospheric buildup of greenhouse gases.

John Holdren, Professor of Energy and Resources at the University of California, was trained as a plasma physicist. His principal research contributions have been in the analysis of energy technologies and their environmental risks. Long before the theme of sustainable development had gained a popular following, Holdren was an advocate of balanced national strategies that were equally concerned with economic development and environmental protection. As a long-time participant in the Pugwash Conferences and Chairman of the Pugwash Executive Committee, Holdren has worked diligently to bridge the perceptual gaps created by the Cold War and develop a new and comprehensive vision of global security. He has long recognized the political linkages between traditional military doctrines, national energy strategies, and the urgent needs for economic development in the Third World. Holdren's audience stretches far beyond the boundaries of academia, to heads of government and leaders of major international corporations. The message of this chapter, for example, draws heavily on a speech given by Holdren to the 1991 ECOTECH Conference, a convocation of business executives from cutting edge companies across a variety of high technology fields.

This chapter is not an argument that we should abandon energy use, choosing instead to freeze in the dark. Rather, it suggests the need for a strategy that simultaneously improves the quality of output from national economies while preserving the ability of natural systems to support human and other forms of life. Such a strategy, according to Holdren, must contain five key elements: (1) policies to encourage increased efficiency of energy supply and use; (2) measures to reduce the environmental impacts of current energy sources; (3) financial incentives coupled with accelerated research and development of sustainable energy options; (4) expanded efforts at international coopera-

tion and development assistance; and (5) policies, programs, and measures to stabilize global population at level not greater than ten billion people.

John Holdren explores the ramifications of these policies in this chapter; other chapters in this section will examine the technological prospects and institutional challenges of changing energy use during the coming decades. Improved use of energy provides one of our best leverage techniques for avoiding the false dichotomy suggested by those who would choose between economic development and environmental protection.

- I. M. M.

1 Energy Problems in Theory

Basically, there are two ways that human societies can get into difficulty with energy: they can suffer from having too little of it; and they can suffer from paying too much for it. (It is even possible, as we shall see, to suffer from both difficulties at the same time.)

Having too little energy can mean failure to meet basic human needs in the form of energy for heating, cooking, pumping and boiling water, and lighting. In more prosperous contexts, it can mean unwanted constraints on the rate of industrialization and economic growth.

As for paying too much, the excessive costs may be monetary, environmental, or sociopolitical. Excessive monetary costs mean that paying for energy diverts too much money from other societal needs, leading to reduced living standards, inflation, recession and debt. Excessive environmental costs mean that the toll exacted by emissions and accidents in the energy system — the toll in disease, lost life expectancy, reduced worker productivity, and loss of ecosystem services — is too high in relation to the energy's benefits.

The sociopolitical costs that can become too high include: constraints on foreign policy arising from overdependence on foreign energy resources; going to war to protect access to those resources; the aggravation of international tension by transboundary pollution; the spread of nuclear weapons capabilities along with nuclear energy technology; adverse impacts of energy choices on the vulnerability of societies to military or terrorist attack; domestic political disputes over the siting of energy facilities; and domestic and international resentments over inequitable distribution of energy benefits and risks.

2 Energy Problems in Practice

In reality, the worst is true: the world is suffering simultaneously from having too little energy (this being so for the majority of the world's people, who are poor) and from paying too much for it (this being so for the poor majority and rich minority alike, although in different ways.) Let me elaborate.

The poorest countries in the world, and the poorest people in the rich countries, have too little energy to meet basic human needs; and the prices they pay for the energy they do have are punishing. The poor are paying high prices in money for commercial energy such as kerosene, gasoline, and electricity. They are paying high prices in the time spent in gathering ever scarcer fuelwood in rural areas. And they are paying high environmental prices in the form of forest degradation from fuelwood harvesting, reduced soil fertility because crop wastes and dung are being burned as fuel, and acute local air pollution from burning dirty fuels with no pollution control.

In the richer countries, the problems in the category of having too little energy are smaller, but the problems in the category of paying too much for energy are larger. The most troublesome of the excessive costs are not the monetary ones — although those are high enough to cause difficulties for some people in some countries. The most troublesome costs of energy supply in the United States and other industrialized nations are the environmental and sociopolitical ones.

Among these environmental and sociopolitical costs, moreover, the most troublesome are not the local ones, however intense those may be, but rather those that reach out across continents and around the world — polluting hazes covering huge regions, radioactive plumes that respect no borders, highly acidic precipitation falling on millions of square kilometres of susceptible soils and watercourses, the global accumulation of greenhouse gases so likely to impose intolerable changes on the world's climate, and of course the contribution of the industrialized world's continuing overdependence on Middle East oil to the chance that yet another catastrophically destructive war will be waged there.

3 Some Relevant Data

To understand our energy problems and prospects in more detail, it is useful to look at a few numbers.

Consider first the current pattern of energy supply. As indicated in Table 1, the inanimate energy used by civilization in 1990 (i.e. excluding food for people and feed for domesticated animals) came 77% from fossil fuels, 12% from the "traditional" fuels of wood, crop wastes, and dung, 6% from hydropower, and 5% from nuclear energy. Of this total supply, about two-thirds went to the 1.2 billion people living in industrialized countries, and one-third went to the 4.1 billion people living in less developed countries. The industrialized countries accounted for 72% of world fossil-fuel use.

If one considers the probable sizes of the remaining resources of these energy forms, one sees no immediate danger of "running out" of energy in a global sense. At the

Table 1. World energy use in 1990
(TW = Terawatts = 10^{12} watts = 10^{12} joules/second = 31.5 x 10^{18} joules/year)

	Industrialized countries	Less developed countries	World total
Oil	3.23 TW	1.27 TW	4.50 TW
Coal	1.97 TW	1.20 TW	3.17 TW
Natural gas	2.16 TW	0.38 TW	2.54 TW
Biomass	0.25 TW	1.27 TW	1.52 TW
Hydropower	0.48 TW	0.29 TW	0.76 TW
Nuclear	0.63 TW	0.03 TW	0.66 TW
	8.72 TW	4.44 TW	13.16 TW

Notes: Figures for oil, coal, gas, hydropower, and nuclear energy from British Petroleum (1991), converted from lower to higher heating values. Biomass figure is author's estimate based on Hughart (1979), Hall et al. (1982), Goldemberg et al. (1987), and Smith (1987).

1990 rates of use, resources of oil and natural gas would last 70 to 100 years. (These are the figures for "conventional" resources of these fuels, not counting heavy oils, oil shales, and unconventional sources of natural gas.) There is at least a 1500-year supply of coal — again, at today's rate of consumption — and quite probably enough affordable uranium resources to run today's types and numbers of nuclear reactors for 500 years or more.

Neither hydropower nor biomass energy use are close to capturing the total energy flows that are theoretically available in these forms. And all this is without considering the further set of energy resources — direct conversion of sunlight to electricity and fuel, windpower, ocean thermal energy, geothermal energy, and fusion — which altogether contribute much less than 1% of civilization's energy supply today, but which might be able to make much larger contributions in the future.

Of course this rather optimistic portrayal of global energy resources looks less rosy if one compares the various potentials, not against today's rates of energy use, but against the much higher rates that will occur in the future if past patterns of growth in energy use continue. And because of the uneven geographic distribution of most energy resources, the picture for some countries and regions looks less rosy than the picture for the world as a whole. But the fact remains that running out of energy resources in any global sense is *not* what the energy problem is about — now, or in the lifespan of anyone reading this book.

4 If We're Not Running Out of Energy, What's the Difficulty?

We may not be running out of energy, but we are running out of a great many other things that matter a great deal.

- *We are running out of the cheapest and most accessible oil and natural gas supplies, and the most convenient and cost-effective hydroelectric sites,* which are the resources that fueled the development of today's industrialized countries and shaped the expectations of the poor ones. The unsurprising reality is that industrializing and industrialized societies found and used the most convenient and least expensive energy resources first: the biggest, richest, shallowest, closest deposits of oil and natural gas, and the closest and most cost-effective hydroelectric sites. Cumulative depletion and rising demand now require that we resort to smaller, leaner, more distant, more difficult (and hence more expensive) resources of these kinds — or to more abundant resources, such as coal and uranium and solar energy, which all happen to be costlier, in terms of capital investment, to convert into the fluid fuels and electricity that industrialized societies require.

- *We are running out of the renewability of biomass energy supplies,* in so far as today's exploitation rates and practices — in connection with the use of wood, crop wastes, and dung as energy sources — are contributing in many regions either to deforestation or to soil depletion. What level of biomass energy use can be made sustainable — as well as compatible with competing uses for the biomass resource for food, fodder, fibre, fertilizer, feedstock, and ecosystem function — will be the key issue in the future of this energy resource.

- *We are running out of the absorptive capacity of the environment for the effluents of energy use.* We have used up and exceeded the local capacity of the atmosphere to dilute and disperse emissions of particulate matter, reactive hydrocarbons, and other toxic effluents in regions of high population density; the fallout of nitric and sulphuric acids formed by the effluents of fossil-fuel combustion has consumed the acid-neutralizing capacity of poorly buffered soils, lakes and streams over still larger regions; and the capacity of the atmosphere to dispose of carbon dioxide from fossil-fuel combustion has been overtaxed worldwide.

- *We are running out of investor and public tolerance for the risks and uncertainties associated with contemporary types of nuclear power plants,* including cost escalation, accident potential, the radioactive waste issue, and the links between nuclear energy technologies and nuclear-weapons capabilities. This depletion of the reserves of tolerance is visible in the results of referenda against the continuation or expansion of nuclear power generation in some countries, and the dearth of orders for new plants nearly everywhere.

- *We are running out of the tolerance of the poor for inequity in the distribution of energy's benefits and of its risks.* The three-quarters of the world's population who live in the less-developed countries of the South have less than a third of the world's total energy use to divide among them — hence some six times less energy use per person than in the industrialized North. Yet they are exposed to the dirtiest air and water (the worst of the air being in Third World village huts where biomass fuels are burned indoors for cooking, space heating, and water heating); and they are more vulnerable than the populations of the North to the consequences of greenhouse-gas-induced climate change being brought about, above all, by the fossil-fuel combustion that so far has been concentrated overwhelmingly in the industrialized countries.

- *We are running out of money to pay for alternatives* to the high-polluting and often low-efficiency energy options on which the world remains mainly dependent. Cleaner and more efficient options are often costlier than the ones they would replace, and even when not costlier in life cycle terms, they still require up-front investments that pose a barrier to their widespread and rapid adoption. Of course it must be said that while "running of out of money" means, in developing countries and in the floundering economies of the former Soviet Bloc, that there is no *ability* to pay, in the richer industrialized countries we are merely running out of *willingness* to pay.

- *We are running out of time to adjust our energy system* before the consequences of the foregoing problems become intolerable. The energy system is too big, too expensive, and too ponderous to be changed very quickly. The global investment in facilities for harvesting, transforming, and transporting energy, valued at replacement cost, is around $8 or 9 trillion; and the operating lifetime of individual facilities is typically 30 to 60 years. Power plants being designed in 1991 to begin operation in the year 2000 will still be running in 2030, and quite possibly in 2050. If we want the energy system in 2030 to be very much different than the one we have today, we must begin to change it *now*.

- *We are running out, finally, of the political will needed to take collective action*. That is partly a matter of lack of public and policy maker understanding of the dimensions and timing of the predicament, and of compartmentalization of understanding and responsibility. It is partly a matter of the interplay of "scientific" uncertainties (about environmental impacts and economic consequences) with a political process over-encumbered with special interests, narrow and short-range conceptions of the public interest, and pork-barrel politics. And it is partly a matter of overestimating the capacity of "the free market" to solve our energy problems.

5 The Marketplace and its Shortcomings

Presidents Reagan and Bush have excused the inattention of their administrations to energy policy — as leaders of many other countries also have done — by asserting that the economic marketplace is the best way to make most energy decisions. But there has not been a free market in energy in the United States — nor in many other places — for the past 100 years (and the United States did not move much closer to one under Reagan and Bush). Instead, an enormously complex structure of regulation, tax incentives, and subsidies continues to make it virtually impossible to tell what the various energy sources really cost in monetary terms, or to say what mix of sources would be used if their true monetary costs governed the choices.

Even if we DID have something more closely resembling a free market, a range of energy policies would still be required to cope with the well-known ways in which markets fall short. Among these market failures, three are of predominant importance:

- the failure to account for externalities and public goods — that is, costs and benefits of particular energy choices that fall on people other than the buyers and sellers of the energy and energy technologies;

- the short time horizons of individual and corporate economic actors, reflected in the high discount rates that lead to neglect of long-term concerns and values; and

- the market's failure to account for the interests of the poor, who are disenfranchised by the circumstance that in a market one votes with one's money.

These particular failures have been crucial in the generation of today's energy/environment/economy predicament, which consists above all of three interacting elements:

- the rising threats to human well-being from the external costs of energy supply;

- the incapacity of the poor to pay for the expansion of energy supplies that would be needed to make them prosperous, if doing so is going to require as much energy as it did in the currently industrialized countries; and

- the likelihood of intolerable local, regional, and global environmental damage if the poor even attempt such energy expansion using a mix of energy supply technologies very much like today's.

6 The Dimensions of the Difficulty

A few more numbers will underline the dimensions of the difficulty. In the 140 years between the middle of the last century and the present day, the population of the world increased five-fold, world use of energy in the aggregate increased 20-fold, and fossil-fuel use increased 100-fold. These increases were major factors in the transformation of civilization, over this century and a half, from a modest and mainly localized disrupter of environmental conditions and processes to a global ecological and geochemical force. Today human activities rival or exceed natural processes as mobilizers of sulphur oxides, nitrogen oxides, hydrocarbons, lead, cadmium, mercury, and suspended particulate matter in the global environment; the actions of humans have increased the global atmospheric burden of carbon dioxide by nearly 30% and that of methane by more than 100 percent, compared to pre-industrial levels; and among all human activities, the technologies of energy supply — above all, fossil-fuel energy technologies — are the dominant sources of most of these global pollutants and significant sources of all of them. (See Table 2.)

Global warming, to which carbon dioxide release from fossil-fuel burning is the largest single contributor, is arguably the most dangerous and intractable of all of the environmental impacts of human activity. It is the most dangerous because climate affects — and climate change can drastically disrupt — most of the other environmental conditions and processes on which the well-being of 5.5 billion people critically depends. These include the magnitude and timing of runoff, frequency and severity of storms, sea level and ocean currents, soil conditions, vegetation patterns, and distribution of pests and pathogens, among others. It is the most intractable because the "greenhouse" gases mainly responsible for the danger of rapid climate change over the next few decades are being released largely by human activities too massive, widespread, and central to the functioning of our societies to be readily altered: carbon dioxide from fossil-fuel combustion and deforestation, methane from rice paddies and cattle guts and the harvesting and transport of oil and natural gas, nitrous oxides from land clearing and fertilizer use and fuel combustion. Altogether some 53% of the global warming potential of contemporary human activities comes from fossil-fuel use, and another 3% or so from the use of fuelwood.

Table 2. Energy's role in global environmental impacts

Human impact	Size of impact compared to natural processes	Energy's share of responsibility for human impact
Emission of lead into the atmosphere	1500%	65%
Spills/leaks of petroleum into the oceans	1000%	60%
Emission of sulphur dioxide into the atmosphere	140%	85%
Accumulation of methane in the atmosphere	100%	25%
Accumulation of carbon dioxide in the atmosphere	27%	80%
Emission of particulate matter into the atmosphere	25%	45%
Emission of non-methane hydrocarbons intc the atmosphere	13%	40%

Notes: Estimates are based on a variety of references and are very approximate. See, e.g. Holdren (1987, 1991), Lashof and Tirpak (1989), Graedel and Crutzen (1989), and IPCC (1990).

These circumstances make clear that the material prosperity of the industrialized countries cannot be maintained and enlarged — and prosperity for the majority of the world's population who are now poor cannot be provided — by straightforward expansion of the energy-supply systems we employ today. For example, to provide today's world population with the average level of energy use per person enjoyed in 1990 in the industrialized countries would require an immediate tripling of world energy supply from its 1990 value of 13 terawatts to 40 terawatts. No serious student of problems of global warming, urban and rural air pollution, acid rain, oil in the oceans, deforestation, soil depletion, and so on thinks this could be managed by multiplying today's energy sources by three.

Yet we must deal not merely with today's population, but with a likely 8.5 billion people by 2025 and 10 billion by 2050, barring either catastrophe or drastic breakthroughs in fertility reduction. To provide 10 billion people with the per capita energy use characteristic of the industrial countries in 1990 would require nearly six times the 1990 world energy supply, or about 75 terawatts; to attempt such a level with anything like today's energy technologies could not fail to destroy, everywhere, the environmental basis of prosperity.

This is not even to mention that funding the corresponding expansion of energy supply technologies in the developing countries would consume an intolerable fraction of the capital that is expected to be available there in this time period for all purposes.

Continuation of energy business as usual, then, is not the solution; it is the heart of the problem.

7 What Should Be Done?

An energy strategy responsive to this predicament would contain five main elements, as follows:

First, *increasing the efficiency of energy end use* is the most obvious response to excessive energy costs, whether those are monetary, environmental, or political. Increasing efficiency refers not to sacrifice but to squeezing more goods and services out of each gigajoule of heat, each litre of fuel, and each kilowatt-hour of electricity. The potential for improvements in energy efficiency is large, as indicated by the nearly 40% increase in the energy efficiency of the US economy (inflation-corrected dollars of GNP per gigajoule of primary energy) in the 17 years following the initial oil-price shock of 1973 — as well as by dozens of detailed engineering-economic studies of the further efficiency improvements waiting to be harnessed in particular sectors of human activity such as transportation, agriculture, housing, and manufacturing. It is quite plausible that the current US standard of living could be provided with about half the current US energy use per capita, using known technologies that would be cost-effective even at today's energy prices (exclusive of environmental costs and other externalities).

In fact, such a living standard probably can be managed with even less energy — perhaps a quarter to a third of the current US figure — given some further technical innovation, a modicum of increase in energy prices (as from internalizing environmental costs), and the sorts of structural and lifestyle changes likely to be brought about by the combination of such price increases and growing environmental concerns. (Such changes might include increased durability of goods, shorter commutes, more attractive public transportation systems, and reduced materials use in packaging.)

In any case, energy-efficiency increases of this order absolutely must be achieved and propagated over the next several decades if there is to be any chance of providing a high standard of living to 8 or 10 billion people without reaching unsustainable levels of energy use.

Second, *reducing the environmental impacts of today's energy sources* must be pursued in parallel with efforts to increase efficiency of energy use. This means, in the short run:

- More widespread use of available control technologies to reduce emissions of sulphur and nitrogen oxides from fossil-fuel combustion, and to reduce emissions of hydrocarbons and particulate matter from fossil and biomass fuels alike;

- greater efforts to reduce the leakage associated with ocean drilling and transport of petroleum;

- increased resources and inspection powers for national and international nuclear-safety and anti-proliferation authorities; and

- in less-developed countries especially, efforts to put fuelwood harvesting on a sustainable basis.

In the medium term, it means:

- Deploying more efficient and cleaner-burning coal technologies (such as integrated-gasification combined-cycle power plants, pressurized fluidized-bed combustors, and fuel cells);

- increased substitution of natural gas for coal and oil, to gain further reductions in emissions of carbon dioxide and other pollutants; and

- replacing direct combustion of biomass fuels in developing countries with biogas and alcohol fuels that burn more cleanly and return some nutrients to the soil.

In the long run, it may be necessary to capture carbon dioxide from the exhaust gases from fossil-fuel combustion and sequester it away from the atmosphere; expanded research on the means and costs of doing this should be undertaken now.

Third, *facilitating the transition to more sustainable energy options* should include financial incentives for renewable energy sources — such as wind power, solar-thermal electricity generation, and biomass-derived alcohol fuels — that are close to economic competitiveness at current prices of conventional alternatives, and that offer significantly smaller environmental or political risks than those alternatives.

It should also include increased support for research and development on such longer-term non-fossil energy options as photovoltaics, solar-thermochemical hydrogen production, ocean-thermal energy conversion, hot-dry-rock geothermal energy, advanced fission reactors, and fusion energy systems. These R&D efforts should pursue not only the attainment of practical, economic ways to harness these possibilities, but also the prospects for minimizing their environmental costs; thus, for example, work on fission should emphasize improved reactor safety, proliferation resistance, and development of an approach to waste disposal acceptable to the public.

It is not yet clear how rapid the transition away from today's heavy fossil-fuel dependence should be, or what non-fossil alternatives ought to play the main roles. The answers will depend both on our growing understanding of the magnitude and timing of the global-warming threat, and on the still incompletely understood costs and impacts of the advanced non-fossil sources. But certainly the investment needed now to ensure that an appropriate mix of non-fossil options is available to choose from, when the need becomes clear, is modest — compared to the potential costs of having a compelling need and no satisfactory choices.

Fourth, *expanding our programs of international cooperation and assistance* is necessary if the elements already mentioned are to be implemented not only in the most advanced industrial nations but also in the industrialized but cash-poor countries of eastern Europe and the former Soviet Union, and in the less developed countries of the South.

This is a proposal not for altruism but for responsibility and self-interest: responsibility, because the countries that are now richest got that way burning cheap fossil fuels and using up the absorptive capacity of the atmosphere for the resulting carbon dioxide, thus depriving the rest of the world of the option of following the same inexpensive path to prosperity; and self-interest, because the rich will not be able to escape the consequences of the environmental catastrophe likely to follow if China, India and the rest try to industrialize with inefficient end use and conversion technologies, and heavy reliance on coal.

Finally, *halting the growth of the world's population at ten billion or fewer* is essential. This achievement would require fertility to fall to the replacement level of 2.1 children per woman (from today's world average of 3.5) by the year 2025. Failing that, increasing numbers of people will offset and overwhelm the gains from all the other elements of the energy strategy described here. Bringing about the needed fertility reduction will itself be an enormous challenge, requiring massive development assistance and other forms of international cooperation. But as difficult as that will be, it will still be easier than coping with the consequences of a world population soaring to 12 billion, 15 billion, or more.

8 Some Concluding Observations

Today we are not doing remotely what is required in any of these aspects of energy strategy. Why not? We are not doing it because, as a society, we are underestimating the danger from global environmental change — above all from climate change — and overestimating our capacity to respond quickly if we decide belatedly to do so. Those of us in industrialized countries are underestimating the linkages between our societal well-being and that of other countries which cannot make it without our help. And we are overestimating the costs of taking constructive action now.

We have succumbed, foolishly, to the siren song that the economic marketplace can be our only energy policy, when in fact the need for policy arises precisely from factors that markets unassisted by policy will always fail to address: environmental and political costs outside the balance sheets of producers and consumers; the uneven distribution of the capacity to participate in markets; and the societal benefits of investments with payback times longer than the time horizons of individual and corporate investors.

Perhaps the biggest indignity in today's energy/environment/economy debate is the persistent assertion that we must choose between the environment and the economy — and the implication that investments in environmental protection make us poorer and diminish our economic competitiveness. To the contrary, intact environmental conditions and processes are essential to a productive economy. Suggesting that we cannot or ought not to invest in the maintenance of our environmental infrastructure is akin to saying the same thing about our transportation infrastructure, or our communications infrastructure, or our educational system. Our economy cannot function without any of these things.

The sad truth, of course, is that we have been neglecting all of these foundations of our economic well-being. It has been a matter of overemphasizing consumption while underemphasizing investment — an approach which makes us seem richer than we really are because some of the prosperity it generates is based on using up our capital. In the environmental arena, that has meant mining our groundwater, consuming our topsoil, logging off our old-growth forests, mucking up the machinery of climate. Whether as a strategy for managing environmental resources, or the national accounts, or social and industrial infrastructure, consuming one's capital is a long-run prescription for bankruptcy.

As for competitiveness, there is much reason to believe that the companies and countries which move out in front in developing technologies and products with low environmental impact — those that provide the goods and services people want while minimizing greenhouse gas emissions and other pollution — will gain the key competitive edge for the 1990s and beyond. These will be the technologies and products that will be in demand in the United States and around the globe, as the imperative of maintaining the world's environmental capital becomes steadily more apparent.

References

British Petroleum, 1991: *BP Statistical Review of World Energy*. BP, London.

Goldemberg, J., T.B. Johansson, A.K.N. Reddy and R.H. Williams, 1987: *Energy for a Sustainable World*. World Resources Institute, Washington, DC.

Graedel, T.E. and P.J. Crutzen, 1989: The changing atmosphere. *Scientific American*, September, 58-68.

Hall, D.O., G.W. Barnard and P.A. Moss, 1982: *Biomass for Energy in Developing Countries*. Pergamon, Oxford.

Holdren, J.P., 1987: Global environmental issues related to energy supply. *Energy*, **12**, 975-992.

Holdren, J.P., 1991: Population and the energy problem. *Population and Environment*, **12**, 231-255.

Hughart, D., 1979: *Prospects for Traditional and Non-conventional Energy Sources in Developing Countries*. World Bank, Washington, DC.

IPCC, 1990: *Climate Change: The IPCC Scientific Assessment*. Cambridge University Press, New York.

Lashoff, D.A. and D.A. Tirpak, (eds.), 1989: *Policy Options for Stabilizing Global Climate*. Environmental Protection Agency, Washington, DC.

Smith, K.R., 1987: *Biofuels, Air Pollution, and Health*. Plenum, New York.

CHAPTER 13

Electricity: Technological Opportunities and Management Challenges to Achieving a Low-Emissions Future

David Jhirad and Irving M. Mintzer

Editor's Introduction

We are moving into a future powered by electricity. In developing countries, electricity use continues to replace the burning of fuels. Extension of the electricity grid to rural and agricultural areas will expand the customer base of electric utilities and broaden the industrialized regions in these countries. In the industrialized world, growing "plug loads" caused by the proliferation of appliances, computers, and other gadgets will combine with the increased penetration of electrically driven heating and cooling systems to raise residential and commercial sector demand for electricity. Advances in process design will cause more industrial operations to shift from direct combustion to electric operations. And in the transportation sector, the combined effects of urban pollution, traffic congestion, and the costs of oil imports will encourage many countries to develop electric-powered railroads, vehicles and mass transit systems. Unless care is taken now in the plans made to meet these demands, emissions of greenhouse gases from the electricity sector will increase rapidly in the decades ahead.

David Jhirad and Irving Mintzer argue that meeting the projected demand for electricity will be very difficult for many developing countries — countries which now are caught in a triple bind. First, these countries do not have (and are unlikely to get) the capital to build all the new power plants suggested by the most recent forecasts of demand. Secondly, many utilities which have traditionally provided electric power in developing countries suffer from declining levels of financial and technical performance. And thirdly, like many other institutions in both developing and industrialized countries, electric utilities in the developing world face severe pressure to address environmental concerns, both local and global.

This chapter offers an overview of the potential strategies which electricity providers, their financiers, and end-users can adopt to overcome this triple bind. The most important opportunities, in the near term, come from energy efficiency — as utilities in industrialized countries are beginning to discover. By applying cost-effective technologies to reduce the energy intensity of key end-uses, environmental impacts can be moderated, financial requirements reduced, and the technical performance of utilities improved. Many of these technologies are available now; others, which utilities will also find essential as they shift away from burning carbon-intensive fossil fuels, are still under development. The most promising of these include advanced combustion systems; solar, wind, biomass, and other renewables; molten-carbonate fuel cells; advanced storage systems; and improvements in transmission and distribution systems.

But Jhirad and Mintzer emphasize that more than technological change will be necessary to overcome the market imperfections and institutional barriers that obstruct the path to a rational energy strategy. Policies to improve the flow of information, to bring the full cost of energy supply and use into the price of fuels, and to reshape the role of electric utilities will be needed to rapidly exploit the new technological opportunities. In the industrialized countries, utilities must be reorganized as energy service companies —enterprises whose purpose is not to sell more "juice," but to provide people with comfort, heat, light, and power. This often means promoting electricity efficiency among their customers, a less risky investment (for the utility) than building new plants. But spreading this practice requires financial and regulatory incentives for efficiency improve-

ments that are in balance with the incentives for increasing energy supply.

In their work with power sector institutions in developing countries, the World Bank and other development assistance agencies are particularly well positioned to encourage the introduction of incentives for improving efficiency. This is not sufficient by itself, however. The development assistance agencies must also promote institutional reforms, and they must encourage co-development of the best of the new technologies. These new technologies must be adapted as necessary to the special circumstances of each receiving country. This combined strategy, if linked with reform of energy subsidies and pricing policies, investments in energy efficiency, and extension of electricity supply into rural areas, could provide the impetus for sustained economic development with a modest increase in greenhouse gas emissions from the power sector in the near-term. In the longer-term, if the full range of cost-effective efficiency improvements and carbon free supply techniques could be implemented in developing countries, and if these could be combined with realistic energy price and subsidy policies, economic development could be expanded while stabilising emissions, all in a cost-effective manner.

David Jhirad, trained as a plasma physicist, is the Senior Energy Advisor in the US Agency for International Development's (USAID) Office of Energy. Jhirad has been the strongest advocate within USAID for the development of comprehensive strategies that link technological advances to institutional and policy reform of the power sector. Irving Mintzer is Coordinator of the Climate and Sustainable Development Programme of the Stockholm Environment Institute and director of its Global Warming Assessment Project. *The authors wish to offer a special acknowledgement to Amulya K.N. Reddy and Dilip Ahuja, whose comprehensive analyses of the energy sector stimulated the development of this chapter.*

- I. M. M.

1 Introduction: Electricity in the Larger Picture

Under all reasonable scenarios, demand for electricity is expected to grow in both industrial and developing countries during the next 30 years. Even if energy use decreases in industrial countries, electricity use will increase, as electrification takes over the burden from direct combustion of fossil fuels. Everywhere in the world, major increases in per capita consumption of electricity will occur - as a result of efforts to accelerate modernization and improve the overall standard of living. But in seeking to establish their electric power priorities, developing countries and nearly all the former Eastern Bloc nations - face a particularly severe triple-bind set of problems:

- First, they cannot mobilize adequate capital with which to develop the transmission, distribution and fuel supply facilities for providing increasing amounts of electric power. Under most "business-as-usual" projections, approximately $100 billion per year will be required for power systems expansion in developing countries in the 1990s. But the combination of traditional internal and external sources of investment funds will be able to provide only about $20-25 billion per year.

- Secondly, the existing electric utility companies suffer from stagnant or deteriorating performance - both financial and technical.

- And thirdly, they are now coming under increasingly severe environmental constraints.

Each of these factors amplifies the severity of the others, and limits the extent to which electrification of these countries can proceed. No single measure will be an effective response to the challenge of this triple bind. Neither policy reforms, nor institutional improvements, nor technological innovation can be adequate in itself. Instead, a multifaceted approach will be necessary in each nation. Successful strategies must include innovative approaches on all fronts (Jhirad, 1990).

This poses a difficult challenge for the governments of developing nations, and the institutions within them — as well as for the international community. When, in the past, the environmental problems of developing countries were perceived as local in nature, these problems could be ignored by outside nations and agencies. But more recently, the increasing recognition of global environmental linkages has accelerated the search for new paths to economic development that are consistent with both local and global environmental goals.

The risks of rapid climate change particularly complicate many of the problems already facing the electric power sector. Yet, many of the most important actions required to address potential global climate problems must be undertaken now, independent of whatever the magnitude and timing of actual global climate changes turn out to be.

Over the long term - 2025 and beyond - a major technological transition to low-carbon or carbon-free electricity generation is possible. As will be demonstrated in this chapter, the benefits of such a transition more than justify the effort of making it. But the transition must be made with realistic expectations. Even with aggressive implementation of efficiency and renewable power sources, fossil fuels will continue to dominate the supply mix for the next 30 years. To prepare for the long-term change, drastic policy and institutional restructuring will be required in the power sector of developing countries, to ensure the continued growth of affordable electricity services within the limits of unavoidable capital and environmental constraints.

2 Facing the Triple Bind

2.1 Capital shortages

Electric utilities in developing countries are finding that their traditional plans for expanding electricity supply capacity will fall short of providing the power services required for economic development. A 1988 report by the United States Agency for International Development (USAID) to the US Congress (USAID, 1988) concludes that developing countries will require an additional 1500 gigawatts (GW) of generating capacity, as well as related new transmission and distribution facilities, to maintain a moderate economic growth rate during the 20-year period from 1988 to 2008. The cost of this added capacity is roughly $2.5 trillion, or an average of $125 billion per year. The World Bank has independently estimated that about $100 billion a year will be required during the 1990s (in current dollars) for the same electrification (Moore and Smith, 1990). Of this annual requirement, $40 billion is needed in hard currency. However, only $7-$10 billion in hard currency is presently available from all external sources, including both bilateral and multilateral agencies, as well as private creditors.

There are also serious local financing gaps. Utilities in developing countries have limited ability to mobilize the local currency required by these energy expansion plans. The average developing country already spends one-quarter of its public budget on the power sector. Other sectors, such as health and education, are competing for the same scarce resources.

Developing countries are quite aware that this power-related financial crisis will become more severe as debt mounts and new loans become more difficult to secure, and that economic disruption and political instability could ensue. Deteriorating utility performance, and the consequent unreliability of delivered power already exact a heavy toll on economic growth. The adverse economic effects of power supply interruptions can equal five to 100 times the average electricity tariff. In India, for example, the value of lost industrial output caused by power shortages is estimated at $6 billion a year (10% of the total annual industrial output).

2.2 Declining institutional performance

The financial performance of many utilities in developing countries is considerably lower than that of typical utilities in industrial countries. Moreover, in recent years, the technical performance of typical utilities in the South has been worsening. Inadequate maintenance, shoddy operations, incomplete accounting, unsystematic billing and unrealistic planning practices limit the ability of utilities to deliver adequate electricity services for sustained growth. Transmission and distribution losses in developing countries typically amount to 20-25% of electricity generated, compared with 7-8% in the industrialized countries. In the industrialized world, system losses are primarily technical, caused by the electrical resistance of transmission and distribution systems. Reactive power losses also occur in electrical networks, but these losses effectively reduce the power delivered from existing system capacity. Thus, resistive losses result in lost electric energy and reactive losses result in reduced system capacity. To developing countries, however, these technical losses are often dwarfed by accounting failures, uncollected revenues, and theft. In many cases, these problems stem largely from a lack of management autonomy and from a systematic inability to provide staff with incentives that are adequate to encourage the highest levels of creativity and performance.

Due to this poor performance, government-owned utilities in developing countries often show a negative return on assets. Indeed, for developing country utilities during the period from 1966 to 1987, the conventional indicators of financial performance declined steadily. The rate of financial return on assets fell from 9.2% to 4.4%; debt service coverage rose from 2.0 to 2.6; and the investment self-financing ratio dropped from 24% to 19% (Munasinghe et al., 1988). Figure 1 shows a steady decline in the financial rate of return over the period 1965 to 1984. The financial performance of utilities has been equally poor when measured by other financial indicators, such as the self-financing ratio, operating ratio, and debt-service ratio.

Inadequate tariffs are a major factor in poor financial performance. Studies undertaken by the World Bank indicate that average tariff yields in developing countries should be about US $0.10 per kilowatt-hour (1989 prices) in order to cover long-run marginal costs of operations. However, in most countries, tariff levels have generally remained well below long-run marginal costs, and even below average operating costs. Historically, developing countries have kept electricity prices low for two reasons: first, the traditional belief that low electricity prices spur industrial growth, and secondly, because low electricity prices are seen as one way to maintain political support from farmers and the urban middle class. In these cases, governments provide partial subsidies to cover the shortfall, and the operational performance of utilities suffers further from the insufficiency of funds.

2.3 Environmental degradation

Increasing concern over environmental degradation complicates the capacity expansion plans of utilities in many developing countries. Electricity production is among the most significant contributors to local air and water pollution, as well as to the emissions of greenhouse gases that may affect climate change. Fossil fuel combustion is thought to be responsible for 65-90% of total carbon dioxide emissions from human activities. Although developing countries account for only a quarter of the global CO_2 emissions from

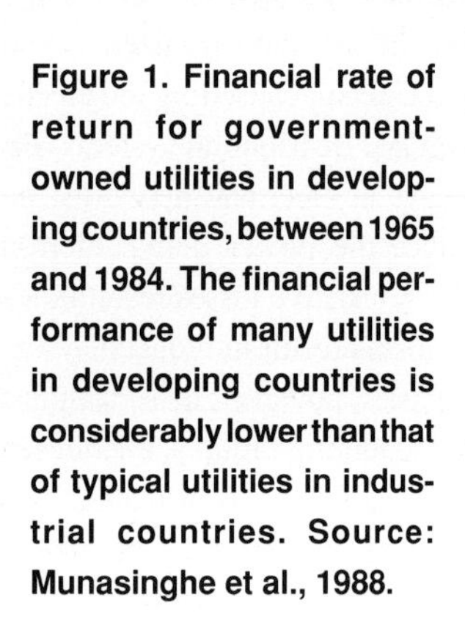

Figure 1. Financial rate of return for government-owned utilities in developing countries, between 1965 and 1984. The financial performance of many utilities in developing countries is considerably lower than that of typical utilities in industrial countries. Source: Munasinghe et al., 1988.

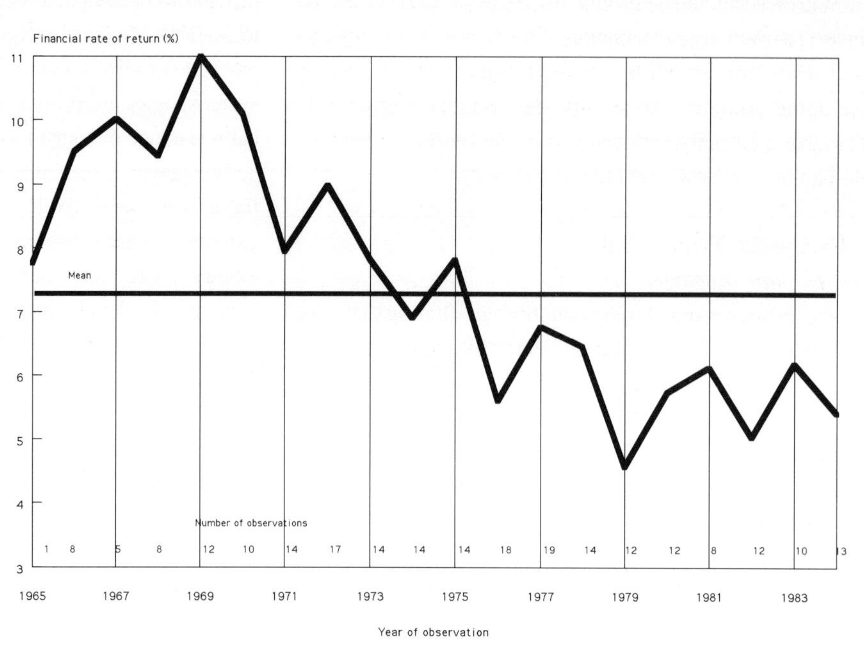

commercial energy use, increased emissions from these countries will accompany the rapid expansion of their electric power sectors. Emissions of SO_x and NO_x compounds and particulates will increase simultaneously, as a result of the same electricity-generating activities, thus contributing to acid precipitation and to a general decline in air quality, particularly in urban areas.

Requiring new environmental control equipment will increase the initial capital cost of new generating equipment, including conventional, pulverized coal-fired power plants. By contrast, when fully commercialized, the greater efficiency of integrated coal gasification/combined cycle technology is expected to compensate for much of their higher capital cost by reducing environmental stresses and damages. While the economic benefits of these environmental protection activities will accrue in other sectors, the power sector will have to bear the brunt of the capital cost for environmental improvement. Innovative and successful solutions to these linked problems will probably involve a combination of new technology (including renewables), a major expansion in natural gas use, and the widespread implementation of energy efficiency in supply and demand.

Figure 2. Energy intensity (primary energy consumption per dollar of Gross Domestic Product) for five industrial countries, and projected energy intensity for an aggregate of Less Developed Countries. (Data are given in metric tons of oil per thousand US dollars of Gross Domestic Product — MTO/$1000 GDP.) The steadily declining energy intensity in industrial countries has resulted in a decline of about 1% annually in the rate of growth in energy demand. Source: Adapted from "Energy for the Developing World", A.K.N. Reddy and J. Goldemberg, Copyright (c) September 1990 by Scientific American, Inc. All rights reserved.

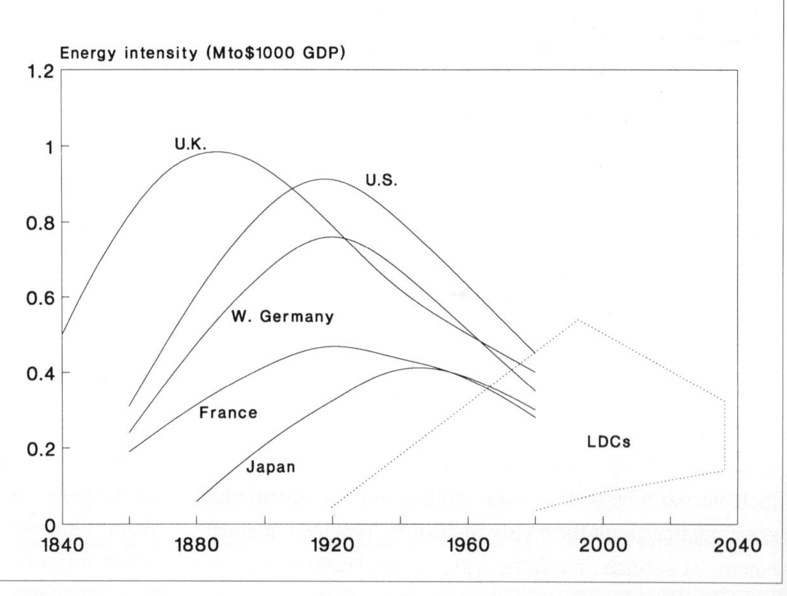

3 Potential Contribution of the Energy Sector to Greenhouse Gas Emissions

Developing countries are emerging as the major growth centres of electricity demand during the 21st Century. In 1970, total primary energy demand in developing countries was 16 million barrels per day of oil equivalent (mbdoe) or 15% of the world total of 104 mbdoe. By 1990 this had increased to 45 mbdoe — or to 27% of the world total (of 165 mbdoe). Assuming average annual growth rates in primary energy demand of 5 to 6% annually, the primary energy demand in developing countries would exceed 100 mbdoe in 20 years, and perhaps exceed 200 mbdoe by 2025. However, even with the highest growth scenario developed by the Inter-Governmental Panel on Climate Change (IPCC), the per capita contributions of developing countries to greenhouse gas emissions in the year 2025 will be at most 10-20% of those of the OECD countries.

Reductions in energy intensity arising from efficiency improvements in the production and uses of energy are an economically desirable way of reducing emissions. But the rate of implementation of such measures has historically been slow. In particular, over the past century, the steadily declining energy intensity in industrial countries has resulted in a decline of about 1% annually in the rate of growth in energy demand (Anderson, 1991). This has occurred in spite of enormous potential gains in efficiency of both energy production and consumption. Figure 2 shows the evolution of energy intensity in the industrialized and developing countries over the last century.

Future greenhouse gas emissions will depend on future rates of population and economic growth, the amount and types of energy consumed, and the rates of adoption of new policies and technologies. Even with gains in energy intensity energy growth rates in developing countries are likely to exceed 4 percent annually (see Figures 3a, 3b, 3c).Most reliable estimates of future emissions recognize that these could vary dramatically, based upon energy management policies. For example, five separate sets of assumptions about future energy use were used in a set of emissions scenarios prepared by the Energy and Industry Subgroup (EIS) and the Agriculture, Forestry, and Other Human Activities Subgroup (AFOS) of the Response Strategies Working Group (RSWG) of the IPCC (Houghton et al., 1990). A summary of these assumptions is provided in Annex 1.

The differences among these scenarios can be illustrated best by comparing the global primary energy mix reflected in each scenario. As shown in Figures 4a and 4b, the two "business-as-usual" scenarios incorporate very rapid increases in energy demand, met primarily by the use of fossil fuels, particularly coal. By contrast, in the scenarios in which policy actions are taken to respond to climate change, worldwide energy demand grows much more slowly, and is increasingly met by non-fossil fuels, resulting in a significantly smaller threat to global climate (see, for example Figures 5a, b, c). Therefore, it is significant to look at the trends involved in electric power use, and whether or not it is reasonable to expect the sorts of changes that could reduce electricity demand.

4 The Power Sector: Short-, Medium- and Long-term Trends

4.1 Electrification and supply trends

4.1.1 Growth of electric supply

One of the most striking trends in the last 40 years has been the rapid growth of electrification, particularly in the developing world. In India, for example, the total power supply capacity has increased 50-fold from about 1300 MWe in 1947 to over 65,000 MWe in 1990. The per capita consumption of electricity has increased about 15-fold over the same period. This trend is expected to continue in the next few decades, moderated only by shortages of capital for power system expansion.

In this context, we note that technological options exist that hold substantial promise for reducing greenhouse gas emissions while meeting capital constraints. Fossil fuels will continue to provide most of the growth in electric power generating capacity over the next 20-25 years, but advanced, gas-fired turbines and integrated coal gasification/combined cycle units will become the preferred fossil-fuel conversion options, because of their much higher conversion efficiency, modular character, and low emissions rates. Advanced gas turbines, molten carbonate fuels cells and electric power transmission, distribution and storage technologies can work well with a variety of primary fuels, including coal, gas, biomass and solar/hydrogen supply options. They may, therefore, represent important building blocks in a transitional strategy to a low carbon or carbon-free power sector in the long-term future.

Global fossil-fired power-generating capacity (i.e. fuelled by coal, oil and gas) produces about 1700 gigawatts (GW) and comprises two-thirds of the world's total generating capacity of 2600 GW (UN Energy Statistics Yearbook, 1987). The developing country share of generating capacity is about 500 GW or one-fifth of the world total. But, electricity use in the developing world is growing at 6-9% annually compared to only 2-3% annually in the developed world. If these growth rates continue, generating capacity in the developing countries will equal that in the industrial world by 2020.

Electricity generating capacity in developing countries is projected to increase in the 1990s, from 471 GW to 855 GW. Electricity demand is expected to grow during this period at an average annual rate of 6.6% in these countries. If demand continues to grow at the projected rates, this would be is substantially lower than the rates of the 1970s (10%) or 1980s (7%). The Asian countries are projected to show the fastest growth rate, averaging about 8% annually. The growth of electricity supply in developing countries continues to lag behind the demand for electricity.

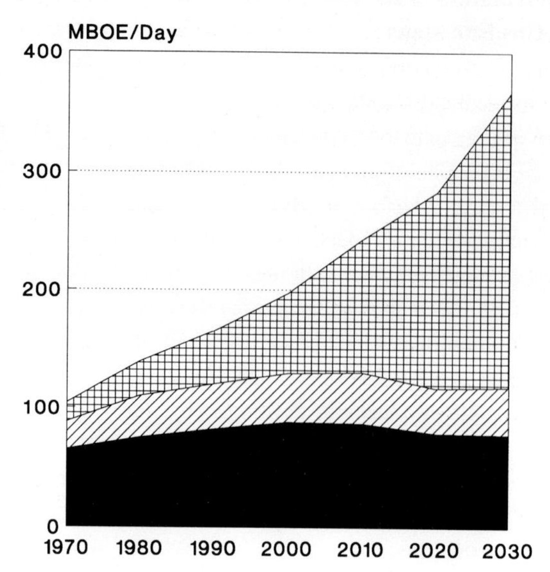

Figure 3a. Expected rates of growth in energy use (million barrels of oil-equivalent per day). In OECD and Eastern European nations, energy use may actually decline, while in developing countries it will grow rapidly.

LDCs
E. Europe/USSR
OECD

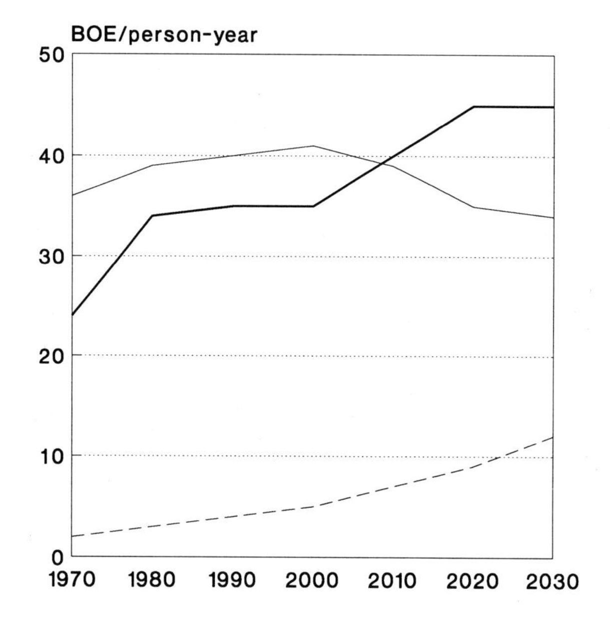

Figure 3b. Expected changes in per capita energy use (barrels of oil-equivalent per person per year).

LDCs
E. Europe/USSR
OECD

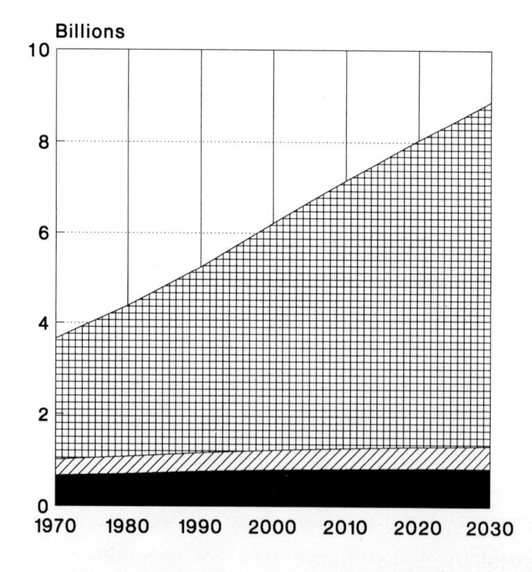

Figure 3c. Expected human population growth, one of the key components in the amount of energy used. Source: Houghton et al., 1990.

LDCs
E. Europe/USSR
OECD

Figure 4. Two "business-as-usual" scenarios for energy use through the year 2100, prepared by the Response Strategies Working Group of the IPCC. In both of these scenarios, government energy policies continue along more-or-less traditional lines, resulting in very rapid increases in energy demand. For the results of scenarios in which energy policy actions are taken to respond to climate change, see Figure 5. Source: Houghton et al., 1990.

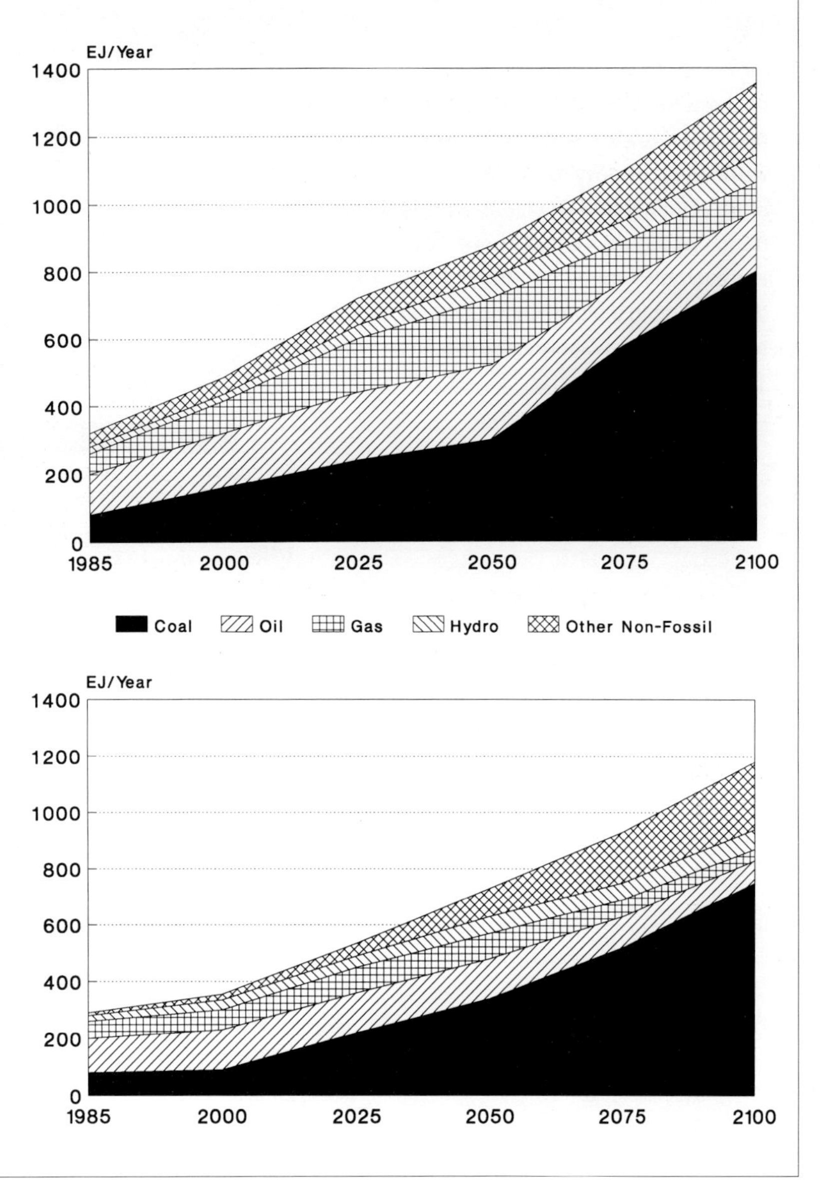

4.1.2 Continued use of coal

Coal burned to produce electricity in developing countries is expected to be an increasing contributor to atmospheric emissions. In a recent World Bank study (Moore and Smith, 1990), the actual 1989 and forecast 1999 breakdown of electricity supply in the developing countries was analysed. (See Table 1.) The data indicate that coal-based thermal supply already provides almost one-half of the electricity produced in developing countries, and the percentage is (according to the World Bank) likely to rise higher still.

Most developing country coal plants have only electrostatic precipitators or fabric filter bag collection systems to limit particulate releases from existing conventional steam plants. Most have no sulphur control equipment. By contrast, in Europe, Japan and the United States, sulphur emissions are being reduced by environmental standards that require desulphurization equipment (on any new steam plants which burn high-sulphur coals) and retrofits on existing plants.

Table 1. 1989 and 1999 electricity supply in the developing countries

	1989		1999	
	TWh	%	TWh	%
Hydro	674	33.2	1207	31.5
Geothermal	11	0.6	29	0.8
Nuclear	80	3.6	212	5.5
Oil thermal	224	11.0	255	6.6
Gas thermal	120	5.9	332	8.6
Coal thermal	907	44.7	1793	46.6
Net imports	14	0.7	16	0.4
Total	2,030	100.0	3,844	100.0

Figure 5. Three scenarios for energy use through the year 2100, in which policy actions are taken to respond to climate change, worldwide energy demand grows more slowly (than in the Business-as-Usual scenarios of Figure 4), and energy demand is increasingly met by non-fossil fuels. All of this results in a significantly smaller threat to global climate. Source: Dennis Anderson, 1991.

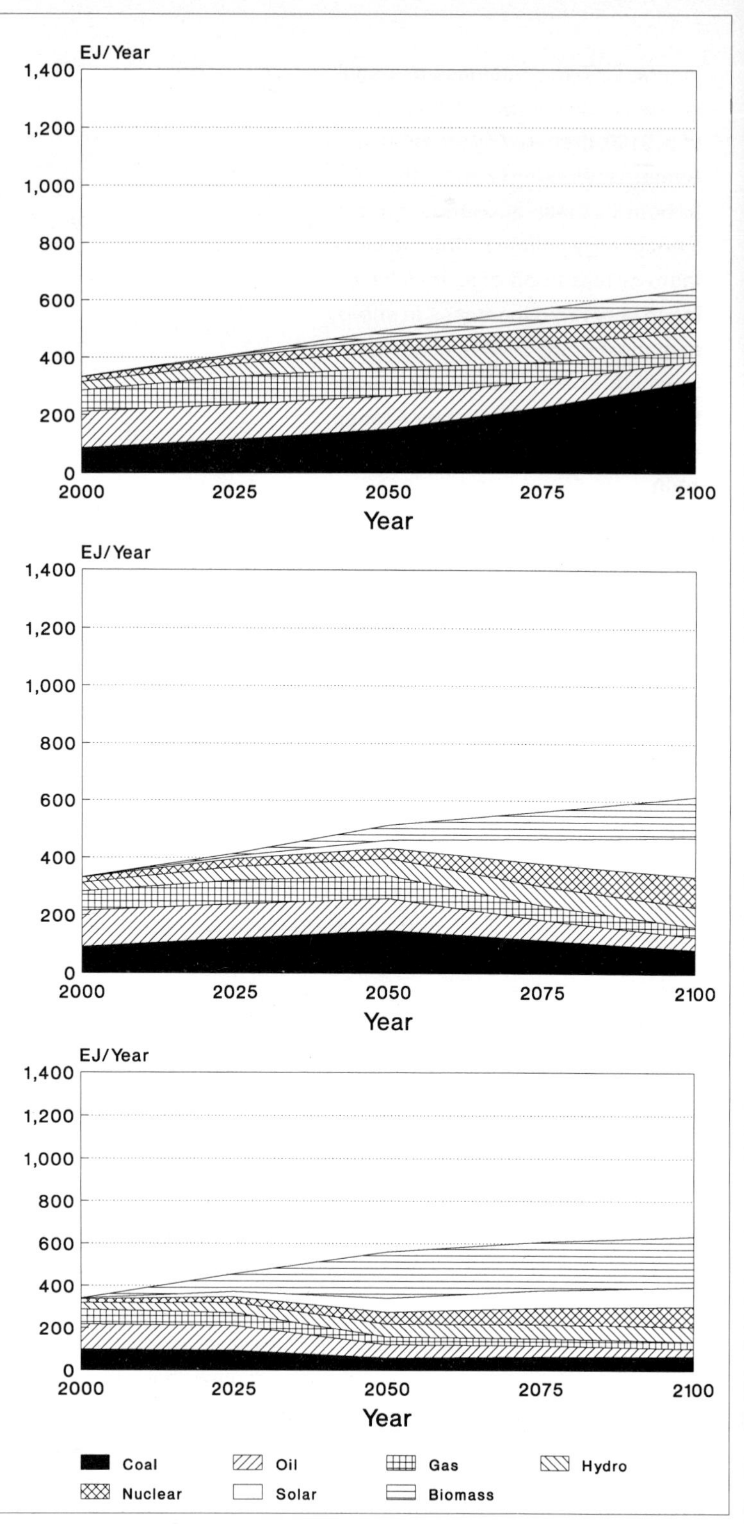

4.1.3 Increased pressure on capital

Over the 1990s, the capital requirements to sustain the projected growth in electric generating capacity total about 1989 US $760 billion. (See Table 2.) About 60% of this amount is for generation, and the rest for transmission, distribution, and other items. On average, approximately 60% of the total is in local currency, principally in India and China. The proportion of foreign currency in total power sector investment is highest for the countries of Sub-Saharan Africa, at about 75%.

Loans for power sector development accounted for about 25% of the total public sector foreign debt among developing countries during the 1980s, and as much as 50% in some countries. Meanwhile, the relative shares of foreign exchange borrowing from official and private sources has changed dramatically over the 1980s. Public and government-guaranteed debt, the traditional sources of foreign exchange for power investment in developing countries, have provided about $7 billion annually since 1983, down from $12-14 billion in 1980. This decline has been greatest

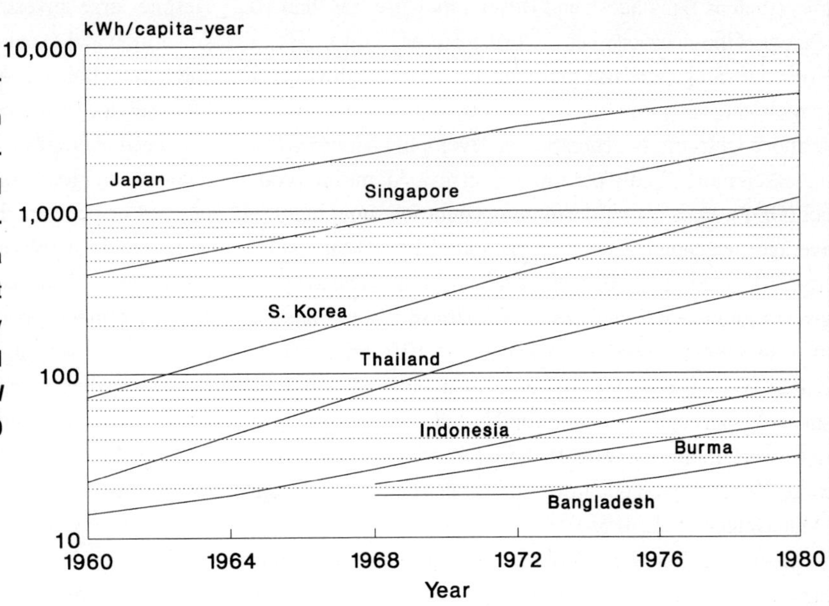

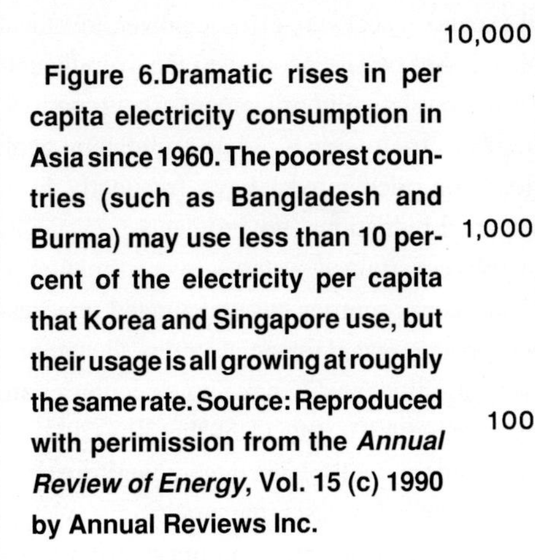
Figure 6.Dramatic rises in per capita electricity consumption in Asia since 1960. The poorest countries (such as Bangladesh and Burma) may use less than 10 percent of the electricity per capita that Korea and Singapore use, but their usage is all growing at roughly the same rate. Source: Reproduced with perimission from the *Annual Review of Energy*, Vol. 15 (c) 1990 by Annual Reviews Inc.

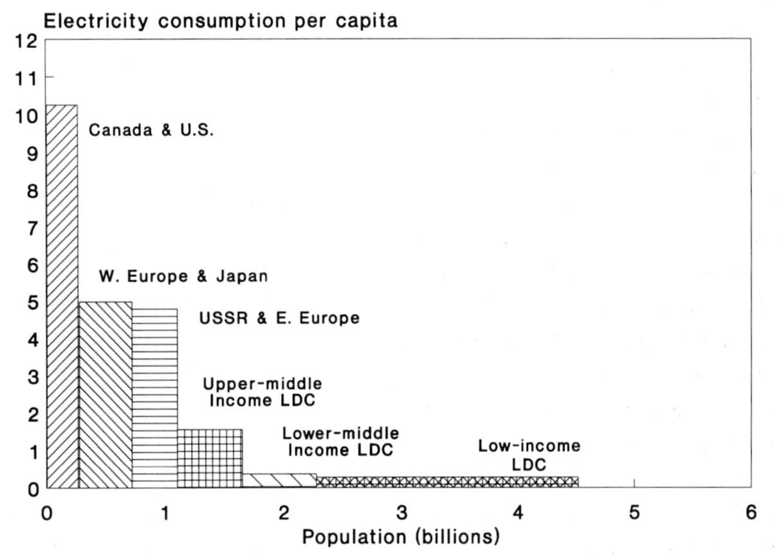

Figure 7. Electricity consumption per capita. This figure shows that, despite the rapid growth rates of electricity consumption in developing nations, the industrial nations still consume far greater amounts of electricity per capita. The average Indian consumes one-fortieth of the electricity used by the average U.S. resident. Source: Jhirad, 1990.

Table 2. Estimated power investment requirements for developing countries during the 1990s (US$ billion in 1989 terms)

Period	1990-1994	1995-1999	Total 1990s	Percent
Power system component				
Generation	193.6	270.1	463.7	61%
Transmission	33.1	46.2	79.3	10%
Distribution	60.9	89.0	149.9	20%
General	26.7	37.9	64.6	9%
TOTAL	**314.3**	**443.2**	**757.5**	**100%**
Of which in:				
Local currency	190.0	272.4	462.4	61%
Foreign exchange	124.3	170.8	295.1	39%

for private creditors, whose lending in 1987 was only one-eighth of the amount in 1981. Official creditors, such as multilateral and bilateral agencies, have maintained a fairly constant level of support for the power sector over the last several decades. If current trends in demand growth continue, as long as financing is available, most experts expect this level of support to be maintained.

4.2 Growth in per capita consumption

Per capita consumption of commercial fuels has grown rapidly over the last 25 years in the developing countries, and this trend is likely to continue. Figure 6, compiled by the Asian Development Bank, illustrates the steady rise in per capita electricity consumption in Asia. The overall trends are similar for all Asian developing countries; the poorest countries (such as Bangladesh and Burma) may use less than 10 percent of the electricity per capita that Korea, Taiwan, Hong Kong and Singapore use, but their usage is all growing at roughly the same rate.

Indeed, electricity generation in developing countries has increased more rapidly per capita than gross domestic product (GDP). Data for 51 developing countries show that the average per capita generation increased by 7% annually, from 196 to 529 kWh between 1968 and 1982. The average growth rate of GDP per capita was almost 2% per annum over the same time period, increasing (in 1980 prices) from US $837 to US $1093. In more than 90% of countries surveyed, per capita generation and installed capacity growth rates were more than double the GDP growth rate, and in more than half of all countries more than three times as high (Munasinghe et al., 1988).

In spite of these relatively rapid growth rates, per capita electricity consumption in developing countries is still only a small fraction of that in the countries of the Organization for Economic Cooperation and Development (OECD). Figure 7 shows the present distribution of per capita electricity consumption worldwide. At 250 kWh per capita, for example, the average Indian consumes only one-fortieth of the electricity consumed by the average US resident.

The authors of an end-use oriented global energy strategy have maintained that a living standard in developing countries comparable to that in Western Europe could be achieved with a per capita electricity consumption of about 2100 kWh — about half of the actual average for Western Europe in 1975 (Goldemberg et al., 1988). This goal, a desirable one, would still require major increases in new generating capacity in countries such as India and China, as well as major improvements in the efficiency of generation, transmission, distribution and end use.

Without a focus on increasing the quality, efficiency, and reliability of electricity services, however, even a tenfold expansion in generating capacity in a developing country would fall far short of providing the standard of living of a modern European nation. Low capacity factors, high transmission and distribution losses, and inefficient end-use equipment will undermine any level of investment in new service; too much of the new electricity would simply not reach the consumers. Thus, simply extrapolating present inefficiencies into the future would be an unaffordable choice, from both an environmental and economic standpoint (Jhirad, 1990).

4.3 Deteriorating utility performance

The financial history of the power sector in developing countries offers no solace to national and international decision makers. At a time when greatly increased capital mobilization is required, the financial performance of developing country utilities has worsened.

A review of about 300 power projects financed by the World Bank between 1965 and 1983 reveals a steady decline in electric sector performance (Munasinghe et al., 1988). Despite large investments providing increased per capita generation and expanded access to electric power, technical and non-technical losses persist. Moreover, the overall quality of service shows no sign of improvement. Costs continue to exceed average electricity prices, while multilateral bank recommendations for price reform have frequently been ignored or postponed by national governments.

Except in supply-constrained systems, overestimation of power demand in many countries has encouraged unnecessary investment and aggravated financial strains. Generating capacity reserve margins in over 70 developing countries are excessive; they average 43% instead of the 20%-30% reserve margins found in well-planned and well-operated systems. Meanwhile, in many cases, the quality of service to existing customers is deteriorating as utilities supply electricity to new areas. There is also a worsening level of maintenance in most developing country utilities.

These problems are not just damaging to service, but financially crippling. There was an excess capacity of 43,000 MW in the developing world in 1989 on a total system load of 331,000 MW, assuming a 30% reserve margin. This excess capacity represented a capital investment of $50 billion. Improved maintenance to increase unit availability and reduce capacity reserve could have saved an equal amount — another $50 billion — initially, and approximately $25 billion annually in future generation investments (World Bank, 1992).

5 Power Technology Options

The costs and performance of supply and demand-side technology options to meet power needs in developed and developing countries have been extensively investigated (Johansson et al., 1989; OTA, 1985; EPRI, 1986; Häfele, 1989; Bemis et al., 1990). No effort is made here to suggest any particular portfolio of options over the next decade. Such an exercise would be highly country-specific, and subject to rapid obsolescence.

But in general, reductions in greenhouse gas emissions per kilowatt hour of electricity produced are possible through:

- switching the mix of fossil fuels toward natural gas and increasing the hydrogen content;
- recycling carbon through biomass utilization;
- using carbon-free sources such as solar, wind or nuclear;
- implementing efficiency in generation, transmission and end use;
- and resorting to carbon dioxide recovery and disposal techniques.

Fossil fuels — and in some countries, conventional hydroelectric power — are projected to provide most of the growth in electric power generating capacity over the next 25 years.

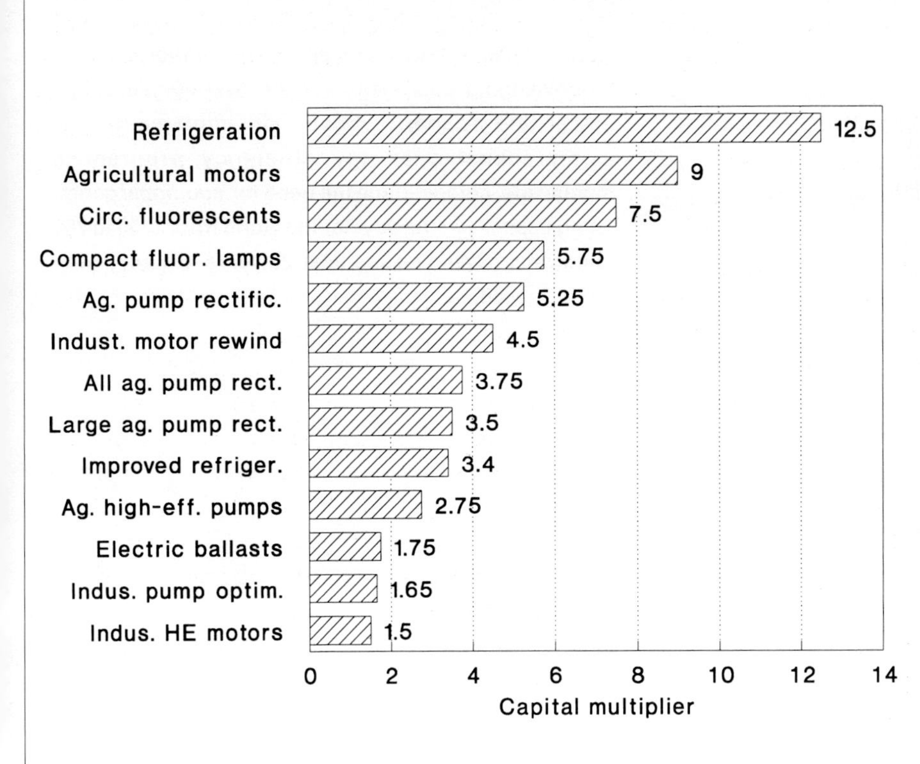

Figure 8. Capital multipliers for 13 key energy conservation measures. For every dollar invested in one of these measures, this figure shows how many dollars would be saved in reduced capital expenditures for construction and operation of new energy supply facilities. For example, a dollar spent subsidizing efficient refrigeration systems would save 12 dollars by reducing the need for new power supplies. Source: After USAID 1991.

Among these, the gas turbine combined cycle technology is emerging as the preferred option. Advanced gas turbines and molten carbonate fuel cells (when fully commercialized) represent vital transitional building blocks to a low-carbon or carbon-free power supply. They are well matched to a variety of fuels: natural gas, coal, distillate fuels, biomass and non-fossil/hydrogen options.

Major technological opportunities exist for efficiency, natural gas development, advanced clean coal technology and renewable power systems. All of these options merit serious consideration by developing country decision makers, especially since they have been seriously underestimated in most investment plans to date.

5.1 End-use efficiency

There are several proven technologies (Johansson et al., 1989) for improving the efficiency of electricity-using devices. These include power factor control systems, variable speed motors, high efficiency lighting, and new industrial processes. However, more important than the obvious savings of electricity they engender, is the role these devices play in reducing the capital investment required to deliver electric power to the end user.

The capital mobilization problem in developing countries demands that capital investments in end-use efficiency be compared with investments in new supply. Of major interest is the capital multiplier: a measure of every dollar saved supply-side per dollar invested in demand-side energy efficiency. A pioneering assessment completed recently by the Government of India, USAID, ODA of Britain, and the World Bank (USAID, 1991a), concluded that implementation of eleven key end-use efficiency measures can dramatically reduce capital by an average multiplier of 5 dollars for every dollar invested. Figure 8 shows the capital multiplier associated with each of thirteen key end-use devices. The best measure, high-efficiency refrigerators, had a capital multiplier of 12.5. Investing in the key measures through a national end-use efficiency programme could deliver a given level of electricity service at a cost of about 20% of the cost of new supply. The assessment, conducted collaboratively by developed and developing country experts, demonstrated the vital importance of end-use energy efficiency to any country seeking to improve its electricity system.

Assessing the potential for more efficient lighting systems, motors, refrigerators, air conditioning systems and industrial processes demands an investigation of how electricity is used in industry and agriculture, as well as in the residential and commercial sector. In India, for example, electric motors are the dominant end-use device in the industrial sector, accounting for nearly 75% of consumption. Other significant end uses are lighting (9%) and aluminum smelting (8%). In the residential sector, the largest end use is fans (33%) folllowed by lighting (27%), refrigerators (16%) and air conditioners and evaporative coolers (11%). In the commercial sector the two major end uses are lighting and the "HVAC" component (heating, ventilation and air conditioning) at 45% each.

Despite the impressive growth of capacity in India, from 31,000 MW to 65,000 MW over the last decade, shortages amount to about 20% of peak power and 10% of energy demands. Power shortages are exacerbated by inefficient end use of electricity, which in turn result from low subsi-

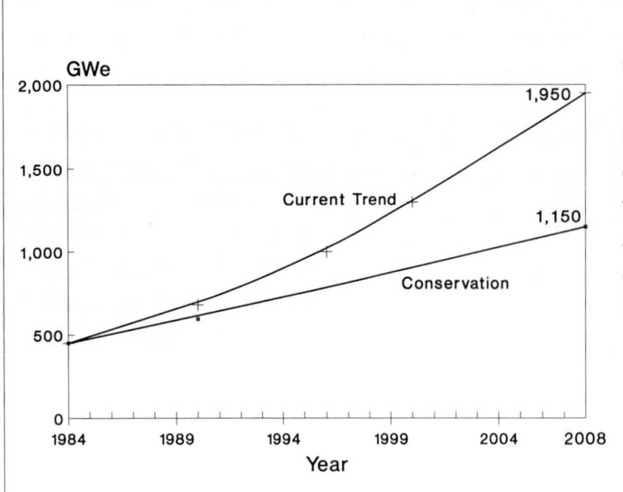

Figure 9. Two scenarios for energy generation capacity in developing countries over the next 20 years. The lower-line scenario, based on implementing a comprehensive energy efficiency programme worldwide, could reduce the need for *additional* generating capacity in the developing world from 1500 to 700 gigawatts (GW) over the next 20 years. Source: After USAID, 1988.

dized tariffs, technological obsolescence of end-use equipment, processes and systems, and inadequate commercial incentives for efficiency.

Despite all of these arguments for efficiency as a substitute for capacity expansion, the Government of India plans to install 110,000 MW of additional capacity by the year 2000 - the largest power expansion undertaken anywhere outside of the United States. This would reportedly cost about $150 to $200 billion (1989), or about 30 to 40% of total public investment over that 8-year period. Implementing such an ambitious programme will strain India's public finances and pose unprecedented managerial and technical challenges.

Instead, power end-use efficiency could potentially save 38,000-61,000 MW in peak generation capacity, and 103-170 billion kWh in energy savings by 2004-05. According to the study, the capital requirements would then be cut by as much as one-fifth, or by $30 billion (USAID, 1991a).

An earlier, generic investigation (USAID, 1988) showed that developing countries can significantly lower the growth rate of generation capacity, without affecting the growth rate of electricity services, if they adopt a variety of efficiency measures in generation, transmission, distribution and end use. These measures involve improving baseload plant capacity factor to over 65%, reducing transmission and distribution losses from the current 25% to 15%, and achieving a 20 to 30% improvement in the end-use efficiency of motors, lighting, and refrigeration systems.

USAID estimated that developing countries can reduce the need for additional generation capacity from 1500 to 700 GW over the next 20 years (Figure 9). This would reduce the need for additional investments in power supply expansion by US $1400 billion over 20 years. However, in spite of this strategically significant potential, the real issue is implementation - creating the right policy and institutional environment to accelerate adoption of supply and demand-side efficiency measures. Realistic implementation rates for energy efficiency measures in developing countries have yet to be established.

5.2 *Carbon-based power generation, transmission, and distribution*

5.2.1 *Increased use of natural gas*

Several reviews of the global natural gas situation (Hansen and Schramm, 1989; World Bank, 1989a) reveal a major opportunity for the greatly expanded use of natural gas in the power sector, far more than the seven percent contribution anticipated in earlier investment plans.

The marginal costs of natural gas delivered to the main markets in developing countries - including exploration, production and transport to the market — is between $3 and $11 per barrel of oil equivalent (Schramm, 1984). Thus, even an oil price which fell to $15 per barrel equivalent in 1989 dollars would justify the development of most onshore and some offshore reserves natural gas in developing countries. The major low-cost gas reserves of the former USSR and Iran could also be developed, via an international pipeline network, to provide natural gas for markets from Western Europe to South Asia.

Table 3. Comparative emissions for gas combined-cycle and coal steam plants

Emissions and waste	Coal steam with scrubber	Gas-fuelled combined-cycle
CO_2, g/KWh	830	380
CO, mg/KWh	75	34
SO_2, mg/KWh	600	0
NO_2, mg/KWh	600	350
UHC, mg/KWh	0	18
Waste water, g/KWh	15	0
Ash, g/KWh	34	0
Rejected heat, MJ/KWh	4.3	2.6

Source: Haupt G. et al. 1990

5.2.2 *Gas turbine combined-cycle technology*

In 1989, developing countries generated about 900 terawatt-hours (TWh) of electricity from coal-based thermal supply; This is expected to increase to about 1800 TWh by 1999. A shift away from coal use in conventional steam plants to gas-fuelled combined-cycle technology, or to integrated coal gasification combined-cycle technology, would be a significant environmental improvement. In addition to sulphur, NO_x and CO_2 emissions can be reduced substantially (by as much as one-half) using gas-fueled combined-cycle plants instead of coal thermal plants with scrubbers (Table 3).

The gas turbine - in scales from 1MW to over 200 MW - will assume a leading role in developing country power generation over the next decade, both in central station and cogeneration applications. Depending on the price of natural gas, gas-turbine combined cycle power plants will be able to provide electricity at lower costs and with less environmental burden than coal or large-scale hydro installations.

In particular, there are multiple economic, environmental, and planning benefits from deploying aeroderivative gas turbines (Williams and Larson, 1989). The modularity of aeroderivative turbines makes it possible to avoid expensive field construction, and rely instead on factory fabricated units, where modern mass production allows quality control and major cost reductions.

Gas turbine combined cycle plants fired with natural gas or distillate fuels can be quickly deployed in field applications, since the lead times for plant siting and construction average about 2-4 years versus 6-10 years for pulverized coal fired thermal plants. Current efficiencies of combined cycles are 45%, with the prospect of reaching 50% or more with steam-injected gas turbines (Johansson et al., 1989). (Steam injected gas turbines take steam not needed for process heat requirements, and inject it back into the combustor for added power and efficiency). These efficiencies greatly exceed the approximately 37% attainable with conventional coal-fired plants. The estimated capital costs of gas turbine combined cycle plants are between $450 and $500 per kilowatt installed (Kamali and Tawney, 1990).

5.2.3 *Clean coal technologies*

The coal-fired steam plant has been the workhorse of the electric utility industry for the last few decades. The efficiency of steam-based (Rankine cycle) power plants has increased steadily for the last 80 years, from about 5% in the 1880s to just under 40% today (Figure 8). However, in the early 1970s, the efficiency of coal fired plants reached a plateau — partly for fundamental thermodynamic reasons, and partly because scrubbers and other pollution control equipment consume 2%-4% of the electricity produced. Using current technology, a pulverized-coal plant with pollution controls to limit both nitrogen and sulphur oxide emissions has a thermal efficiency of about 37%. As has been pointed out (Balzheiser and Yeager, 1987), incorporating flue-gas desulphurization scrubbers into power systems in the US has not been easy or cheap, and the process has taken about 25 years.

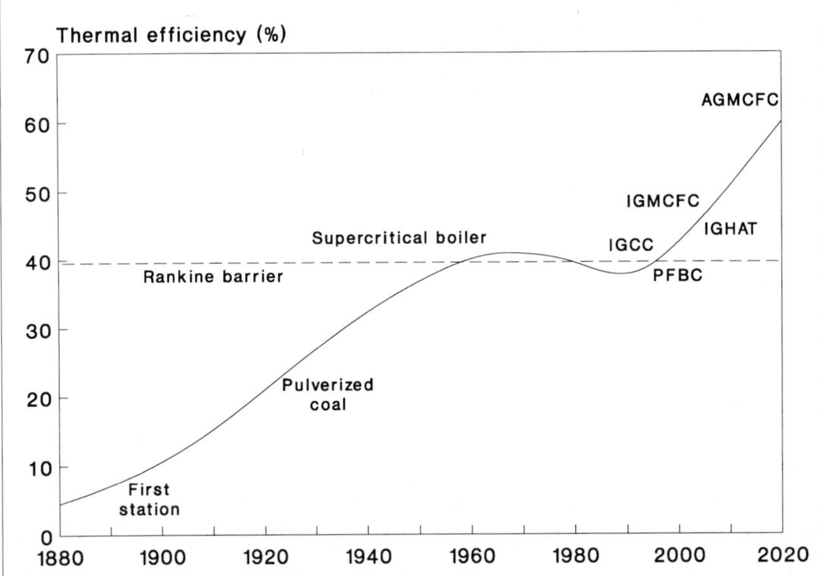

PFBC: Pressurized fluidized-bed combustion
IGCC: Integrated gasification-combined cycle
IGHAT: Integrated gasification-humid air turbine
IGMCFC: Integrated gasification-molten carbonate fuel cell
AGMCFC: Advanced gasification-molten carbonate fuel cell

Figure 10. Evolution of coal-fired power plants. The efficiency of steam-based (Rankine cycle) power plants increased steadily for 80 years, nearing theoretical limits in the 1960s. Since then, efficiency has decreased somewhat, because of the need to use energy to remove pollutants formed during combustion. Breaking the Rankine barrier of practical limitations on efficiency will require innovative approaches based on chemical energy conversion — such as coal gasification and electricity generation by means of advanced combustion turbines and fuel cell technologies. Source: Copyright (c) 1990, Electric Power Research Institute, "Beyond Steam: Breaking Through Performance Limits", *EPRI Journal* , December 1990. Reprinted with permission.

In recent years, power plant designers have preferred to integrate pollution control into the generation process, rather than having to add scrubbers. Atmospheric fluidized bed combustion technology (AFBC) addresses this problem by suspending a mixture of coal and limestone particles in a stream of air that comes from below. Even though efficiency is not improved, the ability to burn lower quality fuels with less environmental impact makes this technology an attractive candidate for use in developing countries and Eastern Europe. AFBC power plants are, however, not effective for reducing carbon dioxide emissions from coal combustion.

Efficiency improvements can be realized in pressurized fluidized-bed combustion (PFBC), in which the steam cycle is supplemented with a gas turbine that runs off the pressurized gas leaving the boiler. PFBC plants are able to achieve thermal efficiencies of 40% or better. The technology is being demonstrated by American Electric Power (AEP) at a 70 MW scale, and there are plans to construct a 330 MW plant by 1996. Two other PFBC units in the 70-80 MW range began commercial operation last year in Europe.

Integrated coal gasification combined cycle plants (IGCC) have major advantages over the current generation of pulverized-coal power plants. At 42 per cent or more efficiency, they can break the Rankine cycle barrier. Their emissions are less than either conventional coal plants or AFBC and PFBC plants, and they operate with either natural gas or coal. In addition, the coal gas which they produce can be used in a variety of products, such as Hydrogen, Methanol, Ammonia, synthetic natural gas, industrial chemicals and gasoline. The Netherlands state power utility is constructing a 250 MW IGCC plant based on Shell gasification technology, for operation in 1993. IGCC technology is also a very attractive candidate to burn high ash coals, such as those found in India (USAID, 1991b).

Further gains in efficiency can be obtained by replacing the combustion turbine with a fuel cell. In a fuel cell, electrodes act as catalysts for the oxidation of a continuous stream of hydrogen-rich gas, releasing and absorbing electrons in the process. The molten carbonate fuel cell (MCFC) is now approaching the utility demonstration stage (EPRI, 1990 and 1991b). Although the first demonstration unit will be fuelled with natural gas, there are plans to test the fuel cell with coal gas as well. Future improvements could lead to a coal-to-electricity efficiency of about 60%, compared with 37% for pulverized coal technology.

The Electric Power Research Institute (EPRI, 1991b) estimates that the US market for two-megawatt MCFC power plants adds up to about 12,000 to 14,000 MW, once commercialization is achieved in the 1990s. Early applications will include urban areas under severe environmental stress, substation upgrades and industrial cogeneration opportunities. The projected price of early production units is about $1500/kW (1989 dollars). Prices of subsequent commercial units are expected to be close to $1000/kW. To spur commercialization, purchasers of early power plants would receive royalties from the sale of future commercial units. In addition, financial commitments by the utility do not become firm unless the prototype demonstration plant performs successfully.

Taken as a whole, the molten carbonate fuel cell combined with coal gasification technology is the cleanest, most efficient, coal plant system now conceivable. Clean coal technologies can be used for both newly constructed power plants and to "repower" existing power plants. In repowering or rehabilitation, the combustion system is replaced with one of the new technologies. Other new components, such as a combined cycle unit, may also be installed. The result is a renovated plant with efficient performance.

5.2.4 *Cogeneration*

Cogeneration is a process which makes use of the fact that steam and electricity are byproducts of many industrial processes. In cogeneration plants, the steam may be used to meet on-site heating needs, while the electricity powers on-site generators, or is sold back through the power grid to other electricity users. Cogeneration has been very popular with large industrial energy users, as one approach for reducing their overall energy costs. Most of the approximately 18 GW of currently operating cogeneration projects in the US, fall into this category.

5.3 *Carbon-free power generation, transmission, and distribution*

In the long term, large-scale displacement of fossil fuels by non-fossil alternatives can stabilize greenhouse gas concentration levels. The principal non-fossil fuel energy options are discussed very briefly here; extensive information on these options is widely available elsewhere (International Energy Agency 1989; Häfele 1989).

5.3.1 *Hydroelectric power*

Hydroelectricity production in 1986 was 661 TWh in the developing world and 1366 TWh in the industrialized world, corresponding to capacities of 154,000 and 413,000 MW respectively. There is a major technical potential to expand the contribution of hydropower to future primary global energy use by fourfold in the developing world. However, as the World Bank (1989a; 1990) has pointed out, environmental and social impacts of large-scale projects must be considered carefully, and these potential impacts will limit hydropower development in many developing countries.

5.3.2 *Biomass energy*

Advanced technologies for conversion of biomass to gaseous and liquid fuels and electricity (especially the use of gas turbines for power generation), combined with measures to increase biological productivity, could allow biomass to provide a much larger share of global power demand, particularly in developing countries.

US and Brazilian experiences suggest that the costs of producing ethanol from biomass is between $75 and $100 per barrel of oil equivalent energy, although advanced processes and higher yield plantations may bring down the costs to $60 per barrel in the long term (Anderson, 1991). For power generation, costs relative to fossil fuels are much more favourable. The US has over 9000 MW of biomass-powered power stations, although the cost is somewhat higher than for fossil fuels.

Recent work (Booth and Elliot, 1990) has shown that gasified biomass used in gas-turbine combined cycle plants has the potential to be competitive with fossil fuels in many countries.

The major disadvantage of biomass utilization, on a significant scale for power generation, is that it requires large land areas, compared to direct solar or fossil options. Efficiencies of conversion of sunlight to electricity via photosynthesis are typically 0.3% or less — about one-fortieth the efficiency of solar photovoltaic cells. Although efficiencies could be improved through higher yield species, competition with land-use for agricultural production will limit the contribution of biomass.

5.3.3 Geothermal energy resources

Such resources are widely distributed, and over 7000 MW of installed generating capacity are in operation world-wide. With improved technology, such as deep drilling capability, further addition of geothermal technology could be significant in specific areas.

5.3.4 Wind energy systems

Wind systems for grid-connected power generation are commercial and competitive with fossil fuels — especially in good wind regimes of the kind that are encountered at Altamont Pass and other locations in California. There are 1500 MW of wind plants installed and operating in that state, with costs for some turbines ranging from 5 to 9 cents per kilowatt hour. Recent technological advances and expanding markets have resulted in cost reductions and performance improvements. In the developing world, countries such as Egypt and India have the potential for several thousand megawatts of grid-connected wind electricity at costs competitive with fossil fuels.

5.3.5 Photovoltaic power systems

Although a diffuse source of energy, the land areas necessary to support commercial energy utilization on a large scale are small in relation to the land available. All of the energy requirements of the United States could be supplied with 1% of the land area, or the amount currently devoted to roads and highways (Jhirad and Hocevar, 1977). In developing countries, assuming 15% conversion efficiency to electricity, less than 0.1% of their land area would be required to meet their entire energy demand from solar.

In 1989, approximately 42 MW of photovoltaic (PV) power systems were produced and shipped as commercial products. Significant technological advances and rapidly increasing production have lowered the costs of electricity production from PV systems from several dollars per kWh to roughly $0.25-0.40 per kWh. In many parts of the developing world, this is competitive with the actual costs of electricity from grid extension or diesel units.

And there is evidence that the costs could drop much further. Technological progress is occurring in production methods for thin-film and multijunction devices, as well as in materials development. Between 1970 and 1990, the costs of photovoltaic units fell from about $200,000 per peak kW of capacity to $6000 per kW in 1989 prices. With the development of large-volume markets and production scale-up, costs could continue dropping at that pace, perhaps reaching $2000 per kW in the next decade and $1000-$1500 in the longer term. This would make solar competitive with fossil fuels for a large fraction of the rural population in devloping countries. In the meantime, there are significant opportunities for governments and international agencies to develop large volume markets for decentralized power systems based on photovoltaics. Bold procurement initiatives that would enable manufacturers to introduce high volume manufacturing techniques would bring down costs substantially, and would make rural power a reality for a majority of the world's population.

While the present applications of PV systems in the developing world focus on small-scale applications of electricity for water pumping, water purification, rural lighting, health clinics, and communications, significant markets for irrigation and rural power services would result from costs in the range of $2000 per kW. Even with existing prices, India has equipped thousands of villages with PV lighting systems, and the goal of the Indian Government is to provide PV-based rural electrification in as many as 100,000 villages over the coming decade.

5.3.6 Solar thermal electric power generation

Conversion of solar heat to electricity is a commercial reality in California, with several hundred MW of grid-connected solar power plants on line. The commercial development of this option, as with wind technology, was accelerated by Federal and California state tax credits, as well as by attractive power purchase contracts implemented under PURPA regulations in California. Unfortunately, for a variety of policy and regulatory reasons, the only US solar thermal technology vendour and project developer has filed for bankruptcy.

While the costs of electricity production from solar thermal electric power systems are still higher (US cents 0.10 / kWh) than for fossil-fuel power plants, this technology offers the potential for relatively near-competitive power generation within a decade or so, provided production scales can reach several hundred MW or more per year.

5.3.7 Nuclear fission

Nuclear power from fission is growing in its contribution to global energy supply, owing to the completion of power plants ordered during the 1970s. While nuclear power provides a large share of the power supply in countries such as

France and Japan, it is also extensively used in a number of developing countries — and wider use is expected over the coming several decades. In 1986 there were 2700 MW of nuclear power generation in China, 1200 MW in India, 1300 MW in Mexico, 626 MW in Brazil, and 125 MW in Pakistan (USAID, 1990).

High capital costs, and concerns about safety, nuclear weapons proliferation, and radioactive waste management have halted orders for new nuclear plants in many countries. Nuclear power could make a contribution to global power supply if these problems are resolved. That would require commercializing a new generation of modular, inherently safer reactors, and implementing socially acceptable methods for radioactive waste management and control of proliferation. Several countries are actively working to develop a new generation of inherently safer reactors, but broad-based societal acceptance of these technical options is uncertain; the results will be decisive in securing a role for nuclear fission in the future.

5.3.8 Fusion power

Nuclear fusion technology is not likely to have an impact on the global energy situation for many decades to come. Once scientific feasibility of controlled fusion for power generation is established, it will have to be followed by technological and commercial feasibility. In addition, first generation fusion reactors are likely to be based on the deuterium-tritium reaction, which releases high neutron fluxes. Such fluxes generate radioactive isotopes in containment vessels, which therefore require replacement and disposal. "Dream fusion" reactions, which release only charged particles and no neutrons, are far more difficult to achieve, and are not the focus of present research. Fusion reactors will also have to be socially acceptable and economically viable.

5.3.9 Transmission and distribution

Loss reduction and improved management of transmission and distribution networks using existing technology are immediate and major priorities for all developing countries (World Bank, 1982; Abraham, 1990). Improvements in transmission and distribution technologies will help developing countriees to address the problem of capital shortages. By reducing energy losses during routine operations, these technical improvements will help to squeeze more useful electricity out of installed generating plants.

Advances in power electronics, automation and control systems are fundamentally altering transmission and distribution technology. Power electronics are part of the "macro-electronics" revolution, now increasingly being teamed with the micro-electronics revolution that has fundamentally altered the computer industry. Innovative transmission, distribution, and storage technology, as well as control and management of national and international power grids, will be central to environmentally sound power development.

Advances in power electronics using metal oxide semiconductors or thyristors (controlled rectifiers), have provided the ability to control blocks of power at multi-megawatt scale. Power electronic devices are also being applied in AC/DC power conversion, in reactive power compensation, and as controllers for flexible high-voltage transmission systems. High voltage DC transmission (HVDC) is the most significant new technology for transferring bulk power over long distances, and power electronic devices will be vital in assuring integration of DC with AC transmission networks. In addition, flexible AC transmission networks (FACS) are being developed that will allow transmission systems to be switched and controlled like telecommunications networks.

5.3.10 Electricity storage

The Electric Power Research Institute (EPRI, 1991c) considers new storage technologies more flexible, and often more cost-effective, than new oil- or gas-fired peaking plants as a means of making use of cheap baseload generating capacity more fully, while meeting a growing demand for peak power in US utilities. The availability of storage can minimize the effect of unexpected changes in fuel costs or load shape, and can increase the reliability of power to consumers. In the US, storage is also viewed by utilities as a means to comply with emissions limits in a timely and cost-effective fashion.

Alabama Electric Cooperative dedicated the first compressed-air storage plant in the US in September 1991. The 110 MW facility compresses air into an underground cavern during periods of low electricity demand. During periods of peak demand, the compressed air can be released and expanded through a turbine to generate electricity. During the generation period, the plant uses one-third of the fuel required by a conventional combusion turbine, thereby generating one-third of the emissions. Several other utilities are seriously considering implementing this technology. The storage unit also allows far better use of plant capacity and capital investment, and therefore addresses the dual need to both reduce emissions and use capital more effectively.

Storage technologies represent only about 3% of US generating capacity. With the exception of the new plant described above, all this capacity resides in pumped-hydro plants, which pump water uphill during off-peak hours and release it downhill at peak times to generate electric power. Large-scale batteries and superconducting magnetic energy storage technologies, currently under development, ultimately promise a holy grail of storage, but such prospects should not obscure the major innovations that are currently occurring in the electric power industry.

Because storage technology can level out the fluctuations in power availability over time, cost-effective electricity storage systems can dramatically increase the value of intermittent renewable technologies to electric utilities. Storage systems can complement solar and wind systems by releasing energy when the wind stops blowing or when clouds obscure the sun. If such storage systems are operated in a centralized mode with the dispatch controlled by the operators of the grid, they can support decentralized, intermittent generators at many locations. Thus, advances in storage systems may make a pivotal contribution in the transition to a low-emissions future.

5.3.11 A world without fossil fuel use

The energy needs of modern societies can be completely provided by electricity and high-quality liquid and gaseous fuels. In principle, nuclear energy and renewable energy resources could provide electricity and chemical fuels on a scale of 10-30 TW(thermal), or enough to power a world of 10-15 billion people. Hydrogen produced through the electrolysis of water could act as an energy carrier and storage medium. A hydrogen economy based on non-fossil primary energy resources would virtually eliminate emissions of greenhouse gases from the power sector, and would also end many of the local and regional air pollution impacts of fossil-fuel use. With the strategic vision and the political will, civilization could be well on the way to a non-fossil advanced power system by the middle of the next century, using only technological options currently within our grasp today. (Jhirad and Hocevar, 1977, 1980; Goldemberg et al. 1988; Weingart, 1979)

6 Barriers to Achieving a Low-Emissions Future in the Electricity Sector

Despite the wide range of lower emissions technologies described in the previous sections, and their general availability, few nations have captured their potential so far. Conventional electricity supply options are still employed far more often than the more effective, less-polluting, and even less expensive alternative approaches. If this pattern of investment continues, increasing electrification in both industrial and developing countries will lead to significantly higher levels of greenhouse gas emissions as well as sulfur dioxide, nitrogen oxides and particulates.

The deployment of the more preferable energy-saving and renewable technologies is hindered by market failures, and other policy and institutional barriers. Such barriers influence decision makers and investors to ignore the lower emissions options and to choose the more traditional methods of meeting electricity demands (Reddy, 1991).

The barriers exist at every level of energy-related activity: they affect individual consumers, energy equipment manufacturers and suppliers, utilities, financial institutions, governments, international funding agencies, and the development assistance agencies of industrialized countries. In the following section, we illustrate several of these sets of obstacles — by no means the full range. Understanding barriers to energy innovation, and implementing measures to overcome them, is the principal challenge to achieving a low-emissions future for the electricity sector. Meeting the challenge will require research — not just on better technologies, but on the reasons why people and institutions behave as they do in pursuing their perceived self-interests.

6.1 Barriers to end-use efficiency: individual consumers

Most people in the industrialized countries, and a growing fraction in the developing world, consume electricity as a fundamental aspect of their daily affairs. But consumers are not a uniform and homogenous group. Their occupation, level of wealth, education, and position in society determine which barriers and market failures have the largest effect on their decisions. We divide the electricity-consuming public into five rough groups, all with members in both industrialized and developing societies.

6.1.1 The unaware

The largest barrier affecting economic choices by the individual consumer is a lack of information and training. In many cases, these individuals do not know that any choice exists among the ways that they might buy and use electricity. They do not know the character or extent of services available from local institutions, and they do not know how to find out. The way to overcome this barrier is through better information, in well designed publicity and demonstration programs. Over the long term, the key is to make available comprehensive education and training programs for the broad consuming public, such as those now being implemented in Singapore. Such programmes can help align private decisions with public objectives.

6.1.2 The first-cost sensitive

Those with limited access to capital for discretionary investments find it very difficult to put these scarce resources into energy-saving measures. Even when investments have very short-payback periods, these consumers typically act as if there were extravagantly high implicit discount rates attached to the investments. As a result, few new opportunities appear sufficiently attractive to push out other alternative uses of their limited funds.

To overcome this barrier, financing mechanisms are required that can convert the high first-cost of such investments into a stream of smaller payments. Some US utilities have begun to address this barrier, by leasing efficiency-improving devices to consumers and charging the rent as part of the monthly bill. The utility structures the lease rate so that the savings from employing a new device (such as a more efficient light bulb) is divided between the consumer and the company. In this "win-win" arrangement, the consumer achieves a reduction of his or her electric bill while the company makes a profit on the lease and avoids the construction of expensive new generating capacity.

6.1.3 The indifferent

For some, the barrier is not a result of having too little money, but too much. There is little incentive to save when energy costs are a minuscule part of their discretionary income. As long as prices are kept low, these individuals will usually choose the modest expense of additional purchases over the inconvenience of changing their habits or equipment. To reach these individuals, governments must set minimum performance standards for key end-use equipment, mandating that all equipment available for a particular appliance meets a certain level of energy efficiency.

6.1.4 The uncertain

Some consumers are risk-averse, and thus unwilling to invest in new, unfamiliar equipment. For them, the barrier is the lack of complete information about the potential costs and benefits from these investments. They may also be unwilling to listen to proponents of any unfamiliar change, for fear of being manipulated or swindled.

6.1.5 The inheritors of energy inefficiency

Some consumers take on energy-using devices without making any choice at all. Renters of housing are often required to use furnaces, lighting, and other appliances that are already in place when they arrive, as part of the rental agreement. No information is provided to them about energy-use characteristics or possible savings; in many cases, the cost of electricity or fuel is set at a fixed level in the monthly bill. Thus, the end user sees no gain from the investment in efficiency.

To address this barrier, performance standards should be set so that the landlords, or equipment owners, must choose relatively efficient devices. The utility may join with the owner in sharing the costs and benefits of improvements - a programme which some utilities, such as Pacific Gas & Electric in California, have found works very successfully.

6.2 Barriers to end-use efficiency

Not all manufacturers choose to build the most efficient equipment that is commercially feasible. Their sales have never depended upon energy efficiency in the past; they do not perceive a market for efficient equipment now. It appears to these firms that their customers are making purchase decisions on the basis of other product attributes, such as style or capacity.

Governments must transmit a message emphasizing the importance to society of energy efficiency and emissions reductions. To accomplish this, pricing policy, regulations, standards, and equipment labelling schemes are all necessary. Equipment financing schemes and special tax credits should be tied to the efficiency of the purchased equipment.

6.3 Producers of electricity

Producers of electricity, even large corporations, face barriers that encourage them to avoid investing in efficiency improvements and renewable technologies. Many regulated utilities in the United States have traditionally been rewarded financially for selling as many kilowatt-hours as possible. Because fuel costs could be passed on directly to end users, these utilities have seen their mission as building generating plants with the lowest first cost, usually burning cheap fossil fuels.

To encourage these utilities to put efficiency options and renewable supplies on an equal footing with cheap fossil fuels, policy and regulatory incentives to implement least cost approaches must be put in place. Furthermore, regulators must evaluate utility performance on the basis of minimizing the cost of service, not just minimizing the cost of electricity supply.

Other utilities avoid investments in efficiency improvements or renewable technologies because of a cultural bias in the corporate culture toward central-station generators and a cultural resistance to innovation. In order to overcome this barrier, the regulatory environment must encourage power utilities to seek the most economical alternative available to meet demands. Here again, the implementation of least-cost solutions is the key.

Education and training must also be increased, so that key decision makers in the bureaucracy can become more aware of and competent to deal with emerging innovations. The education could cover not just technology, but the emerging ideas about how large organizations can effectively "learn"; these might include training in systems dynamics, scenario planning, articulating common goals, and employee involvement. Entrepreneurial attitudes and less risk-averse approaches should be encouraged as new technologies prove their mettle in practical applications.

7 Policy and Institutional Initiatives

Severing the triple bind, and overcoming the barriers to innovation, will demand coherent, integrated policy packages and will require incorporating these elements in both national and international programmes:

7.1 Collaborative mechanisms and actions

Greatly increased collaboration among multilateral development banks, bilateral donor agencies, developing country governments, power utilities and the private sector will be fundamental to implementing energy efficiency and to improving the performance of power-generating institutions.

Resources should be focused on developing countries with a strong commitment to improving power sector performance. These countries should collaborate with donors to perform appraisals of the power sector — its scale, efficiency, mandate, constraints, technologies, and needs. The appraisals should recommend policy packages, resource mobilization strategies, and investment priorities for both the national government and the power utility administrators. System efficiency and demand-side investment options (such as incentives and education to improve end-user efficiency) should be given equivalent priority to supply options — which, in turn, should include a shift to natural gas and renewables. In addition, the appraisals should analyse the policy, institutional, financial and economic constraints to environmentally sound development of the power sector; it should then propose (and prioritize) ways to remove these constraints.

Box 1

Trans-National Policy Instruments:The Global Environment Facility

The ameliorative measures and preventative technologies, necessary to address global environmental problems, are presenting developing countries with incremental costs far above and beyond what the countries expect to gain domestically in intrinsic benefits from applying those measures. Thus, concessionary finance is an important policy instrument, used to allow developing countries to address the new set of global environmental problems.

In 1990, the World Bank, the United Nations Development Programme (UNDP) and the United Nations Environmental Programme (UNEP) became partners in a tripartite agreement to establish the Global Environment Facility (GEF). This fund provides, on a pilot basis, concessionary finance for environmental protection in developing countries, where such protection yields benefits to the global community.

The GEF has been conceived as a truly transnational policy instrument, one that seeks global benefits and distributes the costs of obtaining them on an equitable international basis. Benefits are assigned according to what each country needs to reduce greenhouse gas emissions to below a set of baseline rates. These baseline rates represent levels of gas emissions that would exist if each country independently were to pursue its own economic interests. In other words, the baseline rates already include increases in greenhouse gases due to expected industrialization and growth, as well as some reductions that would result from economically viable programmes of energy conservation and fuel substitution. The costs of obtaining the global benefits are the economic sacrifices made to obtain the additional reductions below the baselines.

Eligible countries for GEF financing are those with a per capita GDP of US $4000 or less in 1989. In this way the costs are borne by industrialized countries (subscribing members of the GEF), and redistributed to developing countries. This is equitable in that it recognizes the historical use made of the atmosphere's carrying capacity for greenhouse gases by the industrialized countries — which, as a consequence, now also have a greater economic capacity to finance the necessary measures to restore the global environment.

Doubts have been expressed about the GEF's purpose, charging that it is a vehicle for further World Bank control over developing countries. (See further Hyder, 1992; this volume.) To forestall these doubts, and to succeed as a policy instrument, the GEF will need to stimulate appropriate investments in addition to existing national programmes, and to garner support from industrialized countries beyond the support already provided for development assistance to poorer countries. But in fact, the GEF, with broad and significant multilateral participation, already has a core fund of over one billion (10^9) US dollars in additional financing for a three year period. This should enable the GEF in its pilot phase to finance a meaningful number of the additional investments needed initially and thereby to demonstrate the effectiveness of this type of policy instrument.

A good example of a collaborative pilot effort to provide concessionary financing for global environmental protection in developing countries is the Global Environmental Facility (see Box 1). New international mechanisms of this kind will help to ensure that the growth of electrification world-wide is compatible with lowered emissions of greenhouse gases.

Implementing innovative approaches to global electrification may well also require an international research network of power utilities, analogous to the Electric Power Research Institute (EPRI) in the United States. Senior executives from EPRI and from utilities in Western Europe, Japan, Canada, and selected developing countries have explored the possibility of forming an International Electric Utility Industry Network, to sponsor collaborative research and technology development related to climate change, and to facilitate the transfer of power technologies among nations. In addition to sponsoring and managing research and development, EPRI has an international reputation for financing power technology commercialization, and for providing objective, independent, and high-quality documentation. EPRI's experience would be invaluable in launching an international network.

Donors could also greatly expand their collaboration with Development Financing Institutions (DFIs), to incorporate energy efficiency investments in major projects in developing countries and in the former USSR/Soviet bloc nations. Financing of industrial modernization by DFIs should include technological advances that result in lower energy intensity and environmentally sound production processes.

7.2 *Policy and institutional reform*

These measures include the development of new policies in developing country administrations — to stimulate private capital mobilization, to reform prices, and to build incentives for the economic efficiency and fiscal soundness of power institutions.

When proper pricing signals are in place, and there is access to adequate financing, utilities, acting as energy service companies (at least in the United States), have shown that energy efficiency can be a profitable business. In the process, these companies find themselves taking on a new role. Their primary purpose shifts from producing electricity to delivering energy services. That means they might become financial entities, underwriting purchases of energy-efficient appliances by customers; or they might become educational organizations; or they may distribute power generated by other companies and individuals. Since "least-cost" is important to these energy service companies, they have discovered that providing extensive information and training to customers is one of their most cost-effective and successful options for improving electricity distribution. But the companies cannot take on these new roles unless regulatory policies are in place to encourage them to do so.

In many countries, major institutional changes are needed, to allow electric power utilities to function as autonomous institutions, instead of political patronage organizations. In countries where governments are both owners and regulators of power utilities, political patronage often prevents utilities from functioning as commercial institutions; there is thus an urgent need for a new social compact between government and utilities.

To promote superior performance, the most important element of such a compact is an independent regulatory

The Programme for the Acceleration of Commercial Energy Research (PACER) **Box 2**
(David Jhirad ,1987)

The Programme for the Acceleration of Commercial Energy Research (PACER) is a $20 million, 6-year project initiated in 1987 by USAID in bilateral collaboration with the Government of India. The purpose of PACER is to foster innovation in the Indian electric power sector, through supporting Indian private firms in technology commercialization, fostering joint ventures with US firms, and creating consortia of R&D laboratories, industrial enterprises (including manufacturers) and end users. The PACER Program also supports two related activities: a competitive research awards programme, and an analysis and outreach programme.

PACER particularly aims at promoting market-driven and environmentally-sound technological innovations in the Indian energy sector. This is achieved through cost- and risk-sharing among the PACER consortia, which each focus their concern on one power project or geographic locale within India. PACER provides up to $3 million in conditional grants to co-finance the costs of technology development and commercialization. If the project succeeds, the promoters repay the grant. If it fails, the grant is written off.

Criteria for eligibility are threefold:

a) The consortium must involve Indian and/or US manufacturers, research institutions and end users;

b) The project must involve the commercialization of innovative and environmentally benign products and processes — and they must be relevant to the India energy sector;

c) The technology must show significant potential for commercialization within a period of 5 years.

PACER is being implemented by the Industrial Credit and Investment Corporation of India Ltd (ICICI), Bombay. An Energy Research Advisory Committee consisting of representatives from government, academia, private sector and electric power utilities provides policy and strategy guidance for the programme. The advisory committee is chaired by the Department of Non-Conventional Energy Sources of the Government of India.

As of July 1991, twelve consortia R&D projects and eight research studies have been approved, involving a total PACER funding of $3.6 million. The approved subprojects include energy efficient product developments, renewable power generation, advanced power systems, and analytical studies of management and policy changes in the Indian power sector.

PACER has gained wide recognition in the Indian and international energy communities as a model for technology cooperation with advanced developing countries. The programme could be replicated on a global scale. Towards that end, international representatives at the PACER conference in New Delhi in 1990 proposed the establishment of a mechanism to raise one billion US dollars annually (Jhirad, 1991). The money would be used to support the development and commercialization of innovative electric power technologies, for use in both industrialized and (particularly) developing countries. The funds could be raised from a global carbon users' fee of about two cents per barrel of oil, or about eight cents a tonne of coal. For India the fee would be about $24 million annually; for the US it would amount to $235 million per year.

framework, clearly separating the roles of government and power supplier. Despite the virtues of the market — as an allocator of capital, raw materials and manpower — regulation will be necessary to safeguard equity, to ensure that environmental externalities are valued, and to protect the interests of future generations. Even in many countries which have a regulatory framework now, there should be substantial changes in the regulatory environment — for example, to encourage more efficient delivery of electricity services to consumers. In all cases, the regulatory process should be independent of patronage, and transparent to the public. These are necessary measures for ensuring that economically efficient, socially equitable, and environmentally sound power options are implemented.

Encouraging the mobilization of private capital, and allowing the private sector to take a role in energy generation, is an important part of this process. Attracting private capital into the power sector in developing countries will require that governments, as well as bilateral and multilateral institutions, provide insurance to private power developers to cover the risks of non-payment by utilities and customers. Pricing reform that allows tariffs to cover costs will be fundamental for the fiscal soundness of both public and private enterprises.

In the meantime, the poor - both individuals and countries - must be assisted through innovative financing and assistance arrangements. For national and international assistance agencies, human resource development and technical and managerial training will remain important, as will concessional financing (the loaning of money at exceedingly low interest rates, below market value and usually below the cost of borrowing for the agency).

7.3 *Innovative financing: catalysing technological advance*

In most developing countries, current financing for technology development and commercialization is insufficient. Unfortunately, most multilateral development banks and international financing institutions are technologically risk-averse. While they play an important role in emphasizing efficiency, institutional restructuring, and policy reform, they do not finance technology research, development and commercialization.

Concessional financing should be provided for joint-venture technology development and commercialization, probably linking firms from industrialized and developing countries. Such mechanisms will reduce the level of risk associated with new product development, and will provide incentives for bringing efficient technology into the marketplace.

The US Agency for International Development (USAID) is promoting market-driven technology development through consortia (Jhirad, 1990). The consortia, which include electric utilities, manufacturers, universities, and national laboratories, establish developing country private and public institutions as partners in innovative technology development programmes. The Programme for the Acceleration of Commercial Energy Research (PACER), a $20 million Indo-US collaborative initiative launched by the Government of India and USAID in 1987, is a good model of such programmes (see Box 2).

8 Near-Term Steps to Address the Issues

Recognizing the urgent need for concerted international action to address the problems of the international electric power sector, a series of consultative technical conferences have been held in Stockholm over the last few years, most recently in November 1991. Called the Stockholm Initiative on Energy, Environment and Sustainable Development (SEED), these meetings have been hosted by the Government of Sweden and cosponsored by the United Nations Commission on Trade and Development (UNCTAD), the United Nations Development Programme (UNDP), and several major bilateral agencies. Participants at the November meeting (including senior decision makers from 21 developing countries) discussed strategies for implementing power sector efficiency and endorsed recommendations for redesigning the roles developing country institutions as well as bilateral and multilateral organizations. In addition, SEED participants recommended that the implementation of energy efficiency be a major objective of Agenda 21 for the UN Conference on Environment and Development to be held in Rio de Janeiro, Brazil in June 1992.

The following set of recommendations were endorsed by donors, lenders and developing country representative attending the SEED meeting. (USAID, 1992) Since these recommendations represent a North-South consensus, they are reproduced here in their entirety in Box 3.

9 Conclusion

As the SEED recommendations make clear, developing countries face massive, daunting problems in meeting the projected increases in electricity demand. Any reasonable solution will entail high financial and environmental costs, but only about $20 billion of the projected requirements of $100 billion per year will be available from internal and external sources. The decline in financial performance of power utilities in developing countries, along with the technologically risk-adverse nature of most multilateral development banks and international financing institutions, will make the mobilization of capital even more difficult than it would otherwise be.

We have shown that neither policy reforms nor institutional strengthening, nor technological innovation alone can suffice. Innovative approaches on all fronts will be necessary. For example, one vitally important set of near-term opportunities requires that developing countries find and use cost-effective technologies to reduce the energy intensity of key end uses. But even if all near-term efficiency improvements are exploited, new and less-polluting electricity supply technologies will still be required after the year 2000. Even then, persistent market failures will undoubtedly occur

Box 3

Recommendations of the Stockholm Initiative on Energy, Environment and Sustainable Development (SEED), November 1991

Developing countries

1. Developing country governments should develop and implement programs for improving power sector efficiency, both in supply and demand. These programs should focus on greatly improved performance compatible with an integrated energy strategy and environmental sustainability.

2. Developing country governments should support efficient alternatives to capacity expansion for utilities through better utilization of existing capabilities, and the development of independent private power facilities. Tariff reforms that make the sector credit worthy should be an integral part of such measures. However, the political, economic, and social conditions in individual countries underscore the need for a country-specific approach in addressing these issues.

3. Developing country governments should strengthen financial mechanisms, institutions, and associated policies and regulations to provide innovative lending in supply and demand side power sector efficiency, including direct lending for private sector initiatives. Sector financing entities, including development financing institutions with portfolios in industrial modernization, agriculture, the environment, and housing, are targets for such institutional reforms.

Bilateral and Multilateral Institutions

1. Bilateral and multilateral institutions should dramatically alter their investment priorities to support end-use efficiency, sustainable and reliable operations and maintenance programs, and private sector initiatives, in addition to traditional investments in supply.

2. Bilateral and multilateral institutions should provide financial and technical support to improve the legal and regulatory framework as well as the management and institutional performance of power utilities.

3. Bilateral and multilateral institutions should expand their financing to cover joint ventures in environmentally sound electric power-related technology cooperation.

4. Bilateral and multilateral institutions should provide insurance for private sector power projects to enable capital mobilization from commercial and other markets.

5. Bilateral and multilateral institutions should commission a study investigating the lack of progress of private sector involvement in developing country power sectors.

6. Bilateral and multilateral institutions should create a fund in specific countries to support the availability and delivery of critical spare parts to ensure high system availability.

Institutional Linkages

1. Bilateral and multilateral institutions should, together with developing countries, perform long-term power and environmental sector appraisals to formulate policy reform packages and investment priorities for public and private entities.

2. Existing bilateral and multilateral networks in energy and environment should be strengthened and expanded to link with developing country financing institutions and recognized centers of excellence.

3. The SEED recommendations should be widely disseminated to relevant agencies, developing country governments, and the private sector. they should also be presented to the refocused World Bank/UNDP Energy Sector Management Assistance Program (ESMAP) programme and its consultative group of donors and developing country representatives.

at crucial points in the investment cycle. Specific policies for the efficient use of capital must be implemented. In addition, policies to improve the flow of information and to bring the full financial, social and environmental costs of energy supply and use into the price of fuels (Jochem and Hohmeyer, 1992; this volume) will be necessary.

Finally, the role of electric utilities must be reshaped, as we have seen, from electricity producers to energy service companies. This will require incentives for companies to profit as much from improved efficiency as from increased supply. In the developing countries, the World Bank and other multilateral development assistance agencies, along with the bilateral aid agencies, must help to implement institutional and financial reforms. These funding agencies must also encourage the codevelopment of the best of the new technologies. Successful codevelopment will require equitable partnerships between companies from industrial countries and their counterparts in the developing world. Such partnerships will be necessary to effectively adapt new energy technologies to the special circumstances of each receiving country.

Fortunately, while the necessary multifaceted approaches pose a difficult challenge for the international community, the will, desire, and technological means to surmount those challenges are more evident now than they have been in the past. A number of promising new supply technologies are viable, or on the horizon of viability, for the first time. All of these reforms can take place in ways that help developing countries to escape the triple bind of capital shortage, declining performance, and environmental risk.

References

Abraham, P., 1990: Strategies for improving the quality of power supply. In *Innovation in The Indian Power Sector*, Tata-McGraw Hill, Delhi.

Anderson, D., 1991: Energy and the environment. *The Wealth of Nations Foundation,* Edinburgh, May.

Balzheiser, R. and K.E. Yeager, 1987: Coal-fired power plants for the future. *Scientific American*, **257**(3), September, 100-107.

Bemis, G.R., A. J. Soinskin, S. Rashkin, S. Jenkin, A. Jenkins, R. L. Johnson and M. Radovich, 1990: *Technology Characterizations, Final Report,* Staff Issue Paper #7R, Docket No 88-ER-8. In the matter of the 1990 Electricity Report (ER90). Energy Resources Conservation and Development Commission, State of California.

Booth, R and P. Elliott, 1990: *Sustainable Biomass Energy*, Shell Staff Selected Papers, Shell Centre, London.

EPRI, (Electric Power Research Institute), 1986: *Technical Assessment Guide,* December.

EPRI (Electric Power Research Institute), 1990: Beyond steam: breaking through performance limits, *EPRI Journal*, December.

EPRI (Electric Power Research Institute), 1991a: On-Site Utility Applications for Photovoltaics, *EPRI Journal*, March.

EPRI 199lb: Fuel cells for urban power, *EPRI Journal*, September.

EPRI 199lc: Alabama cooperative generates power from air, *EPRI Journal*, December.

Goldemberg, J., T.B. Johansson, A.K.N. Reddy and R.H. Williams, 1988: *Energy for a Sustainable World.* Wiley Eastern Limited, New Delhi, India.

Häfele, W., 1989: *Energy Technologies for The First Decades of The Twenty-First Century*, United Nations, Economic and Social Council, Economic Commission for Europe, November 15.

Hansen, S. and G. Schramm, 1989: Natural gas as a domestic energy resource in developing countries issues and potentials. *The World Bank.* April.

Haupt, G., J.S. Joyce and K. Kuenstle, 1990: *Combined-Cycles Permit the Most Environmentally Benign Conversion of Fossil Fuels to Electricity*, Siemens AG-KWU Group, June.

Houghton , J.T., G. Jenkins and J.J. Ephraums, 1990. *Climate Change: the IPCC Scientific Assessment*, Cambridge University Press, Cambridge, UK.

International Energy Agency, (IEA),1989: Organisation for Economic Co-operation and Development (OECD): *Energy Technologies for Reducing Emissions of Greenhouse Gases.* OECD, Paris.

Jhirad, D.J. 1987: *The Program to Accelerate Commercial Energy Research (PACER).* USAID, New Delhi, and Office of Energy, Washington, DC, Aug.

Jhirad, D.J. 1990: Power sector innovation in developing countries: implementing multi-faceted solutions, *Annual Review of Energy,* **15**, 365-398. Inc., Palto Alto, California.

Jhirad, D.J. 1991: A global program for commercial energy research. In *Innovation In The Indian Power Sector*, pp. 429. Tata-McGraw Hill Publishing Co. Delhi.

Jhirad, D.J. and C. Hocevar, 1977: Energy strategies for a solar-powered United States. In *Nuclear Energy and Alternatives*, ed. by O.K. Kadiroglu, A. Perlmutter and L. Scott. Ballinger Publishing Company, Cambridge Massachusetts 1978.

Jhirad, D.J. and C. Hocevar, 1980: In *Energy Strategies: Toward a Solar Future*, ed. H. Kendall and S. Nadis, Ballinger Publishing Company, Cambridge, Mass.

Johansson, T.B., B. Bondlund and R.H. Williams, eds., 1989: *Electricity: Efficient End-Use and New Generation Technologies, and Their Planning Implications,* Lund Univ. Press.

Kamali, K. and R. Tawney, 1990: *Aircraft-Derivate Steam Injected Gas Turbines for Power Plant Application*, Bechtel Corporation, Gaithersburg, Maryland, December.

Moore, E. and G. Smith, 1990: *Capital Expenditures for Electric Power in the Developing Countries in the 1990s*, World Bank Industry and Energy Department Working Paper, Energy Paper No. 21, February.

Munasinghe, M., J. Gilling and M. Mason, 1988: *A Review of World Bank Lending For Electric Power,* World Bank Industry and Energy Department, Energy Series Paper No. 2, March.

OTA (Office of Technology Assessessment), 1985: *New Electric Power Technologies: Problems and Prospects for the 1990's*, Washington, DC, US Congress, Office of Technology Assessment, OTA-E-2466, July.

Reddy, A.K.N. and J. Goldemberg, 1990: Energy for the developing world. In *Scientific American*, September, **263**(3), 110-118.

Reddy, A.K.N., 1991: Barriers to improvements in energy efficiency. *Energy Policy*, **19**, 953-961.

Schramm, G. 1984: The changing world of natural gas utilization. *Natural Resources J.* **24**, 405-36.

UN (United Nations), 1987: *Energy Statistics Yearbook.*

USAID (US Agency for International Development), 1988: *Power Shortages in Developing Countries: Magnitudes, Impacts, Solutions and The Role of The Private Sector,* A Report To Congress, Washington, DC.

USAID (US Agency For International Development), 1990: *Greenhouse Gas Emissions and The Developing Countries: Strategic Options and The USAID Response*, A Report to Congress, July.

USAID (US Agency For International Development), 1992: *Report of The Stockholm Initiative on Energy, Environment and Sustainable Development*, January.

USAID (US Agency for International Development), 1991a: *Opportunities for Improving End-Use Electricity Efficiency in India*, November.

USAID (US Agency for International Development), 1991b: *Feasibility Assessment of Coal Integrated Gasification Combined Cycle (IGCC) Power Technology for India*, June.

Weingart, J., 1979: Global aspects of sunlight as a major energy source. *Energy, The International Journal.* **4**, 775-798.

Williams, R. and E.D. Larson, 1989: Expanding roles for gas turbines in power generation. In *Electricity: Efficient End-Use and New Generation Technologies, and their Planning Implications,* ed. J. B. Johansson, B. Bodline and R. H. Williams. Lund Univ. Press.

World Bank, 1982: *Energy Efficiency and Optimisation of Electric Power Distribution System Losses*, Energy Department Paper No. 6. July.

World Bank, 1989a: *Promoting Natural Gas Investment and Production in Developing Countries*. Energy Series Paper No. 10.

World Bank, 1989b: *The Future Role of Hydropower in Developing Countries*. Industry and Energy Department Working Paper, Energy Series Paper No. 15. World Bank, Washington, DC, April.

World Bank, 1990: *A Review of the Treatment of Environmental Aspects of Bank Energy Projects*. Industry and Energy Department Working Paper, Energy Series Paper No. 24. World Bank, Washington, DC, March.

World Bank, 1992: Gunter Schramm, private communication.

CHAPTER 14

Transportation in Developing Nations: Managing the Institutional and Technological Transition to a Low-Emissions Future

Jayant Sathaye and Michael Walsh

Editor's Introduction

The world is caught in a seemingly insolvable dilemma. On one hand, in both industrialized and developing countries, consumers are increasingly demanding mobility—both for themselves and for goods from far away. The most convenient, and in many cases most desired solution is transport by motor vehicles. But as Jayant Sathaye and Michael Walsh note in this chapter, increasing motor vehicle use inevitably leads to a range of air pollutant emissions to the atmosphere —along with congestion and inefficient use of fuel. These in turn exacerbate both environmental and economic problems. As developing countries become more urbanized and increasingly dependent on cars and trucks, they will confront the linked problems of heavy traffic, urban pollution, high accident rates and low vehicle efficiencies. These developments suggest to Sathaye and Walsh that there is a growing need for new and more sustainable transportation strategies.

Unfortunately, the trend is in the opposite direction. Sathaye and Walsh document how desires for mobility in developing countries have already brought forth a conventional vehicle fleet that is dependent on liquid fuels — principally oil. Expanding industrial markets spur demands for motorized freight transport. As economies grow, freight transport traditionally shifts modes — from rail to roads and air. The gradual movement away from non-motorized transport—which depended on animal carts, bicycles, and people on foot—toward increasing reliance on cars, buses, and trucks has led to multipurpose use of roads with consequent increases in congestion, risks of collisions, and decreases in fuel efficiency.

Jayant Sathaye is the co-leader of the International Studies Group at the Lawrence Berkeley Laboratory in Berkeley, California. Michael Walsh of Washington, DC, is widely recognized internationally as a leading consultant on national transportation strategies and their environmental impacts. For the last 15 years, Sathaye and Walsh have been working on the evolution of transportation management strategies that can deliver high levels of personal mobility while improving vehicle efficiency, minimizing economic costs, and protecting the environment from harmful pollutants. Their work has catalysed successful collaborations among transportation experts in the developing world and the OECD nations.

Sathaye and Walsh emphasize the value of mass transit for developing countries, and the importance of developing effective, efficient, and profitable systems. But they also acknowledge that private automobile ownership will continue to expand in these countries as their economies develop. Grounded in this vision, the authors outline a comprehensive transportation strategy to support sustainable development. The key elements include (1) improving vehicle fuel efficiency, (2) encouraging the introduction of cost-effective technological improvements to control emissions, (3) introducing alternative fuels where these can be competitive with oil, (4) improving fuel quality, and (5) managing urbanization to limit congestion. Ultimately, the solution lies in building institutional capacity for transportation management in developing countries, while facilitating the co-development of advanced transportation systems by combinations of firms in the North and the South. Measures such as these, combined with high standards for new vehicles, inspection and maintenance programs for older vehicles, and land use planning strategies to facilitate the free movement of people and goods (while minimizing the average distance travelled), will help to maintain the momentum of development while limiting the environmental and social costs of transportation.

- I. M. M.

1 Introduction

Satisfying transportation needs — both for economic activity and for the individual needs of citizens — is a critical requirement of socioeconomic development. While few would dispute this, the construction of a complete transportation system can burden both a nation's economy and its environment. In recent years especially, the motorization of developing nations has led to a burgeoning motor vehicle fleet which runs virtually entirely on fossil fuels (primarily oil, although vehicles use alcohol and gaseous fuels in some countries).

The gradual replacement of traditional forms of transport — animal carts, walking and bicycling — by motorized vehicles has led to more multipurpose use of roads, which lowers fuel efficiency and increases congestion. All of this contributes to high levels of local pollution, and increasing emissions of greenhouse gases, including carbon dioxide, volatile organic compounds (VOCs), nitrogen oxides, carbon monoxide (CO) and chlorofluorocarbons (CFCs). Besides contributing to greenhouse warming, most of these gases also add to already high levels of urban air pollution. (Aircraft and ships are also important sources of carbon dioxide and air pollution, but this report addresses only land transport.)

Industries and urban populations are growing rapidly in the developing world. The exceptions to this pattern are in Latin American nations — where large-scale industrial activities have been well-developed for years and the urban population shares are already high. Projections of population and economic trends in most developing countries indicate that a larger proportion of the growing population will move to cities over the next decade (Figure 1). Simultaneously, many developing nations, especially in Asia, are expected to maintain, if not increase, their currently high annual growth rates of GNP and GNP per capita (Figure 2 and Figure 3), As GNP per capita rises, vehicle density also increases (Figure 4). Therefore, if current trends persist, developing world inhabitants will increasingly rely on private transport. This growth will lead to a dramatic rise in transport fuel consumption, and a worsening of today's already serious pollution problems.

The congestion, inefficiency and devastating environmental impacts of today's transport systems all point to the need for more effective transportation strategies. However, particularly for financially constrained developing countries, many potential options for reducing transport emissions (for example, by substituting electricity generated from solar, nuclear, or hydropower sources) prove too expensive. Developing and industrialized nations alike are faced with the challenge of maintaining economic growth rates and satisfying personal mobility desires — less through increasing the roads and truck routes, and more through the more efficient use of resources. Meeting this challenge will require each nation to examine its entire transport system, in search of alternative ways to satisfy the desire for personal mobility, and more efficient means of transporting freight and accomplishing business-related travel.

This paper examines developing world transportation systems in particular. We address the following set of questions:

- How will the changes presently under way — in fuel use, technology and modes of transport — affect environmental and economic resources?

- Can the current trend towards increased motorization and higher vehicle emissions be altered?

- What alternatives may provide the same services? Are these options unique to certain economic, demographic and/or environmental contexts or can they be replicated in other circumstances?

- Which policies are necessary to provide more sustainable transportation services from an emissions and resource use standpoint?

2 The Structure of Transportation: in Transition

An examination of the present trends in developing world transport sheds light on the forces that shape these systems and provides a framework for understanding their future growth.

2.1 Industrialization, urbanization and the growth of transport

The emergence of modern transportation systems in developing countries has followed a pattern similar to that of the industrialized world: economic growth has been accompanied by rapid industrialization and urbanization, with transport structures evolving to keep pace. The expansion of industrial activities and markets spurs the need for motorized freight transport - to ferry goods to the market place — and thus places particular pressure on a nation's transport system. Rail, truck, airplanes, water barges and ships provide a vital service in a country's economic development process. With further economic growth and structural change, the types of commodities and the modes used for their transport shift. The share of the various modes varies over time and across countries.

A good measure of freight transport is *tonne-kilometres* (tonne-km), the distance travelled multiplied by the weight transported. This indicator, coupled with fuel efficiency levels, determines the amount of fuel used and air pollutants emitted from transport activities. The quantity of freight that must be moved depends not only on the level of economic activity in society, but also upon the type of industry: one dollar's worth of steel or cement is much heavier than one dollar's worth of computer. Similarly, the distance travelled between the point of supply and the point of demand varies, depending on each country's geography, and not necessarily on its resource base. A small country (like Singapore) may

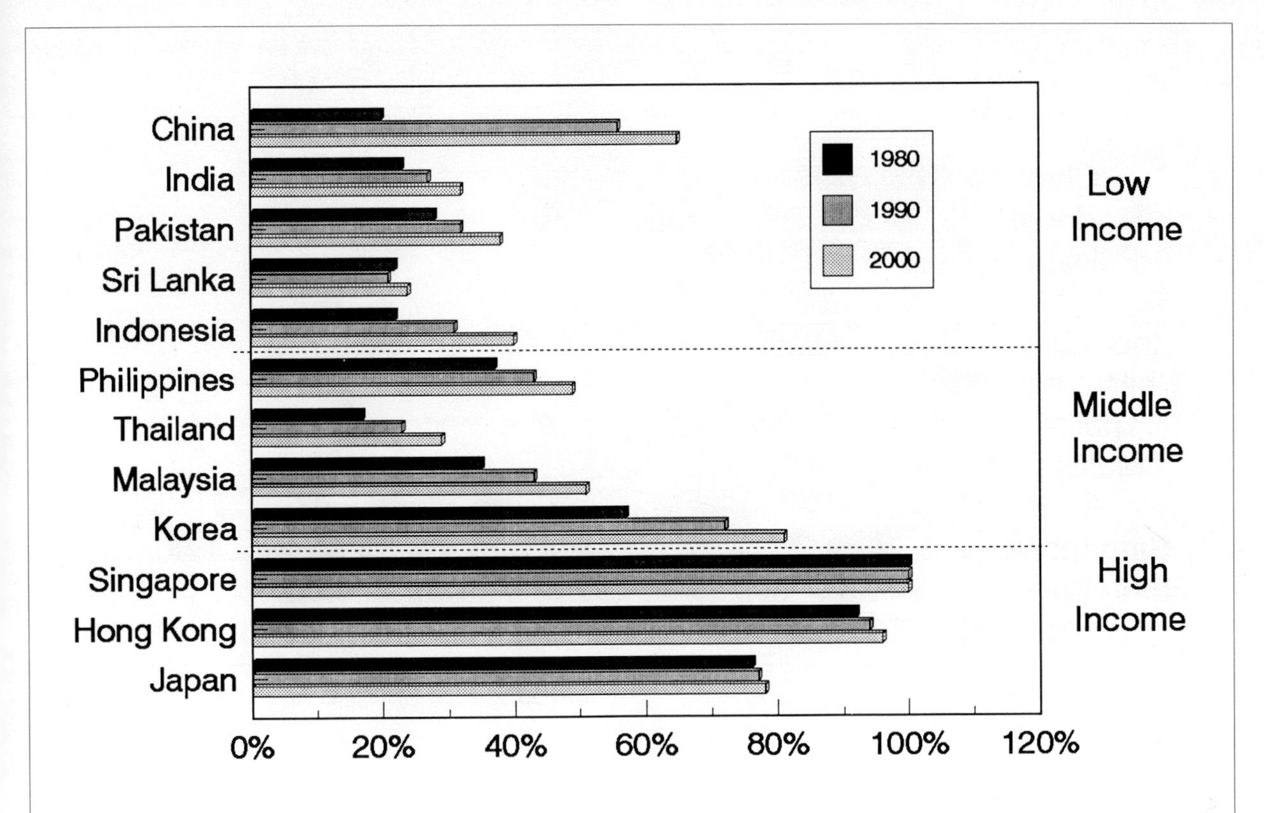

Figure 1. Urban proportion of the populations of 12 Asian nations — past (1980), present (1990), and expected in the near future (2000). Note the extremely rapid urbanization of China, Indonesia, Malaysia, and Korea.

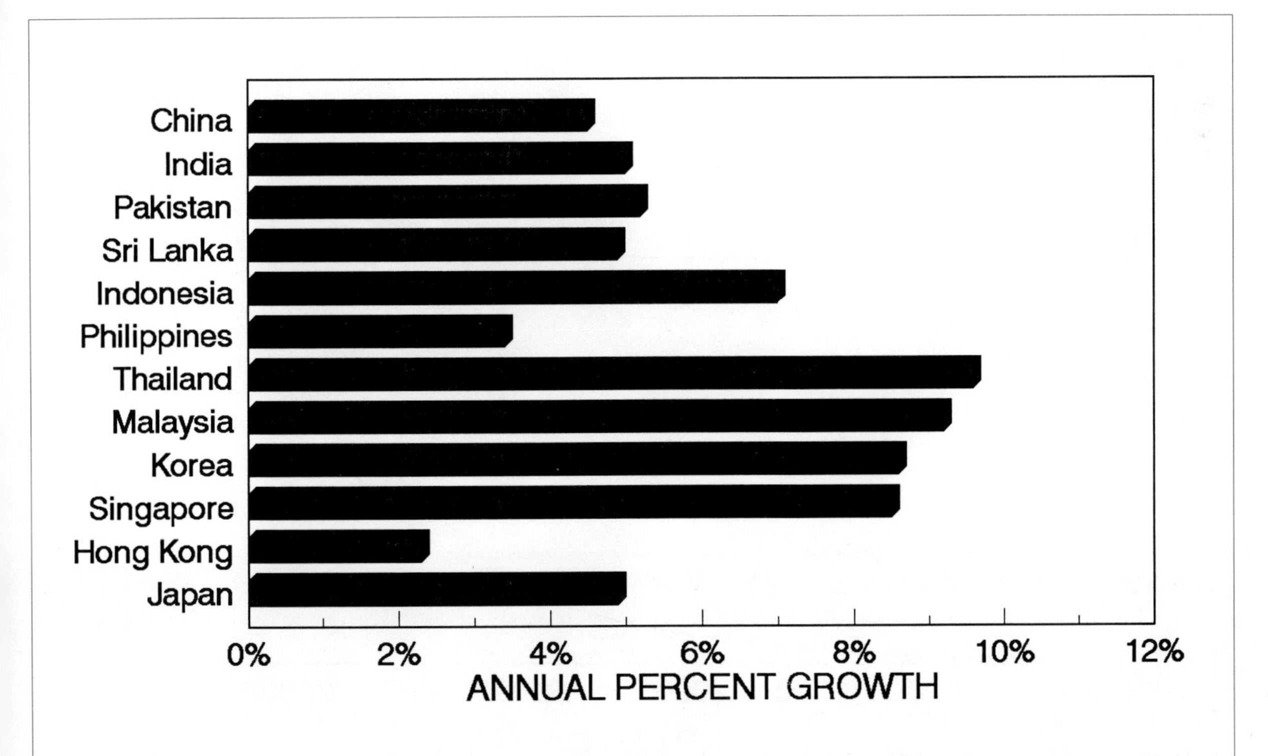

Figure 2. Annual growth of Gross National Product of 12 Asian nations. Growth of GNP is a major factor promoting increased use of fossil-fuel-emitting transportation. Source: Adapted from *The Almanac Asia Week*, 1 February, 1991.

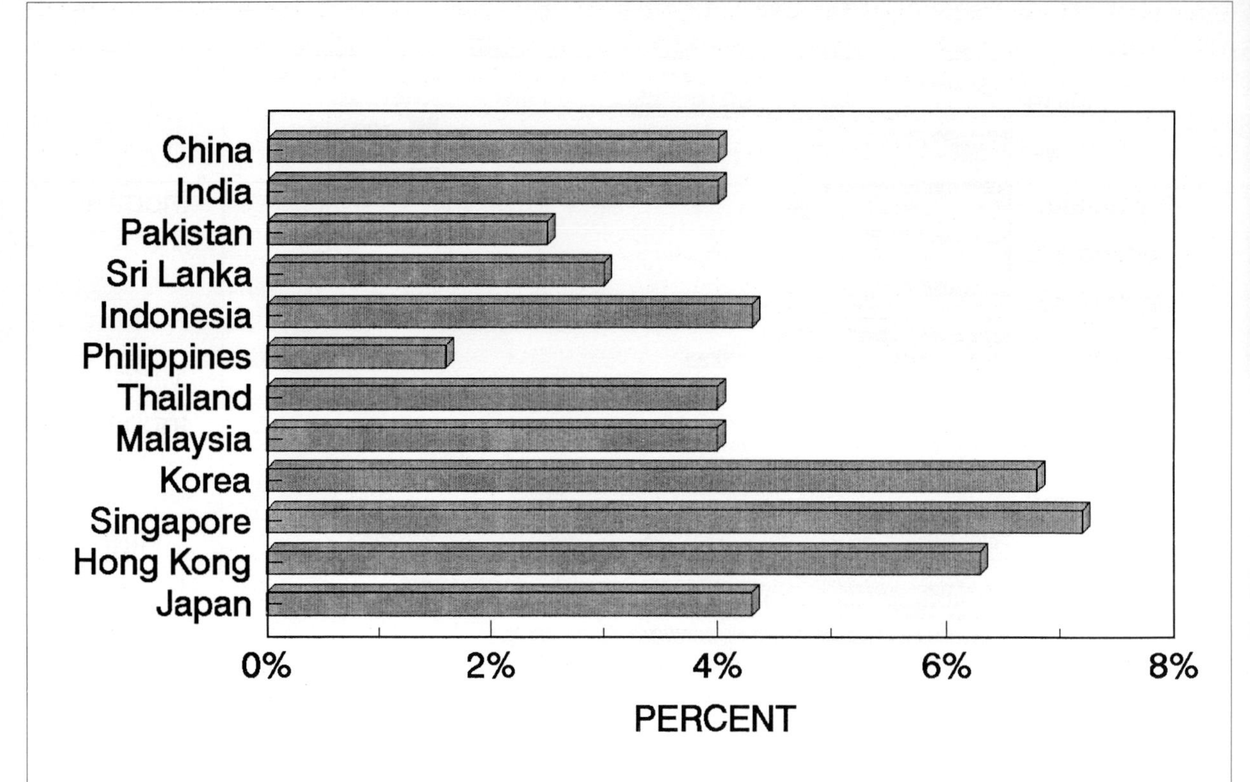

Figure 3. Annual growth of GNP per capita, in 12 Asian nations. Source: Derived from World Bank, *World Development Report 1990*, World Bank, Washington, DC, 1990.

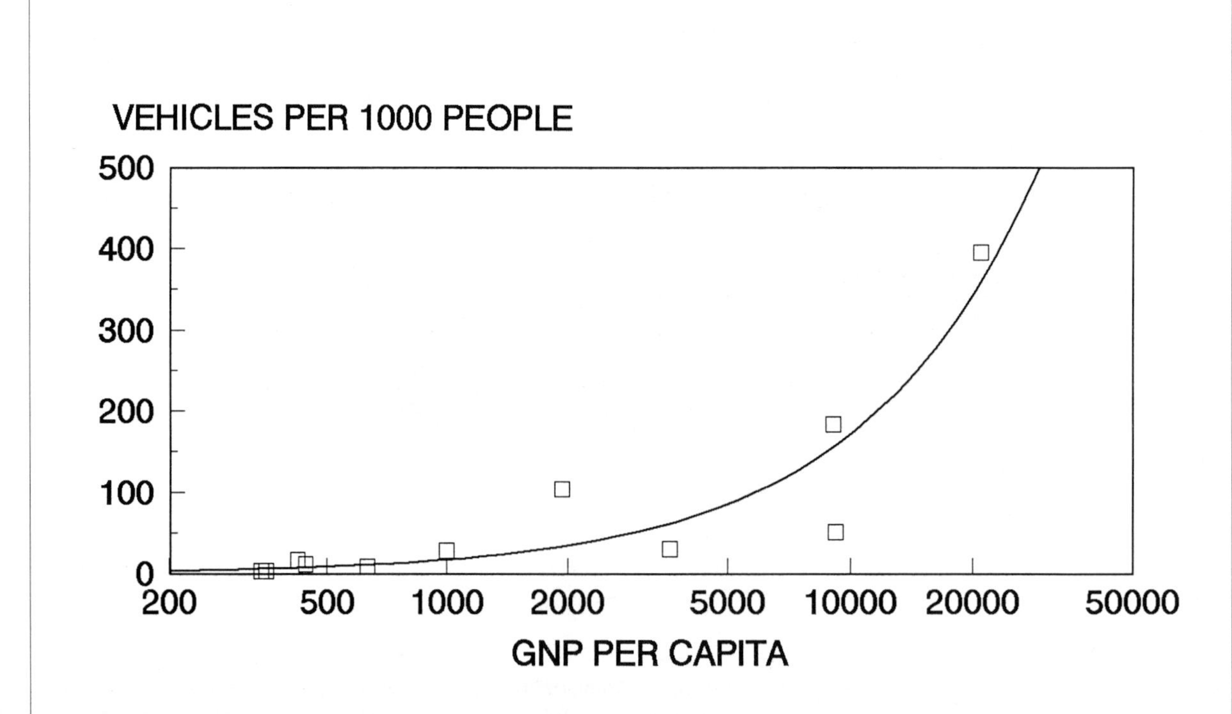

Figure 4. The relationship between vehicle density and GNP per capita, plotted for 1990 in 12 Asian nations. For all the nations, the ratio falls roughly on the same curve. As GNP per capita rises, vehicle density also increases.

import these commodities, incurring much lower tonne-km and consuming less transport fuel for domestic freight traffic than would a large country (like Brazil) which may transport indigenous industrial raw materials over long distances. No rigid relationship exists between freight tonne-km and value-added; the factors which structure this relationship differ from country to country and region to region.

Urbanization is another integral part of the economic development process. The proportion of residents of developing countries who live in cities has grown between 1960 and 1988 — from 17% to 35% in the low-income economies and from 40% to 56% in the lower middle-income countries (World Bank, 1990). As urban areas expand, land which was previously considered unsuitable, at the edges of metropolitan areas, is developed. The growing distance between these new residential suburbs and city centres (or subcentres) increases the need for motorized travel.

Simultaneously, as incomes rise, an increasing proportion of trips shift modes— first to motorcycles, then to private automobiles which, per person carried, are the most polluting option. As we shall see in Section 5 of this chapter, public policies towards land use, housing and transportation also influence this transition to private motorization. But so do ordinary human desires. For example, air conditioning can account for as much as a 15% increase in fuel use per trip (Sathaye and Meyers, 1986a). With rising incomes, more consumers in developing countries are in a position to afford air-conditioned vehicles, and it would be almost impossible to stop these purchases by appealing to ecological concerns. To compensate for the resulting decline in average vehicle efficiency, vehicles and the transport system would have to become just that much more energy efficient.

2.2 *Modal shifts: from rails to roads*

Over time, transport activities grow by shifting from rail to road and air. For freight, as industrialization commences, the share of bulk commodities declines and the weight fraction representing finished goods increases. Because these manufactured items often are intended for delivery to many smaller-scale distributors rather than a few large facilities, the new, lighter, higher value-added goods are more amenable to transport by road and air than rail. Thus, during the last two decades the share of freight carried by trucks has increased from 35% to 51% in Japan; and from 54% to 63% in six major European countries (West Germany, UK, France, Italy, Sweden, and Norway). The share of freight carried by rail declined from 14% to 5% in Japan and from 28% to 18% in Europe.

China (along with India) is one of the few places in the Third World where rail still remains important to the movement of freight. Elsewhere, the long-distance movement of both freight and people between cities is shifting rapidly from rails to roads. Pakistan and South Korea, as illustrated in Table 1, are typical of many nations.

Note that a substantial fraction of freight still moves by ship in China and Korea. The ship fraction in India (not shown in this table) is surprisingly small, perhaps because of inclement monsoon weather; which, if storminess continues to increase (see Mitchell and Ericksen, 1992; this volume), may mean that freight traffic by ship will continue to decline.

The growth of inter-city passenger transport — as shown in Table 2 — has also prompted a transition from rail to road modes. Increasingly, private cars and buses have replaced trains as the primary means for moving passengers from one city to another, and within cities. (As personal incomes rise more people can buy automobiles or use taxis). Only India and China still rely as heavily as they once did on trains for inter-city passenger transport.

The transition from rail to road has a number of implications for energy efficiency, emissions and cost effectiveness. Generally, trucks are more cost effective than rail for trans-

Table 1. Modal shares for freight transport

Country	Road (%)	Rail (%)	Air (%)	Ship (%)	Other (%)	Total (Billion tonne-km)
S.Korea						
1970	11	58	-	32	-	13.4
1987	46	24	6	25	-	54.7
China						
1980	6	48	-	42	4	1202
1987	12	43	-	43	3	2222
Pakistan						
1971	45	54	2	-	-	14.5
1987	71	25	3	-	-	30.7

Source: Korea Energy Economics Institute, 1989; Government of the People's Republic of China, 1989; Asian Development Bank and Economic Development Institute, 1989, pp. 331-380.

Table 2. Modal shares for passenger transport

Country	Road (%)	Rail (%)	Air (%)	Ship (%)	Total (Billion passenger km)
S.Korea					
1970	66	32	1	1	30.4
1987	74	25	1	-	131.6
India					
1970	64	36	-	-	328
1987	78	22	-	-	1150
China					
1970	23	70	-	7	-
1985	37	56	3	4	442

Source: Korea Energy Economics Institute, 1989; Planning Commission, Government of India, 1988; Government of the People's Republic of China, 1989.

porting goods over shorter distances.[1] The distance over which trucks are more cost effective varies by commodity; for example, in transporting food-grains, the break-even distance in India is estimated at 280 km. However, in part because of the subsidized price of diesel fuel, trucks in India transport food-grains over an average distance of 386 km or 38% farther (Reddy, 1991). Similar use of trucks over longer than break-even distances has been noted for China, where a large portion of the freight is trucked beyond its optimal road distances (People's Republic of China, 1989).

Despite these inefficiencies, however, the shift from rail to road is probably an inevitable consequence of economic development. As economies develop, the time required for shipment of goods and flexibility in choice of destination becomes more important. Efforts aimed at reducing fuel use and emissions must consider these concerns. Developing world governments would be hard pressed to reverse the shift from rail to road. They can, however, decelerate the process and lessen the negative repercussions, by (as we shall see) the choice of infrastructure they create and the policies they implement.

2.3 The rise of the urban passenger vehicle

Today, most motorized passenger transport in developing countries occurs in cities. Thus, the greatest problems associated with passenger transport — pollution, traffic, high accident rates, poor efficiencies — tend to manifest themselves most deeply in metropolitan areas (Thailand Development Research Institute, 1990). In Bombay, for example, ownership of cars per capita almost doubled between 1970 and 1982 (from 13.5 to 23.4 cars per thousand people) while increasing only 15% on the national scale (World Bank, 1986). While the proportion of middle- and upper-income households (who can afford cars and motorcycles) is lower in developing and newly industrialized countries, the number of vehicles and levels of congestion are comparable to (or exceed) those of major cities in industrialized countries.

Walking, bicycling, two- and three-wheelers and motor cars are the common forms of personal transport in urban areas. In many developing country cities, animal-driven carts are still in use as well. While bicycles are prevalent, they are primarily used for short journeys. Survey data for Beijing show bicycles used for more than half of all trips of 10 kilometres or less; beyond that distance, the use of buses is more common.

Bicycle ownership in Beijing increased from 1.5 to 2.1 per family between 1980 and 1987 (Liu and Qin, 1991). This high proportion of bicycle and bus use helps check the growth of motor fuel consumption and accompanying air pollution, but it also has exacerbated traffic congestion, thereby reducing traffic speeds, decreasing vehicle fuel efficiencies and heightening bicycle fatalities. The growing congestion in Beijing and lack of cautious driving has reduced bus speeds to an estimated 15 kilometres per hour (km/h) — about the speed travelled by a bicycle. Segregated lanes for bicycle use have helped to reduce these negative repercussions in some cities.

In other cities, the use of cars, motorcycles and taxis is more common. The gradual replacement of bicycles with motorized two-wheelers has become a feature of Asian urban development. A recent survey illustrates how motorized two-wheelers supplanted bicycles during the 1980s in Pune, India (Kulkarni et al., 1991): the share of households owning bicycles declined from 61% in 1982 to 29% in 1989, while the share owning two-wheelers rose from 17 to 41%. Motorized two-wheelers have supplanted bicycles with equal rapidity in other Asian cities: in Bangkok, they increased from 22 per 1000 persons in 1976 to 90 per 1000 persons in 1986. The popularity of two-wheelers has spread from Bangkok to rural Thailand in recent years.

As Table 3 shows, most Third World cities have also seen a substantial increase in the number of automobiles, beginning with the early 1970s (Meyers, 1988). This growth rate in cars has eclipsed that of car growth in cities in Europe and Japan (see Table 4).

The exceptions are Bombay and Calcutta, where the Indian government limited the number of cars manufactured and heavily taxed the imports of automobiles into the country in order to limit automobile ownership and reduce dependence on imported oil.

However, in many developing nations, knowing that automobile manufacture is an engine for industrial growth, policy makers have pursued a strategy of encouraging domestic motor vehicle production with an eye on export markets to earn foreign exchange. South Korea has been a leader in producing motor vehicles for the US market. Malaysia and India are following the same route. Multi

Table 3. Urban car ownership, 1970 and 1980

	1970		1980	
	Population (000)	Cars per '000 persons	Population ('000)	Cars per '000 persons
Seoul	5536	6.3	8366	15
Calcutta	7402	13.0	9500	10
Bombay	5792	13.5	8500	21
Jakarta	4312	18.0	6700	33
Bangkok	3090	49.7	5154	71
Sao Paulo	8400	62.3	12800	151
Mexico City	8600	78.3	15056	105

Source: World Bank, *Urban Transport*, Washington, DC, 1975, 1986.

[1] The advantages of trucks over rail include: shorter time required for transporting goods, less bureaucratic paperwork, less rigid transport schedules, and the convenience of door-to-door delivery. Further, the price of transporting goods encourages the use of trucks over longer distances.

nationals use Mexico and Brazil as a base for export to the United States as well.

2.4 Mass transport: the failure of infrastructure to keep pace

In many cities, the number of registered taxis surpasses the number of buses. For example, Hong Kong and Bangkok each have more than 17,000 taxis versus just over 7000 buses in Hong Kong and about 4000 in Bangkok. However, because of their larger passenger carrying capacity, buses transport far more passengers than do taxis. The wealthier the city, the greater the predominance of taxis; the proportion of taxi to bus passengers is higher in Hong Kong than in Beijing, where much of the taxi ridership is made up of visitors. Surveys in Pakistan show a wide variety of vehicles in comparatively wealthy Karachi, but no registered taxis in Peshawar (Al-Haque, 1983).

Where longer city distances are involved, buses and subways can be the primary mass transport modes. Several developing world cities have built modern subway systems in recent years. These systems have alleviated much of the pressure previously placed on surface road systems. Hong Kong's Mass Transit Railway (MTR) carried about 1.9 million riders daily in 1989, compared to 3.6 million carried by buses. Bus ridership has remained unchanged since the early 1980s, but MTR ridership has doubled since 1982. Similarly, the Light Rail Transit (LRT) in Manila carried 300,000 riders in 1988 (Gimenez, 1989). Traffic congestion along the LRT route has decreased, thus reducing car travel time along the route from 45 to 32 minutes.

Those, unfortunately, are exceptions. In general, the growth of urban mass transportation infrastructure has not kept pace with rising urban incomes. As a result, municipal services — particularly transport services — have been stretched to serve a much larger population than they were originally intended for. Public vehicles are overcrowded, and congestion has increased travel time several-fold in major cities.

Table 4. Annual growth in number of cars, 1970-1980

City	AAGR (%)
Abidjan	10.0
Bangkok	7.9
Bombay	6.1
Buenos Aires	10.0
Cairo	17.0
Calcutta	5.6
Hong Kong	7.4
Jakarta	9.8
London	2.6
Sao Paulo	7.8
Singapore	6.8
Stockholm	3.0
Stuttgart	2.5
Tokyo	2.5

AAGR = Average annual growth rate. Source: World Bank, *Urban Transport*, Washington, DC, 1986.

3 The Deterioration of Air Quality, and Its Links to Transport

In many developing nations, concerns about local emissions take priority over the global accumulation of greenhouse gases, because the repercussions of poor local air quality (smog, health problems, etc.) pose a more immediate and tangible threat. The local ambient air quality in Bangkok, Jakarta, Manila and Singapore often fails to meet these nations' standards. Recent monitoring found one-hour-duration average CO levels at 95 parts per million in Jakarta and 49 ppm in Haad Yai, a city in Southern Thailand. High levels of suspended particulate matters (SPM), carbon monoxide (CO) and lead are largely attributed to motor vehicle use (Kiravanich et al., 1986).

In Thailand, the Office of the National Environment Board has monitored levels of carbon monoxide, particulate matter and lead near major roads in Bangkok since 1984. In areas of the city where traffic is heaviest, all of these exceed the air quality standards on a daily basis (National Environment Board of Thailand, 1990). In Hong Kong, almost 2 million inhabitants are regularly exposed to unacceptably high levels of sulphur dioxide and nitrogen dioxide. Asthma, bronchitis, and lung cancer are common (Hong Kong EPA, 1989). Hundreds of millions of dollars, concluded the Hong Kong Environmental Protection Agency, are "spent every year on combating air pollution or paying to rectify its effects." The expenditures range, they said, from the obvious (maintaining an air control staff) to the indirect (the costs of cleaning buildings and clothes more frequently, replacing corroded materials, and maintaining hospital beds). Even in the United States, where air quality is generally higher, the American Lung Association concluded that auto and truck exhaust causes between $4.4 and $93.5 billion per year in health costs (American Lung Association, 1990).

In the US, the average car consumes about 500 gallons of petrol per year. The carbon released as CO_2 from this annual fuel use weighs approximately 2700 pounds. This is equal to the weight of a typical mid-size American sedan.[2]

Carbon dioxide emissions from motor vehicles in the developing world presently account for about 30% of world wide vehicle releases. Developing nation vehicle emissions levels are expanding at a rate of about 3.5% per year, accounting for about 45% of the global increase (see Tables 5 and 6 for details).

Based on current trends, the share of global carbon emissions that arise from motor vehicles in developing countries will increase further. In addition, as the developing world's vehicle population grows — with only modest, if

[2] This example comes from research conducted under the auspices of the Stockholm Environment Institute.

any, pollution controls — emissions of CO, HC, NOx and other toxic compounds will inevitably increase as well (World Resources Institute, 1988). With the fundamental economic reforms now taking place in Eastern Europe, industrial modernization and economic growth could lead to significant additional carbon dioxide increases there. For the past 15 years, Eastern European motor vehicle oil consumption per capita has averaged less than half of Western European levels. If the consumption there rose to match Western Europe, per capita, then global carbon dioxide emissions would be about 90 million tonnes greater per year — an amount equivalent to about half of all US auto emissions.

4 Technical Opportunities for Improvement

The problems mentioned so far are unnecessarily exacerbated by the low fuel efficiencies typical of most developing world vehicles. Low efficiencies stem from a range of factors: the use of outdated vehicles, poor maintenance, unwise modal choices, high levels of urban congestion and dilapidated or insufficient transport infrastructures. All of these can (and should) be changed.

4.1 Improving vehicle fuel efficiency

The fuel efficiency of vehicles has improved rapidly over the last decade, bolstered in part by the international nature of motor vehicle design and manufacturing. A wide range of technologies that are market-ready, or nearly so, can significantly reduce or even reverse current trends of high vehicle fuel consumption and local air pollution (Bleviss, 1988; World Resources Institute, 1990).

In 1985, the Toyota Corporation tested a lightweight prototype car, dubbed the AXV (for Advanced Experimental Vehicle), that features low aerodynamic drag, a direct-injection diesel engine, a continuously variable transmission and other well understood design features. The car received an EPA fuel economy rating of 98 miles per gallon (mpg) when tested according to the US EPA test procedure (Bleviss, 1988). Similarly, the Volvo LCP 2000 (for Light Component Project), also dating from 1985, gets 65 mpg in mixed driving and up to 100 mpg at a constant 40 miles per hour (mph). Powered by a three-cylinder direct-injected diesel engine, it can accelerate to over 60 mph in 11 seconds. These vehicles are not yet considered ready for commercial production, and may not meet all current US safety and emission requirements, but they demonstrate the feasibility of increasing fuel efficiencies through new technologies.

Government trade policy, by encouraging innovation, can spur fuel efficiency. When India and China began to permit joint ventures with foreign manufacturers, those countries began to produce cars of modern vintage, with a fuel intensity much lower than any cars produced earlier in those countries. Several other Asian car-manufacturing countries have begun to make cars whose fuel efficiency and design approaches the cars of industrialized countries.

Ledbetter and Ross (1990) estimate that if average automobile size and performance were held constant at 1987 levels, new US car fuel economy could be improved to 43.8 miles per gallon, at an average cost of 53 cents per gallon, solely through vehicle design improvements that are readily available. The potential for such improvement is highest in countries where the engines are of older vintage and there is local manufacture and/or assembly of cars. In trucks, switching to more efficient diesel engines and motorcycles to four-stroke engines would significantly reduce transport oil use (Energy and Environmental Analysis, Inc.). However, the potential for improving diesel fuel efficiency is limited, compared to that for petrol engines. This is a particular disadvantage for Asian countries, where a large fraction of the oil is used for diesel trucks and buses.

4.2 Technological improvements in emissions control

The level of tailpipe hydrocarbon emissions is primarily a function of the engine-out emissions and the overall conversion efficiency of the catalyst — both of which depend on the proper functioning of the fuel and ignition systems. There are

Table 5. The extent of carbon dioxide emissions

Average worldwide vehicular carbon dioxide emissions:	
1971:	510 million tonnes carbon
1987:	830 million tonnes carbon
Annual increase, 1971-87:	3%
Current annual increase in carbon dioxide emissions	
OECD nations:	10.4 million tonnes carbon
All other nations:	8.7 million tonnes carbon
Total:	19.0 million tonnes carbon
Relative contribution of motor vehicles to CO_2 emissions worldwide:	
1971:	12%
1985:	14%
Evidence of fuel efficiency increase, from 1971 to 1985:	
Increase in motor vehicle CO_2 emissions:	63%
Increase in motor vehicles:	87%
Rank of groups of nations in amount of CO_2 emissions:	
OECD nations	
Eastern Europe (former Communist bloc)	
Latin America	
Asia	
The Middle East	
Africa	

Source: OECD data contained in the following reports: International Energy Agency, *Energy Balances of OECD Countries, 1970-1985*, Organization for Economic Co-operation and Development, Paris, 1987; International Energy Agency, *World Energy Statistics and Balances, 1971-1987*, Organization for International Co-operation and Development, Paris, 1989; International Energy Agency, *Energy Balances of OECD Countries, 1986-1987*, Organization for International Co-operation and Development, Paris, 1989. For each country, the annual values of oil consumed in road transport were used to calculate direct carbon dioxide emissions from motor vehicles. Indirect emissions, from the production, processing and transport of motor vehicle fuels, are not included here.

Table 6. The impact of motor vehicles on the environment

	Vehicle Impact on Emission
Carbon dioxide (CO_2)	19 pounds into the atmosphere per gallon[a] 300 pounds per 15-gallon fillup 14% of the world's CO_2 emissions from fossil fuel burning from motor vehicles
Tropospheric ozone	Although ozone in the lower atmosphere does not emanate directly from motor vehicles,they are the major source of the ozone precursors: hydrocarbons and nitrogen oxides
Carbon monoxide (CO)	Concentrations in the lower atmosphere increase by 0.8-1.4% per year.[b] 66% of OECD country emissions (78 million tonnes) from motor vehicles in 1980.[c] 67% of US emissions from transportation in 1988[d]
Nitrogen oxides (NO_x)	47% of OECD country emissions (36 million tonnes) from motor vehicles in 1980.[c]
Hydrocarbon compounds (HC)	39% of OECD country emissions (13 million tonnes) from motor vehicles in 1980.[c]
Chlorofluorocarbons (CFCs)	54.1 tonnes consumed by US mobile air conditioners annually[e] 35.6 tonnes consumed in the US annually through leakage, service venting, or accidents[e]
Diesel particulate (Tiny carbon particles, hazardous to respiratory tract, visibility, and as a possible carcinogen)[f]	No overall measurements Diesel engines emit 30-70 times more particulate than petrol-fuelled engines.
Lead	90% of airborne lead from petrol vehicles.
Lead scavengers Additives to remove lead; some (notably ethylene dibromide) may be carcinogenic.[f]	Significant amount emitted.
Aldehydes (incl. formaldehyde)	Exhaust emissions correlate with hydrocarbon (HC) emissions Diesel engines produce a higher percentage
Benzene (identified as carcinogen)	Present in both exhaust and evaporative emissions; 70% of the total benzene emissions in the US come from vehicles
Non-diesel organics	Smaller amount per vehicle, but more mutagenic overall, than diesel particles.
Asbestos	Used in brake linings, clutch facings and automatic transmissions. About 22% of the total asbestos used in the US in 1984 was used in motor vehicles.[f]
Metals	US EPA has identified mobile sources as significant contributors to nationwide metals inventories, including 1.4% of beryllium and 8.0% of nickel. Arsenic, manganese, cadmium and chromium may also be mobile source pollutants. High-risk hexavalent chromium which does not appear to be prevalent in mobile source emissions.

Notes

[a] This figure refers to direct tail-pipe emissions only. Transportation, refining and distribution account for perhaps 15 to 20% of total emissions.

[b] Khalil, M.A.K. and R.A. Rasmussen, "Carbon Monoxide in the Earth's Atmosphere: Indications of a Global Increase," *Nature*, **332**, (245), March 1988.

[c] Organization for Economic Cooperation and Development, *OECD Environmental Data,* Paris, 1987.

[d] US Environmental Protection Agency, *National Air Quality and Emissions Trends Report 1988, op cit*, Ref. 76, p. 56.

[e] US Environmental Protection Agency, *Regulatory Impact Analysis: Protection of Stratospheric Ozone*, Washington, DC, December, 1987.

[f] Carhart, B. and M. Walsh, *Potential Contributions to Ambient Concentrations of Air Toxics by Mobile Sources, Part 1,* paper presented at the 80th Annual Meeting of Air Pollution Control Association, New York, June 24, 1987.

many technological improvements, which are currently becoming widespread or are on the horizon, that make more stringent control of HC and CO feasible. These advances are expected not only to reduce the emissions levels that can be achieved in the certification of new vehicles, but also to reduce the deterioration of vehicle emissions over the life of the car.

First among these advances is the trend toward increased use of fuel injection, spurred in part by better driveability. Fuel injection has several distinct advantages over carburetion: more precise control of fuel metering (which in turn means more efficient burning of petrol), better compatibility with digital electronics, better fuel economy and better cold-start function (much of the HC and CO emitted by vehicles is generated during cold-start, when the fuel system is operating in a rich mode and the catalyst has not yet reached its lightoff temperature). Carburettor choke valves, long considered a target for maladjustment and tampering, are being replaced by more reliable cold-start enrichment systems in fuel-injected vehicles.

Second, improvements to the fuel control and ignition systems — such as maintaining a more balanced air/fuel ratio under all operating conditions and minimizing the occurrence of spark plug misfire — will result in better overall catalyst conversion efficiency and less opportunity for engine failure (which is also a source of extra emissions).

Third, a trend that bodes well for catalyst deterioration rates (and therefore in-use emissions) is the US Environmental Protection Agency's lead phaseout, which reduced the lead content of petrol by about 90% (to 0.1 g/gal) beginning in January 1986. The new Clean Air Act in the United States will actually lead to a complete ban on lead in petrol in a few years.

Finally, there are alternative catalyst configurations that will probably be used in the future to meet lower emission standards. It is likely that dual-bed catalysts will be phased out over time, but a warm-up catalyst could be used for cold-start hydrocarbon control. Additional advances include the optimization of combustion chamber geometry and turbulence levels in an effort to minimize emissions and maximize fuel economy, and the development of new ignition systems (now being tested by SAAB and Nissan), which may virtually eliminate misfires and the need for high-voltage spark plug wires.

4.2.1 Two-stroke engines

Two-stroke engines are common in many of the two-wheeled vehicles so prevalent throughout the developing world. These engines pose a special challenge for emissions control. Generally, most measures for four-stroke engines can be applied to two-strokes. But in addition, since much of the HC emitted from two-strokes comes from their different method of adding lubricating oil, these engines deserve special attention.

Relying on engine oils containing polyisobutylene and maintaining leaner fuel/oil ratios reduces visible smoke. Hence, use of separated lubrication systems helps limit the amount of smoke produced (Sugiura and Kagaya, 1977). Another method is to increase the use of mopeds with catalysts, which have been available in Switzerland and Austria since 1986 (Laimbock and Landerl, 1990). A range of new technologies are now available which can eliminate the historical problems of high smoke and unburned hydrocarbons from two-stroke technologies (Wyczalek, 1991).

4.2.2 Diesel-fuelled vehicles

Diesel trucks and buses have been receiving greater attention as significant sources of particulates and NO_x. Diesel vehicle standards recently adopted in the US will foster technological developments similar to those which have already occurred for petrol cars.

Smoke emissions from diesel engines, composed primarily of unburned carbon particles, usually occur when there is an excess amount of fuel available for combustion. In addition, failing to clean or replace a dirty air cleaner may produce high smoke emissions because it can choke off available air to the engine, resulting in a lower than optimum air/fuel mixture. Vehicle operation also plays an important role: selecting the proper transmission gear (and thereby keeping the engine operating at the most efficient speed), moderate accelerations, lower highway cruising speed changes, and reduced hill-climbing speeds all minimize smoke emissions.

Basic approaches to diesel engine emission control fall into three major categories:

1. *Engine modifications*, including combustion chamber configuration and design, fuel injection timing and pattern, turbocharging and EGR;

2. *Exhaust aftertreatment*, including traps, trap oxidizers and catalysts; and

3. *Fuel modifications*, including control of fuel properties, fuel additives, alternative fuels.

4.3 The lower-case of alternative fuels

Oil (diesel and petrol) fuels almost 100% of road transport, but this is not a result of the innate superiority of this fuel. The type of fuel used by a vehicle - whether petrol, diesel, liquid petroleum gas (LPG), compressed natural gas (CNG) or alcohol - is a function of the availability of fuel and vehicles, the relative fuel price differential and the investment required to convert from one fuel to another.

4.3.1 Natural gas

Clean, cheap and abundant in many parts of the world, natural gas already plays a significant vehicular role in a number of countries (Sathaye et al., 1989). While many qualities of natural gas make it a desirable alternative fuel source, its gaseous form at normal temperatures poses barriers to its widespread application.

Pipeline-quality natural gas is made from a mixture of several different gases; methane typically makes up 90-95% of the total volume. Methane is a nearly ideal fuel for Otto cycle (spark ignition) engines and leads to greater efficiency and power output. On the other hand, its high flame temperature tends to result in high NO_x emissions, unless very clean mixtures are used.

Because of its gaseous form and poor self-ignition qualities, methane is a poor fuel for diesel engines. Since diesels are generally somewhat more efficient than Otto cycle engines, natural gas engines are likely to use about 10% more fuel — albeit cleaner-burning fuel — than the diesel they replace.

4.3.2 Methanol

Methanol has many desirable combustion and emissions characteristics, including lean combustion, low flame temperature (leading to low NO_x emissions) and low photochemical reactivity. As a liquid, methanol can either be burned in an Otto cycle engine or injected into a diesel cylinder. Methanol burns with a sootless flame and contains no heavy hydrocarbons. As a result, particulate emissions from methanol engines are also very low — consisting essentially of a small amount of unburned lubricating oil. Methanol's CO emissions are generally comparable to, or somewhat greater than, those from a diesel engine.

The major pollution problems with methanol engines stem from emissions of unburned fuel and formaldehyde. Formaldehyde is a powerful irritant and suspected carcinogen; it also displays very high photochemical reactivity. While all combustion engines produce some formaldehyde, some early generation methanol engines exhibited greatly increased emissions compared to diesels. The potential for large increases in formaldehyde emissions with the widespread use of methanol vehicles has raised considerable concern about what would otherwise be a very benign fuel from an environmental standpoint.

But this may be a solvable problem. Formaldehyde emissions can be reduced through changes in combustion chamber and injection system design, and are also readily controllable through the use of catalytic converters under warm conditions. Recent efforts to reduce aldehydes by Detroit Diesel have shown dramatic gains (Walsh and Bradow, 1991).

If methanol is made from natural gas, its contribution to global warming is about the same as petroleum; if made from coal, however, the global warming impact could be double that of oil (Walsh, 1989).

4.3.3 Liquefied petroleum gas

Liquefied petroleum gas is already widely used as a vehicle fuel in the US, Canada, the Netherlands and elsewhere. As a fuel for spark ignition engines, it has many of the same advantages as natural gas, with the additional advantage of being easier to carry aboard the vehicle. Its major disadvantage is the limited supply, which would rule out any large-scale conversion to LPG fuel.

Occasionally, environmental concerns, aided by lower LPG prices relative to petrol prices, can encourage the widespread substitution of LPG for petrol. As a result of such a measure implemented by the Seoul government, the number of LPG-fuelled motor vehicles (mostly taxis) in Seoul increased more than four-fold between 1979 to 1983.

4.4 *Improving fuel quality*

The lack of consistent fuel quality is an important restraint on the production of efficient engines. High-quality fuels are necessary to use fuel injection in place of Carburettors. In Brazil, fuel quality is so inconsistent that it precludes manufacturers from producing fuel injected engines.

Improving fuel quality also has the advantage of reducing pollutants from any engine, particularly, lead in petrol and fuel oil fractions in diesel.

4.4.1 Petrol

Throughout much of the industrialized world, unleaded fuel has been the norm for more than a decade. Japan has been the world leader in this regard; for more than a decade now, Japan has maintained a 90% unleaded share in its petrol mix. Developing nations still rely heavily on leaded fuels, however, which leads to heavy emissions of lead. Thus, reducing lead in petrol or using unleaded petrol would lead to a significant improvement in the lead pollution levels in much of the developing world. Where leaded fuel cannot be eliminated, standards should be enacted to require that the lead content does not surpass 0.015 grams per litre.

4.4.2 Diesel fuel

Modifications to diesel fuel composition have also drawn considerable attention as a quick and cost-effective means of reducing emissions from existing vehicles. Two fuel modifications show the most promise. The first, reducing the sulphur content, reduces the indirect formation of sulphate particles from SO_2 in the atmosphere. Recently, the US Environmental Protection Agency (EPA) decided to reduce the sulphur content in diesel fuel to a maximum of 0.05% by weight. This low-sulphur fuel will soon be introduced in Western Europe and Japan. The adoption of similar standards in developing nations could mitigate the direct and indirect adverse effects of sulphur emissions.

The second measure, a reduction in the aromatic hydrocarbon content of diesel fuel, can also help to reduce emissions, especially where fuel aromatic levels are high. For existing diesel engines, a reduction in aromatics from 35% to 20% by volume can reduce transient particulate emissions by 10 to 15% and NO_x emissions by 5 to 10%. HC emissions, and possibly the mutagenic activity of the particulate SO_x, would also be reduced.

4.4.3 Fuel additives

A number of well-controlled studies have demonstrated the ability of detergent additives in diesel fuel to prevent and remove injector tip deposits, thus reducing smoke levels (Walsh and Bradow, 1991). The reduced smoke probably results in reduced particulate matter emissions as well, but this has not been demonstrated as clearly, owing to the great expense of particulate emissions tests on in use vehicles. Cetane-improving additives are also likely to result in some reduction in hydrocarbon (HC) and particulate emissions in marginal fuels.

4.5 Smaller car and engine sizes

Fuel efficiency levels differ substantially from one developing nation to another. Cars and motorcycles tend to be less energy—intensive in the Asian countries than in the Latin American ones. Both car and engine sizes are smaller in China, India and Korea compared to Mexico, Venezuela and the GCC countries. Because of the large size of cars and congestion in Lagos, Nigeria, petrol consumption per car is twice that in Asian countries today. Fuel efficiency is low in Ghana and Sierra Leone because cars of older vintage are imported and maintained for a longer period.

Hence, the potential for savings is large through simply replacing older, larger cars with newer, more efficient ones, particularly if the new ones can be sold inexpensively.

5 The Role of Government Policy

The future of rapidly increasing carbon dioxide emissions (see Box 2, page 214) highlights the urgent need for government policies aimed at mitigating the growth of all types of transport-related emissions. While improved technology can offer opportunities, confronting a problem of this magnitude requires policies aimed both at implementing better technologies, and at controlling the growth of transport-related carbon emissions through other means.

Despite technological advances, many developing countries have been unable so far to make significant progress in reducing vehicle emissions. A central barrier continues to be the lack of necessary government support for carbon-abating measures and policies. In most developing nations, the government wields almost exclusive power, and its objectives and constraints guide the development of transportation. Hence, government policies almost entirely determine the extent of progress in reducing petrol lead content, introducing catalytic converters in petrol-fuelled cars, lowering motorcycle emissions and introducing particulate controls.

In addition, the construction of infrastructure to support transport activity — the building of airports, roads, waterways, national highways and rail tracks — is the purview of local and national governments in each country. Local and national governments will only make strides towards reducing emissions from transport when they recognize its importance and make it a goal of top priority.

5.1 Government priorities and objectives

A wide range of competing priorities often shift the focus of government transport policies away from reducing fuel consumption and emissions. For example, reducing emissions may clash with the perceived need to provide more roads to support more auto and truck traffic. When policy makers consider fuel reduction policies, they often implement these measures on a piecemeal basis, which backfires upon them.

For example, in the late 1970s the Philippine Government increased taxes on petrol fuels, while keeping the price of diesel fuel much lower. As a result, many owners of jeepneys replaced their petrol engines with diesel fuel engines. This shift led to increases in diesel demand, which domestic refineries were unable to meet. In the end, the Philippines experienced an unintended shortage of diesel fuel, which it had to import at high prices, which in turn added unexpectedly to that country's overall fuel cost (Sathaye and Meyers, 1986b).

In many developing nations, equity and employment objectives take precedence over other goals including the protection of environmental quality and the efficient allocation of resources. For example, the development of labour-intensive transport modes, particularly in regions where jobs are scarce, has received higher priority in India. Building roads employs 25 times the number of person-years per rupee as building railroads, because of the labour-intensive manufacture and operation of motorized and animal-driven vehicles, and the heavy manpower requirements of road construction (Sathaye and Bhatia, 1991). The majority of these jobs are in the private sector, which thus provides a way to stimulate the national economy without burdening the government's own resources.

There are many options with which governments may improve the efficiency of fuel use. These range from large-scale policies aimed at relocating communities or changing traffic patterns, to small-scale measures focused on improving the fuel efficiency of individual engines. Emissions per kilometre driven can also be lowered by mandating changes in the driving itself — such as average speed limits or degree of acceleration permitted in vehicles.

An effective overall vehicle pollution control strategy must aim both to reduce emissions per kilometre driven, and reduce the total amount of driving. But the strategy's first priority should be restraining future vehicle growth rates, ideally without restraining mobility or economic growth. Economic measures, physical restrictions and selective policies should each play a role. Recent studies indicate that this effort should also incorporate high taxes on vehicle use and tough controls on land use which, in turn, promote more dense urban development (Pucher, 1988). However, even if overall vehicle growth rates could be constrained to only 5% per year — an optimistic outlook for many rapidly industrializing countries — vehicle emissions would explode over the next 15 years. Thus, in addition to growth restraint, a series of additional measures are necessary.

5.2 *Infrastructure development*

5.2.1 Road quality

Road quality is an important factor in levels of fuel use. In Bogota, Columbia, fuel consumption was reported to be 25% higher for a vehicle moving on a gravel or earth surface than on an asphalt pavement (United Nations, 1986). A poorly maintained asphalt pavement would result in higher fuel consumption with a significant loss to a national economy.

Governments often short-change road maintenance. Generally, maintenance of airports, railways and ports is funded through revenues generated from their operation. Road maintenance is financed by a variety of taxation measures including: fuel taxes, road user fees, and vehicle taxes. Tax avoidance and evasion are commonplace in most developing nations, however fuel taxes appear to be collected in full. For the other taxes and fees, collection typically ranges between 40 and 60% (Heggie and John, 1989).

A widely quoted World Bank study has estimated that in 85 countries with a main road network of 1.8 million kilometres, a quarter of the paved roads outside urban areas need to be rebuilt (World Bank, 1988). The $US 45 billion needed for rebuilding these roads is three to five times greater than the fee would be for maintenance. Furthermore, insufficient road maintenance increases the costs of operating vehicles by 10-40% on paved roads, and by 30-60% on unimproved roads (Heggie and John, 1989). A sound government policy would be to give maintenance a higher priority over new construction by making it more glamorous and attractive. Also, the identification and implementation of the most effective taxation measures can raise needed funding for transport infrastructure, and eliminate more futile fund-raising efforts.

Financing light rail or subway system in developing country cities encounters benefits and costs similar to those in industrialized countries. None of the systems built to date can repay the invested capital through passenger revenues. In many instances, operating expenses are barely recovered through such revenue. On the other hand, these systems reduce air pollution, save time, and increase travel comfort on heavily travelled routes. Such benefits may well outweigh the higher expense of dedicated light rail systems.

5.2.2 Foreign exchange

This is another important consideration in transport infrastructure development. The efficient provision and operation of public transportation is not only affected by general shortages of fiscal revenues, but also by specific shortages of foreign exchange (Heggie and John, 1989). If much of the infrastructure is indigenous, as is likely to be the case with both rail and road, about the same proportion of capital cost will be devoted to foreign exchange. Long lead times required to obtain imported spare parts often idle much too large a fraction of public transport in developing countries. International donor agencies can assist in this regard by focusing on spare parts more than on the acquisition of new vehicles or infrastructure improvement.

5.3 *Fuel efficiency in public vehicles*

Buses are operated by either private or public enterprises; their relative efficiency and quality of service differs accordingly. Generally, publicly owned transport systems tend to be more expensive and less efficient, and in many instances require operational subsidies. On the other hand, private bus operators often maintain profitability with scant regard to labour wage laws and safety. On the same routes, private operators may incur as little as half the expense of their public counterparts.

Fuel costs as a percentage of total operating costs can be surprisingly high for bus transport systems. For the Bangkok Metropolitan Transport Authority (BMTA), the operating total cost was 11.2 baht per kilometre in 1984. Of this total, 2.5 baht, or 22%, was spent on fuel purchases (United Nations, 1986). But in Peshawar, bus operators spent 66% of their expenses on fuel and lubricants, while rickshaw operators spend 81% (Asian Development Bank, 1989). Operators are thus very sensitive to changes in fuel prices.

5.4 *Fuel efficiency legislation*

By deliberately rewarding (or mandating) the purchase and use of highly energy-efficient vehicles, governments could help sustain markets for them. If controlling emissions and the impact of transportation on climate change are taken seriously, then this sort of help may be necessary. Over the past few years, for example, the fuel efficiency of new cars in Japan, Western Europe and the United States has started to decline, in large part because of the remergence of the "horsepower wars" of the 1950s and 1960s. In an age of relatively stable oil prices, performance sells vehicles better than fuel efficiency.

Government measures could include gas-guzzler/gas-sipper taxes on new vehicles, annual registration fees on all motor vehicles graduated according to fuel efficiency, mandatory fuel efficiency standards and carbon taxes on fossil fuels. Broad-based carbon taxes, already under consideration by several European countries, are especially attractive because they begin to make *all* fossil fuel prices reflect the climate risks and other environmental costs associated with each fuel. We have seen, during the 1970s, that high fuel costs can play a significant role in encouraging more fuel-efficient vehicles. A carbon tax would be relatively easy to administer and could be adjusted upward or downward as more information on climate becomes available. It would encourage fuel users in all sectors of the economy, not just commuters and other drivers, to use energy more efficiently. It would also encourage the development and use of non-fossil energy sources. The funds raised could be used to mitigate the impacts of global warming.

A novel idea, which has not been pursued before, would be to regulate the manufacture and import of fuel-efficient vehicles by requiring the preparation of an *energy-efficiency impact statement*. Like an environmental impact statement (as used in the United States), it would require vehicle manufacturers to evaluate energy consumption of alternative vehicle-manufacturing plants, *including the fuel efficiency of the vehicle,* in order to justify that the plant being proposed for construction would be more fuel efficient.[3] (Sathaye and Gadgil, 1992).

5.5 Support for alternative fuels

The well-known use of alcohol cars in Brazil provides an interesting example of government policy towards the use of alternative fuels. Brazil has had a chequered history of alcohol fuel development programmes (Sathaye et al., 1989). The purchase of new alcohol cars, encouraged by government policy, increased from virtually zero in 1979 to almost 100% in the mid-1980s. However, the alcohol shortage that emerged in the late 1980s caused many consumers to shy away from the cars. In 1990, more than 90% of the cars sold in Brazil ran on petrol. Maintaining consumer interest in alternative fuels requires steady government policies — a major challenge in light of the frequent shifts in government priorities.

None the less, such policies could have valuable impact. For the last decade, the possibility of substituting cleaner-burning alternative fuels for diesel fuel has drawn increasing attention. The proposed alternative fuels are natural gas, methanol made from natural gas and, in limited applications, LPG. Potential benefits include conservation of oil products and energy security, as well as the reduction or elimination of particulate emissions and visible smoke.

5.6 Reduction of traffic congestion

Vehicle congestion is an important contributor to inefficient fuel use and higher emissions. A 1984 survey in Thailand, for example, showed that cars with equivalent engine sizes used 10% more fuel in Bangkok than in the rest of the country (Government of Thailand, 1989). Congestion has worsened in Bangkok since then, and average speeds have declined rapidly over the last 3 years. Fuel efficiencies are estimated to have deteriorated further. Similar patterns have been observed in Jakarta, where fuel consumption is 30% higher due to traffic congestion, amounting to $133 million of unnecessary oil use[4] (Tashrif, 1991). In effect, this money was spent on oil instead of on transport alternatives which could have reduced congestion and improved traffic flow.

Why does congestion have such an impact? Many of the findings of an analysis done in Teheran during the 1970s are still valid today (Teheran Development Council Secretariat, 1976). At that time, the Iranian capital had severe traffic congestion caused by inadequate supply of public transport facilities, substantial money-and-time subsidies to car users, inadequate traffic policing, engineering and control, a culture of unbridled driver-and-pedestrian conflict, and bad road surface conditions. The analysis noted a relationship between speeds and fuel intensity (Figure 5): at 15 kilometres per hour, vehicle fuel use equaled 0.20 litres/km, while at 30 km/h it dropped to 0.12 litres/km. The report recommended several measures to reduce congestion: better policing and police training, special parking zones, area-licensing schemes during peak hours, and re-routing buses.

If left alone, congestion will mushroom, because it and the increasing acquisition of private vehicles reinforce each other. Most motorists, given the choice, will eagerly opt to ride in an air-conditioned car rather than a steaming bus which typically takes the same amount of time as a car to reach its destination. Only a few cities have broken this cycle, and then by taking steps before the congestion problem reached crisis proportions.

Those steps usually mean placing mandatory restrictions on traffic flow. Constructing more roads generally has not worked, since the latent demand for road use is so high that any additional space is immediately occupied by more vehicles. A more recent study for Bangkok, for example, concluded that, unless accommodated by effective pollution control policies and demand management efforts, expanding the transport infrastructure in central Bangkok would only serve to spread existing congestion over a larger area and substantially raise existing levels of air pollution (Thailand Development Research Institute, 1990).

5.6.1 Changes in city traffic flow patterns

These improvements have been much more successful at alleviating congestion and increasing traffic speeds. Successful measures include conversion to one-way streets, timed signals, lane controls to segregate traffic, limiting parking spaces, and timed lane entry. Congestion in Singapore and Hong Kong, for example, could be far worse had the transportation system not been tightly managed in these cities by their governments. In Singapore, to reduce congestion in the central business district, the government has adopted the systematic development of housing estates and roads, given priority to public transportation and carefully managed the growth and use of private vehicles through higher taxes, user fees and petrol taxes (Ang, 1989). These measures have held car saturation to 90 cars per 1000 inhabitants. Traffic flows freely in this densely populated city.

In another example, Shanghai has banned motor vehicles registered elsewhere and trucks over eight tonnes from using city roads between 7 a.m. and 7 p.m. Beijing has mandated staggered work hours, to spread traffic over longer periods of the day.

5.6.2 Discouraging private cars

In Hong Kong, the government has applied measures to reduce the use of private cars. The number of private cars licensed had grown at a rate of 14% per year between 1977

[3] For a detailed elaboration of this concept see Sathaye and Gadgil, 1992.

[4] Personal communication with Muhammad Tashrif, Center for Research on Energy, Institute of Technology, Bandung, Indonesia, 6 February 1991.

Figure 5. The relationship between speed and fuel intensity, from an analysis conducted in Teheran during the early 1970s. It suggests that reducing congestion will lead to significant reductions in fuel use. Source: Hansen, 1976.

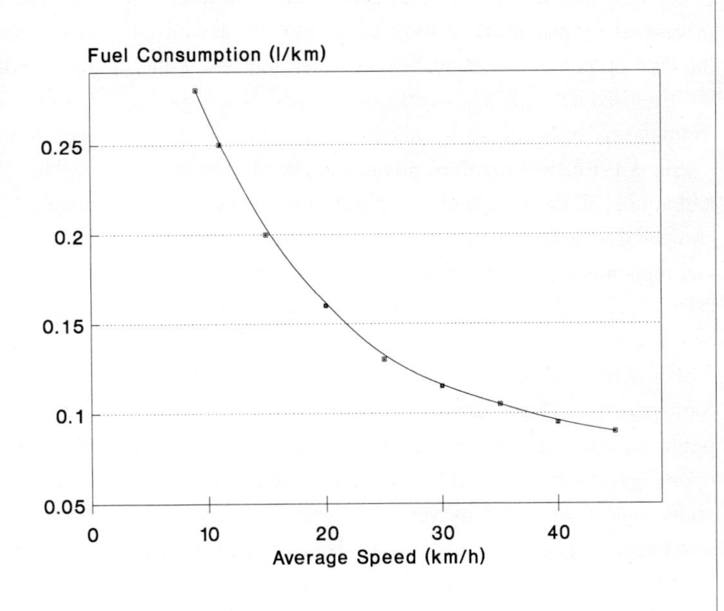

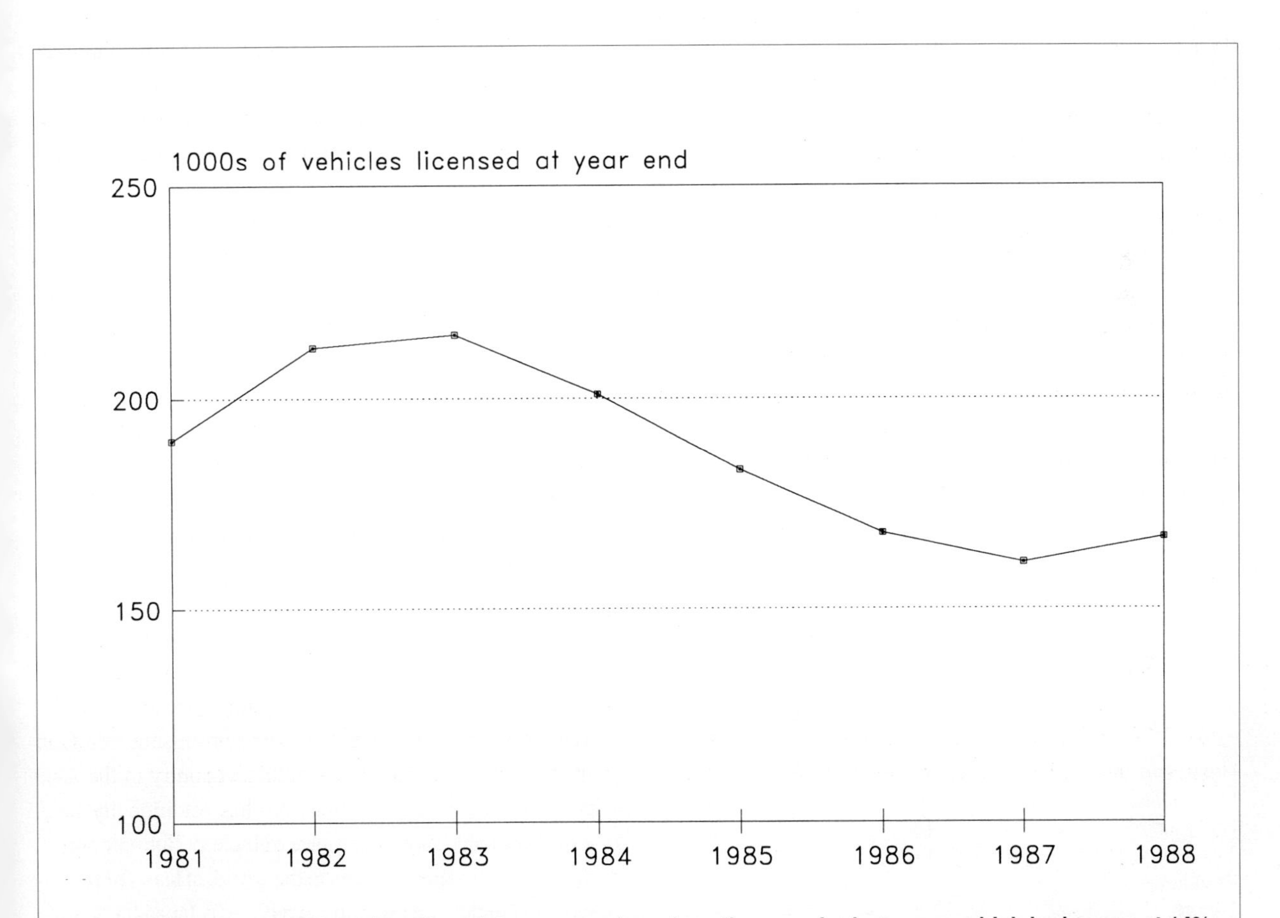

Figure 6. The results of Hong Kong's strict measures to reduce the use of private cars, which had grown at 14% per year between 1977 and 1981. In 1982, to reduce congestion, the Hong Kong government dramatically raised automobile taxes, license fees, and petrol duties. Source: Parker, 1987.

and 1981, leading to growing traffic congestion. In 1982, Hong Kong doubled the first registration tax on new and used private cars; tripled the annual vehicle license fee; and raised the duty on petrol. The annual licence fees and petrol duty were revised upwards again in 1983. The results are shown in Figure 6.

Since 1987, the growth of private cars has begun anew, and almost 30 thousand more cars were registered in 1989 compared to 1986. The government increased various fuel and registration taxes in 1991 to slow the growth of private cars.

5.6.3 *Effective use of mass transit*

Singapore and Hong Kong also offer low-cost, efficient public transport services. Singapore's MRT system has been operational since 1987; half of the city's population lives within one kilometre of its route. In Hong Kong, the auto-restriction measures took place alongside the completion of new metro-rail facilities. Thus, the percentage of passengers carried by metro-rail increased from 13% to 25% between 1980 and 1985, while the percentage using private cars declined from 8% to 5%. (However, as Table 7 shows, because there was an increase in income as well, the percentage of consumers using taxis rose significantly.)

In general, personal motorized vehicles consume several times the amount of fuel, per passenger-kilometre travelled, than buses and subways (Sathaye and Meyers, 1986a). Automobiles consume up to six times the fuel per pass-km as buses. One key feature is efficient intermodal transfer of passengers: easy, accessible, inexpensive points of interchange between foot traffic, bicycles, automobiles, rails, and public transit can help to reduce traffic delays and to smooth the flow of traffic.

5.6.4 *Encouraging shorter journeys*

Increases in the average lengths of journeys and the number of trips taken each year have significantly raised levels of passenger transport fuel consumption in urban areas. Average trip length is affected by the spatial layout of cities, but also increases with the acquisition of motorized vehicles.

Table 7. Daily average passengers carried by mode in Hong Kong (%)

	1980	1985
Heavy buses	44	42
Light buses	20	16
Rail/tram	13	25
Taxis	11	16
Private cars	8	5
Others	4	4
Total passengers ('000s)	7428	9346

Source: Parker L., 1987: *Energy Use in Land Transport in Hong Kong*, Centre of Urban Studies, University of Hong Kong, 1987.

The average car trip in Nairobi was almost twice as long as the average foot or bicycle trip. On the other hand, the municipal government in Curitiba, Brazil, has planned city layout with exquisite attention to easy train-to-bus transfers, one-way expressways, and the creation of commercial zones outside the circular ring surrounding the city.

The Chinese government has also undertaken efforts to abate the growing levels of congestion. Since 1949, Beijing's motorized traffic has increased 100-fold, while the road network expanded only 12-fold. As China's population grew, even a small proportional number of motor vehicles per capita meant heavily congested traffic, leaving little room for future growth. The Beijing roads also had to accommodate the large increase in bicycle traffic; 500,000 bicycles were added to the city in 1989 alone. In China, however, a large gap exists between capital allocated and capital required; the Ministry of Communication wanted $8.1 billion for highway construction in the 1986-90 Seventh Five Year Plan, but received only $54 million. Thus, efforts to limit the already acute congestion have focused on defusing traffic peak periods (as described above) and shortening journeys: encouraging more companies and institutions to move from central areas to the suburbs.

5.7 *More stringent car and motorcycle emission standards*

The adoption of current state-of-the-art emissions standards can substantially lower auto emissions. Developing countries can introduce some of the measures which have been effective in industrialized nations over the past three decades. Many industrialized nations, for example, first introduced initial crankcase HC controls in the early 1960s and exhaust CO and HC standards later that decade. By the early- to mid-1970s, most major industrialized countries had initiated some type of vehicle pollution control programme. California's 1994 emissions standards are the most stringent yet adopted (see Box 1).

In most cases, these measures have had a dramatic impact on emissions. In the United States, cars exhaust less than 20% of the hydrocarbons and carbon monoxide (per mile driven) than uncontrolled vehicles did in the 1960s.

Emissions control approaches differ significantly among nations due to varying types and degrees of air pollution problems, vehicle characteristics, economic conditions and other factors. In choosing the best available measures, each nation will have to consider its own circumstances carefully.

Recent years have witnessed some promising trends in transport policies. Japan, the wealthiest country in the Asia with the largest vehicle population, has traditionally been one of the world leaders in motor vehicle pollution control. It has one of the cleanest fleets in the world, at least for petrol-powered vehicles. Just within the past year, Japan has started to introduce particulate controls on diesel vehicles, boding well for future control of this especially hazardous pollutant. Two rapidly industrializing Asian nations, Taiwan and South Korea, have recently introduced state-of-the-art controls on new petrol cars and are in the process of enacting comprehensive vehicle control programmes. Hong Kong and Singapore have decided to follow suit, and Thailand appears to be on the verge of implementing similar measures as well.[5]

While unleaded and low-leaded petrol have had a limited presence in much of Asia in the past, they are now being (or have been) adopted by Japan, Taipei, Korea, Hong Kong, Singapore, Malaysia, Indonesia and Thailand. Motorcycles in Taipei are subject to the most stringent motorcycle standards in the world — a promising development for other cities in the region with similar two-stroke motorcycle problems.

5.8 Inspection and maintenance (I/M) measures

These programmes have been demonstrated to lower emissions from existing vehicles in two ways:

1) By lowering emissions from vehicles which fail to meet regulations and requiring repairs;

2) By encouraging owners to take proper care of their vehicles, to avoid potential costs of repairing vehicles which have been tampered with or mis-fuelled.

Available data suggest that a well-run I/M programme can bring about very significant emissions reductions — about 25% for HC and CO and 10% for NO_x. The less significant NO_x reductions reflect solely the lower tampering rates from I/M and anti-tampering programmes, since at present there has been no focused effort to design I/M programmes to identify and correct problems that specifically cause excess NO_x emissions.

In a typical I/M programme, the reductions start out slowly and gradually increase over time, because I/M programmes tend to lower the overall rate of fleet emissions deterioration. Thus, maximum I/M benefits are achieved by adopting the programme as early as possible.

[5] The recent military coup raises concerns about whether plans to address vehicle pollution, which were in the late stages of development, will now be implemented. But initial indications are positive.

California's Energy Standards **Box 1**

California's Emerging Standards: Forcing Future Technologies

The technology for petrol vehicles is advancing rapidly. California has been the world leader in fostering this progress. Over the past year, California's Air Resources Board (CARB) developed a sweeping set of standards requirements which, starting in 1994, will dramatically change vehicle technologies during the rest of the century and beyond.

Transitional Low Emission Vehicles (TLEVs)

Must meet a hydrocarbon standard of emitting less than 0.125 grams per mile (gpm) when fuelled with conventional petrol. Compliance with the TLEV standards should not require significant modifications of current engine designs and control technologies. A number of small conventional petrol vehicles have attained TLEV emission levels in certification testing, and a Nissan engine family has achieved emission levels at about 40,000 miles in use.

Low Emission Vehicles (LEVs)

Emit only 0.075 gpm of hydrocarbons. The hydrocarbon and nitrogen oxides standards are intended to be technology forcing, and are based on projected improvements in emission control technology and conventional petrol hardware design. The most promising control technology presently being developed is the electrically heated catalyst. Fuels cleaner than conventional petrol may be able to meet the LEV emission standards without the use of electrically heated catalysts, but considerable effort would need to be expended to ensure the durability of the clean fuel vehicles to 100,000 miles of service.

Ultra Low Emission Vehicles (ULEVs)

Emit no more than 0.04 gpm HC and 0.02 gpm NO_x. The standards for ULEVs are based on the capabilities of electric and compressed natural gas-powered vehicles; and on the projected emission capabilities of vehicles equipped with an electrically heated catalyst fuelled with methanol, ethanol, liquefied petroleum gas or possibly reformulated petrol.

Zero Emission Vehicles (ZEVs)

Sales of these must reach 2% by the year 1998 and 15% by 2003.

6 Directions for Further Research

To restrain fuel use and emissions growth in the long run, several important areas of research need to be investigated — either in each country separately, or internationally.

6.1 The planning process

An economic evaluation of the government's conflicting transport objectives (employment, economic efficiency, oil use, air quality) is necessary. This evaluation should explicitly consider capital requirements and the environment and oil import externalities. Hard currency payments for oil imports account for as much as 70% of export earnings in developing countries. Reduced motorization, improving efficiency of transport oil use, and the use of alternative fuels are the only viable means to restrain the growth of oil imports. Planning is essential to accomplish this goal.

The relationship between economic growth and the need for transport services must be evaluated carefully. Economic growth can be accomplished through reduced motorization, but not with reduced levels of service. The required alternatives necessary to achieve this goal need to be evaluated, but in most countries there is a lack of data for this evaluation.

6.2 Alternative fuels

The use of alternative fuels (particularly CNG and LPG) in place of diesel and petrol must be economically justifiable. Without such cost justifications, including an assessment of energy security and environmental externalities, such programmes have failed in the past.

6.3 Fuel efficiency

As we have seen, many developing countries have made substantial strides in improving automobile efficiencies over the last few years. Efforts to improve trucks, buses and utility vehicles have been less successful. An important research topic is the cost of improving efficiencies of engines and vehicles versus that of fuel supply. Manufacturer perspectives on this issue must also be investigated. The idea, described earlier, of requiring each manufacturer or importer to prepare an *energy-efficiency impact statement* needs to be explored.

6.4 Congestion

Only cities with autocratic governments have succeeded thus far in restraining vehicle growth. Reducing congestion requires better transport policies, improved land use planning and the ability to implement both. The latter measures may prove the most difficult. Thus, further research is needed on changing urban development patterns and effectively implementing improved transport. Packages of alternatives for controlling congestion need to be developed for a wide range of urban topographies and population density patterns.

6.5 Standards for fuel use and emissions

Developing countries must further investigate the possibilities for setting these standards and broadening their implementation. Both economic evaluations and industrial impact studies are required.

6.6 Fiscal considerations and the role of the private sector:

An important consideration is the government's ability to finance road maintenance, state-wide passenger transport development, and other infrastructure improvement. The private sector could play an important role in selected areas. This issue needs to be investigated since improved roads mean higher average speeds and, thus, higher efficiency for business.

7 Conclusion

Motorized transport will remain an essential and growing element of economic development. Increased motorization in congested cities has the potential to increase emissions and to reduce local air quality. Restraining the current rapid growth of transport emissions will necessitate government intervention in several key realms. However, policies have to be able to endure the vagaries of governments over time. The earlier checkered history should serve as a warning to advocates of policies, which can easily become, or already are, uneconomic. The challenge is to draft policies with a long-term perspective without losing sight of the immediate short-term needs to provide adequate services. As these policies are developed, current state of the art technological advances must be introduced to the degree practicable, to ameliorate adverse effects.

References

Al-Haque, M., 1983: Mechanized road transport survey in selected centres of Pakistan, 1981-82, *Quarterly Research Review*, **2**, No. 3.

American Lung Association, 1990: *The Health Costs of Air Pollution: A Survey of Studies Published 1984-89*, New York.

Ang, B., 1989: *Traffic Management Systems and Energy Savings: The Case of Singapore*, Department of Industrial and Systems Engineering, National University of Singapore, Singapore.

Asian Development Bank and Economic Development Institute, 1989: *Transport Policy*, country papers and private sector papers presented at a regional seminar on transport policy, Manila, Philippines, 21-28 February 1989, pp. 331-380.

Bleviss, D.L., 1988: *The New Oil Crisis and Fuel Economy Technologies*, Quorum Books, New York, NY.

Gimenez, A., 1989: *The Manila Light Rail System*, paper presented at Workshop on New Energy Technologies, Transportation and Development, Ottawa, Canada, 20-22 September 1989.

Government of the People's Republic of China, 1989: *Sectoral Energy Demand in China*, UN-ESCAP, Regional Energy Development Programme Report No. RAS/86/136.

Government of Thailand, 1989: *Sectoral Energy Demand in Thailand*, UN-ESCAP, Regional Energy Development Programme Report No. RAS/86/136.

Hansen, S., 1976: *An Analysis of Low-Cost Early Action Local Transport Improvements*, Technical Report No. 5. Plan and Budget Organization, Teheran Development Council Secretariat, Harvard Inst for International Development.

Heggie, I. and M. John, 1989: *Financing Public Transport Infrastructure*, report prepared for the Regional Seminar on Transport Policy, Asian Development Bank, Manila, 21-28 February, 1989.

Hong Kong Environmental Protection Agency, 1989: *White Paper: Pollution in Hong Kong - A Time To Act*, Hong Kong, 5 June 1989.

Kiravanich, P., S. Panich and J. Middleton, 1986: *Air Pollution Problems and Management in Asian Countries*, paper presented at the 79th Annual Meeting of the Air Pollution Control Association, Minneapolis, Minnesota, June 22-27, 1986.

Korea Energy Economics Institute, 1989: *Sectoral Energy Demand in the Republic of Korea: Analysis and Outlook*, United Nations — Economic and Social Commission for Asia and the Pacific (UN-ESCAP), Seoul.

Kulkarni, A., J. Deshpande, G. Sant and J. Krishnayya, 1991: *Urbanization: In Search of Energy*, Systems Research Institute, Pune, India.

Laimbock, F. and C. Landerl, 1990: *50cc Two-Stroke Engines for Mopeds, Chainsaws and Motorcycles with Catalysts*, Report No. 901598, Society of Automotive Engineers, Warrendale, Pennsylvania.

Ledbetter, M. and M. Ross, 1990: *A Supply Curve of Conserved Energy for Automobiles*, draft report, Lawrence Berkeley Laboratory, Berkeley, California.

Liu, L. and Z. Qin, 1991: *Urban Transport and Its Energy Use and Air Pollution in Beijing*, Report No. IDRC 3-P-88-123-02, Institute of Techno-Economic and Energy System Analysis, Tsinghua University, Beijing, China.

Meyers, S., 1988: *Transportation in the LDCs: A Major Area of Growth in World Oil Demand*, Report No. LBL-24198, Lawrence Berkeley Laboratory, Berkeley, California.

National Environment Board of Thailand, 1990: *Air and Noise Pollution in Thailand 1989*, Bangkok.

Parker, L., 1987: *Energy Use in Land Transport in Hong Kong*, Centre of Urban Studies, University of Hong Kong.

Planning Commission, Government of India, 1988: *Report of the Steering Committee*, Delhi.

Pucher, J., 1988: Urban travel behavior as the outcome of public policy: the example of modal-split in Western Europe and North America, *APA Journal*.

Reddy, A.K.N., 1991: *Sustainable Development in India through Reduction of Oil Dependence*, paper presented at the Petrad Seminar on the Role of Petroleum in Sustainable Development, Penang, Malaysia, 7-11 January 1991

Sathaye, J. and R. Bhatia, 1991: *Energy Use in the Transport Sector in India: Policy Options and Research Needs*, report submitted to International Development Research Centre, Ottawa, Canada.

Sathaye, J. and S. Meyers, 1986a: Transport and home energy use in cities of the developing countries: a review, *The Energy Journal*, **8**, LDC Special Issue, October, 85-103.

Sathaye, J. and S. Meyers, 1986b: Changes in oil demand in oil-importing countries: the case of the Philippines, *The Energy Journal*, **7(2)**, 171-179.

Sathaye, J., B. Atkinson and S. Meyers, 1989: Promoting alternative transportation fuels: the role of Government in New Zealand, Brazil, and Canada, *Energy*, **14(10)**, 575-584.

Sathaye, J. and A. Gadgil, 1992: Pursuing Aggressive Cost-Effective Electricity Conservation: Novel Approaches, *Energy Policy*, **20 (2)**, February 1992.

Sugiura, K. and M. Kagaya, 1977: *A Study of Visible Smoke Reduction from a Small Two-Stroke Engine Using Various Engine Lubricants*, Report No. 770623, Society of Automotive Engineers, Warrendale, Pennsylvania.

Tashrif, M., 1991: Center for Research on Energy, Institute of Technology, Bandung, Indonesia. Personal communication, 6 February.

Teheran Development Council Secretariat, 1976: *An Analysis of Low Cost Early Action Local Transport Improvements*, Technical Report No. 5.

Thailand Development Research Institute, 1990: *The 1990 TDRI Year End Conference, Industrializing Thailand And Its Impact On The Environment*, Research Report No. 7, Energy and Environment: Choosing The Right Mix, Bangkok.

United Nations, Economic and Social Commission for Asia and the Pacific, Joint Working Group on the Energy Action Programme in Thailand, 1986: *Rational Use of Energy in Transport*, Report No. ST/ESCAP/433.

Walsh, M., 1989: *Global Warming: The Implications for Alternative Fuels*, Society for Automobile Engineers Technical Paper Series 891114, May 2-4, 1989.

Walsh, M. and R. Bradow, 1991: Diesel particulate control around the world. In *Global Developments in Diesel Particulate Control*, Report No. P-240, Society of Automotive Engineers, Warrendale, Pennsylvania.

World Bank, 1986: *Urban Transport*, Washington, DC.

World Bank, 1988: *Road Deterioration in Developing Countries: Causes and Remedies*, Washington, DC.

World Bank, 1990: *World Development Report 1990*, Oxford University Press, New York.

World Resources Institute, 1988: *Global Trends in Motor Vehicles and Their Use Implications for Climate Modification*, Washington, DC.

World Resources Institute, 1990: *World Resources 1990-91*, Oxford University Press, New York, pp. 148-152.

Wyczalek, F.A., 1991: Two-stroke engine technology in the 1990s. In *Two-Stroke Engine Design and Development*, Report No. SP-849, Society of Automotive Engineers, Warrendale, Pennsylvania.

A Glimpse Into a Future without Pollution Controls: **Box 2**
The Dire Future of Transportation Emissions

In 1990, 17 developing countries were compared in a transportation scenario submitted to the IPCC. This research suggests that, unless there is dramatic change, there will be high growth in the amount of greenhouse gases emitted from motor vehicles in developing nations during the next 15 years.

The countries and their growth: in GDP, GDP per capita, and population

Country	GDP (1985 Bn. US$)		GDP/capita (1985 US$)		Population	
	1985	2025	1985	2025	1985	2025
China	324	2840	310	2000	1045	1420
India	191	1347	250	796	766	1692
Indonesia	84	280	510	1050	165	267
Korea	93	926	2276	18400	41	50
Argentina	70	156	2296	3294	31	47
Brazil	249	891	1879	4163	132	214
Mexico	189[a]	1000	2470[a]	6980	77[a]	143
Venezuela	50[b]	248	2943[a]	7318	17[b]	34
Ghana	4.6	22.8	343	533	13.4	42.7
GCC Countries	163	1084	10200	16400	16.3	66.1
Nigeria	75	373	788	1176	96	317
Sierra Leone	0.9	2.9	247	327	3.5	8.8

[a] 1987 data; [b] 1984 data

Expected changes if business continues as usual:

Freight transport: A continuing transition from rail to truck and air travel throughout most of the Third World, and to ship travel in Indonesia, China and other water-dominated countries. Rail used mainly for moving minerals. China and India will replace coal-fired steam locomotives with more efficient diesel and electric railway systems.

Passenger rail: Several countries (such as Venezuela) have plans to develop urban rail passenger systems.

Truck efficiency: Expected to improve, although the potential improvement for diesel engines should remain limited to around 30-40% in most countries. The fuel transitions for other freight modes should remain minor, resulting in slight increases in the use of natural gas and biomass alcohol.

Personal automobile ownership: continues to grow rapidly in developing countries as incomes rise and government policies facilitate vehicle acquisition:

Country	Car ownership (per 1000 persons)	
	1985	2025
China	0.6	20
India	2.1	16
Indonesia	5.9	45
Korea	11.2	210
Argentina	126	200
Brazil	63	245
Mexico	64	250
Venezuela	87	320
Ghana	3.8	5
GCC Countries	145	232
Nigeria	11	14
Sierra Leone	7.1	10

Box 2 (cont'd)

Congestion and car saturation: increases, in some places (Korea) more than fifteen-fold. Dramatic increase in population of motorcycles in many countries (Brazil) by 2025. Average distance cars are driven per trip declines dramatically, particularly in Asia — but because of the number of new car owners, the total miles driven per capita will increase.

Passenger vehicle fuel efficiency: improves as a consequence of technological improvements, and as new vehicles penetrate the marketplace faster. Larger and more comfortable cars offset the improvement in Korea:

Country	Fuel efficiencies (kilometre/litre)	
	1985	2025
China	10	17
India	11	17
Indonesia	10	20
Korea	10	13
Argentina	8	12
Brazil	7.7	14
Mexico	5.4	15
Venezuela	4.4	8.3
Ghana	4.2	5.9
GCC Countries	10	16.6
Nigeria	5.3	11
Sierra Leone	5.0	6.3

Fuel use: Increases more than four-fold, despite efficiency improvements. Use of oil increases from 2.9 million barrels per day (mbpd) in 1985 to 15 mbpd by 2025, a five-fold increase. Oil demand expands more rapidly in the Asian countries (seven- to ten-fold) than in Latin America or Africa. This creates significant oil demand constraints in developing countries; total oil demand rises from 8.3 mbpd in 1985 to over 35 mbpd by 2025, enough to exhaust the known OPEC excess production capacity. By 2025, China imports 1.8 mbpd and produces another 2.5 mpbd of synthetic oil from coal. India's consumption reaches 5.8 mbpd that year, up from 0.8 mbpd in 1985.

Transportation fuel shares in the study countries

	1985	2025
Energy use (mbpd)	**3.3**	**16.0**
Coal	9%	0%
Oil	87%	92%
Natural gas	—	2%
Biomass	3%	3%
Electricity	—	3%

Foreign exchange: In July 1990, with the oil price at only $15/barrel, the Indian government had to adopt regulations to curb the use of oil, since foreign exchange was inadequate to pay the needed $1.5 billion for oil imports. The estimated level of oil imports for 2025 would require $82 billion. Each country will have to allocate more than twice the share of GDP currently devoted to oil. For oil-importing countries, the heightened demand will require increased payments of hard currency, taking it away from other economic development goals. This will curtain economic growth and may force government resignations or changes in economic structure to encourage more foreign investment.

Emissions: High increases in all types of transport-related emissions, making a substantial contribution to greenhouse gases from all countries surveyed:

Box 2 (cont'd)

Carbon emissions by country, 1985 and 2025[c]

	1985		2025	
	Transport carbon emissions (Mt/C)	Share in total fossil fuel-related emissions (%)	Transport fossil emissions (Mt/C)	Share in total fossil fuel-related emissions (%)
China	29	6%	187	11%
India	18	15%	143	19%
Indonesia	6	24%	58	45%
Korea	5	12%	37	22%
Argentina	8	36%	12	28%
Brazil	21	46%	51	37%
Mexico	21[a]	29%	49	22%
Venezuela	7[b]	29%	14	21%
Ghana	0.38	59%	1.04	23%
GCC Countries	18	26%	116	23%
Nigeria	5	53%	13	22%
Sierra Leone	0.21	65%	0.63	52%
Total	139	—	682	—

[a] = 1987 data; [b] = 1984 data

[c] Sources: **Sathaye, J. and A. Ketoff**, 1991: *CO_2 Emissions from Developing Countries: Better Understanding the Role of Energy in the Long Term, Vol I: Summary,* Report No. LBL-29507, Lawrence Berkeley Laboratory (LBL), Berkeley, USA;
Ketoff, A., J. Sathaye, and N. Goldman (eds), 1991:*op cit: Vol. II: Argentina, Brazil, Mexico and Venezuela,* Report No. 30059, LBL;
Sathaye, J. and N. Goldman (eds), *op cit: Vol. III: China, India, Indonesia and South Korea*, Report No. LBL-30060, LBL;
Sathaye, J. and N. Goldman (eds), *op cit: Vol. IV: Ghana, the GCC Countries, Nigeria and Sierra Leone*, Report No. LBL-30061, LBL.

CHAPTER 15

The Economics of Near-Term Reductions in Greenhouse Gases

Eberhard Jochem and Olav Hohmeyer

Editor's Introduction

Many analysts have suggested that limiting the risks of rapid climate change by reducing the emissions of greenhouse gases will be very costly, especially in advanced industrial economies. Using simple models, such analysts argue that emissions-reducing technologies will increase the costs of production. These costs, they say, will divert investment from more productive opportunities and penalize the companies and countries that impose the most stringent environmental constraints on their domestic activities.

In this chapter, Eberhard Jochem and Olav Hohmeyer demonstrate that just the opposite is true. Economies benefit, even in the short term, from strategies that promote environmental protection through the development of new technologies. To make their point, these authors start with a difficult case: the Federal Republic of Germany, which has already achieved substantial gains in energy efficiency during the last two decades, but where opportunities for further improvement, and even greater economic benefit still exist.

Using a sophisticated macro-economic analysis, Jochem and Hohmeyer show, for the German case, that policies to improve energy efficiency and to shift the energy mix to advanced technologies and less carbon-intensive fuels will generate four important kinds of benefits for the national economy. Such policies will (1) spur overall economic growth, (2) quickly generate a large number of jobs within the country (including the sort of entrepreneurial jobs which encourage a resourceful, self-sufficient, and satisfied work force), (3) increase exports of high technology products, and (4) reduce environmental and social costs of energy use that were previously uncounted in the market transactions for fuel. Taken together, these benefits will work to reduce the social costs paid by the society as a whole to subsidize economic development.

Eberhard Jochem is a leading German economist and systems analyst working at the Fraunhofer Institute for Systems and Innovation Research (ISI). For the last five years, he has been an important member of a board of scientists advising the Enquete Kommission of the German Parliament, studying the linked issues of global warming and ozone depletion. Olav Hohmeyer, also at the Fraunhofer Institute, is an economist and political scientist. To supplement their macroeconomic analysis, Jochem and Hohmeyer have developed a taxonomy of social costs and a methodology for incorporating their consideration into economic analysis. Focusing, in particular, on the social and environmental costs of electricity production, these authors have developed an approach that can quantify the previously unaccounted-for costs that have distorted electric capacity expansion planning decisions for the past several decades.

But Jochem and Hohmeyer make it clear that getting the arithmetic of costs correct is only part of the solution. As the previous chapter on electricity (Jhirad et al.) observed, a series of institutional obstacles and deep-seated market imperfections exist, obstructing the diffusion of a rational energy strategy. Even in a highly educated and technically sophisticated society such as the Federal Republic, Jochem and Hohmeyer found failures of information dissemination and training that keep consumers and investors from making decisions which are in their economic self-interest.

Lack of access to capital, reliance on simplistic and antiquated tools of micro-economic analysis, and adherence to yesterday's financial strategies combine to discourage profitable investments in energy efficiency and new technology. The sociopolitical analysis in this chapter also reveals important non-financial barriers to market penetration. These legal and administrative obstacles to rational energy planning emerge on close inspection of nearly all energy sub-sectors. For example, Germany's tax on car engine size instead of petrol as well as the lack of speed limits on motorways, encourage people to buy and drive powerful, energy-consuming cars. Even worse is the fact that fuel for air transportation with high external costs is not taxed by any country.

Legislation limiting the construction of gas- or oil-fired cogeneration plants also discourages the rational use of energy. These laws make it more attractive to throw away the energy value of waste heat at a power plant, rather than to sell it. Furthermore, the widespread separation of operating budgets from investment budgets in public agencies discourages local government from buying the best, rather than the cheapest equipment. These are only the simplest of examples of ways in which national policies and administrative practices can handicap progress toward rational energy use.

Jochem and Hohmeyer conclude that the key to a rational energy future lies in the evolution of comprehensive national energy policies — a development being bitterly resisted today in many countries. Emphasizing that all aspects of energy policy cannot be designed or implemented successfully by central governments, Jochem and Hohmeyer urge the institution of a policy climate that incorporates the full economic, environmental, and social costs of energy use into the price of fuels to make use of market mechanisms as much as possible. Instead of piecemeal measures, policy makers should consider systematic approaches, that take into account the linked effects of measures implemented separately by disparate sectors of government, trade associations, energy service supplying companies and interested industrial sectors.

A successful strategy might include: (1) "getting the prices right," (2) increasing the level of knowledge about energy options, (3) encouraging investments in renewable energy systems and energy-saving devices (and reaping the secondary environmental and employment benefits), and (4) setting high standards for the performance of energy end-use devices. Jochem and Hohmeyer conclude, that if such comprehensive strategies are implemented, governments will quickly discover that the true economics of renewables and rational energy use are far more attractive than today's market prices suggest. They may be the vehicle for sustaining economic development, even in the most technically advanced market economies.

- I. M. M.

1 The Economic Benefits of Energy Efficiency

The ecologically beneficial role of energy efficiency is widely accepted today. Following an environmental imperative, many OECD member nations have decreased their energy intensity (annual primary energy consumption per dollar of Gross Domestic Product) by 1.5% or more per year since 1973. Behavioural and structural changes (such as the substitution of new products and services) account for only 10% to 25% of this improvement. Most of it stems from *rational energy use:* that is, improved efficiencies in existing technologies and products (Morovic et al. 1989; Howarth/ Schipper, 1991). Rational energy use has thus become a

Table 1. CO_2 reduction potentials of the German Enquete Commission's "energy policy" scenario

Assumptions		
no additional nuclear capacity		
remove obstacles		
modal split changing policy		
Composition of options	**million tonnes CO_2**	%
Rational use of energy		
Additional savings in final energy sectors (including modal split policy)	-75.0	-10.5
Savings in the transformation sector and savings due to cogeneration	-45.0	-6.3
Renewables	-30.0	-4.2
Intra-fossil fuel substitution	-26.5	-3.7
Nuclear power plants at 85%	-25.0	-3.5
Energy conscious behavior	-13.0	-1.8
Total	-215.0	-30.0

widely recognized means, perhaps the means with the most potential, for the near-term reduction of greenhouse gases. A modest set of improvements in energy-efficiency in West-Germany, for example, with no major policies to reduce greenhouse gas emissions in place, are none the less projected to eliminate 165 million tonnes of CO_2 by 2005 and stabilize CO_2 emissions at today's level of some 715 million tonnes despite a 50% increase in GDP by 2005 (Prognos/ISI, 1990).

However, the benevolent *economic* effects of rational energy use are still not well-known. Nor are they accepted in energy decision making. In 1990, the Enquete Commission of the German Bundestag proposed to reduce 120 million tonnes of greenhouse gases (17% of today's total emissions) through improved energy efficiency. This included saving 75 million tonnes in all final end-use energy sectors (residential, transportation, commercial and industry), and 45 million tonnes by doubling today's use of cogeneration in industrial plants, commercial buildings, and district heat generation facilities (see Table 1). At first glance, these reductions seemed over-ambitious; doubts about the plan's feasibility were expressed by speakers at public hearings, by energy suppliers, and by the Federal Ministry of Economics. But these doubts, like many expressions of traditional industrialization policies, reflect a lack of understanding of what has been learned in the last 20 years, about the possibilities and economic impacts of more rational use of energy.

Experiences with energy efficiency show that it serves all four macro-economic goals: full employment, economic growth, well-balanced foreign trade, and price stability. This cannot be said for energy supply policies in many cases. Moreover, when a nation adopts rational energy use, the process in itself brings existing economic problems to light. This occurs because a rational energy policy - reducing greenhouse gas emission at almost no cost by 0.5% or more per year - can exist only if obstacles are removed and market imperfections are alleviated. Finally, the hidden economic costs of traditional ("irrational") energy use are not reflected in energy prices; they are paid, instead, by society as a whole.

The following three sections of this chapter are intended to improve understanding of these complex relationships.

2 Full Employment and Economic Growth

More rational use of energy would alleviate a major problem in OECD member countries today: long-term unemployment, which has increased to 25 million people in the OECD as a whole. Unemployment is expected to become an even more serious issue in the former Comecon Member Countries, as they change from centrally planned to market economies. This emerging burden of unemployment can already be observed in most East European Countries - especially in the former German Democratic Republic, where unemployment rates of some 25% are expected for early 1992, owing to the instantaneous shift to a market economy.

2.1 Energy efficiency spurs economic growth

Rational energy use helps to spur economic growth by cutting waste, and cutting energy costs. Businesses can bring products or services to market with less investment in fuels and heating; they can distribute their wares with less expenses. The private consumer spends less on fuel costs, but more on investment costs or maintenance to improve energy efficiency. But as long as the efficiency investment is economic, the energy consumer saves some net energy costs. These savings can be spent to pay for other goods and services. The economic gains of intensified energy efficiency expenditures are modest, but not negligible.

Quantifying the exact economic growth generated by energy efficiency is difficult, if not impossible, because energy efficiency is part of complex production and consumption processes. In most industrial countries, it does not take much effort to gather the figures (from official statistics) on production, sale and installation of new energy *supply* technology (power plants and transmission lines). But similar data for energy-*saving* technologies is not easy to find. A new or improved industrial process, for instance, may need less energy per unit of production, but it may also be more efficient with regard to labour, capital and new materials; thus, an observer cannot, from statistical sources, distinguish the savings from improved energy use from all other net savings.

But we *can* make that distinction in the case of "dedicated" energy saving investments, when the sole purpose of the new industrial product is to use energy more efficiently than in the past, or to avoid heat losses. Such products have become increasingly successful. The production of 12 of those energy saving products, identifiable in official statistics (burners and gas turbines, high efficient boilers and heat pumps, insulation material, double/triple glazing, heat exchangers, and electrotechnical products) has increased by 4.6% per year since 1982 in West Germany. By comparison, the total production of West German industry has only increased by 2.6% per year on average. The direct impact on employment of this growth in the energy savings industry has been estimated at 40,000 jobs. The secondary employment for planning, installing and maintenance doubles this figure (Jochem et al., 1992).

Germany is not unique. Similar figures have been analysed for France and for specific products in other industrialized countries.

2.2 Analysing the net employment gain

The positive gross employment effects induced by energy efficiency improvements must be considered against the counter-balancing negative employment effects: the decreased energy production and distribution (because consumers need less). Several authors, however, have included these negative impacts in their analyses of employment effects, and still come up with a positive job-creation outlook:

- Hohmeyer et al. (1985) reported (on the basis of an extensive input-output analysis) that the effective implementation of six energy-saving technologies could yield additional net employment of 98,000 jobs by the year 2000. The six technologies were: district heating (in most cases produced by cogeneration plants); insulation in residential buildings, heat exchangers for heat recovery; large gas-engine-driven heat pumps; domestic solar hot-water systems; and agricultural biogas plants. This figure emerged from empirical analysis of only four European Community member countries (Denmark, France, the United Kingdom and West Germany).

  Extrapolating these results to their entire potential in 12 EC member countries, the authors estimate additional net employment of at least 200,000 new jobs. If the full range of energy efficiency technologies is considered, the authors expect an employment effect of about 530,000 jobs by the year 2000 (CEC, 1985), or 600,000 jobs if Spain and Portugal are added.

  For the original technologies, which would yield an estimated 45 million tonnes of oil equivalent (Mtoe) by the year 2000, between 2800 and 8000 jobs per Mtoe will be gained. This figure is based on the labour productivities of the 1980s.

- Dacey et al. (1980) studied the net employment effects of improving fuel efficiency in automobiles in the United States, projecting ahead (at that time) to 1985. Negative employment effects were expected in the iron and steel industry (because of weight reductions of the vehicles), and in the production, wholesale and retail trade of petrol. But the positive employment effects in two plastic sectors and the car manufacturing industry would more than compensate these negative effects.

  One may wonder why the dates of the two studies are before 1985. The obvious reason seems to be the enormous efforts which have to be made for the net analysis, which involves detailed technical analyses and their integration in input/output modelling. New research results will be published in 1992.

  From these findings, however, the following conclusions can be drawn: the net effect of energy efficiency improvements and of renewable energy technologies on production, net employment and income is positive. There is a net employment gain on the order of 3000 to 4000 jobs per Mtoe.

The extent of the employment gain depends on several factors:

- The amount of reduction in energy imports (particularly oil imports). This varies substantially between countries. Denmark imports more than 90% of its energy and Japan more than 80%, while the United Kingdom is an energy exporter.

- The labour-intensity of the industries and non-industrial sectors (crafts or trades) which benefit from the energy savings.

- The degree of reduction of energy costs, which depends on the profitability of the energy efficiency investment.

Apart from a few exceptions, the leaders of OECD member countries do not sufficiently consider these positive net effects in their economic and energy-related strategies. In West Germany, for instance, the 4.1 exajoules of energy savings made between 1973 and 1990 have already produced approximately 400,000 new jobs (Jochem et al., 1992). In other words, the net employment effect in this country has been about 100 jobs per petajoule of saved primary energy. If this figure is taken as a first proxy for other industrialized countries, improved energy efficiency has created a few million new jobs since 1973. The new jobs have been created not only in the manufacturing industry but also in the building sector, in installation, planning, maintenance, consulting and other services.

Along these lines, additional energy efficiency improvements in any country could create new jobs which would reduce today's unemployment and thus cut government spending by US $1 billion per year (more than $5000 per unemployed person annually). Even if there were no threat of global warming, the promise of reducing high unemployment to some extent would be enough to justify energy efficiency investments.

2.3 New jobs are concentrated in the investment period

Unlike many other employment effects, the jobs created by more rational use of energy are not evenly distributed over time. In almost all cases, they are created during the initial period of investment. For example, when a building manager installs wall insulation, the direct employment takes place during installation and no new jobs are created for maintenance. Over the rest of the lifespan of the insulation, the energy-savings may be respent and contribute to other employment, but not directly. A few larger scale investments will distribute the employment more evenly over time; the employment effects of a coal-fired cogeneration plant, for example, will linger for the lifespan of the plant. This is because people are needed to maintain and operate the plant, as well as produce the coal. For smaller-scale technologies, part of the requirement of the energy efficiency industry is to continue spreading from building to building, or venue to venue, to provide its workers with steady base of jobs.

The changing time scale of job creation is apparent in an analysis made of the future of the six key energy efficiency technologies (Hohmeyer et al., 1985). Once again, these technologies are: district heating from cogeneration; insulation in residential buildings; heat exchangers; heat pumps; domestic solar hot-water systems; and biogas plants. The impact on job creation was calculated for four EC member countries, over 35 years with an investment period of 15 years: from 1985 to 2000.

- During the initial investment period, net employment increases from 93,000 to 136,000 new jobs. These are mainly due to substantial additional investment, and to some extent to slowly increasing employment from new operation and maintenance.

- The period of much lower net employment effects, from 2001 to 2020, is due to the fact that no further investment in the six technologies is considered beyond the year 2000 in this calculation, so that the net employment only reflects the net effects of operation and maintenance as well as respending effects of saved energy costs.

The study suggests that the new jobs are concentrated in the investment period. Compared with the employment base of energy consumption, long-term employment in energy efficiency is partly shifted to the near-term period of investment. This pattern coincides with the long-term patterns of employment which show shrinking labour due to demographic patterns in many industrialized countries after 2000.

2.4 *Efficiency-related jobs are more evenly distributed within a country than the jobs depending on energy supply*

Warren (1984) and Tschanz (1985) stress the fact that the regional distribution of net employment will be more equitable. Traditional employment in the energy supply sector is concentrated in areas of heavy industry, manufacturing, energy production, and conversion - all generally taking place in centralized locales. The authors stress the fact that from a local or regional point of view, most fuels and electricity are "imported". Laquatra (1990) argues that a given dollar spent on energy has a smaller impact on local economic growth than a dollar spent on local goods and services - which would include many energy efficiency investments. The difference can be as much as a factor of five; it depends on the type of energy efficiency investment or service (operation and maintenance), the substituted energy and the respending of the saved energy costs.

Warren also points out that energy conservation investments in the residential or public/commercial sector will create jobs for semi-skilled people. A high proportion of those jobs will occur in city areas with high unemployment and deprivation.

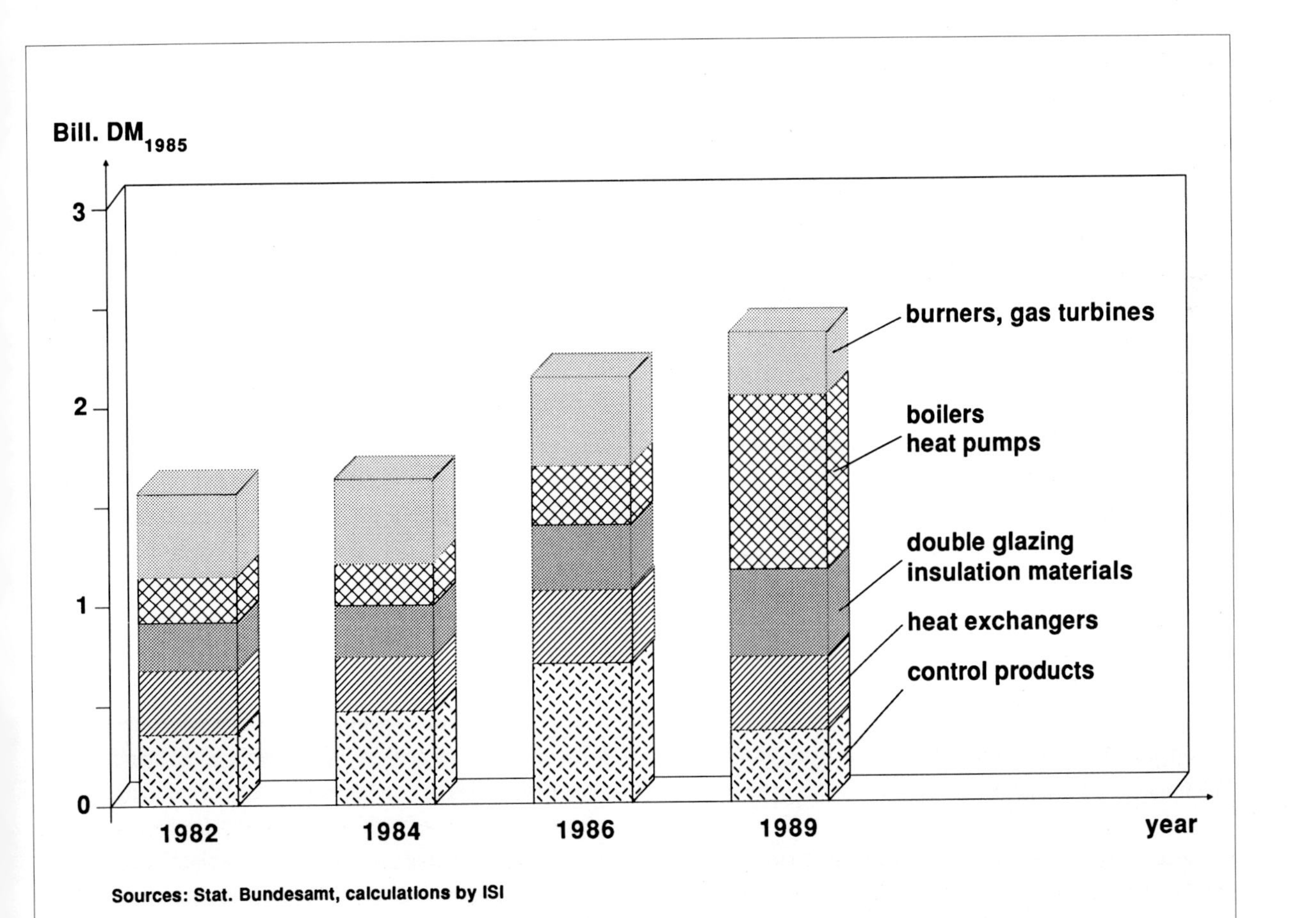

Figure 1. Exports of energy-saving products,West Germany, 1982-1988, in 1985 prices. Production and exports of energy-saving products are increasingly supporting economic growth and world trade, and hence, a sustainable development.

3 Foreign Trade

Most energy conservation products are manufactured domestically; they are all planned, installed, and maintained locally. Thus, improved energy efficiency helps to equilibrate the balance of foreign trade. This is particularly true for countries that are net importers of energy, because they can substitute domestically-produced goods and services for imported energy. All of the benefits from greater balance of trade accrue to countries which practice energy efficiency: reduced inflation, strengthened competitiveness, investment in research and development, and more control over prices (particularly energy prices, which may be stressed as more developing and former Comecon countries become greater net energy importers).

In 1989 Germany reduced its energy imports by almost 10 billion DM, directly through improved energy efficiency, as compared with energy efficiencies of the year 1973. Most of this avoided sum of 10 billion DM — representing one third of the total bill of energy net imports in 1989 — related to avoided imports of crude oil and oil products.

In addition, the exports of energy saving products tend to exceed the average growth for exports as a whole. In West Germany, exports of 12 energy saving products have increased by 8.3% per year since 1982 - about twice as fast as all exports of industrial products, which increased by almost 4% per year (see Figure 1).

Lowering energy cost also enhances the competitiveness of energy-intensive products (Garnreiter, 1982). This is particularly true for products which are easily tradable because of their relatively high value-to-weight ratio: steel, primary aluminium, pulp and paper, and mass polymers. Improving energy efficiency in the production of these energy-intensive products reduces vulnerability to energy price increases. The share of energy cost in gross production of the West German basic materials industries, for instance, dropped almost 30% in seven years - from 6.1% of the total cost of production in 1982 to 4.4% in 1989. The share of energy cost of the cement production dropped from 31.7% in 1981 to 18.8% in 1989, of iron and steel from 13.4% to 9.2%, of pulp and paper from 11.8% to 8.3% and of the chemical industry from 5.9% to 3.7% respectively.

Although many factors determine the competitiveness of the basic materials industries, a continuous improvement of energy efficiency is essential for the survival of energy-intensive industries.

4 Impact on Social Costs

The importance of energy-related social costs, which are not directly included in the energy prices, is largely underestimated. These costs are often difficult to quantify or even monetarize; they include costs of environmental damage and health effects from air pollution and acidity of soils or water. In many cases, social costs do not even seem to have been identified yet - much like the impacts of greenhouse gases, where no effort was made to assign a financial figure until recently. Because the techniques of measuring the social costs and the availability of social cost data are rapidly changing, the estimates vary dramatically; the social costs of energy consumption in West Germany, for instance, are estimated to range between 12 and 40 billion DM per year (Schlomann, 1990), not taking into account the social costs of global warming.

If those figures are accurate, then energy efficiency improvements helped West Germany avoid costs between $4 billion and $14 billion in 1990. This figure is based on assigning a monetary figure to the drop in energy use (measured in exajoules) between 1973 and 1990. (West Germany avoided about 4.2 exajoules in 1990, by this comparison.) It is important to note that this reduction has been achieved at no additional cost; the energy savings achieved were profitable at the microeconomic level.

4.1 The costs of not using renewable energy sources

It is increasingly recognized that conventional energy technologies bear hidden costs, that should be taken into account, when comparing (or setting) energy prices (Hohmeyer and Ottinger, 1991). Neglecting the costs of environmental damages, for example, adds an unrealistic bias against environmentally friendly technologies, such as renewable energy sources. It may lead to a serious underuse of such technologies. This has been shown in detail for wind energy and photovoltaics (Hohmeyer, 1988; 1990a; and 1990b) but it can be assumed equally true for most rational energy technologies. We think that a fair balance sheet will show that the rational use of energy is even more economic than today's market prices would lead you to believe.

In this section of this chapter we examine the cases of wind energy and photovoltaics in Germany. The results on these two technologies, contrasted against the competing conventional technologies for electric power generation, are readily available.

4.2 Internalization

The argument most often heard against more widespread use of renewable energy sources - and against many technologies for the rational use of energy - is that they are too expensive. They do not offer energy (or energy savings) at competitive price levels. But cost comparisons regularly just take into account the so-called "internal" cost elements involved in the production and distribution of a product. Other cost elements are paid for by third parties not involved in the production or consumption of the product; they do not show up in prices and are not considered in standard cost comparisons. These cost elements are referred to as external or social costs. Kapp (1950) suggested the later term, arguing that the term external costs used in neo-classical economic theory, was too narrow for analytical purposes, as it excludes some important cost elements. The term social costs, in the sense of Kapp's definition, is used here to refer to all cost elements handed on to third parties or future generations which do not show up in market prices.

Examples for such social cost elements of energy production and use are the damages done to forests by acid rain, which are paid for by the forest owners or purchasers of wood; the consequences of massive global warming due to antropogenic emissions of greenhouse gases; or the health impacts of major nuclear reactor accidents like Chernobyl. No energy consumer is charged for any of them directly - but all present and future consumers, together, must pay to compensate for their effects. Thus, to analyse the true costs of an energy source (or a product or service), we must *internalize* the social costs - that is, we must bring them in to the same balance sheet on which the internal costs are counted, so that they may be added together for decision making.

When conventional energy sources are compared against renewable energy technologies with a balance sheet that includes social costs, the renewables (as well as most technologies for the rational use of energy) have considerably lower social costs. In other words, the seemingly cheap conventional energy sources can be rather expensive to society. If this is the case, the statements regularly made on the comparative internal costs may be vastly misleading, and investment decisions taken on these grounds may induce substantial losses to society, and even to long-term investors.

The question of relative social costs of electric power has been heavily discussed internationally since 1988, when a first comprehensive report on the subject (Hohmeyer, 1988) was published. We will now summarize the results of this discussion and draw some early conclusions about the relative total costs of renewable energy sources.

4.3 An anatomy of social costs

The following list gives an impression of the range of possible effects which should be considered in any comparison of energy technologies:

- Impacts on human health:
  - short-term impacts like injuries
  - long-term impacts like cancer
  - intergenerational impacts due to genetic damage;
- Environmental damages on:
  - flora, including crops and forests
  - fauna, including cattle and fish
  - global climate
  - materials;
- Long-term costs of resource depletion;
- Structural macroeconomic impacts like employment effects;
- Subsidies like:
  - R&D subsidies
  - investment subsidies
  - operation subsidies
  - subsides in kind for:
    - infrastructure
    - evacuation services in case of accidents;
- Costs of an increased probability of wars due to:
  - securing energy resources (like the Gulf war) prolif eration of nuclear weapons know-how through the spread of "civil" nuclear technology;
- Costs of the radioactive contamination of production equipment and dwellings after major nuclear accidents; and
- Psycho-social costs of:
  - serious illness and death
  - relocation of population due to construction or accidents.

This list is not exhaustive. But it gives an impression of the range of costs which need to be considered before one may conclude that any particular energy technology is too expensive. Although it is relatively easy to enumerate these social cost categories, it is rather difficult to quantify or put monetary values on them. The latest computer runs may allow us to come up with some first quantification of probable global temperature rises, but a sound analysis of the damage costs seems impossible today. We can only guess possible orders of magnitude of such damages. In general, we are like navigators trying to estimate and compare the size of different icebergs ahead of us, while we can only see the tips of these icebergs in the fog. Figure 2 tries to give an impression of this situation.

4.3.1 Empirical evidence on social costs

The empirical evidence presented here is based on one author's research (Hohmeyer, 1988; 1990a; 1991a), taking into account much of the international discussion of the last 3 years.[1] This work was centred around a comparison, applied in the Federal Republic of Germany, between conventional electricity generation (based on fossil and nuclear fuels) and wind energy or photovoltaics. The areas of social costs covered were:

[1] So far most empirical studies of social costs have focused on a few problem areas, mostly on effects on human health and environmental damages like Ottinger et al. (1990) or Barbir et al. (1990). It should be pointed out, however, that there is a growing number of publications in the field addressing different facets of the problem at the theoretical as well as at the empirical level. Two collections of papers exist to date: the special issue of Contemporary Policy Issues (1990) on "Social and Private Costs of Alternative Energy Technologies," containing about 20 papers on the subject; and a report of a German-American workshop on the subject: "External Environmental Costs of Electric Power Production" (Hohmeyer and Ottinger 1991), containing about 30 papers on the topic.

- Environmental effects
- Impacts on human health
- Depletion costs of non-renewable resources
- Structural macroeconomic effects
- Subsidies.

Due to the scarce availability of empirical data and some fundamental problems in monetizing, a number of effects have not been quantified or specified in monetary terms by the authors so far:

- the psycho-social costs of serious illness or deaths, as well as the costs to the health care system;
- the environmental and health effects of the production of intermediate goods for investments in energy systems, and the operation of these systems;
- the environmental effects of all stages of fuel chains or fuel cycles, including those of nuclear energy
- the full costs of man-made climate changes;
- hidden subsidies for energy systems;
- costs of an increased probability of wars due to:
  securing energy resources (like the Gulf War)
  proliferation of nuclear weapons know-how through the spread of "civil" nuclear technology; and
- costs of the radioactive contamination of production equipment and dwellings after major nuclear accidents

Accordingly, one should interpret the results as a preliminary overview. Wherever doubt exists, assumptions have been made favouring conventional energy, counter to the underlying hypothesis that the social costs of systems using renewable energy sources are considerably lower than those of conventional energy. Thus, the authors feel confident that the difference in the real social costs — between the renewables considered and the conventional electricity generation in Germany — is even larger than these results show.

In general, most studies estimate the social costs of environmental and health impacts as damage costs, roughly attributable to specific sources (such as an individual power plant). Some authors favour general estimates of control costs as proxies for the actual damage costs — because these are easier to analyse — while others advocate contingent valuation procedures like "willingness-to-pay" analyses, which survey the extent to which those who capture the benefits of a policy are willing to pay for those benefits, or to compensate others who bear the costs.

For such contingent valuation analyses, a substantial number of individuals are asked about their ranking and valuation of certain things or qualities which are not traded in markets. Through comparisons to traded goods, monetary values can be derived. These types of studies allow the analysts to cover a broader range of impacts as if they were using direct costing. Because the control cost approach allows for a substantial level of arbitrariness, and because the contingent valuation methods result in somewhat less reliable results, these approaches have been chosen by the author for analysis only in rare cases. Control costs have been used for some first estimates on CO_2 emission impacts through global climate change. The figures used in the following are based on an overview of US studies on the subject published by Koomey (1990).

4.3.2 *The costs of traditional energy versus renewable energy*

There is evidence that the prices of non-renewable energy sources (primarily fossil-fuel and nuclear sources) do not reflect long-term scarcity. This is because major aspects of intertemporal allocation, like sustainability and intertemporal justice, are disregarded — in favour of extremely high and wasteful energy consumption today. If energy prices need to steer toward long-term sustainability, then simple models can be drawn up for the calculation of reinvestment costs and appropriate surcharges.[2]

First estimates of such costs are included in the figures quoted in Table 2. This table summarizes the social costs of different means of electricity generation quantified in monetary terms. The macroeconomic effects given are derived from the comparative analysis of differences in production and consumption, resulting from energy scenarios as expressed in terms of changes in GNP and employment. (For a discussion of such approaches, see Hohmeyer, 1991b.)

What, then, is the gross social cost for the production of 1 kilowatt hour of electricity? With a technology based on fossil fuels, the costs are between 0.03 and 0.16 DM_{82} per kilowatt-hour (kWh).[3] For electricity generated in nuclear reactors (not considering fast breeder reactors), gross social costs between 0.1 and 0.7 DM_{82}/kWh result. When one considers the social costs and benefits of electricity generated by wind energy — with the social costs of present electricity generation included as avoided costs — total social net benefits result in the range of 0.05 to 0.28 DM_{82}/kWh. The equivalent for photovoltaic electricity supplied to the public grid lies between 0.06 and 0.35 DM_{82}/kWh.

---

[2] Energy services can be supplied by drawing on non-renewable energy sources (our inherited energy capital) or on our daily energy account from renewable energy sources. Today, the second is the more expensive approach, but it does not diminish future availability of enery services, as the use of non-renewable sources does. To keep this availability constant over time (at a sustainable level) while we are reducing our energy capital, we need to set aside funds for additional future investments in technologies which use renewable energy sources. This is what is meant by the term "reinvestment costs."

Based on the investment cost of a renewable backstop technology and the time span up to the point where the reinvestment will be needed, after the non-renewable energy source is used up, we can calculate the present value of such reinvestment surcharges.

[3] The value of 1 DM at the time of writing is about 0.6 US$ or approximately 0.37 pound sterling.

Table 2. Social costs of electricity generation based on fossil fuels, nuclear energy, wind energy, and photovoltaic solar energy (1982 prices)

	Hohmeyer 1988 (p. 8)	New calculations 1990	
		Fossil power plants 1982	New fossil power plants 1990
a) Gross social costs of electricity generated from fossil fuels (all figures are estimated minimal social costs			
1. Environmental effects	1.14-5.09	2.6-10.67	2.05-7.93
2. Depletion surcharge (1985)	2.29	0.67-4.71	0.67-4.71
3. Goods and services publicly supplied	0.07	0.06	0.06
4. Monetary subsidies (including accelerated depreciation)	0.32	0.30	0.30
5. Public R&D transfers	0.04	0.02	0.02
6. Total	3.86-7.81	3.65-15.76	3.11-13.03
b) Gross social costs of electricity generated in nuclear reactors, excluding breeder reactors (all figures are estimated minimal social costs)			
1. Environmental effects (human health)	1.20-12.00	3.48-21.0	3.48-21.0
2. Depletion surcharge (1985)	5.91-6.23	4.88-47.42	4.88-47.42
3. Goods and services publicly supplied	0.11	0.11	0.11
4. Monetary subsidies	0.14	0.14	0.14
5. Public R&D transfers	2.35	1.46	1.46
6. Total	9.71-20.83	10.06-70.00	10.06-70.00
c) Average gross social costs of the electricity generated in the FRG in 1984			
1. Costs due to electricity from fossil fuels (weighting factor 0.705[1])	2.87-6.56	2.58-11.25	2.19-9.19
2. Costs due to electricity from nuclear energy (weighting factor 0.237[2])	2.48-5.32	2.38-16.62	2.38-16.62
Total (conventional energy)	5.35-11.88	4.96-27.87	4.57-25.81
d) Net social costs of wind energy			
1. Environmental effects (noise)	(-0.01)	(-0.01)	(-0.01)
2. Public R&D transfers (estimate)	(-0.26)-(-0.55)	(-0.16)-(-0.33)	(-0.16)-(-0.33)
3. Economic net effects	0.53-0.94	0.47-0.78	0.47-0.78
4. Avoided social cost of present electricity generation	5.35-11.88	4.96-27.87	4.57-25.81
Total social benefits rounded to two digits	5.6-12.30	5.26-28.32	4.87-26.25
Mean	8.90	16.80	15.60
e) Net social costs of solar energy (photovoltaics)			
1. Environmental effects	(-0.44)	(-0.44)	(-0.44)
2. Public R&D transfers (estimate)	(-0.52)-(-1.04)	(-0.33)-(-0.65)	(-0.33)-(-0.65)
3. Economic net effects (not including 1982 figures)	2.40-6.65	2.35-8.35	2.35-8.35
4. Avoided cost of present electricity generation	5.35-11.88	4.96-27.87	4.57-25.81
Total social benefits rounded to two digits	6.80-17.10	6.54-35.13	6.16-33.07
Mean	11.90	20.80	19.60

[1] Old weighting factor 0.7444

[2] Old weighting factor 0.2556

Again, these probably represent only the minimum social net benefits; all assumptions underlying these figures minimize the advantages of renewable energy sources. This point has been reaffirmed through all national and international discussions on the first results published by the author in 1988 (Hohmeyer, 1988).

Even without including all social costs and even with a deliberate bias against renewable energy sources, the net external benefits in monetary terms of wind and photovoltaic energy are comparable with the basic market prices of conventionally generated electricity. Thus, any statement on the too-high relative costs of renewables must be reconsidered in the light of a full cost analysis, taking into account the substantial differences in social costs between conventional electricity generation and renewables. The way that social costs are handled will have considerable effect on the speed with which these technologies are introduced and diffused; indeed, some nations will probably introduce them quickly (and benefit) while others, daunted by the seeming expensiveness, hold back and lose the advantages.

4.3.3 *Effect of social costs on the date when a technology becomes competitive*

How can one analyse social costs as a factor in the competitive position of a new technology versus an established one? One way is to examine a two-product market, as Figure 2 portrays. The costs of the established technology (P_{ECI}, the lower of the two parallel curves) are increasing gradually — due to rising exploration and mining costs, for example — while the costs of the new technology based on renewable energy sources (P_{ER}) are decreasing considerably over time, as technology becomes less expensive and more capable as illustrated in Figure 3. We have demonstrated the extent of this downward expense curve with wind and solar energy, and the upward curve for most conventional forms of electricity. At the point labelled t_0, the new energy technology reaches cost effectiveness — even if one considers no social costs.

The upper parallel curve (P_{ECS}) in Figure 3 shows the effect of including the net social costs as well. These are defined as the difference between the social costs of the conventional electricity generation and those of the new technology. An illustrative, but static, calculation of the social costs estimated for a representative base year (1988), and ignoring subsequent changes in social costs over time, results in a projection of the costs of conventional electricity that is roughly parallel to the market price curve (P_{ECS}). This results in a new intersection with the renewable energy cost curve, in which the new energy technology reaches cost effectiveness at an earlier point in time. If the social costs reach a sizable order of magnitude, but policies and markets continue to ignore them, then a distorted competitive situa-

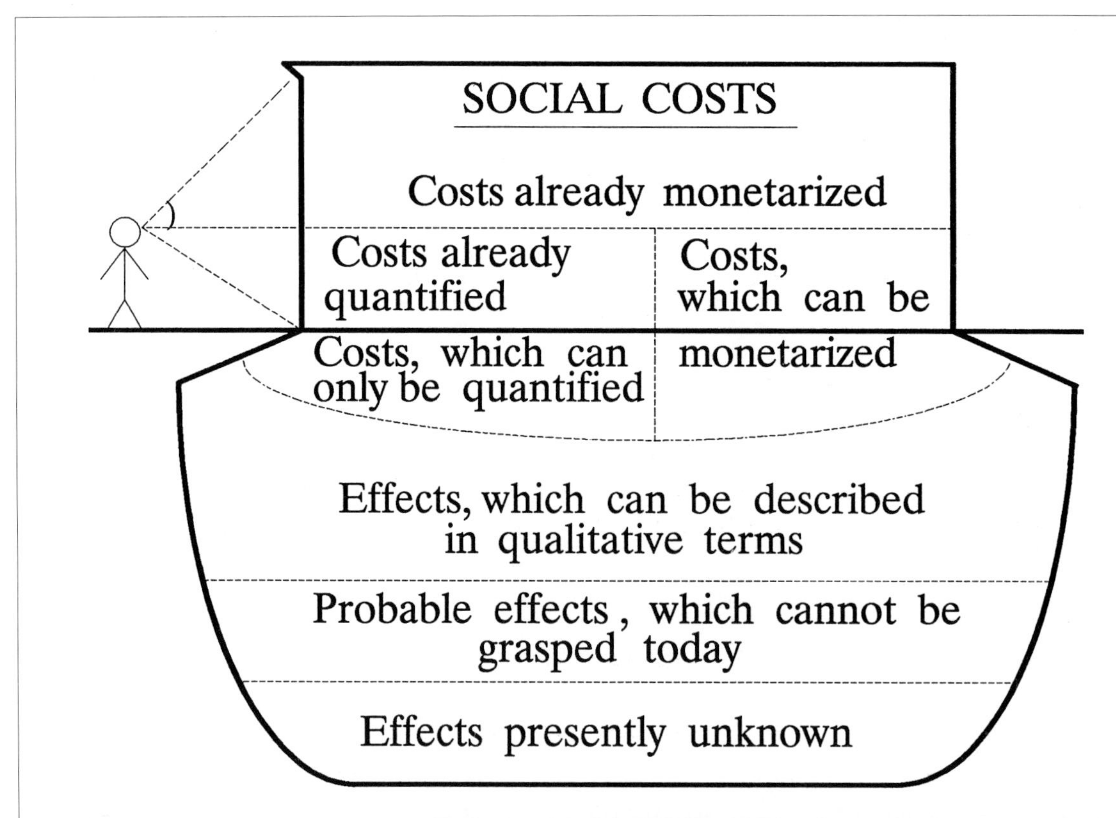

Figure 2. The total social costs of electricity consumption can be compared to an iceberg. From the typical policy or economic view, the costs underneath the water line are hidden.

Figure 3. Analysis of the costs over time of two competing technologies:

P_{ER} represents wind energy, a typical example of renewable energy sources. P_{ECI} represents conventional electricity generation, with only "internal" costs (those that show up on a traditional balance sheet) included. P_{ECS} represents conventional electricity generation, but with social costs considered.

In the top panel, no social costs are considered for the comparison. At first, the conventional electricity is less expensive per user; but over time, wind energy becomes more cost-effective until, at point t_0, it surpasses the conventional electricity generation techniques.

In the middle panel, the point of cost-effectiveness comes sooner. Considering the social costs (such as pollution) makes a renewable source more attractive. This point of cost- effectiveness (t_1) supports wind energy not only sooner, but at a higher price.

The bottom panel suggests the pattern of market penetration for both analyses. Whether or not social costs are not considered, market diffusion of wind energy follows the same rough pattern. If social costs are considered (Q_{ECI}). the penetration still takes place, but at a later date. The delay is represented here as the gap between Q_{ECS} and Q_{ECI}.

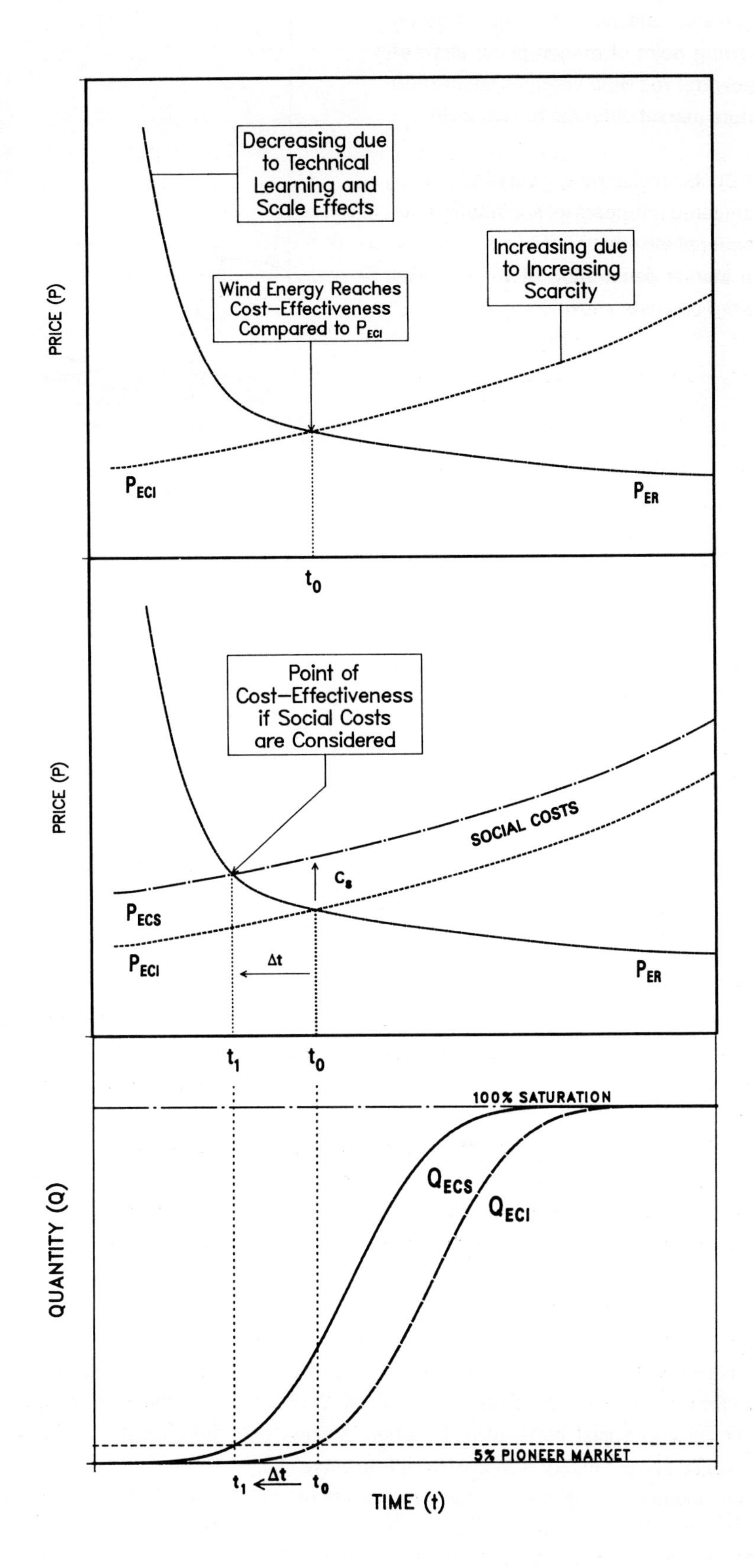

Figure 4. Infleunce of social costs on starting point of market penetration of decentralized wind energy systems and future market diffusion to year 2030

(a) Costs for electricity from wind energy compared with costs for substituted conventional electricity
(b) Market penetration of wind energy based on costs shown above

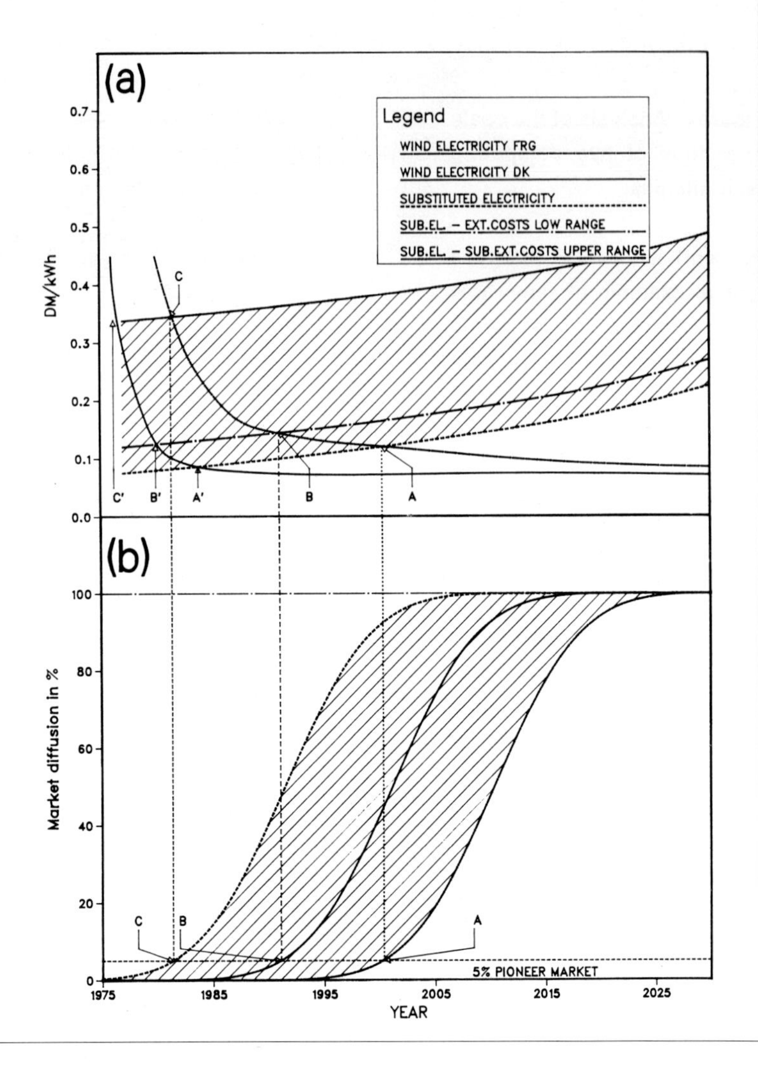

tion results: The wrong price signals are given through the markets to the potential investor for the choice of energy technologies.

4.3.3.1 Wind energy — competitive as early as 1981: Because cost effectiveness does not lead to instant technology substitution — but to a substitution, or market diffusion, process that may easily stretch over 20 or more years — one can picture the impact of not considering social costs entirely in terms of time delay. Figure 3 also shows the curve of market penetration for wind power technologies (as measured in the quantity of electricity sold). The left-hand curve shows the pattern when social costs are understood. But if we do not consider social costs, then the whole process of introducing this technology is delayed, as it is in curve Q_{ECI}, past the optimum introduction schedule for society or customers. (The amount of time delay is shown as Δt.)

The two charts in Figure 4, similarly, show the amount of expense and time lost between now and the year 2025 in West Germany — by not including social costs in investment decisions on wind energy systems. A similar analysis is shown for photovoltaic energy systems in Figure 5.

As we see from Figure 4, the German wind energy cost curve intersects with the market price curve of the electricity around the year 2002 (point A). At this point in time, wind energy produced by a private autoproducer would be competitive with the electricity from the grid at market prices not including social costs. If we add a minimum figure of social costs to this market price curve, the point of competitiveness for wind energy comes much earlier — point B, or 1991. In other words, this point is already upon us.

If we add a higher (but still credible) estimate of social costs, then the cost-effectiveness point came in 1981. If that maximum figure is correct, wind power has been more cost-effective than fossil power plants for eleven years, but this information was hidden by the accounting structure.

4.3.3.2 Photovoltaics — Competitive as early as 2002: Figure 5 illustrates a similar analysis for photovoltaics — competing with electricity from the power grid in the Federal Republic of Germany. The cost curve shown has been estimated on the basis of eight different studies on photovoltaic energy cost developments (Winter et al., 1983; Starr and Palz, 1987; DeMeo and Taylor, 1984 and DeMeo 1987;

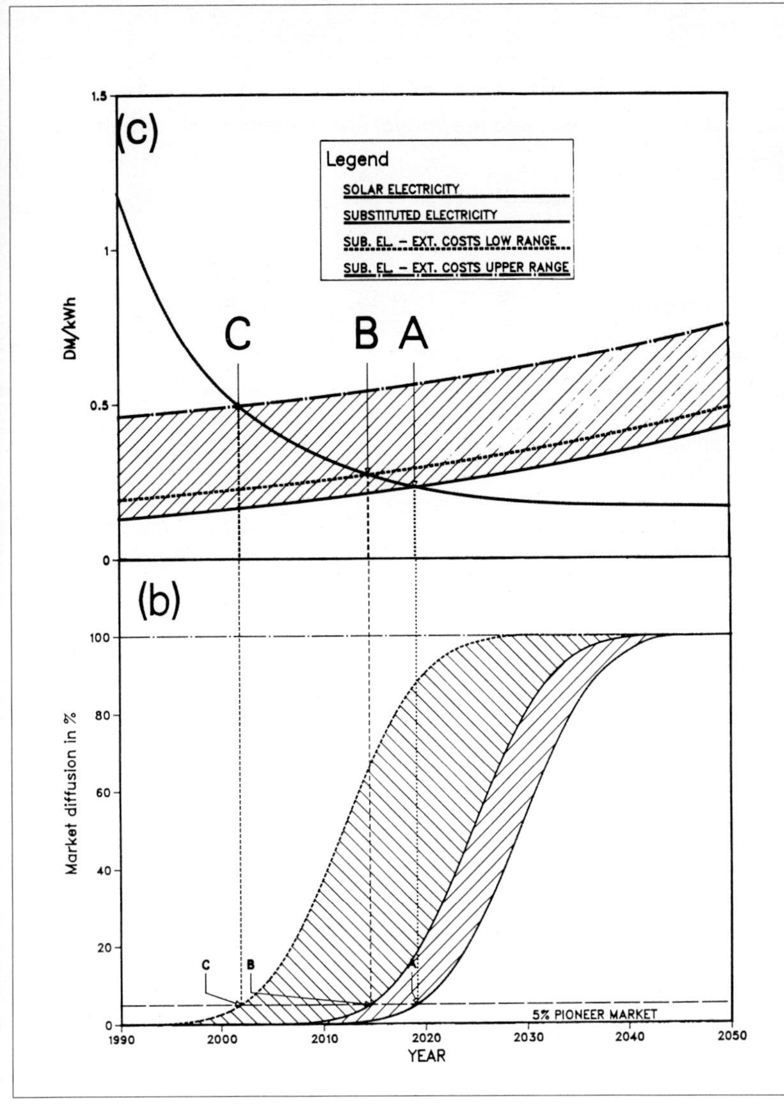

Figure 5. Influence of social costs on starting point of market penetration of decentralized photovoltaic systems and future market diffusion to year 2040.

(a) Costs for photovoltaic electricity compared with costs for substituted conventional electricity
(b) Market penetration of photovoltaics based on costs shown above

Authier, 1987; Nitsch and Räuber, 1987; Wolf, 1987; Stolte, 1982; and Fabre, 1987). Later comparison to other analyses has shown that our estimated cost degression may be conservative. For a more favourable climate like Southern Spain or Southern California, the PV costs can almost be divided by two, simply because of the greater amount of solar radiation per square metre and year in those locales. At the same time, the cost of electricity from diesel generators in isolated locations may be as high as 0.5 to 1.5 DM/kWh depending on the specific transportation costs.

(a) Costs for photovoltaic electricity compared with costs for substituted conventional electricity
(b) Market penetration of photovoltaics based on costs shown above

Once again, the point of intersection shows the time when photovoltaics will be more cost effective than conventional power generation, even without social costs factored in. That moment is the year 2019 (point A). With a minimum estimate of social costs, the date moves ahead 5 years, to point B: 2014. (The estimated minimum social costs for photovoltaics, compared to conventional electricity with new technology for fossil plants, is 0.06 DM_{82} per kilowatt hour.) With the higher estimate of social costs (0.33 DM_{82}/kWh, still including the new fossil plants) photovoltaics reaches its competitive position in the year 2002. Figure 5 illustrates the shifts in market penetration accordingly.

Due to the substantially higher costs of photovoltaics today, the inclusion of social costs will not have an instant effect on its market introduction (as it would in the case of wind energy). But when the future of one or two decades from now is considered, the inclusion of social costs changes the competitive situation and market diffusion of photovoltaics dramatically.

4.3.3.3 The analytical background: These estimates are derived from the social costs given in Table 2. All assumption for this analysis are given in Table 3.

4.3.3.4 The social cost curves of energy conservation and other renewable technologies: A preliminary study is being conducted by Hoffman (1991) at the time of this writing, that compares increased wall insulation in private homes to

Table 3. Assumptions underlying the analysis of social costs and the impact on the competitive situation of wind and photovoltaics (Hohmeyer, 1990)

Assumption	Value
1. General Assumptions	
1.1 Price of Substitutable Conventional Electricity (1982)	25.1 Pf_{82}/kWh
1.2 Working price (62,5%)	15.6 Pf_{82}/kWh
1.3 Payment for Electricity Supplied to the Public Grid	6.5 Pf_{82}/kWh
1.4 Real Price Escalation of Conventionally Produced Electricity	2%/Year
1.5 Real Interest Rate for the Financing of New Investments in Wind and Photovoltaic Installations	5%/Year
1.6 Market Potential for Wind and Photovoltaic Installations	20% TWh/Year
1.7 "Pioneer Market" (5% of the market potential)	1 TWh/Year
1.8 Time Period for the Diffusion Phase (5% to 95%)	20 Years
2. Assumptions About Wind Energy	
2.1 Share of Wind Energy Consumed by Owner	20%
2.2 Share Sold to Utility	80%
2.3a Compound Gain of Wind Electricity (1982)	10.2 Pf_{82}/kWh
2.3b Compound Gain of Wind Electricity Based on Working Price Assumption	8.3 Pf_{82}/kWh
2.4 Life Expectancy of Wind Energy Facilities	15 Years
2.5 Annuity	9.63%/Year
2.6 Operating and Maintenance Cost	1.5%/Year
2.7 Wind Energy Costs in West Germany[a]	
1980	44.8 Pf_{82}/kWh
1986	19.6 Pf_{82}/kWh
1990	15.0 Pf_{82}/kWh
2000	12.1 Pf_{82}/kWh
2010	10.2 Pf_{82}/kWh
2030	8.4 Pf_{82}/kWh
2.8 Wind Energy Costs in Denmark[a]	
1980	12.5 Pf_{82}/kWh
1986	9.1 Pf_{82}/kWh
1990	7.6 Pf_{82}/kWh
2010	7.4 Pf_{82}/kWh
2030	7.0 Pf_{82}/kWh
3. Assumptions About Photovoltaics	
3.1 Share of Photovoltaic Energy Consumed by Owner	50%
3.2 Share Sold to Utility	50%
3.3a Compound Gain of Solar Current	15.8 Pf_{82}/kWh
3.3b Compound Gain of Solar Current Based on Working Price Assumption	11.1 Pf_{82}/kWh
3.4 Life Expectancy of Solar Facilities	20-30 years
3.5 Annuity	8,02-6.505%/a
3.6 Operating and Maintenance Cost	12 Pf_{82}/Wp a
3.7 Solar Energy Costs	
1982	267 Pf_{82}/kWh
1990	122 Pf_{82}/kWh
2000	62 Pf_{82}/kWh
2010	42 Pf_{82}/kWh
2020	26 Pf_{82}/kWh
2050	26 Pf_{82}/kWh

Pf_{82} = Pfennig, 0.01 of a German Deutsche Mark, 1982 prices
TWh = Terawatt hour
DM = Deutsche Mark in 1982 prices
[a] For the electricity costs of small wind energy systems of 50 to 100 kW nominal power, a cost curve has been derived. These estimates are based on the few available German wind energy cost figures for the period 1980-1986 and on well-documented Danish wind energy data for the years 1975-1985.

central gas heating systems and single unit night storage electrical heating systems in Germany. Using process chain analysis and taking account of the impacts from the most important processes in insulation manufacturing and gas and electricity production, Hoffman (1991) arrives at a net figure of approximately 0.05 DM87/kWh final energy — the social costs avoided by increased insulation.

Ignoring these heating-related social costs leads to a severe underestimation of the real costs of energy. In 1987, German households paid about 0.05 DM/kWh for gas and approximately 0.116 DM/kWh for low tarif heating electricity. Thus, the real costs of gas are about double the market price, while heating electricity should cost approximately 140-150% of its present market price. Today, a West German household deciding on investments in home insulation is given the wrong price information, leading to substantial underutilization of rational energy technologies, and far too high energy consumption.

For most other rational energy use technologies (heat exchangers, combined heat and power generation, use of microelectronics for power regulation, fossil-fired heat pumps), similar results can be expected. Like renewables, these technologies are significantly under-valued. Many of them, in addition, could play an important role in coping with the global climate challenge in the short run.

5 Obstacles and Market Imperfections

There is no doubt that energy efficiency improvements have at least the same magnitude of economic potential as conventional energy supply systems, and far more opportunity to reduce greenhouse gas emissions. But to judge from the history of today's power structures, it is very likely that the high potentials of rational energy use will be overlooked or judged as "purely theoretical" or "unfeasible."

Of course, it is not easy to explore the economic potentials of energy efficiency, the "fifth energy resource," because the technologies are decentralized. Instead of a dozen large energy supply companies or engineering companies in any country, millions of energy consumers have to choose independently these new investments and organizational measures. The heterogeneity and diversity of energy consumers and manufacturers of energy-efficient equipment causes a low perception of the high potentials of energy efficiency. But in itself, power structures are not sufficiently developed to keep this technology from being realized; after all, energy supply technologies have flourished during the last four decades.

In theory, given all the benefits of rational energy use, a perfect market performance would optimally invest in, and allocate the rewards from, these new energy technologies and strategies. In practice, however, many obstacles and market imperfections prevent profitable energy-saving potentials from being fully realized (IEA, 1987; Jochem and Gruber, 1990; Hirst, 1991; IEA, 1991; Jhirad and Mintzer, 1992, this volume).

5.1 Lack of knowledge, know-how and technical skills

Private households and car drivers, small and medium-sized companies and small public administrations do not have enough knowledge about possibilities for energy saving or enough technical skills. Lack of information and knowledge is not found only among energy consumers, but also among architects, consulting engineers and installers (Stern, 1985). These groups have a remarkable influence on the investment decisions of builders, small and medium-sized companies, and public authorities. The construction industry and many medium-sized firms in the investment industries face the same problem as small companies on the user's side. Managers, preoccupied with routine business, can only engage themselves in the most immediately important tasks. Rational energy use, with its slow payoffs and learning curve, goes on the back burner.

5.2 Lack of access to capital and historically or socially formed investment patterns

The same small energy consumers, even if they gain knowledge, often face difficulties in raising funds for energy efficiency investments. Usually, their own capital is limited and additional credit is expensive. Especially when the interest rate is high, the firms and private households prefer to accept higher current costs and the risk of rising energy prices instead of taking a postponed energy credit.

In most small and medium-sized companies, all investments — except those for infrastructure — are decided according to payback periods instead of internal rates-of-return calculations. If the lifespan of energy-saving investments (a new condensing boiler or a heat exchanger), is longer than the remaining useful life of existing production plants and machinery, the entrepreneur expects — consciously or unconsciously — a higher profitability from energy-saving investments.

For small and medium-sized local government units in many countries, lack of funds is a severe constraint. Many communities with high rates of unemployment (like Germany) are highly indebted. In addition, energy-saving investments mostly remain invisible and do not contribute to politicians' positive public images.

5.3 Separation of expenditure and benefit

The owner of a building, or a set of energy-consuming equipment, is not always identical with the user. Therefore, two problems exist (IEA, 1987; IEA, 1991):

- For several reasons (lack of market transparency, difference in the financial and operating public authority of a public building) the owner does not receive the complete payoff on energy-saving investments or conversion to another energy source; he may choose a more profitable investment.

- When making energy-saving investments — if he is allowed to do so — the user runs the risk of not getting the payoff for the whole life cycle of the investment, if he terminates his residence or contract earlier.

This obstacle affects the adoption of efficient space heating, air conditioning, ventilation, cooling and lighting equipment in leased buildings and appliances. It is also important in the public sector, where schools, sports halls, hospitals and leased office buildings may have a variety of owners — or where local governments operate and use buildings which state or federal governments own.

5.4 Disparity of profitability expectations of energy supply and energy demand

The lack of energy efficiency-related knowledge among small energy consumers causes a higher risk perception, so that energy consumers and suppliers expect different rates of return on investments. According to available information, energy supply companies in most OECD countries are willing to accept nominal internal rates of return of 8 to 20% after tax for major supply projects (IEA, 1987; Chesshire, 1986). For energy conservation investments, however, energy consumers only demand payback periods between one and five years, which are equivalent to nominal internal rates-of-return of about 15%-50%. This intersectoral disparity in the rate-of-return expectations favours investments in energy supply.

Preliminary estimates suggest that the effect of this intersectoral disparity of profitability expectations is at least a 10 to 15% distortion of energy-saving investments (Jochem and Gruber, 1990).

5.5 The impact of grid-based energy price structures on efficient energy use

The structure of gas, electricity, and district heat tariffs for small consumers, and the level of the load-independent energy charge, are important for energy conservation. Because tariff structures are usually designed in two parts to reflect two services — the potential to obtain a certain amount of capacity at any given time, and the delivered energy — the capacity charge plays an important role in profitability calculations (IEA, 1991).

Among 20 different utilities in OECD member countries, between 53% and 94% of the residential charges for electricity are fixed — they bear no relation to the amount of energy used. In itself, the difference between 53% and 94% means a 75% difference in the profitability of electricity saving investments in the home (Jochem and Gruber, 1990). The effect on profitability may be similar in industry, at least in cases where electricity saving investments do not reduce capacity demand, such as inverters on electric engines. In addition, in most OECD countries utilities do not offer time-of-use or seasonal rates, which would reward consumers for using energy during off-peak hours. Peak demand, a significant cost factor for any power or gas company, is particularly responsive to energy-efficiency improvement.

5.6 Legal and administrative obstacles

There are legal and administrative obstacles in almost all energy sectors. They are mostly country-specific, and often date back to before 1973, when there were declining energy prices and no threat of global warming. Examples are:

- Social housing with upper rental limits, which restrict capital-intensive solutions for refurbishing multi-family buildings or for the installation of highly efficient electrical appliances;

- In West Germany, a price-dependent tax on electric light bulbs discriminates against high-efficiency bulbs;

- In public budget planning, the budgets for operating costs are often kept separate from the investment budget, which means that energy-efficiency investments are not rewarded through operations savings;

- German legislation does not allow the construction of oil- or gas-fired cogeneration plants of a capacity above 10 megawatts, in order to protect domestic hard coal production — but coal-fired plants only become profitable above 40 megawatts;

- Mechanisms to acknowledge decentrally achieved energy savings in public or private administrations are often completely lacking.

6 The Role of a Comprehensive Energy Policy

For every obstacle to achieving the potential energy savings — and hence emission reductions — there is an interrelated measure of energy policy which could remove that obstacle. (Some examples are shown in Figure 6). But the choice of which policies to pursue must be made with care.

During the last 20 years, individually implemented policy instruments — from information to training to grants to energy taxes — have often produced very poor results (Gruber et al., 1982; Bonaiti, 1989). On the other hand, integrated energy demand policies — which consider the interdependence of regulations, consultation "on the spot," training programmes and financial aid — have been very successful. These include residential programmes which have been in place in Sweden and Denmark since 1978, and which have produced impressive results. In Denmark, between 1972 and 1987, annual consumption in space heating dropped from 1.3 gigajoules per square metre (GJ/m^2) by 45%, to almost 0.7 GJ/m^2). In Sweden, residential energy intensity declined by 35% during the past 10 years (Smith-Hansen, 1988).

One should not expect all instruments of energy demand policy to be initiated by governments. Companies, utilities, and industrial associations shall also play a part. Given several obstacles which hinder economic energy-saving potentials from being fully realized, any actor will look for a "central instrument" that could alleviate *all* obstacles. In

Figure 6. Scheme of the interrelationships between market imperfections involving efficient energy use (left column) and policy measures to alleviate them (right column). Given the existing market imperfections, the economic energy saving potentials will not be fully realized. Policy measures, however, may alleviate those obstacles. A set of measures may be more effective in cases of several obstacles in a given sector than individual measures.

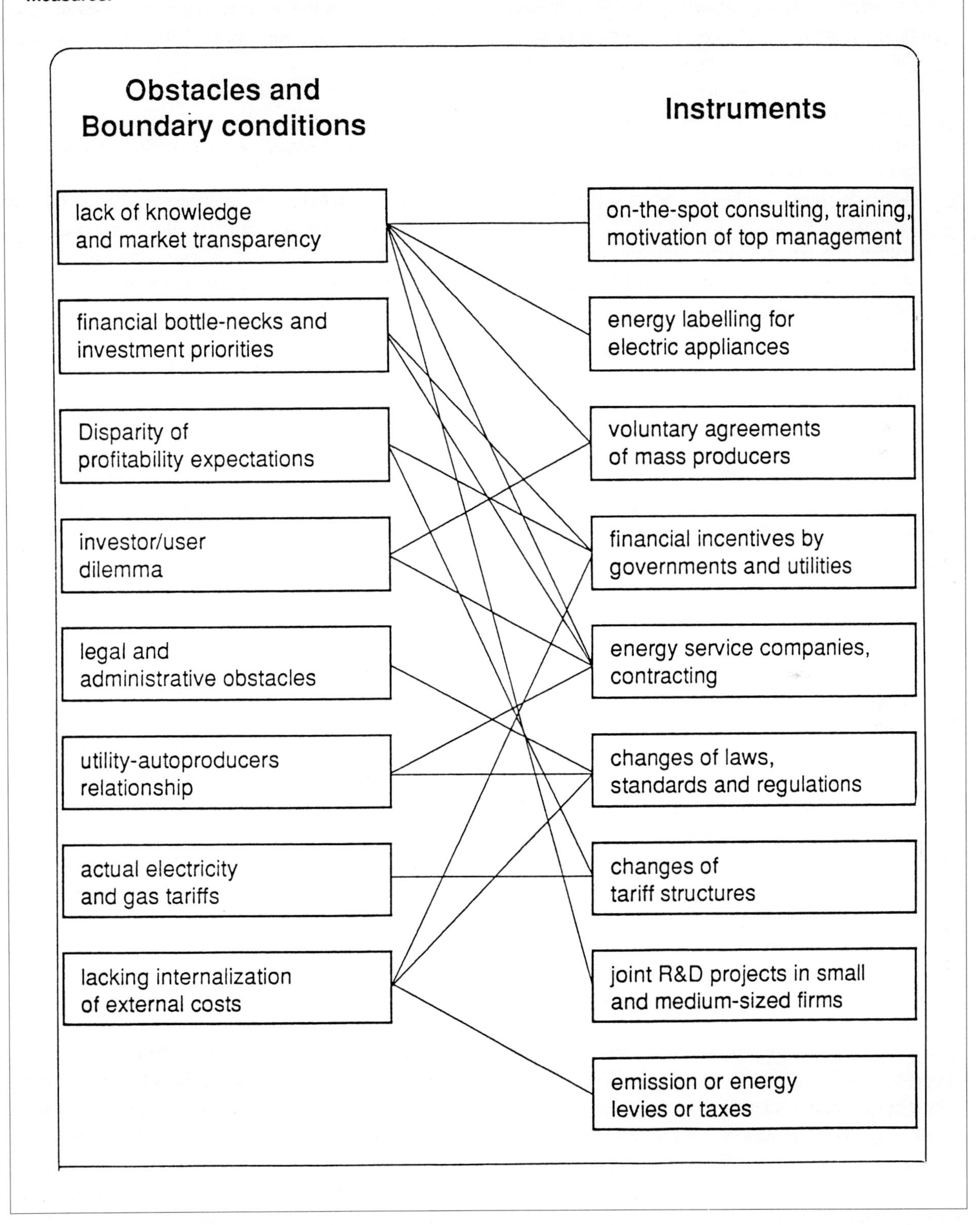

the case of mass products, performance standards are considered to be a very efficient instrument — because standards can be made after some discussions with scientists, engineers, and a few industrial associations. Standards and regulations avoid the need for information dissemination among, consultancy with, and training of, millions of households, car drivers, and small- and medium-sized companies.

But no single, most efficient instrument will be available in all cases of individual energy-efficient solutions (such as the refurbishing of buildings or efficiency improvements in industrial plants). In these cases, a package of policy measures has to be implemented to alleviate several obstacles simultaneously.

A good example is the refurbishing of residential buildings. Homes and apartment houses consume about 20% of the total final energy in the Common Market countries. Refurbishing a building may be primarily an individual event, but its effectiveness depends on such political and social remedies as these:

- Advanced education and training of architects, planners, installers and private home-builders, as carried out in the Swiss "impulse programme," which has had outstanding results since 1978;

- Information and education for landlords and home-owners (particularly on the substitution of energy cost for capital cost);

- Training professional advisors to perform audits and give practicable recommendations;

- A subsidy for these energy audits, which may otherwise be considered too costly by the landlords or home-owners; these subsidies have proven cost-effective (Smith-Hansen, 1988);

- The subsidizing of investments may be bound to a registered energy consultant and a formal heat survey report;

- The government may create a market for these services through a regulation that demands a formal heat survey in the case of a changeover of tenants in dwellings, or a change of ownership of houses and buildings;

- An investment subsidy scheme for specified groups of home-owners or multi-family buildings may be needed to overcome financial bottlenecks. But the cost-effectiveness of this instrument has often been overestimated (Gruber et al., 1982);

- An economically justified minimum thickness of insulation may be secured by new building codes, which should also cover the refurbishing of buildings;

- New rules for the pricing of electricity, gas, and district heat could abolish fixed standing charges for these energies;

- Flexible import taxes on oil may help home-owners and landlords in periods of falling oil and gas prices.

The Danish and Swedish energy-saving programmes owe their success, in large part, to this multi-measure character. Similarly, a comprehensive West German energy conservation policy is expected to play a major role in reducing greenhouse gas emissions between 1987 and 2005 (Enquete Commission, 1991). As shown in Figure 7, the "energy saving scenario," if put into effect, would reduce energy demand 20% by 2005, or 1.2% per year.

Similar results have been calculated and discussed for the Netherlands for the year 2010 by Becht and Van Soest (1989). The sustainable energy policy in that nation is expected to create a 2% annual efficiency improvement over *laissez-faire* development.

7 Bringing Forth the Real Cost of Energy

We have seen that — all costs considered — renewables and home insulation have considerably lower relative costs than market prices would indicate. Moreover, the possibilities for reducing greenhouse gas emissions seem to be high, and achievable at very low cost. In short, we are all — in OECD and developing nations alike — subsidizing our present low market prices of conventional energy sources by not accounting for their true social costs. We are allowing parties who do consume less energy to subsidize those with higher levels of energy consumption. By doing this, we are wasting energy at the expense of future generations.

A variety of energy policies do exist to internalize the social cost of energy into the cost people pay for energy — and thus promote energy saving and sustainable development. This can be done by charging taxes or levies; or, if that does not seem to be feasible in the short run, an increase in the buy-back rate (which utilities must pay for electricity produced from renewable energy sources and pumped back into the power grid) can be a starting point for setting things right.

In Germany the Federal Government has enacted a law, in effect since January 1st, 1991, to increase buy-back rates (for electricity from wind turbines and photovoltaic installations) to 90% of the rate charged to final consumers. This represents, in effect, a doubling of buy-back rates. The same law prescribes rates of 75% for electricity from biogas plants and small hydro installations. It also supports a massive expansion in private applications for building permits for wind energy turbines in the coastal areas of Germany, which have average wind speeds above 5m/s.

Many examples exist of such far-sighted policies. If enough countries follow suit, we will see that some solutions to global climate change-induced problems are a lot less

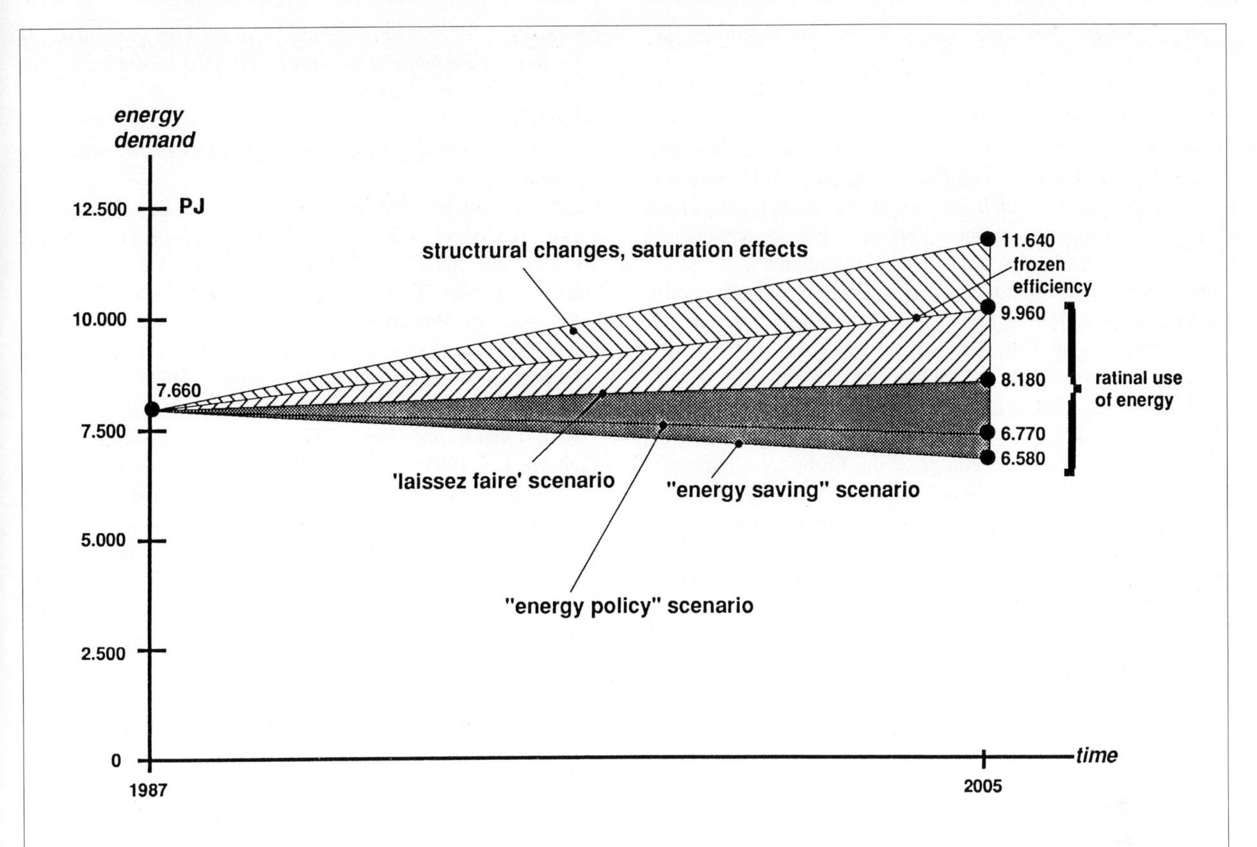

Figure 7. Effects of structural changes and energy efficiency on projected final energy demand, FRG, 1987 to 2005. Structural effects and saturation of energy-intensive products and uses will support the effects of energy efficiency in order to decouple economic growth and energy use as well as greenhouse gas emissions of industrialized countries.

expensive than most people expect. We will also discover that the true economics of renewables and rational energy use are far better than market prices show today, or first glance evidence suggests.

References

Authier, B., 1987: *Kostendegressionspotentiale in der Photovoltaik.* Draft. Burghausen.

Barbir, R., T.N. Veziroglu and H. J. Plass, 1990: Environmental damage due to fossil fuels use. *International Journal on Hydrogen Energy*, **15**, 739-749.

Becht, H.Y. and J.P van Soest, 1989: *Energy Conservation for a long-term, substainable energy policy.* Centrum voor energiebesparing, Delft, The Netherlands.

Bonaiti, J.P., 1989: *Local conservation programmes in France: 1979 - 1986 Methods of diffusion of technical and social innovation.* Manuscript. Inst. Energy Policy and Economics, University of Grenoble.

CEC (Commission of the European Communities), 1985: Links between the rational use of energy and job creation. *Energy in Europe*, **1**(1), 36-38.

Chesshire, J., 1986: *An Energy Efficient Future: A Strategy for the UK.* SPRU, Univ. of Sussex, Brighton.

Contemporary Policy Issues, 1990: *Social and Private Costs of Alternative Energy Technologies.* **VIII**(3) Huntington Beach, California.

Dacy, D.C., R.E. Kuenne and P. McCoy, 1980: Employment impacts of achieving automobile efficiency standards in the United States. *Appl. Economics* **12**(3), 295-312.

DeMeo, E.A., 1987: Photovoltaics for electric utility applications. In *Technical Digest of the International Photovoltaic Solar Energy Conference 3*, held in Tokyo in 1987. pp. 777-783.

DeMeo, E.A. and R.W. Taylor, 1984: Solar photovoltaic power systems: an electric utility and R&D perspective. *Science,* **224**(4646), 245-251.

Enquete-Kommission des Deutschen Bundestages, 1991: *Protecting the Earth - A Status Report with Recommendations for a New Energy Policy.* 2 Volumes, Bonner Universitäts-Buchdrucherei, Bonn.

Fabre, E., 1987: Fabrication and production costs of solar cells and solar generators. In *ISES Solar World Congress 1987, Book of Abstracts*, **1**, Hamburg, pp.1.W.1.03.

Garnreiter, F., 1982: International competitiveness of energy-intensive industries. *Aluminum,* **58**(7), E125-E131.

Gruber, E., F. Garnreiter, E. Jochem, U. Kuntze, W. Mannsbart, F. Meyer-Krahmer and S. Overkott, 1982: Evaluation of Energy Conservation Programmes in the EC Countries, ISI, Karlsruhe.

Hirst, E., 1991: Improving energy efficiency in the USA: the federal role. *Energy Policy*, **19**(6), 567-577.

Hoffman, Christian, 1991: Soziale Kosten des Heizenergieverbrauchs, Nettobilanz der Sozialen Kosten des Heizenergieverbrauchs im Vergleich mit verschiedenen Technologien der Wärmedammung als Beispiele Rationeller Energieverwendung. Universität Karlsruhe, unpublished Diplomarbeit.

Hohmeyer, O., 1988: *Social Costs of Energy*, Springer, Berlin.

Hohmeyer, O., 1990a: Social costs of electricity generation: wind and photovoltaic versus fossil and nuclear. *Contemporary Policy Issues*. **8**(3), 255-282.

Hohmeyer, O., 1990b: *Latest Results of the International Discussion on the Social Costs of Energy - How Does Wind Compare Today?* Proceedings of the European Community Wind Energy Conference held at Madrid, Spain, September 10-14, 1990, H.S. Stephens and Associates, Bedford, pp. 718-724.

Hohmeyer, O., 1991a: Least-Cost Planning und soziale Kosten. Peter Hennicke (ed.): *Least Cost Planning - Ein neues Konzept zur Regulierung, Planung und Optimierung der Energienutzung*. Springer, Berlin, Heidelberg, New York.

Hohmeyer, O., 1991b: *Adequate Berücksichtigung der Erschöpfbarkeit nicht erneuerbarer Ressourcen*, Paper presented at the Seminar Identifizierung und Internalisierung Externer Effekte der Energieversorgung. Freiburg, April 19, 1991.

Hohmeyer, O. and R. L. Ottinger (eds.), 1991: *External Environmental Costs of Electric Power Production*. Springer, Berlin, Heidelberg, New York.

Hohmeyer, O., E. Jochem, F. Garnreiter and W. Mannsbart, 1985: *Employment Effects of Energy Conservation Investments in EC Countries*, CEC, Brussels (EUR 10 199 EN).

Howarth, R.B. and L. Schipper, 1991: Manufacturing energy use in eight OECD countries: trends through 1988. *The Energy Journal*, **12**(4), pp. 15-40.

IEA (International Energy Agency), 1987: *Energy Conservation in IEA Countries*. OECD, Paris.

IEA (International Energy Agency), 1991: *Energy Efficiency and the Environment*. OECD, Paris.

Jochem, E. and E. Gruber, 1990: Obstacles to rational electricity use and measures to alleviate them. *Energy Policy*, **18**(5), 340-350.

Jochem, E., K. Nickel and W. Mannsbart, 1992: *Impacts and Boundary Conditions of Rational Energy Use*. Karlsruhe (to be published in German).

Kapp, W.S., 1979: *Soziale Kosten der Marktwirtschaft*. Fischer Frankfurt.

Koomey, J., 1990: Comparative Analysis of Monetary Estimates of External Costs Associated with Combusion of Fossil Fuels. New England Conference of Public Utilities Commissioners (ed.): *Environmental Externalities Workshop - Papers Presented*. Portsmouth New Hampshire, USA, New England Conference of Public Utilities Commissioners.

Laquatra, J., 1990: *Energy Efficiency in Rental Housing*. Proceedings of the 1987 Socioeconomic Energy Research and Analysis Conference. USDOE, office of Minority Economic Impact. Washington 1990, pp. 718-730.

Morovic, T., G. Gerritse, G. Jaeckel, E. Jochem, W. Mannsbart, H. Poppke and B. Witt, 1989: *Energy Conservation Indicators II*. Springer, Berlin.

Nitsch, J. and Rauber, 1987: *Teilgutachten Erneuerbare Energiequellen für Baden-Württemberg. Möglichkeiten der Umstrukturierung der Energieversorgung Baden-Württenbergs unter besonderer Berücksichtigung der Stromversorgung*. DLR, Stuttgart.

Prognos/ISI, 1990: *The Development of the Energy Economy in the Federal Republic of Germany until 2010*. mi-Poller Verlag, Stuttgart (in German).

Schlomann, B., 1990: Possibilities and limits in the monetarisation of external costs of energy use. *Energy and Environment*, **1**(3), 199-209.

Smith-Hansen, O., 1988: Energy Efficiency Policy - The Danish case. Rockwool A/S Denmark. Discussion paper at the Int. Conference of IAEE in Luxembourg, July 1988.

Starr, M.R. and W. Palz, 1987: *Photovoltaischer Strom für Europa*. Eine Bewertung, Köln.

Stern, P.C. (ed.), 1985: *Energy Efficiency in Buildings: Behavioral Issues*, Nat. Academy Press, Washington, DC.

Stolte, W.J., 1982: *Photovoltaic Balance-of-Systems Assessment*. Final Report, Bechtel Group. San Francisco, June.

Tschanz, J.F., 1985: *Evaluating Potential Employment Effects of Community Energy Programs*. Argonne National Laboratory, Illinois.

Warren, A., 1984: *The Employment Potential of an Expanded U.K. Energy Conservation Programme*. Proceedings of the ACEEE 1984, Summer Study on Energy Efficiency in Buildings. American Council for an Energy Efficient Economy (Eds.), Washington, 1984, pp. J157 - J172.

Winter, C.-J., J. Nitsch and H. Klaiss, 1983: Sonnenenergie - ihr Beitrag zur künftigen Energieverwendung der Bundesrepublik Deutschland. Paper presented to the VDI-GET, 19 May 1983.

Wolf, M., 1987: Actual development of crystalline Si cells and their technical status in production lines. In *ISES Solar World Congress 1987. Book of Abstracts, Band 1*. Hamburg, pp.1.W1.02.

CHAPTER 16

"Wait and See" versus "No Regrets": Comparing the Costs of Economic Strategies

R. K. Pachauri and Mala Damodaran

Editor's Introduction

The international debate on policy responses to the risks of rapid climate change is moving toward a climax. In the process, scientists, economists, environmentalists and decision-makers have sorted themselves loosely into two camps. One group, while cognizant of the remaining uncertainties in the science of climate change, sees critical relationships and synergisms linking this problem to other aspects of the environment/development challenge. This group sees the risks as sufficiently urgent, and the synergisms sufficiently beneficial to argue for bold initiatives now — especially those that sustain or enhance the prospects for economic development while limiting the dangers of environmental damage. This group advocates a policy strategy based on the principle of "No Regrets" — taking steps now that will pay dividends in the future, whether or not the world is on the edge of a major climate change.

The second group emphasizes the fact that scientific knowledge of the climate system is not complete. They note that current knowledge is, for example, insufficient to predict the regional distribution of future climate change with certainty. As a consequence, this group argues that hasty choices made now could impose large and uncompensated costs on human societies — or might even backfire environmentally. This group urges caution and delay — adopting a "Wait and See" strategy in which no deliberate policy actions are taken until scientific certainty about the timing, distribution, and severity of future climate change has been established.

R.K. Pachauri and Mala Damodaran take up the challenge of comparing the economic implications of these two strategies. They assess the potential impacts of global climate change on agriculture, forests and ecosystems, coastal zones, and human health. They then analyse the effects of the principal response options — focusing on forestry measures (particularly afforestation) to enhance biotic uptake of carbon dioxide, and on improvements in the efficiency of energy supply and use.

In the public debate, economic models have been used to estimate the financial impacts of potential measures that address climate change problems. This chapter offers a careful critique of several of the most widely publicized economic modelling tools — including the "Green GNP" model developed by Nordhaus, the "Global 2100" model of Manne and Richels, and the "Carbon Emissions Trajectory Assessment" of Peck and Teisberg. Pachauri and Damodaran summarize the strengths, limitations, and principal assumptions of each model (and, by extension, of similar models of the same structure) for comparisons of the "No Regrets" and "Wait and See" strategies. The authors then evaluate the models in terms of their regional coverage, sectoral scope, time horizons, and key uncertainties.

Their survey has made Pachauri and Damodaran keenly aware of the serious limitations of economic models, when applied to issues as complex as the risks of rapid climate change. They note that models can be a practical aid to decision-making when used to help politicians understand the implications of realistic scenarios of future developments. But they caution against the seductive allure of large, complex computer models, that tend to make non-economists assume they are predictions of the future. No such predictions can be reliably made with the models available today.

What the models can do effectively is to show possibilities and suggest unexpected linkages between policies. For example, Pachauri and Damodaran conclude that excessively draconian policies to counteract global warming are likely to retard economic development. By contrast, their economic analysis suggests that the most important measures within the "No Regrets" strategy — open dissemination of the best information, standards to improve energy efficiency, forestry programs, CFC phase-outs, and efforts to "get the prices right" by incorporating some form of carbon tax — will internalize environmental costs in energy prices and improve the workings of commercial energy markets. Thus, the "No Regrets" strategy will promote economic development and minimize environmental damage whether or not the planet is about to face a major climate change.

R.K. Pachauri, the Director of the Tata Energy Research Institute in New Delhi, and his colleague, Mala Damodaran, bring a deep sensitivity and long experience to the economic analysis of national energy strategies. Pachauri is a political pragmatist, one of the world's leading economists, and chairman of the International Association of Energy Economists. He has written extensively on economic policy, environmental issues, development strategies, and international relations. He is an articulate spokesman for rational energy policies and for comprehensive national strategies that can promote sustainable development. He is also a keen student of international relations, and an advocate of institutional reforms that can encourage efficient and equitable trade regimes.

One conclusion from the models is worthy of special attention. Under either strategy, the costs of a greenhouse warming will not be evenly distributed. Although we do not yet know who will bear the greatest costs and who will capture the most lasting benefits, some will be apparent "winners" in the near term and others will suffer a decline in general welfare. With the "Wait and See" strategy, benefits will be distributed among stakeholders, independent of their own efforts. On the other hand, measures taken following a "No Regrets" approach offer benefits in direct proportion to the extent they are adopted by individual stakeholders and groups.

The uncertain distribution of benefits causes differences between countries and among interest groups to persist, but there is also much on which many can agree. Pachauri and Damodaran suggest that there is one critical element in the challenge of implementing a global strategy to minimize the risks and the damages due to rapid climate change. The critical piece involves establishing an identity of perceived self-interest between those in the North who are principally responsible for current emissions but simultaneously hold the means to undertake mitigation measures and those in the South who may rapidly increase their emissions in the future but stand to suffer the bulk of the disruption and damage if rapid climate change occurs.

- I. M. M.

1 Introduction: The Debate Over Action

International negotiations are in full swing on a possible worldwide effort to avert the threat of global climate change. In this context, it is important to assess the options for judging the economic costs and benefits of specific measures.

There are essentially two schools of thought among scientists, economists, environmentalists, and national leaders. One group feels that the problem of global climate change is linked to other critically important problems of environment and development. The combined risks are serious enough to require urgent and bold initiatives, even if they impose a substantial cost on society. This group also argues that human welfare, by and large, will be enhanced, not jeopardized, through strong efforts to mitigate environmental effects — that taking strong action will lead to a "no regrets" outcome.

The other segment of public opinion — perhaps not as large in numbers, but influential in decision making — prefers the conservative "wait and see" strategy of postponing action. These proponents argue that the scientific evidence for global warming is incomplete; thus, hastily contrived strategies could do more harm than good. At the very least, the costs of these strategies would lead to a loss of human welfare and far exceed any benefits to human society. Their implementation could lead to bureaucracies and laws that would unnecessarily stifle human activity.

Underlying the debate are two distinct theories about the proper behaviour for coping with risk. The "no regrets" strategy favours a policy of "acting, then learning": opting now for measures whose benefits include experimentation, foresight, and cost-effective prevention. The proponents of this view place a high "option value" on the costs of future disasters — meaning that those potential costs should be factored in to investment calculations today. In that sense, the most cost-effective "insurance compensation" to future generations is taking preventive and adaptive measures *now* (Barbier and Pearce, 1990). Moreover, argue the proponents, early actions offer the prospective extra benefit of learning through experience: of gaining better information about the benefits and costs of action through our first steps.

The converse view believes in "learning, then acting". It holds that, since the uncertainties of global warming may themselves be sensitive to the quality and timing of climate research efforts, all uncertainties should be resolved before taking action. With what proponents of this strategy call "perfect information" (meaning complete, correct and exact predictions of the future), the best "no-risks" strategy could be outlined.[1] Thus, the proponents of this viewpoint argue for a sustained commitment to research — to developing new technological options for conservation, energy supply, and reducing climate uncertainty (Manne and Richels, 1990a). As a result, they say, there would be less need for precautionary emission cutbacks and their attendant costs.

Faced with this debate, we cost-benefit analysts have an unenviable task. Ideally, we must suggest a set of measures which would synthesize both schools of thought — or at least work within the assumptions of either one. Our suggestions should anticipate and help prevent the effects of global climate change. But in the event that climate change does *not* take place, they should offer major benefits anyway for the future of the human race, the health of the planet, and the health of human economies. Fortunately, such measures exist; indeed, one of the primary arguments for adopting them is that, given a realistic form of long-term cost-accounting, they are beneficial in any scenario.

2 Economic Impacts of Global Climate Change

In this debate, each side justifies its claim on the relative economics of the case. But estimating the potential costs of abatement is subject to a great deal of controversy, and the attempt to quantify the damages from global climate change is truly a shot in the dark. As Nordhaus put it (1990), it is a virtual shift from the "terra infirma" of climate change to the "terra incognita" of the social and economic impacts of climate change.

Thus, as a first step towards accuracy, it is necessary to list the possible impacts and try and quantify the extent of these impacts, using the Report of Working Group II of the IPCC (1990) as the most comprehensive documentation:

2.1 *Direct climatic change effects:*

- Costs of climatic extremes (the frequency and the extent of divergence from the observed climatic thresholds) are expected to increase in the coming years.

- Temperature increases in the high latitudes could increase the competition for land, and result in a northward retreat of the southern margin of boreal forests.

- A poleward advance of monsoon rainfall is anticipated. However, the warmer climate of the tropics might also induce more precipitation, and hence an increase in humidity.

2.2 *Effects on agriculture:*

- Soil water availability might be reduced. The consequences for agriculture would depend on whether the effect is felt during the growing or the non-growing season. If climate stabilizes in the future, the effects may be highest during the initial period of rapid change, before farming methods have been adapted to new climates. If rapid climate change continues, the high costs of transition may persist into the long-term future.

- The vulnerability of rain-fed agriculture may increase, adding to the strain on regions with poor-quality farming or population pressures.

- A poleward movement of agriculture may take place. Studies suggest that a 1 °C increase in mean annual temperature would advance the thermal limit of cereal cropping in the mid-latitude northern hemisphere by about 150-200 kilometres, and raise the altitudinal limit to arable agriculture by 150-200 metres.

- World trade flows might also be affected in the future — given that climatic conditions, and hence cropping/cultivation patterns, would change. This could lead to land use changes; reduced sugar from sugarcane might be offset by an increase in yield from sugar beet. The ability to exploit this shift largely depends on regional capabilities.

- CO_2 doubling implies an increase in the photosynthetic rate of 30-100%, suggesting agricultural improvements in adverse (saline, water restricted, etc.) environments, These conditions may favour C3 plants (wheat, rice, barley, legumes, and root crops) relative to C4 plants (the world's principal biomass).

- However, increased ozone concentrations may inhibit plant yield.

- If the plant development process is affected by global climate change, growing periods could either increase or decrease. Since current advances in the high yielding varieties have been induced through shorter growing periods, the probability of a proper yield must be investigated.

- Relatively small changes in seasonal rainfall could have major impacts on viability of agriculture, especially in tropical regions.

- Climate-induced change might affect resistance to pests and disease.

- Change in the yields (and hence carrying capacities) of grasslands could affect farm incomes, rural employment, national food production, food security and exports.

2.3 *Impacts on forests and ecosystems*

- Forests would tend to move northwards at a rate of about 100 km for each degree (centigrade) rise in temperature. Developing nations' fuel supplies would be affected, since

[1] We use the term "perfect information" only in the context of economic theory, and not as a practical possibility. In essence, information can never be perfect, but one can talk about deviations from the ideal.

they depend on non-commercial sources for about 40% of their energy input.

- Changes in water balances could accentuate salinity problems and change runoff patterns. Frequent floods would affect species not adapted to low soil oxygen levels. In competition with woody perennials, weeds would be favoured.

- The accelerated and irreplaceable loss of endangered species, ecosystems, and genetic material is a major cause of concern. There is an associated loss of products obtained from nature. The tropical rain forests, currently being replaced by agricultural land, house the greatest concentration of species diversity. Loss of species can have cascading effects if one of the primary producers in the food chain is affected.

2.4 *Effects of sea level rise*

- Coastal natural ecosystems, which contain some of the largest highest degrees of species diversity, would be affected.

- The primary effects of a rising sea level would be increased coastal flooding, erosion, storm surges and wave activity. Large tracts of land would be inundated. There would then necessarily be a shift in productivity, lifestyles and capacity for people to live in these areas. Impacts will vary based on how much each region is coast dependent (with economies built around ports, oil terminals and fish processing); coast preferring (favouring tourism and coastal residential development) or coast independent (without defence and industries located near the sea).

2.5 *Human health and social impacts*

- Substantial numbers of people would have to be relocated, not only within countries but also in new countries. This could create tremendous social and psychological problems.

- In terms of human health, increased rainfall or temperature (especially in the tropics) would lead to an increase in the mortality rate amongst the aged and very young. Waterborne diseases might find larger occurrence, given the increased chances of flooding.

- Warmer working conditions in steel mills, foundries and forges might provoke health problems, or lower productivity. In an extreme case, industries might migrate northwards towards more temperate regions.

- The thermal delay (potential long time lag before ocean warming affects general climate change) has not been incorporated into the IPCC scenarios. Thus, there might be a delay of between a decade and a century before the full warming effect of these cases can be measured (Abrahamson, 1989; Hoffert, 1992, this volume; Crutzen and Golitsyn, 1992, this volume).

3 Economically Viable Initial Measures

Knowing that the economic consequences of no action at all are great, the question still remains as to what action to recommend. Until recently, much of the analysis of the greenhouse effect has been conducted free from the entrails and encumbrances of the economics of each recommended measure. This lack of financial analysis has led to the advocacy and, at times, adoption of techniques with no economic justification. For example, some extreme arguments for renewable energy use have included no regard for capital and operating costs.

When there has been economic analysis, it has led decision makers into an *ad hoc* approach, where as little is done as possible, and solutions are limited to retrofits and revamps. Financial analysts, viewing the greenhouse effect as a burden that firms and organizations might have to bear at a later stage, are naive about the costs of environmental damage. They have assumed that, because the long-term costs are difficult to quantify, they are less significant than short-term costs. (This is similar to the plight of manufacturing directors who cut quality or service costs, but fail to note the resulting cost of lost customers down the road.)

Nor have many financial analysts noted the potential financial benefits from early action. In particular, there exists a lacuna on the economic viability of techniques which exist now, and which are considered viable even in the absence of catastrophic climate change. At least three general categories of such options exist: forestry, energy conservation, and phasing out of CFCs (as in the Montreal Protocol). Because the third option is already under way, and thus less controversial, we focus here on the first two.

3.1 *Biotic options*

Improvements in forestry and agriculture — prevention of deforestation, reforestation of cleared areas, and the long-term stewardship of farmlands — offer significant potential for storing carbon and displacing fossil fuel use. These biotic options, however, are neither as simple nor as inexpensive as some analysts have hoped. (A large part of the land available for afforestation in the tropics is highly degraded; reclaiming it would require large-scale inputs of both nutrients and time.) While forests can help moderate net carbon emissions, increasing tree plantations cannot compensate for the lack of a comprehensive and enlightened energy policy.

3.1.1 *Forest management improvement*

Forests conserve and protect the soil bank. It has been estimated (Ranganathan, 1979, as quoted in Soni et al., 1990) that soil loss from barren land is 800 times greater than from forested land. Studies (Moore, 1986, as quoted in Soni et al., 1990) reveal that areas cleared of all vegetation show a 40% increase in runoff. Also, peak rate flood discharges from small watersheds could be reduced by as much as 60% by good forest management (Mathur et al., 1979, as quoted in Soni et al., 1990). This would save some of the $462 million which Swaminathan, 1990 (as quoted in Soni et al., 1990) estimates is lost in nitrogen-phosphorus-potassium (NPK) nutrients in India, where 6000 million tonnes of topsoil wash away every year.[2]

Because trees absorb CO_2 and release O_2, forests also help in ecological management of local regions. It has been estimated that the *Ficus religiosa* absorbs 2252 kilograms per hectare (kg/ha) of CO_2 per year and releases 1712 kg/ha/yr of O_2/hr (Soni et al., 1990). A band of 500 metres of forest area around factories reduces SO_2 concentration by 70% and N_2O concentrations by 67%, as well as reducing noise pollution (Soni et al., 1990). In addition, forests and related activities provide employment — about 1019 million man-days per annum (Soni et al., 1990) for collecting, processing and marketing major and minor forest produce.

3.1.2 Forestry cost/benefit ambiguities

Mathur and Soni (1986, as quoted in Soni et al.) estimate that the material benefits from forests in India (timber, fodder, and forest produce) account for only 0.06% of the overall advantage of retaining forests. Ecological benefits represent 99.94%, they say, broken down as shown in Table 1.

[2] Admittedly, these estimates are rather random and the range of variation large. Also the estimate of the loss incurred is dubious since flooding in the plain area is not bad per se, since it deposits the rich, fertile soil on the plains, and helps maintain their productivity. However, afforestation of upland areas is required to prevent the excessive movement of solids.

Table 1. Economic benefits of ecological forests Mathur and Soni calculations

Benefits Yields of...	Tropical forest 450 tonnes/hectare	Subtropical decid 410 t/ha	Temperate forest 300 t/ha
1. Production of oxygen	86.54	78.85	57.69
2. Conversion to animal protein	6.92	6.31	4.62
3. Controlling soil erosion & fertility	86.54	78.85	57.69
4. Recycling water/ controlling humidity	103.85	94.62	69.23
5. Shelter to birds, insects, plants	86.54	78.85	57.69
6. Absorption of CO_2	-	-	-
7. Controlling air pollution	173.08	157.69	115.38
8. Preserving genetic diversity	5n	2n	n
9. Benefits (50 yrs)	543.47	495.17	362.30
10. Average annual benefits/ha	10.87	9.90	7.25

But most measurements of the costs and benefits of any forestry decision are ambiguous. No one knows exactly the limits of the area deforested; the amount of carbon released per acre of vegetation and soil; or whether increased CO_2 concentrations actually "fertilize" standing forests, thereby offsetting some of the carbon emitted by deforestation (Trexler, 1991). Quantification must also include such uncertain variables as the life cycle and societal costs of implementing biotic policy options (such as the cost of diverting land from grazing or agriculture), the costs and quality of the land (type of soil, remnants of old vegetation, slope, aspect against the wind), and the recurring costs of planting trees. Additionally, carbon benefits are subject to uncertainty, affected by future natural disasters and an unknown rate of long-term growth. For this reason, Mathur and Soni do not assign the absorption of CO_2 a value in their table.

Thus, estimates of the net costs of large-scale afforestation policies vary significantly. It has been computed for India, for instance, that reforesting recently cut forest land (a relatively cheap option, especially if mechanized methods can be used) will range from $96/hectare to $154/hectare (Chaturvedi, 1991)[3]. Afforestation of land denuded five or more years ago, and allowed to degrade ever since, is far more expensive. The variation is also greater: between $385 and $1539 per hectare (Chaturvedi, 1991). In the United States, Trexler (1991) has estimated the cost of afforesting croplands and pastures at $43 per tonne of carbon saved (marginal cost) and $24 per tonne (average cost).

These cost estimates, however, do not allow for major revenues and feedback processes from the timber and agricultural marketplaces. The secondary market impacts of large-scale tree planting, including the lowering of wood prices, could reduce the long-term carbon benefits by encouraging more timber sales. Unless the timber grown on large-scale areas is kept out of the market (which would increase the price demanded by farmers for the land), such revenue-based afforestation programmes would push out other forestry investments.

3.1.3 Location strategies for afforestation investment

Where, then, is the most leverage in afforestation investment to reduce global warming? One place is cities. Because urban pollution levels are normally higher than those prevalent in rural areas, an urban tree can be up to 15 times as effective at "reducing" atmospheric CO_2 as a tree in a rural area. Trees are also natural air-conditioners and can reduce temperatures by 3-5 °F. For a city like Washington, each °F decrease in temperature can lower cooling costs by $10,000 per hour (Trexler, 1991).

[3] The conversion rate from Indian rupee equivalent to US dollar was taken at Rs. 26 to a dollar.

Also, while it makes no difference to atmospheric CO_2 concentrations whether a tonne of carbon comes from a temperate or tropical country, the carbon released per acre of cut forest is greater in tropical than in temperate regions. This is because the tropical forest crown, unlike that of temperate forests, contains more carbon than the soil. However, though the scale of deforestation in the currently developing countries is high, this argument should not serve as an excuse to place the entire burden of afforestation on them. The most developed nations have already passed through their stages of initial growth, during which they could not stop their own deforestation. Developing nations now face similar economic demands from their people; if the burden of stopping deforestation is imposed inequitably on the developing nations alone, these demands for economic growth must be met some other way.

3.1.4 *Biomass forestry*

Another prospect for leverage is the restoration of forests with biomass potential. While carbon sequestering by forests (the absorption of carbon dioxide and other compounds from the atmosphere through photosynthetic chemical processes) is well publicized as a strategy for offsetting CO_2 emissions, the substitution of biomass for fossil fuels has not received similar attention. But substituting biomass for coal removes carbon compounds from the atmosphere in two ways: by sequestering it in growing trees, and by reducing the amount of CO_2-polluting coal (or other fossil fuels) that are burned (Hall, Mynick and Williams, 1990). In many cases this strategy is as effective as sequestration, per tonne of biomass, in reducing CO_2 emissions. Moreover, fuel substitution can be carried out indefinitely, while carbon sequestration is only effective until the forest reaches maturity. Also, far greater biomass resources can be committed to fossil fuel substitution, because forest growers will tend to seek biomass species with higher annual yields for energy applications.

Finally, biomass energy is less costly than the displaced fossil fuel energy in many circumstances, so that the net cost of displacing CO_2 emissions is often negative. Thus bio-energy strategies have built in economic incentives that make them inherently easier to implement than many alternate strategies for coping with global warming (Pearce and Barbier, 1991).

The cost of sequestration, per tonne of CO_2 emission, is directly related to the cost of growing biomass. According to Moulton and Richards (as quoted in Hall et al.) the average unit costs of a tree-growing programme to offset 56% of US fossil fuel emissions would be $27 per tonne of carbon. (Marginal costs would be $48/t C.) The annual cost of such a large-scale US effort, some $19.5 billion, might be paid for by a carbon tax of $15/t C on all fossil fuels consumed (Hall et al., 1990). According to the proponents, the effect of this would be to increase the cost of coal based electricity generation by 0.4 cents/kWh (a 7% increase) and the cost of gasoline by one cent per litre. If the sequestering rate were half that estimated by Moulton and Richards, the tax required to elicit the same change would have to be twice as large.

3.2 *"No regrets" policy measures to pay for biotic options*

3.2.1 *The energy tax*

While these biotic options (improving forest management, afforestation, and biomass) could theoretically absorb large quantities of carbon annually, the realized potential often shows significantly less effect. Each policy option has a rising cost curve as more of the potential is used up. The marginal cost curve of an individual forestry programme depends on whether societal or private costs are estimated and if the time frame involved is long or short. Market impatience can provide a serious threat to the long-term effectiveness of biotic policy options; this can be prevented through the imposition of an energy tax.

3.2.2 *Debt-for-nature swaps*

In 1984, T.E. Lovejoy proposed an innovative way to link debt reduction with environmental protection measures in heavily indebted third world countries — the well-known "debt-for-nature" swaps (DFNS). In a swap, a non-governmental organization (NGO) would purchase part of a developing country's debt from the lending bank. The NGO would then offer to write off the debt, or donate the debt instrument to that country's central bank, in exchange for certain environmentally beneficial actions taken in that country.

While the DFNS is good for promoting environmental awareness, strengthening NGOs, protecting biodiversity and perhaps incorporating the popular sector into conservation activities, it is far less effective as a source of funding. Unaddressed by DFNS, for example, are the need for general sustainable development efforts country-wide, as well as the need to bring the for-profit private sector into conservation, and sustainable development efforts (Patterson, 1990). The DFNS is but one mechanism in a multitude and cannot be expected to meet all needs.

3.3 *Energy efficiency in industry*

Energy conservation presents industry planners (and industrial policy planners) with their own choice between a "wait and see" policy versus a "no regrets" strategy. The cost comparison is much like the difference between the costs of a new installation versus a retrofit. While a heavy initial investment in energy-efficient production is generally a more expensive option than the "wait and see" initial investment, in the long run it is often far more cost effective.

In another sense, however, the analogy does not apply, because energy conservation measures can be installed at any time as retrofit options themselves. Indeed, industry does not often have the luxury of constructing a new energy-efficient plant from scratch. Hence, there is a need to understand the factors that determine the savings from retrofit energy conservation measures, as compared to new installations.

3.3.1 The benefits of energy conservation

Some of the benefits are tangible and can be quantified, and others can be neither quantified nor (in some cases) identified with ease. None the less, they are significant.

Still other potential energy savings, while achievable in theory, are lost or diluted in the larger system of the company or manufacturing process. There may, for instance, be a mismatch between upstream and downstream equipment that, in effect, cancels out the energy savings which the upstream equipment achieved on its own. Or a retrofit measure might require modifications to existing processes, which are not made.

These "political" factors, rather than technical limits, make an energy conservation retrofit a more expensive and, in all probability, a less beneficial option. The diversity of such factors makes it difficult to evaluate the general potential for savings through retrofit; every case must be examined on its own basis. For an example of the potential issues, consider the case of a retrofit energy conservation measure in a mini-steel plant with an electric arc furnace.

3.3.2 Retrofitting a steel plant

Any small, scrap-based electric arc furnace steel plant essentially consists of an electric furnace (of a 10-15 tonne capacity) where metal scrap is heated to produce molten steel. This molten steel is then passed through soaking pits, rolling mills, and a reheating furnace, to finally be rolled into billets or blooms. Several methods of reducing energy consumption in a mini steel plant can be applied as retrofit options:

- Converting to a *continuous casting process,* where the molten metal is directly converted into billets, would require replacing or adapting most of the plant equipment; but would result in the most significant savings. The investment required would normally be the deterring factor and the economics would vary with the age of the plant: older plants would gain the most from modernization.

- The use of *Ultra High Power (UHP) transformers* would significantly reduce melting time, and thus boost furnace productivity. This saving would outweigh the increased power consumption; savings of about 20 kWh/t can be expected. These transformers, however, require a strong electric power grid which can provide consistent high power without large voltage dips. Since refractory and electrode consumption would simultaneously increase, a low voltage arc could be used to avoid damage to the refractory. Therefore, the probable loss from increased consumption of refractories and electrodes, along with the downtimes needed for their replacement, would have to be studied. In many cases, it may not be possible to economically employ UHP transformers.

- *Recovery of waste heat:* Using the exhaust waste gas emanating from the furnace to preheat the scrap to about 350 °C before charging the furnace, would result in savings of 30-50 kilowatt hours/tonne (kWh/t) — nearly 7% of the input energy. An additional benefit would be a decrease in refractory and electrode consumption. But it would mean slight underutilization of the furnace transformer, due to the lower heat requirement in the furnace. More significantly, a $58,000-$66,000 initial investment would be required.

- *Oxy-fuel burners:* These provide uniform heating to the furnace, eliminating cold spots, reducing the demand from the power grid, and improving furnace productivity. An investment of $300,000 or more produces savings of 10-50 kWh/t. The use of oxy-fuel burners would, however, increase electrode and refractory consumption — especially in furnaces without water-cooled panels.

- *Ladle refining:* This is a process of refining as the molten metal is being tapped from the furnace, with savings of about 10-50 kWh/t. The corresponding investment ranges from $154,000 to $212,000.

- *Continuous charging:* In this method, additions are made through a fourth hole in the roof of the furnace, rather than through the door, reducing heat loss when the door is open.

- *Eccentric bottom-tapping furnaces:* Another technical improvement, eccentric bottom taps result in a saving of about 10 kWh/t. This is an expensive option since it requires the furnace to be changed.

- *Computerized control systems* for melting power, oxygen flare and additives would cost around $204,000 and yield savings between 10 and 15 kWh/t.

Many other energy conserving measures can be considered: the use of foaming slag, substituting pure lime for limestone, integrating civil works with a rolling mill, and others. In every case, the actual savings might vary due to operating conditions, execution of the change, or the many uncertainties that cloud the estimation of the potential of a retrofit. There may also be unforeseen changes in productivity, which cannot be quantified.

4 Economic Factors: Arguments for Immediate Action

Both biotic management and energy efficiency are well suited to a "no regrets" strategy of immediate action. Both undeniably help reduce greenhouse gases, and both pay back the investment required for them. But they are ignored, in part, because they don't fit well into models; neither the investment, nor the financial returns, nor the ecological results, nor the time required for payback, have been adequately quantified so far.

But being unquantified, to date, does not mean that the goals of the future must remain vague. In supporting a view of "sustainable development" — that developing countries' interests lie in a path that minimizes greenhouse emissions — some stress the need to include targets for economic incentive (Nitze, 1991). William Nitze has written that, during the next decade, governments and corporations, by investing in energy and transportation systems, urban infrastructure, agriculture and the development of new technologies, will shape the world economy for the first few decades of the next century (see Nitze, 1992; this volume).

And there is, perhaps, the most compelling argument for the "no regrets" approach: it contains a vision of the future. If the parties to a greenhouse convention commit themselves to initial practices and goals — for improving the efficiency of energy supply and use, reducing deforestation, and deploying the best available technologies — the decisions will have an innate bias towards the reduction of greenhouse gas emissions. If a start is made now, atmospheric concentrations of GHGs could probably be stabilized at their preindustrial levels or less. This could be achieved by stabilizing emissions early in the next century and achieving modest annual reductions thereafter, with a reasonable, and repaid, financial investment. That sort of use of long-term goals would be more successful than a quick, all-out effort to tackle specific emissions (such as methane and nitrous oxide) at excessive cost, if necessary.

If the parties "wait and see", holding out for subsequent protocols and research, the bias will be weaker and less systematic than a "no regrets" strategy. Even with the best intentions and most alert early-warning systems, this sort of policy will tend (for lack of other directions) to safeguard, as long as possible, the entrenched habits of thought which permitted overextended levels of emissions in the first place. The cumulative difference in outcomes over time would be enormous. Whether intended or not to let business as usual continue for another 20 years, that would be the innate direction of a "wait and see" policy. As a result, far more rapid reductions would be eventually required to stabilize concentrations at an ultimately higher level.

This view has been further elabourated by the United States Environmental Protection Agency (Rahman, 1991). The EPA has stated that delays in setting targets would increase local costs of reducing emissions and hence the eventual amount of global warming.

5 Modelling the Costs of the "No Regrets" Strategy

The ability of society to act immediately, however, is restricted not just by policy, but by the lack of available capital. Given the uncertainties attached, governments can not commit enough resources for a suitable response to the anticipated problem. Hence the need for an international approach to marshalling capital. And, despite the uncertainties we have noted, both governments and international bodies need reliable models of the economic effects of the steps they are contemplating taking.

Scientists and policy makers alike have subjected scientific models on climate change to intense scrutiny, in the hope of shaping an appropriate response strategy to the problem. But considerably less attention has been given to models of *economic* consequences. Thus, the next part of this paper is a review and analysis of the most significant economic models which exist so far. In order to conduct that review, we need to look at the general limitations of these models — beginning with the range of problems on which they have focused their attention.

5.1 The range of economic models

5.1.1 Regional scope

Unfortunately, there is a trade-off in current models between regional and sectoral coverage. Despite the fact that climate change is an inherently global problem, relatively few models in use analyse the economic effects of climate change on a global, or even regional, scale. The greater number of single-country models stems in part from the relative ease of handling one nation's data, and from computational limits. But a wider geographic coverage is vital for analysis of the international trade, migration, and human welfare consequences of climate change-oriented policies. Without worldwide models, the far-flung consequences of environmental missteps are all too easily ignored.

5.1.2 Sectoral scope

Hoeller et al. (1991) have surveyed studies which examine the macroeconomic implications of emission reduction costs in quantitative terms. The most complete global model (in terms of modelling the energy sector and its influence on aggregate economic growth) is that of Manne and Richels, 1990b. However, that model does not distinguish among different industrial sectors. And analysis of different sectors is necessary in pinpointing the interrelationship of policies with the industrial structure.

5.1.3 Temporal scope

In addition to wider regional and sectoral scope, economic models need to analyse greenhouse gas emissions (and their effects) over a much longer time horizon. As explained by Crutzen and Golitsyn (1992; this volume), there are lengthy time lags between emissions and their ultimate effects on climate change. These time lags translate into eventual costs and benefits that should appear in more economic models.

5.1.4 Scope of emissions coverage

Most studies conducted thus far focus on fossil fuel related CO_2 emissions. These emissions account for 60% of greenhouse gas emissions and, additionally, are the easiest to link to anthropogenic activities and economic behaviour. But other emissions (such as sulphur dioxide) will also have effects, and the potential costs should be factored into decisions about which measures to tackle first.

5.2 Key uncertainties of economic models

Whatever their scope, all the models must deal with several key uncertainties that will determine aggregate costs (Pearce and Barbier, 1991):

5.2.1 Interfuel substitution

The degree to which nations are willing to substitute one fuel for another is expressed in policy measures such as the carbon tax. (In this form of fuel tax, the level of taxation is determined by the carbon content of each fuel. Users of natural gas pay less than users of petroleum gasoline. For more discussion of the carbon tax ramifications, see Section 8.3 below.)

5.2.2 The aggregate energy-output link

The ratio of overall energy input to overall productive output varies; it depends on the efficiency with which capital is used, goods are produced, and services are provided. In all cases, an increase in the price of energy will shift investment toward other input factors: e.g., capital or labour. For energy-intensive economies, raising fuel prices may seem to be very costly in the short term. The smaller the energy-output ratio turns out to be — in other words, the more output there is per unit of input — the easier it will be for an economy to make the transition to fewer greenhouse gas emissions.

5.2.3 The availability of capital

Similarly, if there are increases in fossil fuel prices, the adjustment costs are likely to be higher as well — at least in the short run — because the retrofitting of installed technology is costly and capital, which is largely sector-specific, will not be equally available to all industries in all parts of the world. The energy-output level would also be affected by changes in relative prices through a feedback mechanism: over the medium- to long-term, price changes will spur more efficient technologies, which in turn affect prices.

5.2.4 Terms of trade

These could affect aggregate income flows. If the need to reduce emissions leads energy-importing countries to switch to higher-value fuels (such as gas), or to finished goods or technology transfer, it could lead to a short-term deterioration in their balance of payments. Global actions could induce a large redistribution of income between regions, since fossil fuels are unevenly distributed over the world and the rates of usage and exhaustion differ.

5.2.5 The industrial burden

The greenhouse effect involves an important economic "externality" — the actions of corporations and businesses, that are not yet appropriately reflected in the market prices for either goods or energy. Emitters do not pay the price for environmental damage. Hence market prices are lower than social cost, which implies that the production of greenhouse gases is likely to be above the efficient level. Taxes and fines will continue to be proposed to correct this, and efforts are also arising within industry itself (Schwartz et al., 1992; this volume). The results will have an important effect on the way the economics of greenhouse gas reduction plays out. Here, too, the more participation and cooperation among government and industry in boosting — rather than stonewalling — early action, the lower the long-term costs will be.

5.3 The Nordhaus model

Nordhaus (1990) has concentrated on efficient strategies to reduce the costs of climate change — in other words, strategies which maximize overall net economic welfare. (This is the "green GNP" model.) At the heart of the Nordhaus model is the relationship between the CO_2 level in the atmosphere and the economic cost needed to stop emissions or deal with damages. The solid U-shaped line in Figure 1 shows this relationship; called the "impact line", it represents the cost to the economy, on any given date, of various levels of CO_2.

Achieving very low levels of human-made CO_2 (at the left-hand side of the chart) means extremely high costs (to develop fossil fuel substitutes, manage forests, or create efficiencies). On the other hand, if no action is taken, CO_2 is high (at the right-hand side of the chart), and costs also soar (to deal with damages). The costs shrink (in Nordhaus's model, almost to zero) only in the middle, at some balance point of increased CO_2 levels. The location of the balance point (indeed, the slope of the entire curve) depends on the constants used to determine what Nordhaus calls the "impact function."

The wavy line represents the "damage function" — the extent of ecological and property damage caused by greenhouse gas effects. This is more or less greater in environments where CO_2 is greater. But since (according to Nordhaus) little is known about the exact mechanism of the damage function, it is drawn as a wavy line to stress its inexactness. One reason for the waviness of the line is that because of the "fertilization" effect of CO_2 or the attractiveness of warm climates, the greenhouse effect may have some economic advantages that offset damage costs, when measured in aggregate. And, of course, the feedbacks described in Chapter 1 would contribute to the waviness of this line, by building in complex, unpredictable chains of cause and effect.

5.3.1 Costs for incremental units of reducing emissions

Using the same function, Nordhaus goes on to produce curves (Figure 2) for the "marginal costs," that is the incremental cost of reducing CO_2 and other greenhouse gases by one unit (one metric tonne of carbon or one unit of the measure used) and "marginal damage," namely the cost of damage to society from any single additional unit of CO_2 or other greenhouse gases.

The resulting curve shows that, the lower the level of human-caused CO_2 becomes, the higher the cost to reduce any unit of CO_2 and other greenhouse gases. In other words, both charts together suggest that the reduction of the first units of CO_2 is virtually free, and the cost rises thereafter. As

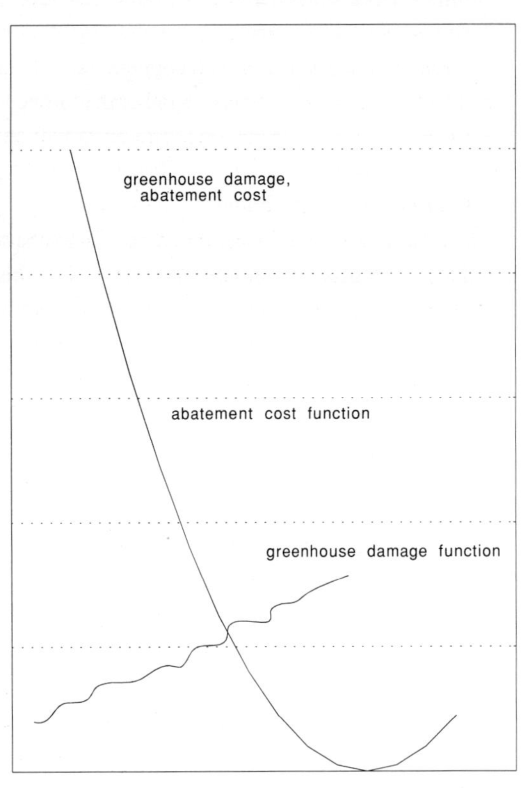

Figure 1. At the heart of the Nordhaus economic model is the relationship between the CO_2 level in the atmosphere and the economic cost needed to stop emissions or deal with damages. This figure demonstrates that relationship. The solid U-shaped line, called the "impact line," represents the cost to the economy, on any given date, of various levels of CO_2. The least cost to society, argues Nordhaus, comes from a CO_2 level in the middle range. If the level is high, the costs of damage repair will be great. If the level is low, that means that (in Nordhaus' view) too much money will have been spent keeping the level of carbon dioxide down.

The wavy line represents the "damage function" — the extent of property and ecological damage caused by greenhouse gas effects. It is drawn as a wavy line to stress its inexactness, but it tends to rise as emissions increase.

As we move leftwards on the chart, reducing CO_2 steadily, we reach some point of diminishing returns, after which it will not be cost effective (or presumably ecologically effective) to reduce greenhouse gases any further. Source: Adopted from Nordhaus, 1990.

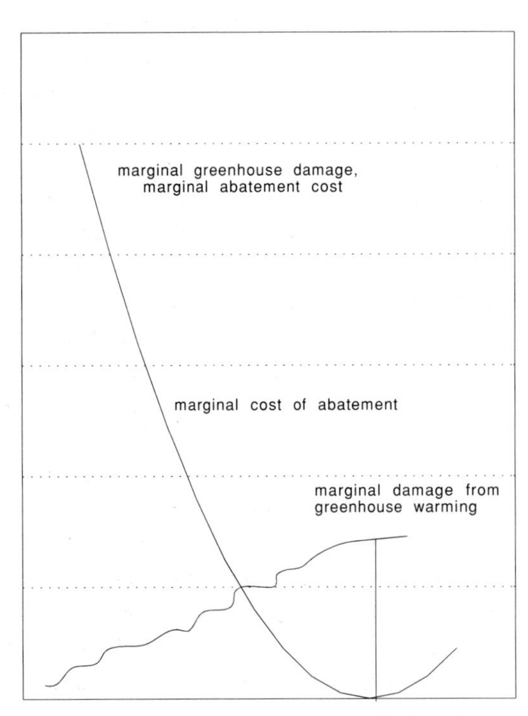

Figure 2. The same analysis as for Figure 1, performed by Nordhaus for "marginal costs" — the incremental costs of reducing CO_2 and other greenhouse gases by one unit. The marginal damage, for example, is the cost of damage to society from any single additional unit of CO_2 or other greenhouse gases. Source: Adopted from Nordhaus, 1990.

we move leftwards on the chart, reducing CO_2 steadily, we will reach some point of diminishing returns (which may or may not be the same as the balance point), after which it will not be cost effective (or presumably ecologically effective) to reduce greenhouse gases any further.

According to Nordhaus, that point of diminishing returns — otherwise thought of as the optimal or most efficient level of control of greenhouse gases — comes at the point where the marginal cost of reduction equals the marginal damage of greenhouse warming. Beyond this point (i.e. to the left of this point of intersection), the incremental gain — reduced damages from greenhouse warming — is outweighed by the incremental cost of additional control measures.

5.3.2 *The Nordhaus model limitations*

The mathematical underpinnings of this model, however, are based on certain assumptions that deserve critique. The estimates of costs and benefits depend upon a 'snapshot' of emissions, concentrations, economic costs, and economic damages at a single point of time. There is no sense of delayed effects, or of initial up-front investments that may pay themselves back later in a variety of ways (as we have shown may happen with both afforestation and energy efficiency).

In an attempt to anticipate the gradual effects of equilibrium warming, the model calculates its impacts in terms of doubled CO_2 concentrations in the atmosphere in the future, while costs are given for reducing CO_2 emission today. However, given the complex economic and climatic dynamics of the situation, those economic calculations are oversimplified: they assume that physical flows of elements through the atmosphere will be constant, instead of unpredictable and varying. They also extrapolate the current mix of industrial sectors in the United States to the rest of the world, overlooking the potential alternative uses of resources, and economic growth efficiencies, that may exist in the developing world.

The calculations also omit other potential market failures due to ozone depletion and air pollution (Nordhaus, 1990). And it assumes that damages would grow at roughly the rate of economic growth, that most emissions will come from energy (rather than industrial or agricultural) sources, that the ratio between energy use and gross national product is more-or-less constant (ignoring energy efficiencies), and that the mix of fuels and energy sources in all countries adds up to the same aggregate mix of emissions.

All of these assumptions are (or can be) incorrect; for example, in developing economies the rate of growth of demand for energy far outstrips the rate of growth of the economy. Even when aggregated into a global calculation, the likelihood of the developed nations counterbalancing this rate of growth is questionable. The study also assumes that the rate of absorption (disappearance of greenhouse gases from the atmosphere) is known; but whenever the emission of GHGs is not known with certainty, the rates of absorption also become indeterminate.

5.3.3 *Nordhaus's policy conclusions: Limited action*

Combining the individual marginal cost curves, Nordhaus arrives at minimum marginal cost estimates if the options are deployed in a cost-effective manner. He concludes that the most effective option is to restrict emissions of CFCs, followed by curbing CO_2 emissions through the imposition of a carbon tax. Forestry options are included in the prescription, but with very little quantitative significance.

Significantly, Nordhaus does not favour consequential and rapid reduction of greenhouse gases. He states that it would be foolhardy to go in for expensive control options, given the uncertainties that cloud the issues. On the basis of the Coolfont workshop, Nordhaus also concludes that CO_2 would have a powerful "fertilization" effect on trees and crops; people do prefer a warmer climate; and technological developments have the potential to offset any harmful effects of climate change. Counter-arguments, such as the influence on plant growth from inadequate rainfall (Gleick, 1992; this volume), are ignored. The appropriate level of control depends, according to Nordhaus, upon three central parametres: the cost of reducing greenhouse gases, the damage to society from greenhouse warming, and the time dynamics reflected in the rate of discount when applied to new investment, with due attention to time lags in the reaction of climate to emissions.

In short, the Nordhaus study estimates that 14% of greenhouse gas emissions can be reduced at extremely low costs, but estimated costs rise sharply with subsequent reduction. The cost estimates that he came up with, at prevailing prices, are presented in Table 2.

Because of the many limitations of the Nordhaus model, however, these figures are probably too pessimistic.

5.4 *The Global 2100 model*

On the basis of their previously developed ETA-MACRO model, Manne and Richels brought forth the Global 2100 model in 1990, incorporating current uncertainties about the impacts of global climate change (Manne and Richels, *Global 2100,* 1990b). This has been done by determining an "expected value of perfect information (EVPI)" as a factor in estimating costs. That value is the value of whatever information is gleaned from a "learn, then act" or "wait and see"

Table 2. Nordhaus ("Green GNP") model
Cost estimates of greenhouse gas emission reduction

Long run marginal costs:		
25 % reduction	$ 38	per tonne of CO_2
50%	119	per tonne
Global costs:		
14 % reduction	$ 3	billion
25%	27	billion
50%	201	billion

strategy. On the basis of their calculations, the expected value (and, consequently, costs) associated with decisions under uncertainty (act, then learn) is 0.167; under a "wait and see" scenario approach (learn, then act) it is 0.124. The EVPI is estimated to be the difference between these two scenario values — that is, 0.042, or about 25% of the expected costs under an "act, then learn" policy.

5.4.1 Assumptions of the Global 2100 model

This model incorporates many of the feedback predictions that were not part of the Nordhaus model. Its projections are based, in part, on the assumption that producers and consumers will be sufficiently farsighted to anticipate both scarcities of energy and environmental restrictions in the coming decades. This is clearly a rather optimistic assumption.

Further assumptions are made about future international oil prices (expected to rise as supplies falter), about regional willingness to import or export oil (increasingly restricted), and about the non-tradeability of emissions quotas. In this model, rising costs and limited supplies of energy retard GNP growth, while rising energy prices lead both to cost-effective efficiency methods, the substitution of labour for technology, and the replacement of oil and gas with electricity or various new energy-supply technologies. The relationship between energy consumption and GNP is not fixed, but depends upon government policy (each nation practises its own independent policy of energy conservation); and the model embodies a dynamic, non-linear optimization approach, in which the levels of supply and demand come into balance during each period, but also affect the supply/demand balance during the following periods. Like many others, this model assumes that natural economies proceed along a smooth, continuous path toward a long-term economic "optimum".

The model is also limited in that it describes only three sectors of the economy: consumer goods, electric energy and delivered non-electric energy. The primary inputs are capital, labour and energy resources, and the model is organized to calculate the strategy, given these inputs, that maximizes the present value of the utility of consumption. (The utility of consumption is a mathematical function based on the assumption that public welfare is directly proportional to the level of consumption in the economy.) The utility function is logarithmic in form and the utility discount rate is chosen so that the rate of return on capital will remain constant as long as the economy realizes its potential growth rate (which, as in most economic models, is determined by the growth of the labour force).

This is a direct outcome of the assumptions, familiar in the economic literature, that are represented by a Cobb-Douglas-type production function. The Cobb-Douglas function suggests that there is a constant elasticity of substitution between the production "inputs" of capital, labour and energy. (In other words, one of these elements may be easily substituted for another to produce the same net value of goods or services; but when any element increases in price, the cost of the output (good or service) will rise. Thus, as energy costs increase and/or environmental standards get stricter, the net amount of return on investment decreases until a less expensive input is substituted for energy. Moreover, according to this assumption, when capital and energy costs are fixed, the rate of growth of the economy depends on the rate of growth of the labour force.

5.4.2 Limitations of the Global 2100 model

The restrictions imposed by the structure of the Cobb-Douglas production function place important limits on the usefulness of the Global 2100 Model. The Global 2100 model is not so much a tool for the scientific analysis of data, as it is a summary of judgements gleaned from many such analyses, organised into a software framework that uses them in "expert system" fashion to perform calculations about the future (Hogan, 1990).

While the overall model design is of value, the Global 2100 model ignores the dislocation costs of rapid climate change and the secondary benefits of investments made to reduce emissions; but it counts the direct costs of those emissions-reducing investments. Thus, the model skews its judgments toward the "wait and see" approach.

Manne and Richels have essentially ignored three key issues in their analysis. First, one main economic issue in global warming is the absence of a market value for the impacts of greenhouse gases. As a result of wide separations over time and space, there is no existing market mechanism by which the beneficiaries of uncontrolled greenhouse emissions can identify (let alone compensate) those who bear the most costly impacts of rapid climate change. Second, the model fails to recognize that reducing CO_2 emissions will allow us to avoid the costs of some of these damages. Finally, the model overlooks the factor of environmental accounting: that costs of pollution are unpaid by, or unincorporated within, the costs of production today.

The non-inclusion of these three points in any part of the analysis is a serious shortfall in the model. Hogan (1990), however, defends this on two counts: First, that emission reduction will probably take place on a slower, lower scale than what is proposed or hoped for. In this case, the benefits of emissions reductions may be trivial. Thus, even achieving a 20-% emission reduction may be a liberal, not a conservative, goal. Secondly, the optimal time path of controls (such as a carbon tax) would follow from the model itself, so that they would not take place so quickly that they overwhelmed the balance between labour and energy.

5.5 *The Peck and Tiesberg model*

In the development of a new model, Peck and Tiesburg (1991) attempted to actually state the damage function in mathematical terms: cost of coping with damage as a function of the amount of emissions. Their model, called the Carbon Emissions Trajectory Assessment, reflects a further effort based on the works of both Nordhaus and Manne and Richels.

Peck and Tiesberg state that some of the costs of global warming are related to the level of warming; these include sea level rise, and effects on animal reproduction. Others, such as effects on storm frequency and plant productivity, rise and fall with the rate of warming. And for the costs related to the rate of warming, they feel that a "recovery effect" is likely — in which the costs of a rapid temperature change will be dissipated over time, as the affected systems adapt to the change. They then assume a damage function in this form:

$$D = \alpha[(\Delta T)^{\lambda}_{-0} + (1-\mu)(\Delta T)^{\lambda}_{-1} + (1-\mu)^2(\Delta T)^{\lambda}_{-2} + ...]$$

where
D = annual damage costs in the current decade (in dollars)
$[T]_{-t}$ = temperature change, t decades earlier (in °C)
λ = power of the damage function
μ = damage recovery rate
α = scaling effect.[4]

They assume that m= 0.1 — which means that there is approximately a 65% recovery after one century, and a 90% recovery after about 220 years. Other values would reflect alternative assumptions about the irreversibility of climate-induced damages. In this formulation, marginal damage costs are variable; high marginal costs of temperature change can be obtained by setting l at around 3. The scaling constant is set by calibrating to Nordhaus's warming damage estimate: a 3 °C warming.

The results obtained using the rate-related damage function were not found to be sensitive to the assumption of the recovery rate parameter, mu. The results, however, are sensitive to the decadal rate of temperature change used to calibrate damage to Nordhaus's benchmark.

6 The Use of Economic Models for Policy Decisions

The serious limitations of these three models are typical of all economic models of global warming measures to date. In their decisions about what to count, based in part upon what is easily quantified, they all reveal a subtle, probably unintentional bias towards a "wait and see" or "learn, then act" strategy.

Most tend to emphasize the cost of measures to deal with global warming; potential benefits of those measures go unrecorded or vaguely assumed. The complexity and diversity of the environmental response to global warming is oversimplified, and the question of intertemporal fairness (fairness to later generations) unresolved. The severity of regional level impacts is glossed over or lost in aggregate global figures. This is tragic, in some cases, because the areas most liable to inundation from sea level rise are the most constrained by a paucity of funds to deflect or repair the damages (Warrick and Rahman, 1992; this volume).

Most models exacerbate these problems by assuming that *laissez-faire* and perfect markets will exist throughout the world, along with the free flow of technology and unbridled technological innovation. Nor do the models take into account the current achievements of some countries (notably Japan and West Germany) in achieving enviable standards of energy efficiency while reducing both energy intensity and greenhouse gas emissions.

Even the cost estimates, which have been so well stressed, are not accurate; the most widely used models are based on economic data from 1958 to 1979 and 1973 to 1985. Those years, unfortunately, included the oil shocks of 1973-74 and 1979-80 — the effects of which may well have been anomalous. In both cases, the abrupt nature of the oil price increase might have had a greater impact on the economy than the absolute amount of the price change (Miller, Mintzer, and Brown, 1990). In addition, models that predict human responses over the next 100 years are subject to a high level of unreliability.

However, despite their limitations, models can contribute to decision making when they help to illuminate scenarios — devices to help policy makers consider the relationships between environmental acts and economic growth. Problems arise when policy makers regard these models as precise forecasts; but when used as guides for investigating policy directions and research needs, and highlighting least-cost methods of cutting down greenhouse gas emissions, the usefulness of these models is not in doubt. For example, they have great value in highlighting the possible ramifications of the carbon tax.

6.1 The carbon tax

Both the Nordhaus model and the Global 2100 model had assumed that emissions taxes would be an integral part of efforts to control carbon emissions. The question was, what would be the economic effects of such a tax? How high should the tax go (in other words, what extent of emissions reduction should it provoke)? And how rapidly should that tax come in?

Nordhaus had published on this question as early as 1982 ("How Quickly Should We Graze The Global Commons?"), when he described an optimal path for emission control and emission taxes. While his results depend critically on oversimplified assumptions about the costs of climate change and the lack of other economically important benefits from pollution control, he was able to demonstrate the importance of gradually escalating rates, so that early levels of control and tax are much less than the eventual targets. This result has important political implications: a slowly increasing tax may be much easier to accept than the swift implementation of a seemingly draconian measure. The gradual increase is also necessary to spur research on energy efficiency, which helps make further increases in the tax palatable.

[4] The formula is then rewritten in a simpler form as

$$D = [\delta T]_{-0} + (1-\mu)D_{-1}$$

Where

D_{-1} = annual damage in previous period.

In Global 2100, Manne and Richels (Hogan, 1990) also assume a gradually increasing carbon tax, though at a faster rate than that assumed by Nordhaus. The tax begins around $29 US per tonne of carbon in 2000, then rises sharply as emission limits/environmental standards are tightened. To discourage the use of high-carbon based fuels, in the absence of low carbon alternatives, the tax should reach a sufficiently high level. By the mid-21st Century, Manne and Richels suggest, the tax would be around $250 per tonne of carbon emitted into the air.

The effectiveness of a carbon tax depends on the relative proportions of the use of each fossil (carbon) fuel, and on the availability of substitutes for them. In the Global 2100 model, an effective long-run equilibrium tax could be calculated by factoring in the cost and emission coefficients of synthetic fuels and non-electric backstop supply technologies such as hydrogen fuels.

The models also reveal what would take place if a carbon tax does not occur. We have seen that substitution possibilities may depend on the investments possible from a carbon tax; and if substitution possibilities are limited and cheap backstop technologies not available, then greenhouse gas reductions will not occur through alternative energy technologies, but via the more expensive route (to society) of reduced energy use, mandated not by the marketplace but by policy makers.

6.2 Research initiatives and technology transfer

There are at present numerous proposals for an immediate cutback in CO_2 emissions ("act then learn"), in order to hedge against unfavourably rapid changes in climate. Manne and Richels, however, conclude that the size of the hedge is quite sensitive to quality and timing of climate research results *(Global 2100,* 1990b). If, as the "learn then act" proponents suggest, we can believe that technologies for solving global warming problems exist, then the hedge will be quite large, and we can feel safe in depending on it; in terms of the models, the constrained and unconstrained emission paths lie close to each other. And the incentive for immediate abatement measures would be much less. But in the absence of faith in near-term research accuracy, the need for precautionary action would be greater.

An optimistic scenario of energy futures could draw upon the possibility of abundant low-cost, carbon-free supply alternatives (such as photovoltaic power) and highly efficient end-use technologies. But in a pessimistic scenario, there is a heavy dependence on carbon intensive synthetic fuels and less autonomous conservation. Which scenario should policy makers use to base their decisions upon? We feel that the near term policy implications are quite clear: if a sustained commitment is made to reducing climate change uncertainties and developing new supply and conservation options, it will serve us well in either future. Better information reduces the hedge and improved supply and conservation technologies enhance the capability to deal with such a future, if it occurs. But this policy also presupposes the unconstrained flow of information and technology between nations - something that has not occurred in the past and is unlikely to take place in the future.

7 Conclusions

What makes the choice of public policies to respond to the risks of rapid climate change difficult is the problem of balancing the expected present and future costs of various options against the uncertain backdrop of consequences and impacts of global warming. Excessively strong measures to counteract global warming could retard growth and deprive present and future generations of several forms of welfare, considered essential in maintaining current standards of living.

In any event, there will be some apparent winners in the short term; others will seem to be losers when pollution prevention options are implemented. Without doubt the impact of global warming would not be uniform round the globe, and it is clear that some are likely to suffer more than others. It is equally clear that some communities and nations which are likely to be adversely affected are the ones who have the least resources to direct towards solutions to the problem. Unless an identity of interest can be established between those who are responsible for causing the problem but simultaneously possess the means to undertake mitigation measures rapidly and ambitiously, as well as those who do not have the means but may be responsible for future growth of emissions, solutions will be difficult to identify — much less implement.

Efforts to understand the impacts of both global warming, and of human intervention, need to be intensified and extended as quickly as possible. Even then, the dilemma will remain: whether or not to delay action till a certain prerequisite level of research output is available. The models to date give us one answer: that we do not need to resolve this dilemma yet. There are several cost-effective measures that can be implemented immediately, without any doubt regarding their positive benefit cost ratios. These include measures to increase energy efficiency, extend forestry programs, and rapidly phase out the use of dangerous CFCs.

Several efforts need to be made in the research arena to ensure that economic models point the way toward efficient and equitable public policies. In the field of economics, a far more disaggregated approach to modelling would be required, with specific models being developed for specific regions and countries. Some of the sweeping assumptions underlying a number of existing models must also be replaced with more plausible parametres and realities. This clearly implies that institutional capabilities to analyse this entire subject area need to be developed all over the world. This would be important, also, for ensuring that leaders, communities and groups are sensitized to the realities of the problem through an analysis carried out from a local perspective. Global appeals and models would have limited value in persuading a small country in Africa, for instance, to give up a biomass-based lifestyle, purely because the global community concludes that it is destroying a part of the global wealth of forests.

This takes us to the heart of the problem and the core of the existing challenge. What the scientific and academic community should aim at is the evolution of a consensus for a global decision-making process. Information and analysis are a vital resource that the academic must peddle vigorously and judiciously for the benefit of policy makers. While there are several differences between countries and interest groups, there is also much on which agreement seems likely. The important thing is to see that areas of agreement are quickly identified, based on whatever scientific evidence exists today, and on mutual long-term economic and environmental self-interest.

This self-interest would not be difficult to establish, even with the uncertainties of global warming. Those uncertainties must, after all, be projected against the enormous damage that may be inflicted on this planet and life existing on it. Closer interaction between researchers and policy makers has never been more important than it is in this field and at this instant of time.

References

Abrahamson, D.E., 1989: Global warming: the issue, impacts, responses. In *The Challenge of Global Warming* ed. D.E. Abrahamson, Island Press.

Barbier, E.B. and D.W. Pearce, 1990: Thinking economically about climate change. *Energy Policy,* **18**(1).

Chaturvedi, A.N., 1991: Personal communication.

Hall, D.O., H.E. Mynick and R.H. Williams, 1990: *Carbon Sequestering versus Fossil Fuel Substitution - Alternative Roles for Biomass in Coping with Greenhouse Warming*, The Centre for Energy and Environmental Studies, Princeton University, PU/CEES Report No. 255, November.

Hoeller, P., A. Dean and J. Nicolaison, 1991: *Macroeconomic Implications of Reducing Greenhouse Gas Emissions : A Survey of Empirical Studies*, OECD Economic Studies, No 16, Spring 1991.

Hogan, W.W., 1990: Comments on Manne and Richels : CO_2 emission limits : an economic analysis for the USA. *The Energy Journal*, **11**(2).

IPCC (Intergovernmental Panel and Climate Change), 1990: *Climate Change: The IPCC Impacts Assessment*, Report prepared for IPCC by Working Group II by W.J. McG. Tegart, G.W. Sheldon and D.C. Griffiths, sponsored by WMO and UNEP, Commonwealth of Australia.

Manne, A.S. and R.G. Richels, 1990a: CO_2 Emission Limits : an economic cost analysis for the USA. *The Energy Journal*, **11**(2).

Manne, A.S., and R.G. Richels, 1990b: Buying Greenhouse Insurance, preliminary draft of chapter in forthcoming monograph "Global 2100, The Economic Costs of CO_2 Emission Limits", November.

Miller, A., I. Mintzer and P.G. Brown, 1990: Rethinking the economics of global warming. *Issues in Science and Technology.*

Nitze, W.A., 1991: *The Greenhouse Effect : Formulating a Convention, Energy and Environment Programme*, Royal Institute of International Affairs, London.

Nordhaus, W.D., 1990: To Slow or Not To Slow: The Economics of the Greenhouse effect. *The Economics Journal*, **101**, 920-927.

Patterson, A., 1990: Debt-for-nature swaps and the need for alternatives. *Environment*, **32**(10).

Pearce, D.W. and E. Barbier, 1991: The greenhouse effect : a view from Europe. *The Energy Journal*, **12**(1).

Peck, S.C. and T.J. Tiesberg, 1991: *Temperature Change Related Damage Functions: A Further analysis with CETA.*

Rahman, A.A., 1991: *Possible Criteria for Burden Sharing under a Global Ceiling : Experiences and Suggestions*, Presented at the International Conference on a Comprehensive Approach to Climate Change Policy, July 1-3, 1991, CICERO, Oslo.

Soni, P., M. Negi and H.B. Vasistha, 1990: Ecological impacts: reclamation of derelict lands. *Economics Times*, April 5.

Trexler, M.C., 1991: *Minding the Carbon Store: Weighing US Forestry Strategies to Slow Global Warming*, World Resources Institute, Washington, DC.

USEPA (United States Environmental Protection Agency), 1990: *Policy options for Stabilizing Global Climate*.December.

CHAPTER 17

International Organisations in a Warming World: Building a Global Climate Regime

Kilaparti Ramakrishna and Oran R. Young

Editor's Introduction

In 1972, the United Nations Conference on the Human Environment (UNCHE) was held in Stockholm, Sweden. This meeting — the first global conference to focus on negative human impacts on the environment — led directly to the formation of a new member of the UN family, the United Nations Environment Programme. In the 20 years since, many meetings have been held to address environmental issues; extended negotiations have led to several important international agreements designed to protect environmental quality, and a number of international organisations have emerged as forums for consideration of ecological problems that span national boundaries.

But as the efforts evolved into negotiations over an international framework convention on climate change, a variety of special challenges emerged for traditional institutions trying to establish a cooperative international regime. Among the complicating factors are: (1) the presence of persistent scientific uncertainties, including a pervasive and continuing ignorance about the timing, severity and regional distribution of effects of climate change; (2) a wide separation both in time and space between those who benefit from industrial activity, and those who must absorb the damages; and (3) a protracted lag of unknown duration between the time at which a response strategy is implemented and the expected observation of its salutary effects.

In addition, because the response strategies could affect such a wide variety of economically important activities (both in industrial and developing countries), it is not possible to work out a simple and quiet arrangement among a few key interest groups familiar with the problem. Rather, achieving a practical and equitable agreement will require the participation of many different stakeholder groups.

In this chapter, Ramakrishna and Young review international negotiations on the climate issue, and trace the evolution of the involvement of specialized and subsidiary bodies of the United Nations. They evaluate the performance of these various institutions, which must both cooperate and compete for resources and influence in a complex arena. A process has evolved to separate the functions of agenda formation (dominated first by the United Nations Environment Programme and the World Meteorological Organisation, and more recently by the Intergovernmental Panel on Climate Change) from institutional bargaining (now in the hands of the UN General Assembly and its designee, the Intergovernmental Negotiating Committee for a Framework Convention on Climate Change). As Ramakrishna and Young observe, many interrelated factors will determine if the proximate goal of this process — to prepare a workable agreement in time for signature at the UN Conference on Environment and Development planned for June 1992 in Rio de Janeiro, Brazil — can be achieved. Even if it is achieved, a larger challenge remains: to implement a flexible, cooperative, long-term regime to manage global emissions of greenhouse gases and to minimize the damages due to rapid climate change.

Kilaparti Ramakrishna is a Senior Associate and Director, Program on Science in Public Affairs at the Woods Hole Research Center and a world-renowned international lawyer with widespread experience in multi-lateral negotiations. He played a prominent role in the negotiations leading to a number of international environmental agreements and is now Special Advisor to the Intergovernmental Negotiating Committee for a Framework Convention on Climate Change. Oran R. Young is a Professor of International Relations and Director, Institute of Arctic Studies at

Dartmouth College in New Hampshire, and an expert on cooperative regimes to manage international environmental problems. Ramakrishna, an Indian, and Young, an American, bring a unique hands-on perspective, a deep sensitivity to the needs of various stakeholders, and a valuable balance to the debate on the evolving international climate regime. Studying historical precedents and the current situation, Ramakrishna and Young conclude that a new institution will be necessary to promote and maintain a cooperative international regime addressing the challenges of changing climate. The demands placed on such an institution will be interdisciplinary. It must blend scientific competence and credibility with the administrative skill to monitor compliance with the new regime, the economic authority to distribute compensatory funds, and the political mandate to revise and renegotiate the regime as circumstances evolve over time. Ramakrishna and Young are not unduly sanguine about the prospects for achieving an agreement on a new International Climate Authority, either quickly or efficiently — but as this chapter shows, much depends on our ability to create a sustainable, well-managed institution eventually.

- I. M. M.

1 Introduction: The Issues Facing International Organisations

International organisations figure prominently in efforts to solve transboundary environmental problems, including the threat of global climate change. In this chapter, we seek to illuminate the roles played by international organisations in the ongoing effort to create a global climate regime, to evaluate the performance of these organisations, and to make suggestions for improvements in the future.

The first section provides a narrative account of the process to date; its emphasis is on the activities of the Intergovernmental Panel on Climate Change (IPCC) during the phase of agenda formation, and on the events leading up to the onset of institutional bargaining and the creation of the Intergovernmental Negotiating Committee for a Framework Convention on Climate Change (INC). Looming large in the background are the activities of the United Nations General Assembly (UNGA), which adopted a resolution in 1990 calling for negotiations to reach agreement on the terms of an effective framework convention on climate change. The UNGA also established a Preparatory Committee (PrepCom) to lay the groundwork for the United Nations Conference on Environment and Development (UNCED), to be held in Brazil during June 1992 at which the framework convention will be signed.

The second section of the chapter analyses the performance of international organisations working on the problem of climate change (listed in Table 1). The analysis is made with an eye toward identifying both sources of strength and sources of weakness. In the process, we make a number of suggestions regarding organisational reforms intended to improve the capacity of international organisations to cope effectively with the prospect of global climate change.

Finally, those engaged in international negotiations often foresee a need for establishing or adapting organisations to implement and administer the provisions of the new environmental regimes being created. These regimes are a set of agreed-upon principles, norms, rules and decision-making procedures that govern the interaction of actors in a specific issue area. The character of the organisations to be established typically becomes a major concern. In some cases, the same organisations figure in both capacities, serving initially as instruments of regime formation and subsequently assuming roles in the management and administration of the regimes they help to establish. In the case of global warming, the United Nations Environment Programme (UNEP) and the World Meteorological Organization (WMO) are the obvious candidates to assume this dual role.

2 The Evolution of a Global Climate Regime

Historically, environmental agreements have been concluded under the auspices of several types of organisations, including subsidiary and specialized agencies of the United Nations, and regional international organisations. They may also be simply concluded at the initiative of a group of countries. Never before, however, has the UNGA decided to conduct negotiations on an environmental issue directly under its own auspices.[1]

This departure is attributable, in part, to the fact that the proposed climate convention is not just another environmental agreement. It is more complex than any environmental agreement attempted previously. Moreover, given the linkages between stabilizing the composition of the Earth's atmosphere and economic development, some would go so far as to argue that the climate convention will not be foremost an environmental agreement, but an agreement covering many aspects of economic and environmental policy.

Interest in the role of international organisations in dealing with climate change has risen rapidly since 1988. Before that time, beginning with the First World Climate Conference of 1979 (WMO, 1979), several meetings had discussed the potential effects of global warming, as well as policies to

[1] Negotiations launched by the General Assembly on the Partial Test Ban Treaty (PTBT) and the Law of the Sea (LOS) Convention, may at first blush appear as environmental agreements concluded under the auspices of the General Assembly. While it is true that they did deal with and produce legal regimes for the protection of environment, the principal purposes of PTBT and LOS were security and access to ocean wealth, respectively.

Table 1. International organisations concerned with global climate change

GEF	Global Environment Facility	Funding body intended to promote environmentally sound and sustainable economic development, created in a tripartite agreement between the World Bank, UNEP, and UNDP.
INC	Intergovernmental Negotiating Committee for a Framework Convention on Climate Change	Formed by UNGA to negotiate the terms of an international climate convention and to play significant role in the processes of agenda formation and institutional bargaining giving rise to a climate regime.
IPCC	Intergovernmental Panel on Climate Change	*Ad hoc* organisation, established by UNEP and WMO in November 1988 to meet the need for a balanced and integrated scientific assessment of the risks and impacts of climate change and the potential policy responses to it.
PrepCom	UNCED Preparatory Committee	Established to lay the groundwork for the United Nations Conference on Environment and Development (UNCED), to be held in Brazil during June 1992.
SCDC	Special Committee on the Participation of Developing Countries	Committee formed by the IPCC to consider issues related to concerns and needs of developing countries.
UNDP	United Nations Development Programme	Subsidiary organisation and programme of the United Nations. World's largest channel for multilateral technical assistance and pre-investment cooperation. Active in more than 150 developing countries and territories.
UNEP	United Nations Environment Programm	Subsidiary organisation and programme of the United Nations, formed after the United Nations Conference on the Human Environment (UNCHE) in 1972.
UNGA	United Nations General Assembly	One of the six principal organs of the United Nations. Discusses questions arising within the scope of the United Nations Charter.
WMO	World Meteorological Organisation	Specialized agency of the United Nations, concerned with weather and climate.
The World Bank	The International Bank for Reconstruction and Development	Specialized agency of the UN established to promote international flow of capital for productive purposes and to finance reform programmes leading to economic growth in its less developed member countries.

Note: A subsidiary organisation receives its funding from and reports to one of the six United Nations principal organs. A specialized agency has its own constitution, becomes part of the United Nations system by a special agreement between it and one of the principal organs of the UN, and is funded by its signatory member countries.

prevent, mitigate, and adapt to it. Other meetings of importance were held in Villach, Austria, in 1980, 1983, 1985, and 1987.[2] However, it was not until the November, 1987 meeting in Bellagio, Italy, and then at the June 1988 Toronto Conference on "The Changing Atmosphere: Implications for Global Security," that a conference devoted its full attention to developing policy options for responding to climate change.

Even so, when the idea arose — initially at the Toronto Conference — of adopting an international convention on the atmosphere by 1992, many states (including the United States) expressed the view that 1992 was extremely early and, therefore, an implausible target date for completion of a climate convention. But by the time the Second World Climate Conference concluded in 1991 (Jager and Ferguson, 1991), it was beyond question that negotiations on climate

[2] For a survey of both scientific literature and some of these conferences, see Kellogg (1987).

change would begin with the objective of concluding a framework convention by 1992. Obviously, the political climate has changed markedly since the Toronto Conference. In the face of such rapid change, it is easy to lose perspective, and to overlook some of the institutional innovations that played key roles in international efforts to meet the goal set at Toronto.

2.1 The contribution of the IPCC

Deserving special attention, in this context, is the Intergovernmental Panel on Climate Change (IPCC), which UNEP and WMO launched in November 1988 (WMO/UNEP, 1988) to meet the need for a balanced international scientific assessment of the risks and impacts of rapid climate change, and of potential responses to it.[3] The IPCC's work derived, in part, from the WMO's experience at assisting 160 UN member countries in standardizing, collecting, and disseminating atmospheric data. Despite this emphasis on science, the secretary-general spoke at the opening session of the Panel of global warming as one of the most important long-term challenges facing humanity. He said that WMO could not be simply a bystander during the consideration of the consequences of the scientific studies. The executive director of UNEP concurred, and emphasized the unprecedented global cooperation which made the launching of the IPCC possible.

Several distinctive features set the IPCC apart from similar initiatives that international organisations have undertaken in the past. Though the Terms of Reference (the charter under which the IPCC operates) stress the process of peer review and state that the activities of the working groups and subgroups are intended to be technical in nature, the structure is designed to allow the organisation to address policy concerns effectively. Most participants have either been approved by their governments, or were appointed because of their work with government agencies in the member countries. And at its first session, the IPCC established three working groups. Only one dealt with scientific analysis; the other two focused respectively on socioeconomic impacts and policy responses. According to the Terms of Reference of the working groups, their reports are to be written in such a way that they address the needs of policy makers and the concerns of non-specialists.

To ensure coordination among the working groups, the IPCC created a bureau composed of the IPCC chairman, the panel's vice-chairman and rapporteur, the chairmen of the three working groups, and the vice-chairmen of the working groups (two each from Working Groups I and II and five from Working Group III). As it soon became apparent that this arrangement was not sufficient to ensure adequate participation by the developing countries, the IPCC also formed a Special Committee on the Participation of Developing Countries.[4] IPCC's *First Assessment Report,* adopted at the fourth plenary session in August 1990, sets forth the conclusions of the working groups on science, impacts, and response strategies as well as the recommendations of the Special Committee on Developing Countries (IPCC, 1990).[5]

2.2 The movement towards open-ended negotiation, leading to the creation of the INC

At the same time that the IPCC was beginning to conduct its scientific and policy investigations, an impetus had begun to establish open-ended negotiations on a multinational framework convention dealing with climate change. By mid-1990, UNEP and WMO had passed resolutions authorizing the first open-ended negotiating session and initiated informal inter-governmental discussion about how to organize negotiations. Participants from over 70 countries attended the initial preparatory session, held in Geneva in September 1990. Significantly, an overwhelming number of those attending came from the developing countries, suggesting that the perspectives of the developing countries will receive adequate attention in climate change negotiations and that any measures adopted to stabilize the composition of the Earth's atmosphere will have the support of both the North and South.

At the September 1990 meeting, new emphasis was placed on urgency and structure. The executive- director of UNEP stressed that both he and the secretary-general of WMO had been requested by their governing bodies (see Table 1) "to prepare for negotiations *now.*" The UNEP/WMO preparatory documents set out a number of topics to be considered in the negotiating process (UNEP/WMO, 1990):

- identification of gases that should be considered in atmospheric emission reduction and stabilization targets;

[3] The objectives for the IPCC set up by the governing bodies of WMO and UNEP were summarized in the WMO Executive Council Resolution (EC-XL):
(i) Assessing the scientific information that is related to the various components of the climate change issue, such as emissions of major greenhouse gases and modification of the Earth's radiation balance resulting therefrom, and that is needed to enable the environmental and socio-economic consequences of climate change to be evaluated;
(ii) Formulating realistic response strategies for the management of the climate change issue.

[4] This Group was chaired by Jean Ripert of France. Mr Ripert was subsequently elected as the Chairman of the Intergovernmental Negotiating Committee at its First Session in Chantilly, Virginia, US (See Box 1.)

[5] In accordance with WMO Executive Council resolution 4 (EC-XLII) of 22 June 1990 and UNEP Governing Council decision SS.II/3 Climate B of 3 August 1990 concerning IPCC, the UNEP and WMO are jointly arranging for the continuation of the Panel (WMO, 1990 and UNEP, 1990).

- proposed dates and base years for emission reduction targets (important elements in negotiations, because they can significantly affect the size of each country's "license to pollute");

- and criteria for calculating emission levels - according to population, GNP or GDP, the area of a country, its climatic conditions, the size of its natural carbon sinks, its energy consumption per unit of production, or some combination of all these criteria (see Grubb et al.,1992; this volume).

Prodded by some participants' growing concern about the potential risks of climate change and others' recognition of the emerging opportunity to establish an equitable regime governing North/South relations, the September 1990 meeting resolved, as a target, that a meaningful legal act should be developed for adoption by the United Nations in 1992. It was less clear, however, whether this goal should be pursued even if it meant agreeing to a legal instrument of only a declaratory character (i.e. one which did not impose specific obligations on the signatories).

The resolve, which the heads of UNEP and WMO had expressed at the September meeting, was endorsed formally two months later, at the Second World Climate Conference (SWCC), also held in Geneva. The SWCC, whose sponsors included a consortium of international agencies, received and took note of the IPCC report. The Ministerial Declaration adopted at the close of the SWCC called for negotiations concerning a framework convention on climate change to begin immediately, following a decision of the United Nations General Assembly on ways, means, and modalities for organizing these negotiations.

The UNGA took up this issue in December 1990 and adopted Resolution 45/212, calling for the initiation of negotiations on a framework convention - including appropriate commitments and any related instruments agreed upon by consensus - to be signed during the United Nations Conference on Environment and Development (UNCED).[6] The UNGA decided to entrust the climate negotiations to a newly created body, the Intergovernmental Negotiating Committee for a Framework Convention on Climate Change (INC), assisted by an *ad hoc* secretariat located in Geneva.

In parallel to the INC process, the UNCED Preparatory Committee (Prepcom) has accorded high priority to the development of institutional arrangements that will integrate environment and development concerns. The INC negotiations will test the ability of the international community to integrate these concerns at all levels of policy making within existing and new institutions, and will provide the first indications of what form these institutional arrangements will take.

3 Goals and Performance of International Organisations Today

Because the process of forming a global climate regime is still under way, we cannot now provide a definitive evaluation of the performance of international organisations in the effort to come to terms with climate change at the international level. None the less, it is not premature to initiate a review of the performance of the various components of the United Nations system that have become involved in this enterprise.

How well have these international organisations performed in orchestrating a response to the prospect of climate change? What are their strengths and weaknesses in this regard? What sorts of organisations will be needed to implement a climate change regime? What reforms in the existing system of organisations active in this area are both feasible and desirable? To address these questions, we take up, in turn, three vital stages of the overall process: agenda formation[7], institutional bargaining[8], and regime implementation[9].

3.1 Agenda formation

The work of the IPCC - the principal vehicle for agenda formation in the area of climate change - has been, for the most part, a success story. The Panel has played a key role in monitoring the state of scientific knowledge relating to climate change, and in bringing this knowledge to bear in a policy-relevant fashion. It has effectively promoted the concept of climate change as a self-contained issue, rather than as a component of some larger effort to devise an overarching law of the atmosphere.

In bringing science to bear on policy, the IPCC has benefited from a unique and largely fortuitous combination of circumstances, not of its own making. UNEP and WMO established the panel as a step toward staking out roles for themselves in the climate change issue area and, at least in the minds of some of their leaders, assuming a position of leadership in devising a global climate regime. Yet the UNGA soon intervened to limit the role of the two organisations in this area — creating the INC and, as a result, ensuring that the process of institutional bargaining relating to climate change would unfold under its own auspices, rather than under the auspices of UNEP and WMO. The separation of the roles of agenda formation and institutional bargaining has allowed the IPCC to draw on the scientific expertise and

[6] The UNCED is a special meeting convened by the UNGA per resolution 44/228 to be held in June 1992, in Brazil. The principal purpose of the meeting is to draw together current concerns about economic development and environmental protection. Brazilian authorities expect between 30,000-60,000 people to attend this meeting and the associated activities (UNGA, 1990).

[7] Agenda formation is the process through which issues requiring actions come to the attention of political decision makers.

[8] Negotiation over the content of a constitutional contract setting forth the provisions of regimes.

[9] The process of turning a regime, as spelled out in the provisions of a constitutional contract, into a social practice.

Climate Change Negotiations: the Interplay of State Actors **Box 1**

The first session of the Intergovernmental Negotiating Committee for a Framework Convention on Climate Change (INC) was held in Washington, DC during February, 1991 at the invitation of the government of the United States. The INC, since then, concluded three more meetings in Geneva (June 1991), Nairobi (September 1991), and Geneva (December 1991). The fifth session was scheduled to take place in New York in February 1992, with the possibility of a resumed fifth session in April 1992. Thus, within just 14 months, the INC will have completed negotiations for a framework convention for adoption at the UNCED in June 1992.

The climate negotiations are both complex and extremely significant for the future of international relations. In his statement at the first session of the INC, the UN secretary general (S.G.) characterized the significance of the climate negotiations by saying that a parallel exists between the San Francisco Conference that created the United Nations in 1945 and the process being set in motion at the INC meeting. Similarly, in terms of complexity of the subject matter, many commentators have drawn a parallel between the climate negotiations and the United Nations conference on Law of the Sea (UNCLOS), which took 9 years to negotiate and adopt the Law of the Sea convention (United Nations, 1983). Even a casual observer of the negotiations so far will attest to both the significance and inherent complexity of the task at hand for the INC.

At its first session, the INC elected as Chairman of the Committee Mr. Jean Ripert, special advisor to the Minister of Foreign Affairs of France. Mr. Ripert was Chairman of IPCC's Special Committee on the Participation of Developing Countries, and was intimately familiar with the concerns of the developing countries and the work of the IPCC, leading to the production of the *First Assessment Report.* The Secretary General of the United Nations appointed Mr. Michael Zammit Cutajar as Executive Secretary of the INC, and Director of its *ad hoc* secretariat. A career international civil servant and Maltese national, Mr Zammit Cutajar brings to the Secretariat years of experience gained at UNEP and UNCTAD, and his extensive knowledge of the United Nations system.

On the substantive side, progress at the first session, particularly on the organisation of the work, was less than satisfactory. Many countries stressed the importance of addressing the issue of global climate change in an integrated and comprehensive manner, and taking full account of the special circumstances and needs of developing countries. Yet there was great opposition to considering emissions reductions, preservation and expansion of sinks, and financial and technical assistance separately, divided among various working groups. It was clear, however, that all of these topics could not be addressed in the plenary. After intensive discussions, agreement was reached on the establishment of two working groups: one to deal with commitments and the other with mechanisms. The progress made by the working groups will be integrated by the plenary, and the final text will be treated as one package.

The second session began with the primary task of electing bureau members for the working groups. And it continued a process begun in the first session, where delegations submitted texts ranging from specific elements to be contained in the convention, to an entire draft convention. After the Bureau and Secretariat compiled these submissions, they were discussed with a view to streamlining the text in accordance with the general structure of international agreements.

Negotiations on the text began in earnest in the third and fourth sessions. One element that retarded progress in the third session was the attempt by delegations to ensure that the texts they had submitted were incorporated, verbatim, into the documents produced by the Bureau. It required considerable diplomacy, on the part of the INC Bureaus and the working groups, to overcome delegates' insistence on maintaining their own special wording of proposals. Several new proposals contained language such as: "pledge and review," "targets and timetables," "joint implementation," "emissions trading," and the need for specific protocols. Discussions on these proposals have been inconclusive. Ultimately, it was decided that, if delegates agree on the substance of protocols, they should be reflected in the framework convention itself, rather than in supplementary documents signed by fewer countries.

When the negotiations began, many delegations feared that the developing countries would form a blocking coalition and impede progress. In the negotiations so far, however, many divergent views exist among the developing countries. Three distinct groups have emerged within the G-77: The Alliance of Small Island States (AOSIS), the oil-producing states, and the other developing countries. It is unlikely that the G-77 will form a united blocking coalition. Yet, it is unclear at this time how these groups of nations will interact with each other, and with the various coalitions among the industrialized countries. The ultimate role and position of these groupings will have become more evident during the fifth session.

At the end of the fourth session, the INC produced a "Consolidated Working Document" (CWD) containing the results of the work of Working Groups I and II. This CWD still reflects a divergence of positions on the specificity of

cont'd

Box 1 cont'd

the text regarding emissions reductions, carbon sinks, and financial and technological assistance. Despite the existence of the CWD, the content of the final text is still open to substantial change and, as a result, so is the shape of the institutions to be entrusted with the task of implementing the convention. Delegations expressed significant resolve at the conclusion of the fourth session to overcome the remaining obstacles. It is most likely that during the fifth negotiating session in New York, substantial progress on the CWD will have been made, and the INC will have a draft convention for adoption at UNCED '92.

credibility associated with UNEP/WMO enterprises, while largely avoiding the politicization that would be inevitable if the Panel were to try to serve in both roles.

However, while the reports of the IPCC have undoubtedly helped direct public attention toward climate change, the panel's efforts in themselves have hardly engendered the sense of urgency needed in climate change negotiations. Instead, UNCED leaders have driven the negotiations, working alongside the rising public expectation that some international treaty or legal instrument dealing with climate change will be ready for signature when the UNCED convenes in June 1992. Meanwhile, the IPCC has taken on a role as technical advisor, focusing on such relatively constrained matters as emission scenarios and technological control systems, instead of on broader underlying concerns. This may well have been a prerequisite of success in framing the issues and involving key participants.

But as a result, no group has yet been willing to tackle directly the larger questions of interregional and intergenerational equity that lurk beneath the surface in efforts to deal with climate change. Admittedly, there is much to be said for the proposition that the IPCC should concentrate on technical, rather than political, matters in the next phase of institutional bargaining concerning climate change. This is a point of some significance; if the threat of climate change turns out to be as serious as some projections suggest, it will be essential to have an unimpeachable source of scientific information on effective responses to climate change-related problems.

Yet we cannot help noting that the issue of climate change is profoundly political. The underlying problem is, in large measure, a product of the activities of advanced industrial societies in the North. But the drive to come to terms with climate change is likely to generate the perception of mounting pressure on the developing countries of the South, to place limits on their efforts to industrialize and modernize. By its nature, the IPCC is not well equipped to be responsive to these larger concerns. However, these concerns are bound to surface in any serious effort to cope with the threat of climate change at the international level.

Even if the INC produces a draft convention on climate change in time for consideration by the UNCED, the work of the IPCC will not be finished. At best, a climate convention ready for signature in 1992 will cover only the initial phase of an evolving global climate regime. Accordingly, there will be a need for an ongoing mechanism to evaluate advances in scientific understanding of climate change with an eye toward determining what additional steps should be taken at the international level, and what the timing of these steps should be.

The most important adjustment required in this area, we believe, involves institutionalizing the work of the IPCC, so that its subsequent efforts are less ad hoc or fortuitous while, at the same time, avoiding the stultifying effects of both bureaucratization and politicization. There are no panaceas in this realm. But as we argue later in this essay, it may make sense to fold the IPCC into a new organisation, chartered by the United Nations to implement and administer a global climate regime.

3.2 Institutional bargaining

In establishing the INC, the United Nations General Assembly acted to ensure that the negotiation of a global climate regime would unfold in the setting of a legislative conference. It is fashionable in some circles to criticize the resultant conference diplomacy for its deliberate pace and its alleged inability to generate timely adjustments to changing physical, political and social conditions. Yet it is far from clear that these criticisms are persuasive in the case at hand. Given the fact that the UNCED will take place in June 1992, and that there is a widespread expectation that some climate convention will be ready for signature at that time, the INC will experience strong pressure to produce results. In fact, the pace at which this issue has proceeded already, from the stage of agenda formation to institutional bargaining at the international level, has been remarkably swift by any standards.[10]

What is more, we are now learning how to build flexibility into international regimes to allow them to adapt to changing circumstances. One device worthy of consideration is provision for shifts in membership, which has proved effective in the international whaling regime and the Antarctic Treaty System. Another device is the review conference, empowered to amend a regime's substantive provisions, as in the

[10] A study conducted by UNITAR concluded that on average a multilateral agreement takes about 12 years for its adoption and entry into force.

case of the rapidly evolving ozone regime. Such devices will be valuable as the INC continues its work.

None the less, there are other reasons to be concerned about the performance of the INC. There is, to begin with, the problem of substantive content in contrast to timing. Because of the tight schedule imposed by the timetable of the UNCED, those negotiating a climate convention are under great pressure to produce results. But the issues at stake are both highly complex and politically sensitive. The danger, then, is that the INC will produce a draft convention that has little substantive content or, to be more concrete, a document resembling the 1985 Vienna Convention on the Protection of the Ozone Layer, which without the Montreal Protocol could hardly have accomplished any reductions in the use of chlorofluorocarbons. To be successful, we believe the INC should seek to produce a climate convention that encompasses one or more substantive protocols to accompany a set of framework provisions - as the 1976 Barcelona Convention for the Protection of the Mediterranean Sea Against Pollution did.

Yet there is no point in adopting a dogmatic stance on this issue; we are mindful of the fact that the 1987 Montreal Protocol on Substances that Deplete the Ozone Layer and the 1990 London Amendments to this protocol have introduced substantive points into the ozone debate, despite the somewhat inauspicious start of the Vienna Convention.

3.2.1 The value of leadership

The evidence suggests that a major determinant of success or failure in institutional bargaining is the presence of one or more individuals capable of providing effective leadership. These leaders may be representatives of states participating in conference diplomacy, such as Tommy Koh in the case of the United Nations Conference on the Law of the Sea. Or they may represent international organisations, such as UNEP's Mostafa Tolba in the cases of ozone and the international movement of hazardous wastes. In exceptional cases, they may be talented individuals who surface from time to time in different contexts, such as Maurice Strong, who was the secretary general in 1972 for the UN Conference on the Human Environment (UNCHE), set up a dynamic preparatory process, and went on to become the first executive director of UNEP. He is now secretary general of the UNCED PrepCom. Our perception is that Bert Bolin's leadership has been an important factor in the success of the IPCC.

With respect to the negotiations organized by the INC, it is surely premature to render a judgment regarding this matter. But it is a cause for some concern that we have yet to see clear evidence of the emergence of effective leadership in this forum. This observation takes on even greater significance because the record of dealing with North/South disagreements over greenhouse gases through conference diplomacy is not an auspicious one. Such efforts as the law of the sea negotiations in the 1970s and 1980s, and the various rounds of trade negotiations conducted under the auspices of the General Agreement on Tariffs and Trade (GATT) — for example, the current Uruguay Round — have been hampered both by an insufficiency of creative ideas relating to institutional design, and by a lack of political will to take them seriously. So far as we can see, the INC has yet to set a new standard of performance in this area.

3.2.2 The value of the diplomatic model

There is no basis for assuming that the problems we have raised regarding the process of institutional bargaining would disappear even if it were possible to replace conference diplomacy with the sort of legislative politics more familiar in domestic settings. When the issues at stake involve core interests of powerful groups in society, legislatures are typically just as slow to act and just as likely to respond unimaginatively as international conferences. We do not believe, therefore, that efforts to reform the process of establishing international regimes for issues like climate change should take the form of calls for the replacement of conference diplomacy with some other legislative procedure.

Rather, the current need in the case of climate change is to encourage innovation in thinking about institutional arrangements and to identify effective leaders willing and able to take steps to incorporate new thinking into the provisions of a climate convention, rather than settling for empty formulas simply to meet the demands of the UNCED timetable.

3.3 *Regime implementation: the most challenging problem*

Whatever the terms of an international climate regime, primary responsibility for their implementation will reside with national governments. Not only do international organisations lack the resources to tackle the problem of implementing the substantive provisions of a climate regime, they also do not possess the authority needed to regulate or channel the behaviour of the myriad individuals and organisations operating in domestic arenas. This is true whether the issue at stake is reducing the energy intensity of transportation systems, protecting moist tropical forests from destruction, or controlling emissions of nitrous oxide and methane in connection with agricultural production.

What is more, those who advocate the establishment of new organisations to administer international regimes should bear the burden of proof that these organisations are necessary. Not only are organisations costly to operate in purely fiscal terms, but they are also intrusive, potential sources of bureaucratic rigidities, and targets for manipulation by powerful interest groups. Having said this, however, we hasten to add that there are compelling reasons to expect that international organisational arrangements of some complexity will be needed to deal effectively with the problem of climate change.

While the science of climate change is evolving rapidly, there remain major uncertainties in our understanding of both the physical systems involved (Hoffert, 1992; Wigley et al., 1992; this volume), and the driving social forces giving rise to climate change (Holdren, 1992; Sathaye and Walsh, 1992; Jhirad, and Mintzer, 1992; this volume; Stern et al., 1992). There is no escaping the fact, therefore, that a climate regime will need a sophisticated capability for monitoring the progress of science in this area. The administrators of such a regime will need to make balanced judgments about remaining uncertainties relating to climate change, and to evaluate the extent to which new knowledge suggests changes in the provisions of the regime. As we have already suggested, this function may be met by incorporating the IPCC or a revised version of this Panel into an evolving climate regime.

The problem of regime implementation may well emerge as the most challenging issue facing those endeavouring to come to terms with climate change at the international level. Agenda formation and institutional bargaining may be difficult and complex, but they can none the less be handled through the operation of well-established international practices. Regime implementation, by contrast, will take us into uncharted waters. It is therefore to be expected that the process of forming a climate regime will involve hard bargaining over provisions relating to the character of organisations to be created to manage and administer the regime.

One of the most perplexing issues facing leaders of the INC over the coming months will undoubtedly centre on tradeoffs likely to arise between the need to reach agreement in time for signature during the UNCED, and the need for time to solve problems associated with developing an organisational infrastructure. Though we recognize the need to act promptly, we would hope that leaders in the INC negotiations will resist the temptation to gloss over key issues merely to foster the illusion of agreement.

4 Implementation of the Global Climate Regime and International Organisations

This brief review of the process of building a climate regime indicates that the approach the international community adopts in addressing global warming is likely to be different from any attempted before. The success of this approach will depend on how and when governments agree on binding legal agreements, specifying clear commitments and defining national responsibilities, explicit targets and schedules, and workable organisational mechanisms. An assessment of the existing United Nations system suggests that a new international organisation will be needed to address all these matters effectively.

It is hard to see how a climate regime can function effectively in the absence of arrangements that amount to an International Climate Authority. The IPCC's current activities are hardly the extent of the organisational arrangements that will be required to administer a climate regime. Depending upon the precise nature of the regime's substantive provisions, there will be a need to monitor compliance on the part of individual countries and their subjects with the regime's rules, to make decisions about matters such as the distribution of compensation funds and requests for temporary exemptions from regulatory requirements, to manage technology transfers mandated under the regime's provisions, and handle the process of refining or revising the terms of the regime itself in the light of changes in the science of climate change. Because a number of these functions can only be carried out with secure funding, there will be a need as well to organize the raising and disbursing of revenues on some predictable basis.

Experiences to date with international organisational requirements of this scope and complexity are not encouraging. Two of the most ambitious arrangements - the International Seabed Authority outlined in Part XI of the 1982 Law of the Sea Convention and the institutional arrangements envisioned under Chapter II of the 1988 Convention on the Regulation of Antarctic Mineral Resource Activities - have failed to materialize. These failures were due, at least in part, to controversies over the authority and powers of the proposed organisations.

There are, however, an array of precedents that are worthy of consideration in any discussion of appropriate organisations for the implementation of a global climate regime. These include the GATT secretariat in the realm of international trade, the International Monetary Fund (IMF) in the field of monetary arrangements and foreign indebtedness, the International Telecommunications Union (ITU) in the area of communication flows, and the commissions set up in conjunction with the international regimes for whaling and the conservation of Antarctic marine living resources.

Fundamentally, the options regarding implementation of a climate regime boil down to two:

(1) relying on the services of an existing organisation, such as UNEP — which has, for example, provided administrative support (at least during an initial period) for the regimes dealing with pollution control in the Mediterranean Basin and the control of transboundary movements of hazardous wastes

or

(2) establishing some new organisation(s) designed specially for the regime in question, such as the arrangements referred to above.

Here, too, there are no panaceas. The implementation of a climate regime will involve both administrative responsibilities and political issues that could easily overtax the organisational resources of UNEP or WMO, undermining the ability of these organisations to handle their current functions in the process. Any effort to saddle UNEP or WMO with the tasks involved in operating a climate regime would almost certainly swamp these organisations. In addition, existing institutions — whether environmentally oriented or development oriented — have necessarily evolved specialized competence and orientation, which is not well fitted to the multi disciplinary character of the environment development problem. On the other hand, a decision to establish new organisational structures to administer a climate regime runs the risk of falling prey to continuing political manoeuvres, with the result that the system of organisational arrangements is established in name only, or that paralysis sets in shortly after the arrangements are set in motion.

On balance, we are convinced that the time has come to face this issue squarely and to move forward with the creation of an International Climate Authority (ICA) as part of a global climate regime. While the proposed ICA should have its own mandate and source of revenues, it should also operate as a component of the United Nations system. As to scope, the ICA should be carefully limited to those activities clearly needed to cope with climate change effectively, on the understanding that it is always preferable to opt for handling the relevant functions in the least intrusive and least complicated fashion possible.

However, while many recent environmental agreements establish commissions and secretariats to monitor progress under individual conventions, a number of countries (including the United States) are reluctant to engage in innovative thinking about institutional arrangements. The American position, outlined at an informal meeting of UNCED's Working Group III, is that:

> *"[On] institutions, better coordination and management are the principal requirements for improved action on the environment in the UN system."*

Though this position does not rule out the establishment of new mechanisms, it clearly states that they ought not to be considered until "their necessity and feasibility have been clearly demonstrated."

For its part, the UNCED Prepcom document on institutions states that the "question of institutional arrangements will arise primarily from the needs identified in consideration of the substantive issues and related financial and other cross-sectoral measures..." Though this is surely a reasonable position to take, it hardly constitutes a new approach to institutional questions. Similar sentiments were expressed 20 years ago during the preparatory process leading to the United Conference on the Human Environment (UNCHE) in 1972.

The UNCHE resulted in the establishment of UNEP. And despite reservations such as those outlined above, new environmental agreements have regularly brought into existence new international organisations. This observation may lead some to conclude that, regardless of the positions individual countries adopt at the outset concerning institutional questions, negotiations on the terms of a climate change convention will eventually lead to the establishment of appropriate institutional arrangements to manage and administer a global climate regime. But given the complexity of the climate change issue, the importance of integrating environment and development interests, and the opportunity afforded by the UNGA's involvement in the negotiating process, it is critical that questions relating to institutional arrangements in the area of climate change be investigated systematically as early in the process as possible.

Both the strengths and the weaknesses of this approach are well illustrated by the Scientific Committee established under the terms of the international whaling regime and the Scientific Committee for the Conservation of Antarctic Marine Living Resources. As is often the case, the trick in this instance will be to devise an ongoing mechanism capable of bringing scientific knowledge to bear on a well-defined set of issues in a timely manner, without simply creating another arena for the interplay of political interests.

5 Conclusion

The perspective we have adopted in this article is reformist rather than revolutionary. We assume, for example, that the essential character of international society will remain unchanged over the foreseeable future, so that the problem we face is one of enhancing the ability of international organisations to come to terms with climate change in a decentralized social setting.

Yet parts of our argument may seem unduly optimistic to some, even as other parts of the argument seem insufficiently critical to others. For instance, we take the view that conference diplomacy in international society is not as inadequate as many analysts believe. As processes for consensus-oriented debate and discussion continue to grow more sophisticated, they will be a crucial vehicle for insuring the integrity and effectiveness of these international negotiations, where no single actor has sovereignty over any other.

At the same time, we argue that a successful climate regime will require the services of an International Climate Authority of some complexity. One of its mandates would have to be the establishment, and management, of a specialized climate fund. This would meet the need for new and additional funding, not just for the enforcement of environmental agreements and the administration of negotiations, but also for subsidizing the transfer of efficiency-improvement and low-emissions technologies from industrialized to developing nations. The Authority, in managing the fund, could adopt a role similar to that of the Global Environmental Facility (GEF), (see Box 2), but modified so that control was shared equitably, on both policy and operational matters, between donor and recipient countries. Voting rights could accrue equally to each nation, not weighted (as they are with the World Bank) according to a nation's economic status. Most importantly, the Authority would have to operate under principles mutually agreed upon by both groups of participant nations. This type of Authority, itself unprecedented in design, is necessary to manage the response to concerns about poverty, equity, development, and climate change, so that the needs of an increasingly interdependent world can be reasonably addressed.

The Climate Fund and the Global Environmental Facility **Box 2**

In a number of non-papers circulated during the second and third sessions of the INC, several developing countries (independently, and in a collective G-77 position), called for "new, additional, and adequate" funds with which to meet their obligations under a climate convention. Many representatives of developing countries have stated clearly that a successful conclusion of a meaningful climate change convention depends upon realizing access to these funds. In addition, many have argued for the disbursement of such funds without placing conditions upon the recipient countries. Also of significant interest, and of great concern to many developing countries, is the mechanism that will disburse the funds. Developing countries have stated over and over again that the climate fund should be controlled by the parties to the convention, and not by existing mechanisms or institutions.

In contrast, several industrialized countries recommended the use of the Global Environmental Facility (GEF), the result of a tripartite agreement between the World Bank, the United Nations Development Programme (UNDP), and the United Nations Environment Programme (UNEP). GEF is situated in the World Bank. Its operating budget comes from the contributions of member countries to the Global Environment Trust (GET) Fund. GEF's purpose is to assist in the protection of the global environment, and in promoting environmentally sound and sustainable economic development. It was established in order to maximize the comparative advantage of each agency which participated in its founding. Implicit in this arrangement was the understanding that no new coordinating structure would be created, and that only modest modifications would be made to existing institutional and organisational structures. In other words, for industrialized countries — particularly those that opposed setting up new institutions — GEF represented an opportunity to address environmental funding concerns within existing global financial mechanisms.

Precisely for those reasons, the developing countries are sceptical of GEF, and of its ability to address their environmental and developmental concerns in an equitable manner. Many developing countries view the GEF as dominated by the World Bank. They believe that any attempt to entrust GEF as the mechanism for implementation of a climate fund will only result in the operations of the World Bank as usual, in a new guise. Much of this scepticism emerged after the initial discussions leading to the formation of GEF, which took place only between participating agencies and contributing countries, and which did not involve developing countries. It is perhaps fair to say that much of the debate concerning GEF's role in disbursing the proposed climate fund has not necessarily been well informed, or has not included reference to specifics.

But there is a specific case worthy of discussion. Many actors in the climate change negotiations are intimately familiar with, and sometimes have first-hand knowledge of, the Vienna Convention for the Protection of the Ozone Layer, and the Montreal Protocol on Substances that Deplete the Ozone Layer. Many of these delegates see similarities between the attempts to protect the ozone layer then, and the effort to stabilize the composition of greenhouse gases now. Likewise, many delegations desire that the framework convention on climate change should closely resemble the Vienna Convention. It is therefore understandable that they have experience with the implementation of the "Ozone Fund" (the Interim Multilateral Fund for the Implementation of the Montreal Protocol).

In the case of the Ozone Fund, the development of project eligibility, the review of projects, and the approval of projects were assigned to the Executive Committee of the Parties to the Montreal Protocol, not to the GEF — despite the arrangements made by participating agencies in the GEF, and the coordinating role performed by the GEF in administering the Ozone Fund. This is obviously due to specific concessions, sought and extracted by the Executive Committee of the Parties to the Montreal Protocol. If a climate fund were created under the climate change convention, the signatories to the convention could similarly negotiate a method with GEF that adequately meets their concerns. The need here, therefore, is not necessarily creating a new institution, but having a clear sense of what the needed institution should do — and determining whether that could be accomplished by existing mechanisms.

In the words of Tariq Osman Hyder of Pakistan, speaking for the developing countries at the Global Environmental Facility Meeting in Geneva (December 1991), the GEF "is a new hybrid situated in a *terra incognita* between the World Bank and the more usual UN bodies and organisations." This hybrid status could seriously handicap the effectiveness of the GEF (also see Hyder, 1992; this volume.) It is up to the countries that oppose the creation of new funding agencies to make a case for GEF, and to show how it could adequately meet the concerns of developing countries.

References

IPCC, 1990: *Climate Change Reports of Working Group I, II, and III.* WMO/UNEP, Geneva, Switzerland.

Jäger, J. and H.L. Ferguson, (eds.), 1991: *Climate Change: Science, Impacts and Policy; Proceedings of the Second World Climate Conference.* Cambridge University Press, Cambridge.

Kellogg, W., 1987: Mankind's impact on climate: the evolution of an awareness. *Climatic Change*, **10**, 111-136.

Stern, P., O. R. Young and D. Drukman (eds.), 1992: *Global Environmental change: Understanding the Human Dimension,* National Academy Press, Washington, DC.

UNEP, 1990: *UNEP Governing Council decision SS.II/3 Climate B of 3 August 1990.* (Concerning the continuation of the IPCC Panel). Geneva, Switzerland.

UNEP/WMO, 1990: *Ad Hoc Working Group of Government Representatives to Prepare for Negotiations on a Framework Convention on Climate Change.* UNEP/WMO Prep./FCCC/ L.1/Report, Geneva, pp. 24-26.

UNGA (United Nations General Assembly), 1990: *Resolution 44/228, United Nations Conference on Environment and Development*, in which the General Assembly decided to convene the United Nations Conference on Environment and Development in 1992 in Brazil and, *inter alia*, to establish the Preparatory Committee for the Conference, open to all States Members of the United Nations or members of the specialized agencies, with the participation of observers, in accordance with the established practice of the General Assembly. Geneva, Switzerland.

United Nations, 1983: *The Law of the Sea: United Nations Convention on the Law of the Sea with Index and Final Act of the Third United Nations Conference on the Law of the Sea.* United Nations, New York, 224 pp.

WMO (World Meteorological Organization), 1979: *World Climate Conference; Declaration and Supporting Documents*, WMO, Geneva, Switzerland.

WMO, 1990: *WMO Executive Council Resolution 4 (EC-XLII) of 22 June 1990,* (Concerning the continuation of the IPCC Panel). Geneva, Switzerland.

WMO/UNEP, 1988: *Report of the First Session of the WMO/ UNEP Intergovernmental Panel on Climate Change (IPCC), Geneva, 9-11 November 1988.* WMO/UNEP TD - No. 267.

CHAPTER 18

Modifying the Mandate of Existing Institutions: NGOs

Navroz K. Dubash and Michael Oppenheimer

Editor's Introduction

For the last 20 years, non-governmental organisations (NGOs) have played a key role in bringing environment and development issues to the attention of policy makers and business leaders. Many of today's environmental NGOs trace their roots to the international movement which formed at the beginning of this century to protect wildlife and wildlands. But over the decades, the purview and interests of these organisations, and their successors, expanded. Today, environmental NGOs are active in more than 100 countries, working on a broad range of issues affecting human health, natural ecosystems, and sustainable development. They concern themselves, for example, with toxic wastes, pesticides, water quality, air pollution, and the environmental impacts of energy supply and use.

At the same time, development-oriented NGOs are active in many Third World countries, and in some industrialized countries as well. They principally focus their attention on the rate, quality, and equity of economic development in the South; and on the economic and foreign policy dynamics of North-South relations. In the last several years, with positive effect, development NGOs have made common cause with environmental organisations to address common issues. The risks of rapid climate change, which endanger natural systems and threaten to undermine the prospects for economic growth, are often the starting point for this new commonality of purpose.

In this chapter, Dubash and Oppenheimer trace the historical development of NGOs, and the linkages between them, in industrialized and developing countries. Using case studies of the Narmada movement in India, the Green Forum in the Philippines, and the Kenya Energy and Development Organisations (KENGO), as well as specific examples from popular campaigns around environmental issues in industrialized countries, Dubash and Oppenheimer illustrate the range of challenges faced by NGOs in different economic, political, and social milieux.

Michael Oppenheimer was trained as a chemical physicist at MIT and the University of Chicago. He subsequently held the position of physicist at the Harvard-Smithsonian Astrophysical Observatory. Oppenheimer has worked for 15 years on atmospheric issues, including acid deposition, ozone depletion, and global warming. Navroz Dubash received his early education in India and subsequently obtained an A.B. in Public Policy from the Woodrow Wilson School at Princeton University. He has worked extensively on environmental issues in the US, India, Bangladesh, and throughout South Asia. Dubash and Oppenheimer have been actively involved in the international negotiations on climate change and ozone depletion. They bring to this analysis a sensitivity to North-South issues and a hands-on understanding of the strengths, weaknesses, and synergisms that exist within the international NGO community.

It is clear, as Dubash and Oppenheimer demonstrate, that the future role NGOs play in international deliberations and governance will be significant. And that the nature of their role is still evolving. But how can they best be prepared to deal with the scientific complexities, the long lag times between issue definition and the solution of environmental problems, and the inherently difficult issues of distributional equity? The authors conclude that successful strategies must contain several elements. Among these are (1) measures to move from a reactive mode providing criticism to a proactive mode developing solutions to environmental problems; (2) campaigns to develop interactive links between policy-oriented "think tank"-type NGOs and grass-roots organisations; and (3) programmes to enhance the technical and scientific capabilities of NGOs worldwide, but especially in the developing world. Dubash and Oppenheimer believe that NGOs are especially well-situated to represent easy-to-overlook aspects of environment and development issues in the twin crucibles of public attention and political policy making. These organisations can be an important force for constructive social change, helping to overcome social inertia and bureaucratic resistance to necessary reforms.

- I. M. M.

1 Introduction

Human activity is altering the biosphere faster than our understanding of it is accumulating. Humans have spun a complex web of social, economic and political interactions across the globe which threaten the biosphere; yet even now, humanity remains dependent upon the continued viability of the biosphere. As humankind struggles to devise measures to protect the biosphere, we ask what role non-governmental organisations or NGOs can play in these deliberations. While NGOs have been successful in stimulating action on local and national environmental problems, what are the limits and benefits of their expansion into the international arena? Given citizen power and participation that are the source of NGO energy and power, what is the appropriate role for NGOs in addressing global environmental problems? How can NGOs assist, interact with and influence policy makers on these questions?

For the purposes of this chapter, we use the term NGO to refer to non-profit organisations, whose actions are substantially independent of governments or the private for-profit sector — and whose activities are directed toward a clearly defined set of goals that they perceive to be in the interest of society as a whole. Thus, NGOs may choose to interact with private sector organisations, the state, political parties, and legislatures, reserving the right to change their agenda and/or their manner of interaction with these other institutions.[1] For this chapter, we limit the discussion further, to those NGOs that define themselves as concerned with issues of environment and development.

NGOs have a unique role to play in addressing global environmental problems. Alone amongst the various actors on the world scene, they are relatively free from the narrow, short-term political interests within which many governments operate, and from the profit motive which drives the corporate world. Consequently, while other actors approach these issues from the standpoint of what they feel to be in their particular self-interest, combined with what they feel to be politically acceptable, environment- and development-oriented NGOs approach global environmental issues primarily from the perspective of their perception of the greatest social good.

However, different NGOs, depending on their philosophical foundations, organisational goals and the nature of the society within which they work, may define "the social good" differently. The result is not an immediate consensus on action, but rather, as a first step, an alternative framework within which the scientific, social, political and economic issues surrounding global environmental problems can be debated and discussed, with a view to outlining a solution. The role of NGOs in global environmental discussions is to work toward consensus on a just course of action, and then to exercise their influence to direct the bounds of political acceptability toward their vision of a solution.[2]

Despite the relatively recent recognition of global environmental problems such as depletion of the ozone layer and global warming, NGOs have already exercised their influence on these issues. In particular, the discussions leading to the establishment of an Intergovernmental Negotiating Committee (INC) on climate change were driven by a coalition of advocacy- and research-oriented NGOs, scientists and international organisations. NGO activities — of which the most visible were two meetings organized by the Stockholm Environment Institute at Villach and Bellagio — provoked a reaction from governments, who first established the Intergovernmental Panel on Climate Change (IPCC) and subsequently the INC. In the words of a former US official involved in the process:

> *".. the two workshops, the meetings of the Advisory Group on Greenhouse Gases and other activities ... indeed played a significant catalytic role in establishing the IPCC ... governments could no longer permit ... NGOs to drive the agenda on the emerging climate issue."* [3]

While NGOs can count this early success as well as earlier influence on the development of the Montreal Protocol to their credit, global environmental problems pose novel complications that NGOs have not had to address in dealing with localized environment or development issues. These characteristics are: persistent scientific uncertainties, which decrease the likelihood of consensus on policy action; a long lag time between implementation of policies and their results; and a hazy distinction between "winners" and "losers" from global environmental problems. In addition, many NGOs in developing countries and some Northern NGOs view global environmental problems as a consequence of the "development as industrialization" world view. From this

[1] The factors that influence the degree of independence which NGOs have, in reality, are interesting and worth pursuing, but are beyond the scope of this study.

[2] We are indebted to Shekhar Singh, Indian Institute of Public Administration, for this clear formulation of the role of NGOs.

[3] Personal letter from William A. Nitze, former Deputy Assistant Secretary for Environment to Michael Oppenheimer, April 4, 1991.

perspective, consumption patterns and the international economic order that perpetuates them are the root cause of these problems. The three characteristics described above and the broad scope of these problems will provide the background for our discussion on the role of NGOs in addressing global environmental problems.

2 The Nature of NGO Activities

NGO activities are typically stimulated by an innovative idea, a specific policy challenge to be met, a particular opportunity to be exploited or some combination of these. The result is a bewildering diversity of organisational structures, ideologies and tactics. While this is a strength of the non-governmental sector, it can also be confusing for those who deal with NGOs. This section examines trends in the local and regional work being undertaken by NGOs so as to identify the main elements of their concerns and strategies. Without seeking to be comprehensive, the discussion below demonstrates these various missions that shape the discussion of global environmental problems. The discussion highlights a broad distinction between the NGOs in the industrialized North and those in the developing countries of the South that recognizes their differing socio-economic circumstances.

2.1 Trends in developing countries

During the last 15 years, the number of NGOs in developing countries has increased dramatically. While work in the non-governmental sector has, in the past, been dominated by Northern charity organisations, this emphasis has been shifting. It is giving way to a broader definition of the problems faced by developing country populations, as indigenous Southern organisations take a larger hand in defining and acting on local problems. This transition may be characterized as a three-step process:

- from damage control — a short-term relief orientation;

- to development — a project focused approach, targeting individuals or communities;

- to empowerment — an approach which seeks to address the broader causes of impoverishment (Elliot, 1987). In this section, we develop the theme of empowerment as a driving motivation for Southern NGOs and show how it relates to some dominant characteristics of NGO work. The case studies that follow this discussion provide an illustration of these characteristics.

Many of the new NGOs in the South are small, local groups working to ensure a better quality of life for the local people. They have their roots in struggles by indigenous peoples and rural populations for access to and control over natural resources. Their work has taken many forms: fighting against human rights abuses by local elites or by the state; struggling for control over local natural resources; and facilitating income enhancement projects and rural micro enterprises. The common thread that runs through all these issues, however, is that they involve a substantial element of community organizing as a means of empowering the citizenry. A consensus is emerging among NGOs that lasting change cannot be achieved if "development" is viewed simplistically as building wells, hospitals and the like. Rather, to be successful, development programs must involve people in all stages of the process: identification of needs, conceptualization and implementation.

An important result of a long-term NGO emphasis on the need for empowerment is a wariness of technological fixes. This suspicion of technology reflects bitter experiences of recent history. In the past, the South has generally obtained largely outdated, pollution-causing and dependency-creating technology from the North. It is now generally acknowledged that even "Green Revolution" technologies — biotechnological advances in the production of such crops such as rice, corn and wheat — while increasing yields per hectare, have caused a considerable amount of ecological damage in developing countries. Crop resistance to disease has been diminished due to decreased diversity; multiple cropping has depleted soil nutrient levels; and extensive use of irrigation has created a demand for large irrigation projects, which have displaced populations and flooded farm and forest lands. In addition, there is good reason to believe that, while total production has increased, income disparities have been enhanced by the Green Revolution — with most of the benefits accruing to large farmers who are able to afford the cost of agricultural inputs.

Southern NGOs feel that "Northern" technologies do not address localized needs, but rather reflect availability and commercial interests of the industrialized countries, and often encourage dependence of the South on the North. The implications for environmental problems of a global nature are clear: Southern NGOs are likely to be very suspicious of Northern promises — by both Northern governments and NGOs — of massive infusions of efficient, sophisticated technology. From the point of view of Southern NGOs, new technologies will have to meet more stringent criteria than in the past. Not only will they have to be environmentally benign, but their social implications will have to be examined closely, since Southern NGOs view environmental protection as an intrinsic part of socio-political development and concurrent technological choices. Thus, while there are many promising technologies available — such as improved cook-stoves, biogas digesters and photovoltaics — the implementation of these technologies will have to be carefully undertaken, keeping in mind the needs, desires and capabilities of the local communities.

The overwhelming lesson is that many Southern NGOs are driven by a longer term vision of empowerment, and that global environmental problems and their proposed solutions must be viewed through this lens. In practical terms, this implies the need for global environmental issues to be successfully related to broader development and human rights issues, national and international patterns of resource control, use and distribution and the appropriate role of technology in making the transition to sustainable development paths.

2.1.1 The "Save Narmada Movement" of India

The Save Narmada Movement had its origins in the struggle by rural populations, predominantly indigenous peoples, to oppose the Narmada Valley Project, an Indian government effort to construct a large multi-purpose hydro-power facility. One component of this project, the Sardar Sarovar dam, threatens to displace over 100,000 people from their lands (Dubash, 1990). Social activists from within the region, and from other parts of India, have been involved with the Save Narmada Movement from its early days — bringing with them organisational skills and a range of contacts with other activist groups, both local- and urban-based. The emphasis has been on organizing these groups and on amplifying their voices. A major element of the campaign has been the insistence of the movement, based on documented evidence, that the state governments involved have withheld information from the affected persons, and consistently distorted information on the project so as to present the project more positively than is warranted.

From its inception as a localized opposition movement, the anti-Narmada stir has expanded to include development groups, human rights advocates and environmentalists. In so doing, the movement has acquired a high degree of visibility within India, and has become the rallying cry for many NGOs opposing similar destructive development projects throughout the country. Equally significant, it has provoked a broader debate on the dominant model of industrialized development broadly advocated by the Indian state. From the perspective of the environmentalists of the Save Narmada Movement and similar efforts, an environmentally sustainable society should allow local people access to and control over local natural resources (Kothari and Bakshi, 1989). Therefore, rather than framing environmental questions in terms of productive versus protective uses of natural resources, these NGOs weigh alternative productive uses of natural resources against each other (Guha, 1988).

It is noteworthy that the Narmada activists have mobilized an impressive array of supporters from sectors within Indian society, including prominent academics, journalists, lawyers and labour unions. This has lent credence to the movement, generated additional publicity and has also brought greater technical competence to the movement. The movement has also used a wide range of creative strategies — from direct action, to letter campaigns by its prominent supporters, to forging international links (discussed later in this chapter) to bring pressure to bear on the state and national governments. Volunteers have undertaken economic and scientific analyses for the project, which have been used to refute official studies. The considerable impact achieved by independent studies has led the movement to consider drawing on this expertise, to develop detailed alternatives to the project. For example, the movement has considered outlining a strategy to meet the energy demands of one of the beneficiary states, Gujarat, using energy efficiency improvements and renewable energy technologies. These new sorts of activities would signal a shift from a reactive mode to a pro-active one.

Finally, attempts to structure the spontaneous stirrings sparked by the Narmada Movement have led to the formation of the "Movement for Peoples' Development," a network of local groups of activists working on the spectrum of issues that fall under the general environment and development rubric. This group intends to translate the experiences learned by grass-roots movements, such as the Narmada movement, into concrete policy changes at various levels of government. In its early days, as well as in its more aggregated forms such as the Movement for Peoples' Development, the Narmada case illustrates the strength of the movement: a strong grass-roots base that provides legitimacy, and a definition of environmentalism that successfully unifies environment, development and human rights concerns.

2.1.2 The "Green Forum" of the Philippines

The Green Forum shares many characteristics with the Narmada Movement. In advocating a holistic approach to questions of environment and development, it has attracted an impressively broad cross section of Philippine society. The membership includes social organisations concerned with urban and rural poverty, environmental degradation, womens', youth and tribal rights, religious groups and academic institutions (Green Forum, 1990).

The Green Forum views the Philippines as being composed of "diverse ecosystems or bio-regions each with a community that needs to find a symbiosis with its ecosystem" (Green Forum, 1990). A society based on this vision would differ from advanced industrial society in treating natural units (ecosystems) at par with the needs of social units, in determining policy formulation. The various components of this vision — all based on the community within an ecosystem as the central unit — include maintenance of cultural integrity, the integration of spiritual and ecological values, democratization of ownership, control and management of natural resources, and decentralized, community based development aimed at equity rather than growth.

For the Green Forum, "the main function of NGOs is the empowerment of people and communities toward self-development and self-government" (1990). This consensus on the work of the Green Forum emerged from eight workshop consultations around the country, that were attended by a total of 1251 participants from 934 different organisations. While these views represent the legitimate desire of local communities to retain control over the conditions of their existence (a trend visible in industrialized countries as well), it is unclear how they envision moving beyond the social structures currently in place.

An extraordinary aspect of the Green Forum's work so far is the extent to which agreement has been forged on a common set of goals by representatives of many sectors of Philippine society. As with the Narmada movement, this agreement testifies to the political importance of making internal NGO policies be consistent with the reforms they advocate — toward decentralization and democratic decision making in this case. In addition, the importance placed on development of the platform itself, rather than careful exposition of strategies to achieve those plans, testifies to the stress on process, and the empowerment component on which every action and strategy must be based.

2.1.3 Kenya Energy and Environment Organisations (KENGO).[4]

The Kenya Energy and Environment Organisation has been operating since 1981, working in two areas: biomass and energy conservation, and trees and soil conservation. KENGO is different from the other two cases illustrated here in that it has its roots in specific micro programmes. But it shares the emphasis on participatory development rooted in direct contact with grass-roots groups that characterize the Narmada Movement and the Green Forum.

KENGO's approach to the issue of domestic resource degradation is founded on the understanding that programmes to improve the economic well-being of local communities are inextricably linked with programmes to undertake environmental protection. Accordingly, KENGO researchers and field extension staff work with local populations in the spirit of "harambee" or self help — to identify ways in which indigenous trees and plants are beneficial to the local economy, and subsequently to disseminate such information. They also work with local community groups to put tree planting, soil conservation and energy conservation programmes into practice. Close interaction with the community is essential to ensuring that programmes are appropriate to the needs and skills of implementing communities, and to the resources available to them.

As an example, the ceramic wood stove developed by KENGO in collaboration with other Kenyan NGOs is designed to be constructed by groups of local artisans from local materials, yet its use results in substantial improvements in energy efficiency. Working as a link between local communities and scientific establishments, KENGO brings expertise from universities and research institutes to bear on problems of environmental degradation. In developing its indigenous trees and plants programme, KENGO carried out a joint project with nutritionists and agronomists from Kenyatta University, effectively providing the link between indigenous scientific capability and an understanding of ground level realities.

A contributing factor to KENGO's effectiveness is the relationship it has built with the government. In particular, KENGO has managed to further its goals without losing sight of the constraints an NGO faces in the prevailing political climate in Kenya. To achieve this, it phrases its goals as common with those of the government: improving the quality of life for Kenya's people. The group has backed up this broad statement of common purpose by cooperating with government agencies, to the extent of being a member of government-organized District Development Committees in areas where KENGO has a field presence. This relationship of trust has allowed KENGO to influence government programs and attitudes from the inside, in all likelihood with far greater success than it would have done had it relied on an attitude of confrontation.

For example, KENGO has organized "travelling workshops" for policy makers, media executives and scientists, in order to expose them to ground level problems, and to introduce them to the idea of participatory community development. Similarly, KENGO publications are distributed to government agencies, and the organisation has been given access to government-controlled radio broadcasts to air programmes on trees and tree planting.

2.1.4 Lessons from developing country NGOs

The case studies above have included a nascent national movement, born out of protest against a specific project (India); a well-organized attempt to build a national consensus on the elements of a sustainable development future (Philippines); and a network of groups working to conduct research, disseminate information and implement specific ideas of sustainable development (Kenya). All share a belief in the need for a long-term view of development, one that links environment and development by the notion of sustainable management of natural resources. The means by which this long term end is achieved is by development of a constituency — those that most directly use resources — that understands both the imperative of sustainable development, and their potential for empowerment through participatory processes to sustain their natural resource base.

Meeting these challenges requires an analysis of political economy; this approach contests control over resources and the manner in which they are used. In this sense, the longer-term emphasis on empowerment can lead NGOs to increased conflict with those who currently control resources: governments, private corporations, and wealthy individuals. The risk of political conflict depends on the socio-political matrix within which an NGO must operate.

As we have seen in these case studies, the packaging of NGO actions reflects concerns of ideology, as well as tactical considerations which incorporate a shrewd assessment of socio-political realities. But in many cases, the drive for empowerment manifests itself as a struggle for control over, and access to, resources. Their vision of decentralization leads many NGOs to be wary of international negotiations between nations, which they perceive as being a further trend toward centralization of decision making and control over resources.

[4] Information on KENGO is taken from Arum and Awori (1992) and Arum (1991).

2.2 *Trends in industrialized countries*

In order to comprehend the current functioning of Northern NGO groups, it is worth considering their history. In the late 19th Century and the early 20th Century, several environmental advocacy groups were established in the US and in the UK. Their primary objective was the protection of natural areas or the prevention of natural resource development. Although the young science of resource management was itself primitive, these organisations relied on non-technical political activism, and they generally targeted specific resources or particular locations.

Concern over air pollution in urban areas is centuries old, but it did not coalesce into large-scale citizen activism until the 1950s, when clean air legislation was first adopted in the U.K. and in certain US localities. Concerns over air and water quality (from an aesthetic as well as a health perspective) merged with a growing awareness of the potential toxicity of industrial and agricultural chemical releases to create the modern environmental movement. These groups subsequently added an interest in energy during the oil shortages and energy-related crises of the 1970s. During the 1980s, acid rain, energy, ozone depletion and greenhouse issues merged to yield the current global issue perspective held by several of these organisations.

These groups are characterized by a variety of philosophical stances. Some, particularly the large, national groups, are militantly non-partisan and non-ideological. Environment *per se* is their central concern. Organisational interest in broader socio-economic issues is incidental. These concerns shape institutional policy only when they prove to be obstacles to environmental goals. At the other end of the political spectrum, some grass-roots groups regard environment, economic equity, and, in the US, racial issues to be inextricably linked.

During the last 15 years, NGOs have shifted their attention outward, from local to regional to the global scale. At the same time, they have increased their level of technical expertise. Originally, the groups had included primarily citizen activists, and a handful of individuals with expertise in resource management. With the shift in direction of the 1960s came biologists and other scientists, as well as attorneys. Economists joined the ranks during the 1970s. Generally, these trends appeared first among US groups. With the incorporation of such expertise came the ability and the tendency to propose alternatives, rather than just oppose projects. Since environmental protection now covered greater and more abstract ground, the tactic of merely opposing projects was insufficient — there was frequently no particular target. Employing this new expertise, NGOs have contributed greatly to the force of the energy efficiency and renewable energy movement in recent years, and also to the drive toward adoption of pollution prevention technologies.

In North America, NGOs have effectively combined the original grass-roots approach targeted on specific projects, like waste dumps or wilderness protection, with the national-level policy approach, to strongly influence events. Think-tank style analysis, litigation, lobbying, and political organisation are not often combined within a single group. Groups specialize and cooperate to attain the appropriate mixture of talents and approaches.

2.2.1 *The Clean Air Campaign (United States)*

The long legislative battle to strengthen the US Clean Air Act provides a relevant example. US groups — characterizing the full spectrum of skills, sizes and geographic coverage — joined together to wage a decade-long campaign beginning in 1981. The coalition developed a Washington-based lobbying strategy to influence Congress and the White House. Coordination was achieved through the National Clean Air Coalition. The Coalition was made up of national environmental, health, and union groups. The Coalition members were of two types: those with large memberships and strong local chapters throughout the nation; and smaller groups valued for their legal and technical skills. The latter were deployed to sustain the ongoing scientific debate with industrial opponents of the legislation, as well as to assure that the various legislative proposals were designed to achieve their stated goals.

The Coalition provided financial and technical support for various state and local groups, which organized grass-roots campaigns and lobbied their Congressmen on the issue. In some instances they engaged in activities aimed at influencing electoral politics. The national groups remained closely tied to the local groups throughout a decade of activity. This combination proved to be far more effective than a national campaign alone would have been. We note the electoral element here as an exception to the rule that US environmental groups generally avoid strategies which involve electoral politics.[5]

Ironically, a public education campaign, linked to the 1984 presidential election, did play a critical role in the ultimate strengthening of the Clean Air Act. Acid rain was of local concern to citizens of New Hampshire due to the vulnerability of their forests and lakes. The national groups joined to raise the issue during the pivotal New Hampshire primary election campaign, by focusing information about acid rain on the media, and by holding "citizen hearings" to air the issue. This effort forced each of the major candidates running in New Hampshire to stake out a stronger position on the issue than several of them had originally intended (except President Reagan, who ran unopposed in the Republican primary in that state).

[5] However, they do have a political arm, the League of Conservation Voters, which supports particular candidates, but not the parties per se. In contrast to the League, most of the national groups have tax-exempt status, which prohibits attempts to influence elections. This tax-exemption provision, and the philosophical non-partisanship of the movement as a whole, have led to a tendency to not lean heavily on electoral approaches.

The landslide nature of Reagan's re-election precluded the possibility that the acid rain issue would receive an immediate impetus from this activity. But an important seed had been planted. When George Bush ran for President in 1988, he had to first run the very same gauntlet of the New Hampshire primary. The electorate, educated on the issue by the effort of the environmental groups 4 years before, wrung a strong commitment on acid rain from Bush. Upon his election, he had to live up to the commitment, and strengthening of the Clean Air Act thus became inevitable.

This history supports the efficacy of combined national and grass-roots campaigns, involving groups with both organizing and technical skills. The efficacy of raising issues at election time in particular has led to development of a similar effort around the global warming issue, ahead of the Florida primary election in 1992. As in New Hampshire, this effort remains non-partisan and not aimed at supporting particular candidates. Whether such an even-handed approach can succeed more than once has yet to be seen.

US groups have proven effective at blocking anti-environmental initiatives, but somewhat less effective at initiating solution-oriented programs. Of course, this is true of any so-called "interest group"; it is inherent in a non-parliamentary system. Recent examples include the defeat of a proposal to drill for oil in the Arctic National Wildlife Refuge, and the apparent reversal of President Bush's attempt to redefine wetlands — a move that would have led to the elimination of many of them. Perhaps even more impressive was the successful fight in the early 1980s to prevent weakening of the Clean Air Act, in face of a strong push by a newly elected and very popular president (Reagan). Ultimately, positive and strong changes were made in the Act, but only after a decade of struggle (see above).

Nevertheless, environmental groups have had a number of successes, in addition to the Clean Air Act, which illustrate a capability of marshalling a variety of forces to achieve their goals. The very salience of the global warming issue, and the convening of the INC provide a case in point. Here, an issue which could not have been predicted to assume immediate importance to the "man on the street" was converted into a concern at the highest level of government — as well as for the general public in the US and Europe — by a campaign spearheaded by scientists and environmentalists. The primary vehicle was a thorough airing of the issue in the media, which included an education campaign to raise the general level of awareness of the global warming issue.

2.2.2 *Forging joint solutions with the private sector: McDonald's and the Environmental Defense Fund (US)*

A more complex episode involved the decision by the McDonald's Corporation to abandon the use of foam plastic disposable containers. A grass-roots effort and an associated consumer boycott brought the company under economic pressure. At about the same time, a cooperative analysis arrangement was developed between the Environmental Defense Fund (EDF) and McDonald's to study solid waste disposal and packaging options. The combination of grass roots action, bad publicity, and sound analysis and specific proposals by an outside actor (EDF) with access to top management provided the company with both the motivation to change packaging and a reasonable set of alternatives. This set of circumstances also bolstered the position of those in the company who supported more environmentally sound solid waste policies, and encouraged them to advocate these policies from within the corporation. Consequently, McDonald's discontinued foam packaging.

This episode may provide a model for future progress, but some aspects of the situation were unique. In this case, an aggressive grass-roots campaign helped to create a middle ground where the technical group had room to operate. Not all corporations are as vulnerable to consumer preference as McDonald's, or as willing to entertain outside suggestions from environmentalists who are perceived to be reasonable by top management. On the other hand, the "inside-outside" strategy between EDF and the grass roots groups was uncoordinated and unintentional. This lack of coordination probably had benefits in this case, but in other cases it could hinder a good outcome if the "inside" technical group pursues goals at odds with the grass roots, or if the grass roots demands do violence to the realm of the possible.

2.2.3 *The Green Party of Germany*

In Europe, the evolution of environmentalism has been more partisan. Technical resources, while existing within national and international groups there, are thinner, and litigation is virtually non-existent. On the other hand, the strong and vocal membership base of European NGOs has enabled them to mobilize public opinion around environmental issues, and the coalition nature of governments has allowed Green parties, building on public support, to exert substantial influence on mainstream politics. A prime example involves the rising influence of the Green Party of the Federal Republic of Germany during the middle 1980s. A variety of environmental and disarmament issues — including ozone depletion and forest decline (waldsterben) related to air pollution — were publicly raised through the grass-roots contacts of environmental groups, and fuelled the strength of the party beginning early in the decade. By the late 1980s, it had garnered 9% of the national parliamentary vote, and appeared likely to hold a critical position in parliamentary elections then planned for 1990. A so-called "Red-Green" coalition with the Social Democrats was envisioned.

Events moved in a different direction. Dominant parties co-opted the Greens on environmental issues by adopting elements of their positions. At the same time, the reunification thrust Green Party issues off centre stage. Consequently, Green Party electoral strength declined. Nevertheless, the current strong position of the German government (relative to the US for instance) on CFCs, acid rain, and global warming bear witness both to the power of vocal grass-roots action and mobilization of public opinion, and to the success of the electoral, partisan approach in the European context.

2.2.4 *Lessons from the industrialized nations*

Therein lie both the strengths and the limitations of the Northern groups. North American groups are very successful when targeting particular projects through the grass roots, and at influencing some policies through national groups which analyse, lobby, litigate and even negotiate directly with industry. Coalitions of organisations combine these strategies effectively on an issue-by-issue basis. But they lack a political base. Alliances with labour, church, human rights, and consumer organisations are routinely carried on issue by issue. But the latter organisations themselves have only limited political power. Therefore, it remains in doubt whether they can muster the strength to influence underlying phenomena, like the economic decisions which affect global warming so strongly. There is also a certain non-ideological characteristic of American politics, which furthers consensus on environment while obscuring these underlying concerns. Political economy and socio-cultural criticism play little role in North American environmentalism. They surface only intermittently, as in arguments over water supply and water rights.

In Europe, grass-roots pressure and Green political power abound. These characteristics leave European environmentalists in a better position to address broader socio-economic issues than their colleagues in the US. Thus, European NGOs are more likely to view local as well as global environmental issues as connected to international trade patterns, the levels and form of official development assistance, the role of transnational corporations, the debt burden many Southern countries bear, and the whole host of factors that feed into North-South relations. Since Southern perspectives on environmental issues are fundamentally linked to the poverty prevalent in many Southern countries, this willingness to address economic issues in conjunction with environmental problems marks an important step toward a global perspective on the problems at hand.

An encouraging sign of this emerging trend is the serious role of environmental NGOs in free trade talks. Similarly, European groups are trying to play a role in continental unification issues with respect to 1992. If environment becomes a concern on a par with trade, economy, and arms, and if the linkages between all these issues are made explicit, then it may be possible to address the structural issues of both North and South. Whether the non-partisan, non-ideological approach of US groups is up to this task remains in doubt.

3 Action on Global Environmental Problems

3.1 *The expanding international influence and activities of NGOs*

The ability of various NGOs to have an impact on negotiations and international processes is growing. Battles are being fought and won, and precedents set on the extent to which NGOs are able to observe and influence international processes — such as the preparatory committees for the United Nations Conference on Environment and Development.[6] More significantly, NGOs — particularly those with an academic and research bent — are being requested to contribute to, and comment on, the preparation of government positions. As NGOs strengthen their technical and policy capabilities, they are granted greater access, albeit often informally, to the decision-making levels of international processes, redressing, in part, the dominant influence industry experts have had on these processes in the past.

Increasing NGO strength and influence at the national level also has ramifications for the international negotiations — as researchers and policy experts begin to evaluate the role NGOs can play in the functioning of international agreements. As an example, one proposed mechanism to limit emissions of greenhouse gases suggests that enforcement for Northern countries would be based on ensuring full and open access to information, with the expectation that NGOs would enforce country commitments through domestic political processes (Chayes, 1990).[7] This represents a significant shift from the traditional role of NGOs acting to make a system work more effectively to NGOs acting as an integral part of an enforcement system, and is a potent illustration of the expanding impact of NGOs.

Despite these favourable indications, NGO participation at the international level has been largely restricted to Northern NGOs. This is rapidly changing. Northern NGOs have generally had better interpersonal contacts with officials at intergovernmental organisations and with their own governments due to the nature of their national operations. In addition, the "revolving door" between the government and non-profit sectors in some Northern countries have allowed NGOs to recruit former government personnel with extensive international contacts and vice versa. Southern NGOs have not, thus far, had the resources — financial or human — with which to address these larger scale problems, particularly since many pressing local symptoms of these global issues occupy all available time and resources. However, international events such as UNCED have mobilized and activated large numbers of Southern NGOs, who have come to the conclusion that it is not sufficient to fight their battles on a national level alone, but that they must also attend to the international dynamics of environment and development issues. As a result, many Southern NGOs have begun extrapolating their domestic agenda to the international stage,

[6] While the UN General Assembly resolution calling for UNCED only invited participation from NGOs with consultative status with the UN Economic and Social Council, NGO observer status has since been extended to numerous other groups, many of whom are actively participating in the process with mixed levels of success (note United Nations General Assembly resolution A/C.2/44/L.86).

[7] We add, however, that Chayes proposes enforcement for developing countries through conditions on access to funds, potentially a far stronger and potent weapon for enforcement.

and have presented some incisive critiques of the political economy of international environmental negotiations.[8]

Having established that NGOs have recently managed to obtain a greater voice at the international level, how best ought they make use of these new opportunities? We have seen that the strengths of Northern NGOs are highly developed technical capabilities and effective strategies using political institutions, media, and public pressure to monitor and influence the activities of the state. However, the effectiveness of Northern NGOs' advocacy — at local and national levels — is limited by the extent to which an issue can be focused and a constituency can be mobilized. Southern NGOs have, by contrast, been innovative and successful at developing and implementing specific projects at the local level that have served to provoke and stimulate a debate on policy direction. However, they have limited technical capability to draw on and less experience than Northern NGOs at policy advocacy, although their strength in these areas is rapidly growing.

3.2 Developing a consensus on sustainable development

Action on global environmental problems is complicated by a need to define and choose among sustainable development paths. The Brundtland Commission defined sustainable development as "development that meets the needs of the present without compromising the ability of future generations to meet their own needs" (WCED, 1987). Daly and Cobb (1989) suggests several related questions: how are needs defined; how are we to define what will, and will not, jeopardize the prospects of our children; and does this definition, by itself, constitute an adequate criterion for sustainable development.[9] While these questions are not answered here, they are relevant to our subject. We believe that governments, the private sector, and NGOs in the North should confront issues of trade, debt, and related issues of economic sovereignty in the context of negotiations on global environmental problems.

As a generalization, developing countries and industrialized countries face drastically differing socio-economic circumstances. Industrialized countries tend to have urban populations with low growth rates that rely on manufacturing and services for the bulk of their economic livelihood. By contrast, most developing countries — with the notable exception of Latin America — are made up of rural, agrarian populations in the midst of a demographic transition (World Bank, 1991). The picture is further complicated by massive rural to urban flows.

These factors, in addition to the NGO emphasis on decentralization, empowerment and control over resources demonstrated by the three case studies above lead to the conclusion: nations must not be restricted by international factors in their choice of policy mix. Rather, they must be directed by the aspirations of their citizens and their social and ecological base. How free are countries, in reality, to define their practice of sustainability, and to what extent are they constrained by the exigencies of forces such as world trade? The drastic fall in prices of primary commodities over the last decade is a forceful example of the impact of world trade patterns on the domestic situation of a country. Similarly, the shift in emphasis from projects to "structural adjustment" policies by the World Bank, starting in the early 1980s, are of great significance in determining the development path a country adopts. In contrast to project specific lending, structural adjustment programmes advocate a two part "fixed menu" of macro-economic stabilisation followed by structural adjustments emphasizing the liberalisation of trade and privatisation of industry and agriculture. Consequently, development in the South is greatly influenced by events in the North and by the patterns and pressures of the international economic order. This is not to deny the benefits of global economic integration, but the extent to which there is a trade off between these benefits and economic sovereignty must be examined and factored into decisions.

In addressing issues of economic sovereignty, the question of patterns of resource use and their link between lifestyle patterns in North and South as well as a host of associated issues will likely be raised. For example, it is possible to conceive of widely differing opinions on the extent to which appropriate technological choices are a solution for resource scarcity. Thus, the socio-political issues inherent in questions of control and use of resources are likely to be sharpened. So far, these issues have been blurred by the seeming consensus on the need for "sustainable development," but it is likely that as the various interpretations of this phrase and their implications become clear, the consensus will be threatened. For our purposes it is useful to highlight these issues as possible sources of disagreement, and to emphasize the need for discussion of these broader issues between governments, the private sector and NGOs, many of whom are substantively engaged in addressing these questions.

Having stated the case for addressing broader questions of sustainable development, we note that the need to address these issues should complement, rather than substitute for action on specific global environmental problems. As governments become convinced that concrete measures must be undertaken to address specific environmental issues, the resultant discussions and negotiations open up many opportunities for NGOs. NGOs need to simultaneously take advantage of specific opportunities that present themselves, while emphasizing other international socio-economic links that affect the development of sustainable development policies. In particular, Northern NGOs need to pay more

---

[8] See, for example, the recent study on climate change by the Centre for Science and Environment, an Indian NGO (Agarwal and Narain, 1991).

[9] Daly and Cobb also raises these questions, and rightly points out that the Brundtland Commission wisely opted to avoid a detailed discussion in order to gain political acceptability for the term "sustainable development."

attention to the latter, and Southern NGOs must begin to consider ways of developing and using opportunities that global discussions present.

3.3 Information sharing and action networks

One constructive mechanism by which NGOs continue to refine the notion of sustainable development is by working together in action oriented networks and coalitions. Networks of NGOs working together serve multiple purposes: they provide a concrete context for discussion of the broader issues of sustainable development aimed at, in the long run, reaching shared understandings; they allow NGOs to multiply their technical capabilities by pooling resources; they provide a coherent vehicle for dialogue with governments and the private sector. Most important, these networks can achieve concrete change on various subsets of global environmental problems and, by extension, on issues of sustainable development.

There are various examples of such NGO networks that are already in existence — some of them country specific, and others that focus on a specific issue. We will examine each type in order to understand what such networks can achieve and the manner in which they do so.

3.3.1 International NGO Forum on Indonesia (INGI)

The INGI was set up as a mechanism by which to influence lending policies to Indonesia as orchestrated by the Intergovernmental Group on Indonesia (IGGI) — the coordinating mechanism for lending agencies who operate in Indonesia. With this basic objective, INGI's mandate allows for further steps that promote the objective of strengthening and broadening participatory development processes in Indonesia. INGI is structured around two steering committees — one of Indonesian NGOs and the other of non-Indonesian groups comprised of NGOs from the IGGI member countries — largely OECD countries. This mechanism facilitates communication between the Northern and Southern partners and allows each group to keep the other informed on events within Indonesia, particularly at the grass-roots level, and internationally on the policies of the lending institutions. Thus, INGI members share information, for example, on an incipient project or programme loan, evaluate it, and communicate their reactions and concerns to the Indonesian government and the lending institutions.

Each annual INGI conference has produced a set of recommendations for the IGGI. These transmit the viewpoint of NGOs with considerable grass-roots experience to lending institutions in relevant, easily digestible form. These recommendations have covered questions of resettlement, women and development, international issues such as debt and protectionism as well as various environmental topics at both the project and policy levels. In addition, the Forum has recommended to IGGI that a socio-cultural impact assessment and environmental impact assessment be undertaken for all official development projects. Domestically, Indonesian NGOs have begun a dialogue with the Government of Indonesia during which INGI concerns have been raised.

We believe that INGI provides a valuable perspective to the members of IGGI, one that is not reflected in their main source of information, the World Bank's annual report on lending in Indonesia. The link between Northern and Southern groups has resulted in a more complete ability to monitor the foreign assistance flows to Indonesia from both the Indonesian and donor ends, and has built up trust and an understanding of the functioning of NGOs on both sides of the relationship. Also noteworthy is the extent to which access to foreign NGOs and to donors has increased the leverage Indonesian NGOs have with their own governments. While this causal relationship cannot be easily proven, it does seem likely that the formation of INGI has given Indonesian NGOs a louder voice and better information with which to act within Indonesia. The INGI model provides a framework within which NGO ideas and input is channelled through to decision- making bodies — both of the Indonesian government and of international lending agencies — in a useful and relevant form.

3.3.2 Multilateral Development Bank (MDB) Campaign[10]

The MDB campaign has built a strategy for reform of the environmental policies of the MDBs based largely on close and equal ties between NGOs in industrialized and developing countries. Tactics employed by the campaign have relied on detailed information flow from Southern NGOs to the North on specific instances of environmentally and socially unsound projects being funded by the World Bank and other MDBs. Northern NGOs have served as a conduit, to communicate these concerns to the funding institutions, and to generate public pressure for reform in the North. The approach of Northern NGOs has been to question the use of taxpayers' money to support environmentally and socially damaging projects in the South, and, from this base, to place conditions on the financial contributions to the MDBs from the industrialized countries.

While there are still considerable problems with the World Bank's policies from an environment and development perspective, the MDB campaign appears to have been remarkably successful in leveraging change within the MDBs. For example, since 1983 the World Bank has increased its environmental staff manyfold, has commissioned environmental issue papers and action plans, has committed to financing specific environmental programs and has called for greater participation from local Southern and Northern NGOs (Rich, 1990). While there is no way of attributing these changes directly to the campaign, its impact on specific projects has been considerable. As an example, a strong international campaign in Japan led to the withdrawal of Japanese funds for the Narmada Valley Project, and forced

[10] Much of the information on the MDB campaign included in this section is taken from Rich (1990).

the World Bank to halt disbursements on its funds subject to the report of an extensive independent review.

The campaign thus meets the objectives of the Northern NGOs, while benefiting the Southern NGOs both through policies that better incorporate their concerns as well as by strengthening their struggle against specific projects. The campaign has also forced Southern governments to take their NGOs more seriously since Southern NGOs now have an effective strategy for holding their governments accountable. In many cases the MDBs — due to pressure from the campaign in the North — have urged Southern governments to give their NGOs access to information and to involve them in project design and management, albeit with limited success.[11]

The MDB campaign illustrates the benefits of pooling knowledge and building on the comparative advantages of different NGOs. It also demonstrates a viable mechanism for ensuring that grass-roots opinions and perceptions are fed into the very highest levels of policy dialogue. In addition, while Northern and Southern partners each operate within their respective social contexts, the process of defining shared goals and strategies is a significant one, and one that causes many elements of disagreement to be aired. The lessons learned by the partners in the campaign about each others' priorities and visions is a valuable body of knowledge, and one that would be well worth sharing. Finally, from its initial basis as an effective vehicle for damage limitation, the MDB campaign — largely by forcing the door open through effectively directed public pressure — has cast itself in a proactive role, where the participant organisations have a chance to comment on policy development, rather than simply the impact of policies.

3.3.3 Climate Action Network (CAN)

From the very earliest stages of discussion of the climate change issue, CAN attempted to provide a coherent NGO voice. Participants in the network, acting prior to its establishment as an entity, played a role in stimulating governments to take action on the threat of climate change. While the network was originally established by NGOs from the industrialized countries of Western Europe, Australia and North America, it is in the process of expanding to include NGOs from Asia, Africa, Latin America and Eastern Europe.

The major focus of CAN has been on the contribution of industrialized countries to greenhouse gas emissions. This represents the original composition and has coloured the priorities of the network, resulting in a narrow focus on the climate issue and the climate negotiations in particular. As a result, the Climate Action Network also provides an example of the difficulties of international NGO collaboration on a complex issue. The network characterizes the goal oriented focus and strategic imperative of industrialized country NGOs, which leaves little scope for discussion of broader socio-economic issues. This contrasts with the breadth of developing country concerns, which must be translated into terms that match the specific opportunities the climate negotiations present. Finally, CAN demonstrates the need for scientific capability in order to effectively address complex issues such as climate change. INGI and the MDB campaign have had relatively greater success in defining concerns shared between North and South and in developing mutually acceptable strategies. An explanation is that INGI and the MDB campaign do not have an issue focus, but rather focus on a country and a mechanism respectively, allowing the participants greater flexibility in interpreting and developing the agenda.

CAN has approached this problem by facilitating the creation of regional climate networks. These networks, initiated and developed by local NGOs in the region, are tasked with translating the climate issue into terms that match local concerns, and ultimately creating the appropriate context within which climate change can be understood and responded to by all NGOs. Achieving this goal requires action at several different levels: information on the science and policy aspects of the issue is disseminated to local NGOs, with the long-term goal of providing two-way linkages for exchange of information and development of ideas; research on local impacts of climate change and on policy issues of particular relevance to the region are undertaken, often in collaboration with academic or research institutes; links are established with governments and regional intergovernmental bodies with the goal of developing credibility and a channel for input of positions and informed opinions. CAN confronts the problem of creating relevance and providing a context for discussion of global environmental problems. This is a problem that is likely to be repeated; the early efforts of the Climate Action Network are an ongoing attempt to overcome this hurdle to effective NGO action.

The collective face presented by CAN has been favourably received by governments, largely due to the benefits gained by pooling resources and expertise. This acceptance and credibility has been earned as a result of well thought out and effective technical and policy research contributions to deliberations, a demonstrated understanding of the international negotiating system, and innovative methods such as production and dissemination of an informative and thoughtful advocacy newsletter *(ECO)* on a daily basis during

[11] It is only fair to note that reservations have been expressed in some quarters about the MDB campaign. These concerns centre on the argument that such a transnational campaign gives Northern NGOs a say in the governance of Southern countries without any reciprocal power for Southern NGOs. Further, critics argue that the ability of Northern NGOs to influence events in the South is liable to undermine democratic institutions in the South and to damage relationships between Southern NGOs and their governments. In response, proponents of the MDB Campaign insist that transnational campaigns of this sort are an alliance between peoples, irrespective of national boundaries, that allows citizens to have a more effective say in the activities of their states and states' creations such as the MDBs.

climate meetings.[12] The result is an ability to influence government positions even as they are being formulated, during the negotiations as well as between sessions. In addition, CAN as a whole possesses a sufficiently detailed understanding of the process to address and provide input on technical details which often are the most significant determinants of the outcome of such a process.

3.3.4 *Lessons from international NGO activities*

All three collaborative attempts described above are examples of NGOs forging links across national boundaries, expanding the NGO role of monitoring and advocacy to include institutions formed or actions initiated as a result of collaboration between countries. In the first two cases the Southern partners had a large part in setting the agenda, and the strategy focused on overcoming the restraints Southern NGOs face in their own countries as a result of international linkages between states. CAN represents an attempt to place a problem that is largely Northern in cause as of today, but which has global effects, in the context of global sustainable development issues. While one network is country specific, another is mechanism specific and the third is issue specific, they are all predicated on the same notion: effective accountability across international boundaries is based on cooperation and exchange of information between NGOs that allows each to function more effectively in their own setting. Many such networks seek to address the underlying causes of global environmental problems. By fostering common attitudes and understandings between NGOs in different countries on specific issues, they also work towards building mutual understandings of broader notions such as sustainable development, that must be the cornerstone of long term NGO efforts to address the root causes of global environmental problems.

It is worth keeping in mind that each of these networks, in particular INGI and the MDB campaign, work on aspects of the sustainable development issue that have indirect impacts on global environmental problems. In order to maximize the impact of such networks, it is important to ensure that they are provided information on, and included in discussions of overarching global environmental problems such as climate change. Thus, providing the MDB campaign, the Tropical Forestry Action Plan campaign, and various energy efficiency initiatives with information on incipient environmental problems at a global level will enable NGOs to work more effectively within their issue or country specific networks, keeping the broader picture in mind. As we have seen with the INGI and MDB campaign, such a system will also allow NGOs to transform their reactive approach to a preventive one.

[12] "ECO" has been produced by environmental groups at various environmental meetings since the Stockholm Conference on Environment held in 1972. This particular set of ECOs has been jointly produced by the Climate Action Network.

4 The Future and Potential of NGOs

If NGOs are to stake out a position as legitimate players in the resolution of global environmental problems, they must develop the technical capabilities which will enable them to influence development of policies. In addition, creative approaches to the private sector could yield productive results. However, new strategies must not be developed at the expense of continued NGO credibility as actors free from short-term political and economic interests, and with close links to their constituency. Finally, the three examples above, by dint of their ability to present clear, well thought out technically competent positions backed by their collective credibility and their independence from official processes, offer the most promising model for constructive input to official government and intergovernmental processes.

4.1 *Tapping the potential of the private sector*

We have seen that many NGOs have developed a useful degree of political clout within their own countries, have worked to influence governments and intergovernmental actors by pooling their efforts, and have succeeded in raising public consciousness of environmental issues. However, as was discussed earlier, the particular characteristics of global environmental issues limits the utility of hitherto successful strategies. How does one create a specific constituency for policy action when the degree of impact on various "losers" and even the existence of "winners" are in doubt? How does an NGO generate public pressure for change through media and information campaigns when the public is bombarded by contradictory information at a high level of scientific abstraction? Global environmental problems do not provide the public with a simple, easily understood message that relates to their daily life.

One possible alternative approach is contained in some of the innovative strategies NGOs are beginning to consider in their dealings with the private sector. The key lies in combinations of tactics. In the case of global environmental problems external pressure may not be sufficient by itself to effect policy or corporate change. Individual approaches to members of the private sector may be insufficient. NGOs using an "inside-outside" strategy, as we have seen with the case of the McDonald's Corporation, can stimulate change.

We note that such a strategy does not require *all* NGOs to adopt a policy of dialogue with industry; such a position would weaken NGO leverage. As with the McDonald's case, an optimal policy is one that simultaneously creates incentive for change through external pressure, and provides a mechanism for responding to that pressure. In the case of global warming, NGO actions to stimulate international negotiations on climate change provide a mechanism for pressure; the private sector is forced to adapt or work to delay agreement on an effective convention. In order to decrease the chances of a unified opposition forming within the private sector. NGOs must demonstrate the advantages that companies can gain by factoring implementation of an effective climate convention into their research and development programs, and planned investments. In addition, NGOs should highlight the commercial opportunities for innovative technologies that will emerge within the framework of an international convention. A positive attitude toward action on global environmental problems from the private sector would also help break down resistance to political action. This linkage was strikingly demonstrated during negotiation of the Montreal Protocol. During the negotiations, the withdrawal of opposition to a phase-out of ozone depleting chemicals by some major chemical manufacturers facilitated political agreement on the most dangerous chlorofluorocarbons and halons.

4.2 NGO links with the grass roots

Most environment and development NGOs perceive the democratization of local, national and international decision-making processes as an essential part of their role. The emphasis many NGOs have placed on empowerment and participatory processes is intimately linked to their demands for full and open access to information and to decision-making structures. In order to be consistent, NGOs place much emphasis on a loose and open internal administrative structure. As groups have expanded their staff, budget and scope of their agenda, an element of tension has been introduced between the perceived imperative of an open process, and the need for managerial structures that would allow rapid, more centralized decision making.

However, the close contact many environment and development NGOs maintain with their constituencies provides a great deal of their legitimacy as representatives of the interests of society as a whole. It gives NGOs the right to represent what they feel to be in the greater interest of society but it also obligates NGOs to be accountable to their constituency. Global environmental problems represent a challenge to NGOs in terms of adequate links — between groups and within groups — to ensure accountability. It is relatively simple for an NGO representative to explain the action a group is taking with respect to a local toxic waste dump, or a local company that threatens to strip forest land. It is more difficult for grass roots groups or active members of an NGO to relate to the dynamics and subtle progression of international environmental negotiations. However, if NGOs float free of their grass-roots base, for reasons of convenience and expediency, their legitimacy will be severely eroded. In order to avoid this, it is necessary to ensure that adequate and appropriate information flows are established from NGOs at the national level to regional and local NGOs. Similarly, feedback on global issues must bubble up to the national level, so as to ensure legitimate advocacy at the international level, even if it does slow down NGO reaction time.

In this context, NGOs could put their grass-roots contact to effective use by serving to remind negotiating parties of the human scale of the problems. In particular, some NGOs with grass-roots contacts are in a position to undertake detailed studies of potential impacts of global environmental problems using their in-depth knowledge of communities and ecosystems. This would serve as a stimulant to the negotiating process and ratchet up the degree of public pressure and awareness by sharpening the distinction between winners and losers. An example is provided by a detailed study of the disastrous effects of large-scale resettlement of island communities prepared by the Association of South Pacific Environmental Institutions, which was distributed to the delegates at a session of the climate negotiations (CAN, 1991). Thus, even if the direct impacts of global environmental problems defy quantification, NGOs, backed by the necessary scientific capability and case studies, can vocalize the concerns of the powerless.

4.3 Enhancing scientific and technical capabilities

The role of environment and development NGOs in global environmental problems will largely be determined and driven by those best able to address the scientific and political complexities that characterize these problems. NGOs, if they wish to increase their credibility with governments and the private sector on these issues, will have to expand their scientific and technical knowledge. NGOs in the South that are technically capable tend to be exclusively research organisations with few, if any, links to the advocacy and action community. Greater indigenous technical capability is needed to translate global issues into local ones and to develop local mitigation and adaptation strategies based on an adequate understanding of the global technical issues. While Northern NGOs have begun this process, their capabilities could be further strengthened by forging stronger links with the research community, and by developing institutional structures that will allow them access to greater resources. This section will explore ways in which NGO capabilities, particularly in the South, can be built so as to adequately address global environmental problems.

As the current discussions on global warming show, developing countries are severely hampered by their lack of scientific capability. Industrialized countries are in a position to manipulate the scientific evidence so as to serve their interests. An example is the current debate over the global warming potential ascribed to the different gases. These weights can be assigned on a variety of criteria, and could shift the relative importance of the various gases to future warming, shifting the burden of mitigation measures among nations. Global NGO coalitions, acting as informed observers if not researchers, could play a significant role in balancing the unequal access to information and research which allows industrialized countries a virtual monopoly over the complex scientific knowledge that underlies these problems. As a caveat, however, we add that such cooperation between NGOs is predicated on a relationship of mutual trust and acceptance of common goals between industrialized and developing country NGO partners. This point relates to the earlier discussion on the need to examine the underlying social issues and longer term visions of sustainable development.

While there is a great deal of activity in the environment and development sectors in the South, most of this activity, understandably, is centred on increasing the quality of life in local communities. While this work must form the core of action on global environmental problems, there is a need to increase the number of skilled, technically competent individuals from the South who understand community level issues, and are able to make the international linkages. Despite the current surge in financial resources available for work on global environmental problems, results have been limited. Those working on global environmental problems are stretched thin, leaving little or no resources for training a new pool of individuals.

In order to break out of the cycle of scarce human resources leading to inadequate coverage of an issue, a commitment by Southern NGOs to allow individuals or groups the time to devote to specific issues is necessary. Capability building can then be enhanced by means of small training workshops organized by region, complemented by internship opportunities and involvement in international activities. Internship opportunities with other NGOs, both in the North as well as with other Southern NGOs are useful in establishing personal links, sharing ideas and broadening the resource base. International activities will enable Southern NGOs to acquire skills necessary to articulate local perspectives on global issues.

Southern NGOs could be strengthened by expanding their efforts to build contacts with Southern universities and research institutes. There is great potential for a symbiotic relationship between action and advocacy NGOs and research organisations. In addition, Southern NGOs could pool their resources in order to establish issue specific information centres on a national or even regional basis. While technical information flows from North to South are heavily emphasized, there are several Southern initiatives that may be more compatible with the needs and circumstances of developing countries.

Several regional information centres on energy use have recently been established. We consider a recent proposal for a regional centre on energy efficiency, which would contribute greatly to developing an understanding of energy efficiency issues. The centre would: collect information on the uses and implementation of relevant technologies, both from the South and North; serve as a research centre where NGOs working on energy efficiency issues could develop their ideas as short-term fellows; compile a resource database of available experts in the field of energy efficiency; and disseminate the accumulated information in suitable packages to advocacy NGOs, government officials, the private sector and the public.[13] The attractions of such a centre are great. It would provide a context and an issue focus within which to build capability, and would additionally constitute a centrally compiled body of information. The fellows programme would allow individuals from NGOs to further develop their ideas, and focus on enhancing their capabilities and skills. The centre would serve as an intermediary between research organisations and advocacy groups, uniting technical capability with advocacy skills and political savvy. Finally, the centre would provide a mechanism for links with other sectors of society including governments and the private sector. This would allow NGOs to educate themselves of the needs of these sectors and the opportunities available for cooperation.

5 Conclusion

Non-governmental organisations have blossomed in the last 15 years. While continuing to serve as a source of accountability — keeping governments and corporations honest — they have expanded their scope to proposing and developing alternatives to policies they dislike. As we have seen, in the last few years they have emerged as significant actors on the global scene, and as their technical capabilities grow, are likely to increasingly command the attention of states and the private sector.

Global environmental problems present humankind with a unique set of difficulties. In order to solve these problems, it may be necessary to subject our entire pattern of resource consumption and the international system which maintains that pattern to scrutiny; and it may be necessary to commit to immediate mitigation steps. NGOs have begun this process by attempting to define the notion of equitable and sustainable development relatively free from the restraints of political or corporate interests. Governments can employ this information, if it is presented in a relevant and useful form, to ensure that their policies are truly representatives of the desires and needs of the people they serve. The International NGO Forum on Indonesia demonstrates one mechanism by which people's perceptions and opinions are factored into decision making in a positive manner; the multilateral development bank campaign has shown the power of NGO monitoring across national boundaries, and the Climate Action Network has proved a constructive model for interaction between governments and NGOs in finding creative yet immediate solutions to global warming.

Active measures on the range of human activities that contribute to global environmental problems must first overcome a great deal of social inertia. NGOs are well placed to bring these issues to public attention, and mobilize opinion around them. A large factor in overcoming this inertia is the attitude of the private sector. In addition to undertaking information dissemination, NGOs should help ensure a responsible and progressive private sector by offering the carrot of commercial opportunities and wielding the stick of consumer pressure.

[13] The authors are grateful to Agus P. Sari from WALHI, Indonesia, for sharing these ideas with us.

References

Agarwal, A. and S. Narain, 1991: *Global Warming in an Unequal World: A Case of Environmental Colonialism*, Centre for Science and Environment.

Arum, G., 1991: Indigenous tree planting takes root in Kenya. In *Earth Summit: Conversations with Architects of an Ecologically Sustainable Future*, ed. S. Lerner. Common Knowledge Press, Bolinas, CA.

Arum, G. and A. Awori, 1992: *Natural Resources Development: KENGO's Case Experience*, Draft.

Chayes, A., 1990: Managing a transition to a global warming regime or what to do till the treaty comes. In *Greenhouse Warming: Negotiating a Global Regime*, ed. Jessica Matthews, World Resources Institute.

Daly, H. and J.B. Cobb Jr., 1989: *For the Common Good*, Massachusetts, Beacon Press.

Dubash, N., 1990: The Birth of an Environmental Movement: The Narmada Valley as Seed-Bed for Civil Society in India, unpublished paper.

Climate Action Network (CAN), 1991: *ECO*, **LXXVIII**(4), p 3.

Elliot, C., 1987: Some aspects of relations between the North and the South in the NGO Sector. In *The Challenge for NGOs, World Development*, ed. Drabek, 15 Supplement, Pergamon Press, Oxford.

Guha, R., 1988: Ideological trends in Indian environmentalism, *Economic and Political Weekly*, p. 2579.

Green Forum, 1990: *Creating a Common Future: Philippine NGO Initiatives for Sustainable Development.*

Rich, B., 1990: The Emperor's new clothes: the World Bank and environmental reform. *World Policy*, pp. 305-329.

Kothari, S. and R. Bakshi, 1989: *Lokayan Bulletin 7*, (editorial), pp. 4-5.

World Bank, 1991: *World Development Report 1991*, Oxford University Press, Oxford.

World Commission on Environment and Development, 1987: *Our Common Future*, Oxford University Press, Oxford.

CHAPTER 19

Modifying the Mandate of Existing Institutions: Corporations

Peter Schwartz, Napier Collyns, Ken Hamik and Joseph Henri

Editor's Introduction

In the last 30 years, corporations have taken on an increasingly important role in the world economy. Investment decisions made by corporations shape the development patterns of large regions, the goals and means of national economies and the direction of international trade. In many cases, the technological choices made in conjunction with these decisions will determine for decades to come the timing, character, and volume of greenhouse gas emissions.

But the relationship between corporations and the environment is more than a one-way street. Few people, either inside or outside corporations yet realize how much these entities will be affected by the policy decisions made in response to the perceived risks of rapid climate change. Initiatives taken by governments, lending policy decisions made by multi-lateral banks and other development assistance agencies, and campaigns led by activists will all contribute to the climate of financial risks in which strategic planning and investment decisions by corporations play out. And, of course, the lands held by corporations, and the workforces that they employ, are as subject to the forces of climate as any other lands or workforces.

Many see the political forces of the climate change debate as unequivocally hostile to the interests of the corporation, and its need to preserve profitability in a changing world, but Peter Schwartz, Napier Collyns, Ken Hamik and Joe Henri have a broader and more sophisticated understanding of the challenges that environmental issues pose for the modern corporation. Schwartz et al. observe that corporate concern with the environment is not a form of altruism, but a practical response to the concerns of both shareholders and regulators. They identify important links between aggressive, forward-looking corporate environmentalism and the sorts of strategies, including periodic self-examination and the development of new and advanced technologies, that are necessary to thrive in the current global business community.

Pursued systematically, the programme outlined here is nothing less than a market-driven, long-run plan for sustainable management of resources and emissions. It is the best basis for hedging against the risks of an uncertain future, sharpening the competitive edge of the company in its principle markets, improving the quality of products and services, developing customer loyalty, and increasing the profitability of the firm.

Schwartz, Collyns, Hamik and Henri will be considered utopians by some — on one hand for saying corporations will be compelled to change, and on the other, for daring to include corporations as part of the solution at all, instead of as inexorable villains. But the conclusions of these authors are based on direct experience and practical analysis, not on wishful thinking. As former planners in the influential Group Planning branch of the Royal Dutch/Shell group of corporations, Peter Schwartz and Napier Collyns helped develop Shell's widely respected methods of scenario-based strategic planning. (Schwartz' book, The Art of the Long View, describes these methods in detail.) Schwartz and Collyns are now engaged, along with energy analysts Hamik and Henri, in spreading this disciplined approach to other companies and industries via the multi-media channels of their cooperative company, the Global Business Network.

Schwartz et al. argue that businesses willing to confront the challenges of environmental uncertainty can capture multiple benefits. This demonstration of leadership will establish credibility with their clientele, confidence among their shareholders, legitimacy with the public at large, and their own long-term vision of evolutionary change.

There is no one-shot miracle cure for reaching these goals, however. Making a commitment to internal dynamism requires more than good public relations and advertising copy. It requires instilling a process of continuous self-auditing from the top, allowing constructive self-criticism of the firm from the bottom and from outsiders, and making an on-going commitment to constructive change, openness with information, and honesty. This latter commitment is as important to management success as choosing the hot new product idea or developing the next manufacturing improvement. But as Schwartz et al. argue persuasively, companies that can be courageous today will be competitive tomorrow and will remain profitably in business for the long term.

Woven intimately with any effort towards corporate environmentalism is the Quality Movement — a statistics-based approach to management that is achieving unprecedented successes in Asia, North America, and Europe. Art Kleiner, a writer who is part of the editorial staff of this book, describes the link between the Quality Movement and efforts to reduce greenhouse gas emissions on page 292. Kleiner has written on similar subjects for the Harvard Business Review, *the* Whole Earth Review, *the environmental journal* Garbage Magazine, *and in a forthcoming book called* The Age of Heretics.

- I. M. M.

1 Corporate Response to Environmental Issues

Traditionally, corporate environmental policies have been made in response to government regulations. But today, there are more sweeping economic forces driving corporate environmentalism: the globalization of the marketplace, technological innovation, and the emerging value of institutional legitimacy. The most forward-looking corporations now consider environmental issues to be among the most potent forces affecting business strategy. They recognize that business success and environmental improvement are not mutually exclusive — that in fact businesses have much to gain from environmental practice. They can and should do more than comply with existing regulations.

In order to stay in business, a firm must be able to anticipate and respond to customer needs. One of the most important customer needs emerging around the world is for sustainable economic development, that protects both public health and environmental resources. Markets and public policy will ultimately determine the scope of corporate environmental responsibility, but a firm that intends to ensure its own continued success, or even its continued existence, must anticipate a new environmental business ethic. It must meet the environmental and health needs of its customers.

As the world struggles to protect the global environment and nations create or modify policies to control environmental impacts, corporations find it is increasingly difficult to continue operations that have been successful in the past. The world is defining a new business ethic based on corporate environmental responsibility, an ethic that is fundamentally altering business strategies. But climate change and environmental considerations are not the only factors in this shift. "Business as usual" is a fading memory in a world where relentless innovation and international competition have shaken even the largest economies to their core. The increasing interconnection of world markets has forced businesses around the world to innovate and develop new, more efficient production processes and delivery systems. The resulting drive to innovate and produce durable, high-quality products may also prove to be the means by which business helps to save the environment.

1.1 Premises underlying corporate environmentalism

The purpose of this chapter is to describe how environmental goals can be incorporated into business strategy to make a company more successful. The argument is based on a few simple premises:

- *Environmental responsibility isn't simply altruism.* Businesses are increasingly expected to adhere to a high level of ethical conduct, and judged on the basis of how well they fulfil their responsibilities to shareholders and other stakeholders. They will also be judged on their fulfilment of environmental responsibilities.

- *Highly efficient production processes and continual experimentation with cleaner products and production methods benefit the environment* by reducing energy consumption, waste and raw material inputs. Over the long run, this leads to increased competitiveness for the corporation.

- *Defining an environmentally sound business strategy requires articulation of long-term goals* that can unite an organization, make it a stronger competitor and lead to long-term success.

- *Consistent, rational and clearly defined environmental regulations* (that focus on market incentives rather than specific technologies) *benefit business* by providing a level playing field and allowing for maximum innovation. Businesses have the responsibility to more actively participate in the search for effective environmental policies.

- *An aggressive environmental strategy can lend legitimacy and make a company a leader in its industry.* In contrast, companies with poor reputations are frequently magnets for lawsuits and increased government scrutiny. A clearly articulated environmental strategy can pre-empt onerous regulations.

- *Communication and understanding with various stakeholder groups is a critical factor* in successfully addressing environmental issues and achieving other long-term corporate goals.

Our focus is on the larger corporations because they are generally the leaders who initiate change in any particular industry, but we acknowledge the important roles of smaller firms and by no means exclude them.

2 An Anatomy of Corporate Environmentalism

2.1 What is needed?

Although many corporate managers recognize the need for environmentally responsible economic development, they are hampered by a business decision making system that has traditionally ignored environmental issues. Often environmental issues can't be translated into the language of business decision making, because they deal with things that have always been externalities or were previously assumed to be free goods.

Global climate change is a particularly frustrating business issue for a number of reasons. It is characterized by a lack of scientific certainty, extensive media coverage and a surplus of opinion. As is the case with many environmental issues, the problem has been detected after a long delay (CO_2 levels in the atmosphere began rising with the advent of the industrial revolution) and there are no clear means of connecting the actions of a particular business with changes in the earth's climate. Stakeholders in this issue, as with most environmental issues, are a large, disparate group separated both in space and time. Any solution to the problem of global warming is certain to entail difficult economic choices and will be debated by a host of vested interests.

A corporate strategy designed to respond exclusively to a single environmental concern, even one as broad as global warming, is inadequate. Global climate change is not an isolated problem. Ozone depletion and deforestation of tropical rain forests share many similar characteristics. In each of these situations the summation of many incremental actions has resulted over time in an acute environmental problem. The individual corporation makes only a small contribution to the problem and their action alone will result, at best, in only an incremental improvement. Worse, if unilateral action raises the corporation's costs to the point where its products are no longer competitive, the corporation might be eliminated by its competitors. But businesses are already being forced to deal with these issues.

The solution to this dilemma must include a combination of systematic responses on the part of individual corporations and structural changes on the part of industries and the regulators who oversee them. The goal of these changes must be sustainable, environmentally sound, long-term development. Sustainable development ". . .is development that meets the needs of the present without compromising the ability of future generations to meet their own needs." (World Commission on Environment and Development, 1987)

2.2 What are the obstacles?

There are many impediments to environmental reforms but the principal obstacles, particularly in the United States, are psychological and philosophical rather than economic. While European corporations tend to treat environmental regulations as a part of the operating environment, US business leaders seem to share a residual feeling that environmental regulations are part of an anti-business, anti-progress political agenda. Richard Darman, Director of the Federal Office of Management and Budget, articulated this paranoia when he argued that the goal of US policy was not to "make the world safe for green vegetables." This prejudice — coupled with the short-term orientation of the investment community and a naturally adversarial business environment — has resulted in suspicion towards environmental issues.

Another serious issue in the United States is the legacy of an adversarial "winner-take-all" system of jurisprudence, that focuses on narrow disputes rather than the long-term good of society. Environmentalists and corporations have frequently resorted to the courts to settle their differences, often resulting in an ingrained and mutual set of suspicions. Environmentalists feel that corporations must be compelled to action, while corporate decision makers feel they don't have to act until they have lost all legal avenues of appeal. The best solution, one that protects the environment and meets corporate concerns, won't be found through adversarial clashes in the courtroom.

Although compliance costs are quickly calculated and debated by the business community, regulators and environmental organizations, rarely is there comment on the costs of an adversarial system of justice. The costs of poor business practices and fighting compliance requirements may in fact be higher than compliance and a sound business strategy.

Corporate suspicion of the intent behind environmental regulation is gradually fading, but other, more tangible, issues remain. Accounting rules profoundly influence business practices. While current accounting standards recognize all the "costs" associated with environmental issues, they unfortunately are not designed to deal with any of the external "benefits" associated with clean air or species diversity.

Accounting practices also require that corporations expense environmental costs in the period in which they occur, rather than capitalizing the expense over the life of the environmental improvement. Given just these two problems with the current system of accounting, it's not surprising that environmental issues typically appear in an unfavourable light in corporate cost/benefit analyses.

2.3 Is there a better strategy for business?

There is a different way of looking at the challenges raised by the environment, a systematic strategy called corporate environmentalism. Corporate environmentalism is a business strategy based on collaborative efforts that join the interests of stakeholder groups, increases mutual understanding, minimizes conflict, and lowers the heretofore hidden costs of litigation and lobbying. Corporate environmentalism promotes innovation and competitiveness, by encouraging companies to understand their own business and become more aware of the broadening array of customer needs. Individual companies can flourish by applying this strategy, but achieving broad societal success requires that government regulators realign the incentives faced by corporate managers, so that the incentives innately support public goals.

There are significant parallels between corporate environmentalism and Total Quality Management (see Kleiner, in the box at the end of this chapter). Examining the broadening array of customer desires leads to a better, more insightful understanding of what the customer actually needs. Improved customer knowledge allows the company to question whether the products it offers truly meet the needs of its customers. Thus, it helps the company to develop stronger ties to customer markets. By producing more efficiently and in an environmentally sensitive manner, companies can expand their markets and produce environmental gains.

There is no ideological agenda inherent in corporate environmentalism, but it invites — even requires — companies to think differently about how they do things. It does not require a full ethical resolution of environmental issues — a decision to ferret out the "right" or "wrong" of any environmental dilemma. But it requires a dedication to understanding customer needs; the willingness to harness innovation to improve production efficiencies; and the resources to continually search for ways to eliminate pollution at its source — the production process.

Certainly, this approach has risks. Frequently, an environmentally responsible action raises costs in the short run. Volvo has recently finished construction on a new paint spraying facility, that dramatically reduces air pollution caused by the painting process. By choosing to build this plant in Sweden, and designing the equipment to exceed the already stringent Swedish clean air regulations, the high cost of the new facility will raise the cost of Volvo's products. But Volvo went ahead because the company believes it will gain in the long run by building the cleanest facility possible — rather than just meeting standards. Volvo believes that other companies may eventually "catch up," but, in the meantime, they will have established a competitive advantage and increased brand loyalty among the educated consumers who are the company's target market.

3 Elements of a Corporate Environmental Strategy

3.1 Broadening corporate legitimacy and aspirations

Corporate environmentalism makes corporations a central part of society's effort to ensure environmental quality and sustainable development. Thus, it broadens the basis of corporate legitimacy, and increases the sphere of activity to which a corporation might reasonably aspire.

In the United States, corporations derive their legitimacy in part from the fact that they offer a limited liability investment vehicle and provide returns on the investments of its owners. However, this model does not necessarily define the German or Japanese corporation. Nor does a single goal such as maximizing return on equity define all US corporate strategies. Corporate goals range over a variety of profitability and market share goals. They certainly include long-term survival.

Corporations' views of their own goals can change rapidly. Even companies in the grip of recession, or mature industries whose profit margins seem too slim to support reductions in environmental impact, may find their time and industry life cycles effectively reset to zero by the juggernaut of innovation, and by the global economy. The US shoe industry in the early 1970s would have fitted most definitions of a mature industry. Italian designers dominated fashion shoes, Adidas seemed to be the only significant specialty athletic shoe, and no one predicted significant growth in the market. But in retrospect, the industry was poised at the edge of tremendous change. The advent of Nike and Reebok reset the industry life cycle clock and sparked a wave of innovation and competition. Today's athletic shoe industry would be unrecognizable to an industry executive in the early 1970s.

A mature industry is an industry waiting for innovation to bring about fundamental change. The US automobile and steel industries were surprised to find themselves so quickly removed from world leadership. Many other industries around the world are potentially on the brink of a new wave of innovation and competition. Environmental innovations and management techniques may well be the forces that set off these new waves.

3.2 Articulating corporate self-interest

Whatever a corporation's goals, or however its legitimacy may be defined, it is reasonable to assume that no individual or entity will act against its own self-interest. This simple assumption leads to an important point — corporations will work to protect the environment if environmental goals and corporate self interest are aligned.

Definitions of self-interest vary between corporations. Consistently successful businesses, however, take a long-term view of success, and put a high value on their ability to deal with innovation. They continually reassess their market, their position in that market and their assumptions about what they are trying to achieve.

An important requirement of corporate environmentalism, and perhaps one of the most challenging, is for corporations to define their own interests clearly and honestly. To be useful, this must go beyond a defence of the status quo. The Chief Executive Officer of Volvo recently stated the obvious: that his company's products were dirty, smelly and damaging to the environment. He made this admission in the context of redefining Volvo's scope of responsibilities and as part of an ambitious environmental strategy. In other words, he recognized that Volvo's long-term interest lay in innovating beyond being a polluter; in the long run, there was little value in defending an old process or product, when innovation would quickly render that process or product obsolete.

Profit maximization is an essential part of every corporation's goals but it should not be considered the only, or even the principal, goal. It is a single theme within a complex melody. While profitability is essential, so are the long-term survival of the company, growth in shareholder value, the corporation's reputation, responsibility to employees and the interests of other stakeholders. A preoccupation with profit maximization is often the result of confusion between the goals of management and those of the corporation. Management is compensated on the basis of profits, therefore they emphasize profits. Continued profitability is a prerequisite of long-term survival but it is misleading to suggest that corporate interests extend no further than the bottom line.

Another source of confusion arises when managers perceive environmental regulations as constraints on their prerogatives as active decision makers. Good managers are expected to make decisions and take actions; they typically want to move ahead, grow, expand. They tend to oppose environmental regulation, in part, because it seems to constrain their action. The value of corporate environmentalism is that it encourages, even demands action, but align incentives in such a way that management can follow its prerogative for action in an environmentally sensitive way.

3.3 *Clarifying the interests of society*

In order to achieve society's goal of slowing the rate of global climate change, some way must be found to align corporate self-interest with the needs and aspirations of society. An important step in this process is for society to articulate clearly its goals and provide incentives for corporations to meet those goals. The design of regulations determines whether they will enhance competitiveness or simply stifle it. Regulations that seek outcomes rather than specific solutions will spur innovation, while still achieving society's goal of a cleaner environment.

History is rife with examples of societal inconsistency when it comes to environmental regulation. The US regulatory strategies for dealing with asbestos and dioxin show in one case the wrong strategy for dealing with a problem and in the other a fundamental misunderstanding. Regulations requiring the removal of asbestos from public buildings arguably increased public exposure to the material. Recent work on the toxicity of dioxin, one of the most potent carcinogens known, suggests that the removal of entire communities away from contaminated areas was not necessary. The business community is cautious, and wonders if global climate change will be addressed any more carefully or rationally by regulators.

Regulatory policy frequently ignores the environmental implications of the business incentives it creates. Tax policy, accounting rules, and regulations written by a variety of agencies direct corporate actions along various, frequently conflicting paths. There must be an understanding of environmental implications and coordination among the regulators in order to provide clear and unambiguous signals to the business community. There is as much incentive for regulators to pursue collaborative solutions to regulatory issues as there is for business.

Besides regulators and environmentalists, there are other groups who are taking the initiative to define environmental goals. A variety of investment fund managers solicit and invest funds in companies who behave in an environmentally sound manner. The Valdez Principles, a code of conduct calling for environmentally sound corporate behaviour, was drafted by a group that included the California Public Employees Retirement System, a pension fund that invests billions of dollars in publicly traded stocks.

3.4 *Aligning the interests of better business and a better environment*

Working with a number of corporations on environmental strategy, we have found that many industry leaders actually welcome environmental regulation. Not because they desire restrictions on their operations, but because the costs and uncertainty of working without a clear environmental policy can be very high. Investment in certain abatement technologies or cleaner and more efficient production processes may be delayed until the direction of policy is clear.

A level playing field and carefully designed regulations stimulate innovation. A "command-and-control" regulatory structure, mandating "the best available current technologies", stifles creative solutions to environmental problems. Giving corporations standards, and establishing clear expectations, allows each company to find individual solutions that meet both corporate and public policy goals.

Unfortunately, policy isn't made on a purely rational basis. Special interest groups also alter the course of environmental debate. Electrical utilities in the eastern US burn coal produced locally, helping to provide jobs for thousands of unionized coal miners. Those same utilities have been forced to install expensive pollution abatement equipment, to remove the contaminants found in the high sulphur eastern coal.

A cheaper alternative for clean air would have been to buy coal from the western US, where the coal typically has a much lower sulphur content. In part to satisfy union interests, air quality regulations were written in such a way as to require particular abatement technologies rather than simply setting air quality standards. This type of regulation preserves jobs, but chokes off market responses and innovation. Better organized information leads to better business decisions and environmental improvement. In the US, the Superfund Amendment and Reauthorization Act of 1986 required companies to monitor and report emissions of certain chemicals. The Act did not establish limits, nor did it mandate any penalties for emissions. It was generally opposed by US industry; but a study by a Tufts University research group found that "the mere gathering of information promoted mutual technical assistance in the company, the transfer of good practices from division to division, and increased contact with customers and suppliers". (Baram et al., 1990; Kleiner, 1991). Regulations requiring that information be collected systematically help to clarify environmental issues and catalyze decisions.

3.5 Minimizing environmental damage

The key to minimizing environmental impact lies in addressing pollution at its source — the production process — rather than in attempting to abate the pollution problem once it has been produced. Successful companies have accomplished this by looking afresh at the services their customers require, the products they produce, and the processes they employ. For example, major chemical companies, such as DuPont and Dow, have reassessed both their product mix and their production processes — in the latter case, striving to make them more efficient and reduce pollutants. The 3M company began a programme called "Pollution Prevention Pays" in 1975, to reduce their use of inputs and either recycle or find new uses for waste products. The programme has saved more than $482 million, helped to eliminate more than 500,000 tonnes of waste and resulted in $650 million in energy conservation savings (Porter, 1991).

Electrical generation also demonstrates the business and environmental advantages of efficient production processes. Dennis Anderson, in *Energy and the Environment* (1991) describes how advanced coal combustion techniques, developed in response to environmental regulations, have broadened the generation and environmental compliance options for electrical utilities. Innovations in processing and combustion techniques reduce fly ash, sulphur and NO_x emissions. The processes are significantly more efficient than traditional generation techniques and require less fuel to produce the same amount of electricity. For an electrical utility consuming thousands of tonnes of coal and spending millions of dollars on pollution abatement, a technology that allows it to reduce both inputs and costs without reducing output is very attractive.

Environmental benefits need not be measured in tonnes of avoided emissions to be worthwhile, nor are they possible only for the largest corporations. Small changes add up to big improvements. For example, transmitting documents via a facsimile machines is seven times more energy-efficient than using an overnight courier service (Gellings, 1991). Newer photocopying machines allow copies to be made on both sides of a sheet of paper, rather than only single-sided copies. The actions of small companies and individuals working systematically to increase their efficiency can result in larger environmental benefits when extended to the rest of the population.

The challenge of corporate environmentalism and the secret to using it successfully as business strategy lies in looking at traditional corporate activities in non-traditional ways. Hexamethyleneimine (HMI) is a byproduct of DuPont's nylon production process. Traditionally, it had been considered a waste product of zero or negative value. But when DuPont looked at HMI as a potential input for other processes, an exciting discovery was made. DuPont found more demand for HMI in the pharmaceutical and coatings industries than could be met through byproduct production. Additional capacity was built to produce solely HMI — maintaining the improvement in the quality of the overall production process while eliminating an environmental hazard.

4 The Multiple Payoffs of a Corporate Environmentalism Strategy

4.1 Profitability

As the examples of minimizing environmental damage demonstrate, corporate environmentalism (when properly implemented) plays an important role in improving profitability. Innovative technology is not free, and re-engineering production processes to eliminate wastes can be expensive, but those higher costs are frequently more than offset by savings in the cost of inputs, more efficient production and avoiding the expenses associated with pollution abatement or waste disposal.

Corporate environmentalism can also be a form of insurance against the high costs associated with environmental surprises or disasters. Rather than spend a relatively small amount to insure against environmental damage, corporations frequently employ a "just say no" strategy. For instance, the oil industry has opposed government regulations calling for double-hulled tankers, claiming the benefits don't justify the expense. In the wake of the Exxon Valdez spill, one wonders if the costs of the safeguards might not be outweighed by the costs of an accident. Exxon recently settled with the State of Alaska for $1.1 billion in addition to the amount already spent trying to repair the damage caused by the spill.

A "just say no" strategy that precludes saying "yes" to prudent insurance measures is not a sound business strategy. Double hulls on tankers may not be the best solution to the problem of transporting petroleum products; in fact, some evidence suggests that double hulls may create other safety problems. But a smart, environmentally responsible company should actively promote efforts to find the most effective way of preventing environmental damage. The insurance premium of environmental responsibility may be quite small when compared to the increasing penalties for environmental accidents and poor public relations.

Public perception is becoming an increasingly valuable commodity as consumers seek more than simple need gratification with their purchases. In affluent markets, such as the US, Japan and Germany, consumers increasingly demand high-quality products that reinforce their self-perception, real or imagined. Surveys of US consumers have found that over 70% of consumers claim to base their purchase decisions in part on environmental considerations. The significance of environmentally motivated purchasing remains to be proven, but there is clearly a growing, world-wide consumer preference for durable, clean products. Volvo markets its automobiles, in part, on their safety record by appealing to buyers who have a strong sense of responsibility for their families. More recently, Volvo has begun to incorporate environmental responsibility into their advertising, seeking to attract customers who are concerned about the environment.

Durability and high-quality benefit both the environment and the manufacturer. Durability means that fewer units are produced, and efficient production processes reduce both energy use and waste. Volvo cars last 15 years or more, and new ones are designed so that 70% of the vehicle can be recycled. Even without a directly environmental-oriented sales approach, the Japanese auto industry offers another example of environment-enhancing values: production efficiency and consumer preference for quality. Japanese cars have proved to be very durable. Highly efficient production processes have allowed manufacturers to keep down waste, reduce expensive (imported) inputs and price their products competitively.

4.2 Competitiveness

Constant change and the need for innovation are the only constants in the modern business environment. Corporate environmentalism can lead to a long term competitive advantage based on the ability to deliver a better product or service with fewer inputs and less pollution.

Michael Porter, in an essay published in Scientific American (1991), made the important point that environmental regulations don't necessarily reduce a nation's competitive advantage. Quite to the contrary, Porter asserts that competitive advantage grows out of an ability to continually improve, innovate, and upgrade. (The first version of the essay was titled, "Environmental Protection and Economic Competitiveness: The False Dichotomy.") Challenging, rational and fair government- or industry-sponsored environmental standards can foster increased competitiveness for the nation, industry and individual firm. The frequently voiced concern that restrictive environmental regulations are a net cost to the economy is based on a narrow perception of competitive advantage and the benefits of environmental quality.

Japan is an interesting example of how short-term sacrifice leads to long-term competitive advantage. In 1974 and 1975, Japan suffered through a severe recession, as the nation made infrastructure investments designed to lower energy consumption. It was a decision made in the wake of the OPEC oil crisis, and, although there were environmental benefits in lowering energy consumption, the investments were made primarily for economic and national security reasons. In 1976, Japan emerged in a very strong competitive position with a much more energy-efficient national infrastructure. Recently, after many years of neglect towards pollution control and environmental standards, Japan has put aggressive standards in place which have led to growing exports of pollution control equipment. It's ironic that many US businesses claim they can't afford to comply with environmental regulations because of competition from Japan, yet Japan has surged ahead of the US in certain areas of environmental regulation with no discernible loss of competitiveness.

Members of the European Community are also finding that environmental protection leads to competitive advantage. Strict stationary air pollution standards have given Germany a wide lead in patenting air pollution and other environmental technologies (Porter, 1990). Similarly, Sweden and Denmark have substantial export industries in water pollution control equipment due to their strict domestic standards. As environmental regulations in the rest of the industrialized world catch up to these strict standards, the producers who have already learned to operate under strict environmental standards will have an advantage over their less regulated competitors.

Developing countries have less developed infrastructures than the industrialized nations, but this may prove to be an advantage for corporate environmentalism. These nations also lack many of the established constraints that prevent countries such as the US from adopting an aggressive, environmentally sound infrastructure policy that would lead to a long-term competitive advantage in world markets. World markets are putting much more importance on domestic environmental policies (e.g. Denmark has banned the import of tropical hardwoods, the US has banned the import of various animal products) and the environment may be an area where developing economies overtake the industrialized nations.

4.3 Leadership and legitimacy

Two of the potentially most valuable aspects of corporate environmentalism are intangible. It isn't possible to put a price tag on the benefits of being perceived as an industry leader, but the benefits can be substantial. In addition to the fascination that people feel for a leader, industry leaders can more easily define the competitive agenda so they can stay ahead of their competitors. The value of leadership is evident from the efforts of those companies who have lost a leadership position to regain it — as IBM has tried unsuccessfully to do in the personal computer market.

The perception of leadership, particularly in environmental matters, can offer significant advantages. Government officials around the world seek the opinion of the "industry leader" when it comes to drafting regulations. The corporation that is perceived as a leader in environmental issues will find that its opinions and concerns carry more weight among its peers and the general public. Investors are motivated primarily by returns, and also by the company's "story". The growing number of environmental investment funds is evidence that issues other than return on investment are important to investors. Employees and customers are also attracted by the prestige of an industry leader. The brightest employees and best customers are motivated by more than salary or price; they want to associate with the industry leader. Corporate environmentalism is one way of building that image.

Leadership and legitimacy are based on a variety of factors, the most important being accountability and responsibility. The irresponsibility and lack of accountability in the US savings and loan industry led finally to its collapse. Whatever good the industry was able to provide the economy has been lost in a tidal wave of public disdain. Companies such as Volvo, Dow, DuPont and 3M, that hold themselves accountable and widen their scope of responsibilities, are more than simply leaders. They are considered to be an important part of the fabric of society.

5 Auditing and Defining a Long-term Corporate Vision

For any company seeking a strategy for dealing with environmental issues, the first fruitful step is to seek an understanding of itself. That means asking what the company will be doing 10, 20 or even 30 years into the future. The process of articulating its goals and self-image is a challenging exercise for any company, but one which is essential for designing a long-term strategy that is sensitive to the environment. Corporate environmentalism essentially calls for a series of audits: of corporate goals, customer needs, decision-making processes, production processes and marketing strategies.

Simply thinking about the company's goals over the long term provides multiple benefits. Corporate environmentalism requires a company to ask itself the basic questions that any well run business should be asking itself continually. This process, in itself, leads to a vision which can unite the organization. If the vision is communicated effectively to the rest of world, it will help to make the company a stronger competitor. What are customers really buying? Farmers don't choose a pesticide because of the intrinsic beauty of its chemical formulation. They have needs that certain products are designed to meet. A company that thinks of itself only as a manufacturer of pesticides may find itself in a much less competitive position compared to a company that considers itself to be in the pest management business. A growing consumer demand for organic or pesticide-free agricultural products, and increasingly stringent regulations, suggest that limits may exist to growth in the traditional pesticide business. A forward-looking company should be concerned about continued strength of the traditional business, and should ask itself if its portfolio of products should be expanded or adjusted.

Identifying stakeholders, planning for the long term, and designing ways of incorporating environmental costs into the decision-making process are important steps on the path to an environmental strategy. A company might find there are no corporate stakeholders who are concerned about the company's environmental strategy. Or, more likely, the top leaders may discover unexpectedly that a significant group of stakeholders are intensely concerned about environmental issues. Pacific Gas & Electric Company was surprised when a survey of its employees, shareholders and customers revealed that employees were the most concerned about environmental issues, followed by shareholders and finally, customers.

Planning for the long term and incorporating environmental considerations into corporate decision making will be challenging even for the largest and best-managed corporation. In a struggling small (or large) business, the chief executive might recognize the value of long term planning but find it difficult to accomplish in the face of day-to-day crises. However, the ability of a business to successfully overcome its difficulties will depend, in part, on its ability to define a long term vision.

If marginal or struggling companies are a significant source of environmental damage then it behooves industry leaders to define an industry-wide vision — and thus take steps to prevent government regulation and a loss of legitimacy for the entire industry. Sharing important environmental lessons with marginal companies, or pressuring those companies to act in an environmentally sensitive manner, is merely enlightened self-interest. In California's Silicon Valley, manufacturers meet regularly to exchange ideas and information on environmentally sound manufacturing techniques. Naturally, these businesses don't share everything. No one gives away information that provides a significant competitive advantage, but they have found benefits through the exchange of ideas. Though altruism isn't the principal motivation for many of the companies, it is clearly in their self-interest to pre-empt onerous regulations. An entire industry can suffer for the weakness of a few of its members, no matter the individual responsibility. And by the same token, an entire industry can benefit from just a little cooperation.

6 Meeting the Challenges

An effective strategic vision will challenge a company to accomplish more than ever before. If it is easy to achieve, it serves no useful purpose. A company's environmental strategy should be bold, yet cognizant of the external and internal challenges associated with designing an effective strategy. Recognizing and using the challenges as a source of motivation is one way of overcoming them.

6.1 External challenges

These include the impatience of capital markets and the investment community, public opinion and the influence of special interest groups, and the oftentimes environmentally inappropriate preferences of customers.

6.1.1 The impatience of the US capital markets

This is a well-documented phenomenon often blamed for the short-term outlook of many US corporations. The managers of the major investment funds are evaluated on their quarterly performance. In turn, they are quick to divest from stocks that don't show gains each quarter. This drives the managers of those corporations to produce better quarterly results rather than plan for the long term. The costs of implementing an environmentally sound business strategy might cause short term declines but lead to better long term performance; unfortunately, investors aren't often willing to wait.

One might argue that there is little that one company can do to change the market, but a thoughtfully designed long-term strategy and effective communication can overcome the short term outlook. In the early 1980s, the stock of Royal Dutch/Shell was consistently underpriced by securities analysts, because the company wasn't following the herd and "drilling on Wall Street", or growing through acquisitions. Shell's management was convinced that their long-term strategy would be effective, and they were proved right. By drilling its own wells rather than acquiring overpriced oil fields, Shell was much better positioned than its competitors when oil prices fell in the latter part of the decade. Long-term investors who held the company's stock through the 1980s were rewarded with the highest returns in the industry and Royal Dutch/Shell is now one of the most profitable oil companies in the world.

The first and most important step in trying to educate the capital markets is to have a clearly defined business and environmental strategy. The second step is to communicate, as effectively as possible, with investors, regulators and other stakeholders about the benefits of supporting the company in its long-term efforts. There are indicators (such as the growth of "green" investment funds) that the investment community is beginning to understand and value long-term environmental goals. It's conceivable that the tables may soon be turned, and instead of the market penalizing a business for pursuing a long term environmental strategy, the market will penalize businesses who lack one.

Shell's experience also underscores the important internal challenge of distinguishing the interests of the corporation from those of its managers. If executives are compensated on the basis of stock price performance, they must work extra hard to convince the investment community that they have an effective long-term strategy - or they can focus on providing short-term results. A serious commitment to corporate environmentalism requires that the external goals of environmentally sound business practices be reflected in properly designed internal incentives.

6.1.2 Public opinion and the efforts of special interest groups

These are also significant external challenges. McDonald's Corporation continually experiments with different ways of reducing inputs and waste, ranging from reducing its use of wooden delivery pallets to developing starch-based cutlery that would decompose naturally in landfills. Recently, the company switched from polystyrene to paper packaging for its food products, a move that has been hailed and assailed by a variety of groups. The switch has been portrayed by some as "caving in" to public interest groups because the polystyrene containers were recyclable, and the special paper and foil wrappers that replaced them are not. The plastics industry in particular opposed the move to the paper and foil wrappers. McDonald's move was actually based on advice from the Environmental Defense Fund, an environmental organization that suggested a strategy of reducing total materials usage rather than attempting to collect and recycle materials both on and off McDonald's premises. A careful life cycle analysis indicated that although the plastic clamshells were recyclable and better for the environment than, for instance, the cardboard boxes used by Burger King, the paper and foil wrappers used less material and took up less space in landfills.

6.1.3 Consumers

They also play an important role in a corporation's environmental strategy. Like investors, they need to be educated. Dow Chemical's market research led the company to produce two products which were potentially harmful for the environment. One was a disposable cleaning rag called "Spiffits" that came with a pre-measured amount of cleaner; the other a carpet stain remover called K2R. Spiffits are still available to consumers, but Dow has been criticized for introducing a non-biodegradable, chemical-soaked paper product into the nation's landfills. K2R is not available because an environmental organization threatened to start a legal battle over the solvents used in the product, causing Dow to withdraw the product. Dow's effort to pursue corporate environmentalism stumbled against consumer demand for convenience; educating and refocusing consumer demand for products that damage the environment is an important and difficult challenge.

6.1.4 New political challenges

These will also develop as industry and environmentalists pursue their sometimes divergent goals. The Clean Air Act of 1990 allows US electric utilities to trade "pollution credits," which will give them the right to produce a certain amount of air pollution. Companies with plants whose emissions levels exceed air quality standards will be able to buy pollution credits from companies whose emissions levels are below the threshold. Companies will thus have increased financial incentives to reduce their emissions. This free market approach allows for maximum ingenuity on the part of electrical utilities to find the best business solution for improving air quality.

It also allows for ingenuity by other parties. Pollution credits may one day be traded on a public commodity exchange. This would provide access for non-utilities to purchase the credits. It doesn't require a great leap of imagination to imagine an environmental organization sponsoring a charity concert ("Green-Aid"?) or other efforts to raise funds for the purpose of buying and retiring the credits. This would force utilities to lower their emissions even further. The free market would thus create much more realistic pricing of environmental quality.

Confronted with a number of conflicting stakeholder demands, and needing to find innovative solutions to problems often lying outside their areas of expertise, many corporations have engaged in collaborative efforts to resolve environmental issues. While McDonald's was engaging in its collaborative process with the Environmental Defense Fund, PG&E was participating in multilateral process with consumer groups, regulators and the Natural Resources Defense Council (an environmental organization), to design an incentive programme that allows PG&E to profit by encouraging energy conservation.

When the parties enjoy mutual respect and share a desire to achieve a fair solution, collaboration is an effective way of crafting innovative, reasonable and lasting understandings. An agreement reached through collaboration is more likely to withstand criticism and changing economic conditions than a unilateral corporate programme, no matter how well intended. Collaboration helps to prevent criticism in a number of ways: potential critics have a better understanding of a corporation's objectives and have been able to judge for themselves the sincerity of the corporate commitment. Critics are much less likely to speak against programmes which they helped develop. The public legitimacy of intransigent organizations who disdained the original invitation to participate in the collaborative process is somewhat suspect if they didn't manage to get themselves included in the original effort. Finally, collaboration allows companies to avail themselves of expert advice, often at very low cost.

6.2 Internal challenges

The external challenges of corporate environmentalism are mirrored by significant internal challenges. These include the difficulty of incorporating environmental considerations into decision making, the significant short-term costs associated with redesigning production processes, and the fact that changing procedures can be unsettling for an organization.

6.2.1 Difficulties in decision making

Corporations aren't monolithic entities. They are groups of people united in an enterprise. But observers, for the sake of brevity and clarity, must usually discuss corporate actions as we have in this essay — ignoring the important roles played by individuals within the firm.

The strategies and policies designed by management must be implemented by employees if the strategy is to be effective. Total Quality Management works only when every level of the organization and every person in the organization is aware of the company's objectives and believes they have a role to play in achieving that objective. Similarly, corporate environmentalism can succeed only when environmental issues have the commitment of top management, and are also considered in every decision at every level of the company. One of the best means of achieving this goal is to make a portion of each person's compensation dependent on their knowledge and effort in achieving environmental goals. Aligning incentives within the organization is just as important as aligning external incentives for the corporation as a whole.

Robert O. Anderson, the former chairman of ARCO, brought concern for the environment to one of the world's major petroleum companies. His well-publicized concern for the environment encouraged the people within his organization to discuss environmental issues and seek environmentally sound solutions to business problems. This internal dedication helped to give ARCO a good name among environmentalists and made the company an industry leader on environmental issues. ARCO's good reputation has persisted even after Anderson's departure, indicating that sustainable image advantages are certainly possible.

6.2.2 Cost considerations

The rules of generally accepted accounting principles (GAAP), together with each corporation's own investment criteria, are the DNA of business decision making. Just as the pattern of DNA chromosomes determines the form and affects the behaviour of a human being, this collection of rules and codes provides the fundamental blueprint for a company's structure and actions. Unfortunately, accounting rules governing the capitalization and allocation of expenses are not designed properly to encourage environmental investments. By including a weighted measure for environmental (and other) externalities in the formulas for return on investment (ROI), a new set of GAAP or internal accounting criteria could "genetically alter" a corporation's methods of managing current projects and making future investments. This would build in a bias against ecologically (and thus economically) unsound projects, and hard-code the assessment of environmental impacts into individual project proposals.

Even without changing the accounting formulas, new methods of judging costs and benefits are necessary. For example, under current rules the expense of changing a production process must be borne almost immediately, while the long-term benefits may be intangible or realized only after many years have passed. The environment reveals mistakes slowly, often after a harmful practice has become widely accepted as a standard technique. A short-term view can lead to long-term problems. Incorporating environmental issues into decision making requires that managers look carefully not only at the accounting reports, but also at the intangibles that won't appear on the bottom line. The corporation must instill an appreciation for those intangibles and articulate rules of thumb for dealing with the shortcomings of the accounting system.

In the financial statements of Homestake Mining, one of the oldest gold mining companies in the US, appears a note explaining that the company has been required to reserve funds against the clean up of a Colorado River. Homestake was one of many mining companies who mined along the river and used mercury to refine gold from the ore. Most of the companies are long gone and now the inhabitants of the river valley are exposed to potentially dangerous mercury levels. A hundred years ago, or even 50 years ago, mining companies didn't stop to wonder if there would be any harmful long-term consequences of their actions. Homestake has learned to ask these questions and has lengthened its view. Homestake's environmental efforts at a mine in Northern California brought praise from environmental organizations, but the company still has to correct for the myopia caused by accounting and short-term thinking. Due to the accounting principle of "joint liability," Homestake may be held liable for damages beyond the scope of its own activities, indeed for damages caused by other (now defunct) companies.

By contrast, Volvo has begun to price its cars according to a life cycle analysis that assumes Volvo will take on a lifetime responsibility for its products, from use of raw materials to final disposal of unusable parts. Volvo has collaborated with Saab and the Swedish government to design an "environmental load value (ELV)" measurement index that evaluates the environmental cost of materials used in the manufacturing process. This ELV is used by Volvo as one factor in the selection among alternative component designs or construction materials.

6.2.3 Internal change

The unsettling nature of change is an internal challenge which shouldn't be underestimated. The way an organization reacts to change will determine how far and how fast a strategy of corporate environmentalism can go. Managers and employees may find new ways of thinking difficult, but the constantly changing and innovating business world leaves little room for complacency. Corporations who are seeking innovation or are ready to harness innovation when it arrives will be the survivors over the long term.

7 Towards the Right Incentives

Few examples exist, so far, of companies that have found an ideal strategy for dealing with critics and regulators. Those that succeed know that, at the heart of their strategy is continual self-assessment, experimentation, and a commitment to using environmental issues as sources of opportunity rather than obstacles to be avoided.

As more is learned about the causes and effects of global climate change there will be increased regulatory efforts to mitigate its effects. Corporations that resist public policy efforts and fight to maintain the status quo might prevail temporarily, but they will soon find themselves at the disadvantage of competitors from countries with more aggressive environmental regulations. An effective, proactive strategy calls for corporations to become actively involved in designing regulations that meet environmental concerns and position domestic industries to maintain their competitiveness. It is in the interests of corporations and the general public to design regulations that focus on standards and end results rather than specific abatement technologies or procedures.

Communication is one of the most essential elements of corporate environmentalism. Continuing education of consumers and the investment community is essential. Investors who understand and respect a corporation's long term strategy will be loyal, long term investors. Consumers who understand the environmental consequences of their purchases are more likely to be repeat customers. Corporate environmentalism simply encourages a company to be better at what it should be doing — continually searching for better ways to serve its customers.

Acknowledgments

The authors would like to thank all the people who contributed their ideas and criticism to this work. In particular: James Butcher of Global Business Network, author of *The Environmental Movement: Threat or Strategic Opportunity?*; Lee Schipper of Lawrence Berkeley Labs (Schipper, 1991), author of *Improved Energy Efficiency in the Industrialized Countries*; Tony Lent of ABT Associates; Hardin Tibbs of Global Business Network, author of *Industrial Ecology: An Environmental Agenda for Industry*; Yana Valachovich of Global Business Network; and Seth Zuckerman, educator and author of *Saving an Ancient Forest.*

References

Anderson, D., 1991: *Energy and the Environment, An Economic Perspective on Recent Technical Development and Policies,* Edinburgh: The Wealth of Nations Foundation.

Baram, M.S., P. Dillion and B. Ruffle, 1990: *Managing Chemical Risks: Corporate Response to SARA Title III,* Medford, MA: The Center for Environmental Management, Tufts University.

Gellings, C.W., 1991: Saving Energy with Electricity, paper given at August 12, 1991 Electric Power Research Institute's Advisory Council Seminar.

Kleiner, A., 1991: What does it mean to be green? *Harvard Business Review,* July-August, 38-47.

Porter, M.E., 1990: *Environmental Protection and Economic Competitiveness: The False Dichotomy,* unpublished draft: December, 1990.

Porter, M.E., 1991: America's green strategy. *Scientific American,* April, p. 168.

Schipper, L. 1991: Improved energy efficiency in the industrialized counties, Past Achievements, CO_2 Emission Prospects. *Energy Policy,* March, 127-137.

WCED (World Commission on Environment and Development), 1987: *Our Common Future,* Oxford: Oxford University Press.

The Lesson of Continuous Improvement: A vehicle for combining environmental quality with economic growth

Art Kleiner

Ten years ago, environmentalist groups and multinational corporations battled regularly over the concept of zero emissions. Environmentalists argued that industry should eliminate all toxic emissions from every factory or refinery. Industrial leaders protested that it would be too costly, because it would require an armada of scrubbers, cleansers, and monitors. It would probably be impossible, because the technological ability to detect emissions continues to improve, and today's "zero emissions" might be tomorrow's "one part per quadrillion" of toxins. Fortunately, this stalemate no longer exists — at least in principle. The most capable industrialists have begun to understand that reducing emissions towards zero can be a boon to profitablity. The reason has much to do with an industrial renaissance loosely known under the terms "total quality" and "lean production." With roots primarily in Japan and America, this movement has dramatic implications for pollution control everywhere in the world.

The movement toward total quality management (also referred to as the "quality movement"), is based on ideas most prominently articulated by Dr W. Edwards Deming and Kaoru Ishikawa. They include the principle that an entire system can and should be continuously improved (a process known in Japanese as *kaizen*). A practitioner of total quality management (who might be anyone in an industrial plant) will continuously analyse the flow of work, seeking ways to make the process more consistent, less variable, less wasteful, and more serviceable than it was before. This is done by finding the most significant measurements of quality, and charting their progress over time — a process which gradually reveals where the most leverage lies for beneficial change. Teams of such employees, meeting across their specialties and levels of education, work together to put the necessary changes into practice.

Because disparate parts of a system may affect each other, the practice of total quality management (or just "quality") means that traditional turf battles over jurisdiction and authority must give way to a greater unity of purpose. Everyone in a "lean production" enterprise is aware of the ultimate goal of the operation: profiting over the long run through service to customers. A worker on an assembly line may get involved in the process of ordering supplies, because the worker's system includes the need for better-quality materials. In such operations, everyone is trained in tools for statistical analysis and measurement — so that they may distinguish, for example, ordinary ups and downs (about which nothing need be done) versus shocks to the entire operation.

Quality-oriented enterprises have discovered that their total costs do *not* inexorably increase. Their production costs go down over time. These enterprises spend less money on inspection, because precision is designed in to the process. They spend less money on warehouses, because supplies are ordered and shipped when they are needed. They spend less money ferreting out unnecessary data, or working with inadequate supplies. And, it turns out, they spend much less money on environmental cleanup.

Wherever it has taken hold, the quality movement has — almost without conscious effort — promoted environmental quality. In an Imperial Chemical Industries (ICI) dye plant in France, for example, workers involved in a late-1980s quality effort succeeded in eliminating toxic discharges into a stream. In part, because their neighbours included fire-fighters, they wanted to eliminate the risk of anyone being hurt in a chemical fire.

The disciplines of "pollution prevention" and "waste reduction" are mirror-images to the quality process, and equally powerful: they show that environmental quality, rather than costing more, will significantly reduce industrial expenses. Rather than attaching scrubbers and sensors to a smokestack or discharge pipe, the pollution prevention practitioner reduces waste before it reaches the pipe. Landfills and garbage dumps are treated as warehouses — expenses that must be reduced. The goal of zero discharge is, essentially, the effluent equivalent to the goal of zero defects in manufacturing.

Sometimes the necessary changes are obvious in retrospect, or remarkable for their simplicity: redesigning the shape of a catalytic converter, so that heat spreads more evenly through a chemical reaction. Or not allowing metals to sit unused for too long, so that they corrode and must be dumped. In one Westinghouse metal-finishing factory in Puerto Rico, the company cut dragout — the contamination accidentally carried as chemicals flow from one tank to another — by 75% simply by shaking the tank to remove solids before shunting the chemicals on to the next tank.

But there is a catch. To be effective, like the quality movement, pollution prevention will require industrial and political leaders to move deliberately (albeit quietly, if necessary) away from procedural rigidity and old habits of authority. These efforts cannot be instituted without involvement by the top leaders of any enterprise. But equally important, they cannot be dictated from above. They depend on willing participation, involvement, and openness of information.

An engineer seeking to improve a pipeline's flow, for instance, will seek information wherever it is available — including from a competitor's plant (or even a plant in an antagonistic nation). Deming's suggested goal of "optimizing

cont'd

cont'd

a system," in the deepest sense, means that it is not possible to view competitors as enemies. Here is the apparent conundrum: each facility thrives by making others thrive alongside. While profits remain a proprietary asset, corporations which habitually keep information secret, particularly data about toxic effects of chemicals or wastes, must now learn to reveal that data — in part as evidence of good faith, but more importantly because effective waste reduction (and cleanup) depend on the data being not just available, but actively shared.

Decisions in a quality-oriented business take place differently than in the normal chain of command. They are not necessarily made democratically, but they are made with clear discussion, few or no secrets, and far less pressure to achieve immediate "bottom-line" results. Steady, deliberate progress becomes a way of life; each small change means fewer materials are required, more precision is introduced, and coordination is improved. Leaders see their role as articulating common goals, and inspiring people to follow them. Workers are trusted to observe the results of their efforts over time. Their suggestions are not just heeded for morale's sake, but because the improvement of the operation depends upon them.

To an outsider, the quality movement can seem hopelessly idealistic. Yet it is credited as the source (for example) of the success of the Japanese automobile. It allowed industrialists to consider a previously unthinkable possibility: what if zero defects, though never achievable by law, became for all intents and purposes a practical reality? Now, industrialists and policy makers may take the same attitude about zero discharge: impossible to ensure by edict, but a reasonable target of aspiration. This would make other aspirations possible as well: factories placed nearer to residential areas, so that people can walk to their jobs. Clean cities. More profitable, more efficient industries. A healthier workforce. And, perhaps, a significant slowdown in the production of greenhouse gases and other causes of global climate shock.

References

Crawford-Mason, C. and L. Dobins, 1991: *Quality or Else,* The Revolution in World Business, Boston, Houghton Mifflin Company.

Deming, W.E., 1986: *Out of the Crisis,* Cambridge, Massachusetts Institute of Technology Center for Advanced Engineering Study.

Ishikawa, K., (English translation by David Lu), 1985: *What is Total Quality Control? The Japanese Way,* New York: Prentice-Hall.

Stephenson, Jean, ed.: *Pollution Prevention Review* (A journal of ongoing discussion of waste reduction and life-cycle analysis efforts), New York.

Womack, J.P., D.T. Jones, and D. Roos, 1990: *The Machine That Changed The World,* New York: Rawson Associates/Collier/Maxwell Macmillan.

CHAPTER 20

International Trade, Technology Transfer and Climate Change

Konrad von Moltke

Editor's Introduction

As the world emerges from the Cold War into a new multipolar era, international trade relationships will be realigned in fundamental ways. The Uruguay Round of talks on the General Agreement on Tariffs and Trade will set many new precedents which can affect intellectual property rights, mechanisms for technology transfer, and financial flows between North and South. The concurrent discussions leading to a framework convention on climate change will touch on many of the same issues.

In this chapter, Konrad von Moltke explores the linkages between the policy dialogue on climate change and the current negotiations on international trade regimes. His analysis begins with a review of the historical separation between these discussions, and between the communities that participate in them. He then evaluates the potential implications that these complex debates may have on each other; and identifies ways in which sound trade policies can support the principal goal of the climate negotiations — establishing a cooperative regime to reduce the risks and the ultimate damages due to rapid climate change.

von Moltke's analysis suggests that it is possible to achieve simultaneously the separate objectives of these two negotiations. But achieving both objectives will require an act of courage — to supplement the current international trade goal of economic efficiency with an added criterion. The design of trade incentives and barriers should reward actions which benefit all countries: increased efficiency in the use of resources, especially the resources that are key to the risks of climate change. This will require several important and far-reaching changes in the calculus of economic growth — internalizing the environmental costs of energy use into the price of fuels, and rationalizing the use and management of subsidies.

For the last two decades, Konrad von Moltke, Senior Fellow of the World Wildlife Fund (WWF) and Professor of Environmental Studies at Dartmouth College, has been a leading advocate in the search for structures which link environmental management with international economic policy: trade, structural adjustment, development assistance, debt rescheduling and technology transfer. Before coming to Dartmouth and to WWF, von Moltke founded the Institute for European Environmental Policy in Bonn, FRG. By opening regional offices in Paris and London, von Moltke established a network of collaborators who, working together, contributed to the intellectual foundation for the current environmental policy of the European Community. Today, von Moltke continues this process, expanding the collaborative process to new issues and, reaching beyond Western Europe, to disparate stakeholder groups in the US, Eastern Europe, South America, and Asia.

von Moltke notes that, if the process of negotiation is poorly managed, the goals of trade policy can undermine international efforts at environmental protection. The converse is also true. To limit these risks and to improve the chances of reaching mutually reinforcing outcomes in the separate debates, von Moltke suggests a new structure in which the two regimes are kept separated, but issues at the interface of trade and environment problems are handled by specialized institutions in each domain. He proposes, for example, that trade disputes involving environmental issues be passed to a strong, independent, international environmental institution for resolution.

This process may involve important structural changes in international trade regimes, but it will not require radical or revolutionary policies. By encouraging the gradual evolution of institutions with these goals uppermost, it may be possible to protect the environment while facilitating full disclosure of environmental risks, to protect intellectual property rights while encouraging technology transfer, and to promote free and equitable trade regimes that can sustain the prospects for economic development worldwide.

- I. M. M.

1 Introduction

The linkages between global climate change and economic activities are manifest. But while some consideration has already been given to the general economic implications of adjusting to climate change (IPCC, 1990; OECD, 1991a), too little attention is paid to the implications for international economic policy. The linkages between international trade policy and climate change have not yet been discussed in even a preliminary manner; the linkages between technology transfer and climate change have indeed been discussed, but without due regard to the economic policy dimension of technology transfer.

The ultimate objectives of international economic policy are presumably the same as for domestic policy: to increase human welfare. Traditionally, that has meant increasing the availability of goods and services. Now, it may also mean increasing their quality. In either case, the criteria and instruments of economic policy are quite different at the international level than within nations. Originally the exclusive domain of governments, international economic policy is now frequently determined by the demands of rapidly internationalizing economies, where private interests (such as multinational corporations) predominate.

These participants, so far, have focused on international financial flows and imbalances, development assistance, technology transfer and trade — without regard to the possible environmental consequences. Confronted with rapidly internationalizing environmental policies, they will increasingly need to include these in their calculations. This is particularly true for climate change, an environmental issue for which the international level is the level of original jurisdiction. A policy of successful response to climate change policies must originate internationally.

This paper will identify several trade policy aspects of climate change, and highlight the role of technology transfer. It is worth keeping in mind that these policies are closely related to the other major dimensions of international economic relations.

2 Trade and Technology Transfer Policies: Historically Unrelated

Trade policy has a long history, reaching back to the very origins of public economic policy. Mercantilist theories of public economic policy, widespread in the 18th Century, focused primarily on trade as a means of achieving domestic economic growth.

Technology transfer, on the other hand, is the newest branch of international economic policy. Its attractiveness arises from the increasingly central role that technology plays in modern economies. Technological innovation is frequently the driving force behind structural economic change, in many instances the defining competitive advantage or disadvantage of a nation or private enterprise.

When there is no government policy regulating technology transfer, it tends to occur copiously and continuously — within international corporations and through the operation of international markets. An effective technology transfer policy seeks to ensure that these processes are equitable, and that they do not conflict with overriding goals of public policy. It recognizes the crucial need for countries to gain access to new technologies appropriate to their specific needs. Indeed, the issues of technology transfer have often been discussed apart from their economic dimension, but it is important to recognize this economic dimension if technology transfer policies are to be successful.

The advent of the possibility of climate change has brought forward the interrelationship of trade and technology transfer policies, and has made it clear that they cannot be legislated independently. Nor can they ignore the implications of climate change considerations, which have already begun to affect them.

3 Trade and the Climate Change Policy

The relationships between trade liberalization and environmental policy have only recently received widespread attention (ODC, 1991, OECD, 1991a; *The Ecologist,* 1990; von Moltke, 1991a; Reinstein, 1991). However, significant linkages to climate change response exist in virtually every dimension of trade policy. The linkages will require continuous attention from policy makers in the coming years. Trade policy can serve to reinforce and support climate change response policy, and vice versa; and the policies for both climate change response and trade will need to be modified to accommodate each others' priorities.

To begin with, an effective response to climate change will induce a process of structural adjustment in domestic economies. Managing the consequences of such adjustments lies at the heart of trade policy. Countries which resist or lag in adjusting to new geopolitical realities (of which climate change is one) typically use trade measures to protect themselves from the immediate consequences. However, this practice threatens the integrity of the international trade regimes. On the other hand, well-functioning trade regimes can serve to universalize structural change within the nations that belong to them; they will act to support climate change policy if such change corresponds to their needs. Thus, while climate change policy poses a potential challenge to trade regimes, trade policy may enhance the effectiveness of climate change response activities.

Neither positive nor negative outcomes are automatic. Trade regimes currently do not, and to a certain extent cannot, take into account environmental concerns. Should this situation persist, the most likely outcome will be broadly negative for trade and climate change regimes alike. As some countries move towards introducing climate policies, others will resist them, unbalancing and upsetting the integrity of the trade regimes. At the same time, the rules governing trade will indiscriminately help and hinder efforts to establish climate change policies. Similar results may be expected from a failure to institute effective international climate change policies.

Conflicts not resolved in direct environmental negotiations risk being carried into trade regimes, again threatening their very structure. A recent dispute between Mexico and the United States over "dolphin-safe" tuna provides a vivid example of what might happen. The lack of an effective international regime to protect dolphins led to an attempt to use trade controls as a surrogate measure to force other countries to comply with the United States approach to dolphin protection. The resulting dispute created a situation where all parties lost, dolphin protection was threatened, and the General Agreement on Tariffs and Trade (GATT) was seen to be utterly insensitive to the needs of conservation.

The evolving debate about trade and the environment will cover a large number of issues, including the need for harmonization of environmental policies, institutional mechanisms, non-tariff barriers and subsidies. The principal fora for this debate are the General Agreement on Tariffs and Trade (GATT), the Organization for Economic Cooperation and Development (OECD) and regional or bilateral fora such as the European Community (EC) or the proposed North American Free Trade Agreement (NAFTA) which could emerge from continuing negotiations between Canada, the United States and Mexico, White House, 1991).

These trade policies may create incentives or disincentives in relation to climate change policies. Unless trade policies are explicitly designed to address the policy issues raised by climate change, the nature of such outcomes will be left to chance. Thus, one of the first priorities for international policy makers should be to ensure the compatibility of trade and climate change policies.

Trade policies may prove indispensable in other ways to achieving adequate climate change measures, because of issues of international equity and the absence of multilateral enforcement mechanisms for environmental agreements. Beyond these complex, interactive interrelationships, it is necessary to consider the inescapable linkage between trade policy and transport, one of the economic areas most sensitive to climate change concerns.

4 How Can Trade Priorities Support the Goals of Climate Change Policy?

4.1 Advantages and availability of "no-regrets" policies

A major dimension of climate change policy is the development of more efficient economic structures, which make less wasteful use of the energy and materials which impact climate change: fossil fuels, forest products, many industrial chemicals, and most food items, to name just a few. Policies which support a stable or growing level of human welfare, while reducing the impacts of climate change, often turn out to be more economically efficient, even if the most narrow and traditional of economic standards are applied. Numerous declarations emphasize the availability of such "no regrets" policies, in particular the ministerial declaration of the Second World Climate Conference (WCC, 1990).

Indeed, some countries have already indicated that they view the initial measures which they are taking in the area of climate change as yielding significant economic benefits, perhaps even greater than those offered by more traditional investments (von Moltke 1991b). This suggests that investments in such "no regrets" measures may become the basis of significant competitive advantage in international trade, whether international climate change policies are implemented or not. In principle, trade policy should be designed to allow countries to benefit from competitive advantages such as these.

However, this may lead to measures to protect such advantages, which would in turn create impediments to technology transfer. From a planetary perspective, it may be more desirable that countries act to secure the advantages available from early action on climate change. These advantages frequently involve technological innovations which would require a balance between being protected (to encourage innovation at the host country) and being disseminated widely.

4.2 Putting environment in the definition of economic growth

Economic efficiency is also a proximate goal of trade policy, but not all advances in traditional economic efficiency will yield climate change benefits. Some will tend to exact a climate change cost. For example, the traditional economic criteria of efficiency have caused a massive increase in the use of certain chlorofluorocarbons — whose negative environmental impacts, including climate change impacts, are by now well known.

In effect, climate change policy introduces an additional qualitative criterion to the principles of economic growth: i.e. efficiency in terms of the use of certain resources relevant to climate change. This new criterion does not necessarily conflict with traditional economic criteria, provided economic policy is appropriately developed to take it into account.

Economic "growth," as traditionally conceived, is indiscriminate and relies for public support on a very simple premise. If total resources available for consumption increase, only serious mismanagement of the distribution of these resources will impede a significant number of individuals from obtaining a larger share for their use — a phenomenon widely identified with an increase in public welfare. The converse is also true: it is widely believed that an increase in welfare requires economic "growth." This fallacy assumes that the current distribution of resources — between individuals, countries or regions — cannot be optimized without increasing their total supply. Such redistribution is politically difficult; there is none the less ample scope for it.

In short, in economies where growth is defined in quantitative and qualitative terms, trade policy can support climate change policy objectives, but will not automatically do so. This will require a careful and deliberate structuring and matching of both policy arenas to achieve the simultaneous benefits which are available. The key is likely to be found in measures which adequately internalize environmental costs, particularly those of concern for climate change (OECD, 1991b; Kosmo, 1987). Internalization involves a process which causes cost factors which were previously not calculated (such as the value of a degraded environment) or broadly distributed (such as health costs associated with pollution) to be attributed to specific activities. In practice, this is equivalent to introducing the polluter pays principle (which is after all a principle of cost attribution) into the trade structure (OECD, 1989, 1991b). The issue of subsidies can illustrate the manner in which internalizing environmental costs can link to trade policy.

4.3 The use and management of subsidies

Subsidies have played a role in the early phases of environmental policy in most countries. In many countries, subsidies have been viewed as essential to accelerating the adoption of health, safety and environmental protection measures by public authorities and private industries (Hartkopf and Bohne, 1983). They have also been widely used to support research and development of environmentally sound technologies. There is hardly a country which has not used subsidies to support investments in environmental infrastructure, such as publicly owned wastewater treatment facilities. Users of these facilities typically have to pay primarily for ongoing operating expenses, since capital costs are derived from general revenue.

While subsidies may indeed achieve the desired goal, they often do so only at the expense of continuing economic distortion — which ultimately may exact a much higher price from the economy than the original subsidy or grant. Moreover, subsidies once granted prove notoriously difficult to withdraw. Examples include agricultural subsidies, and programmes that permit the cutting and sale of public forest timber at unrealistically low prices. Thus the use of subsidies, even for the best of purposes, always entails an element of economic risk.

Trade negotiators have long sought to restrain a tendency of most countries to use subsidies to achieve policy outcomes (Hufbauer and Erb, 1984). But they do not care about subsidies which are not relevant to trade. For example, subsidized pesticide prices in developing countries are acceptable under trade policy, so long as domestic and imported pesticides are equally subsidized.

The European Community confronted the issue of subsidies in environmental policy at an early stage. Article 92 of the EC Treaties specifically outlaws aid "which distorts or threatens to distort competition" (a broader criterion than that used in most trade regimes) although certain exceptions are made. None the less, in 1975, the EC Commission decided that subsidies for environmental protection were allowable up to 15% of the net investment (Commission of the European Communities 1975). As a rule, subsidies have been directed at environmental investments and have not carried over to operating costs. The range of such subsidies is quite remarkable, including direct payments, tax rebates, and accelerated depreciation allowances. In the end, the subsidies allowed by the European Community have presumably contributed to better and quicker environmental protection in the member states than could have been achieved otherwise (Rehbinder and Stewart, 1985).

The arguments for and against subsidies are complex. Indeed, most trade agreements do not outlaw all subsidies, only those which discriminate against imports (that is, favour domestic producers over importers.) Other than the EC, most trade regimes have not yet begun to consider how subsidies for environmental purposes should be handled. Presumably limited subsidies of the kind permitted by the European Community — that do not distort competition — would be more widely acceptable in trade regimes which are concerned only with competition across borders.

In general, trade policy will be limited in its strictures; it will not seek to stop governments from adopting poor economic policies, provided these are neutral in their impact on trade. Pushing governments towards desired policy goals is the domain of structural adjustment programs and policies, and — to a more limited extent — of development assistance. Both of these have environmental implications, as well, covered later in this chapter and in Nitze et al. (1992, this volume).

4.4 Opposing the environment's hidden subsidy

The definition of subsidies differs quite significantly depending on the regime in question. The Code on Subsidies and Countervailing Duties of the General Agreement on Tariffs and Trade (GATT) focuses on injuries inflicted on another party by subsidies (Hufbauer and Erb, 1984). The European Community uses a wider definition of subsidy, which can include the selective enforcement of certain rules and regulations when this has a discriminatory impact and is not limited to trade between countries or injury. Climate change policy may require an even broader definition of subsidy, including a tacit form which is only now becoming recognized as a subsidy: the lack of measures to ensure full internalization of environmental costs.

In practice, when an industry or public facility does not internalize its environmental costs — that is, does not include them in the calculations of how much it costs to produce a product or service — then this is fully equivalent to a subsidy. Producers and consumers are being provided with an incentive to transfer everyday costs of production and use to the general environment — for example, by substituting emissions for responsible waste disposal. This causes overconsumption of scarce environmental goods — water, air, the capacity of the atmosphere to absorb greenhouse gases — an activity which could well be uneconomical if costs were properly accounted for. In other words, a large subsidy is currently being granted to producers and consumers by the environment.

Indeed, the current beneficiaries of this subsidy act like beneficiaries of any other subsidy, and vigorously resist efforts to remove it. We have seen this problem arise as the pressure grows to internalize environmental costs. Producers attempt to bargain with the fact that they have not been asked to internalize those costs in the past and say that they depend (in effect) upon the government's tacit approval. If they are held financially responsible for their emissions, they say, they might go out of business.

At present, environmental costs represent a limited factor in the production of most goods. But with increased internalization, environmental costs are liable to become increasingly prominent. Hence, there will be more overt pressure for extravagant environmental subsidies, with damaging economic and environmental effect. In so far as trade regimes outlaw the lack of internalization of environmental costs, as representing an unacceptable subsidy, they can serve to universalize the process of internalization and avoid granting any industry or country special treatment.

If the trade-distorting impacts of these subsidies are recognized by trade negotiators, additional measures will be necessary to ensure that environmental costs are properly internalized. This can be achieved by a variety of means, such as comparable environmental standards and enforcement; or the introduction of appropriate pricing and other economic incentives; or both. In practice, such measures require viable international environmental regimes.

Under such circumstances, trade policy would powerfully reinforce climate change policy, primarily by providing governments with a legitimate international enforcement mechanism for agreed policy measures. Trade policy is singular among international policy areas in offering internationally recognized domestic measures to redress harm inflicted by certain trading practices outlawed by international agreement. To the extent that climate change policy incorporates internationally agreed economic instruments, these will be in conformity with GATT — and GATT provides a means for implementing them.

4.5 Subsidies as a vehicle for technology transfer

The issue of subsidies is closely related to the goals of technology transfer. A consensus is emerging among governments that international assistance to developing countries (DCs) is necessary in order to elevate the priority placed on achieving globally desirable environmental goals. This is particularly true with respect to climate change.

The Financial Mechanism created in the framework of the Montreal Protocol process provides one precedent for this view; the Global Environmental Facility represents another (Helland-Hansen, 1991; Reed, 1991). In both instances, international financial mechanisms are being used to subsidize environmental policy in selected developing countries. Such subsidies are desirable, even essential, to accelerate the process of environmental adjustment, which in some instances may be unattainable by other means. Nevertheless it is important to keep in mind that these are subsidies — and have all the potential liabilities of subsidies. They allow the beneficiaries to escape the process of internalizing environmental costs, and once granted they may prove difficult to withdraw. Consequently these international financial mechanisms must be constructed in such a manner as to avoid, or at least minimize the distorting effects of subsidies (Reed, 1991).

The present limited size of these two international instruments — a total of only approximately $1.5 billion spread over a period of years and a large number of countries — is probably their best protection against distorting effects. Amounts this small will not create distortions that are significant enough to be concerned about. Nevertheless, the funds will presumably be enlarged, and could indeed grow quite substantial. Demands articulated by developing countries suggest resources at least ten times as large may be necessary.

The level at which distortions are likely to become serious is difficult to calculate; yet sums of the order of $10 billion annually can be distorting, even if spread over many countries. A number of additional criteria should be taken into account: subsidies should be devoted primarily to costs which are less project-specific — for example, monitoring, assessment, research and development, and to facilitate the necessary capital investment associated directly with relevant projects. Priority should be given to small projects with large leverage. Subsidies should focus on supporting the essential infrastructure for technology transfer, including "soft" technology transfer — that is know-how, training, and information resources. They must be strictly limited in time.

5 Where Do Trade Priorities and the Goals of Climate Policy Conflict?

As outlined above, good climate change policy can be good trade policy. Unfortunately, good trade policy is not necessarily good climate policy — since the requirements of climate change introduce additional, qualitative criteria to a trading system, which is mainly oriented towards quantitative growth.

5.1 Trade priorities with potentially damaging environmental effects

For example, from the perspective of trade, the source of electricity — solar, hydropower, fossil or nuclear — is indistinguishable. But to reduce the risks of rapid climate change, it is essential to distinguish between sources of electricity. It is also essential to allow electricity importers and exporters to identify the mix in generation which stands behind electricity transfers. Without such measures, carbon dioxide budgets will be difficult to develop as a basis for climate change policy.

Many of these issues are currently being debated in relation to the importation of electricity to the United States from the James Bay hydropower complex. A huge power generation system is planned for Northern Quebec, and some of the anticipated surplus power is to be exported to the United States. Public debate focuses on the rights of indigenous peoples (the Cree Indians in this instance) and the adequacy of assessments of environmental impacts which have been undertaken thus far. Opponents in the United States are seeking to block electricity imports from James Bay because of the environmental consequences of impounding the water behind the dams. While the issue here is not climate change, the dispute raises many of the issues which are likely to arise in relation to trade and climate change.

The Montreal Protocol uses trade control measures to create an incentive for countries to accede, and to ensure compliance with the treaty. It specifically targets trade in controlled chlorofluorocarbons, as well as trade in products containing these substances or produced with them. This introduces the concept of implied environmental impacts of products entering international trade. Presumably this concept will require further elaboration in the ozone regime, and may also be applied to other emissions which affect climate. In both instances, it creates far-reaching conflicts with trade regimes, which heretofore have resisted distinctions between "like" products because these can easily be the source of discriminatory trade actions.

5.2 The opposing roles of trade and environmental regulation

Free trade is an abstraction not unlike the free market. To be viable, free trade requires regulation. This is particularly evident with regard to the environment. Without regulation, free markets do not internalize environmental costs; indeed, they efficiently externalize them. For example, because international markets are largely unregulated, the prices of commodities traded internationally do not reflect the environmental damage caused by their production. In the case of agriculture, conservation therefore requires subsidies (but subsidies do not guarantee conservation); in the case of non-energy minerals, extraction has all but ceased in most developed countries, except in locations with dramatic natural advantages.

However, once public policy has determined that environmental costs must be internalized, free trade becomes the instrument of choice to ensure the decisions are broadly disseminated and implemented. In the absence of international environmental measures, actions by individual countries or groups of countries — probably the industrialized countries, seeking to mitigate their contribution to climate change — entails significant risk of conflict with trade regimes in two ways. These countries are likely to seek to protect their producers from competitive disdvantages with measures which may not be permissible under trade rules; and they may use trade measures to pressure other countries to act as they do. The tuna/dolphin dispute may be a harbinger of future trends in this respect.

Thus, among the most important issues facing climate change negotiators is the delimitation of roles between the trade and the environmental regimes. Despite the vulnerability of trade regimes to environmental concerns, and thus the inevitable temptation to bolster the regimes, the response should not be to transfer environmental responsibilities to trade regimes — which are manifestly incapable of discharging such a role. Instead, the first priority should be the creation of broadly effective international environmental regimes. This still leaves the delimitation of the boundaries between trade and environmental regimes as an important, long-term concern. Whatever the outcome, more complex trade regimes will emerge, involving more complex and more substantive environmental disputes than heretofore.

5.3 The potential conflict over simplicity

In many respects, the increased complexity of potential disputes is an issue of real concern to trade negotiators, who have long sought to transform a range of complex trade policy issues into a simplified format so as to render them more manageable. The increasing use of tariffication is an excellent example of this approach. Tariffication is a process which transforms policies with trade-distorting effects (such as domestic agriculture subsidies), and calculates the border tariff required to achieve an equivalent level of trade protection. It enables negotiators to compare quite different domestic policy practices by translating their effects into terms of tariffs. Countries are then pressed to replace their subsidies or other practices by equivalent tariffs, which can then be lowered. In other words, this process uses tariffs as a common "currency" of trade negotiation. Like any currency, it tends to eliminate the individual, qualitative aspects of a trade dispute. Indeed, this characteristic is the basis of its success as a negotiating technique.

In trade negotiations, a fair amount of effort has been devoted to transforming substantive issues into relatively more simple procedural or formal ones. This has generally facilitated the settlement of disputes and contributed to international processes which are relatively expeditious. By their very nature, trade disputes often require rapid settlement. If certain practices, such as dumping or subsidies, continue for any length of time, the structural damage to the relevant sector of a country's economy can already have been done before any remedies become available — and is then often irreversible.

The focus of environmental policy is quite different. It is driven by environmental conditions, a harsh taskmaster which cares little for formal compliance with procedural rules. The only thing that counts is environmental quality. Several countries have discovered that it is possible to have the most progressive of environmental policies and still have unacceptable environmental quality, not because of inadequate compliance but because the environment was still overburdened. This emphasis on certain substantive results tends to make environmental policy difficult to integrate with other policy concerns, which seek to create "a level playing field" a notion which implies that all parameters can be controlled and that all "players" can abide by the rules. The environment is not a "player" in this sense because its rules are laws of nature, not open to negotiation.

5.4 *Separation of the two regimes: the "cut-out" process*

The institutional solution to this dilemma is again the separation of environmental management and trade disputes as far as possible, for example by the creation of strong international environmental institutions. The trade/environment interface would then be reflected in the institutional interface between trade and environment institutions. Trade disputes involving environmental policy issues would be subject to a "cut-out" process, which transfers them from the trade to the environmental institution for settlement.

A precedent exists for such an approach. Taxes and fiscal measures have a similar relationship to trade policy: they are not contradictory in principle, although in practice problems can arise. Most trade regimes do not cover taxes and fiscal measures, leaving resolution of these issues to international financial institutions.

But climate change issues are actually more difficult to manage than tax and fiscal concerns, because of the element of scientific uncertainty, and the factors outside human control. Thus the characteristics of the institutional arrangements required to manage an environmental "cut-out" process are more stringent. Environmental issues will require a high level of scientific fact-finding and assessment. Appropriate institutions do not exist at present; they would need to be developed slowly so as to establish their credibility and authority through performance.

One arena where we may expect some cross-over is transport — where trade liberalization has an important indirect impact on climate change. It is almost axiomatic that increased trade will engender increased transportation of people and goods. The transport sector is, however, a principal target of climate change policy. In terms of environmental effect, measures to increase the energy efficiency of transport can be overwhelmed by increased volume of transport services. Trade liberalization is an important source of this increase. At the very least, trade liberalization will increase the pressure for increased energy efficiency of transport. Conversely, it is likely that transport costs will rise on account of climate change policy; such an increase would in turn affect the economics of international trade and theoretically lead to a reduction in trade volume.

6 Trade Measures as Incentives or Disincentives to Achieve Compliance with Climate Change Policy

6.1 *Conditionality*

There is a fine line between the use of trade regimes to ensure that climate change policy does not create unacceptable economic distortions, and the use of the same regimes to create incentives or disincentives to achieve compliance with climate change policy. Both approaches will probably be needed in climate change policy. The former is an extension of current trade regimes; the latter is likely to be seen as a misuse by those involved in trade, since it threatens trade restrictions through a structure designed to reduce and eliminate them.

This threat is known under the name "conditionality." It refers to the increasingly widespread practice of attaching conditions to the granting of international economic concessions — making access to certain markets dependent on specified actions. For example, a market may be opened to a producer only upon the signing of a climate change response treaty, or an international agreement to protect dolphins. Other "conditions" might include development assistance, structural adjustment lending, or technology transfer.

In principle, the use of conditionality in trade relationships is problematic under GATT rules. However, such an approach has already been used on occasion by the United States — for example, in attempting to encourage greater freedom of immigration from the former Soviet Union (which is, of course, not a GATT member). Similarly, the United States has introduced conditions linked to its Generalized System of Preferences (GSP). Countries benefiting from GSP are required to comply with internationally recognized standards of worker protection. Theoretically, such provisions could be extended to include "internationally recognized standards of environmental protection" — whatever climate change convention is negotiated. However, GSP represents an exception to the general rules of GATT concerning most favoured nation treatment (MFN). It covers a portion of the exports from the poorest countries, which would all need assistance to comply with any climate change requirements. Thus, this precedent cannot be generalized.

On the other hand, the Montreal Protocol includes quite explicit language concerning trade, language which is presumed to be consistent with GATT under its Article XX (b). "These provisions were critical since they constituted, in effect, the only enforcement mechanism in the Protocol" (Benedick, 1991). It effectively limits the export of controlled substances from contracting parties in bulk (by counting such exports against a country's use). It provides for import bans from countries which are not parties. These bans cover controlled substances (certain chlorofluorocarbons) in bulk and in products as well as, to a lesser extent, products manufactured with controlled substances. This regime is still subject to further elaboration as various provisions of the Protocol are implemented.

Similar provisions might be considered for a climate change regime. Yet it is difficult to envision an approach which would be both effective and acceptable from the perspective of trade negotiators. The difficulty arises from the fact that only a limited number of greenhouse gases are actually traded internationally, in bulk or in products — with CFCs, the most prominent among them, already subject to the controls of the Montreal Protocol.

Thus the principal area of action would need to concern the difficult issue of production processes and their climate change impact. This proved controversial in the Montreal process, and is likely to prove unmanageable within climate negotiations, since the production of virtually every product traded internationally entails some climate change impact. The resultant regime would need to be so generic and vague as to be unmanageable. The limits of using trade regimes for the enforcement of climate change policy presumably can be no broader than the adoption of cost attribution — the "polluter pays" principle — and the elimination of the subsidies of non-internalized environmental costs (as described earlier in this chapter).

6.2 Disclosure requirements

An alternative approach may develop from new methods of end-user information and labelling, which incorporate the principles of life-cycle accounting. As such approaches gain greater currency, the climate change impact of particular production processes can be incorporated, requiring manufacturers to provide relevant information if they wish to gain access to certain markets. Full-scale product life-cycle management has not yet been introduced in any country, but its elements are already becoming visible. Numerous countries, following Germany's lead, have adopted or are adopting labelling for environmentally sound products. International controls of toxic substances are slowly moving towards cooperative approaches between countries, which are liable to give trade advantages to participants as compared to non-participants.

Germany has recently adopted waste control regulations which will dramatically alter practices regarding packaging and recycling, with many potential impacts on its trading partners. Such practices could be acceptable under existing trade regulations within limits defined by regulations covering non-tariff barriers to trade (NTBs). While these limits can presumably be renegotiated to accommodate climate change imperatives, this is liable to prove difficult. What is possibly a test case is currently before GATT, with the Mexican complaint against the United States concerning US regulations governing the protection of dolphins in tuna fisheries.

6.3 Trade in emission rights

Finally, there have been proposals to allow international trade in emission rights — that is, to permit some countries to emit larger quantities of greenhouse gases than might otherwise be allowed under international agreements, if other countries agree to reduce theirs below the permissible level. Such a regime could indeed provide incentives for climate change measures, if these are economically attractive in their own right and can be "sold" to other users. However, such a regime is likely to prove difficult to administer, requiring levels of government control which go far beyond those currently acceptable in many countries (BMU, 1990).

6.4 Intellectual property concerns

One of the most complex issues under consideration, in both the Uruguay Round of the GATT and the North American negotiations concerns international trade in intellectual property and its domestic protection. "Intellectual property" includes primarily copyright law and patent protection. The latter issue relates directly to technology transfer. This is a crucial issue, in particular, between the United States and Mexico. The outcome is liable to be ambivalent from the perspective of climate change policy. On the one hand, increased protection of intellectual property rights will increase the willingness of private owners of these rights to make them available on a broader basis (Touche Ross, 1991). On the other hand, the inevitable outcome will be to render the acquisition of these rights more expensive than they might otherwise have been — provided they were made available — and thus create additional financial barriers to transfers which are desirable from the point of view of climate change policy. Confronting this ambivalence will require a delicate balancing of financial and institutional incentives.

7 Technology Transfer

Environmental policy makers have long faced a conundrum: the inappropriate use of technologies is the root cause of much environmental devastation, yet the answer appears to lie in the development of more technologies, and in ensuring that these technologies are made widely and rapidly available. Certainly all other policy alternatives appear significantly more disruptive, requiring a change in public and private priorities in all developed countries. As a recent report put it: "Radical changes in economic structure or consumer behaviour are not necessary to introduce appropriate technologies" (Touche Ross, 1991). It remains to be seen whether this is actually the case, but technological innovation is certainly the least risky policy approach to climate change.

Consequently, environmental policy makers have increasingly advocated the development of international structures for technology transfer to promote the availability of low and non-waste technologies (Heaton et al. 1991).

In parallel with this development in environmental policy, there has been a transformation of the economic role of technology. Increasingly, access to new technologies — including low and non-waste technologies — has become a major source of competitive advantage. In other words, innovative technologies have a crucial economic value which transcends the issue of intellectual property rights, forming part of the strategic planning of many major corporations. In some instances, control of technological innovation ranks with the traditional factors of production — capital, raw materials and labour — even in extractive and productive industries. That is why the transfer of intellectual property rights has taken on such importance in trade negotiations. Clearly, the changing economic realities run counter to the environmental need for rapid diffusion of new technologies. Climate change policy will need to confront this dilemma.

7.1 Appropriate technology transfer policies

If innovative technologies are viewed as an item of trade, as ultimately they are bound to be, two crucial questions emerge: How can the "balance of trade" in intellectual property rights be rendered stable? What can governments do to facilitate the desirable process?

Most discussions of appropriate technologies assume that the developed industrialized countries hold the solutions; therefore the need is for a structure which allows the one-way transfer of these solutions to less developed countries (Touche Ross, 1991). This is a dangerous assumption to make. The developed world is confronting the problems of industrialized societies, virtually all of which are physically located in the temperate zone. Some of the solutions which are developed will prove appropriate to less developed countries, located for the most part in different climatic zones. Others will not be appropriate or may require adaptation which is only possible in the region. Countries with large subsistence economies have specific problems to confront, which developed countries have had no economic incentive to address. Moreover, any one-way transfer of information and resources is an inherently unstable system with the risk of creating unacceptable dependencies.

The "Green Revolution" provides an example of the kinds of relationships which may alternatively evolve — a useful model, if only in its decentralized design. The fundamental research was supported by US foundations, undertaken by researchers from many countries in Mexico and the Philippines, and distributed widely from these countries throughout the world. For environmental quality-enhancing technologies, the advanced developing countries in particular — India, Mexico or Brazil, for example — are technologically capable, and can support significant development activities aimed at other countries within their region.

The overriding priority of public policy must be the strengthening of the innovative forces and the technology assessment capability of less developed countries and those outside the temperate climate zone. Presumably this implies that technology transfer policy must concentrate on creating the conditions for such transfer, rather than executing transfers themselves, while emphasizing the need for "North-South", "South-North" and "South-South" transfer. In such a scheme, advanced developing countries play a crucial role as mediators between the rest of the developing world and industrialized countries. Moreover, appropriate regional institutions capable of reflecting the specific environmental, social and economic conditions of countries more accurately than global ones, will be needed to drive this process.

Technology transfer clearly plays a central role in managing climate change. It will be increasingly difficult to influence, as access to advanced technologies becomes a source of competitive advantage. Public policy must concentrate on creating conditions which support the development and transfer of appropriate technologies rather than seeking to actually fund and execute such transfers. These will need to occur on a scale far transcending the capability of public institutions to finance or to manage.

8 Conclusion: Structural, But Not Radical, Change

The trade and technology transfer aspects of climate change policy are part of a larger pattern of international economic relations.

To reflect the priority of climate change policy, all of the key elements of the larger pattern of international economic relations will need to be adjusted. In most instances, this implies structural, but not radical change. The introduction of priorities reflecting the climate change imperative is not necessarily in fundamental conflict with traditional economic criteria.

Both trade and technology constitute central elements of any system of incentives and disincentives to support international agreements on climate change. Using these tools of international economic policy requires an exceptionally careful approach if serious distortions and conflicting priorities are to be avoided.

References

Benedick, R., 1991: *Ozone Diplomacy, New Directions in Safeguarding the Planet*. Cambridge: Harvard University Press.

BMU (Bundesminister fuer Umwelt, Naturschutz und Reaktorsicherheit) 1990: *Bericht des Bundesministers fuer Umwelt, Naturschutz und Reaktorsicherheit zur Reduzierung des CO2-Emissionen in der Bundesrepublik Deutschland zum Jahr 2005* (Report of the Federal Minister for Environmental, Nature Protection and Nuclear Safety on the Reduction of CO_2 Emissions in the Federal Republic of Germany to the year 2005), Bonn, BMU.

CEC (Commission of the European Communities), 1975: *Commission Communication Regarding Cost Allocation and Action by Public Authorities on Environmental Matter.* (75,436/EEC, OJ No. L 194, 25 July).

Ecologist, (eds.), 1990: Gunboat diplomacy and the Uruguay round. *The Ecologist,* **20**(6), Special GATT Issue, November/ December.

Hartkopf, G. and E. Bohne, 1983: *Umweltpolitik,* Band I, Grundlagen, Analysen und Perspektiven. Opladen, Westdeutscher Verlag.

Heaton, G., R. Sobin and R. Repetto, 1991: *Transforming Technology: An Agenda for Environmentally Stable Growth in the 21st Century,* Washington, DC, World Resources Institute.

Helland-Hansen, E., 1991: The Global Environment Facility. *Int. Environ. Affairs,* **3**(2), 137-144.

Hufbauer, G. C. and J. Erb, 1984: *Subsidies in International Trade,* Washington, DC, Institute for International Economics.

IPCC (Intergovermental Panel on Climate Change), 1990: *Climate Change.* The IPCC Response Strategies, Washington, DC, Island Press.

Kosmo, M., 1987: *Money to Burn? The High Costs of Energy Subsidies,* Washington, DC, World Resources Institute.

von Moltke, K., 1991a: International track and the environment: Friends or foes? unpublished manuscript.

von Moltke, K., 1991b: Three reports on German environmental policy. *Environment* (**33**), 25-29.

ODC (Overseas Development Council and World Wildlife Fund), 1991: *Environmental Challenges to International Trade Policy,* a joint conference sponsored by the Overseas Development Council and World Wildlife Fund. Washington, DC.

OECD (Organization for Economic Cooperation and Development), 1989: *Economic Instruments for Environmental Protection,* Paris, OECD.

OECD (Organization for Economic Cooperation and Development), 1991a *Guidelines for the Application of Economic Instruments in Environmental Policy,* Environment Committee Meeting at Ministerial Level, January 30-31.

OECD (Organisation for Economic Cooperation and Development), 1991b: *Environment and Trade: Major Environmental Issues,* note by the Secretariat, Paris: OECD, (draft).

Reed, D., 1991: *The Global Environmental Facility, Sharing Responsibility for the Biosphere,* Washington, DC, WWF-International Multilateral Development Bank Program.

Rehbinder, E. and R. Stewart, 1985: Environmental Policy, in *Integration Through Law. Europe and the American Federal Experience, vol. 2,* Mauro Cappelletti, et al., eds., Berlin, Walter de Hruyter.

Reinstein, R. A., 1991: *Trade and the Environment,* manuscript.

Touche Ross (Management Consultants), Global Climate Change, 1991: *The Role of Technology Transfer,* A Report for the United Nations Conference on Environment and Development, prepared by Touche Ross Management Consultants, Financed by the United Kingdom Department of Trade and Industry and Overseas Development Administration. London, Touche Ross.

White House, 1991: *Response of the Administration to Issues Raised in Connection with the Negotiation of a North American Free Trade Agreement,* Executive Office of the President: Transmitted to Congress by the President on May 1, 1991.

WCC (World Climate Conference, 1990: *Ministerial Declaration of the Second World Climate Conference,* November 7, 1990.

CHAPTER 21

Sharing the Burden

Michael Grubb, James Sebenius, Antonio Magalhaes and Susan Subak

Editor's Introduction

The world is moving toward a negotiated international agreement to limit the risks of rapid climate change and to minimize the associated damages. In the process, several nagging and complex questions emerge repeatedly. These questions cannot be resolved through any scientific experiment or computer simulation, but they are all critical to any effective new bargain. How will responsibility for the buildup of greenhouse gases in the atmosphere be established among countries? How will the targets for future emissions reductions and the (limited) rights to increases in future emissions be allocated? And how will the burden of future costs be distributed equitably?

Michael Grubb, James Sebenius, Antonio Magalhaes and Susan Subak offer a sound and cogent basis for addressing these questions. They begin by reviewing "the facts": the available data on past and current emissions of the most important greenhouse gases. Noting the weaknesses and limitations of these data, Grubb et al. observe that industrial countries have contributed disproportionately to past emissions, both in aggregate and per capita. This is especially true for emissions of carbon dioxide and chlorofluorocarbons — the most important anthropogenic contributions to the risks of rapid climate change. The pattern continues to the present day; emissions from industrial nations account for about 75% of the fossil-fuel-derived carbon dioxide in the atmosphere. But the increasing emissions from agricultural and forestry activities, and the pressure of growing population, suggest that the South cannot evade future responsibility for the problem, any more than the North can. How, then, can nations share the burden equitably?

Grubb et al. catalogue, evaluate and compare a wide range of rationales for dividing emission rights among the members of the world community. These include allocation schemes based on land area, GNP, population, willingness to pay, the status quo, "reasonable" emissions and comparable burdens, and the need to help most those who are least well off. Each implies a different distribution of burdens and benefits, all articulated in succession within this chapter.

Rarely has such a diverse and distinguished group of people taken on such a task collaboratively. Michael Grubb, a Senior Fellow at the Royal Institute of International Affairs, is a highly reputed analyst especially concerning the international economic and political aspects of climate change. He also serves as an advisor to the Executive Director of the United Nations Environment Programme, and to policy makers in both the industrialized and developing worlds. James Sebenius is a Professor of International Relations at Harvard University, specializing in the dynamics of complex, multi-party negotiations. He has made major contributions to the most constructive aspects of the Law of the Sea negotiations and to several other important international agreements. Antonio Magalhaes, an economist, ecologist, and climate scientist with the Fundacion Esquel in Brazil, has made a series of important intellectual contributions to the debate on rational adaptations to climate change. Focusing on the particular problems of dry regions such as Northeastern Brazil, Magalhaes has developed strategies for policy-making under uncertainty, designed to minimize the damages from climate variability and climate change while promoting the prospects for balanced and sustainable economic development. Susan Subak is a Senior Analyst at the Stockholm Environment Institute's Boston Center. She developed and maintains the best country-level database of greenhouse gas emissions that is now available. Her research focuses on evaluating methods for sharing targets for future emissions reductions.

Grubb, Sebenius, Magalhaes and Subak do not have a magic formula or a secret plan for sharing the burdens of future emissions limits. They do not promise a quick or simple fix. But they offer a market-oriented approach that promotes equity, efficiency, and cost-effectiveness. Using a combination of allocation criteria, they propose a mixed, flexible, and evolving system based primarily on population and current emissions. Such a system, incorporating tradeable emissions rights with finite "lives," could provide several benefits. These include (1) a real incentive for all participants to limit future emissions growth; (2) sufficient resource transfers to enable poor countries to adopt more efficient technologies than they otherwise might and (3) a practical, effective, and manageable scheme which could be modified periodically to reflect advances in scientific understanding, technological advances, or changes in the perceived international balance of risks. This thoughtful and provocative approach may help disentangle one of the most significant, troubling political knots of modern times.

-I.M.M.

1 Introduction

Coping with global warming could have a substantial impact on human affairs. Reducing emissions could require major industrial restructuring, especially of the energy sector, incurring costs which are uncertain but which could amount to several per cent of annual GNP. The feared impacts of greenhouse warming and the costs of adapting to them may likewise become significant. And should issues of "liability" for damages become salient, compensation costs may become a contentious political issue.

Who can and should bear the burden of these costs? This chapter explores the question of their distribution, which will become increasingly unavoidable in national and international negotiations, particularly concerning emissions abatement over the next few decades. Because the greenhouse warming problem is both global and long term, perceptions of the fairness of any proposed regime will assume special importance internationally. Not only are equity considerations of intrinsic interest, but they will also affect the decisions of countries to join and abide by any greenhouse control agreements.

Clearly, the participation of large, industrialized countries such as the United States is of great importance. Industrialized countries are responsible for the bulk of the increase in global greenhouse gas concentrations since the Industrial Revolution. But the participation of key developing countries in any control regime is also essential to its long-term effectiveness. With the projected population and economic growth of the developing world, the source of the greenhouse problem will shift over time to those nations, especially if India and China choose their apparent least-cost development paths, relying on their vast coal resources. China, for example, plans to expand its coal consumption fivefold by the year 2020, which would add nearly 50% to current worldwide fossil carbon emissions. Anti-global warming steps taken alone by the industrialized world could be heavily offset over time by inaction in the developing countries. Thus, any effective response must attract and enable the meaningful participation of the South as well as the North. As discussed below, this means that resource transfers are likely to be an unavoidable component in a long-term regime.

The question of fairness in global greenhouse burden-sharing is a tricky one practically, philosophically, and politically. The most useful way to begin exploring these dimensions is with a clearer understanding of the global emissions profile. An examination of various approaches to emissions accounting puts the problem in current and historical perspective, offers a sense of the value and the limits of the data, indicates who might be called on to make reductions and by how much, and facilitates useful comparisons across countries and regions. The first part of this chapter undertakes this background task, while the second part examines the equity question philosophically and politically in order to suggest the most promising approaches.

1.1 Sources and distribution of greenhouse gas emissions

Contributions to climate change span many different gases and sources. Before discussing more detailed aspects of emissions and accounting for responsibility, we therefore start by considering the relative significance of other sources of greenhouse gas emissions.

One major difficulty concerns the data. Of all potential substances which might affect climate change, the best data and most analysis concern emissions of CO_2, especially from fossil fuels. In most cases, fossil fuel emissions and CFC releases are known to within ±5% (giving an uncertainty range of 10%). By contrast, CO_2 emissions from deforestation and other land use changes are much more uncertain, with recent estimates varying by well over ±50% — for individual countries and for the world in aggregate. The buildup of methane is well documented with atmospheric observations, but emissions from different sources and countries are rarely known with better certainty than ±30%, and in some cases the uncertainty range is greater than ±50%. The unquantified effects on the rate of increase in methane concentration that result from carbon monoxide-induced declines in the atmospheric concentration of hydroxyl radicals add to the uncertainty surrounding methane emissions estimates.

Contributions from other greenhouse gases (especially N_2O and tropospheric ozone), and various other gases involved in different aspects of atmospheric chemistry are still more uncertain. For some particular sources (e.g. small-scale biomass burning and cars without catalytic convertors), the impact of "other" gases may be as important as that of CO_2; but the global impact of these other emissions probably totals less than 10% of that caused by the buildup of the three major greenhouse gases — CO_2, CFCs and methane.

The only available estimates of national emissions from different sources in all major regions are those from the World Resources Institute (WRI, 1990) and the Stockholm Environment Institute (SEI, 1992). The WRI data has attracted considerable critical debate (Agarwal and Narain, 1991; Hammond et al., 1991, and responses). We use the SEI data for CO_2, CH_4 and N_2O, whilst recognising that all data sources still contain substantial uncertainties as noted above (see box for further details of this data).

Additional difficulties are introduced in attempts to compare the radiative impact of different gases because the gases have very different lifetimes and rather complex decay patterns in the atmosphere. As a result, their relative impact varies with the time horizon considered. The Intergovernmental Panel on Climate Change (Houghton et al., 1990) estimated the equivalent global warming potentials of different gases on different timescales. These figures may be revised as estimates improve, and they do not reflect the relative climatic impact of different gases precisely, but give a reasonable basis for comparing the importance of major emissions sources and the relative contributions of different countries.

Emissions from industrialized countries (defined here as members of the Organization for Economic Cooperation and Development, OECD, and former members of the COMECON group) accounted for at least half the total warming impact of anthropogenic emissions in the late 1980s. Activities in these countries have been the principal sources of fossil CO_2, CFCs, and fossil methane. By comparison, developing countries were the principal source of CO_2 from deforestation and land-use changes during the 1980s; the best available data suggest that these sources produced about one-fifth of total CO_2 emissions. In developing countries, agriculture and forestry activities are the principal sources of methane and nitrous oxide emissions, with contributions of these gases greater on an absolute (though not per capita) basis. The exact numbers depend not only upon uncertainties in the data but upon the timescale considered. (See Figure 1.)

Because the industrialized countries represent less than 25% of the global population, the regional differences in relative contribution to the warming effect are greater when expressed in per capita terms. In general, per capita greenhouse gas emissions in industrialized countries are several times those in most developing countries. But, for particular sources, some developing countries have higher per capita emissions rates. In terms of fossil CO_2 alone (which is the

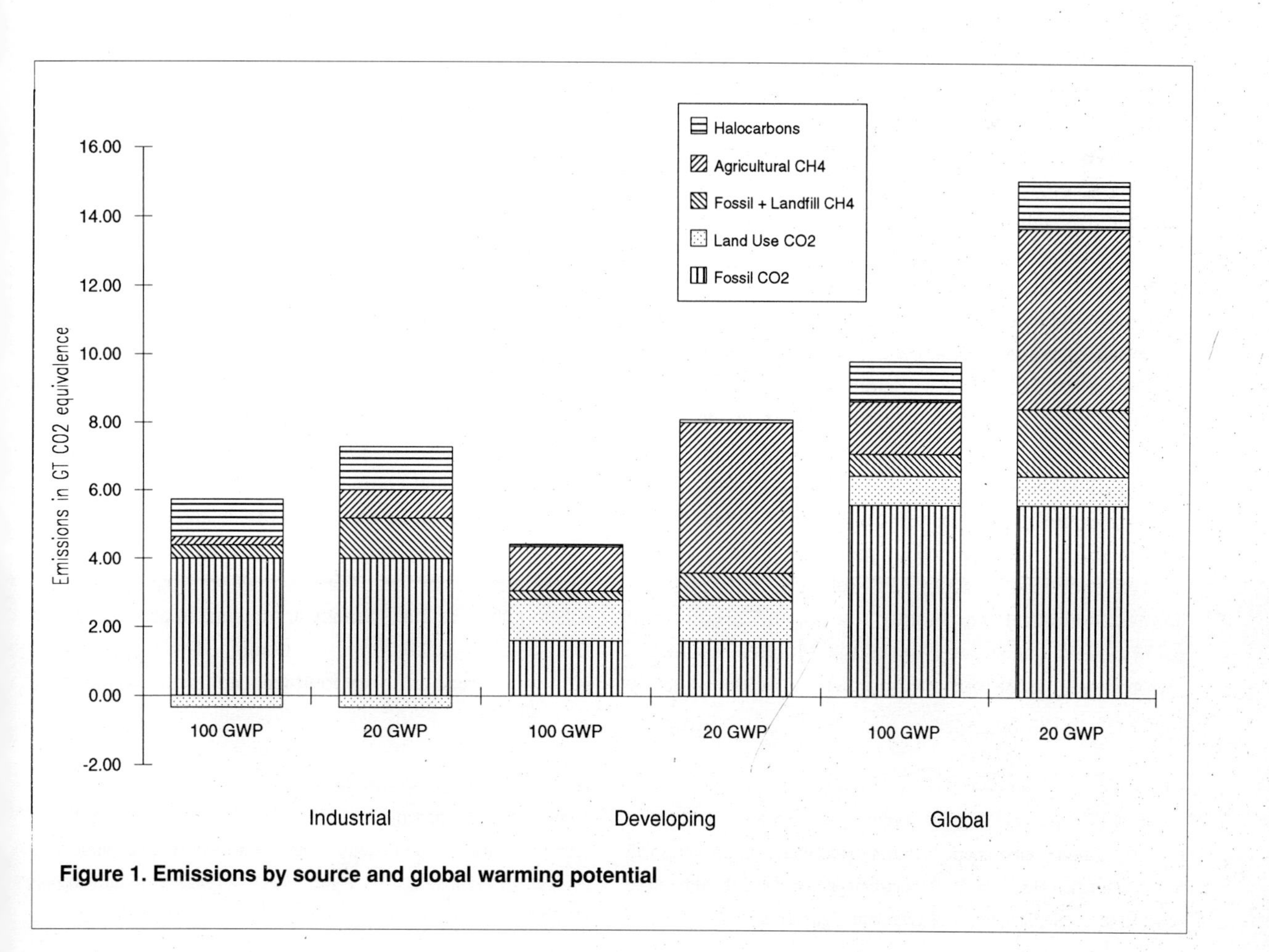

Figure 1. Emissions by source and global warming potential

largest single source by far), per capita emissions from industrialized countries are on average about 8 times those of the developing world.

None the less, their much larger population makes the total emissions from developing countries quite significant despite the much lower per capita levels. Total emissions of greenhouse gases from these countries are growing rapidly as a result of economic development and population growth.

Taking a global and long-term view reveals the real immensity of the problem. Ten billion people is a typical estimate of the levels at which world population might approach stabilization during the next century. If per capita emissions of fossil CO_2 alone from industrialized countries do not decrease, and those from developing countries were to approach similar per capita levels over that period, global fossil CO_2 emissions would be increased to up to seven times the current levels — compared to the 60%+ cut required to stabilize atmospheric concentrations. It is apparent that the development path taken by developing countries over the next few decades will be crucial in determining the level of future global emissions.

In any effort to abate greenhouse gases, it is clearly desirable to consider all gases as far as possible. This is not in itself an equity issue. However, expanding the sources from the easily measured industrial emissions (primarily of fossil-derived CO_2 and CFCs) to the more uncertain sources (e.g. CO_2 emissions from deforestation, energy-related and agricultural emissions of methane and nitrous oxide, etc.) places more emphasis on the contributions from developing countries. The perceived contribution of some countries is extremely sensitive to the range of sources considered. For example, only a few Latin American countries release fossil fuel-related CO_2 at rates above the global per capita average, but, if the other greenhouse gases and non-energy sources are considered as well, a dozen Latin American countries appear to be emitting at above the average levels. In practice many of the non-industrial sources are very difficult to control, resulting in part from basic subsistence activities.

Notwithstanding the theoretical desirability of a "comprehensive" approach, the difficulties of measurement and control of many of the non-industrial gases make it clearly impractical to include all sources. Any quantified abatement efforts will initially, at least, have to focus on the major and measurable greenhouse gas sources that are well known and can be readily documented at present. Of these, fossil-derived emissions of CO_2, CFC releases, and (perhaps) deforestation stand out. Furthermore, of these, CFCs are being phased out anyway under the Montreal Protocol[1]. Deforestation, apart from measurement difficulties already noted, has declined rapidly in several areas because of its many other damaging consequences. We emphasize the following conclusion: no reasonable change in the assumptions on which the analysis is based can change the prime role of industrial country emissions in contributing to the current (and near-term future) risks of rapid climate change. As we will shortly discuss, however, the complexities and ambiguities of the data will severely hamper the prospects of a number of possible burden-sharing approaches in the process of negotiation. (See Box, p. 310)

1.2 Equity and realpolitik

To attract widespread participation, any agreement will not only have to be effective, but seen as fair. The question of fairness in global greenhouse burden-sharing is a tricky one. Practically, it could raise demands for financial resources from countries that cannot or will not provide them. Politically, it raises challenges to existing national traditions and priorities. Philosophically, it demands the resolution of contradictory notions about responsibility, self-interest, co-operation, and distributive justice.

All systems with global impact will impose costs on participants. The perceived fairness of the manner in which any greenhouse gas control regime allocates costs — of adaptation, mitigation, and/or compensation — will be central to its political acceptability. Allocation will pose considerable, some might say insuperable, political difficulties. Yet the allocation problem is unavoidable, whatever control system is adopted — whether it be national emissions targets, carbon taxes, a tradeable permits system, or another approach.

Burden allocation will be affected by differing views of what is equitable, including perceptions of responsibility for past and current contributions to the problem. Arguments about the equity of each cost-sharing plan will clearly be important as tactical instruments in negotiations, but they will also have an intrinsic role and an impact on the structure of the agreements. Though cynics tend to regard equity arguments as mere public relations cover for raw interests or underlying power relationships, in practice some degree of equity considerations have appeared as a consistent feature in a large number of previously negotiated international environmental and natural resource regimes.[2]

Thus, there will be a mix of intrinsic and tactical aspects to the use of equity arguments in greenhouse negotiations. Considerations of ideal fairness may well have a stronger role to play here than in many arguments, in part because of the economic stakes involved, combined with the need to gain widespread and very long-term adherence to a control

---

[1] Many CFC replacements are also active greenhouse gases, but most have much shorter atmospheric lifetimes so that their overall impact will be much smaller. However, CFC replacements could and should also be considered within the context of greenhouse gas emissions.

[2] On equity components of environmental agreements, see Young (1989); for discussion of equity issues in the context of negotiations in general, see Gulliver (1970); Fisher and Ury (1982); Lax and Sebenius (1986).

regime. But inevitably, the negotiations and outcome will also be strongly tempered by the perceived balance of political power, and the practical economic implications of the chosen solution.

1.3 Framing the burden-sharing issue in the context of tradeable permits

Although we could look at the equity issue in the abstract, or applied to a variety of greenhouse control approaches, it is clearer to examine its implications in the context of a single approach. We believe the most appropriate context is to consider the equity issues raised in seeking to allocate initial emission permits in an international tradeable permits system. (The equity issues involved are in fact very similar to those raised in considering how to redistribute revenues from a global greenhouse gas tax, but debate has focused more on tradeable permits, for a variety of reasons beyond the scope of this chapter.)

Let us briefly review the basic workings of such a tradeable permit system:

First, an *overall* global target for emissions would be negotiated. This target might be fixed, or it might be subject to renegotiation over time as scientific understanding and political factors evolved.

Then, (tradeable) emission permits would be allocated to different countries, such that total permits issued would correspond to the overall target. These permits might be allocated once and for all, or they might have finite lives and be periodically reallocated.

Countries would then have to ensure that they held sufficient permits to cover actual greenhouse gas emissions. If their initial allocations exceeded their actual emissions, they could sell the surplus permits to others, who were short of permits and who valued them more highly than the original owners. The parties could then negotiate, bilaterally, the terms of an exchange, and a market for permits could arise. Obviously, there are many important detailed questions about the practical operation of such a system, many of which are examined by Grubb and Sebenius (1992) and others in the same OECD volume. But issues of implementation and control are not central to the discussion of burden-sharing, which depends principally upon the initial allocation.

Thus, if a country such as China received more permits than it needed, while the US had a permit deficit, the US could choose to buy permits from a surplus country such as China. The resulting resource flow to China could be quite substantial, depending on overall permit scarcity (i.e. the stringency of the *global* emissions target), the extent of the US permit deficit, and the Chinese surplus. If the permit "market" works well, then countries (or other actors) would have to face the trade-offs between limiting emissions domestically, and paying to fund abatement elsewhere. Ultimately this would minimize the total costs involved in meeting any given emissions target.

A permit system thus separates the question of "who pays" from that of "where should abatement be undertaken for minimum cost." In doing so, it both enables an efficient outcome, and separates issues of efficiency and implementation from those of equity - which is embodied in the permit allocation. The permit allocation question becomes a *political* issue — to be influenced by equity considerations — while the location of the reductions that are actually made are the result of a decentralized *market* process. Thus, examining which countries get what share of the total emissions permits — the allocation issue — is an ideal vehicle by which to examine equity considerations generally. A surplus of permits to a given country implies the possibility of its receiving significant resource transfers from others who need permits, while a deficit of permits relative to another country's current greenhouse gas emissions implies its need to buy from others. For this reason, we frame the following burden-sharing discussion in the context of the permit allocation issue in a tradeable permits system. But it is easy to translate the implications of this analysis for systems involving targets, carbon taxes, or other schemes.

2 Equity Rationales

In discussing permit allocation, a clear distinction needs to be drawn between accountability for emissions, and allocation of abatement efforts and/or permits. Allocation involves a wide range of considerations including fairness, resources, operational feasibility, and *realpolitik* — whereas accounting is essentially grounded in statistics and definitions. Some suggestions for allocation (of which we will discuss a few) are based on essentially ad hoc rationales, and have little ethical underpinning. By contrast, a number of other, more "pure" rationales have well-developed philosophical bases, which we will briefly describe before assessing how such principles might fare in practice.

2.1 Some **ad hoc** *allocation schemes*

A variety of criteria for burden allocation have been proposed that are attractive on one or another basis. Undoubtedly, many other such candidates wait in the wings to be put forward by inventive proponents. We will assess a few of these proposals before turning to other approaches that are grounded in ethical principles of greater power, depth, and generality.

2.1.1 Land area

One such criterion, suggested by Westing (1989), would allocate emissions permits on the basis of countries' land area; the greater the area the higher the permit allocation. The advantages include the fact that area is easily defined and measured. Arguably, this approach would offer a disincentive to high population densities. But the absence of any link to human activity makes it impractical. Some sparsely populated countries would see very large allowances. The Antarctic, with 10% of land area, would presumably be excluded, but the former Soviet Union, Canada, Brazil, Australia and

Relative Emissions from Developing and Industrialized Countries*

Estimates of national emissions are likely to be the subject of intense debate over the accuracy of national data, and how that data should be interpreted.

For example, a general consensus exists that, during 1988, almost three-quarters of the CO_2 from fossil fuel combustion was released in industrialized countries. But when non-industrial sources are included (e.g. burning of forests and other land-use changes) the contribution of industrialized countries was about 56%. Including all sources of CO_2, the average per-capita contribution of developing countries was about one quarter that of the industrialised world; but given wide variations between individual countries, some particular countries with high deforestation rates in Africa and Latin America had amongst the higher per-capita emissions in the world. Uncertainties in the data are unlikely to change the broad conclusions but are sufficient to make disputes over which sources to include, and which estimates to use, highly contentious.

Theoretically, an allocation system for reducing the risks of rapid climate change should consider all greenhouse gases. Today, an estimated 55% of greenhouse gases are released from industrialized countries. This comprehensive estimate includes halocarbons and CFCs, methane (CH_4), nitrous oxide (N_2O), and carbon monoxide (CO), in addition to CO_2. The impact of CFCs and HCFCs is uncertain because of indirect atmospheric effects (notably ozone depletion), and whilst uncertainties in emission levels are greater for the minor trace gases, of which (as compared with the industrial gases) a higher proportion arise from developing countries.

Figure 1 illustrates carbon dioxide and methane emissions from various activities in industrialized and developing countries. Most CH_4 emission measurements are uncertain by +30-50%, but agricultural sources of CH_4 (e.g. rice paddies, livestock, and biomass burning) are three or four times the level of fossil fuel-related sources of CH_4. In total, methane (CH_4) contributes about one-fifth of the projected warming effect due to greenhouse gas buildup, although this fraction depends somewhat on the time horizon used: because methane has a much shorter lifetime in the atmosphere than does CO_2, its importance (and that of developing country emissions) increases when a shorter timeframe is used. For a 100-year time horizon the developing country contribution to total emissions in 1988 was about 45%; for a 20-year horizon, it was around half (51% on this data) of the total.

Again, individual countries vary greatly in their absolute and relative total and per-capita emissions. Also, the fact that most biotic methane emissions arise from basic subsistence activities makes suggestions of control, and indeed their inclusion at all alongside industrial emissions, another contentious issue. Agarwal and others have also observed that the perceived pattern of "net" emissions changes if "sinks" are explicitly included as equivalent to negative emissions and allocated equally per-capita, though this is primarily an observation about the inequalities in per-capita emissions noted here.

Figure 2 illustrates how the ratio of developing country emissions to average global emissions increases for non-energy sources. This pattern is revealed in the upward slope of the bars in the developing country regions, and the corresponding downward slope in the industrialized regions, where per capita emissions from the biotic and agricultural sources are proportionately (though not necessarily absolutely) lower.

Figure 3 shows the ratio of fossil CO_2 emissions that emanate from each major region. The figure illustrates the regional fraction per capita, per adult, and per unit of GDP expressed in conventional terms and adjusted on the basis of purchasing power parity. Unlike the per capita emissions pattern, emissions per GDP are typically greater in developing countries.

There are still many differences of opinion about how to measure and judge historical emissions, and what time frames to assume for the analysis. The accuracy of estimates of even the most well-documented greenhouse gas source — fossil fuel production — is limited because of spotty data on fossil fuel trading and fuel type and because of changes in political borders. Measurement problems are, of course, much greater for land use changes: agricultural and forest surveys tend to be infrequent and incomplete. It is unlikely that the historic data on biotic emissions will ever be accurate at the country level.

Analysis of the available data suggests that the historical fossil-fuel related emissions from developing countries represent only about 14 percent of the global total, as compared to 28 percent of current fossil-derived CO_2 emissions. The eventual removal from the atmosphere of CO_2 released further back in time lessens the importance of some of the historical emissions, but the overall pattern of higher emissions in industrialized countries is clear and indicative for all nine regions as seen in Figure 2.

*based on Stockholm Environment Institute, 1992

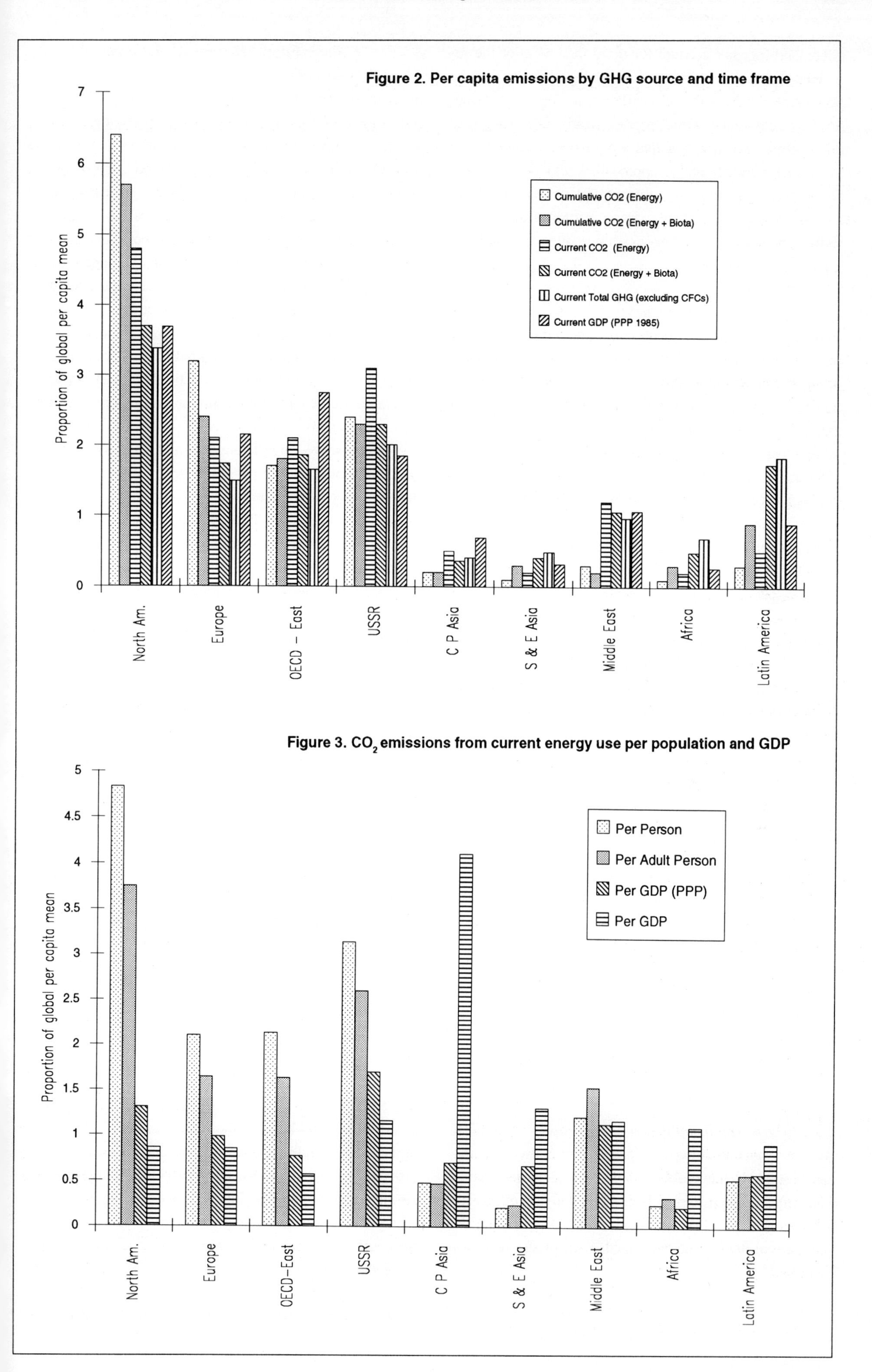
Figure 2. Per capita emissions by GHG source and time frame
Proportion of global per capita mean
Cumulative CO2 (Energy)
Cumulative CO2 (Energy + Biota)
Current CO2 (Energy)
Current CO2 (Energy + Biota)
Current Total GHG (excluding CFCs)
Current GDP (PPP 1985)
North Am.
Europe
OECD - East
USSR
C P Asia
S & E Asia
Middle East
Africa
Latin America
Figure 3. CO2 emissions from current energy use per population and GDP
Per Person
Per Adult Person
Per GDP (PPP)
Per GDP
OECD-East

Greenland together account for more than 40% of the remaining land area, while representing less than 10% of the population and 20% of global carbon emissions. The former Soviet Union alone would receive nearly 20% of global allowances (and much of that would be associated with Siberian tundra). Such an approach would reward those countries that already have the greatest natural resources, and might even exacerbate existing territorial disputes. In short, its ad hoc nature and perverse consequences in practice rule out this criterion.

2.1.2 *Gross national product*

A second such ad hoc allocation scheme would give more permits to each country that emitted less carbon for each unit of income or wealth it produced. Such a "per GNP" allocation scheme may be superficially appealing, on the grounds that carbon emissions are tied to economic activity, and the system's objective should be to encourage maximum carbon efficiency of economic production. But this faces the objection that it is a system which directly rewards riches, and thus perpetuates inequity.

Moreover, the practical difficulties may be insurmountable. There are basic problems of measurement and conversion; GNP is a very crude measure of economic activity since it measures only registered commercial transactions. Especially in developing countries, the underground, "informal" economy can be quite sizeable. International GNP comparisons, moreover, are subject to enormous difficulties given vagaries of exchange rate differentials in purchasing power. Even if such distortions could be avoided, there is still the fundamental fact that less developed countries often use far more energy per unit of GNP than do industrialized ones. Strict GNP allocations could, for example, require India to finance carbon abatement programs in the United States - an untenable proposition.

Many other ad hoc measures (e.g. based on energy efficiency or conservation achieved) have been proposed; it is infeasible to assess them all. Fortunately, it is not necessary, since the problem of equity in burden-sharing has been extensively studied in more general terms (Bergesen, 1991; Ghosh, 1991; Hayes, 1992; Rose, 1990; Shue, 1991; Young and Wolf, 1992.). The underlying "pure" equity rationales that have emerged from these investigations which are of possible significance for the greenhouse negotiations appear to be relatively few in number.

2.2 *"Pure" (philosophically-based) equity rationales*

At least seven normative principles — plus many variants — can be applied to the problem of greenhouse burden-sharing. They often crop up, explicitly and implicitly, in discussions of the subject. As a first step, it is useful to offer an uncritical summary of these principles together with their main justifications.

2.2.1 *"Polluter Pays": Historical and/or current responsibility*

In its most general form, this is simply the principle that countries should in some way pay for pollution that they generate or have generated — an "international polluter pays principle." The OECD formally adopted a form of the polluter pays principle ('PPP') in 1974 as a guide to environmental policy in stating that, if measures are adopted to reduce pollution, the costs "should be reflected in the cost of goods and services which cause pollution in production and/or consumption" (OECD Council, 1974). This principle is essentially concerned that polluters should bear the costs of abatement without subsidy. It is most easily conceived as a pollution tax, but in the international context this raises the question of how revenues would be disbursed, which is equivalent to the permit allocation question. The PPP itself does not necessarily resolve the equity/allocation issue though a common-sense interpretation would suggest that large polluters should end up paying more, by implications with transfers to lesser polluters.

One version of the PPP for greenhouse gases would place the onus on historical contributors to current atmospheric concentrations.

Since the Industrial Revolution, industrialized countries have "used up" more than their fair share of the atmospheric resource, leaving less for the developing countries. In this interpretation, industrialized countries have incurred a large "natural debt," which should be compensated accordingly with proportionately smaller emission permit allocations to those countries that, cumulatively, have contributed more to the problem. (Smith, 1991, developed more fully in Smith et al., 1991). A complementary - though logically quite distinct - justification for this approach relies on the fact that those living in the industrialized countries now benefit from their history of past emissions. They are richer in part *because* they have used up a disproportionate amount of the atmospheric resource, and to the extent that poorer countries are denied this opportunity, the rich who have exploited it first should compensate the poor by granting them greater emission permit allocations.

2.2.2 *Equal entitlements*

An egalitarian doctrine suggests that all human beings should be entitled to an equal share in the atmospheric commons. Agarwal and Narain (1991) and Bertram (1992) argue that the entire world's population should be allowed shares of permissible greenhouse emissions. Grubb (1989) argues for a modified form of per capita allocation. Fujii (1990), cogently supported by Ghosh (1991), urges a much stronger version for this principle: that everyone should have an equal right to an identical emissions quota regardless of the country he or she is born into, or the generation of which he or she is a part. Thus, all people, regardless of nationality — and regardless of whether they lived in the past, are alive today, or are yet to be born — would get an equal share under this scheme, for which Fujii develops a detailed system of accounts.

The principles of historical responsibility and equal entitlements come together naturally in calculations of "natural debt." For example, by setting a past baseline year and treating the atmosphere from that point on as an unallocated resource, the analyst would simply apportion emissions rights on a per-capita basis to see who has exceeded "their" allocation and who has fallen short. "Debtors" would be allocated proportionately fewer tradeable emissions permits.

2.2.3 *Willingness to pay*

A principle independent of historical responsibility or equal entitlements is simply that countries should bear costs based on what they are — or objectively "should" be — willing to pay, given the potential damages they face from greenhouse warming. According to this principle, contributions should be determined by a combination of ability to pay (reflecting current wealth, and hence welfare impacts of a given payment), national benefits gained in terms of reduced climatic stresses, and level of general concern about the climate change issue.

This approach reflects a time-honoured principle of welfare economics which suggests that subjectively perceived benefits ought to affect the distribution of burdens. Dorfman (1991) draws the important distinction that, for a common property resource, willingness to pay should be based not on how a country values a given reduction in its *own* emissions, but instead, how much — given its level of development and other circumstances — it would value a reduction in worldwide aggregate emissions. Presumably, those who more greatly value such reductions should be willing to pay more for abatement — and hence would be allocated fewer permits. (They would need to purchase the balance from others.)

2.2.4 *Comparable burdens*

A simpler rationale for sharing global costs is to find an allocation that, in an appropriate sense, affects all countries similarly or imposes "comparable burdens" (Grubb, 1989). Commonly articulated phrases that reflect this conception of equity include the views that countries should "make the same effort," "do their fair share," and "feel the same pain." In a purely monetary sense, this would suggest sharing the costs according to each nation's ability to pay. If one looks beyond income and wealth distribution, a range of other factors, discussed below, might go into an assessment of how the burdens of mitigation or adaptation could be rendered comparable across countries.

2.2.5 *Broader distributional implications*

Another possibility is to ignore all the above explicit criteria that depend on responsibility, population, or willingness to pay, and instead just focus on the broader distributional implications of any scheme. Such an approach goes beyond "comparable burdens." The question to ask of a proposed allocation of emissions permits would be simply whether it would mitigate or worsen the existing international distribution of wealth. (To the extent that the "willingness to pay" principle depends on ability to pay, it can overlap with this broader criterion; both ability to pay and comparable burdens are less expansive versions of such broader distributional criteria.)

One standard philosophical approach to assessing such distributional issues was developed by John Rawls (1971). For an interested observer to evaluate the fairness of any proposed allocation of permits among countries, that observer should imagine a "veil of ignorance" that obscures the country to which the observer would be assigned after the allocations would be adopted. In other words, *prior* to learning his or her citizenship, what allocation system would that observer decide was the best?[3] For many, this leads to the principle that allocation should seek to affect the international distribution of wealth, to most improve the position of the *least* well-off. At a minimum this would suggest that developing countries be left at least as well-off under an emission control regime as they would be in its absence. If possible, their economic lot should be improved. Henry Shue (1991) provides a more precise and memorable argument for this criterion:

> Even in an emergency one pawns the jewelry before selling the blankets... Whatever justice may positively require, it does not permit that poor nations be told to sell their blankets [compromise their development strategies] in order that the rich nations keep their jewelry [continue their unsustainable lifestyles].

2.2.6 *Status quo*

Still another principle, described for example by Young and Wolf (1992), asserts not only that past emitters should be held harmless but that their current rate of emissions has come to constitute a status quo right now established by past usage and custom especially since past carbon emissions were not generally regarded as damaging. According to this line of ethical reasoning, industrial countries are entitled to their higher rates of emission by analogy with the common law doctrine of "adverse possession," which can legitimate the claim of "squatter's rights." An alternative practical justification derives from the fact that, if agreement is not reached, the status quo pattern of emissions will presumably continue in force. This principle has found some application, albeit contested, in emerging international regimes for allocation of the radio spectrum and for geosynchronous satellite orbits.

[3] It is similarly possible to use an intertemporal version of this Rawlsian principle in which the veil obscures not only the country to which one might belong, but also which present or future generation. We do not attempt to grapple with the magnified practical and ethical difficulties which this introduces.

One implication of the status quo principle has played a consistently important practical role in international agreements: the status quo sets a prominent *baseline* from which adjustments can be made up or down. Agreements such as the Montreal Protocol to control CFCs frequently contain national targets expressed in terms of required reductions from existing levels in a base year. Each country's required reduction from a status quo baseline could be an equal amount, a fixed percentage, or set by a progressive schedule with the greatest percentage cuts being imposed on those with the highest emissions rates or most advanced level of economic development. In line with the distributional criteria discussed above, changes from a baseline level might be calculated to impose "comparable burdens" on all countries in a similar class. (Of course, developing countries might be allocated rights to increase their emissions from their status quo — typically low — baseline up to some level, while industrialized countries were reducing their emissions from their — typically high — status quo baselines.)

2.2.7 "Reasonable" emissions

An alternative criterion suggests setting a "reasonable" level of emissions for each country — a level that would support a consistent, modest standard of living, given the national climatic and other conditions. Permits would be granted for emissions at this level, but not for those "luxury" emissions in excess of this amount. (One could, of course, choose to set the target standard of living near a survival level, but the more common view would aim for a non-extravagant level somewhere near the lower range of currently industrial countries.) This justification for this principle is related to the "basic human needs" school of thought. As with the status quo principle, the "reasonable emissions" principle also sets a baseline — albeit much lower for industrialized countries — from which variations could be calculated.

3 "Pure" Equity, Politics, and Negotiations over Burden-Sharing

Having uncritically enumerated this list of contending burden-sharing criteria — polluter pays, equal entitlements, willingness to pay, comparable burdens, broader distributional implications, status quo, and reasonable emissions — we might proceed in several directions. One option would be to embark on a detailed intellectual assessment of the ethical merits of these principles — to determine in the abstract which one(s) should exert the greatest moral force on allocation choices. A number of authors have implicitly or explicitly made such an assessment, have decided on the "right" principle(s), and have constructed a "mapping" from general principle(s) to specific emissions allocation (Agarwal and Narain, 1991; Fujii, 1990; Hayes, 1992; Krause, Koomey and Bach, 1989; Smith, 1991.) Another option would be simply to note the sizeable number of available general ethical principles — evidently with varying degrees of intellectual respectability, legitimacy, and political appeal to different groups — that are likely to be drawn upon in allocation negotiations. One might then conclude, perhaps a bit cynically, that the ethical aspects of the negotiations will be inconclusive at best, and more likely, irrelevant, in the face of state interests and power.

In our view, however, a combination of these approaches is more appropriate. Debates over the merits of various ethical principles and political considerations will interact in ways that make certain approaches more promising than others. By "promising," we mean both ethically appealing and politically feasible. To ferret out the most promising, we will consider several related aspects of the principles listed above — concluding with an argument for an evolving mix of the radically different "status quo" and "equal entitlements" doctrines.

3.1 Baselines, targets, reasonable needs, and comparable burdens

The dominant theme in international discussions over a greenhouse control regime has been the search for national "targets," or reductions from current level that will collectively exist within an overall world reduction target. Many discussions have focused on the options for equal or progressive percentage reductions from a current baseline. It is often argued that target emissions allocations should reflect equity considerations by being set at levels that impose "comparable burdens" on all participants or at levels that correspond to "reasonable" needs (as noted above).

However, there are very serious drawbacks to any effort to reflect equity in a programme of reasonable needs, comparable burdens, or fair reductions from a baseline. No matter how these concepts are defined, they imply grounding the basis for emissions (and hence allocations) in the real physical and economic conditions that affect the energy and emissions which each country "needs." One approach is to seek an index of "reasonable" carbon emissions that can form the basis for permit allocations. But what factors might arguably affect this "reasonable" level? We note that the same factors that will enter arguments over "reasonable" needs will affect negotiations over "comparable burdens." More generally, these factors will come to dominate negotiations over individual country-by-country allocations. Thus, while these concepts — baselines, targets, comparable burdens, and reasonable needs — are distinct, our analysis of their severe drawbacks is based on the same line of argument.

Clearly, population and the level of economic development are very important to each of these approaches. But per capita consumption varies widely, even between countries of comparable economic development. Nor are emissions consistent against those factors. For example, Figure 3 (see Box) shows the level of CO_2 emissions, by continent, against several indices relating to population growth and economic development. North America, Europe, and Latin America all produce roughly the same aggregate levels of CO_2 emissions per unit of Gross Domestic Product; but North America emits more than twice as much fossil CO_2 per capita as

Europe, and almost ten times as much per capita as Latin America.

Many other factors would claim roles, but most seem quite ambiguous. Cold countries require energy for heating; many hot countries use it for air conditioning; and those in between may use it for both, according to season. Large, sparsely populated countries may require more energy for travel (more densely populated countries might seek an allowance for the resulting congestion!), but they are also likely to have relatively more non-fossil energy sources, such as hydro power, biomass and other renewable sources. Corrections might be attempted for the availability of non-fossil energy sources, but it is impossible to quantify these to any useful degree of accuracy, and countries already vary greatly in the extent to which they can or do exploit the resources available to them.

This list of possibilities and difficulties may be extended almost indefinitely. Reaching a consensus on the formula used to estimate a "reasonable" level of energy consumption for emissions (or on what constitutes a "comparable" burden) will probably lead to a dead end. In fact, ad hoc analysis of this sort tends to be founded on the incorrect assumption that energy consumption primarily reflects questions of geography and economic development. Detailed international energy studies, such as those supervised by Grubb (1991) make it abundantly clear that this is simply not the case. The relationship between economic growth and energy consumption is also not a simple one. And it seems likely to weaken further in the future. Apart from population and the level of overall economic development, the dominant factors that determine energy demand and emissions cannot be readily quantified. They concern the energy policy, institutions, and overall "energy culture" of countries. Not only does finding a rational basis for "reasonable" carbon emissions seem impossible, but any approximate formula will be far from satisfying some countries' political interpretation of "fairness" — especially as "fairness" refers to equality of burden.

Consequently, the process of trying to define "reasonable needs" or "comparable burdens" relative to baselines would in practice probably become a piecemeal negotiating process, on a country-by-country basis. Although this sort of negotiation would probably not be as difficult (because of the greater flexibility involved) as allocating hard emissions abatement targets for each country, it is still likely to be extremely difficult. Compared with any formula-based approach, piecemeal, country-by-country allocations would probably amplify the negotiating difficulties greatly, for three major reasons.

First, it opens the gates for every country to plead its special case for a higher permit allowance, drawing on its own interpretation of the ambiguous factors (noted above) which arguably affect energy and, more broadly, emissions requirements. The second drawback would amplify this: the pressure in such negotiations is always to make excessive "business as usual" emission projections, and to highlight the difficulty of abatement, so as to get a larger allowance. The difficulties in limiting emissions will be great enough without the diplomatic prizes (and permits) going to those who can amplify them the most. Thirdly, circumstances will change with emission paths, population changes, technical developments, and resource discoveries, none of which can be well predicted. It is hard to see piecemeal allocations giving any long-term basis for allocation; the whole miserable process would have to be repeated every few years, replaying processes of exaggeration, special pleading, and brinkmanship. In consequence, such an approach would clearly risk huge negotiating inefficiencies; even if an initial allocation could eventually be agreed, no feasible long-term basis for permit allocation would be produced. An allocation formula is far preferable, and on a simpler basis than that of "equal burdens" or "reasonable needs."

The search for equitable targets has been inspired by the successful experience of the Montreal Protocol, an undertaking that was far simpler than the larger greenhouse problem that involves many more nations, activities, and costs. Much more likely in the greenhouse case is a scenario like that of the European Community's negotiations over the Large Combustion Plant Directive, in which it took 12 relatively homogeneous nations 5 years of intensive, often twice-weekly bargaining sessions to arrive at a range of required reductions. More recently — and ominously — the European Community agreed to stabilize its *overall* greenhouse emissions at 1990 levels by the year 2000. But negotiations over which member states would be required to make what reductions (the "target-sharing" problem) broke down, and attention shifted to a carbon tax. In the words of Grubb (1989), "these goals are very unlikely to be achieved by the countries of the world sitting around a table and agreeing on who should reduce by how much. The idea that a protocol on limiting carbon emissions will be like the Montreal Protocol writ large is an illusion best dispersed before it leads us irretrievably down a blind alley."

3.2 Willingness to pay, ability to pay and distributional implications

We next turn to the misleadingly named "willingness to pay" criterion, which conflates the level of concern about the greenhouse problem, likely damages to be suffered, and relative ability to pay. This latter element closely links a willingness to pay with the pure "distributional" concept — urging that the greenhouse regime should at least not worsen, and ideally should improve, the international distribution of wealth.

At present, an intractable difficulty on the "damage" side of the willingness to pay equation is lack of knowledge about the regional impacts of climate change. Without mutually agreed-upon regional assessments, a "willingness to pay" approach would encourage piecemeal negotiations, in which each country had an incentive to underestimate the potential damages from climate change (on account of the willingness-to-pay rationale), and overestimate the costs of abatement (in line with the "comparable burdens" criterion), so as to obtain a lax target (a large emissions allocation) for itself. In the extreme, those who genuinely expected to be greenhouse "winners" due to more favourable climate or abundant crops - along with those who might cynically adopt such a position for tactical purposes - might even demand compensation from those wishing to prevent climate change.

Further, willingness to pay (to avoid negative greenhouse impacts) would likely become the political equivalent of a "victim pays" principle. This would seem egregious, given that the greatest damage from climate change will most likely occur in the impoverished developing world, which has generally contributed the least to the problem and has the fewest resources to adapt to the consequences.

To the extent that "willingness to pay" is driven by a country's *ability* to pay, one should expect very modest contributions from the developing world — whether as a matter of fairness, political possibility, or necessity. In part, this is because capital constraints and poverty combine to create a very short-time outlook in much of the developing world. Subsistence living focuses people on immediate needs, with little energy or resources for long-term issues. Government policy becomes dominated by short-term aspirations, the external pressures of debt and macroeconomic imbalance, especially the realpolitik of creditor demands. A situation of almost institutionalized crisis management makes it almost impossible to divert resources from the most pressing needs, even when this could yield longer term benefits.

Even if such countries agreed to attempt significant abatement efforts, it could be politically and managerially impossible for them to follow through on the required policies without direct short-term incentives and financial assistance. Their insufficient base of research capability and coordination, combined with education and resource constraints, reinforce the low priority of long-term environmental issues, especially those of a global nature.

The IPCC Special Committee on the participation of developing countries (1991) noted many of these issues, and others, and proposed ameliorative measures; likewise the Brundtland Commission explored the pressures and constraints of poverty (WCED, 1987). An effective global climate regime for the long term can not be developed without taking these constraints into account. By default, the industrialized world is the only conceivable source of many of the resources, and most of the technology, needed in a global climate regime.

In short, a damage-based *willingness* to pay criterion suffers from severe informational and incentive problems. Whether one looks at it as a matter of fairness or simply as the product of necessity, differential *ability* to pay, however, will have to govern much of the burden sharing involved in an effective regime. This might be made operational by way of an existing agreement such as the United Nations Scale of Assessments.

However, the question of the broader distributional implications of an emissions allocation scheme also arises. As a minimal statement of what an allocation system should *not* do — namely, worsen the international distribution of wealth — general distributional criteria have some force, and could be compatible with many other approaches. Yet a number of analysts, as well as developing country diplomats, have seen the greenhouse issue as a moral claim and bargaining lever to advance an entire "Southern" agenda: debt relief, commodity price stabilization, and international institutional reform. (In part, Southern leverage derives from the fact that without developing country participation, the long-term effectiveness of an anti-greenhouse regime would be severely compromised.)

Such redistributional arguments, if too expensive and pressed too aggressively, could inextricably link the climate problem with innumerable other problems and conflicts, quite possibly preventing progress on the climate or other linked issues (Sebenius, 1990). At the same time, industrialized country efforts to treat the carbon emissions problem as completely independent of broader distributional inequity questions are doomed to sharply limited effectiveness at best. In sum, the middle ground — endorsing a more sharply defined ability to pay criterion — holds greater promise than either a damage-based approach or a full-scale effort to create a 1990s version of the New International Economic Order.

3.3 A historic "polluter pays" principle and natural debt

The historic "polluter pays" principle is based on the observation that industrialized countries have preemptively "used up" more than their fair share of the atmospheric resource during the last 300 years, leaving less for the developing countries. Moreover, those living in the industrialized countries became richer in part *because* of this apparently costless usage over the centuries. To the extent that poorer countries are denied this opportunity, the rich who have exploited it first should compensate them. To many involved in the greenhouse debate, these arguments are taken as self-evident and consequential.

There are, however, several potential objections to the principle of historical accountability:

First, it would make current generations pay by virtue of their geographical location, for the activities of past generations who had no idea of potential costs of their actions and no incentive to limit their emissions.

Secondly, the issue of who has benefited from historical emissions can be contentious — given that the patterns of production, trade, consumption, and migration shift over time and are intricately interwoven. Boundary changes could create major difficulties for allocating past emissions to current states — the break-up of the former Soviet Union provides a telling example.

Thirdly, it will be argued that though development has generally bequeathed the greatest benefits to descendants in the same country, important benefits have spread much more widely. In particular, it is easier for later developing countries to gain the same benefits with lower emissions, by drawing on the better technologies developed elsewhere. Thus, although development clearly has negative externalities (such as pollution), it also has positive ones (such as medicine and environmentally friendly technologies), and both can be transmitted internationally and intertemporally.

We also note that while the issue of historical responsibility can be cast in the relatively neutral language of "responsibility" for past emissions as a factual matter, it is easy for the negotiations to take on the morally laden qualities of "blame" and "guilt." This can lead to an escalation of rhetoric and hardening of positions. (In the words of one commentator: "Why should I feel guilty because my grandfather rode a coal-burning train?"). In short, though the (historical) polluter pays principle has some force, the philosophical argument risks becoming tangled.

As the debates move from principles to concrete implementation, other issues arise. In most forms the historical polluter pays principle implies an allocation based (inversely) on "natural debt". It is not clear how such historical accountability would be persuasively operationalized. For example, Krause et al. (1989) suggested a cumulative limit on CO_2 emissions of 300 Gt of C, arbitrarily divided equally between industrial and developing countries; industrial countries have already emitted much of their share on this scheme and would have to cut emissions very rapidly if this were a hard target. The rationale for the Krause et al. allocation is unclear; while various other allocations could be suggested, most are subject to serious underlying data problems — uncertainties, limitations, interpretations — that will be exacerbated in the opportunistic environment of diplomatic negotiation. With respect to definitively assigning responsibility, the data analysis of Section 1.1 of this chapter suggests the range of options about which reasonable arguments can be had: which gases and sources to include, the cutoff period, the global warming potential factor used, and so on.

Even fossil carbon dioxide (CO_2) emissions are somewhat imprecise, especially going back several decades - while historical releases from all other sources except CFCs seem unmanageably uncertain (see Box). Insisting on strict historical accountability could thus, in practice, torpedo attempts to include non-fossil carbon sources and sinks, irrespective of how knowledge about current emissions were to develop. The decay of the different gases partially eases this by implying that less weight should be placed on emissions further back in history, which are presumably more uncertain; but it also introduces some difficulties, arising from uncertainties in estimating the decay of different gases. It could also raise debates about the impact of "latent" warming, for example the delay in temperature rises induced by the oceans.

Given a political will it might be possible to reach compromise estimates, but such uncertainties and complexities could prove important obstacles. Overall, such factors muddy the water of any clearly quantified equity argument concerning past emissions, though in general terms it remains clear that the industrialized countries have been the main source of past greenhouse gas emissions, and their current generations are the major beneficiaries of that process.

With sufficient agreement on the data, the atmospheric "stock" could be reflected through a given number of tradeable permits, which would be divided on a per capita or other basis. Since the industrialized countries have already emitted much more than developing countries, their remaining holding would be correspondingly lowered. On most such historically driven allocation schemes, industrialized countries would have already used up much of their "entitlement," so that permits for future emissions would be primarily "bought" from developing countries. The huge scale of the implied transfers could represent a serious political problem to this approach, a problem to which we shall return shortly in connection with per capita criteria.

In all, whilst the dimension of historical responsibility should not be forgotten, and the concept of natural debt is ethically an important notion, as a direct component in emissions and allocation formulae it could raise both technical and political obstacles to the point of insuperability. Return for a moment to a central justification for the principle of historical responsibility: the wealth accumulated by industrial countries in part due to their long-term free use of the atmospheric sink. This rationale is appealing — but has the curious feature that the primary justification for considering *historical* emissions is based on *current* wealth. And it is intended to compensate *future* increased energy costs of developing countries, if they are constrained in the use of fossil fuels.

Tentatively, we suggest that the ethical divide between historical criteria and those focused on current emissions and current relative ability to pay may not be as large as it appears. In practice, current emissions (as shown in Table 1), are somewhat related to past emissions. Including historical

Table 1. Total CO_2 emissions relative to world average by different criteria

	Current (average 1980-1986)			Cumulative (1860-1986)		
	Country	Capita	Land	Country	Capita	Land
MDCs	2.39	2.62	1.37	2.92	3.18	1.65
LDCs	0.54	0.52	0.78	0.36	0.35	0.49

LDCs are defined as those with GDP of less than $4,000 per capita in 1986 (World Bank, 1988). Source: Subak and Clark, 1990.

emissions would certainly not change the direction of North-South resource flows implied by most underlying allocation criteria (e.g. per capita), and might have little impact on the overall ranking of many different countries.

Given all of these factors, it appears wiser to concentrate the search for a viable allocation formula on contemporary indicators, without formulae linked to some explicit quantification of historical "natural debt." In our judgment, as noted earlier, the most promising allocations would depend on radically different criteria, namely variants of the status quo principle (current emissions) and equal entitlements (per capita).

3.4 "Status quo" (current emissions)

Current emissions offer a natural starting (or "focal") point for negotiations over reductions, in part because they represent the "disagreement point" — what many countries would emit in the event of complete negotiation failure. Moreover, many environmental agreements, such as the Montreal Protocol and acid rain negotiations in both Europe and North America, are couched in terms of departures from current levels.

A strong version of the status quo would simply allocate permits in proportion to current emissions. For example, if a 10% cut were required, current emitters would receive permits equal to 90% of their emissions — and would have to reduce by this amount or purchase the balance from elsewhere. Under many other allocation schemes, industrialized country emitters would receive permits for a much smaller fraction of their carbon output. As a result, the status quo approach is likely to arouse the least amount of political opposition in industrialized countries, a factor of potentially considerable importance in the acceptability of the regime. Another attraction of the status quo approach, to some industrialized countries, is its treatment of the carbon emissions problem first and foremost as a carbon problem — rather than as a broader environmental problem entangled with all sorts of other issues, especially large-scale resource transfers.

Weaker variants of the status quo principle might take current emissions as the baseline and introduce some elements of "progressivity" on the basis of ability to pay, stage of development, or extent of energy produced from non-fossil (i.e. hydro, solar, wind, biomass, or nuclear) sources. This might be thought of as a "modified" status quo approach. In practice, as discussed above, it might face considerable definitional difficulties, as countries would argue their cases for different abatement rates on the basis of "comparable burden" arguments. However, if such complexities could be overcome, industrialized countries (such as the United States) that are uneasy about large resource transfers incident to a greenhouse agreement, might over time become relatively comfortable with some variant of this approach.

Even if a current emissions approach were only applied to a subset of industrialized countries, other industrialized countries might also find emission permit allocations framed in this way to be attractive. Superficially, these countries would be treated equally. Each country would be directly or indirectly responsible for emission reductions proportional to its own carbon output. Resource transfers would be the product of relative reductions — not wealth or development per se.

As an ethical principle applied to pollution in general, however, the strict status quo principle appears questionable: it implies that past pollution should be rewarded by the right to continue indefinitely. Moreover, as noted, such an allocation would largely attack the carbon emission problem as *separable* from the range of North-South inter-dependencies. Developing countries would resist strong forms of this approach vehemently as neither providing them the resources necessary to undertake costlier (greenhouse-friendly) development paths, nor recognizing responsibility for the current problem. They would also object that such a system did not reflect the fundamental importance of differential stages of development in determining abatement prospects.

In short, a status quo-based approach by itself would not attract key developing countries unless it were independently accompanied by sizeable commitments to transfer financial and technological resources. The developing world would largely go its own way. Over time, this would sharply limit the effectiveness of the regime.

3.5 Equal entitlements (absolute and modified per capita allocations)

For much of the international community involved with greenhouse negotiations, per capita allocations probably have the strongest claim to an ethical basis, and several analysts have pointed to this and also to desirable features of a population-based allocation (Grubb, 1989; Epstein and Gupta, 1990; Bertram, 1992). The "equal entitlements" doctrine implies that initial permit allocations should be directly proportional to national population. In its most basic presupposition — that each human being should have an equal right to an as yet unallocated, scarce global common — this principle has considerable appeal. Of course, one can argue that global equality of the right to emit (or to receive equivalent compensation), need not be the same as equality of such a system's *effect* on broader measures of well-being; equal emissions need not be the same as equal benefits from emissions. But objections of this sort are unlikely to carry much weight.

In practice, the per capita approach would reflect some measure of historical emissions, because countries with higher per capita emissions tend to have emitted the most in the past. Moreover, many of the practical consequences of a per capita allocation — though not all — are also appealing. The key feature would be that all the industrialized countries, with higher-than-average per capita emissions, would have to buy permits from developing countries at least for fossil CO_2; thus the resource transfers essential to developing country action would be a feature of this approach. This would also be true for a system with broader coverage that included most other sources and sinks; deforestation is the only source with generally higher per capita emissions in poorer countries.

One practical difficulty with per capita allocation is that, in its simple form, it rewards population growth, which itself is an important component of the problem. The magnitude of this incentive is debatable; it may be insignificant in relation to other determinants of population growth and policies, but it certainly could raise concerns. However, various ways of breaking this link are available (Grubb, 1989). For example, allocation could focus on population above a given age, or population lagged many years behind the issue of the permits (which achieves much the same result). It could use a fixed historical date for population data, or include an explicit term related inversely to population growth rates.

Of these variants, the first and last increase measurement problems because they require some knowledge of the age structure; the second could also raise measurement and definitional problems associated with infant mortality; the third raises practical difficulties (associated with, for example, migration) and introduces some arbitrariness concerning the base date. All of these modifications also weaken the simple ethical basis of an equal entitlements doctrine. Nevertheless, these obstacles are not insuperable. In addition to breaking the incentive to population growth, they could also reduce the scale of implied resource transfers — together with the associated political objections in major industrialized countries — given the greater population growth rates in most developing countries.

The major political obstacles arise from the potential scale of the transfers implied by per capita allocations. Table 2 illustrates the scale of possible transfers which might be implied by such an allocation, with a global target resulting in an effective permit price of $20 per tonne of carbon ($20/tC, sufficient to add perhaps 30-50% to the international traded price of coal). Note that this is a lower price than many studies of abatement costs would imply, and transfers at $20/tC would increase with permit price. At unaltered 1988 emission and population levels, the transfers would amount

Table 2. International transfers with pure per-capita permit allocations

a) Based on 1988 data

	Population M	**Fossil CO_2 emission**	**MtC Allocation**	**Transfer at 20$/tC $bn**	**%GNP**	**/ODA**
NAM	271	1414	298	-22	-0.42%	-1.79
OEC	556	1220	611	-12	-0.16%	-0.34
SUE	427	1420	469	-19	-0.81%	
CHI	1174	608	1289	14	4.34%	6.81
ROW	2653	918	2913	40	1.20%	1.25
Global	**5082**	**5580**	**5580**			

b) Based on 2025 projections with abatement. Fossil CO_2 abatement from 1995 annually at 2.00%/yr below baseline projections

	Population M	**Fossil CO_2 emission**	**MtC Allocation**	**Transfer at 20$/tC $bn**	**%GNP**
NAM	332	1293	282	-50	-0.43%
OEC	613	987	520	-23	-0.40%
SUE	551	1511	468	-53	-0.70%
CHI	1563	982	1327	18	1.05%
ROW	4920	2002	4178	110	0.73%
Global	**7979**	**6775**	**6775**		

NAM - North America
OEC - Other Economically Developed Countries
SUE - Soviet Union and Eastern Europe
CHI - China
ROW - Rest of World

Sources: *Population and emission projections* are given as in the IPCC Energy and Industry Subgroup Report.
GNP is derived from World Development Report, 1990; except SUE and CHI, derived from CIA data. Projected growth rates as in IPCC EIS report. This data reflects conventional exchange rates. Thus, real GNP figures, on a purchasing power parity basis, would be higher for ROW and especially for SUE and CHI. The relative scale of receipts are thus overstated.
Overseas Development Assistance 1988 data: World Development Report, 1990.

Note: Most modifications to per capita allocations to discourage population growth would also reduce the scale of transfers. The figure of $20/t C is probably a lower bound for achieving significant impacts; some studies project an effective marginal cost for substantial abatement efforts up to ten times this figure. For the stated emission and population levels, the transfers would be directly proportional to the permit price.

to about 0.42% of GNP in the US and twice this for the former Soviet Union and Eastern Europe (SUE) with GNP estimated at conventional exchange rates.[4]

These estimates reflect, respectively, the high per capita emissions of the US and the high emissions from the SUE countries relative to economic output; the average transfers from the rest of the OECD countries would be around 0.16% of GNP. The boost to developing countries would be relatively larger than the negative impacts of OECD countries, amounting to 4.3% of GNP in China (though much less on a purchasing parity basis) and an average of 1.2% for the rest of the world. The calculations here are deliberately simplified, but illustrate potential magnitudes of exchanges. Some studies have used global energy models to examine the transfers implied by different targets and allocation systems, though these inevitably also contain a range of (necessarily) arbitrary assumptions,[5] and most result in much higher costs and (for per-capita allocations) transfers.

Under these conditions, based on 1988 data, the net North-South flow would roughly offset the current international flows from South to North exacted in debt service. Overall, these transfers would be similar in magnitude to current official overseas development assistance (ODA). It would require more from the US than that country's current ODA levels, and about a third of the current ODA from the rest of the OECD countries.

Over time, as emissions, population and permit prices change, the transfers would change in proportion to the permit price. Table 2(b) shows possible transfers as projected for population and emission profiles in 2025, at an increased permit price of $20/tC, if abatement reduces emissions at a rate of 2%/yr below the baseline projected by the 1990 IPCC's Energy and Industry Subgroup. Total emissions are greater, but the difference between industrialized and developing countries is less. The transfers are large — in absolute terms roughly in proportion to the price increase — but they are similar as a fraction of industrialized country GNPs, and somewhat less in relation to the more greatly expanded GNPs of the developing countries.

Naturally, actual transfers would depend on prices and responses. Few analysts suggest that a major impact on CO_2 growth can be obtained with financial incentives much lower than the ones discussed above. Modifications to the per capita formula to discourage population growth would reduce the scale of transfers; a higher effective permit price would obviously increase it.

Such transfers raise varied and major political and practical questions. Most observers in OECD countries would consider it politically incredible that the US could or would agree to a system of CO_2 control which involved international transfers exceeding total ODA, perhaps much more, under foreseeable circumstances. Still more problematic is the situation of the former USSR and East European countries where the implied transfers appear unmanageable given the state of their economies, and politically out of the question in the current conditions, though the current industrial collapse and consequent sharp decline in emissions may do much to reduce this anomaly. But for countries in such special conditions to become part of such a system there would presumably have to be other inducements or major special provisions.

Given these observations, pure per capita allocations appear infeasible. Even if this judgment is correct, however, the simplicity and attractive features of per capita allocation suggest that they deserve strong consideration at least as a significant component and a long-term goal towards which a permit system could evolve.

4 A Mixed, Evolving Allocation System

The above discussion has considered various mechanisms, and suggested that even what many consider the most attractive single approach, allocation on a modified per capita basis, may be politically infeasible at least for several decades. The alternative is to search for a mixed formula which carries the right properties but which may be politically more feasible.

It is in fact impossible for independent analysts to determine what is a feasible allocation: that will arise from the process of negotiation between sovereign states. What can be done is to search for approaches which help to keep negotiations as simple as possible, and yet result in an effective and efficient agreement.

In this spirit, are there attractive approaches? Starting from the observation that a tradeable permit system would be of limited effectiveness unless it involves resource exchanges between participating countries, a key requirement is for a formula which ensures that the general direction of transfers is from industrialized economies to developing ones rather than vice versa. The magnitude of the exchange must be (a) sufficient to create a real incentive for all participants to limit emissions; (b) sufficient to provide the resources to enable poorer countries to adopt more efficient technologies; while (c) not so large as to be politically impossible, in which case, however attractive the formula on paper, it will be of no practical use. A key concern of developing countries, in particular, is avoiding any risk of leading to reverse transfers, which could constrain their development aspirations.

This implies an allocation which has many of the broad features of a per capita system, but which carries the flexibility for negotiations to affect the scale of transfers implied by the allocation - without compromising the environmental target. One simple way of doing this is to allocate permits in proportion to a mixed weighing of current emissions and population. If appropriate, allocation could be qualified (as

[4] The fraction for SUE may be similar to the US if purchasing power parity estimates for Soviet GNP were used.

[5] For a review of modelling studies, including some which offer estimates of international transfers under various modelling and allocation assumptions, see Hoeller, Dean and Nicolaisen (1990).

already discussed) to avoid incentives to population growth.[6] Under this approach, the direction of transfers would be preserved, but the lower the relative weighting accorded to population, the lower the transfers, and vice versa. In this simple scheme, the allocation negotiations would focus upon only two parameters: the total permit issuance (defining the overall target emissions), and the relative weight between population and emissions in the allocation formula.[7]

If permit allocations did increase directly with higher contemporaneous emissions, this would weaken the incentive for limiting emissions and slower population growth. An important variant would be to "lag" this formula by a few years — for example, to express it as a moving average of the previous decade's emissions (which would also help to even out annual fluctuations). This would introduce a dynamic element to the incentives, since payments by industrialized countries would reduce in proportion to the rate at which they could reduce emissions, while the receipts by developing countries would be increased to the extent that they could restrain the rate of population and emissions growth. This adds another dimension for meeting the varying demands of the system.

Finally, we note that there is no reason why an allocation formula and its coefficients should be immutable. Indeed, it could and probably should evolve over time. In the above scheme, the allocation could evolve from one which is sufficiently tilted toward current emissions in the permit allocation — to make it politically feasible that the major economic powers would participate in the early stages — towards a much stronger per capita weighting over a period of decades that would offer a powerful inducement to major developing countries.

Alternatively, it is possible to imagine a multi-tier system with differential allocation criteria applied to developing and industrialized countries. As was the case with the Montreal Protocol, permits might be allocated to imply steady reductions in industrialized countries from current levels, while developing countries might receive allocations enabling them to increase emissions up to a per capita limit — from which allocations would be constant or progressively reduced. The path taken could be specified in advance or left to subsequent negotiations — though the incentive on all countries to prepare long-term abatement strategies would be increased if it were clearly understood that the weighting accorded to contemporary emissions would fall steadily over time.

People will hold differing views about what relative weightings would be fair and feasible, and the extent to which some weighting towards the criterion of current emissions can be justified on a strictly ethical basis. We also recognize the immense political obstacles which stand in the path of such an ultimate per capita based formulation, arising both from detailed issues of implementation and from the political and intellectual objections to such an unprecedented approach to international negotiations. (After all, in prominent past examples of international "resource" division through negotiation — of the benefit from deep seabed mineral resources (the "common heritage of mankind"), the electromagnetic spectra, geosynchronous satellite orbits, and even Antarctica — a strict per capita allocation has yet to result.) But this is an unprecedented issue. Success can never be guaranteed, and, for all approaches, will depend upon the degree of political will and good faith.

In the approach presented here, at least the parties would be focused on some relatively simple, if important tradeoffs. They would be striving for a system which, if achieved, could yield both incentives for all parties to limit global emissions in the least costly way, while providing the resources to enable poorer states to do so.

---

[6] Young and Wolf (1992) have investigated and insightfully discussed a variant of such a mixed system. Hayes (1992) has also offered a mixed system, but based on willingness to pay (as reflected in the UN scale of assessments) combined with contribution to the greenhouse problem (as reflected in natural debt calculations per Smith (1991). In our view, the lack of a term reflecting current emissions as well as the reliance on historical emission data and allocation of responsibility both imply significant political problems with this approach, despite some ethical appeal.

[7] To formalize this, if T is the target level and p is the proportion of population in the allocation formula, a country with (modified) population and emissions — which are respectively X% and Y% of the corresponding total qualifying population and emissions of the participating countries — would receive an allocation of $(pX + (1\text{-}p)Y) \times T$ permits (each of which would carry an economic value roughly defined by the severity of the abatement target T): Barret (1992) examines this proposal amongst many others and concludes that appears to be one of the few allocation formulae which could be flexible enough to allow a formal solution even on the grounds of strict and narrowly-defined economic self interest by all parties (by using a very low population weighting).

References

Agarwal, A. and S. Narain, 1991: *Global Warming in an Unequal World.* Centre for Science and Environment, New Delhi.

Barret, S., 1992: Acceptable allocation of tradeable carbon emission entitlements in a global warming treaty. In *Tradeable Entitlements for Carbon Emission Abatement.* UNCAD, INT/91/A29, Geneva.

Bergesen, H., 1991: A legitimate social order in a "Greenhouse" world: some basic requirements. *International Challenges,* **11**(2).

Bertram, G., 1992: Tradeable emissions permits and the control of greenhouse gases. *Journal of Development Studies,* **28**(3).

Dorfman, R. 1991: Protecting the Transnational Commons, unpublished manuscript, Harvard University, June.

Epstein, J.M. and J. Gupta, 1990: *Controlling the Greenhouse Effect: Five Global Regimes Compared.* The Brookings Institution, Washington, DC.

Fisher, R. and W. Ury, 1982: *Getting to Yes.* Houghton-Mifflin, Boston, MA.

Fujii, Y., 1990: An assessment of the responsibility for the increase in the CO_2 concentrations and inter-generational carbon accounts. IIASA Working Paper WP-05-55.

Ghosh, P., 1991: *Structuring the equity issue in climate change.* Tata Energy Research Institute, New Delhi.

Grubb, M.J., 1989: *The Greenhouse Effect: Negotiating Targets.* Royal Institute of International Affairs, London.

Grubb, M.J., 1991: *Energy Policies and the Greenhouse Effect, Volume 2: Country Studies and Technical Options.* Royal Institute of International Affairs, Dartmouth, Aldershot, UK.

Grubb, M.J. and J. Sebenius, (forthcoming, 1992): Participation, allocation and adaptability in international tradeable emission permit systems for greenhouse gas control. In *Tradeable Permits for Abating Greenhouse Gases,* OECD, Paris.

Gulliver, P.H., 1970: *Disputes and Negotiations: A Cross-Cultural Perspective.* Academic Press, New York.

Hammond, A.L., E. Rodenberg and W.R. Moomaw, 1991: Calculating national accountability for climate change. In *Environment,* **33**(1).

Hayes, P., (forthcoming 1992): *Greenhouse reduction regimes: evaluating the costs.* UN University, Tokyo, Japan.

Hoeller, O., A. Dean and J. Nicolaisen, 1990: A survey of studies of the costs of reducing greenhouse gas emissions. OECD Department of Economics and Statistics, Working Paper no. 89, OECD, Paris.

Houghton, J.T., G.J. Jenkins and J.J. Ephraums (eds.), 1990: *Climate Change: The Scientific Assessment of the Intergovernmental Panel on Climate Change*, Cambridge University Press, Cambridge, UK.

Krause, F., J. Koomey and W. Bach, 1989: *Energy Policy in the Greenhouse: Volume I.* International Project for Sustainable Energy Paths, El Cerrito, California.

Lax, D. and J.K. Sebenius, 1986: *The Manager as Negotiator.* The Free Press, New York.

OECD Council, 1974: Communication C(1974), 223, Paris, 14 November.

Rawls, J., 1971: *A Theory of Justice.* Harvard University Press, Cambridge, MA.

Rose, A., 1990: Reducing conflict in global warming policy - the potential of equity as a unifying principle. *Energy Policy,* December.

Sebenius, J.K., 1990: Crafting a winning coalition. In *Climate Change: Negotiating a Global Regime,* World Resources Institute, Washington, DC.

Sebenius, J.K., 1991: Designing negotiations toward a new regime: the case of global warming. *International Security,* **15**(4), 110-148.

Shue, H., 1991: Personal communication, July 17.

Smith, K.R., 1991: Allocating responsibility for global warming: The Natural Debt index. *Ambio,* **20**(2), 95-96.

Smith, K.R., J.N. Swisher, , R. Ranter and D.R. Ahuja, 1991: Indices for a greenhouse gas control regime that incorporate both efficiency and equity goals. World Bank Working Paper, Washington, DC.

Stockholm Environment Institute, 1992: *National Greenhouse Gas Accounts: Current Anthropogenic Sources and Sinks.* S. Subak, P. Raskin and D. von Hippel. Stockholm Environment Institute, Boston, MA.

Subak, S. and W.C. Clark, 1990: Accounts for greenhouse gases: towards the design of fair assessments. In *Usable Knowledge for Managing Global Climate Change*, edited by W.C. Clark, Stockholm Environment Institute, Stockholm, Sweden.

Westing, A.H. 1989: A law of the atmosphere. *Environment,* **31**(3).

World Bank, 1988: *World Development Report 1988*, Washington, DC.

World Commission on Environment and Development (WCED), 1987: *Our Common Future.* Center for Our Common Future, Geneva, Switzerland

World Resources Institute, 1990: *World Resources 1990/91.* Chapter 2, and Table 24.1, Washington DC.

Young, H.P. and A. Wolf, (forthcoming 1992): Global warming negotiations: does fairness count? *The Brookings Review,* **10**(2).

Young, O., 1989: The politics of international regime formation: managing natural resources and the environment. *International Organization,* **43**(3), 349-375.

CHAPTER 22

Climate Negotiations: The North/South Perspective

Tariq Osman Hyder

Editor's Introduction

Among the objectives of the UN Conference on Environment and Development, three stand out particularly strongly for this chapter. One goal involves preparing conventions on climate change, biodiversity, and forestry in time for signature in Rio de Janeiro in June 1992. Second, commitments will be sought for new and additional resources to help developing countries integrate environmental issues in their development plans. And, third, new financial mechanisms and institutions must evolve to manage those resources efficiently and effectively. Realizing all three of these goals will require complex, sensitive, and far-seeing negotiations. In that context, the climate negotiations now being carried out under the Intergovernmental Negotiating Committee for a Framework Convention on Climate Change (INC) are a significant model for the entire UNCED process.

Tariq Osman Hyder, the Director General for Economic Coordination of the Pakistani Foreign Ministry, is a keen political observer of, and an active participant in, the international negotiations of the INC and other important climate-change-oriented deliberations. An experienced civil servant with a long and successful career in the Foreign Service, he headed his country's delegation to both the INC and the management meetings of the Global Environmental Facility (GEF), a new financial mechanism jointly administered by the World Bank, the United Nations Development Programme, and the UN Environment Programme. At the December 1991 meeting of the GEF Management Group, Hyder was asked to present the joint statement of the G-77, the common position of the developing countries. He has thus become one of the world's leading spokesmen for the position of the nations of the South.

None the less, Hyder is known for his ability to understand and meet all sides of these complex negotiations. His analysis in this chapter is grounded in his perception of the key players: (1) the United States, (2) the rest of the industrialized countries, and (3) the developing countries. Despite their apparent differences over short-term positions and tactics, Hyder views the long-term interests of the US and those of the rest of the industrialized countries as inexorably linked. He argues that climate negotiations must, therefore, be viewed fundamentally as a North/South issue.

The current global economic situation — a continuing recession in which productivity and productive capacity is growing more rapidly than demand — puts the most industrialized countries (the G-7 nations) in the driver's seat for these and many other international negotiations. Taking full advantage of this situation, the North continues to impose conditionalities on its support for and cooperation with the South. These conditionalities have in the past included demands for democratic governance, increased recognition of human rights, free-market economic policies, political empowerment of women, and action against narcotics. To these, new conditions concerning environmental protection have recently been added.

The South, on the other hand, seeks to evolve an integrated approach to issues of mutual concern, according to Hyder. He notes that all high-priority issues can be resolved only in the context of strong, broad-based development in the South. This means, he cautions, attacking not just the symptoms of current problems but also their underlying causes.

Hyder recognizes that environmental issues, including the risks of rapid climate change, have now taken their important and appropriate place on the policy agenda of the North. The South, however, continues to focus on questions of greater primacy to their national economies. These include: (1) stable and assured access to world markets, (2) the opportunity to attract new capital for investment, (3) the integration of recent scientific and technological advances into national development plans, and (4) the acquisition of the additional finances necessary to promote sustainable development.

As Hyder makes clear, the climate negotiations provide important and timely opportunities for promoting international cooperation in an increasingly interdependent world. As a prerequisite for this, industrial countries must explicitly recognize their historical responsibility for emissions of greenhouse gases. These negotiations, in Hyder's view, provide an opportunity for the United States and other industrial countries to commit to stabilizing and ultimately reducing their own domestic emissions of the most dangerous gases. For their part, the developing nations can develop national inventories of greenhouse gas emissions and emissions management plans to limit the rate of growth of future emissions.

The key step for the negotiations in the near-term may well be the joint acceptance of a set of common principles to guide the negotiations. Hyder, reflecting the views of many in the developing countries, suggests that these principles might include: (1) equity, (2) sovereignty, (3) the right to development, (4) the need for sustainable development, (5) recognition of the special circumstances of developing countries, and (6) a commitment to international cooperation on these issues. If the obstacles to agreement during the climate negotiations on these (or a similar set of) principles can be overcome through the continued goodwill of all the parties, then the prospects for achieving a successful, effective, and practical climate convention look bright indeed.

- I. M. M.

1 Objectives of the UNCED Process

The first phase of the UNCED process will culminate in June 1992 at the Earth Summit in Rio de Janeiro, Brazil. This phase of the process in Rio has seven broad objectives. The first objective in this phase is to prepare Conventions on Climate Change and on Biodiversity for signature at the Rio meeting.

The second objective is to present and agree on a new "Earth Charter". This joint declaration will give the principles by which people should conduct themselves in relation to each other and to the environment, so as to ensure their common future in both environmental and developmental terms.

Third, the conference will discuss "Agenda 21", a programme of action for implementing the principles in the Earth Charter. This programme of action will indicate what governments must do to bring the peoples of this planet into the 21st Century on a sustainable path, coordinating the process by a series of non-binding statements of goals and objectives. This compilation of goals and objectives is linked to a list of strategies and actions to be taken in the coming decades by governments and non-government organizations.

The fourth objective, one of special importance to developing countries, is to address the issue of Financial Resources. The developing countries seek a commitment from the industrialized nations of sufficient resources to allow and encourage developing countries to integrate international environmental issues into their own priorities for national development.

The need to respond to international environmental problems imposes significant additonal expenses on the already pressed resources of many developing countries. Integrating these emerging and increasingly important environmental issues into the national development plans of the South will require the provision of new and additional financing from the industrial economies of the North. Additional resources are already needed for the development process in general but, in particular, they must be enough to cover the incremental costs that these countries will have to bear as they respond to international environmental concerns. The UNCED process must advance the discussion of practical measures to meet these very real needs for new and additonal financial resources.

The fifth objective of this phase is to develop the necessary Financial and Institutional Mechanisms for equitably and efficiently managing these Financial Resources.

The sixth objective is to assure the transfer of environmentally sound and economically attractive technologies to the developing countries. Such transfers will provide the means to integrate economic development in the South with global environmental protection.

All of these are closely coupled with the seventh objective: the orderly management of Institutional Change — i.e. how best to strengthen and reform existing international institutions so that they may play a more effective role in the whole process linking environment and development.

To a large extent, the current negotiations in the Intergovernmental Negotiating Committee for a Framework Convention on Climate Change constitute a model reflecting all the main issues in the UNCED process. These negotiations also provide a model for understanding the complex of powerful forces and varied perceptions which energize global negotiations and which will shape the end results.

2 Principal Parties in the Climate Negotiations

In the Climate Change negotiations there are three main actors, without which there can be no Convention. These are (1) the United States; (2) the rest of the industrialized countries; and (3) the developing countries. While there are certain differences between the USA and its OECD partners, their long-term interests are similar. By contrast, the long-term interests of the industrialized North are quite different from those of the developing South.

It is therefore essential to recognize that the over-arching issue in Climate Change negotiations, as indeed in the entire UNCED process, is of a North/South nature. A realistic and clear perception of this fact would contribute towards the identification and attainment of realistic objectives for the Climate Convention. In our increasingly interdependent world, a clear understanding of the environment/development linkage in a North-South context, rather than rhetoric founded on purely environmental grounds, is likely to bring the chief actors closer to meaningful dialogue and, ultimately, to practical solutions of these important international problems.

A historical survey of the North/South issue illustrates that many international initiatives have been launched towards bridging the gap between North and South. These initiatives have rallied advocates under many names: from development cooperation to the New International Economic Order. Many of these initiatives arose at the United Nations level, led by eminent statesmen and scholars, and coordinated through regional, non-aligned, and Group of 77 discussions. The end result was a series of UN resolutions and discussions, as well as many valuable studies. Most of the developed countries accepted the rationale behind these initiatives, studies, resolutions, and declarations, but did not put them into practice. Today, the current UNCED process shows signs of a disturbingly similar character.

The world has reached a point where human progress holds out the promise of rectifying many of mankind's material, physical, and socio-economic problems. It is, therefore, even more tragic that at the same time the majority of mankind continues to suffer from illiteracy, disease, malnourishment, unemployment, poverty, and without sufficient hope.

3 The Economic Context of the Climate Negotiations

Today we have a recessionary global situation where productivity is rising faster than demand. Hence, like it or not, we are moving towards a managed world economy, despite lip service to the ideal of national and international free market mechanisms. The developed countries, particularly the G-7, are in the driving seat in this new management mode. It remains, however, in their interest to promote global collaboration and compromise, without insisting on terms that risk potential polarization and conflict, economic or otherwise.

At the same time, the end of the Cold War, and the growing request for economic assistance by the former Soviet Union and her allies in Eastern and Central Europe, have put new pressure on the limited pool of global savings. These new requests have further marginalized the developing countries, both politically and economically.

Just at a time when the former Soviet Union was declining, the Gulf War demonstrated the political will and technological capability, on the part of the West, to fill the vacuum that had been created. The Southern world is therefore faced with a situation in which an ascendant North is assertively trying to lay down the law about how the Southern World should behave if it is to be found worthy of being integrated into the world system.

Increasingly, new conditionalities are being imposed on the South by the North and by the international organizations and financial institutions that they dominate. These conditionalities include: responsive or democratic governance, respect for human rights, free market economies maximizing the role of the private sector, reduced defence expenditure, environmental responsibility, political empowerment of women in development, and action against narcotics. At the same time, on the institutional level, the North has been active in working towards strengthening and creating international institutions concerned with these issues. It began, for example, on the humanitarian side, with mechanisms for human rights, and has now been extended to a stronger and more centralized role for the Humanitarian Coordinator of the UN.

In the South it is felt that, individually, each of these issues deserves adequate attention at the national level. It is also self-evident that it is in the interest of all governments in the South to have broad-based support from a happy, healthy, and prosperous constituency. None the less, conditionalities imposed from without, whatever the motives, are bound to be resisted by sovereign states.

The basic objective of the South is to engage the North, in consultation with the South, in an integrated approach to issues of mutual concern. An integrated approach is preferable to any single issue getting isolated attention, or to priority being given selectively to certain targeted issues. All responsible leaders would agree that it is good to worry about such issues, but this concern should be grounded within a broader framework.

4 The Linkage Between Environment and Development

None of these linked issues can be resolved unless and until there is broad-based development in the South. Only such broad-based development can provide the foundation of international security. The Northern approach is to attack the symptoms, with a residual emphasis on poverty eradication. But the international community must insist on addressing the underlying causes for concern. Development, environmental protection, peace, and security are indivisible.

The environment is increasingly taking pride of place on the policy agenda of the North. But while the North wants to confine its focus to what its political leaders consider the key issues, the South has a different agenda. The developing countries want to focus on structural questions: on greater access to world markets, attracting more capital and investment, resolving the debt problem, achieving access to science and technology, and acquiring additional finances for technology development. These, many believe, are the first necessary steps towards sustainable development.

The North, which has already industrialized, wants the whole world to immediately work towards sustainable development. The South feels that one has to reach a basic level of economic development before going to the next stage of sustainable development. Can one realistically tell a poor peasant not to cut wood to make a small fire to feed his hungry child? Today, the two sides remain divided by a dichotomy of perspective and of intent.

5 Implications of the Climate Negotiations

Having outlined the context of the negotiations, we may now examine the main issues. A series of questions emerge: What are the implications of a potential Framework Convention on Climate Change? What is the relevance of the scientific context, and the scope of the Convention now being negotiated? How do the different actors perceive each others' intentions? What are the concrete formulations put forward by the two main sides, the North and the South, on the basic elements that would constitute the Convention? What are the principal stumbling blocks obstructing the path to agreement? How are we to proceed, and with what confidence-building steps? What are we likely to end up with and what lessons can we draw from this process?

The Framework Convention on Climate Change has potential implications for national governments at many levels. For the future of all mankind, it is obviously worthwhile to work together and to stabilize the present climate. In the process, we must minimize the risks of adverse change, and adapt wherever and whenever necessary. If the negotiations are successful, the convention could provide a concrete model for cooperation in a truly interdependent world. A successful convention could, perhaps for the first time, encourage concerted international cooperation at many levels on an issue which touches all nations and all peoples.

At the same time, when we discuss future limits on greenhouse gas (GHG) emissions, we are negotiating changes and potential limits on many economically important activities. These include: energy production and consumption, the utilization of land, added costs, and technological choices. In practice, we are negotiating on the pace and path of future economic development for all peoples and countries. Naturally, therefore, as we get further into this process, countries will become yet more cautious in the negotiations. National governments recognize increasingly that what they eventually agree to on climate change could have a greater impact on their future than any other single development since the industrial revolution. This responsibility is sometimes not sufficiently taken into account by many NGOs and public interest groups, as they grow impatient with the seemingly slow pace of the talks. For the sake of comparison we note that the Law of the Sea convention took more than 10 years to negotiate and has yet to receive sufficient ratifications.

6 Scientific Context of the Negotiations

The scientific context is also relevant at a number of levels. All the Governments involved in negotiating the Convention are conversant with the underlying scientific research — its preliminary results, recommendations, and uncertainties. Thus, the well-intended efforts of many NGOs (and indeed some INC Bureau members) to educate the negotiators further on certain issues misses the point about the real prerequisites to progress. The main building blocks for the Convention will come from the political assessment and the will of the concerned States to enter into genuine collaboration as equals.

On the issue of whether or not there will be winners or losers from climate change, I believe that not a single country is basing its negotiating strategy on the assessment that it might be a winner. Indeed, the scientific uncertainty, by itself, weighs against such a possibility.

With due respect to the work of the IPCC, it must be observed that the developing world is not completely at ease with the contributions of that organization. This is only natural, given the fact that the IPCC reflects the prevailing dominance of the Western countries in the field of scientific observation and study. Western scientists with access to adequate finances and communications are often on the same wavelength, due to frequent intercommunications and meetings. Scientists from developing countries lack these facilities and opportunities. Hence, we hope that the future work of the IPCC will reflect a more balanced composition.

Many people in the South are sceptical about some of the environmental studies pertaining to developing country emissions that have been prepared by various private organizations in the North. We note that, as soon as a developed world study appeared to show the potential historic linkage between sunspot cycles and global temperature, the results were quickly shot down so as not to interfere with the scientific premise behind the climate change negotiations. This, however, was unnecessary given the commitment by all States to work together towards considering the environmental future of mankind, without waiting for all scientific uncertainties to be resolved.

7 Policy Options and Strategies

If the main choice is seen between States choosing a "no regrets" strategy over a "wait and see" policy (Pachauri, 1992; this volume), it would be an injustice to all States to think that they see the two as equally valid options. Each State, by entering this negotiation, has demonstrated its preference for a no regrets policy. What is at issue is how to put it into practice, the time table involved, and how to pay for the necessary measures.

7.1 The role of the United States

Even after the conclusion of the fourth session of the INC on a Framework Convention on Climate Change (held in Geneva from 9-20 December 1991), the scope of the potential Convention is not yet clear. The main difficulty is the difference of approach between the USA and its OECD partners. The USA still wants a flexible instrument, since it is not yet willing to commit itself to limiting its emissions. The EC has proclaimed a shared objective of stabilizing CO_2 emissions at 1990 levels by the year 2000. The Community hopes to follow this with attempts to reduce emissions further after 2000.

At the third INC session in Nairobi, the US representative observed in his statement on "commitments on sources and sinks" that:

> *"In the United States we have examined a number of possible approaches to addressing the goal of reducing net greenhouse gas (GHG) emissions, including the proposal to stabilize CO_2 emissions at 1990 levels by the year 2000 and to further reduce these emissions by 20 percent by 2010. Quite frankly, Mr. Chairman, we have indications, based on our analysis completed to date, that the costs of achieving these targets may be quite high for the United States, given our high dependence on our domestic coal resources for basic energy supply."*

The US approach remained basically the same during the fourth INC meeting in Geneva. During this meeting, in its official statement on General Commitments, the US delegate observed:

> *"What kind of a convention are we working toward here? Although we have many different ideas, broadly speaking we can say they seem to point in either of two directions. Either toward a rather specific, quantitative, legally binding convention, with obligations that have rather major political and economic implications, or, on the other hand, toward a convention of a more general nature, a framework convention that is a flexible instrument establishing an evolutionary process for developing the most comprehensive and effective global response over the longer term... We favor the latter approach..."*

Because of these fundamental differences in approach, the objective and scope of the Convention remains unclear at this time. There is a feeling that the final US position will only be resolved during the US Presidential election process, although this would be far later in the day than many negotiators would like. Of course, without the US there can be no meaningful Convention, not just because the US is the world's most powerful economy, but also because it is the world's greatest emitter of CO_2.

The US may, in time, harmonize its positions with its EC partners, at least with regard to specific targets and timetables. However, it is clear that the main US concerns are to not deepen the effects of the recession on its people, and to not degrade its competitive position with respect to other developed countries. The US seems particularly concerned about the economically and politically assertive (some would call it aggressive) behavior of the EC on climate-related issues. In this particular respect, its concern is similar to that of the developing countries.

One important factor outside the INC conference rooms is the growing strength of the public environmental movement in the US and other developed countries. In fact, some suggest that the governmental responses of the developed countries are in large part due to the pressure exerted on them by the "green" lobbies among their own people. In this sense, the environmental lobby is one of the most important engines driving the industrial countries towards a meaningful Convention, encouraging these governments to make specific commitments on their own part. It may become increasingly difficult for industrial country governments to maintain the support of their own people unless they can balance citizen demands for environmental protection with other demands for continued economic growth.

7.2 North/South perceptual differences

Coming back to the basic North/South issue, the different perceptions of each side regarding the intentions of the other are quite relevant. Some in the industrial countries feel that the developing countries are not serious about taking on any meaningful emission limitations, now or in the future, and are using this issue to pressure the industrial countries for more financial assistance. In these quarters, the belief exists that the developing countries want to have almost unlimited access into the treasuries of the rich North, at the expense of the already hard-hit taxpayers of Northern countries.

In spite of these extreme fantasies of environmental blackmail by the South, most industrial countries realize that very little can be expected from the developing countries (either at this stage or subsequently) unless their development efforts are supported and sustained.

In the South, some individuals also question the motives of those in the North. These individuals believe that the Climate Convention, whatever its stated objectives, represents a clever plan by the North to exercise control over the development plans, strategies, and projects of the South. These individuals also often believe that those in the North want to dictate the direction and pace of growth in the South.

Each of these extreme perceptions tends to grossly exaggerate the existing realities. But, fantasies aside, the fact remains that the developing countries are in a relatively weaker position (because of their pre-existing debts). These countries must be extremely careful as they proceed in the negotiations because they have the most to lose if an inappropriate Convention is negotiated. A common stance would strengthen the negotiating position of the developing countries.

Of course, any Convention which is not acceptable to the majority of the developing countries, including the major actors amongst them, would be quite empty. It would be a setback to all other global negotiations, affecting both environment and development issues.

Fundamentally, the assessment of the industrialized countries appears to be that, since their own highly fossil fuel-based growth has used up so much of the global emissions bank, that the remaining reserve for future emissions cannot be equitably divided. Hence, developing countries should not be permitted to utilize fossil fuels to the same extent for their own economic progress, lest the global environment be irreparably harmed. For the sake of climate stabilization, and to permit the industrialized countries to maintain their higher emissions levels, the developing countries must increasingly base their development on non-fossil energy, leap-frogging to that state of development; or else these countries should accept slower economic growth until the next generation of energy sources becomes technologically feasible and financially attractive.

Thus, in the view of some in the North, climate stabilization is seen as a zero-sum-game, a concept taken from mathematics to strategic military games, and now ripe to enter the environmental debate. This theoretical analysis is emerging in the political arena just as the East/West rivalry has been overtaken by the North's new priority on global environmental concerns. If the developing countries aspire to reach the per capita emission levels now sustaining the high-consumption lifestyles of the industrial world, it can only be at the cost of the already high per capita emissions, and the still higher per capita emissions projected for the industrial countries in the future by the Energy and Industry Sub-group of the IPCC.

Of course, the conclusion of this Northern scientific and political assessment is not presented to the developing countries quite so baldly. It is put in a more diplomatic way. The industrialized countries have made many mistakes in their march to development (and dominance). It is in the interests of the developing countries themselves not to repeat the same mistakes. But meanwhile, the North still shows too little evidence of willingness to change its own behaviour.

8 Formulating an Alternative Vision

As the negotiations on a climate convention proceed, it is important to realize that the question of equity is uppermost in the minds of leaders from the developing countries. If one looks at the global environment as a bank belonging to mankind, the natural capital in this bank is the absorptive capacity of the atmosphere. One can readily see that, since the industrial revolution, the cumulative emissions that have arisen from activities of the relatively small populations of the industrialized countries have used up around 11 times as much of the shared capital as have the activities of the citizens of the developing countries in the same period. Even today, the activities of the 75% of mankind living in the South release only 25% of the total global emissions of greenhouse gases, while the 25% living in the North emit the remaining 75%.

Because of this continuing imbalance, the developing countries insist that the industrial countries admit their responsibility for creating the present problem. The current problem would never have arisen if the industrial countries of the North had developed along the path followed in the South, and maintained the per capita emission levels similar to those in developing countries. Thus, the industrial countries must accept their responsibility to compensate the developing countries for incremental costs that these countries will now incur, both for mitigation and for adaption measures. It has been estimated that the global environmental costs for which the industrialized countries are responsible to the developing countries are 3000 times as large as the current total debt of the developing nations.

Imagine a situation in which a person went to a bank, took out a million dollar loan, and refused to pay it back. The borrower then turned around to other small (and more prudent) depositors, insisting that the small depositors should make up the difference in order to preserve the liquidity of the bank on which all depended. This, in effect, is what the industrialized countries would like the developing countries to do.

The quality of life is different from person to person within countries, and among countries. But the value of life and the right of each individual to aspire to and work for a better life can not be denied nor legislated against through any international Convention.

We come back, therefore, to the complaint of the South that the North tends to concentrate on selective issues without taking into account matters of equal weight to other states. When human rights was higher in the industrial world's agenda than the environment is now, selective human rights problems were taken up. While many of them deserved the attention they received, the industrialized countries ignored the equally deserving human right to development, and the right to free movement and emigration in search of a better life. It would be unfortunate if the environment debate proceeded down a similarly unbalanced and ultimately unsuccessful track.

9 Elements of a Climate Convention

The essential elements of the Convention lie in defining through collective agreement the objectives of the convention, the principles on which it will be based, the commitments entered into by the industrial and developing States in terms of emissions, finances, and technology transfers; and the institutional mechanisms necessary to put these commitments into practice.

There is broad agreement on the objective of stabilizing greenhouse gas concentrations in the atmosphere at a level which would prevent rapid and dangerous changes in climate. There is broad agreement that such a level should be reached within a time sufficient to allow ecosystems to adapt smoothly to climate change, to ensure that food production is not disrupted, and to sustain economic development in an environmentally sound manner.

Some developing countries support the inclusion of the concept of equity within the definition of the Convention's objective. In this context, equity is understood to mean a convergence toward equal per capita emissions of GHG, as soon as possible. France also proposed the concept of convergence.

10 Building a New Consensus

The Chairman of the G-77 (the Chief Delegate from Ghana) presented the joint formulations in Table 1 to the INC plenary. He stressed that the issues of economic development and equity are vital for the conclusion of a Convention in which the developing countries can participate. Prospects for arriving at a consensus depend upon these factors being given the importance they deserve. The developing countries cannot be expected to accept any formulation that would institutionalize or help to perpetuate the present

Table 1. Principles of the G-77 and China on the climate convention

The Group of 77, which represented over 120 developing countries during the fourth INC session, jointly proposed the following formulations as the basis for negotiating this section:

Sovereignty
(a) The principle of the sovereignty of States shall be adhered to and strictly respected in all fields of international cooperation, including that for protection of the global climate.

Right to Development
(b) The right to development is an inalienable human right. All peoples have an equal right in matters relating to living standards. Economic development is the prerequisite for adopting measures to address climate change. The net emissions of developing countries must grow to meet their social and economic development needs.

Sustainable Development
(c) Protection of the global climate against human-induced change should proceed in an integrated manner with economic development in the light of the specific conditions of each country, without prejudice to the socio-economic development of developing countries. Measures to guard against climate change should be integrated into national development programmes taking into account that environmental standards valid for developed countries may have inappropriate and unwarranted social and economic costs in developing countries.

Equity and Common but Differentiated Responsibility
(d) All States have an obligation to protect the climate for the benefit of present and future generations of mankind on the basis of intragenerational as well as intergenerational equity. This obligation shall be carried out with different time frames for implementation in accordance with common but differentiated responsibilities and capabilities between developing and developed countries and taking fully into account that the largest part of emissions of greenhouse gases have been originating from developed countries and those countries have the main responsibility in combating climate change and the adverse effects thereof.

Special Circumstances
(e) The parties shall give full consideration to the specific needs and special circumstances of developing country parties, especially those developing countries which are particularly vulnerable to the adverse consequences of climate change and also those developing countries which would have to bear a disproportionate or abnormal burden under the Convention.

Precautionary Principles
(f) In order to achieve sustainable development in all countries and to meet the needs of present and future generations, precautionary measures to meet the climate challenge must anticipate, prevent, attack or minimize the causes of, and mitigate the adverse consequences of, environmental degradation that might result from climate change. Where there are threats of serious or irreversible damage, lack of full scientific certainty should not be used as a reason for postponing cost-effective measures to prevent such environmental degradation. The measures adopted should take into account different socio-economic contexts.

International Cooperation
(g) The need to improve the international economic environment for the developing countries and to promote their sustained economic development are prerequisites for enabling developing countries to participate effectively in the international efforts to protect the global environment including climate protection.

In this context, new, adequate and additional financial resources and transfers of and access to environmentally safe and sound technology on most favourable, concessional and preferential terms shall have to be channelled to developing countries in order to enable their full participation in global efforts for the protection of the climate.

States shall promote an open and balanced multilateral trading system. Except on the basis of a decision by the Conference of Parties no country or group of countries shall introduce barriers to trade on the basis of claims related to climate change.

economic disparity amongst the members of the international community.

At the December INC meeting in Geneva, the industrialized countries agreed that these formulations on Principles constituted a balanced and integrated basis for further negotiations. The US reiterated its contention that the principles could only be decided upon once the actual commitments were negotiated. The experienced and distinguished Vice Chairman from Mauritania observed at this point that one necessarily had to begin at some point.

Significantly, some industrialized countries conveyed their disquiet with the concept that all peoples have an equal right in matters relating to living standards. Even within their own countries, they pointed out, living standards differed. They feared that a legally binding treaty including such a formulation could commit them to providing an equal living standard for all, a goal which was both untenable and impractical.

Representatives of the G-77 explained that this was not their intention. Within their own countries, the industrial countries could hardly fault such a national formulation supporting the right of each citizen to strive for a better living standard. In the international context there could not be a double standard.

The heart of the potential Convention lies in the Commitments that will be entered into by State parties. The original Bureau working draft in Geneva was unnecessarily long. The thrust of the document was dominated by the concerns of the industrialized countries.

This draft was split into three sections. The first was on Commitments on Sources and Sinks, but subdivided into three subsections on: Common Commitments, Other Commitments, and Specific Commitments for Stabilization and Reduction of Emissions. The second section was on Commitments of Financial Resources and Technology Transfer, subdivided into two parts. The first subpart dealt with the Provision of Financial Resources, and the second with the Transfer of Technology. The third section concerned Commitments regarding the special circumstances and needs of those developing countries already on the cutting edge of potentially adverse climate change.

Many in the developing countries felt that the draft commitments presented at the beginning of the fourth session by the INC Bureau lacked clarity, did not flow from the Principles that had been proposed, and failed to differentiate the commitments and responsibilities of the industrial countries from those of the developing countries. By making the entire section so diffuse, the Bureau draft had, in effect, watered down the commitments expected from the industrial countries while increasing the commitments of the developing countries. Delegates from developing countries felt the same way about the Bureau drafts presented in the section on the mechanisms of implementation which flow from the commitments. There was a general feeling that an unfavourable result had been arrived at, prejudiced from the start toward the industrial countries of the North, at the expense of the developing countries of the South.

The Group of 77 again tried to formulate a common proposal on Commitments. Given the large number of developing countries and, in particular, the many which have still to reach a national position on Commitments, it was generally agreed to be impossible for all developing countries to arrive at a common or joint position. Several observers noted that some in the industrial countries were relieved that no such common position could be arrived at by the full Group of 77 in Geneva.

The Group of 77 came remarkably close to presenting a common position, thanks to the remarkable efforts and goodwill displayed by all members of the Group. Countries such as Saudi Arabia, Kuwait, Iran, Venezuela, Argentina, and the Small Island States stood out for their outstanding negotiating efforts, and for their willingness to accept common formulations for the sake of a joint declaration. The lack of intersessional consultations among the developing countries made it difficult to reach agreement on such complex issues. But the main obstacle was the limited time available to reach consensus within the over-rapid pace of official negotiations at Geneva.

Despite these difficulties, extremely valuable work was done. As time ran out to finalize a joint position, it was widely felt that all the work should not be lost, and that as much work as possible should be presented to the Conference for incorporation in the consolidated working draft. Such a document will form the basis for the next round of negotiations, which will take place in New York in February 1992.

Although a comprehensive position was not agreed upon in the deliberations of the developing countries at Geneva, a consensus had been reached that the section on Commitments should be concisely divided into only two sections. The first would outline the general Commitments. The second would present the specific Commitments in two subsections, differentiating between those applying to industrialized countries and those pertaining to developing countries.

In the first section on general Commitments, it was agreed that the developing countries would not at this time consent to submitting their national programmes to some international process of review and approval.

As to the second section, it was agreed that, until the OECD countries had come to a common position with the US on emission limitations, there would be no reason for developing countries to make proposals regarding industrialized country emissions in advance. If the developing countries had agreed to accept the EC policy measures, there would have been no pressure on the industrialized countries to do as much as they should — and as much as their own people wanted - to reduce the risks of rapid climate change. On the other hand, if the developing countries had pressed for 50% reductions in CO_2 emissions by industrial countries (that many leading scientific and political figures have suggested are necessary to stabilize the atmospheric concentration of

Table 2. Proposed commitments for states party to a climate convention: A 44 developing countries proposal

1. *In pursuance of the objective of this Convention, the Parties shall, in accordance with the Principles stipulated:*

(a) Develop and periodically update national inventories of sources and sinks of greenhouse gases;

(b) Cooperate in systematic observation, exchange of information and research and development on the potential effects of human activities on climate, and the environmental and socio-economic impact of climate change and related responses;

(c) Undertake, in accordance with their special conditions, measures within their national plans, priorities and programmes that contribute both to their economic development and to their efforts with regard to both sources and sinks in order to combat the adverse consequences of climate change;

(d) Participate as appropriate in the programmes of international organizations to meet the objective of the Convention;

(e) Promote public education and awareness of the environmental and socio-economic impacts of greenhouse gas emissions, the importance of the role of sinks and reservoirs and of climate change, and cooperate therein;

(f) Cooperate in the development and application of requisite technologies and practices, including in the improvement of energy efficiency, in the development of safe renewable energy sources, as well as in the protection and enhancement of sinks and reservoirs of greenhouse gases, and in the training of personnel;

(g) Conduct timely, nationally formulated and determined, project-related socio-economic and environmental impact assessment of actions proposed for the purpose of addressing climate change.

2. *Developed country Parties* shall, in accordance with the Objective and Principles stipulated, as immediate measures:

(a) (We are awaiting an agreed pronouncement from the developed countries on how they plan to address greenhouse gases in climate change, before we can formulate a response);

(b) Through their assessed contributions to the International Climate Fund, specific to this Convention, expeditiously mobilize and provide on a grant basis new, adequate and additional financial resources to meet the full incremental costs of developing country Parties to take measures provided for in this Convention; to cover the costs to developing countries ofadaptation and mitigation measures that may be needed as a result of the adverse consequences of climate change and, the direct and indirect social and economic costs to developing countries that may result from the implementation of the Convention. The International Climate Fund shall operate under the authority of the Conference of Parties and shallbe distinct and independent from other funds and international financial institutions;

(c) Transfer and provide assured access to technologies and know-how required for compliance with this Convention on concessional, preferential and most favourable terms, to developing country Parties;

(d) Ensure that the protection of intellectual property rights does not hinder the transfer of technology to developing countries in compliance with the Convention;

(e) Support developing country Parties in their efforts to create and develop their endogenous capabilities in scientific and technological research and development aimed at combating the adverse consequences of human-induced climate change.

In addition, the developed country Parties may undertake further nationally determined measures.

3. *Developing country Parties* shall, in accordance with the Objective and Principles stipulated, and in accordance with their national development plans, priorities, objectives and specific country conditions, consider taking feasible measures to address climate change, provided that the full incremental costs involved are met by the provision of new, adequate and additional financial resources from the developed country Parties.

Developing country Parties may, on a strictly voluntary basis, take additional nationally developed measures.

CO_2), they would have been labelled as unrealistic by the industrial countries.

The G-77 reached consensus on what the Commitments by the industrial countries should be in regard to their obligations to the developing countries under the Convention. These delegations also agreed that an International Climate Fund should be set up as a distinct and independent entity.

Regarding the Commitments of the developing countries, the representatives of the G-77 agreed that the burden of

Commitments under the Convention should fall first on the industrial countries. Those Commitments which the developing countries could consider entering into would be determined by themselves and completely dependent on prior delivery of their obligations to the developing countries by the industrial countries of the North. Some developing countries felt that, individually, they had already advanced to a stage where, on a strictly voluntary basis, they might be able to take on additional, nationally determined commitments.

As a result, before the time for the meeting expired, the following countries presented a proposal to redraft the basis for the entire section on Commitments:

Algeria, Bangladesh, Benin, Bhutan, Brazil, Cameroon, Chad, Chile, China, Colombia, Congo, Cuba, Dominican Republic, Gambia, Ghana, Guinea, Guinea Bissau, India, Indonesia, Kenya, Lesotho, Madagascar, Malawi, Malaysia, Mali, Mexico, Morocco, Mozambique, Namibia, Niger, Nigeria, Pakistan, Peru, Philippines, Sao Tome and Principe, Sierra Leone, Sudan, United Republic of Tanzania, Thailand, Tunisia, Uganda, Viet Nam, and Zimbabwe.

In view of the large number and broad distribution of countries that agreed to this proposal, and noting the inclusion of most of the large developing countries, this proposal had an important impact. It seems likely that this proposal will significantly influence the negotiations that follow. The text of this 44 developing countries proposal is given in Table 2.

Many other countries would have signed this proposal if there had been more time to discuss it. Others could have gone along with a G-77 consensus, but needed instructions from their capitals before signing a document which had not yet gained the final drafting approval of all the countries. (This is important because the Group works through the principle of complete consensus.) To a large extent, the process was more important than the product it produced. The developing countries had arrived at a common approach, even if time was lacking for a common formulation. This achievement puts them in a better position during the next round of negotiations.

The small Island States also presented a valuable proposal on the elements which should constitute the section on Commitments. It will provide source material, both for the developing countries to unify their position, and for the Conference in general to work towards a final agreement.

Norway came up with an interesting proposal regarding the setting up of a clearing-house mechanism, a variant of the tradeable emission permits concept. The objective of the clearing-house mechanism is to provide a cost-effective approach towards joint implementation of the Commitments to curb climate change. Industrial countries could offset some of their emission reduction obligations by funding emissions-limiting projects in developing countries. Under this proposal, the selection and implementation of these projects would be coordinated through a clearing house. Germany made a similar proposal on joint implementation of Commitments. Certainly these concepts merit further consideration and refinement, despite the objections that have already been voiced by some NGOs from industrial countries.

The Commitments that most developing countries want the industrial countries to enter into are quite clear. These should be set within the legally binding framework of the Convention; they are also defined in the Principles proposed by the G-77 countries. But it is not necessary, at this stage, to go into detail concerning the institutional mechanisms that will flow from the actual Commitments until the Commitments themselves are agreed.

None the less, attention should be given to the most important question of the funding mechanism that has to be set up under the Convention. Most of the industrial world would like to confine the funding mechanism for this Convention and for parallel Conventions (as well as for all other strictly environmental global concerns) to the recently established (and still experimental) Global Environmental Facility. This new "lending window," (often referred to by its acronym, GEF) is jointly administered by the World Bank, UNEP, and UNDP. But, to date, the GEF has been dominated by the World Bank. Many industrial countries feel that only the World Bank and, in particular, the GEF, have the expertise to carry out the proposed tasks while avoiding the cost of unnecessary administrative duplication.

For a variety of reasons, the developing countries have expressed both in these negotiations, in the negotiations on Biodiversity, and during the negotiations in the Preparatory Committee Meetings of UNCED, their clear preference for stand-alone funds to implement and support the Conventions on Climate Change and Biodiversity. These stand-alone funds are in addition to a separate "Green Fund" (a Fund for the Promotion of Sustainable Development) that will be needed to support the equally important but separate category of broad needs for sustainable development, including those specifically outlined under Agenda 21.

If the developing countries sign a Convention on Climate Change or Biodiversity with specific obligations, they cannot be expected to depend upon a separate institution not responsible to the Conference of the Parties for the reciprocal financial Commitments that go along with the Commitments and are clearly due to them. Furthermore, the World Bank — like the IMF and unlike UN Organizations — is responsive to a weighted system of voting and control. Equity does not count in the discussions of the World Bank or in its decision-making. As a consequence, applications to the Global Environment Facility are limited not only to four subject areas, but also by certain technical criteria.

During the GEF meeting held in Geneva from 3-6 December 1991, a joint statement on the future evolution of GEF was made by Pakistan on behalf of the developing countries attending the meeting. There were nine participants and seven observers from developing countries at this meeting. The joint statement noted that the future evolution of GEF may have impacts on the still-evolving debate in the UNCED fora regarding financial mechanisms. In several of these fora, preference has been expressed for separate funding mechanisms. The main operational issues included the criteria for assessing projects, the intersectoral targets for allocating resources, the methods of project selection, and the procedure for sanctioning funds for projects. The overarching issue for the developing countries attending the meeting was where the ultimate responsibility would lay for setting policies and for sanctioning expenditures.

The joint statement delivered by Pakistan on behalf of the developing countries to the GEF meeting outlined five essential elements that should be taken into account in deciding the future direction of the GEF. The first is an operational recognition within the GEF process that meeting the development needs and addressing the issue of poverty are central to the success of cooperative efforts to protect the global environment. GEF projects should, therefore, reflect developmental concerns.

Second, the fundamental responsibility for deciding the scope and case-by-case applicability of project criteria, intersectoral expenditure targets, and the sanctioning authority for release of GEF funds rests squarely and equally with the participating countries.

Third, since the number of participating countries is still relatively small at this stage, more meetings should be held to manage effectively the administrative responsibilities of the GEF.

Fourth, all countries should be invited and feel free to attend GEF sessions as observers — to encourage other countries to join as participants, to widen awareness of the GEF process, and for potential recipients to present and follow their own projects.

Fifth, while appreciating the present and competent Chairman, in the future the appointment of the Chairman of the Facility should be subject to the approval of the participating parties.

The statement of the developing countries concluded that GEF is a new hybrid situated in a *terra incognita* between the World Bank and more traditional UN bodies and organisations. The participating countries must give the GEF a certain coherence and sense of direction, if not structural adjustment, so that the facility merits wider support and is therefore sustainable.

The joint presentation of the developing countries was considered balanced and constructive by the industrial country delegations attending the December meeting of the GEF. A number of industrial countries supported some of the ideas contained in this statement, and there was general agreement that the administration of the GEF had to become more transparent, efficient, and certainly more responsive to the wishes of the participants, if it were to be extended beyond its pilot stage.

As a compromise between the position on GEF of the developing countries, and that of most industrial countries, some have suggested that while separate Funds are set up under the supervision and control of the Conference of the Parties, a functional link with GEF is possible. The parties would set the criteria and release funds. The financial assets could be held as distinct "windows" under GEF auspices. GEF could also be used in the context of supervised project selection and execution. This might satisfy both sides to some extent, but not as things now stand.

The issue of financial resources is crucial to the success of the entire UNCED process. While the additional financial resources required by the developing countries can not estimated be with precision at this time, various proposals regarding the sources of potential funding are being examined in an academic way. These proposals include debt relief, funding linked to Special Drawing Rights (a form which the US would probably not prefer), charges for use of the commons, a private, environmentally sound investment fund, earth stamps (an interesting recent NGO proposal), allocative measures with a revenue potential, tradeable permits, and a non-renewable energy or "carbon tax".

A variety of possibilities have been raised in the debate over funding mechanisms. Should funding needs be met on a case-by-case basis, or through the establishment of a single "sourcing" fund? Should funding be voluntary or mandatory, given the substantial increases in the magnitude of resource flows required to meet all the principal requirements of the developing countries? While voluntary funding would always be welcome, the size of the task requires that a major portion must come from a negotiated burden-sharing arrangement.

In the current progress of the negotiations, the various parties are now in a position to identify the "hot spots" (i.e. the principal obstacles to agreement) in the negotiating process. These hot spots are now more clearly defined in the proposals on the table — either from individual countries, or from groups of countries. They include the differences in approach, already described, between the USA and its OECD allies. Then there are the differences of degree between the OECD member countries themselves. The concept of "pledge and review" also carries different meanings for different countries, both among the developed countries who have proposed it and the developing countries. In fact, this concept continues to raise its head at every INC meeting, to be struck down by some participant. Then, phoenix-like, it is presented again in another incarnation at the next meeting.

Serious differences remain between the developing countries, on the one hand, and the developed countries on the other. It is difficult for most of the developing countries to accept the proposition that they should enter into commitments which would adversely bind them, either now or later on, for the sake of a problem caused by the developed countries — who neither wish to equitably share the remaining emission reserves in the atmosphere, nor to share (even in a small way) the benefits and resources that they have built up by plundering the world's greenhouse gas reservoir capacity.

Among the developing countries themselves, there are different terrains, differing resource bases, a variety of emission records and potentials, and a spectrum of population sizes. These differences make the task of arriving at a common negotiating position rather difficult — but not impossible, as recent experience has shown.

By the same token, many industrialized countries would be happy to see the developing countries split up into various groups and blocs, with which they could negotiate trade-offs from a position of relatively greater strength. Some countries would be approached in the context of potential free trade agreements, or other trade and commercial linkages. Other countries, more at risk from climate change, could be promised some compensatory financing from large and contiguous industrial countries. The fossil fuel resource-rich states would be warned of the dangers of a global recession, and the possibility that (among their customers) increasing resources would be devoted to alternative and renewable energy sources.

Many levers of influence are available to the industrial world. However, the basic premise of each developing country must be that while, in the short term, each one of them can possibly obtain a favourable tradeoff from the industrial countries, in the long run their interests and those of their people can be better assured and protected through collective strength.

11 Orchestrating the Process of Agreement

How, then, are we now to proceed, and with what confidence-building steps? By the time this is published, the fifth INC session in New York in February 1992 will be over. The high-level INC meeting proposed for April 1992 will be ready to take place. Therefore, in part, these modest suggestions may have already been overtaken by events.

I would expect that at the next INC session, the Bureau will have put together a consolidated working document, which clearly outlines the areas of agreement and divergence. The negotiating States can then get down to even more serious work. The OECD countries can try to harmonize their position further. The developing countries, through the G-77, will also undertake the same exercise. The concepts are now clear; the treaty language options are on the table.

Once the main groups and the principal actors have more clearly defined their positions, contact groups with a clear mandate from their constituencies should be set up to facilitate, through intensive negotiations, the task of reaching a consensus.

While the various positions of different States will still make a consensus difficult to arrive at, there are some general and specific factors which have a bearing on how successful the negotiations will be, and what we are likely to end up with.

The main factor will remain the sincerity of the North in its claims that environmental concerns have truly made us an interdependent world.

Then there is the tremendous concern and goodwill which has been developed on the whole environmental issue, among the peoples of both industrial and developing countries, which surely must have an eventual impact on the course of global negotiations that seek to integrate environmental and developmental concerns.

12 Debt Relief and Environmental Practice

Furthermore, a number of industrial and developing countries have developed a tradition of putting forward very constructive and practical ideas in the INC process. The Nordic countries, as usual, stand out for the comprehensiveness and idealism of their approach.

Since the issue of new, adequate, and additional financial flows is central to the position of developing countries for entering into a new global pact on environmental concerns, the current economic position of these countries and the evolving response of the industrial countries has to be assessed. At this moment, with regard to resource transfers, the fact stands out that development cooperation has failed. Resource transfers from the North to the South have not only remained stagnant but in the 1980s, because of accumulated debt and declining terms of trade, the net resource transfer has been reversed. Now the overall position is negative, with the developing countries giving more financial resources to the industrial countries than they receive. Such a state of affairs obviously has a very negative impact on the already slow development of the poor countries, and cannot help but affect their sincere efforts towards sustainable development.

The total debt of the developing countries is approximately US $1.28 trillion. Whatever the composition of the debt, the fact remains that debt write-off and relief would free all those resources in the developing countries for more economic, environmentally friendly, and increasingly sustainable development. The fact that so many proposals have been floated to deal with the debt problem demonstrates the general recognition that the problem is far from resolution and more needs to be done. However, the action taken recently on Poland and Egypt brings out the fact that the developed countries only move on those debt cases where they have the political will, and refuse to accept those debt write-off cases as precedents for action elsewhere.

Proposals that link debt relief to environmental protection, while idealistic and worthwhile in themselves, have unnecessarily restricted themselves to stand-alone debt-for-nature swaps. What is needed is specific action on debt relief, freeing the already limited resources of the South for sustainable development. If debt is broadly divided into four categories — government debt owed to other governments, government debt owed to commercial banks, debt owed to the international financial institutions, and sovereign debt owed by governments to institutions and individuals through international capital markets — a priority order of action can be established. Debt write-off in the order of priority listed is feasible, leaving aside only the last category.

A balanced and early conclusion of the Uruguay Round would help the developing countries to help themselves. It has been estimated that a truly free global trade regime, allowing the developing countries to export without restriction, would earn them eventually ten times as much as is now received through existing Official Development Assistance. The present tendency, however, is to move towards stronger economic groupings in North America, the EC, and the Pacific Rim, which would further marginalise the developing countries economically.

In the international context, a significant policy statement was issued by the OECD Ministers Meeting on Environment and Development of 3rd December 1991. The statement recognizes sustainable development as a shared and common objective. The statement pledged support to ensure additional (but unquantified) external resources for developing countries so that achieving this objective could be assured. It was made clear that the OECD countries expected the developing countries to provide the major part of the resources required for this effort, and that OECD assistance was dependent on a number of conditionalities. These included good governance, the effective integration of economic, social and environmental policies, and participatory approaches which involve accountability and broad-based involvement in the formulation and implementation of government policies. Market-based economies were stressed. Slowing down population growth was strongly recommended. In return, the OECD countries offered to help the developing countries to strengthen their scientific, technological, and administrative capacities.

At this meeting the OECD — including countries (such as the US, France, and Germany) with significant power production from nuclear energy — bravely tackled the "wild card" of the entire environmental debate. The OECD leadership called for the operation of nuclear power-generating facilities only under the highest available standards of safety. Yet, even with such assurances, it remains to be seen how widely this capital-intensive technology (which many believe has the capacity to be "environmentally neutral" with regard to global warming) will be shared.

At the EC Maastricht summit on 9th and 10th December 1991, it was significant that the industrial countries have only accepted the premise that sustainable development has a legitimate claim for additionality on their resources in a more limited and self-serving context. The Agreement stipulated that additional economic assistance would be given to EC countries, such as Spain and Portugal, in order to permit them to merge developmental and environmental concerns.

The evolving response of the industrial countries remains selective and with an ideological rationale far more sophisticated than that applied by the developing countries. The industrialized countries have already launched an integrated and comprehensive defensive position against the demands of the developing countries. It is this defensive position that underlies many of their new conditionalities on even present financial resources. The industrial countries seem to be saying:

> *"If the developing countries want more resources, we can help them to find them. All they have to do is to restructure their internal economies. They must make more use of market mechanisms to allocate resources (if market mechanisms and structural adjustment reduce the social entitlements of the poor and the weak, especially the children, that is unfortunate, but some costs must be incurred for long-term progress.) The developing countries will have to be more efficient in managing available resources and government expenditures."*

This leads to the culminating argument of the industrial countries. They seem to say:

> *"If you lack the vision to see how this can be done we can help you to do so on the ground. Begin by cutting your military expenditures; that will free a significant percentage of what you could get from our giving you debt relief."*

In principle this sounds fine. However, there are no supporting proposals from industrial countries for international action or machinery to resolve the long-standing regional issues that underlie much of this military expenditure. Security also has a price, and when inadequate attention is paid to it, the cost of rectifying that mistake may be still greater than is realized at first — as Kuwait and the international community have found to their regret.

It would appear, at this stage, that the expectations (or rather the justified demands) of the developing countries — in terms of new, adequate and additional financial flows and technology transfers — are regarded as completely unrealistic by the major industrial countries.

The furthest the industrial countries are willing to go at this stage is to help with some funding and with the provision of some relevant technology. This help depends on the willingness of developing countries to accept a process in which all countries prepare and present an inventory of their emissions and stabilization strategies to an international forum.

The industrial countries feel that, in such a system, any developing country would be as free to criticize (for instance) some aspect of US fiscal policy dealing with energy pricing as the US would be to critique a similar policy in some developing country. However, the developing countries know that in practice, if there is any dispute, they have no leverage against the major industrial countries. On the other hand, the industrial countries could use this process to impose strict conditionalities against weaker developing countries.

At best, the developing countries might accept a situation in which they would permit a review of those climate-related environmental projects for which all additional costs had been paid by the developed countries.

The current response from the industrial countries remains disappointing, and the prognosis for change is no better.

In this particular situation, another important factor is the question of what kind of a Convention the US would like to live with. Thus, it is probable that only a hollow Framework Convention could be agreed which is without legally binding commitments, at least in regard to emissions reductions from either developed or developing countries. In such a case, it would be left for each country to decide on its own targets, to declare the commitment to those targets either within or outside the framework of the Convention, and then to review its performance itself. Subsequent Protocols to the Framework Convention might provide the mechanism and opportunity for legally binding commitments, once sufficient political will had been generated.

If this were to happen should we be disappointed? I do not believe so. I repeat my belief that in such negotiations, one is tempted to overlook the notion that the process is as important, if not more so at times, than the product. The countries of the world have now met together to decide how to save the world from the potential risks of rapid climate change. That by itself is a great achievement. Countries now understand each others' positions and specific problems; a foundation has been laid for operational activism within and outside the framework of the Convention, both in national terms and in terms of international cooperation.

13 Lessons for the Future

What lessons can we draw from this ongoing process and where do we go from here? I would confine myself to two observations. The developed world must stop dividing environmental action into that which has supposedly local implications and that which has, in the opinion of the developed world, global climate, or global environmental, impact and implications.

The fact of the matter is that, in addressing such problems, we have to begin at the individual, at the community, and at the local level. Only then can we move on to reach the national level, let alone the global level. In this way, we can change societal habits and attitude, encourage positive environmental action across the board, and reduce the risks of rapid climate change. The artificial distinctions between local and global impacts must now be abandoned.

Irrespective of the conclusion of Conventions dealing with Climate Change and Biodiversity, and irrespective also of the future directions of the entire international environmental debate, the developing countries are as committed as the industrial countries are to giving the greatest priority to sustainable development for their peoples. The importance of clean air, drinkable water, and the abatement of pollution from all sources, is being given increasing importance by them in terms of planning and in terms of action on the ground. However, the scope and pace of these national efforts depend on the availability of scarce resources that face many competing demands. Poverty constitutes the greatest threat to the environment.

To give but one example, clean drinking water is a priority and should be a basic human entitlement. However, for a large portion of mankind the main problem is a lack of access to assured and adequate sources and supplies of water. That is why, in respect to the Global Environmental Facility, the developing countries wonder why water scarcity and land use problems are given so low a priority in terms of project criteria and the target allocation of resources. Ignoring the issue of water supply indicates a lack of foresight in the still-evolving perspective of the industrial countries, and suggests a spurious division between national and global consequences for the environment.

14 The Interdependent World

At this juncture, the industrialized countries must realize that the rules of the game in the North/South dialogue have changed. In the past, the concept of an interdependent world was mainly a humanitarian ideal expressed in terms of rhetoric rather than practice. The Southern world was seen in terms of "lifeboat" and "triage" theories where the weak might have to be left behind.

It is now clear that, at least in terms of the global atmosphere and environment, we are all in the same life boat. If developing countries are not given the trade opportunities, debt relief, credit facilities, financial assistance and the technology flows that they require for their development, we will all eventually pay the price.

Each person has an equal right to the global reservoir — of both oxygen and emission banks. It is true that the developing countries must try hard to curb their populations. However, while the developed world continues to stress the supremacy of the free-market capitalist system, with its emphasis on the free and unhindered movement of capital, goods, services, and technology across borders, it has forgotten about the other equally important component of that system — the free movement of labour.

Since 1492, Europe has expanded beyond its borders and across the world — into vast areas and continents which either belonged to the peoples of the South or lay within their natural path of expansion. Eventually, in the final trade-off on the environment, development, resource transfers, and population, the right to emigration abroad must also be included to some degree. The industrial countries should seriously consider compensating the developing countries through vastly increased immigration opportunities. This would not only be equitable, but would lead to a truly interdependent world.

CHAPTER 23

Shaping Institutions to Build New Partnerships: Lessons from the Past and a Vision for the Future

William A. Nitze, Alan S. Miller and Peter H. Sand

Editor's Introduction

Pollutant emissions from human activities are changing the composition and behaviour of the atmosphere. In the process, they threaten to disrupt the Earth systems upon which all life depends. To minimize these risks and to limit the damages from changes which are now unavoidable, it will be necessary to re-tool existing institutions to meet the new and emerging challenges.

William Nitze, Alan S. Miller and Peter H. Sand begin with an assessment of the major trends that have contributed to the current conditions. They observe that the traditional development path followed historically by the advanced industrial countries has produced the great bulk of atmospheric pollutant emissions. However, the continuing trend toward dematerialization of these economies — a process that leads to continuous reductions in the quantity of material required to produce a unit of output — has also been spreading steadily in recent years. This trend in engineering design is reinforced by the management theories of statistical quality control and the principles of continuous improvement. It is strengthened further by the use of market-oriented policy mechanisms and the movement toward full social cost pricing.

At the same time, the ability of the world community to tackle environmental problems has also evolved. Nitze, Miller and Sand regard the processes leading to the Montreal Protocol on Substances that Deplete the Ozone Layer and the London Guidelines on Toxic Substances as demonstrations that cooperative action between scientists, governments, and non-governmental organizations can be practical and fruitful. In both cases, the combined research of devoted scientists in many countries provided a shared information base, allowing industry representatives and members of environmental groups to make common cause, joining with official government delegations to evolve an efficient, flexible regime.

Other international dialogues also provide useful precedents. The work of the International Atomic Energy Agency (IAEA) shows that a UN agency can operate in an impartial manner, without partisan political pressure, to monitor and enforce collective supervision of a complex technology distributed around the globe. The formation of the Global Change System for Analysis, Research and Training (START) under the auspices of the International Council of Scientific Unions (ICSU) demonstrates the potential benefits of experts and officials in many countries working together to implement cooperative regional strategies for addressing global problems.

Nitze, Miller and Sand have an unmatched record of successful experience in negotiating international agreements on complex environmental issues. Their work, taken together, shows that good lawyers can play a critical and constructive role in world affairs. Bill Nitze was the former US Deputy Assistant Secretary of State for Environment, Health and Natural Resources; he is now President of the Alliance to Save Energy. In both roles, he has been a key protagonist in the global warming negotiations. Alan S. Miller, Executive Director of the Center for Global Change at the University of Maryland, was a major figure in US efforts to bring industry and environmental groups together in a productive dialogue leading to the resolution of the stratospheric ozone depletion problem. Peter H. Sand, formerly Chief Counsel for the UN Economic Commission for Europe and now Senior Legal Advisor to the UN Conference on Environment and Development, made major intellectual and practical contributions to the negotiations and implementation of the UN/ECE Treaty on Long-Range Transboundary Air Pollution (LRTAP).

What gives this chapter depth is the authors' clear vision of a cooperative approach to the climate problem. Nitze et al. view the Conference of the Parties on a Climate Convention as a vehicle for strengthening existing institutions and, if necessary, forming new ones. They envision a convention that incorporates short-term global goals for stabilization of atmospheric concentrations of greenhouse gases, leading to a comprehensive protocol for managing the reductions of greenhouse gas emissions. They note that achieving this goal will involve the development of national strategies and action plans — a process requiring unprecedented cooperation among governmental and non-governmental organizations at the national, regional, and international levels. In the view of these hard-nosed international lawyers, this process, if carefully nurtured and prudently managed, (1) could stimulate necessary investments in energy efficiency, (2) catalyse large-scale resource transfers to support the development and implementation of new technologies and (3) improve the performance and the allocation of resources among existing institutions. If diligence, discipline and the spirit of cooperation hold out, the process might also foster the development of benign new institutional practices, based on the principles of decentralized decision making and global collaboration.

- I. M. M.

1 The Challenge of Sustainable Development: the Need for New Partnerships

Mankind has entered a new era in which human activities threaten to disrupt the Earth systems upon which we depend. To protect those systems from irrevocable damage, we must reshape and tie together a broad array of existing institutions to build new partnerships. Nowhere is this more true than with respect to human emissions of CO_2 and other greenhouse gases.

Important as it is, however, the greenhouse problem should be viewed as part of a broader global challenge. Growing numbers of people all over the world have been using up the Earth's store of natural resources and environmental amenities at unsustainable rates, as they pursue a better livelihood for themselves and their children. We must now break the link between the quest of people the world over for a better life and the environmental degradation that has been a consequence of that quest. In order to achieve what the World Commission on Environment and Development (1987) called "sustainable development," we must preserve the ability of future generations to meet their needs and aspirations as we struggle to meet our own.

Indeed, sustainable development means development which does not diminish the prospects for future generations to enjoy a quality of life at least as good as our own. Development, itself, is inevitable; in any possible scenario, the combination of population growth and efforts to increase living standards will lead to increased consumption of water, minerals, plants and other natural resources, as well as associated changes in land use. We must, however, curb the exponentially increasing depletion of natural resources and environmental pollution, if we are to avoid a collapse in the natural systems which support human life. Curbing these trends will require a coordinated international strategy to reduce population growth, develop and deploy new technologies and policy mechanisms, and change patterns of production and consumption in all countries. Only through such a strategy can we achieve the more stable relationship — between human activities and the rest of the natural world — that is the essence of sustainable development.

1.1 The transition to a less resource intensive economy

The wealthy industrialized nations are largely responsible for creating this global challenge — since they developed the exploitative attitudes towards nature, the development patterns, and the industrial technologies that have now spread across the globe. These countries have a corresponding obligation to take the lead in meeting the challenge of sustainable development. Fortunately, the wealthy countries are in the process of creating many of the technologies, industrial management techniques, policy instruments, and economic concepts necessary to achieve sustainable development.

This process is associated with an ongoing reduction in the amount of energy and other material inputs consumed in producing a given amount of value, a phenomenon sometimes called "dematerialization" (Herman et al, 1989). Two reinforcing trends are contributing to dematerialization in the advanced industrial countries. The first is a decrease in the share of GDP represented by production of more energy- and natural resource-intensive products, such as steel, nonferrous metals, and chemicals; and a corresponding increase in the share of information-intensive products, such as pharmaceuticals, financial services, and telecommunications.

The second is a reduction energy and materials use by energy- and materials-intensive sectors through optimization of existing processes, process improvements, and deployment of wholly new methods of manufacture. For example, the yearly energy improvements in the process that produces ethylene from ethane have averaged 3% since 1960, resulting in a cumulative 60% decline in the amount of energy required (Ross and Steinmeyer, 1990).

Both of these trends should be replicated in developing countries as part of the coordinated international strategy. The greatest short-term benefits will come from bringing the energy supply and energy-intensive manufacturing sectors in developing countries up to (or above) current Western standards. Chinese coal burning power stations have an average combustion efficiency of less than 20%; US coal-fired plants have an average efficiency of approximately

36%. Similarly, Egyptian refineries consume approximately 20% of their crude oil input in process energy; US refineries consume about 8%. Upgrading the technical performance of these energy conversion facilities will allow the operators to squeeze a greater quantity of product from each unit of resource input.

Simultaneous efforts must be made to ensure that new facilities constructed in developing countries incorporate the most up-to-date technology. A number of developing countries have demonstrated that their workforces are capable of producing internationally competitive products. The Ford automobile assembly plant in Hermosillo, Mexico, for example, had the best assembly plant quality in an international sample surveyed by MIT researchers — better than the best Japanese and North American plants (Womack et al., 1990). The rapidly-growing Bangladeshi textile industry has had such success penetrating Western markets that it has provoked trade sanctions. The key constraint is providing the capital to cover the incremental costs. To achieve similarly rapid gains in efficiency, OECD countries and international lending institutions will initially have to provide much of this capital.

Finally, developing countries will have to make their own transition to less material-intensive economies. This will take more time, because of the need to upgrade their workforces and gain greater access to international markets. The most important step the OECD countries can take to aid this process is to accelerate the transition of their own economies to higher value-added production. Japan has led the way in this effort; the US and Europe will have to follow.

1.2 The rise of lean manufacturing: implications for emissions reduction

This revolution is in turn being accelerated by the rise of "lean manufacturing" based on the techniques of statistical quality control and the principle of continuous improvement. A basic tenet of statistical quality control is that a robust product is created by "designing in" quality through a manufacturing process in which variations are reduced to a minimum rather than by inspecting for defects at the end of the line. Developed by American statistician W. Edwards Deming during the Second World War, statistical quality control and continuous improvement have been adopted throughout Japanese industry and will soon be adopted by every other manufacturing firm that hopes to be competitive on world markets.

Statistical quality control and continuous improvement have far-reaching applications to reducing greenhouse gas emissions. They suggest that the best way to prevent pollution is by creating a process in which harmful emissions have been "designed out" of conversion processes, not by catching effluents at the end of a pipe or smokestack. Just as companies have adopted "zero defects" as a goal towards which they strive through continuous improvement, they should adopt the goal of "zero discharges" to be pursued in the same way (Kleiner, 1991).

Pollution prevention can have benefits for both the company and the environment. As described by Schwartz et al. (1992, this volume), individual firms such as 3M and Dow Chemical have already made significant progress in implementing this pollution prevention approach (Kleiner, 1991). The energy supply sector has made similar progress in reducing emissions, largely through more efficient combustion processes and recycling of heat (e.g. using cogeneration technologies or techniques for "cascading" heat through various stages of the manufacturing process) that would otherwise be lost to the atmosphere. Combined-cycle electricity generating systems using modern gas turbines can now achieve conversion efficiencies of 50% or more, a substantial improvement over the 30-40% efficiencies achieved by the best conventional plants.

1.3 Using policy instruments to harness the market

The OECD countries are also developing a variety of policy tools for encouraging energy efficiency improvements, reducing pollution, and encouraging economic development that will reinforce the trends described above. These run the gamut from planning techniques to taxes to tradeable permits, promoted by the US as a device for using market forces to reduce the cost of achieving environmental goals (TFCACC, 1991). Under a tradeable permits scheme, each unit (country, state or province, individual company or power plant) is allocated entitlements to emit a certain quantity of a pollutant (such as sulphur dioxide or carbon dioxide) during a specific time period. If it does not plan to release the full amount allowed, that unit may sell its excess entitlements to another unit which desires to emit more.

The US has taken the lead in developing demand side management (DSM) — a system whereby publicly regulated utilities meet customer demands for electricity by improving the efficiency of their customers' facilities, rather than increasing energy supply. Such utilities are permitted to recoup part of the value of the energy savings from their DSM investments, as an incentive for making those investments. The US also innovated Integrated Resource Planning (IRP), in which regulated utilities are required (or given incentives) to find the most cost-effective means for providing energy services, before they can obtain approval to build new plants or acquire new energy supplies. Both DSM and IRP can be applied, as well, in non-regulated contexts.

The spread of DSM and IRP programmes in California, New England and other states has already produced significant increases in energy efficiency investments, demonstrating that investment in energy efficiency is usually the lowest cost option. The Europeans and Japanese have used high energy taxes and efficiency regulations to encourage more efficient use of energy and the development of new energy technologies. They have also taken the lead in requiring manufacturers to assume responsibility for the full life cycle of their products, including disposition at the end of their useful lives. A German requirement that German automobile manufacturers recycle their old cars, for example, is already causing the manufacturers to redesign their new cars for easier disassembly and recycling.

1.4 Full social cost pricing

All of these policy initiatives represent initial steps towards the development of a new economic accounting system based on the concept of full social cost pricing. The current system of balance-sheet- and income-based accounting, in wide use around the world, does not incorporate environmental costs and benefits into its calculations of national assets or income. But these are genuine costs. Thus it is possible for a country to appear to achieve high rates of economic growth by depleting its natural resources and poisoning its air, oil and water. This is precisely what has happened in Eastern Europe since the Second World War. Even in the United States, it has been estimated that, if US GNP between 1950 and 1986 is adjusted for the remediation costs of environmental damage, the "adjusted" annual GNP per capita exhibits hardly any growth for that period (Daly and Cobb, 1989). In many Eastern European and developing countries, "adjusted" per capita GNP has actually fallen over the same period.

We must develop internationally accepted measures to incorporate such adjustments into national income accounts — by reflecting depletion or enhancement of natural resources, along with the costs of incremental pollution and cleanup. If the long-term environmental costs (of deforestation or emission of various industrial pollutants) and the long-term environmental benefits (of reforestation and pollution prevention) were incorporated into official accounts of national assets and income, politicians and other decision makers would have an incentive to reduce the costs and increase those benefits.

But the greatest obstacle to incorporating those costs and benefits is the difficulty in agreeing on their respective values. There are a number of techniques for arriving at those values, including polling people on their willingness to pay, establishing a market in permits to deplete or to pollute, or aggregating the estimated damage costs of specific impacts. Initially we will probably have to make very conservative estimates of costs and benefits that can be used until better ones have been agreed upon. This approach has been taken by public utility commissions in several US States in incorporating environmental externalities into the prices of energy supply options. For example, in a recently initiated IRP proposal in Massachusetts, the state Department of Public Utilities and the largest power company (Massachusetts Electric) have developed separate monetary values for eight different air emissions, to assess the total social cost of energy resources. Once the two groups reach agreement, these values will be used in formulating the energy plan ultimately approved by the commission.

1.5 The critical challenge: getting people to use the tools already available

Much progress has recently been made in industrialized countries toward the development of the tools described in the preceding sections. Developing countries also possess a number of valuable technologies and approaches, particularly in agriculture, forestry, traditional medicine and other non-industrial sectors of their economies. There is ongoing work to be done in refining and integrating these tools, but our greatest task is to motivate, train and organize people to use them. This is a challenge in the US and other rich countries, where many people fear and resist the changes required. It is a much greater challenge in the developing world, where the infrastructure for making those changes often does not exist.

The magnitude of the challenge will be in part determined by population growth rates, particularly in developing countries. The World Bank currently estimates that population will stabilize at approximately 11.5 billion people, shortly after 2100 (Bulato et al., 1990). The task of increasing per capita incomes without further environmental degradation will be hard enough if world population stabilizes at 10 billion people. It will be much harder, if not impossible, if world population stabilizes at 14 billion. John Holdren estimates that "(one) third of current growth in world use of industrial energy forms continues to come directly from population growth, and much of the energy growth in the decades immediately ahead resides in the already huge and still rapidly growing populations of the less developed countries, where energy use per person today is very low and development prospects hinge on its getting larger (Holdren, 1991)." Therefore it is important that the US and other countries support the United Nations Fund for Population Activities and other international family planning initiatives as part of an overall sustainable development programme.

To meet the challenge, institutions at all levels will have to reshape themselves as agents for change. This will in turn require them to form new partnerships with a wide range of public and private sector organizations that are essential to helping build that infrastructure and providing the technology, knowledge and other resources necessary to carry out the required changes.

2 Reshaping Existing Institutions

2.1 The existing institutional mix

We already have a broad array of public and private institutions dedicated to some aspect of sustainable development. On the development side, we have a collection of organizations belonging to or affiliated with the United Nations:

the World Bank;
the United Nations Development Programme (UNDP);
the Food and Agriculture Organization of the United Nations (FAO);
the United Nations Industrial Development Organization (UNIDO);

the UN regional economic commissions.

These are supplemented by:

the regional development banks;
the Organization for Economic Cooperation and Development (OECD);
overseas development assistance organizations established by individual OECD governments, such as the US Agency for International Development (AID).

On the environment side, we have:

the United Nations Environment Programme (UNEP);
International Union for the Conservation of Nature and Natural Resources (IUCN);
government ministries, agencies and organizations devoted to environmental protection and conservation.

These are in turn supplemented by:

the European Community;
the Association of Southeast Asian Nations (ASEAN);
the Organization of African Unity (OAU);
the Organization of American States (OAS);
the United Nations regional commissions;
other regional organizations that have neither a strictly developmental nor environmental focus but are involved in both.

Finally, there are a host of non-governmental organizations ranging from scientific associations to environmental groups focusing on one or more elements of the environment and development agenda.

The challenge lies not in adding to this already rich tapestry, but in refocusing the missions of many of these institutions and better allocating and coordinating responsibilities among them.

2.2 *Rethinking the role of development assistance*

This is particularly true for the major international and national development institutions such as the World Bank, UNDP and overseas development assistance agencies. Over the last 40 years, these institutions have financed a series of projects in the developing world that have arguably done more harm than good. Certainly many of the recipient countries, particularly in Africa and Latin America, have had disappointing economic records. Even where assistance generated a long-term increase in economic activity, the environmental costs may ultimately outweigh the economic benefits. The Aswan Dam, for example, is now seen as threatening the very people it was designed to serve.

The economic and social costs of the Aswan Dam include making the Egyptian economy dependent on an electricity source derived from unreliable flows from the Blue Nile, depriving downstream riparian and delta areas of the silt required to preserve agricultural productivity and avoid subsidence, and displacement of or damage to historical monuments. It is arguable that these costs already outweigh the benefits from the dam. In time it may be necessary to destroy the dam if the Egyptian economy and society are to survive.

The only clear beneficiaries from many of the internationally financed mega-projects have been the elites in recipient countries through whose hands the financial assistance flowed, the large and well-compensated bureaucracies administering the aid, and the myriad of consultants, contractors, and suppliers actually receiving and spending it. By contrast, those countries in East Asia and elsewhere that have received the least development assistance have turned in the best economic performance.

The reasons for this contrast are not hard to find. Korea, Taiwan, Singapore, Hong Kong and other countries that have achieved sustained growth have a number of common characteristics. They all have placed great emphasis on education and training. They all have kept taxes and inflation low and provided other incentives for private capital formation. Government bureaucracies have been kept lean and public sector investment has been focused on providing the roads, ports, telecommunications and other infrastructure essential to economic development. They have given priority to development of the agricultural sector. Prices for food, energy and other essentials have been left to market forces with little intervention to keep prices low for urban consumers. Their public and private sectors have collaborated to become competitive in world markets by continually upgrading technology, manufacturing facilities and work force quality. Their trade policies have been outwardly directed towards export-led growth rather than inwardly directed towards import substitution.

By contrast, many of the countries receiving development assistance have had insufficient investment in public education, high taxes and inflation, extensive government regulation of industrial and commercial activity, control of food prices for urban consumers, policies designed to insulate their economies from world markets, and little overall appreciation of the role of the private sector in economic development.

Market-oriented reforms and trade liberalization in many developing countries during the 1980s have reduced this contrast, but further basic political and economic reform in Eastern Europe and developing countries is still needed. This sort of reform is a precondition; sustainable development cannot exist without it. The reformers' main theme should be decentralization of decision making to the extent possible. Decentralized decision making by local groups — such as rubber-tappers in the Amazon and the Chipko "tree-huggers" movement in the foothills of the Himalayas in India — has helped to protect forest and other natural systems from destruction. The alternative — centralized decision making, with little understanding of local conditions — has contributed to environmental disasters such as the shrinking of the

Aral Sea and the salinization of the surrounding countryside (a result of diverting water to irrigate cotton).

Governments do need to supplement the operation of the marketplace by internalizing environmental costs and benefits into the decision calculus through fees, taxes, and regulatory regimes. This is as true for a country like Taiwan, which has experienced tremendous environmental degradation from its rapid industrialization, as it is for Bolivia or Ghana. But our unhappy experience with top-down planning in Eastern Europe and many developing countries suggests that specific decisions on what crops to plant, how to manage forests, and what businesses to invest in should be left to the people directly involved. Rather, development assistance should support stable money, open access to international markets, education, training, and incentives that reflect overall societal goals. Government funding should provide for the extension of electricity, telecommunications and other infrastructure — not subsidies for the cultivation of particular cash crops or the manufacture of specific products. With this type of outside support, local people can get on with the job of building better lives for themselves and their children (World Bank, 1991).

2.3 A decentralized approach to designing and implementing development projects

This cannot take place without a basic change in the way development projects are designed and implemented. The World Bank, the regional development banks, development assistance agencies, and other institutions devoted to economic development have focused their resources on large mega-projects such as hydroelectric dams, mineral extraction facilities, and large-scale irrigation projects. This focus has encouraged a centralized approach to project management. Project managers, as well as technical experts (from funding organizations and their contractors) work directly with governing elites in recipient countries in designing and implementing projects. Neither group has had much incentive to spend time in the field consulting with the local people directly affected by the project, or to ensure that local concerns are addressed.

The result has been projects that help move large quantities of development assistance funding and temporarily enhance the power and prestige of local politicians — but disrupt the lives of local inhabitants and cause long-term damage to the environment. Among the best examples are large hydroelectric projects such as the Narmada Dam in India. Such projects displace many thousands of people from their homes, flood millions of acres of forest, and produce large amounts of electricity — to meet needs that could be met cost effectively by more environmentally sustainable means.

To avoid such outcomes in the future, economic development institutions must change their approach for selecting and implementing development projects. Rather than simply processing project proposals put forward by central governments in recipient countries, these institutions should require recipient countries to implement a bottom-up, participatory decision-making process under which project proposals either originate from the local groups most directly affected, or are presented to these groups for discussion and approval before being submitted for funding. A critical element in such a process is full disclosure of detailed information about the project to all interested parties, as soon as possible. The World Bank, or other institution receiving a project proposal, should satisfy itself that such a participatory process has been followed, and should make its own decision process as transparent as possible.

In addition to making detailed information about the project available to the local groups most directly affected, both recipient country governments and potential providers of funding should make it available to non-governmental organizations (NGOs) and the public at large. Local and international NGOs may be able to help the local people identify potential problems that they might not otherwise recognize.

To make such a decentralized decision-making process work, developmental assistance organizations will initially have to channel a large part of their resources into providing the education, training and infrastructure necessary for local communities to make intelligent decisions about their own future. This task will, in turn, require these organizations to form new partnerships with industry, environmental groups, universities, and other non-governmental organizations that have much of the knowledge and specialized resources required. USAID and other development assistance organizations are already forming partnerships with the World-wide Fund for Nature, Conservation International and other environmental NGOs to design and implement conservation projects in the field, in cooperation with affected communities and local NGOs. Eventually, NGOs and local communities themselves will become a primary source of development project proposals.

Decentralization of the process for formulating and implementing development projects will require both governments and development assistance organizations to change their own roles. Both will have to delegate increasing amounts of authority to local missions, offices or government agencies, which will in turn increasingly work with NGOs. Central governments will have to change their role from making detailed project decisions to providing an overall fiscal and regulatory framework and supporting infrastructure. Central offices of development assistance organizations will have to change theirs from approving and funding specific projects to providing operating guidelines, technical support and inspection services to local offices. These changes may meet resistance from central officials anxious to preserve their power and authority, but they are essential changes if we wish to avoid repeating the mistakes of the past.

2.4 *The importance of structural reform: the Brazil example*

The above analysis suggests that central governments should focus their efforts on providing an overall fiscal and regulatory framework and supporting infrastructure (such as education and training, transportation systems and telecommunications networks). To fulfil this role effectively, many governments will have to undergo structural reform. Brazil is a case in point. The hyper-inflation, unrestrained foreign borrowing, currency overvaluation, and massive state intervention in the economy that characterized Brazilian economic policy in the 1970s led the nation into an economic black hole in the 1980s. Public and private sources of investment capital dried up. Job opportunities (in a nation where 50% of the labour force is below the age of 25) disappeared. Denied other opportunities, an army of unemployed Brazilians, desperate for land, set out to colonize the Amazon. They followed a new, all-weather highway built with World Bank assistance.

Using farming techniques inappropriate to the region, these settlers rapidly exhausted the soil, forcing them to move on to new sites to begin the cycle again. The end result was an ecological devastation of immense proportions. The root cause of this devastation, David Hopper reminds us, was not the lack of ecological consciousness by the settlers themselves or even the specific subsidies for clearing the forest, but "the blundering macroeconomic policies of the Brazilian government that, ultimately, failed to provide the investable resources for new jobs in urban, out-of-the Amazon occupations (Hopper, 1991)".

Instead, the Brazilian government could have reformed its fiscal policies to reduce inflation and provide incentives for greater private investment and reduced central government intervention in the private sector. It could have focused its own resources on providing education, training, and other basic tools to help people get jobs where they were currently living.

2.5 *The need for a new international process*

We must now create a new international process for addressing global climate change that gives various elements of the UN system, other international organizations, national and local governments, and different non-governmental organizations appropriate roles in bringing about the changes necessary to achieve sustainable development. We already have several precedents for such a process.

3 Lessons from the Past

The following precedents have been selected to illustrate how an overall regime to address climate change might be structured, and how the elements of that structure might operate. The "Montreal Protocol" on Substances that Deplete the Ozone Layer is the only example of an overall regime established to address an environmental problem (like global climate change) whose consequences would not be fully apparent for many years. The history of the Protocol also illustrates the potential of new types of collaboration among government and non-governmental organizations.

The London Guidelines for the Transboundary shipment of Hazardous Wastes, the International Registry of Potentially Toxic Chemicals, and the recent US experiences with disclosure requirements for chemical discharges all illustrate the importance of accurate information about environmental impacts of different substances, and about the actual discharges of these substances, to any climate regime.

Finally, the UN/ECE Long Range Transboundary Air Pollution Treaty and the Mediterranean Action Plan — and various non-governmental initiatives — provide precedents for the regional and non-governmental elements of such a regime.

3.1 *Ozone: the Montreal Protocol*

The most recent model is the Montreal Protocol on Substances that Deplete the Ozone Layer (United Nations Environmental Programme, 1990). The Protocol established initial targets and timetables for cutting production and consumption of ozone depleting chemicals and an iterative process for periodically reviewing the targets and timetables in light of new knowledge about environmental impacts, the availability of substitutes, and the costs and benefits of different phase-out schedules. This process led to an agreement by the parties at their London meeting in June 1990 to phase out production and consumption of most CFCs by 2000. These international agreements have in turn led the US and other key countries to develop national plans for fulfilling their obligations under the Protocol. These plans include specific implementation measures — such as production permits or taxes — that are designed to implement the phase-out in the most cost-effective manner.

The negotiating process leading up to the Protocol was characterized by a high degree of cooperation. No government or international body dominated the conferences. Individual scientists, research institutions, industrial groups such as the CFC Alliance, and environmental groups were all deeply involved in negotiating the agreements, developing substitutes, conducting the assessment process, advising governments about the most effective implementation strategies, and educating the public. Industry has also cooperated closely with US EPA and other government agencies in helping specific developing countries formulate and implement specific strategies for phasing out CFCs at minimum cost. Finally UNEP and its Executive Director provided overall direction and leadership to the negotiation and implementation of the Protocol, working directly with industry, environmental groups, and other non-governmental organizations, as well as with governments.

Three features of the Montreal Protocol process have particular relevance to the international process for addressing climate change. The first is the role of national leadership. Early unilateral action by the US and several European countries to ban the use of CFC-based aerosol propellants in the late 1970s led the way in phasing out ozone-depleting

CFCs and provided US industry with an incentive to support an international agreement. Moreover, US industry may have gained some competitive advantage from early research on substitutes that was conducted in anticipation of possible further domestic regulation. Leadership by Western Europe and Japan in committing to stabilize CO_2 emissions is playing a similar role during the climate negotiations. This time it is European and Japanese firms that may be gaining a competitive edge from research and development, in anticipation of changes in government policy to implement emissions reduction commitments.

The second feature is the forming of a community of scientists and experts "from many nations, committed to scientific objectivity, [that] developed through their research an interest in protecting the planet's ozone layer that transcended divergent national interests (Benedick, 1991)." The critical role of an independent group of experts has also been noted in the context of the Mediterranean Action Plan formulated under the Barcelona Convention, a regional seas agreement also negotiated under UNEP auspices. "Through the process of negotiating the Med Plan, new groups were introduced to national decision making who succeeded in redefining the national interest, so that states, particularly the weaker ones, came to support the Med Plan despite strong structural incentives to cease their collaboration (Haas, 1990)." The greater complexity of the climate negotiations — and the resulting need to involve a larger number of people with a wider range of perspectives — will make creation of a similar network of experts with shared views more difficult, but it remains an important objective.

The third feature is the role of the private sector in assisting developing countries to make the transition to more environmentally friendly technologies at minimum cost. In the US, industry is already working with EPA and other government agencies to help certain developing countries make the transition to non-ozone depleting technologies. Private firms are now building on this experience by taking independent initiatives to assist developing countries. The Industry Consortium on Ozone Layer Protection, a group of major electronics companies, has established a computerized data base — accessible without charge or royalties — on substitutes for solvents. One company, Northern Telecom, has a special relationship with Mexico to support that country's decision to comply with [the] Montreal Protocol timetable for industrialized countries.

Effective strategies that enable developing countries to address the risks of global warming will similarly require extensive cooperation and support from industry, but for the most part few large companies — even those with technology likely to benefit from policies designed to reduce greenhouse gas emissions — have been willing publicly to express support for such reductions. The climate convention and subsequent protocols should therefore contain provisions designed to encourage and reward early innovators as much as possible. The most direct way of doing this is to reward countries that take steps to reduce their greenhouse gas emissions after a certain date by giving them credit for those reductions in meeting their emissions reduction obligations under future international agreements.

3.2 *Hazardous substances: The London Guidelines and the IRPTC*

Another useful set of precedents emerged from the process leading to the London Guidelines for transboundary shipments of hazardous substances and the International Registry of Potentially Toxic Chemicals (IRPTC). Here again UNEP member governments negotiated a special regime for providing countries importing or using potentially toxic pesticides and other dangerous chemicals with information about regulatory and other actions taken by exporting countries with respect to those chemicals. The chemical industry and environmental groups played an important role in drafting the guidelines in a manner that gave developing countries timely information to make decisions on importing or using the chemicals without unduly burdening or discriminating against exporting companies.

3.3 *The value of feedback: The toxic release substances control act*

The major environmental risk from using dangerous chemicals does not come from use of small quantities in specific applications, but from the cumulative impact of discharging large quantities of different chemicals into the environment over long periods of time. Providing better information on the specific physical properties and effects of individual chemicals is important, but it cannot substitute for better data on actual uses and discharges into the soil, water and air. The importance of such data has been demonstrated by Title III of the US Superfund amendment and Reauthorization Act. Also called the Toxic Release Substances Control Act, it requires US companies to publish annual reports of their discharges of specified chemicals. The data gathered and published in compliance with the Act have already led Monsanto, Dow and other major chemical companies to commit to substantial reductions in their discharges (Baram et al., 1990).

Similarly, accurate data on annual emissions of different greenhouse gases is important for developing an effective international strategy to address climate change. Without such data, it will be difficult to negotiate overall emissions reduction goals, to allocate responsibility for achieving those goals among specific countries, or to ensure that individual countries are fulfilling their commitments under international agreements. It will be particularly difficult to design and implement systems for enabling one country to get credit for either emissions reduction investments in another country or for emissions trading. Public disclosure of data on greenhouse emissions from specific sources will, moreover, provide an important incentive for the emitters to take voluntary steps to reduce those emissions.

3.4 *Nuclear energy: the International Atomic Energy Agency*

The International Atomic Energy Agency (IAEA) provides an example of a specialized UN agency with authority to monitor and supervise a particular technology. The principal agreement guiding this activity is the Nuclear Non-Proliferation Treaty. It shares many of the characteristics of the IRPTC with respect to interacting with industrial, environmental and other non-governmental organizations as well as with governments and other international organizations. With the help of a highly trained technical staff, the IAEA gathers detailed data on the world nuclear industry, reports on developments in nuclear technology and other industry trends, provides technical advice and assistance to member countries and their nuclear industries, develops and recommends operating procedures and safety standards for nuclear power stations, and monitors compliance with the NPT. All of these functions are relevant to implementing international agreements with respect to global climate change.

There may be merit in expanding the IAEA's scope to cover renewables and other non-fossil technologies in addition to nuclear energy. The IAEA performs data gathering, information exchange, technology transfer, inspection and regulatory functions that will be required under a climate convention and subsequent protocols. If its scope was expanded to cover other non-fossil technologies and it was given authority to gather data about direct or indirect greenhouse gas emission from all energy sources, then it could play a valuable role in promoting transfer of non-CO_2 emitting energy technologies and in providing data under climate agreements. The IAEA's power to conduct inspections of parties' nuclear facilities will need to be strengthened in the light of disclosures concerning Iraq's nuclear programme. Such strengthened authority could be a valuable tool in encouraging compliance with international commitments concerning greenhouse gas emissions.

3.5 *Regional pollution regimes: LRTAP and Mediterranean Regional Seas*

The UN/ECE Long Range Transboundary Air Pollution Treaty (LRTAP) provides an example of a successful regional treaty encompassing all of Europe, the US and Canada that has established an iterative process for negotiating successful emissions reduction protocols for sulphur, nitrogen oxides, and volatile organic compounds, based on an ongoing assessment of the sources, impacts, and reduction possibilities for the specific emissions in question. The EC parties to LRTAP have developed their own compliance strategy through the Large-Scale Combustion Plant Directive, as have the US and Canada through their respective clean air legislation and the recently concluded bilateral air quality accord.

The Mediterranean Regional Seas Plan is another example of a regional regime that has acted as a catalyst for action by member states to reduce their emissions into an environmental commons. The Med Plan, similar programmes for other regional seas, and regional agreements for managing interior seas and river basin are all precedents for possible regional initiatives for reducing greenhouse gas emissions. Such a regional initiative is currently being negotiated within the European Community and it may be followed by a similar initiative for North America (Canada, US, Mexico). Broader regional initiatives involving countries whose economies span the development spectrum (West and East Europe; North and South America; Japan and Northeast Asia) could be important building blocks towards a more global strategy.

3.6 *Leadership by NGOs*

Finally, there are precedents for non-governmental organizations taking the lead in addressing an important development or environmental issue. The Rockefeller and Ford Foundations played such a role in bringing about the green revolution in Mexico, India and other developing countries by funding a research programme on new crop varieties and creating the Consultative Group on International Agricultural Research (CGIAR). This effort has resulted in a huge increase in agricultural productivity in these countries and has created institutions such as the International Rice Research Institute that could play a critical role in developing dry rice farming and other agricultural technologies to address climate change. The success of the CGIAR system in meeting this challenge will depend upon its ability to shift its focus from merely increasing yields to improving the ratio of yields to total resource inputs (particularly water and chemicals). It will also be important for these institutions to develop crop systems that are more resilient to the effects of climate change (particularly by reversing the trend towards genetic simplification by promoting genetic diversification).

An important example of NGO leadership is the Global Change System for Analysis, Research and Training (START), which was established by the ICSU's International Geosphere/Biosphere Program (IGBP, 1991). START aims to establish a series of regional research centres covering different geographic areas (e.g. the Amazon Basin, the Arctic) that would conduct research, analysis and training on global issues particularly relevant to the region in question. Each centre would combine different disciplines (natural sciences, anthropology, engineering, economics, law) and would be the hub of a network of universities, training centres, and existing institutions in the region. The North African Centre might, for example, focus on different aspects of the desertification problem, ranging from physical processes at the desert/savannah boundary to social patterns of nomadic peoples, to public policy options for promoting better range management. One of the greatest benefits of such regional research and training centres, particularly in developing countries, is their potential for building what Peter Haas calls "epistemic communities" (Haas, 1990) of experts and officials in different countries, who can work together in implementing cooperative regional strategies for addressing global issues.

4 A Vision of the Future

4.1 A new institution in the making: the Climate Convention Conference of the Parties

As we look towards the future, we must heed the conclusion of the UNCED Report, *Our Common Future* (World Commission on Environment and Development, 1987), that actions to protect the environment and actions to promote economic development must be integrated to an extent never seen before. Nowhere is this conclusion more apparent than in addressing global climate change. Climate change is only one of many environmental challenges facing the world, and in most countries less immediate than others, but it requires us to deal with a broader range of human impacts on the environment than any other issue. The international convention on climate change being negotiated for signature at UNCED will also set a pattern for addressing the overall environment and development agenda. For these reasons, we have focused our vision of the future on key elements of a climate convention, and on institutional arrangements for dealing with the climate issue.

The international community is already in the process of creating a new institution that will act as a catalyst in leading existing institutions to form new partnerships and decentralize decision making. This institution is the conference of the parties to the convention on climate change being developed for signature at UNCED in 1992. The new institution will incorporate not only the conference of the parties *per se*, but also a number of subsidiary bodies including a strong secretariat, working groups and task forces. The provisions of the convention will not be finalized until just before UNCED 1992, but it will almost certainly contain a number of elements with far-reaching institutional implications.

4.2 Three key elements of a Convention

4.2.1 Global goals

The first of these elements is a set of global goals for reducing greenhouse gas emissions: a short-term goal of stabilizing worldwide greenhouse gas emissions at a certain level by 2000 or shortly thereafter and a long-term goal of substantially reducing those emissions by 2025-2050. These goals will not initially create binding legal obligations on individual parties, but will force all parties to think about their potential contributions to achieving the global goals. Industrial countries, for example, will have to recognize the need to stabilize their non-CFC emissions by 2000 or shortly thereafter in order to make the benefits of the CFC phase-out available to offset the inevitable growth in developing country emissions.

4.2.2 A comprehensive protocol

This process could be carried a step further by a second element that may be included in the convention: provision for negotiation of a comprehensive greenhouse gas emissions reduction protocol. Developing countries will probably insist on specific commitments by industrialized countries to reduce their own emissions, and to provide technological and financial assistance as conditions to any commitments to reduce their own emissions. The first of these conditions might be met by a commitment of all OECD countries to stabilize their CO_2 and possibly other non-CFC emissions at current levels by 2000 or shortly thereafter. Given the continuing US refusal to join Western Europe and Japan in making such a commitment, however, specific OECD emissions reduction commitments may not be incorporated into an international agreement until after the parties have gone through an initial process of preparing national strategies. These strategies could contain unilateral "pledges" of specific national policy initiatives to reduce emissions. Then the groups would be able to analyse the projected results of those pledges, and engage in an escalating competition to achieve greenhouse gas reductions, a sort of "emissions cutback race" with new pledges following each round of response.

4.2.3 National strategies

This brings us to the third and most critical element likely to be included in the convention: a requirement that each party prepare and furnish to the secretariat and each other party a national strategy for addressing greenhouse warming, within a short period after signature of the convention. These national strategies will contain baseline data on current greenhouse gas emissions levels and projections for the future; a detailed description of new policy measures sector by sector (utilities, buildings, transportation, industry, agriculture, forestry); a projection of the greenhouse gas emission reductions to be achieved by each policy and a statement of the assumptions used in calculating such projections; and particularly in the case of East European and developing countries, an estimate of the incremental financial and other resource requirements for implementing the strategy.

Preparation of these strategies will require a degree of cooperation among governmental and non-governmental organizations at the national, regional and international levels that is unprecedented in the history of international negotiations. As mentioned above, a number of OECD countries have already made considerable progress in preparing such strategies, but most countries do not have the human and technical infrastructure to carry out such a task. Thus, developing countries will need financial, technical and logistical support to prepare their strategies. Ideally, the process will also require all parties to involve state and local governments, non-governmental organizations and citizens' groups in preparing their strategies.

Even without such a requirement, the comprehensive scope of any national or regional strategy to reduce greenhouse gas emissions will make it impossible to formulate such a strategy without broad, bottom-up participation from affected groups. In the energy sector, for example, state and local governments, electric and gas utilities, energy users in the industrial, commercial and residential sectors, and manufacturers of automobiles and other energy-using equipment will all have to cooperate in devising incentive structures that encourage efficient and non-polluting energy use. The same is true in the agricultural sector, where farmers, extension services, research centres, fertilizer and pesticide manufacturers, and distributors and purchasers of agricultural products will all have to cooperate on a local level to come up with more sustainable crop strategies. Environmental and other non-profit organizations have important contributions to make as well. This type of collaboration is beginning to produce results in the US and other OECD countries, but must be greatly expanded.

4.3 National plans

At the purely national level, a number of OECD countries are in the process of formulating national plans for addressing climate change. The Scandinavian countries, the Netherlands and Germany are furthest advanced in this effort, having established specific national targets and timetable for emissions reductions and identified specific implementation measures including higher energy taxes, efficiency regulations, and development of renewable energy sources. These national plans are an example of how individual countries, working closely with their own industries and environmental groups, lead the international community by taking unilateral steps that can then be reflected in an international agreement.

4.4 Precedents for sectoral strategies

There are precedents for the type of collaboration required. Energy cooperatives among public service commissions, individual utilities, and environmental and consumer groups have been formed in California, Massachusetts and other US states. These collaboratives have negotiated the outline of demand-side management programs that have eliminated the need for many electric power plants through efficiency improvements. In California, for example, the two largest utilities plan to meet over 70% of their demand growth in the 1990s through conservation and most of the remainder through renewables, thereby avoiding the need to build any conventional power plants during the decade.

A second precedent is the work of IUCN, the Worldwide Fund for Nature (WWF), and local environmental groups in working with national governments to create national wildlife conservation strategies for Costa Rica, Mozambique, Nepal, Zambia, and other developing countries. In Mozambique, for example, the government and WWF have developed a strategy that includes development of local village infrastructure, improved agricultural technique, and rural industries recognizing that poor villagers have to be given alternatives to destroying more forest.

Another group of precedents for achieving the required level of cooperation exists in the record of recently concluded arms-control negotiations, and the coordination of efforts to prevent the deployment of nuclear weapons by Iraq and North Korea, and their dispersion from the Republics of the former Soviet Union. The US government has recognized the urgency of halting the spread of nuclear weapons and is coordinating an international campaign to achieve that objective, backed by a substantial commitment of financial and other resources. The US and other OECD countries will ultimately have to make a similar commitment in curbing worldwide greenhouse gas emissions.

5 The Need for Policy Coordination

It is critical that all parties during and after the convention — particularly major developing countries — propose both specific policy changes and the accompanying investments in human and physical infrastructure that will make the policy changes effective. In the energy sector, for example, many developing countries will need to phase out subsidies on fossil fuel production and increase the price of electricity. These changes will not be possible, however, unless poor people in those countries have access to alternative sources of energy services for their basic needs that enable them to maintain or improve their living standards. The most important such source is energy efficiency.

Only if villagers or urban workers are provided with energy efficient lighting, stoves, motors, and appliances and are trained in their use will they readily accept large increases in electricity prices. The same is true in the transportation sector, where increase in the prices paid for fuels must be accompanied by increases in vehicle efficiency and better public transportation. These improved products and services will only be available at reasonable cost if local manufacturers and energy companies are given incentives to encourage their use. Finally, at least in the initial phase, barriers to the import of energy efficient products, such as India's 250% tariff on compact fluorescent light bulbs, will have to be removed.

5.1 The importance of energy efficiency as a means to overcome capital constraints

This focus on energy efficiency is necessary for another reason. The World Bank expects the developing world to spend about $1 trillion on electricity generating and transmission facilities in the 1990s. India is expected to account for over $200 billion of this amount and China almost $160 billion. The Bank estimates that about 38% of the total, or $380 billion, will have to be raised from foreign borrowing. The resulting average annual requirement of $38 billion is roughly 15 times the Bank's current lending for power projects (Jhirad, 1989). It is highly unlikely that sums of this size will be forthcoming given the other competing demands on the world's scarce capital during the decade. Even if a substantial portion of the required sums are raised, they will have the effect of crowding out other investments that are critical for sustainable development. Realistically, therefore, developing countries face a choice between holding back their development due to inadequate energy supplies or using energy efficiency to provide the needed energy services with a much lower level of capital investment.

5.2 *Meeting obligations abroad: a catalyst for resource transfers*

An efficiency-based development strategy will require an unprecedented amount of foreign assistance from governmental and non-governmental organizations at all levels. This assistance should be linked to an element of the climate convention — a provision that would allow industrialized countries to meet some of their obligations to reduce greenhouse gas emissions by efforts made outside their borders. If Japan, Germany, or the US found it more cost effective to meet some of their obligations by investing in China, Poland, or Brazil rather than at home, they would be permitted to do so provided they entered into an agreement with each recipient country spelling out how credit for the resulting emissions reductions should be shared. Since any comprehensive protocol allocating global emissions reduction targets among specific countries will require the industrialized countries to make the bulk of the initial reductions, this option could prove very attractive. It would be further encouraged if the allocation scheme evolved into a tradeable permit system.

5.3 *Enhancing transfers: debt swaps, trade liberalization, and investment incentives*

The effectiveness of resource transfers to meet emissions reduction obligations under a climate convention could be greatly enhanced by targeted debt swaps, reduction in trade barriers, and additional incentive for foreign investment in developing countries. All three will require new partnerships among existing sets of institutions. Debt swaps will involve banks, environmental groups, finance ministries and other government agencies. We have some positive experience with debt-for-nature swaps in the habitat conservation area, but this concept needs to be greatly expanded to include deployment of environmentally friendly technologies and practices in industry and other key sectors.

Reduction in trade barriers will involve the GATT, trade ministries, and a variety of non-governmental organizations. Developing countries will have to lower tariffs and other barriers to the import of environmentally friendly goods and services. Industrialized countries will in turn have to remove the quotas, "orderly marketing agreements" and other barriers to the developing country exports necessary to pay for these environment-friendly goods and services. Providing additional incentives for foreign investment will involve international financial institutions, government agencies providing political risk insurance and credit guarantees, and private firms. (See von Moltke, 1992; this volume.)

In essence, we need a bargain between industrialized and developing countries whereby the former provide credit, guarantees and insurance against non-commercial risks and the latter provide favourable investment regimes enabling the foreign investor to control his investment, protect his intellectual property, and remit his capital and (within reasonable limits) profits.

6 The Need for New Institutions

Achieving these goals will require better allocation of responsibilities among, and coordination of, existing institutions than has been the case to date. Each type of organization has, within the whole, a significant role to play:

- UNEP, UNDP, the UN regional commissions, UN affiliated private organizations such as the Business Council for Sustainable Development, and their counterparts at the national level will have to do more systematic jobs of diffusing information about appropriate policy instruments, operating practices, and technologies.

- The World Bank, the regional development banks, development assistance agencies, and UNDP should give priority to building the local capacity to understand, implement and manage those instruments, practices and technologies.

- National governments, working closely with non-governmental organizations, should give priority to providing the fiscal and pricing policies, regulatory frameworks, and investment incentives to encourage greater domestic and foreign investment in sustainable technologies consistent with the country's national strategy.

- Non-governmental organizations and citizens groups should give priority to building popular support for needed policy changes and working with governments to develop national strategies and specific projects consistent with those strategies.

6.1 *Integrating environment and development in the UN System*

One of the central themes of this paper has been the need to utilize existing institutions more effectively rather than to create new ones. However, there are at least two partial exceptions to this generalization. First, a mechanism for integrating environment and development within the UN system must be created. Neither UNEP nor UNDP, as they are organized now, can perform this task. Their charters are too limited. UNEP should be given additional resources to expand its monitoring, data gathering and analysis, and dissemination functions, as well as its role as a catalyst in creating and supporting new regimes to address specific environmental problems such as transboundary shipments of hazardous waste and loss of biodiversity. Its environment focus makes it unsuitable, however, to act as the central decision-making point on sustainable development for the UN system.

The same is true in reverse for UNDP, which should focus on technical assistance, pre-feasibility studies, and helping countries to implement national development strategies . One possible solution would be to merge the governing bodies of these two organizations or at least combine their meetings in a manner that would permit joint decision making. Another solution would be to create a new sustainable development commission; its mission would be to continue the work of UNCED and its composition would include key UN agencies, governments and non-governmental organizations.

In the meantime, there are other steps that could be taken. In the short term, the UN should strengthen coordination among its existing subsidiary bodies by creating an Environment and Development Coordination Board reporting to the Secretary General. (To be effective, this Board will require better policy coordination among client agencies at the national level.)

6.2 Regional centres

The second exception concerns the desirability of creating regional centres for conducting climate change research along the lines of the START initiative described above. Frequently, developing country policy makers point out that their efforts to participate fully in international negotiations on climate change and to create effective national strategies are hampered by the shortage of trained professionals in their country. Regional climate research centres will help each developing country to train a cadre of scientists, engineers, economists and government officials who would be familiar with the issues and capable of devising local responses. A series of regional research and training centres in Asia, Africa and Latin America would act as the nodes in regional networks tying together existing universities, research institutions, and government agencies. One major goal of this network could be to facilitate the exchange of experts between developing and industrialized countries so that the views and knowledge of each could be broadened in the process.

The creation of such centres was one of the central recommendations of the Second World Climate Conference. However they are established, it is critical that they also promote interaction between disciplines — among physical scientists, social scientists, and policy makers.

6.3 The relationship between climate- and non-climate-related institutions

There remains one over-arching institutional problem that we have not yet addressed. Global climate change is an issue with implications for such a broad array of human activities that it is sometimes treated as a proxy for the whole sustainable development agenda. UNCED will deal with a number of issues — ranging from land based sources of ocean pollution to protection of biodiversity — that are not directly related to climate change *per se*.

These issues will have to be addressed in national, regional and international strategies for achieving sustainable development. Probably the only way of handling the inevitable overlap between strategies designed to address climate change and strategies designed to address non-climate related issues is to prepare them together and then integrate them into an overall sustainable development strategy.

Most of the obvious steps to reduce greenhouse gas emissions — improved energy efficiency, development of renewable energy sources, reforestation, development of drought-resistant crops — also serve other environmental objectives. Where conflicts between objectives or funding priorities arise, they will have to be resolved at the local and national levels before national strategies are finalized. For example, one of the six major items on the UNCED agenda is approval of a worldwide action plan (titled "Agenda 21") for achieving sustainable development in the 21st Century. The World Bank, the Global Environment Facility, and whatever UN institution is charged with following up on the UNCED Agenda 21 should focus on helping this happen. To encourage coordination at the international level, the review body established under the climate convention should seek the assistance of more specialized agencies such as the International Energy Agency or the FAO in reviewing the emissions and other projections in national climate strategies.

7 Conclusion

The type of institutional network for addressing climate change and other sustainable development issues discussed above will require a combination of decentralized decision making and global collaboration that appears inconsistent with the traditional role of the nation state. It can only be achieved with broad and open dissemination of information in all countries and the empowerment of non-governmental organizations, citizens' groups and individuals to participate in making informed decisions based on that information.

The nation state is not about to wither away as the central decision-making unit within the international community. But it can become part of a broader decision-making framework extending downwards into local communities and upwards into multinational institutions. The catalyst for making this happen is the spread of participatory democracy, which in turn is the key to sustainable development.

In concluding, we do not wish to leave the impression that it is possible or desirable to create a unified structure of institutions dedicated to addressing global climate change and achieving sustainable development. These institutions will have to evolve organically in a necessarily somewhat chaotic fashion. One metaphor for successful international organizations and agreements is an anthill, functioning as a dynamic aggregation rather than a unified system of many environmental regimes. It is probably unrealistic to create a coordinated grand design for this aggregation, but we can give its members better tools and organizational links to achieve their common purposes.

References

Baram, M.S., P. Dillion and B. Ruffle, 1990: *Managing Chemical Risks: Corporate Response to SARA Title III,* Medford, MA, The Center for Environmental Management, Tufts University.

Benedick, R.E., 1991: *Ozone Diplomacy,* Cambridge, Harvard University Press.

Bulato, R.A., E. Bos, P.W. Stephens and M.T. Vu, 1990: *World Population Projections,* 1989-90 Edition, published for the World Bank by the Johns Hopkins University Press.

Daly, H.E. and J.B. Cobb, 1989: *For the Common Good,* Boston, Beacon Press.

Haas, P.H., 1990: *Saving the Mediterranean: The Politics of International Environmental Cooperation,* New York, Columbia University Press.

Herman, R., S.A. Ardekani and J.H. Ausubel,1989: Dematerialization: *Technology and Environment,* 50-69, Washington, DC, National Academy Press.

Holdren, J.P., 1991: Population and the energy problem. *Population and Environment: A Journal of Interdisciplinary Studies,* **12**(3), 249 Spring 1991.

Hopper, W.D., 1991: *Some Thoughts on the Financing of Environmental Programs,* Paper prepared for Aspen Institute Environmental Policy Workshop, Aspen, Colorado, July 1991.

IGBP, 1991: Global Change System for Analysis, Research and Training (START), *Global Change, Report No. 15,* Report of a Meeting at Bellagio, December 3-7, 1990, Boulder 1991.

Jhirad, D., 1989: *Annual Review of Energy, Electricity.*

Kleiner, A., 1991: What does it mean to be green? *Harvard Business Review,* **69**(4), July-August 1991.

Ross, M.H. and D. Steinmeyer, 1990: Energy for industry. *Scientific American,* September 1990, p. 94.

TFCACC (Task Force on the Comprehensive Approach to Climate Change), 1991: *A Comprehensive Approach to Addressing Potential Climate Change,* Report to US Department of Justice, Washington, DC, February, 1991.

UNEP (United Nations Environment Programme), *Montreal Protocol on Substances that Deplete the Ozone Layer,* September 19, 1990.

Womack, J.P., D.T. Jones and D. Roos, 1990: *The Machine That Changed the World,* New York, Rawson Associates.

World Bank, 1991: *The Challenge of Development,* World Development Report 1991, Oxford University Press.

WCED (World Commission on Environment and Development), 1987: *Our Common Future,* London and New York, Oxford University Press.

Annex I

Emissions Scenarios from Working Group III of the Intergovernmental Panel on Climate Change[1]

The Steering Group of the Response Strategies Working Group requested the USA and the Netherlands to develop emissions scenarios for evaluation by the IPCC Working Group I. The scenarios cover the emissions of carbon dioxide (CO_2), methane (CH_4), nitrous oxide (N_2O), chlorofluorocarbons (CFCs), carbon monoxide (CO) and nitrogen oxides (NO_x) from the present up to the year 2100. Growth of the economy and population was taken common for all scenarios. Population was assumed to approach 10.5 billion in the second half of the next century. Economic growth was assumed to be 2-3% annually in the coming decade in the OECD countries and 3-5% in the Eastern European and developing countries. The economic growth levels were assumed do decrease thereafter. In order to reach the required targets, levels of technological development and environmental controls were varied.

In the **Business-as-Usual scenario** (Scenario A) the energy supply is coal intensive and on the demand side only modest efficiency increases are achieved. Carbon monoxide controls are modest, deforestation continues until the tropical forests are depleted and agricultural emissions of methane and nitrous oxide are uncontrolled. For CFCs the Montreal Protocol is implemented albeit with only partial participation. Note that the aggregation of national projections by IPCC Working Group III gives higher emissions (10-20%) of carbon dioxide and methane by 2025.

In **Scenario B** the energy supply mix shifts towards lower carbon fuels, notably natural gas. Large efficiency increases are achieved. Carbon monoxide controls are stringent, deforestation is reversed and the Montreal Protocol implemented with full participation.

In **Scenario C** a shift towards renewables and nuclear energy takes place in the second half of the next century. CFCs are now phased out and agricultural emissions limited.

For **Scenario D** a shift to renewables and nuclear in the first half of the next century reduces the emissions of carbon dioxide, initially more or less stabilizing emissions in the industrialized countries. The scenario shows that stringent controls in industrialized countries combined with moderated growth of emissions in developing countries could stabilize atmospheric concentrations. Carbon dioxide emissions are reduced to 50% of 1985 levels by the middle of the next century.

[1] From: Houghton, J.T., G.J. Jenkins and J.J. Ephraums, 1990: *Climate Change: the IPCC Scientific Assessment*, Cambridge University Press, Cambridge, UK.

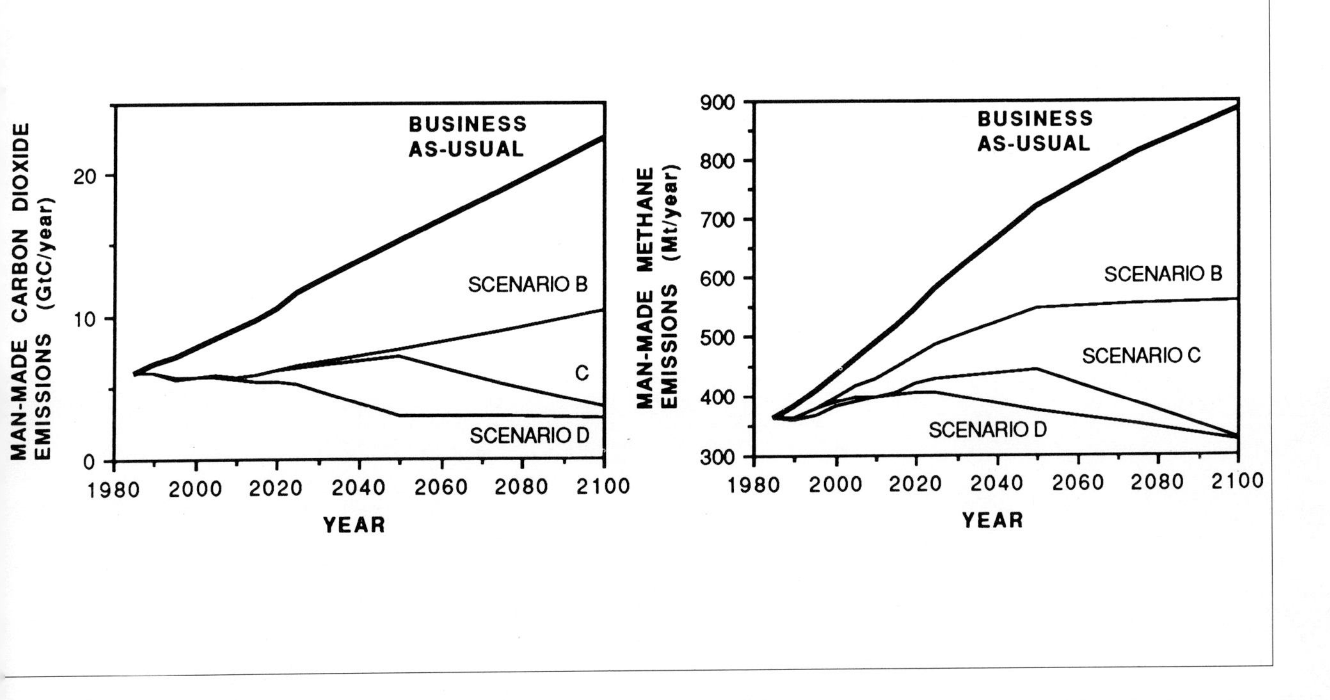

Man-made emissions of carbon dioxide and methane (as examples) to the year 2100, in the fours scenarios developed by IPCC Working Group III.

Annex II

IPCC Scenarios for Greenhouse Gas Emissions from the Energy Sector[1]

SCENARIO	Energy Supply(b)	Energy Demand(c)	Level of Greenhouse Gas Emission Controls(d)	CFC Status(e)	Deforestation(f)	Agriculture(g)
EIS/AFOS Reference Case(a)	Carbon-Intensive	Business as usual	Modest; business as usual	Protocol/low compliance	ModerateCurrent trends	
2030 Emissions	Carbon-Intensive	Business as usual	Modest; business as usual	Protocol/ low compliance	Rapid/ Moderate	Current trends
2060 Emissions	Major switch to natural	High degree of energy efficiency	Stringent	Protocol/full compliance	Major reforestation	Current Trends
Control Policies	Major phase-In of non-fossil source	High degree of energy efficiency	Stringent	Total phaseout	Major reforestation	Declining factors
Accelerated Policies	Major phase-in of non-fossil sources	High degree of energy efficiency	Stringent	Total phaseout	Major reforestation	Declining factors

(a) Because the EIS/AFOS scenario was constructed with the aid of experts from many countries, the assumptions as defined here may not exactly match the specific descriptions provided below. However, the assumptions are qualitatively very similar to these specific descriptions.

(b) For the Carbon-Intensive cases, energy supply is dominated by fossil fuel technologies. For the 2060 Emissions case, fossil fuels continue to play a major role, but the natural gas share of the energy supply market is increased. For the Control Policies case, non-fossil technologies become economic in the latter part of the next century and supply most primary energy needs after 2075. For the Accelerated Policies case, non-fossil technologies play a much larger role starting in the early part of the next century.

(c) In the Business as Usual cases, the annual rate of Improvement in energy intensity (or primary energy use per dollar of GNP) decreases from an initial value between 1.0% and 1.5% to an average value of 0.7% to 1.2% for the years 2075 to 2100. The average annual rate of improvement in energy

(d) In the Modest Controls cases, current emission control technologies are assumed. In the Stringent Controls cases, the following controls are assumed: more stringent NOx and CO controls on mobile and stationary sources, including all gas vehicles using three-way catalysts (in OECD countries by 2000 and in the rest of the world by 2025); from 2000 to 2025 conventional coal burners used for electricity generation are retrofit with low NOx burners, with 85% retrofit in the developed countries and 40% in developing countries; starting in 2000 all new combustors used for electricity generation and all new industrial boilers require selective catalytic reduction in the developed countries and low NOx burners in the developing countries, and after 2025 all new combustors of these types require selective catalytic reduction; other new industrial non-boiler combustors such as kilns and dryers require low NOx burners after 2000.

(e) In the Protocol/low Compliance cases, the Montreal Protocol is assumed to come into force and supply through 2100, with 100% participation by the U.S. and developed countries and 85% participation by developing countries. The assumptions are the same in the Protocol/Full Compliance cases except that all countries participate. In the Phaseout cases, CFCs and halons are completely phased out by 2000, and production of CCI4 and CH3CC1 are frozen with 100% participation.

(f) Tropical deforestation increases gradually in the Moderate case, from 11 million ha/yr in 1985 to 15 million ha/yr by about 2100. Tropical deforestation increases exponentially in the Rapid case, reaching 34 million ha/yr in about 2050, with almost complete tropical forest deforestation by about 2075. In the Reforestation cases, deforestation stops by 2025 and about 1,000 million ha are reforested by 2100.

(g) Levels of activity for the major agricultural activities were not varied amoung scenarios, because population levels were not varied. Emission factors, however, were changed amoung scenarios. Current estimates of average emission factors were used in the Current Trends cases, while reduced factors (assuming technological and management improvements) were used in the Declining Factors cases.

[1] Source: **USAID** (US Agency for International Development), 1990: *Greenhouse Gas Emissions and the Developing Countries: Strategic Options and the USAID Response,* A Report to Congress, July.

Glossary

Absorption - *1.* The taking up and storage of energy in matter, a basic physical process with several implications for global climate change. Solar radiation and heat may be absorbed and held for long periods of time in the oceans, thus delaying the impact of greenhouse gas emissions on global temperature change. Radiation is also absorbed by gaseous molecules and aerosol particles in the atmosphere. This type of absorption can be selective by wavelength. Some polyatomic trace gases are transparent to incoming short-wave solar radiation but absorb and re-emit the long-wave, infra-red (thermal) radiation emitted from the Earth's surface. These gases are commonly referred to as "greenhouse gases." *2.* The term absorption also refers to the taking up and storage of gases through the respiration of living plants and animals. CO_2 may be absorbed, for example, by forests or ocean plants, which convert them to oxygen. In terms of climate change, the plants become sinks, removing greenhouse gases and thus reducing the gases' potential effects.

Acid deposition (Acid rain) - The processes by which acids are delivered to the Earth's surface from the atmosphere. The deposition may take place in either wet form (as aqueous acids) or in dry form (as dry salt particles). Wet deposition is through precipitation (drizzle, fog, sleet, snow, and rain) containing principally sulfuric and nitric acids. These acids are produced by the chemical combination of water vapor with sulfur dioxide and nitrogen oxide gases. These gases are emitted into the atmosphere from the smokestacks of electricity generating stations and smelters, and from automobile exhaust pipes. The harmful effects of acid rain include damaging buildings and metals, killing fish and aquatic plants and microorganisms, weakening and killing trees, and aggravating or causing human health problems.

Aerosols - Particles, other than water or ice, suspended in the atmosphere. They range in radius from one hundredth to one ten-millionth of a centimeter — or 10^2 to 10^{-3} microns (m). Aerosols are important as nuclei for the condensation of water droplets and ice crystals, and as participants in various atmospheric chemical reactions. Perhaps most significantly, they absorb solar radiation, then emit and scatter it. Thus, they influence the radiation budget of the earth-atmosphere system, which in turn influences the climate on the surface of the earth. Aerosols from volcanic eruptions can lead to a cooling at the Earth's surface, which may delay greenhouse warming for a few years following a major eruption.

Afforestation - The establishment of forest cover on land not previously forested. Afforestation may be necessary to increase the net capacity of Earth's forests to absorb CO_2 and other greenhouse gases.

Agenda formation - In international negotiations, the formal process through which issues requiring action are brought to the attention of political decision-makers.

Albedo - Reflectivity. When radiation strikes a body, albedo is a measure of the fraction reflected, either back towards the source or anywhere away from the body. In climate considerations, albedo refers to the percentage of solar radiation reflected back from the Earth into space (about 30% on average). The albedo of any locale depends upon many factors, including the color and roughness of the terrain, the extent of forest or agriculture cover, and the amount of cloud and snow cover. Clouds, ice and snow reflect a greater proportion of radiation than do land and ocean surfaces.

Aldehydes - organic compounds with a structure resembling a chain of carbon, hydrogen, and oxygen atoms. Aldehydes such as formaldehyde are made commercially, through reactions that include the catalytic removal of hydrogen from alcohols. But aldehydes are also present in fossil fuel engine exhausts. Formaldehyde, for example, is found typically at 29-43 parts per million in automobile exhaust, and at a higher percentage in diesel engines. Aldehydes have been linked to cancer risk in humans, and formaldehyde was listed as a hazardous air pollutant in the 1990 United States Clean Air Act.

Algae - Simple rootless plants that grow in sunlit waters in relative proportion to the amounts of nutrients available. They are food for fish and small aquatic animals, and a factor in *eutrophication.*

Alluvial flood plain - Plain created by the deposit of sediment by flowing water, typically during a seasonal flooding. This makes any alluvial flood plain vulnerable to storms, floods, or other disasters.

Anoxic - oxygen-free. See *methane.*

Anthropogenic - Made by people or resulting from human activities. Usually used in the context of emissions that are produced as a result of human activities.

Atmosphere - The envelope of air surrounding the Earth, and bound to it by the Earth's gravitational attraction.

Autotrophic - Able to produce food from inorganic substances.

B.P. - Before the Present Day. 10,000 years B.P. means roughly ten thousand years ago.

Barrier islands - Beaches, reefs, or islands, habitable in places, that provide a measure of protection for the mainland during hurricanes, tidal waves, and other maritime disasters. The presence of barrier islands will probably affect the reconfiguration of coastlines in the event of significant sea level rise.

Biolimiting - Factors determining or restricting the growth of a particular life form. It has been hypothesized that in the oceans near Antarctica, shortage of iron has been a biolimiting factor, restricting the growth of phytoplankton which would otherwise increase in population. According to some theories, if iron was more plentiful in these waters, the phytoplankton would sequester more CO_2 from the atmosphere.

Biomass - All the living matter in a given area. Can also refer to stored energy in the form of organic matter: wood, charcoal, agricultural wastes, and animal dung. When burned, this is the primary source of energy for nearly half the world — about 2.5 billion people in developing countries. Biomass is considered renewable because plants and animals reproduce; but the biomass stock may also be consumed faster than the rate of renewal. Biomass burning is thought to be a significant source of atmospheric methane and carbon dioxide. In energy accounting, the term "biomass" may refer to the potential stored energy content of living organisms (such as forests or fuel crops) present at a specific time in a defined unit (community, ecosystem, crop, etc.) of the earth's surface.

Bioregions - Localities defined by natural ecological systems, as in a river watershed, where the indigenous plants, animals, and native human population manifest the characterisitics of an integrated system.

Biosphere - The system of earth and its atmosphere that supports life. In the global carbon cycle, the biosphere serves as a sink (reservoir); carbon is stored and preserved in living organisms (plants and animals) and life-derived organic matter (litter, detritus). The biosphere controls the magnitude of the fluxes of several greenhouse gases, including CO_2 and methane, between the atmosphere, oceans, and land. The terrestrial biosphere includes living biota (plants and animals) and the litter and soil organic matter on land. The marine biosphere includes the flora, fauna and detritus in the oceans.

Biota - The animal and plant life (flora and fauna) of a given area.

Brundtland Commission - Informal name for the World Commission on Environment and Development (WCED), established in 1984 by the U.N. General Assembly, and chaired by Norwegian Prime Minister Gro Harlem Brundtland. In its report *Our Common Future* (1987), the commission popularized the idea of sustainable development through global cooperation.

C3 plants - Plants (such as soybean, wheat, and cotton) whose carbon-fixation products have three carbon atoms per molecule. Compared with C4 plants, C3 plants show a greater increase in photosynthesis when CO_2 concentration is doubled

C4 plants - Plants (such as maize and sorghum) whose carbon fixation products have four carbon atoms per molecule. Compared with C3 plants, C4 plants show little photosynthetic response to increase CO_2 concentrations above 340 ppm. They show a decrease in stomatal conductance, which results in an increase in photosynthetic water-use efficiency.

Capital multiplier - An index showing the amount of money saved from foregoing investments in energy production, and investing instead in energy demand-side efficiency improvements. In one study, eleven key end-use energy efficiency measures produced a capital multiplier of five: that is, five dollars would be saved in energy supply costs for every dollar invested in these eleven key improvements.

Carbon balance - See *net primary production.*

Carbon sequestration - the storage of carbon, including anthropogenic carbon dioxide that would otherwise affect global climate change, in large-scale carbon sinks, such as forests.

Carbon tax - A financial tax imposed on the sales of fossil fuels. In this form of fuel tax, the level of taxation is determined by the carbon content of each fuel. Users of natural gas would pay less than users of petroleum gasoline.

Catalytic converter - A device fitted to motor vehicle exhausts to reduce emissions of dangerous pollutants. Such devices use catalytic agents (minute quantities of metals such as platinum) to stimulate chemical reactions that convert carbon monoxide and hydrocarbons into carbon dioxide and water vapor. In some forms of catalytic converters, nitrogen oxides are reduced as well. Catalytic converters designed for automobiles, buses, and trucks are contaminated by lead, so vehicles fitted with them must be run on unleaded fuel.

Cetane - a colorless liquid hydrocarbon found in petroleum, used as an indicator of fuel efficiency for diesel oils. The "cetane number" (similar to the "octane" number for gasoline) indicates the ability of a diesel engine fuel to ignite quickly after being injected into the cylinder. Cetane-improving additives improve fuel efficiency and may reduce hydrocarbon and particulate emissions.

Chlorofluorocarbons (CFCs) - A family of inert, nontoxic and easily liquified chemicals, implicated in two major environmental problems: ozone-layer depletion and global warming. CFCs are used as coolants in refrigerators and air-conditioners, as propellants in aerosol cans, as solvents, and as blowing-agents that inflate flexible foams. It takes about 15 years for a CFC molecule to drift into the upper atmosphere, where it can last 100 years or more, destroying an estimated 10,000 ozone molecules over that time. All conventional CFC uses are covered by the Montreal Protocol, and are likely to be halted in industrial countries by 2000 at the latest.

Climate - The long-term statistical average of weather conditions. Global climate represents the long-term behavior of such parameters as temperature, air pressure, precipitation, soil moisture, runoff, cloudiness, storm activitiy, winds, and ocean currents, integrated over the full surface of the globe. Regional climates, analogously, are the long-term averages for geographically limited domains on the Earth's surface.

Climate sensitivity - The Earth climate's responsiveness to increases or decreases in radiation, such as those caused by the greenhouse effect. Sensitivity is an index that predicts how rapidly and strongly the climate will react to changes in radiative forcing, after adjusting its own internal balance (coming to equilibrium). This index is arrived at by estimating the eventual change in surface air temperature, relative to the pre-industrial level, that would occur if carbon dioxide emissions were to double and the concentrations of all other greenhouse gases remained at their natural background (i.e. pre-industrial) levels. Climate model estimates of Earth's sensitivity range from a temperature increase of 1.5 to 4.5 °C above the pre-industrial level, due to doubling the carbon dioxide concentration in the atmosphere.

Climate system - All the interrelated processes and components of earth, air, land, water and life that control or influence what we conventionally think of as the weather. These processes and components include the oceans, the ice masses (from small glaciers to large ice sheets), some aspects of the biosphere, and the changing chemical composition of the Earth's atmosphere-climate system. The properties that characterize the climate are thermal (temperatures of the surface air, water, land, and ice), kinetic (wind and ocean currents, together with associated vertical motions and the motions of air masses, aqueous humidity, cloudiness and cloud water content, groundwater, lake lands, and water content of snow on land and sea ice), and static (pressure and salinity of the oceans, and the geometric boundaries and physical constants of the system). These properties are interconnected by the various physical processes such as precipitation, evaporation, infrared radiation, convection, advection, and turbulence.

Cloud-radiative feedback - a process in which greenhouse gas emissions lead not just to rising temperatures, but also to changes in the distribution of clouds over the planet. The changes may affect average cloud height, average cloud temperature, average cloud cover, or the distribution of these variables over time. These changes, in turn, affect the amount of radiation reflected back to space (see albedo), changing the effects of further greenhouse gas emissions on future temperature change.

Cogeneration - Recycling of heat generated during manufacturing or industrial power production. Typically, the energy created by cogeneration is used to make steam. That steam is used to substitute for other fuel combustion, or to create electricity. Otherwise, the heat would be lost to the atmosphere.

Some utility companies allow "cogenerated" electricity to be sold back to them, and distributed to other customers over the power grid.

Combustion chamber - the space in an internal combustion engine cylinder where most of the burning of fuel takes place. Adjusting the shape or turbulence levels of the combustion chamber can help to increase engine efficiency and reduce emissions.

Commitment - Not yet observable changes in climate that are inevitable (though not necessarily known in advance), because of historical changes in greenhouse gas concentrations. When applied to temperature changes, this is called the "commitment to future warming." Lags in the climate system, such as slow reactions of oceans and land ice, mean that a century or more may elapse between a climatic "cause" and its ultimate effect on temperature change or sea level rise. Nonetheless, the Earth is "committed" to climate effects from emissions already released into the air; the effects have not taken place, but they probably cannot be avoided.

Concessional financing - The loaning of money at below market interest rates — and usually below the cost of borrowing for the lending agency — in order to promote a specific goal such as energy efficiency.

Conditionality - The increasingly widespread (but controversial) practice of attaching conditions upon international economic concessions or development assistance funds, or to the opening of international markets. A condition might involve signing an international climate change policy agreement, or an agreement to protect dolphins or forests.

Conflate - To bring together separate elements into one text.

Conversion - The change of mass into energy. For example, a fuel (such as gasoline, coal, or biomass) may be converted into heat or electricity. Conversion efficiencies describe the degree to which this conversion takes place without waste.

Convolution - A mathematical process, using a form of Fourier analysis, which allows a researcher to calculate the amount of carbon that has moved from the atmosphere into the ocean over time. The inputs to this use of convolution include the levels of atmospheric CO_2 concentration at various points in time and the rates of historical emissions. The inverse of this process, deconvolution, allows us to calculate the levels of CO_2 release in prehistoric and recent times, based upon amounts of CO_2 deduced from the isotope records of ice cores.

Cretaceous (66,400,000-144,000,000 y BP) - An era in which mean global temperatures may have been 10 °C warmer than they are today, and deep water temperatures may have been 18 °C warmer. The last period of the Mesozoic era, the Cretaceous was marked by the rapid rise and spread of deciduous trees, by shallow seas submerging most of the Earth's present land surface, and by dinosaurs — which became extinct after this period. Today's recoverable fossil fuel reserve contains enough carbon to raise the Earth to mid-Cretaceous temperatures, an event which could be considered a "worst-case" scenario.

Crown - The upper part of a tree, including the leaves and living branches. In tropical forests, but not in temperate forests, the crown contains more carbon than the soil — a factor in considerations of tropical forest protection.

Cryosphere - The portion of the climate system consisting of the world's ice masses, sea ice, glaciers, and snow deposits. Snow cover on land is largely seasonal and related to atmospheric circulation. Glaciers and ice sheets are tied to global water cycles and variations of sea level, and change over periods from hundreds to millions of years. The ice sheets of Greenland and the Antarctic, which can be considered quasi-permanent topographic features, contain 80% of the existing fresh water on the globe, thereby acting as a long-term reservoir in the hydrological cycle. Any change in their size will therefore influence the global sea level. The breadth of the cryosphere also influences the amount of radiation reflected from the Earth's surface back into space (see *albedo).*

Deconvolution - A mathematical process, the inverse of convolution, which allows us to calculate levels of atmospheric CO_2 release or uptake in prehistoric and recent times, based upon amounts of CO_2 deduced from the isotope records of ice cores.

Deforestation - Loss of forest. At least eleven million hectares of tropical forest are lost every year. Although the causes vary by region, one estimate indicates that slash-and-burn agriculture and scavenging for wood-fuel, often in the wake of commercial road-building, accounts worldwide for 40-50% of deforestation. Grazing accounts for 10%, commercial agriculture for 10-20%, forestry and plantations for 5-10%, and forest fires for 1-15%. The adverse effects of deforestation include the loss of habitat for countless animal and plant species, many as yet undiscovered by scientists; the destruction of the homes and livelihoods of native tribes; the denuding of mountainsides, providing an easy path for soils to wash away and rain to flood valleys below; and the addition of carbon dioxide to the atmosphere. Deforestation destroys what may be an important sink for excess CO_2 in the atmosphere, and the oxidation of organic matter releases CO_2 to the atmosphere, potentially adding to regional air pollution or global climate change.

Demand-side management (DSM) - A system in which publicly regulated power utilities meet customer demands for electricity by improving the efficiency of their customers' facilities, rather than by increasing energy supply. Where it has been tried, demand-side management has mitigated the financial risk in building new power plants, or has eliminated the need for them.

Dematerialization - Ongoing reduction in the amount of energy and materials consumed in producing a given amount of value. The "lean manufacturing" industrial trend involves continuous dematerialization and quality improvement.

Dendrochronology - The dating of past events and variations in the environment and climate by studying the annual growth rates of trees. *(Dendroclimatology* is the use of tree growth rings as proxy indicators of past climates.) The approximate age of a temperate forest tree can be determined by calculating the annual growth rings in the lower part of the trunk. The width of these rings suggests the climatic conditions during the period of growth. Wide rings signify favorable growing conditions, absence of disease and pests, and favorable climatic conditions. Narrow rings indicate unfavorable growing conditions or climate. Tree rings record responses to a wider range of climatic variables, over a larger part of the Earth, than any other type of annually dated proxy record.

Deuterium - See isotope.

Diesel engine - See *internal combustion engine.*

Ecosystem - The basic functional unit in ecology: the interacting system of a biological community and its non-living environmental surroundings. These are inseparable and act upon each other. The term was coined by the Oxford ecologist A.G. Tansley in 1934, though the concept is much older.

Ecumene - All the parts of the world inhabited (or habitable) by people.

Eemian (125,000-130,000 y BP) - An interglacial optimum period, part of the Pleistocene, and warmer than today. During the Eemian period, the Caspian Sea was several meters higher than it is now. Causes of the warmth may have included eccentricities in the Earth's orbit, giving markedly more radiation during the Northern hemisphere summer.

Effective radiation height - The height at which incoming radiation from the sun to Earth (absorbed solar energy) is equal to outgoing radiation from the Earth to space (black body radiation). The temperature at the radiation height is coupled to the Earth's surface temperature; as one changes, so does the other. The coupling between the two temperatures allows us to make a first estimate of climate sensitivity.

Effluent - anything that flows out of a pipe, smokestack, storage tank, channel, or sewer. Waste reduction is the process of eliminating potential greenhouse-gas emissions or other pollutant releases before they are emitted to the atmosphere, soil, or surface waters.

Eighteen-ninety (1890) - The date at which meaningful records of temperature began to be kept in enough locales to provide an overall record of mean global temperatures.

El Niño - An irregular, quasi-biennial variation of ocean current that, from January to March, flows off the west coast of South America, carrying warm, low-salinity, nutrient-poor water to the South. It does not usually extend farther than a few degrees past the equator, but occasionally it penetrates beyond 12° S, displacing the relatively cold Peru Current. (Generally this occurs around Christmastime; hence the name "El Niño.") The effects of this phenomenon are generally short-lived, but occasionally (most recently in 1982-1983), the effects are major and prolonged. Under these conditions, sea surface temperatures rise along the coast of Peru and in the equatorial eastern Pacific Ocean, and may remain high for more than a year, with disastrous effects on marine life and fishing. Excessive rainfall and flooding occur in the normally dry coastal area of western tropical South America. One reason for confidence in atmospheric General Circulation Models is their generally satisfactory portrayal of atmospheric response to sea surface temperature anomalies associated with El Niño.

Electrode - A conductor (not necessarily metallic) used to establish electrical contact with a non-metallic portion of a circuit. Electrodes would be an integral part of ultra-high-power (UHP) transformers in a steel plant.

Energy (fuel) intensity - A measure of the quantity of energy services derived from a given amount of energy or fuel. In economics, "energy intensity" refers to the annual primary energy consumption per dollar of Gross Domestic Product.

Energy efficiency - The amount of fuel needed to sustain a particular level of production or consumption, in an industrial or domestic enterprise. Energy efficiency measures are designed to reduce the amount of fuel consumed, either through greater insulation, less waste, or improved mechanical efficiencies, without losing any of the value of the product or process. Improving energy efficiency is a technological means to reduce emissions of greenhouse gases without increasing production costs.

Eocene (37,000,000-58,000,000 y BP) - An epoch, part of the Tertiary era. Like the Cretaceous era, the Eocene epoch was significantly warmer than the present day. Seas expanded far beyond their present boundaries; palm trees grew where London and southern Alaska are today.

EPRI (Electric Power Research Institute) - A consortium of power utilities in the United States, organized to conduct research on electric power generation, transportation, and storage; a potential model for international power utility research networks.

Equilibrium altitude - Synonym for *effective radiation height.*

Equilibrium response - The forecast of a change in climate resulting from a single, one-time change in radiative forcing. Because most such changes accrue gradually, a pure equilibrium response (where the climate reaches an equilibrium balance before any further changes take place) is not realistic; but these responses are much easier to estimate through climate models than their more transient counterparts, called *time-dependent responses.*

Equilibrium sensitivity - Synonym for *climate sensitivity.*

Equilibrium warming commitment (or equilibrium warming) - For a specified year, the equilibrium warming commitment is the eventual increase in temperature that would result, at some point in the future, if atmospheric concentrations of greenhouse gases remained constant at the levels of that year. See *commitment.*

Equinox - The time of year when the sun crosses the plane of the Earth's equator, making the length of night equal to that of day. This occurs twice a year, once at the beginning of Spring and once at the beginning of Fall.

Eutrophication - the aging process of a lake, pond, or slow-moving stream, in which organic material (from plants) accumulates and slowly replaces oxygen. Eventually, the body of water fills in and becomes dry land. In recent years, this process has been accelerated by plant or algae growth in many bodies of water, encouraged by environmental pollution from such sources as detergents containing phosposrus, the leaching of fertilizers, sewage and toxic dumping, and heated water from the cooling systems of power plants and other industries. There is concern that greater atmospheric concentrations of CO_2 will also accelerate eutrophication.

Evapotranspiration - The process of water vapor transfer from vegetated land surfaces into the atmosphere; an essential part of the global hydrologic cycle. Evapotranspiration includes *evaporation* (the change of liquid water, from bodies of water and wet soil, into water vapor) and *transpiration* (in which which water is drawn from the soil into plant roots, transported through the plant, and then evaporated from leaves and other plant surfaces into the air).

Exajoule - 10^{18} joules, approximately equal to a quadrillion BTUs or one Quad..

External forcing - Synonym for *radiative forcing.*

Feedback mechanism - a process of system dynamics in which a system reacts to amplify or suppress the effect of a force which is acting upon it. For example, in the climate system, the force of warmer temperatures may melt snow and ice cover, revealing the darker land surface underneath. The darker surface absorbs more solar energy, causing further temperature increases, and warming temperatures further — thus melting even more snow and ice cover. This is positive feedback, in which warming reinforces itself. In negative feedback, a force ultimately reduces its own effect. For example, when the Earth's surface grows warmer, more water evaporates, forming more clouds. If the clouds which form are extensive and widely distributed, covering large areas of the surface, they will tend to reflect more solar radiation back into space than the dark ground underneath would, cooling the Earth's surface — and reducing the force of warmer temperatures. Often incompletely understood, feedback mechanisms are one reason why a hasty or symptom-oriented "fix" to the climate problem often has a different effect than what was intended. Seemingly unrelated aspects of climate processes and human activities are sometimes found to be linked through feedback mechanisms.

Fertilization effect - Feedback mechanism in which increased atmospheric CO_2 increases the growth of plants, enhancing their abilities to absorb CO_2. This mechanism compensates somewhat for the addition of carbon dioxide to the atmosphere, but it may have a limited capacity.

Financial indicators - Measurements of the health and success of a company's finances and operations, used by investors to judge the prospects of a business. These include rate of return on investment, the proportion of income used to service debt, and various operating ratios — such as income compared to sales revenues, earnings per share, and profits per dollar of sales. During the last ten years, most financial indicators have pointed up the inefficiencies of conventional utility operations — both in developing and industrialized countries.

Flux - Flow per unit of area per unit of time. The term flux may be used to describe a flow of energy (usually measured in watts per square meter) or a flow of gas or liquid (sometimes measured in grams per square meter per day).

Forcing - See *Radiative forcing*.

Function - A mathematical formula establishing an unambiguous relationship between two variables. For every value entered (input), the process of performing the mathematics will produce a different value (output) as a result. To say, for example, that "population density is a function of soil fertility" is to say that every measure of soil fertility can be entered into a formula (or a hypothetical formula) to produce a value of population density. As soil fertility changes, population density will change as well. Many functions are plotted as graphs, with the input as the X-axis and the output as the Y-axis; they also become the basis for formulas used in computer models, such as models of global climate processes.

Gaia hypothesis - Put forward by James Lovelock and Lynn Margulis, this hypothesis holds that living organisms on Earth (including microorganisms) actively regulate atmospheric composition and climate, helping provide climate stability in the face of challenges like the increasing luminosity of the sun, or increasing anthropogenic greenhouse gas emissions. The mechanisms which "Gaia" uses to regulate itself may be unexpected, and may range from comforting to disastrous for people.

General Agreement on Tariffs and Trade (GATT) - An international body devoted to eliminating import quotas, lowering tariffs, and otherwise promoting free, nondiscriminatory international trade. Begun as an agreement among 22 nations at the Geneva Trade Conference of 1947, it has sponsored negotiations and agreements among many nations, and affected thousands of tariff concessions. In recent years, GATT-sponsored agreements have clashed with environmental protection concerns, GATT itself has become a forum for debate over integrating economic and environmental needs. Some have begun to see GATT agreements as a source of potential leverage in achieving sustainable development and solutions to problems involving global climate change.

General Circulation Models (GCMs) - A computational model or representation of the earth's climate, used to forecast changes in climate or weather. Most GCMs concentrate on the circulation of the ocean or atmosphere (the latter are often called "atmospheric general circulation models"). Atmospheric GCMS consist of equations that describe the atmosphere's basic dynamics, and inculude descriptions of its physical processes. Functions represent the conservation of energy, momentum, and mass, and calculate the distributions of wind, temperature, precipitation, and other indicators of climate as a result of emissions from human and natural sources. More elaborate climate models couple the atmospheric equations to others which describe the structure and dynamics of the ocean, and to other components of the climate system (land surface and ice). The most advanced models, three-dimensional GCMs with coupled representation of atmospheric and oceanic processes, can be run on only the largest and fastest supercomputers. Typically, GCMs are used to determine and describe potential climate changes that would result from a particular set of prescribed boundary conditions, after equilibrium is reached. It is more difficult to analyze dynamic results with these types of models, in which the boundary conditions are (more realistically) changing over time.

Genotype - A group or class of organisms with the same overall genetic constitution. Rice and wheat are each plant genotypes; but there is considerable variability within those genotypes for such characteristics as resistance to climate-related stresses.

Geomorphology - The study of the characteristic, origin, and development of land forms.

Gigajoule - One billion (10^9) joules.

Gigatonne (Gt) - A measure of mass, used for large quantities (such as the amount of carbon in the earth's atmosphere). A billion metric tonnes, 1 Gt = 10^{15} grams.

Glacial epochs - Periods during the history of the Earth when there were larger ice sheets (continental-size) and mountain glaciers than today. The most recent glacial epoch, the Pleistocene, encompassed most of the last 3,000,000 years. In overall occurrence, all the glacial epochs that have ever occurred occupy only 5% to 10% of all geologic time. During major glacial epochs, which seem to recur at intervals of 200,000,000 to 250,000,000 years, great ice sheets form in the high latitudes and spread out to cover as much as 40% of the Earth's land surface. Accompanying drops in temperature during some glacial epochs may have been as much as 14° C in the mid-latitudes. During a glacial epoch, major glaciations are relatively short-lived, each lasting less than 10,000 years. The periods between these glaciations, the interglacials, persist for only about 10,000 years, so that for most of an epoch, the ice sheets are either growing or diminishing in size. The Pleistocene epoch has been distinguished by seven or eight glacial advances within the last 700,000 years. Its last glaciation ended about 9,000 years ago.

Global climate model - Often used as a synonym for *General Circulation Model,* but can also refer to simpler classes of global models.

Global Warming Potential (GWP) - the estimated warming effect over a period of time, resulting from a hypothetical instantaneous release of one kilogram of a given greenhouse gas in today's atmosphere. Each gas (CH_4, CFC-11, etc.) has a different GWP. Indexed against the warming effect of 1 kg of carbon dioxide (CO_2), the GWP allows climate modellers to compare the relative radiative heating of various gases, taking into account the differing times that gases remain in the atmosphere and the degree to which their characteristic absorption bands are already saturated.

Green Revolution - An organized effort beginning in the 1960s, sponsored by the United Nations Food and Agricultural Organization (FAO), to increase world food production by introducing high-yield cereal varieties developed in the Phillipines and Mexico. The Green Revolution efforts created an infrastructure of agricultural research and development, and very impressive yields of grain on limited land. But the new strains of plants required large quantities of fertilizer, pesticides, and water (which has helped justify massive dam-building programmes with terrible ecological consequences). The adverse effects include the evolution of pesticide-resistant pests, the destruction of fish stocks by pesticides, a decline in food production due to soil destruction, and the encouragement of agricultural costs beyond the means of many small, independent farmers.

Greenhouse effect - An atmospheric process in which the concentration of atmospheric trace gases (greenhouse gases) affects the amount of radiation that escapes directly into space from the lower atmosphere. Short-wave solar radiation can pass through the clear atmosphere relatively unimpeded. But long-wave terrestrial radiation, emitted by the warm surface of the Earth, is partially absorbed and then re-emitted by certain polyatomic trace gases. On average, the outgoing long-wave radiation balances the incoming solar radiation. Thus, a reduction in this outgoing radiation (because of an increase in greenhouse gases) means that both the atmosphere and the surface will be warmer than they would be without the greenhouse gases.

Greenhouse gases - The trace gases which contribute to the greenhouse effect. The main greenhouse gases are not the major constitutents of the atmosphere — nitrogen and oxygen — but water vapour (the biggest contributor), carbon dioxide, methane, nitrous oxide, and (in recent years) chlorofluorocarbons. Increases in concentrations of the latter four gases have been linked to emissions from human activity.

Group of 77 (G-77) - An organization of non-industrial and developing countries, negotiating collectively on some international economic issues and political concerns about climate change and development. The name is a deliberate response to the international economic clout of the "Group of Seven."

Group of Seven (G-7) - The seven largest industrial countries, whose leaders meet regularly in economic summit meetings. The countries are: Canada, France, Italy, Japan, the United Kingdom, the United States, and West Germany.

Holocene (1,000-10,000 BP) - A relatively warm epoch, marked by several short-lived particularly warm periods. The most recent, from 6,200-5,300 y BP, is called the Holocene optimum. It was characterized by increased precipitation and higher lake levels.

Homeostasis - The state of sustained equilibrium in which all cells, and all life forms, exist. (According to the "strong" Gaia hypothesis, the atmosphere/biosphere system also exists in homeostasis.) An organism in homeostasis adapts to changed environmental conditions by adjusting its own internal state; for example, cold-blooded animals - and warm-blooded animals that hibernate - adjust to colder temperatures by changing their own internal temperature, so that their entire system may remain in homeostasis.

Hydrologic cycle - The process of evaporation, transport of water vapor, condensation, preciptation, and the flow of water from continents to oceans. This cycle is a major factor in determining climate; it influences surface vegetation, the clouds, snow and ice, and soil moisture. The hydrologic cycle is responsible for 25-30% of the mid-latitudes' heat transport from the equatorial to polar regions.

Hydrologic models - Models of the interrelationships between geography, soil, and water in specific river basins.

Hydrology - The science dealing with the properties, distribution, and circulation of water, including the study of water in soil, in living things, and in the air.

Hydroxyl - The atmosphere's primary oxidizing agent, this is an ionic (electrically charged) molecule composed of one oxygen and one hydrogen atom (OH). Present as a gas in the atmosphere, hydroxyl reacts easily with methane; this reaction leads to the production of CO and CO_2. Because it reacts in this way with several different compounds, including CFCs, hydroxyl (which is only present as a gas, in minute quantities in the atmosphere) is dubbed "the detergent of the atmosphere." Since individual OH molecules exist only temporarily, during the intermediate stages of chemical reactions, they cannot be measured directly. Thus, the concentration and variability of concentration of these ions are virtually unknown. OH is the main sink for methane, and for all of the hydrogenated halocarbons (including the CFCs); and it is an important aspect of the chemistry of tropospheric ozone. It is therefore crucial for determining future changes in these substances.

Insolation (from INcoming SOLar radiATION) - The solar radiation received at any particular "sunfall" — any specified area of the earth's surface. Sunfall varies from region to region depending on latitude and weather. More insolation is received, per hectare, in sunny equatorial climates than in the industrial north.

Institutional Bargaining - Negotiation over the content of constitutional contracts or institutional mandates, setting forth the provisions of regimes.

Integrated resource planning - A policy in which regulated utilities are required (or given incentives) to find the most cost-effective means for providing energy services — including the promotion of end-user energy efficiency — before they can obtain approval to build new plants or acquire new energy supplies.

Interglacial period - See *glacial epoch.*

Internal combustion engine - Engines where fuel is burned inside the engine. By contrast, in steam engines and other engine forms, fuel is burned in a separate furnace. Most internal combustion engines are gasoline-fueled, in which the gasoline propels pistons which in turn propel the engine. But gas turbines, rocket engines, rotary engines, and turbine propulsion engines all use internal combustion. The earliest, and still dominant, form of the internal combustion engine is the Otto cycle engine — named for Nikolaus Otto, who built the first engine to use this cycle successfully in 1876. The Otto cycle was originally a four-stroke engine, requiring two revolutions of the crankshaft; it has since been refined to two strokes, requiring one crankshaft revolution. The primary alternative form is the diesel engine, demonstrated in 1896, which uses air compression instead of a spark to ignite the fuel, and which differs from the Otto cycle engine in many significant ways. Internal combustion engines release more carbon monoxide, reactive hydrocarbons and nitrogen oxides into the North American atmosphere than all other urban and industrial sources combined.

Internalization - The process of including in financial balance sheets the cost factors which were previously not calculated (such as the value of a degraded environment) or whose brunt was borne by other elements of society (such as the health costs associated with pollution). Advocates of internalization argue that these costs should realistically be assigned to industrial enterprises, because they are an inherent outcome of the production process. Once such costs as environmental degradation or health effects are internalized, the relative cost-benefit ratios of investment decisions change, creating a much stronger incentive for sustainable development and waste reduction.

Isostatic rebound - Uplift of the Earth's crust following the last glaciation, due to the disappearance of the large continental ice sheets.

Isotopes - Atoms of a single element (with the same number of protons) that have different masses (because they have a different number of neutrons). Isotopes are labelled with the approximate mass preceding the symbol of the element: ^{18}O, for example, denotes an Oxygen isotope with an atomic mass of 18, instead of the mass of 16 which oxygen has under ordinary conditions (as ^{16}O). Some isotopes have characteristics making them useful for analyzing chemical history; for example, they release electrons or other sub-atomic particles, allowing their presence to be detected, and they decay (gradually change into another element) at a steady, measurable speed. Carbon-14 (^{14}C) decays into ^{14}N, with a half-life of approximately 5700 years. When detected in samples of sediment or ice, isotopes can be analyzed to compile records of past climate characteristics. From ^{14}C isotopes in core samples frozen in ice, the carbon content of the atmosphere over time can be derived, along with the atmospheric concentration of carbon-based molecules such as methane (CH_4). Deuterium, an isotope of hydrogen (^2H), provides a history of average air temperatures. The ratio between isotopes of oxygen (^{16}O and ^{18}O) can be analyzed, providing a record of past surface temperatures.

Joule - a unit of energy. One joule is equal to the work done when a current of one ampere is passed through a resistance of one ohm for one second. One joule = 10^7 ergs = 9.48×10^{-4} BTUs.

A 100-watt light bulb uses 100 joules every second. Measuring joules allows the comparison of energy needs, capacities, and efficiencies. For example, all of the world's humanity used 31.5 x 10^{18} joules of electrical, mechanical, fossil fuel and heat energy in 1990.

Kerogen - A mixture of hydrocarbon compounds that can be extracted from oil shale rock to produce shale oil. The net energy yield of shale oil is less than that of conventional oil because of the energy used to extract, process, upgrade, and refineshale oil. The conversion of kerogen to shale oil and its burning release more carbon dioxide per unit of energy than conventional oil.

Last Glacial Maximum (18,000 y BP) - The last prolonged period of Ice Age cold climate before the present day.

Lean production - A term coined by John Krafcik, of the MIT International Motor Vehicle Program, describing a form of manufacturing that uses less of everything, compared with mass production: less human effort in the factory, manufacturing space, investment in tools, engineering hours, and inventory warehouse. Lean production is also based on the principle of continuously improving products and processes, rather than meeting preordained specifications; and on sharing involvement in quality production with everyone on the organizational ladder.

Little Ice Age (A.D. 1550-A.D. 1850) - A cold period in Europe, North America, and Asia, marked by rapid expansion of mountain glaciers, especially in the Alps, Norway, Ireland, and Alaska. The end of the Little Ice Age coincided with the beginning of the industrial revolution and a rapid rise in emissions of greenhouse gases; however, a definitive correlation between the two has not been proven.

Loess - a loose surface sediment commonly thought to have been formed by wind action during the Pleistocene Epoch. Loess deposits may indicate formerly dry, continental climates having moderate-to-strong prevailing winds.

Mean sea level (MSL) - The average height of the sea surface, based on hourly observation of the tide height on the open coast, or in adjacent waters that have free access to the sea. In the United States, MSL is defined as the average height of the sea surface for all stages of the tide over a nineteen-year period.

Metabolic activity - The chemical changes that occur in a living animal or plant. In animals, these include synthetic reactions (such as the manufacture of proteins and fats), and destructive reactions (such as the breakdown of sugars into carbon dioxide and water). In plants, they include photosynthesis - a metabolic process which is boosted by increased amounts of CO_2 in the atmosphere.

Methane - A hydrocarbon with four Hydrogen atoms attached to each Carbon atom (CH_4). Methane is called "swamp gas" because it is produced by bacteria when organic matter decomposes under anoxic (oxygen-free) conditions, as in swampy land. Methane-producing bacteria are sensitive to oxygen, but they exist in habits such as animal digestive tracts, sanitary landfills, swamps, sludge, and other decaying organic matter, where the oxygen has already been removed by other bacteria. Methane can be used as a fuel, and as a gas is significant as a contributor to the greenhouse effect — the only major greenhouse gas produced in greater amounts by developing countries than industrialized countries.

Methanol - Wood alcohol. The simplest form of alcohol, it can also be made from coal and natural gas. As a fuel, it blends easily with gasoline (to produce gasohol), burns with a sootless flame and contains no heavy hydrocarbons. Its greatest potential polluting emission is formaldehyde, whose emissions can be controlled technologically. This makes methanol a potentially less harmful alternative to petroleum products in transportation.

Modelling - An investigative technique that uses a mathematical or physical representation of a system — such as a global physico-chemico system, global physico-chemico-biologic system, an agricultural operation, a petroleum refinery, or a temperate forest ecosystem — to explore its potential characteristics and limits. Many models (or system representations) are programmed as computer simulations of causes and effects within the system. These models are often used to test the effects of a change in system components on the overall performance of the system. For example, models of global or regional climate may be used to attempt to simulate the effects on temperature of a change in the concentration of greenhouse gas emissions.

Negative feedback - *See Feedback mechanism.*

Negative feedbacks - In climate processes, negative feedbacks promote climate stability by diminishing the effect of warming or cooling forces. See *Feedback mechanism.*

Net Primary Production (NPP) - Plants both take in and emit carbon dioxide. NPP is the net amount of CO_2 taken in by vegetation in a particular area. It is an important element in the balance of carbon exchange between the Earth and the atmosphere. Two main processes are involved: *Photosynthesis* is the fundamental energy-gathering process of life: sunlight + carbon dioxide + water are transformed into organic carbon + oxygen. This occurs mainly in the leaves of terrestrial plants and in microscopic blue green algae in the ocean. *Photorespiration* (autotrophic respiration) takes place simultaneously, when plants are exposed to light; the plants take up oxygen from the air and release carbon dioxide. It takes place primarily when plants are exposed to light. In an unperturbed world, the balance between these two processes produces a net loss of carbon dioxide — approximately enough to balance the carbon which is formed into soils and peat, plus the amount consumed in *heterotrophic respiration* (respiration by microbes, converting organic matter back into atmospheric CO_2). The carbon balance can be changed considerably by human activities and land use changes, and by climate changes. Since the pools and fluxes are large (NPP 50-60 GtC per year), any perturbations that affect photosynthesis or photorespiration can have a significant effect on the atmospheric concentration of CO_2.

Nitrification and denitrification - The essential components of the nitrogen cycle are these two microbiological processes. Nitrogen is an essential component of amino acids, proteins and nucleosides; the capture of nitrogen from the air is carried out by bacteria which live symbiotically in association with leguminous plants. They fix nitrogen into the form of highly soluble nitrates (salts containing NO_3). Nitrates in soil are taken up by plants, which convert them into useful organic compounds, such as proteins. Animals in turn get most of their nitrogen-containing nutrients by eating plants or other animals that have eaten plants. Denitrification takes place when bacteria and fungi decompose protein and other nitrogen-containing compounds from plant and animal waste back into nitrogen gases, which are released to the atmosphere to begin the cycle again. (Some of this nitrogen is converted directly back to nitrates, without returning to the atmosphere.) In the stratosphere, the gas nitric oxide (NO) a product of the breakdown of N_2O, serves as a catalyst in ozone destruction reactions and thus regulates the amount of ozone in the stratosphere. Humans intervene in the nitrogen cycles in various ways. Land use disturbances, agricultural activity, fossil fuel burning and nitrogen fertilizer application all influence the rate of release of nitrogen-containing gases. Any activity that adjusts the concentration of O_2, nitrates, or carbon in the soil will affect denitrification, since these substances all regulate that process.

"No-regrets" policy - Policy based on the idea that the problem of global climate change is linked to other critically important problems of environment and development. The combined risks are serious enough, and the eventual benefits of action great enough, to require urgent and bold initiatives, even if they impose a substantial immediate cost. Advocates of this policy argue that strong action will lead to a "no regrets" outcome, even if climate change turns out to be an exaggerated fear. The benefits will include experimentation, foresight, and cost-effective prevention. Moreover, say the proponents, early actions offer the prospective extra benefit of learning through experience: of gaining better information about the benefits and costs of action through our first steps.

Noise (statistical noise) - In models, the omnipresent and apparently meaningless background of natural climatic variability, from which the significant data (the "signal") must be distinguished.

Non-Governmental Organizations (NGOs) - A catch-all term that covers what are known in various nations as scientific institutions, citizens' groups, regional associations, professional groups, community groups, public-interest groups, voluntary groups, environmental organizations, and religious organizations. All of these are playing increasing roles in climate change-related deliberations. Trade unions are by definition non-governmental, but the term NGO is not generally applied to them.

Otto cycle engine - See *internal combustion engine.*

Ozone - A molecule made up of three atoms of oxygen (O_3. In the stratosphere, it occurs naturally and provides a protective layer shielding the earth from ultraviolet radiation and subsequent harmful health effects on humans and the environment. In the troposphere, it is a chemicial oxidant and major component of photochemical smog.

Paleoclimatology - The study of past climates, throughout geological history, and the causes of the variations among them.

Particulates - very small pieces of solid or liquid matter such as particles of soot, dust, fumes, mists or aerosols. The physical characteristics of particles, and how they combine with other particles, are part of the feedback mechanisms of the atmosphere.

Perturbation - Synonym for *radiative forcing.*

Petajoule - 10^{15} joules.

Photolysis - Chemical decomposition by the action of radiant electromagnetic energy, especially light. Triggered by solar radiation, photolysis removes both nitrous oxide and ozone in the stratosphere.

Photorespiration - See *net primary production.*

Photosynthesis - See *net primary production.*

Photovoltaic (PV) - Technology in which a semiconductor electrical junction device converts the radiant energy of sunlight directly and efficiently into electrical energy. Most solar cells are made from single-crystal silicon.

Phytoplankton - Microscopic marine organisms (mostly algae and diatoms) which are responsible for most of the photosynthetic activity in the oceans.

Pleistocene (10,000-1,600,000 y BP) - Also called the Quarternary. This ephoch was characterized by numerous (at least 17) worldwide changes of climate, cycling between glacial (cool) and interglacial (warmer) periods, with periodicities of 100,000, 41,000 and 23,000 years. During the last 18,000 years, still nominally within the Pleistocene, the Earth has been consistently growing warmer.

Pliocene (1,600,000-5,300,000 y BP) - A warm epoch with only limited glaciation. The mid-Pliocene was the last time that comparable temperatures existed to those predicted by General Circulation Models to take place within 300 years. Preciptation was greater than at present, including in the arid regions of Middle Asia and Northern Africa, where temperatures were lower than at present in summer.

"Polluter Pays" principle (PPP) - the principle that countries should in some way recompense the rest of the world for the effects of pollution that they (or their citizens) generate or have generated.

Pollution prevention (Waste reduction) - Eliminating the production of hazardous wastes and greenhouse gases at their source, within the production process. This can often be achieved through a variety of relatively simple strategies, including minor changes in manufacturing processes, substitution of non-polluting products for polluting products, and simplification of packaging. Companies practicing waste reduction have saved hundreds of millions of dollars, and used it to catalyze employee involvement and eliminate the need for expensive end-of-the-pipe filtering.

Polyatomic - Molecules composed of more than one atom. Only polyatomic gases that absorb and re-emit solar radiation.

Positive feedback - see feedback mechanism.

Productivity - The output of any production process, per unit of input. To increase productivity means to produce more with less. In vegetation, productivity is the ability to produce life: to create carbon compounds from atmospheric carbon dioxide through photosynthesis. (See *net primary production.)* In factories and corporations, productivity is a measure of the ability to create goods and services from a given amount of labor, capital, materials, land, resources, knowledge, time, or any combination of those. Since capital goods tend to decline in value and wear out, most economists distinguish between *gross capital productivity* (total yield) and *net capital productivity,* which dicounts depreciation.

Proxy climate indicator - Dateable evidence of a biological or geological phenomenon whose condition, at least in part, is attributable to climatic conditions at the time of its formation. Proxy data provide an indirect measure of climate; they include evidence of crop yields, harvest dates, glacier movements, tree rings, glaciers and snow lines, insect remains, pollen remains, marine microfauna, and isotope measurements (see *isotopes).* Tree rings, pollen deposits from varved lakes, and ice cores are the most promising proxy data sources for reconstructing the climate of the last five millenia, because the dating are precise on an annual basis, while other proxy data may only be precise to ±100 years.

Radiative balance - The net amount and direction of radiation, taking into account the sum of all radiation, transferred in all directions, through the Earth's atmosphere and to and from space. The radiative balance is one of the primary forces controlling Earth's temperature, rainfall, and (less directly) sea level.

Radiative forcing (external forcing, perturbation) - A change imposed upon the climate system which modifies the radiative balance of that system. The causes of such a change may include changes in the sun, clouds, ice, greenhouse gases, volcanic activity, and other agents. Convention lumps all these together as agents of radiative forcing. Radiative forcing is often spcified as the net change in energy flux at the troposphere (watts per square meter). Many climate models seek to quantify the ultimate change in Earth's temperature, rainfall, and sea level from a specified change in radiative forcing.

Radiatively active trace gases - Gases, present in minute quantities in the atmsophere, that absorb incoming solar radiation or outgoing infrared radiation, thus affecting the vertical temperature profile of the atmosphere. These gases include water vapor, carbon dioxide, methane, nitrous oxide, chlorofluorocarbons, and ozone.

Rankine cycle - An ideal thermodynamic cycle, used as a standard against which to judge the performance of heat-engine and heat-pump installations (such as a steam plant).

Refractory - A ceramic material used in high-temperature structures and equipment; potentially vulnerable to damage when used in ultra-high-power transformers in steel plants.

Regime - A set of agreed-upon principles, norms, rules and decision-making procedures, which govern the negotiations, establishment of agencies, and other interactions by international participants in the area of a specific political issue.

Regime implementation - The process of turning a regime, as spelled out in the provisions of a constitutional contract, into a social practice.

Regolith - The layer of loose rock and mineral material that covers almost all land surfaces. As it becomes mixed with water and organic matter, it is gradually transformed into soil.

Reinvestment costs - The cost of setting aside additional funds, besides those used to provide energy services, to invest in technologies based on renewable energy sources, so that a power system may be eventually switched over to those. Reinvestment costs are an example of an investment which cannot save a utility or government money in the short run, but will do so over the long term.

Relative sea level (RSL) - Sea level, measured in comparison to the nearest coast line. RSL is affected by mean sea level, and also by vertical movements of the land along the adjacent coast.

Replacement fertility rate - The rate at which only enough new children are born to replace people who die, thus keeping the population constant.

Respiration - See *Net primary production.*

Riparian - Pertaining to rivers, or dwelling on the bank of a river or other body of water.

Salinity - The degree of salt in water. The rise in sea level due to global warming would result in increased salinity of rivers, bays and aquifers. This would affect drinking water, agriculture and wildlife.

Sensitivity - See *Climate sensitivity.*

Seventeen-sixty-five (1765) - Because it is close to the beginnings of the industrial revolution, many climate models use this date as a starting point, from which to assess the impact of greenhouse gas emissions on climate. Since pre-industrial times (circa 1765), radiative forcing has increased by about 2.5 Watts per square meter. That much more radiation reaches the lower layers of the atmosphere than in 1765 - equivalent to the effect of an increase in the Sun's output of a little more than one per cent. Some negotiation strategies over climate-change-related policies seek to make industrialized nations responsible for historical emissions, generally going back to this date.

Signal-to-noise ratio (SNR) - A quantitative measure of the statistical detectability of a signal: showing how reliable and easy it is to detect a significant pattern of data against the background of statistical irrelevancies (noise). In climate modelling, the lower the signal-to-noise ratio, the more difficult it is to detect that an enhanced greenhouse effect has had an impact on global temperature change or change in some other climatic condition.

Sink - Vehicle for removal of a chemical or gas from the atmosphere-biosphere-ocean system, in which the substance is absorbed into a permanent or semi-permanent repository, or else transformed into another substance. A carbon sink, for example, might be the ocean (which absorbs and holds carbon from other parts of the carbon cycle) or photosynthesis (which converts atmospheric carbon into plant material). Sinks are a fundamental factor in the ongoing balance which determines the concentration of every greenhouse gas in the atmosphere. If the sink is greater than the sources of a gas, its concentration in the atmosphere will decrease; if the source is greater than the sink, the concentration will increase. Sinks for methane include reaction with *hydroxyls* (OH) in the troposphere, and storage in soils.The primary sink for nitrous oxide is photochemical decomposition in the stratosphere.While no sinks for fully halogenated hydrocarbons (most CFCs) exist in the troposphere, these gases are eventually broken down by ultraviolet radiation and converted to reservoir species in the stratosphere. The hydroxyl radical is a sink for non-fully halogenated hydrocarbons (HCFCs and HFCs). It is also a sink for carbon monoxide.

Solar luminosity - The total amount of energy emitted by the sun per a particular unit of time.

Solar radiation - The amount of radiation or energy received from the sun at any given point. The "solar constant" is a measure of solar radiation: 0.140 watts per square centimeter, at a point just outside the Earth's atmosphere, located on a surface that is perpendicular to the line of radiation, and measured when the Earth is at its mean orbital distance from the sun.

Statistical process control (SPC) - The use of statistical analysis to improve quality by reducing unwanted variation in industrial and management processes. A fundamental part of successful pollution prevention programs, SPC involves the use of analytic charts for diagnosing the root causes of problems and identifying systemically valid solutions.

Stochastic - Random. The term is used to describe natural events with a substantial random component to them, such as rainfall, runoff, and storms.

Stomates (stomata) - Pores or apertures in the outer cell layer of the aerial parts of leaves, stems, and flowers, through which water is lost by transpiration and gas exchange takes place. This exchange of carbon dioxide gas between the plant and the atmosphere is called "stomatal conductance"; its rate is largely governed by the diameter of the stomatal pores.

Stratosphere - The upper atmosphere. It extends from the tropopause to about 50 kilometers above the earth's surface, and has a comparatively low water vapor content compared to the troposphere. The ozone layer which shields the Earth from ultraviolet radiation is located in the stratosphere.

Sulphate - Ions containing sulphur and oxygen (SO_4) which react easily with hydrogen to become sulphuric acid (H_2SO_4). In the atmosphere, sulphate particles have increased markedly over the past 50-100 years, due to industrial emissions of sulphur dioxide (SO_2). Sulphate aerosols are key components of several atmospheric feedback processes. For example, they cool the Earth both directly (by reflecting away incoming solar radiation under clear sky conditions) and indirectly (by making maritime clouds more reflective).

Sustainable development - According to the WCED, this is "development that meets the needs of the present without compromising the ability of future generations to meet their own needs." Sustainable development implies economic growth together with the protection of environmental quality, each reinforcing the other. The essence of this form of development is a stable relationship between human activities and the natural world, which does not diminish the prospects for future generations to enjoy a quality of life at least as good as our own. Many observers believe that participatory democracy, undominated by vested interests, is a prerequisite for achieving sustainable development.

Technology transfer - The practice of making technological information and aid available at low or no cost to agencies in developing countries. Although it may conflict with patent considerations, technology transfer is an effective means of ensuring the spread of energy-efficient, greenhouse-gas-diminishing industrial capabilities. The term also refers to the co-development of new or advanced systems through partnerships between enterprises in different countries.

Teragram (Tg) - One trillion (10^{12}) grams or one million metric tonnes. The total CH_4 load in the atmosphere is on the order of 4000 Tg (or 4 gigatonnes).

Terawatt (TW) - One trillion (10^{12}) watts (joules/second). Total world energy use in 1990 was 13.16 TW.

Thermal delay - potential long time lag before ocean warming affects general climate change (see *absorption*)..

Time-dependent response - the forecast of a change in climate resulting from a change in radiative forcing which takes place slowly, on a realistic time-scale. These responses are more difficult to estimate through climate models, but more realistic, than their *equilibrium response* counterparts.

Ton-kilometers (ton-km) - The distance traveled multiplied by the weight transported. This is a good measure of the quantity of freight transport. Coupled with fuel efficiency levels, it allows us to compare different transport activities in terms of the amount of fuel used and air pollutants emitted.

Total Quality management (TQ, "quality movement") - Industrial practice based on the principle that an entire system can and should be continuously improved (a process known in Japanese as Kaizen). A practitioner of total quality management (who might be anyone in an industrial plant) will continuously analyze the flow of work, seeking ways to make the process more consistent, less variable, less wasteful, and more serviceable than it was before. *Statistical process control* is an integral part of this practice.

Trace gas - A minor constituent of the atmosphere. Trace gases include water vapor, carbon dioxide, ozone, methane, ammonia, nitric acid, nitrous oxide, ethylene, sulfur dioxide, nitric oxide, dichlorofluoromethane (Freon 12), trichlorofluoromethane (Freon 11), methyl chloride, carbon monoxide, and carbon tetrachloride. All of these contribute to the greenhouse effect.

Transient response - see Time-dependent response.

Transpiration - See evapotranspiration.

Tropopause - The boundary between the troposphere and the stratosphere.

Troposphere - The lower atmosphere, from the ground to an altitude of about 8 kms at the poles, about 12 kms in midlatitudes, and about 16 kms in the tropics. Clouds and weather systems, as experienced by people, take place in the troposphere.

Urban warming - The heat and activity of cities, not considered a major direct factor in global climate change.

Validation - Comparing a climate model's predictions with observations of the real climate, in order to test the reliability and accuracy of the model. The most obvious way to test a climate model is to use it to analyze past events, and then see whether its simulated prediction "came true," or how close it was to being correct.

Volcanism - The phenomena of volcanic activity. Large volcanic eruptions spew massive amounts of ash into the atmosphere that absorb solar radiation, thus potentially generating a cooling effect on planetary temperatures. At the same time, volcanoes release carbon dioxide and sulfur dioxide, and decrease stratospheric concentrations of ozone, thus producing a potential warming effect.

"Wait and see" policy - A conservative strategy for dealing with the risks of climate change, aimed at postponing action until complete, correct, and exact predictions of the future are available. Proponents of this strategy argue that the scientific evidence for global warming is incomplete; thus, hastily contrived strategies could do more harm than good. At the very least, the costs of these strategies would lead to a loss of human welfare and far exceed any benefits to human society. Their implementation could lead to bureaucracies and laws that would unnecessarily stifle human activity.

Waste reduction - See *pollution prevention.*

Watt - Unit of electric power as it is changed into heat and light energy, equal to one joule per second. A 100-watt light bulb uses 100 units of electrical energy every second.

Weather - Atmospheric conditions as experienced by people, on timescales ranging from seconds to months. Most extreme natural events that threaten society fall into this category, including lightning strikes, tornadoes, and tropical cyclones.

Younger Dryas (10,500 BP) - A sudden, abrupt cold episode, lasting about 500 years, occurring in the midst of a longer warming trend which has otherwise taken place continually since the Last Glacial Maximum.

References

Bernard, D. P., T. Lekstrum, W. A. Kurz and N. C. Sonntag, 1990: Glossary, *Briefing Book for the SEI Policy Exercise on Global Climate Change,* prepared for J. Jäger, Stockholm Environment Institute, Stockholm.

Brown, L. R., C. Flavin and S. Postel, 1992: *Saving the Planet: How to shape an environmentally sustainable global economy,* Washington/New York, Worldwatch Institute/W.W. Norton.

CDIAC (Carbon Dioxide Information Analysis Center), 1990: *Carbon Dioxide and Climate,* glossary prepared by the Carbon Dioxide Information Analysis Center, Oak Ridge National Laboratory, operated by Martin Marietta Energy Systems, Inc., for the U.S. Department of Energy, Environmental Sciences Division publication # 3532.

DOE-UK (Department of the Environment, United Kingdom), 1991: "The Potential Effects of Climate Change in the United Kingdom," glossary.

Goldsmith, E. and N. Hildyard, 1990: *Earth Report 2,* Mitchell Beazley Publishers, London.

Harte, J., C. Holdren, R. Schneider and C. Shirley, 1991: *Toxics A to Z: A Guide to Everyday Pollution Hazards,* University of California Press, Berkeley, CA.

Houghton, J.T., G.J. Jenkins and J.J. Ephraums, editors, 1991: *Climate Change: The IPCC Scientific Assessment,* Cambridge University Press, Cambridge, England.

Parker, S. B., ed., 1989: *McGraw-Hill Concise Encyclopedia of Science and Technology,* McGraw-Hill, New York.

PIGW (Policy Implications of Greenhouse Warming, Report of the Mitigation Panel), glossary, 1992 (unpublished draft); includes some entries from Sharp and Leftwich, 1988: *Economics of Social Issues,* Business Publications, Inc., Plano, Texas.

SEI (Stockholm Environment Institute), 1990: "Policy Exercise on Global Climate Change Proceedings," prepared for meeting on 2-7 September 1990, Austria.

Starke, L., 1990: *Signs of Hope: Working Towards Our Common Future,* Oxford University Press, Oxford, England.

This glossary was composed by Art Kleiner, with research by: Gwendolyn Andersen, Paul Carroll, Janis Dutton, Vince Schaper, and Anu Sud. Definitions have been adapted from the chapters of this volume and the above listed references.

Index

R

S